People, Land and Water

People, Land and Water

Participatory Development Communication for Natural Resource Management

Edited by
Guy Bessette

International Development Research Centre
Ottawa • Cairo • Dakar • Montevideo • Nairobi • New Delhi • Singapore

London • Sterling, VA

First published in the UK, USA and Canada in 2006
by Earthscan and the International Development Research Centre (IDRC)

Copyright © 2006, IDRC

ISBN: 1-84407-343-2   hardback
        978-1-84407-343-6 hardback

Typesetting by JS Typesetting Ltd, Porthcawl, Mid Glamorgan
Printed and bound in the UK by TJ International Ltd, Padstow
Cover design by Mike Fell

For a full list of publications please contact:

Earthscan
8–12 Camden High Street
London, NW1 0JH, UK
Tel: +44 (0)20 7387 8558
Fax: +44 (0)20 7387 8998
Email: earthinfo@earthscan.co.uk
Web: www.earthscan.co.uk

22883 Quicksilver Drive, Sterling, VA 20166-2012, USA

Earthscan publishes in association with the International Institute for Environment and Development

IDRC publishes an e-book edition of People, Land and Water
(ISBN: 1-55250-224-4)

For further information, please contact:
International Development Research Centre
PO Box 8500
Ottawa, ON
Canada K1G 3H9
Email: pub@idrc.ca
Web: www.idrc.ca

IDRC is a Canadian public corporation that works in close collaboration with researchers from the developing world with the aim of building healthier, more equitable and more prosperous societies.

A catalogue record for this book is available from the British Library

Library of Congress Cataloging-in-Publication Data

People, land, and water : participatory development communication for
natural resource management / edited by Guy Bessette.
      p. cm.
   Includes bibliographical references.
   ISBN-13: 978-1-84407-343-6 (hardback)
   ISBN-10: 1-84407-343-2 (hardback)
  1. Natural resources–Management. 2. Communication in economic
development. I. Bessette, Guy, 1952–
   HC85.P46 2006
   333.701'4–dc22

                                             2006002676

The paper used for the text of this book is FSC certified.
FSC (The Forest Stewardship Council) is an international network
to promote responsible management of the world's forests.

Printed on totally chlorine-free paper

FSC

Mixed Sources

Product group from well-managed
forests and other controlled sources

Cert no. SGS-COC-2482
www.fsc.org
© 1996 Forest Stewardship Council

Contents

List of Figures, Tables and Boxes

Figures

Tables

Boxes

List of Contributors

Mario Acunzo is a communication for development officer at the Extension, Education and Communication Service of the United Nations Food and Agriculture Organization (FAO) in Rome.

Claude Adandedjan is a senior education fellow at the International Centre for Research in Agroforestry (ICRAF), Sahel programme.

Awa Adjibade is a sociologist currently based in Lomé, Togo.

Madeline Baguio Quiamco is an assistant professor at the College for Development Communication of the University of the Philippines, Los Baños.

Guy Bessette is a senior programme specialist at the International Development Research Centre in Ottawa.

S. T. Kwame Boafo is Chief, Executive Office, Communication and Information Sector of the United Nations Educational, Scientific and Cultural Organization (UNESCO).

Maria Celeste Cadiz is Dean and Associate Professor at the College of Development Communication of the University of the Philippines, Los Baños.

Rawya El Dabi works for the International Development Research Centre's Partnership and Business Development Division in Cairo.

N'Golo Diarra is a researcher and trainer at the Centre de Services de Production Audiovisuelle (CESPA) in Mali.

Corinne Dick is at the American University of Beirut in Lebanon.

Vicenta P. de Guzman is the Executive Director of the Legal Assistance Centre for Indigenous Filipinos (PANLIPO) in the Philippines.

Waad El Hadidy is a programme manager at the Centre for Development Services in Egypt.

Mona Haidar is a researcher at the Environment and Sustainable Development Unit at the American University of Beirut in Lebanon.

Shadi Hamadeh is a professor of animal sciences and is currently leading the Environment and Sustainable Development Unit at the American University of Beirut in Lebanon.

Amri Jahi is a senior lecturer and researcher at Bogor Agricultural University in Indonesia.

Meya Kalindekafe is a senior lecturer in ecology at the University of Malawi.

Chris Kamlongera is the Director of the SADC Centre of Communication for Development.

Jones Kaumba is a senior communication for development trainer at the SADC Centre of Communication for Development.

Lun Kimhy is the Deputy Provincial Programme Adviser of the Partnership for Local Governance in Ratanakiri, Cambodia.

Yacouba Konaté passed away in 2003. At the time of his death, he was the coordinator of a participatory development communication project at the Permanent Inter-State Committee for Drought Control in the Sahel (CILSS).

Kofi Larweh is the Station Coordinator of Radio Ada, in Ghana.

Lourdes Margarita A. Caballero is a research associate at the College of Development Communication of the University of the Philippines, Los Baños.

Luningning A. Matulac is a professor of educational communication at the College of Development Communication of the University of the Philippines, Los Baños.

Pierre Mumbu is a researcher and lecturer at the Institut Supérieur de Développement Rural of Bukavu, in the Democratic Republic of Congo and also works as a consultant in community radio and development communication.

Amadou Niang is the Director of the Millennium Development Goals Centre for West Africa, in Bamako.

Michelle Obeid is a researcher at the Environment and Sustainable Development Unit at the American University of Beirut in Lebanon.

Nora Naiboka Odoi is a development communication specialist at the Kawanda Agricultural Institute in Uganda.

Kadiatou Ouattara is a journalist and works with Journalistes en Afrique pour le Développement (JADE) in Burkina Faso.

Souleymane Ouattara is a journalist and the Coordinator of Journalistes en Afrique pour le Développement (JADE) in Burkina Faso.

Rosalie Ouoba is a sociologist, based in Burkina Faso. She is actively involved with the Union of Rural Women of West Africa and Chad.

Sours Pinreak is an adviser to the Land Rights Extension team in Cambodia.

Nora Cruz Quebral is recognized as the founder of the discipline of development communication. She currently heads the Nora Cruz Quebral Foundation for Development Communication and is still associated with the College of Development Communication of the University of the Philippines, Los Baños.

C. V. Rajasunderam, now retired, has worked as a consultant in development communication.

Chin Saik Yoon is the Publisher and Managing Director of Southbound in Malaysia.

Ahmadou Sankaré is the coordinator of a participatory development communication project at the Permanent Inter-State Committee for Drought Control in the Sahel (CILSS) in Burkina Faso.

Karidia Sanon is a researcher and lecturer at the Université de Ouagadougou, Burkina Faso.

Fatoumata Sow is a journalist currently working at UNESCO's regional office in Senegal.

Diabodo Jacques Thiamobiga is an agronomist and sociologist working in the field of rural development in Burkina Faso.

Jakob S. Thompson is an associate professional officer at the United Nations Food and Agriculture Organization (FAO) in Rome.

Cleofe S. Torres is an associate professor at the Centre for Development Communication of the University of the Philippines, Los Baños.

Le Van An heads the Department of Science and International Relations at the Hue University of Agriculture and Forestry.

Maria Theresa H. Velasco is an associate professor and Chair of the Department of Science Communication at the College of Development Communication of the University of the Philippines, Los Baños.

Rami Zurayk is a researcher at the Environment and Sustainable Development Unit at the Ameircan University of Beirut, in Lebanon.

Foreword

Nora Cruz Quebral

For many communication professionals who have made development a personal commitment, shuttling from theory to practice and then from application to re-conceptualization is what their calling is about. For this breed, the 'field' is the main laboratory and proving ground. Working there can be frustrating, uplifting and chastening, but never dull, as the contributors to this book can attest, particularly so when one is questioning seminal concepts and trying out new ones in their place, which is what participatory development communication (PDC) does in this book.

Development communicators, being new kids on both the development and communication blocks, may not quite have gained full acceptance among their social science peers. They are a dynamic lot, nonetheless, who stake their alternate positions with passion or with studious persistence. A sign of their independence is the several names that they have given their specialty. In addition to PDC, this book cites participatory communication for development, participatory communication or communication for social change. Participation and dialogue are givens in all the variants.

There is none better systematized than PDC, however. This book offers a methodology and a terminology honed in the field. Based on a succession of 'writeshops', workshops and round-table conferences attended by action researchers and outreach workers from Asia and Africa, the collective experience was in itself an experiment in cross-continental dialogue. The linguistic and cultural differences were not trivial. They were a macrocosm of the divides that practitioners and field researchers normally encounter in their local communities.

The group effort recounts diverse interventions in which development communication concepts were tested, modified or found to be a fit. The overall results should encourage other fieldworkers to enlarge the discussion further with their own experiments. For the contributors to the book, re-reading the cases at leisure will still be part of their learning experience. They can better appreciate the conceptual differences between communication approaches and match up an approach with the appropriate methods and terms.

The friendships and increased understanding of each other's ways that ensued among the contributors were equally valuable outcomes of their PDC adventure. They bode well for a global PDC network, which is probably what the organizers had in mind all along.

As an endnote, the words 'participatory development communication' – singly or joined – connote certain values that define PDC as much as its strategies, tools and techniques. They are what makes the term different from other types of communication and are at the core of its guiding philosophy. PDC professionals are reminded to let those values shape their practice and to remember, as well, that sustainable natural resource management is but a facet of the larger goal that is human development.

Manila
April 2006

Preface

This book presents conceptual and methodological issues related to the use of communication in order to facilitate participation among stakeholders in natural resource management (NRM) initiatives. It also presents a collection of chapters that focus on participatory development communication and NRM, particularly in Asia and Africa.

There are many approaches and practices in development communication, and most of them have been implemented in the field of environment and natural resource management. But, even when considering participatory approaches in NRM, communication is often limited to information dissemination activities that mainly use printed materials, radio programmes and educational videos to send messages, explain technologies or illustrate activities. These approaches, with their strengths and weaknesses, have been well documented.

Participatory development communication takes another perspective. This form of communication facilitates participation in a development initiative identified and selected by a community, with or without the external assistance of other stakeholders. The terminology has been used in the past by a number of scholars[1] to stress the participatory approach of communication in contrast with its more traditional diffusion approach. Others refer to similar approaches as participatory communication for development, participatory communication or communication for social change.

In this publication, participatory development communication is considered to be a planned activity that is based on participatory processes and on media and interpersonal communication. This communication facilitates dialogue among different stakeholders around a common development problem or goal. The objective is to develop and implement a set of activities that contribute to a solution to the problem or the realization of a goal, and which support and accompany this initiative.[2]

This kind of communication requires moving from a focus on information and persuasion to facilitating exchanges between different stakeholders to address a common problem, to develop a concrete initiative for experimenting with possible solutions, and to identify the partnerships, knowledge and materials needed to support these solutions.

This book situates the concept and its methodological issues. It has been produced through a three-step process. First, practitioners from Asia and Africa were invited to submit chapters that offer examples and illustrations of applying participatory development communication to natural resource management. Second, a peer-review workshop was organized in Perugia, Italy, in September 2004, in preparation for the Roundtable on Development

Communication organized by the United Nations Food and Agriculture Organization (FAO) to discuss and review these chapters. Third, during the roundtable, the first chapter of this volume was presented and introduced to the participants in order to orient the discussions of the working group on communication and natural resource management.

These steps led to the preparation of this book, which we hope will play a role in both promoting participatory approaches to development communication in the field of environment and NRM, and in sharing the viewpoints of practitioners from Asia and Africa.

Guy Bessette
April 2006

Notes

1   See, in particular, White et al (1994) and Servaes et al (1996).
2   See Bessette (2004).

References

Bessette, G. (2004) *Involving the Community: A Facilitator's Guide to Participatory Development Communication*, IDRC, Ottawa, Canada, and Southbound, Penang

Servaes, J., Jacobson, T. L. and White, S. A. (1996) *Participatory Communication and Social Change*, Sage Publications, London

White, S. A., Sadanandan Nair, K. and Ascroft, J. (1994) *Participatory Communication, Working for Change and Development*, Thousand Oaks, New Delhi and Sage Publications, London

Acknowledgements

I would like to thank the contributors, who were willing to share their experiences and reflections with regard to participatory development communication and natural resource management.

My thanks also go to Manon Hogue, who very patiently and with a lot of enthusiasm revised the different chapters and accompanied them with a note of introduction. Without her work and support, there is no way this volume could have been finalized for publication. Her commitment and skills brought to this book much more than a technical contribution. I and the other authors of this publication salute you and thank you, Manon.

List of Acronyms and Abbreviations

AIDS	acquired immune deficiency syndrome
ARDA	Association for Rural Development in Arsaal (Lebanon)
BACDI	Bayagong Association for Community Development Inc (the Philippines)
CBCRM	community-based coastal resource management
CBNRM	community-based natural resource management
CCD	Convention to Combat Desertification
CDC	College of Development Communication (University of the Philippines)
CDS	Centre for Development Studies
CEDRES	Centre d'Études pour le Développement Économique et Social (Burkina Faso)
CESAO	Centre d'Études Économiques et Sociales de l'Afrique de l'Ouest
CESPA	Centre de Services de Production Audiovisuelle (Mali)
CIERRO	Centre for Rural Radio Development (Burkina Faso)
CILSS	Permanent Interstate Committee on Drought Control in the Sahel
ComDev	classical communication for development
DAES	Department of Agricultural Extension Services (Malawi)
DRC	Democratic Republic of Congo
FAO	United Nations Food and Agriculture Organization
GIS	geographical information system
GUCRE	Gestion des Usages Conflictuels des Ressources en Eau (Management of Conflicting Uses of Water Resources) project
ha	hectare
HIV	human immunodeficiency virus
IB-ESA	Isang Bagsak East and Southern Africa
ICRAF	International Centre for Research in Agroforestry
ICT	information and communication technology
IDRC	International Development Research Centre
IKS	indigenous knowledge system
IPD	Institut Panafricain pour le Développement (Burkina Faso)
JADE	Journalistes en Afrique pour le Développement (Journalists in Africa for Development)
KARI	Kawanda Agricultural Research Institute (Uganda)
km	kilometre
KVIP	Kumasi ventilated improved pit toilet
m	metre

mm	millimetre
NAADS	Integration of Natural Resource Management in National Agricultural Advisory Services (Uganda)
NARO	National Agricultural Research Organization (Uganda)
NEF	Near-East Foundation
NEPAD	New Partnership for Africa's Development
NGO	non-governmental organization
NRM	natural resource management
PANLIPI	Legal Assistance Centre for Indigenous Filipinos
PDC	participatory development communication
PLA	participatory learning and action
PLG	Partnership for Local Governance (Cambodia)
PNKB	Kahuzi-Biega National Park
PRA	participatory rapid appraisal
PRA	participatory reflection and analysis
PRA	participatory rural appraisal
PRCA	participatory rural communication appraisal
RIA–3	Research Institute for Aquaculture – Region 3
SADC-CCD	Southern Africa Development Community Centre of Communication for Development (Zimbabwe)
UFROAT	Union of Rural Women of West Africa and Chad
UK	United Kingdom
UNDP	United Nations Development Programme
UNESCO	United Nations Educational, Scientific and Cultural Organization
UNICEF	United Nations Children's Fund
US	United States

I
Introduction

Facilitating Dialogue, Learning and Participation in Natural Resource Management

Guy Bessette

Poverty alleviation, food security and environmental sustainability: The contribution of participatory communication

Poverty alleviation, food security and environmental sustainability are closely linked and represent major development challenges for all actors involved in the field of natural resource management. Poverty alleviation requires sustained economic growth, but it is also necessary to ensure that the poor benefit from that growth. Efforts must also be made to improve food security, not only through an increase in productivity, but also by providing conditions of access and proper utilization by the poor.

Promoting environmental sustainability includes challenging goals such as fighting land degradation (especially desertification), halting deforestation, promoting proper management of water resources through irrigation schemes and protecting biodiversity. All these activities must be designed and implemented with the active participation of those families and communities who are struggling to ensure their livelihood in changing and unfavourable environments. But they must also include other stakeholders who are playing or can play a role in these changes: government technical services, non-governmental organizations (NGOs), development projects, rural media, community organizations and research teams. Finally, local and national authorities, policy-makers and service providers must also be involved in shaping the regulatory environment in which the required changes will take place.

Effectively addressing the three interlinked development challenges of poverty alleviation, food security and environmental sustainability requires that development practitioners work actively with all stakeholders with a view to facilitating dialogue, learning and active participation in natural resource management initiatives.

Best practices in natural resource management research and development point to situations in which community members, research or development team members and other stakeholders jointly identify research or development parameters and participate in the decision-making process. This process goes beyond community consultation and participation in activities identified by researchers or programme managers. In best case scenarios, the research or development process itself generates a situation of empowerment in which participants transform their view of reality and are able to take effective action.

Participatory development communication reinforces this process. It empowers local communities to discuss and address natural resource management practices and problems and to engage other stakeholders in building an improved policy environment.

But what about the issues involved in applying participatory development communication to natural resource management practices and research? What are the challenges and the difficulties associated with this approach? What insights and lessons can be drawn from our practices in the field? This chapter offers a reflection on these issues and suggests orientations to further reinforce natural resource management practices and research through participation and communication.

Moving from information dissemination towards community participation

Traditionally, in the context of environment and natural resources management, many communication efforts used to focus on the dissemination of technical packages towards the end-users who were expected to adopt them. Researchers wanted to 'push' their products to communities and development practitioners in order to receive community commitment to their development initiatives.

Not only did these practices have little impact, but they also ignored the need to address conflicts or policies.

Participatory development communication takes a different approach. It suggests a shift in focus from informing people with a view to changing their behaviours or attitudes to facilitating exchanges between various stakeholders. These exchanges help the stakeholders to address a common problem or implement a joint development initiative in order to experiment with various solutions and identify the required partnerships, knowledge and material conditions.

The focus is not on information to be disseminated by experts to end-users. Rather, it is on horizontal communication processes that enable local communities to identify their development needs and the specific actions that could help to fulfil those needs, while establishing an ongoing dialogue with the other stakeholders involved (e.g. extension workers, researchers and decision-makers). The main objective is to ensure that the end-users gather enough information and knowledge to carry out their own development initiatives, evaluate their actions and recognize the resulting benefits.

Such a communication process pursues objectives related to increasing the community knowledge base (both indigenous and modern); modifying or reinforcing common practices related to water use and soil productivity so that natural resources can be managed more efficiently; building and reinforcing community assets; and approaching local and national authorities, policy-makers and service providers. Appropriate communication approaches should also be set up to implement the required initiatives, as well as to monitor and evaluate their impact, while planning for future action.

With participatory development communication, researchers and practitioners become facilitators in a process that involves local communities and other stakeholders in the resolution of a problem or the achievement of a common goal. This, of course, requires a change in attitude. Learning to act as a facilitator does not happen overnight. One must learn to listen to people, to help them express their views and to assist them in building consensus for action. For many natural resource management researchers and practitioners, this is a new role for which they have not been prepared. How can they initiate the process of using communication to facilitate participation and the sharing of knowledge?

Some of the chapters presented here describe this process in action. In Part III, in 'From Rio to the Sahel: Combating Desertification', Sankaré and Konaté describe how such an approach was developed in the context of desertification. Communication strategies were used to stress information dissemination, mobilization and persuasion; but they had little impact. There was a need to try out and implement other approaches. An experiment in participatory communication was used to support various local initiatives designed to fight desertification in the Sahel and to facilitate community participation.

The process led community members and local development actors to identify the problems facing them with regard to desertification, to express their needs and to decide on local solutions and concrete initiatives to experiment with. The project used communication tools such as practical demonstrations, and radio and community discussions, as well as traditional songs and poems to support and accompany the initiatives.

The process included four main phases: training, planning, experimentation and evaluation. Training and planning were the foundation because they mobilized all actors (e.g. community members, project leaders and communication facilitators from the locality) to discuss the process of the action research and how communication would be used to facilitate participation. Not only did this process facilitate community participation, but it also contributed to creating synergy between various development structures.

These initiatives were successful because people were involved in the decision-making process and were not simply invited to participate in specific activities. The project also demonstrated that halting desertification, like other development challenges, demands community participation and synergy between different development actors. It cannot be programmed in a top-down way.

In 'Growing Bananas in Uganda: Reaping the Fruit of Participatory Development Communication' Odoi tells the story of how the shift was made to implement communication for participation in the context of action research with banana growers. The banana research programme of Uganda's National Agricultural Research Organization (NARO) wanted to develop a two-way communication strategy to enhance farmer participation in experiments with different technologies in order to improve banana production and foster farmer-to-farmer training using communication tools developed in a participatory manner. This research used participatory development communication as a tool to foster the active participation of the community in identifying and solving their natural resource management problems.

Researchers encouraged farmers to form farmers' groups. They then helped the farmers' groups' representatives to identify and prioritize their natural resource management problems within their banana gardens, as well as to find the causes and potential solutions to these problems. The researchers also worked with the farmers to identify their communication needs and objectives regarding the identified problems, the activities that could be undertaken to alleviate the problems and the communication tools that could assist the farmers in sharing their new knowledge with their farmers' groups.

During this process, the researchers discovered that some farmers already had the appropriate knowledge concerning the natural resource management concerns that were raised, but that this knowledge could be reinforced. They also noted that farmers did not have a forum within which they could share information with each other. Thus, there was a need for communication tools.

As a result of the research activities, plots of land that farmers had previously abandoned started yielding good bananas. Farmers also grew confident enough to show their plots to other farmers and to share their knowledge with other farmers in their community. They learned to use communication tools such as photographs, posters, brochures, songs and dances.

After they appreciated the power of belonging to a group, they created a formal farmers' association through which they could search, access and share relevant information and services about community problems. As a result of these activities, the farmers have become proactive instead of passively waiting for external assistance.

A research action project in the basin of the Nakanbe River in Burkina Faso (see 'Water: A Source of Conflict, a Source of Cohesion in Burkina Faso' by Sanon and Ouattara) is another example of a participatory communication approach that brought all the stakeholders together to manage community conflicts related to water.

Approaches to water resource management are often centralized and allow for little participation of the local populations that are actually affected by water issues. Field research conducted in this basin revealed that 50 per cent of modern water sources (hand pumps and modern wells) that had been established by different projects were non-functional as a result of the lack of involvement and ownership by the beneficiaries. The participatory communication approach used by the research team favoured the use of two-way communication and emphasized dialogue among the different stakeholders with regard to water use. The approach also focused on capacity-building at the local level in terms of organization, participation and decision-making in water resource management and conflict resolution, and in implementing or reinforcing local water management committees.

Once again, participatory communication was helpful in identifying solutions to conflict situations in the villages and in setting up or reinforcing social institutions, such as the water management committee. However, it also built community members' confidence in their capacity to address their problems and to seek their own solutions, rather than wait for external assistance. In this case, the central role played by women in the management of water resources in the villages was also recognized.

Another case from Viet Nam (see 'Engaging the Most Disadvantaged Groups in Local Development: A Case from Viet Nam' by Le Van An) describes how a participatory communication approach was used to reinforce community-based natural resource management (CBNRM) research with upland communities. The research started after new policies were put in place by the government to protect forests in the uplands. However, following these measures, only 1 per cent of the land was left available for agricultural production. Thus, local communities who used to practise swidden agriculture had to change their practices and move to sedentary farming. This research initiative tried to help them improve their livelihood in this new context.

Due to these forced changes in their farming system and low access to assets and natural resources, production was low and there were few opportunities for income generation. Participatory communication was used to foster the participation of these local communities in identifying their needs and priorities, and to discuss ways of improving their livelihood. This approach introduced a change from the traditional ways of intervening in a given community from the outside. For the first time, groups of farmers who shared common characteristics and interests were asked what problems they wanted to start working on and what solutions they wanted to experiment with.

The question of reaching the poor and the most disadvantaged groups in the community was a major preoccupation because these people had few opportunities to participate in research or development programmes. Emphasis was put on the participation of poor farmers and of women. Improving the capacity of the commune's leaders and organizations also helped them to apply such participatory approaches with community members so that they could contribute to community plans and activities.

The natural resource management practitioner as a communication actor and facilitator

Establishing relationships

As soon as a researcher or natural resource management (NRM) practitioner first contacts a local community to establish a working relationship, that person becomes a communication actor. The way in which the researcher or NRM practitioner approaches the local community, understands and discusses the issues, and collects and shares the information involves methods of establishing communication with people. The way in which communication is established and nurtured affects how people feel involved in the issues and how they participate, or not, in the research or development initiatives at issue.

Within this framework, it seems important to promote a multidirectional communication process. The research team or the development workers approach the community through community leaders and community groups. The community groups define their relationship with the new resource people, with other stakeholders and with other community groups.

Many researchers still perceive community members as beneficiaries and future end-users of research results. Even if most people recognize that the one-way delivery of technologies to end-users simply has little impact, the shift in attitudes and practices is not easy. For this shift to happen, one must recognize that community members are stakeholders in the research and development process. Therefore, approaching a community also means involving people and thinking in terms of stakeholder participation in the different phases of the research or development process as a whole. Building mutual trust and understanding is a major challenge at this stage and will continue to be so during the entire period of interaction between researchers or practitioners and the community.

Negotiating mandate

Researchers do not come to a community without their own mandate and agenda. At the same time, communities also want their needs and problems to be addressed by resource people who approach them. Most of the time, they will not distinguish between NRM problems, difficulties in obtaining credit or health issues because these are all part of their reality.

Because they cannot address all these issues, researchers and practitioners should explain and discuss the scope and limitations of their mandate with community members. In some cases, compromises can be found. For example, it may be possible to involve other resource organizations that could help to resolve problems which are outside the mandate of the researchers or practitioners. This can often be the case with the issue of credit facilities.

Power relations and gender roles

The management of natural resources is clearly linked to the distribution of power in a community and to its socio-political environment. It is also closely associated with gender roles. This is why social and gender analyses are useful tools for examining the distribution of power in a community. Failure to use these tools may turn the participatory process into a manipulation process or make it selective of only a few individuals or groups in the community.

The chapter concerning communication and sustainable development in Part IV (see 'From Information to Communication in Burkina Faso: The Brave New World of Radio' by Ouattara and Ouattara) refers to a situation where a traditional healer had an unquestionable authority over everything that concerned the community, and used the participatory communication process to reinforce his authority over the community. The members of the intervention team, who were not used to such behaviour, were *de facto* manipulated by the situation. What kind of participation was then possible?

This situation is not exceptional and can only be prevented by identifying the main actors in a community and understanding their roles and relationships before any process is launched. Social analysis, gender analysis and identification of local communication systems, tools and channels should take place before any intervention that involves people in identifying problems and solutions.

Understanding the local setting: Collecting data or co-producing knowledge?

This attitude change has its corollary in methodology. Researchers have been trained in data collection, which emphasizes an extractive mode that does not facilitate participation. Participatory development communication (PDC), however, suggests that researchers or practitioners collaborate with community members and other stakeholders to assemble and share baseline information. This points to a process of co-producing knowledge that draws on the strengths of the different stakeholders.

Participatory rapid appraisal (PRA) and related techniques have been widely used in the field of natural resource management to assemble baseline information in record time with the participation of community members. However, we often find situations where techniques such as collective mapping of the area, transect walks, problem ranking and development of a timeline are still used in an extractive mode. The information is principally used for the researcher's or the project designer's benefits, and little consideration is given to the information needs of the community or to any restitution activity that would ensure the sharing of results.

In these cases, even with the 'participatory' label, these techniques can reinforce a process guided from the outside. PDC stresses the need to adapt attitudes as well as techniques. Co-producing knowledge differs from simply collecting data, and it can play an essential role in facilitating participation in the decision-making processes involved in a research or development project.

Understanding the communication context

Who are the different groups that comprise the local community? What are the main customs and beliefs regarding the management of land and water, and how do people communicate among themselves on these issues? What are the effective interpersonal channels of communication? What views are expressed by opinion leaders or exchanged by people in specific places? What local associations and institutions do people use to exchange information and points of views? What modern and traditional media does the community use?

Here again, we find value in integrating the biophysical, social and communication aspects within an integrated effort to understand the local setting. In the same way that they collect general information and conduct PRA activities to gather more specific information, researchers and development practitioners should seek to understand, with the help of the community, its communication channels, tools and global context.

Identifying and using local knowledge

Identifying the local knowledge associated with natural resource management practices is part of the process of co-producing knowledge. It should also be linked with two other issues: the validation of that knowledge and the identification of modern and scientific knowledge that could reinforce it.

Specific local knowledge or practices may be well suited to certain contexts. In other contexts, it may be incomplete or have little real value. Sometimes, specific practices may have been appropriate for previous conditions, but these conditions may have changed. This emphasizes the importance of validating common local knowledge against scientific evidence and through discussions with local experts or elders, as well as community members. It may also prove useful to combine and blend modern knowledge with local practices to render the latter more effective or more suited to local needs.

Another point is worth noting with regard to the use of local knowledge within a participatory communication approach. The process should not be conducted in an extractive mode by people outside the community. It should be a decision made by the community which is searching for some solutions to a given problem. Two chapters discuss issues related to participatory communication and local knowledge.

In the research conducted by Ouattara and Ouattara on communication and sustainable development in Part IV ('From Information to Communication in Burkina Faso: The Brave New World of Radio'), women from the community were trained as facilitators, and separate meetings were conducted with men and women. The facilitators always explained to the women the importance of their knowledge in the search for solutions to a specific problem.

A modern solution to a given problem will also have more chance of being adopted if a similar practice already exists in the community. For example, in the Sahel, the use of rocks to protect fields against erosion found easy acceptance because people already used dead branches to stop water from invading their fields.

In 'The Old Woman and the Martins: Participatory Communication and Local Knowledge in Mali' in Part III, Diarra reports on a case from Mali where ancient knowledge was used to improve agricultural production and the well-being of the community. An old woman in the village could predict good rain years and drought years and orient farmers to cultivate either on the high tablelands (during years of good rain) or by the side of the river (during years of drought). For this reason, each family had two plots of land, one by the riverside and the other one in the tablelands. Her well-protected secret was that she could make these predictions by observing the height at which martins built their nests in the trees near the river.

After her death, and with the permission of the village authorities, her story was told to the villagers in order to motivate the community to protect the shallow river from too much bank erosion. The villagers agreed to participate in such activities to protect the birds and the knowledge that they brought with them each year. This story tells us how local knowledge may be used in day-to-day lives and also motivate people to better manage their resources.

Involving the local community in diagnosis and planning

Participatory development communication also requires that the local community be involved in identifying a development problem (or a common goal), discovering its many dimensions, identifying potential solutions (or a set of actions) and making a decision concerning which ones to experiment with or implement. It also means facilitating interaction and collaborative action with other stakeholders who should be part of the process.

Traditionally, many researchers and practitioners used to identify a problem in a community and try out possible solutions with the collaboration of local people. With participatory development communication, the researcher or development practitioner becomes a facilitator of a process that involves local communities and other stakeholders in identifying and resolving a problem or achieving a common goal.

The communication process should help people to identify a specific problem that they want to address; discuss and understand the causes of the problem; and identify possible solutions and decide on a set of activities to experiment with. It is useful to stress that this does not happen during the course of a single meeting with community representatives. Time is needed for this process to mature.

In some cases, the departure point is not a specific problem but a common goal that a community sets for itself. As with the problem-oriented process, the community will decide on a set of actions to try to achieve that goal. At the end of both processes, the community decides on a concrete set of actions.

Ideally, development and research objectives should be identified at this point to strengthen and accompany the chosen community initiative. However, generally speaking, these objectives have already been identified in a research and development proposal that was conceived before the process was undertaken with the community. One solution to this problem is to plan a revision of the initial objectives with the community at the start of the research or development project. But, ideally, the administrative rules of donor organizations and the methodological habits of practitioners should be modified to facilitate community participation at the identification phase of what could become a research or development initiative.

Developing partnerships at the local level

The concept of developing partnerships between all development stakeholders involved with local communities is central to participatory development communication. We often find situations where a research or development initiative is conducted with a local community without considering other initiatives that may be trying to engage the same community in other participatory processes. This situation leads to a lot of strain in the communities and can also result in an overdose of participation. Identifying other ongoing initiatives, developing a communication link with them and looking for opportunities for synergy or collaboration should be part of the methodology.

These activities with a local community also allow researchers and practitioners to identify possible partners who could be involved in the research or development process. It could be a rural radio, a theatre group, or an NGO working with the same community. By establishing contacts from the outset, these groups will feel that they can play a useful role in designing the research initiative instead of perceiving themselves as mere service providers.

Local communities interact with governmental technical services, NGOs, development projects, rural media, community organizations and research organizations. All these organizations come to them with their own perspectives; in many cases, there is no link between the various development projects. To maximize the impact of the various local initiatives, it seems important to develop partnerships and build synergies at the community level.

This issue of collaboration is not an easy one. 'From Information to Communication in Burkina Faso: The Brave New World of Radio' by Ouattara and Ouattara raises the issue of collaboration with technicians from governmental services, and, more specifically, the problem of cohabitation of participatory and non-participatory approaches. Technicians are accustomed to executing and implementing programmes already identified by government authorities. Their mandate often consists in making people adopt their recommendations and participate in their programmes, which contradicts the participatory approaches that we want to implement. Therefore, there is a need to provide training in participatory development communication for the partners with whom we want to collaborate.

Constraints and challenges

If the foregoing steps are to be achieved, certain conditions must be met. In Part III, El Dabi gives an example from Egypt where participatory communication could not be introduced (see 'Introducing Participatory Development Communication within Existing Initiatives: A Case from Egypt'). This initiative aimed to develop implementation mechanisms for a strategic development plan in southern Egypt. The barriers that needed to be considered had to be identified, and realistic modifications that would enhance the potential participation of public, private and civil society actors in local development were to be proposed.

Local authorities were to be trained in participatory planning and participatory development communication, which was to be introduced by undertaking an assessment of the communication problems, channels and materials of all stakeholders; by designing a training programme for the immediate stakeholders to understand and apply the methodology in their communities; and by providing assistance to the immediate stakeholders to develop a strategy for their communities' development plans.

However, several obstacles hindered the implementation of this plan. First, participation was perceived as a process to allow stakeholders to voice their problems, not as a mechanism for them to look for ways of overcoming these problems. Second, the project did not allocate sufficient time to perform communication assessments or to conduct the training in a participatory way. Third, but not least, insufficient resources were allocated for the institutionalization of participatory approaches. As a result, it was not possible to introduce participatory communication in this particular context.

The chapter by Sow and Adjibade, also in Part III, provides examples of some of the practical difficulties we face when implementing participatory communication, particularly in a rural context (see 'Experimenting with Participatory Development Communication in West Africa'). This chapter also outlines some of the conditions that must be met. The authors raise the importance of prior knowledge of the local language and of the communication channels and tools used in the community; of negotiating with the men in a community to identify the conditions under which women can participate in specific activities; of time and distance considerations; of the development of partnerships with local organizations; of considering local authorities (traditional, administrative and family); and of harmonizing the understanding of participatory communication among the facilitators, decision-makers and participants involved. The chapter also reminds us that more time must be allocated to implement participatory communication processes than that usually planned in development projects.

Sow and Adjibade also remind us that participatory communication activities usually lead to the expression of the need for material and financial support to implement the solution(s) identified during the process. Provision must be made somewhere to answer these needs, whether as part of the initiative itself or through partnerships, otherwise the process stops where it should begin. The chapter shows that it is not useful to separate participatory communication activities from development activities and that resources must be planned to support these two complementary aspects.

Another chapter in Part III, 'Strategic Communication in Community-Based Fisheries and Forestry: A Case from Cambodia', presents the experience of introducing communication within a participatory natural resource management project in the Tonle Sap region of Cambodia. The initiative, as described by Thompson, emphasized communication as an integral part of its activities. It applied a wide range of tools and methodologies to inform, educate and promote participation. However, in the absence of a global communication plan, these efforts remained limited. Participatory development communication approaches can identify the best-suited community interventions and the management options for each community to ensure community-based natural resource management. However, the different communication activities must be integrated within a systemic and strategic plan to achieve their potential effectiveness.

Supporting natural resource management through communication strategies and tools

With PDC, communication strategies are developed around an initiative that has been identified by the community in order to tackle a specific problem or to achieve a common goal.

After community members have gone through the process of identifying a concrete initiative that they want to carry out, the next step is to identify both the various categories of people who are most affected by this NRM problem and the groups who might be able to contribute to the solution. They may be either specific community groups or other stakeholders who are, or could be, involved.

Addressing a general audience such as 'the community' or 'the farmers' does not really help to involve people in communication. Every group who makes up the community, in terms of age, gender, ethnic origin, language, occupation, and social and economic conditions, has its own characteristics, its own way of perceiving a problem and its solution, and its own way of taking actions. Likewise, communication needs will vary considerably within each specific community group or stakeholder category.

In all cases, it is important to pay particular attention to the question of gender. In every setting, the needs, social roles and responsibilities of men and women are different. The same is true of the degree of access to resources, participation in decision-making processes and the way in which they will perceive a common problem or potential solutions. The same is true for young people. There is often a sharp distinction between the roles and needs of girls and older women, as well as between the perceptions of older men and young people in the face of the same problem. Consequently, their interests are different, their needs are different, the ways in which they see things are different, and their contribution to the research or development initiative will also be different.

Communication needs and objectives

Development needs can be categorized broadly into material needs and communication needs. Any given development problem, as well as the attempt to resolve it, will present needs related to material resources and to the necessary conditions to acquire and manage these resources. However, there are complementary needs that involve communication in order to share information, influence policies, mediate conflicts, raise awareness, facilitate learning, and support decision-making and collaborative action. Clearly, these material and communication aspects should go hand in hand and be addressed in a systemic way by any research or development effort.

Nevertheless, participatory communication puts a greater focus on the second category of needs and ensures that they are addressed, together with the material needs that the research or development effort is concentrating on.

Communication objectives are based on the communication needs of each specific group concerned with the identified problem or set of activities that will be part of the project. These communication needs are put together by all the stakeholders involved and go through a selection process. The choices can be made on the basis of the needs that are most urgent or those that are most susceptible to action. These needs are then translated into a series of actions that should be accomplished to address each requirement.

Generally, in the context of NRM, these actions are linked to one or another of the following communication functions: raising awareness; sharing information; facilitating learning; supporting participation, decision-making and collaborative action; mediating conflicts; and influencing the policy environment.

Using communication tools in a participatory way

We often find situations where researchers or practitioners who want to use communication in their activities intend to produce a video, a radio programme or a play without first trying to identify how it will contribute to the research or development initiative. The expression 'communication tools' in itself implies that they are not the 'product' or the 'output' of the communication activities.

Participatory development communication takes another perspective. It leads participants through a planning process, which starts with the identification of the specific groups as well as their communication needs and objectives. The research or development team, together with community members and other stakeholders, then identifies the appropriate communication activities and tools that are needed to reach these objectives. It is a collective and consensus-building process, not a strategy developed outside the social dynamic.

PDC also puts traditional or modern media on the same level as interpersonal communication and learning experiences, like field visits or farmers' schools. The importance of using these communication tools in a way that will support multiple-way communication must, of course, be clearly stated at the outset of the project.

We have to consider two situations regarding communication tools. We often discuss this issue with the perspective of research and development teams using communication tools to support their activities. However, community members must also be able to use these communication tools for their own purpose.

Three criteria seem particularly useful in selecting communication tools: their current use in the community, the cost and constraints of their use, and the versatility of their uses. Whenever possible, we should first rely on the communication tools already in use in the local community for exchanging information and viewpoints, or the tools people are most comfortable with. Considerations of cost and sustainability and of different kinds of use should also be examined before making a decision.

In terms of the tools used by NRM practitioners, the chapters in this volume highlight various mixes of interpersonal communication and community media: community thematic discussions, participatory theatre, radio and participatory communication, farmers' field schools, video, photography, illustrations and community meetings.

Community thematic discussions

Almost everybody considers community thematic discussions as an important communication tool. But these discussions also imply a process and specific attitudes on the part of the facilitator.

Thiamobiga's chapter 'How the Parley Is Saving Villages in Burkina Faso' in Part IV gives us two examples of facilitators and the processes that are at work when using this tool. The chapter also describes a specific case in which community discussions were instrumental in managing bushfires and preserving the natural environment. Thiamobiga stresses the link between participatory communication and the '*parley*' – a traditional way of addressing issues and problems at the community level.

Participatory theatre

Participatory theatre also appears to be a favorite communication tool. In 'Burkina Faso: When Farm Wives Take to the Stage', also in Part IV, Thiamobiga discusses the experiences of a theatre of women farmers and explains the process of using theatre debate as a participatory communication tool. Theatre debate, in which a discussion follows the play and some parts of the theatre are played repeatedly following comments, was used to tackle soil fertility problems and was employed by women as a form of empowerment. The idea, at first, was to use this tool to help women voice their concerns and to illustrate causes and potential solutions associated with the problems. But the process initiated an empowerment process through which the women decided that they would play themselves.

There is a traditional ceremony performed in times of drought, when women are allowed to dress up as men to call for rain. The participating women wanted to refer to that ceremony so that they could bring forward topics that could be addressed directly by the men of the community without the risk that they would be offended (during the ceremony, men do not have the right to take offence).

By participating in the discussion to identify the problems related to soil fertility and by learning to express themselves as actors in a play, the women not only put their plots' soil fertility problem on the community agenda, but also gained self-confidence and became more assertive. The impact was also stronger because, in this case, community members were addressing other community members about common issues, rather than development actors from the outside promoting solutions.

At the same time, such involvement from community members, in this case women farmers, raised expectations that could not be met after the completion of the intervention. There was no direct follow-up, and although the experience was empowering for the participants, there was little impact at a broader level. This issue addresses the importance of planning for scaling up a specific intervention at the very beginning of the planning phase.

Radio and participatory communication

Ouattara and Ouattara's chapter 'From Information to Communication in Burkina Faso: The Brave New World of Radio' in Part IV reminds us not only that radio is the most common media in rural Africa, but also that it is still underdeveloped as a participatory communication tool. The research first started to use radio to promote the involvement of community members, together with a communication strategy based on 'endogenous communicators'. The programmes were designed on the basis of interviews and discussions conducted with community members and a communication team that included a radio producer, a representative of the farmers and a representative of a development structure active in the region (e.g. a development initiative or NGO). The development representatives were trained to prepare the field activities, to participate in the production of programmes and to collect feedback following the broadcast.

Other activities were then introduced to complement the media approach and to reinforce community participation. NRM problems and potential solutions were identified through discussion groups consisting of women, young people and adult men. In each locality, a committee which included local development actors was set up to define activities that could respond to prioritized needs. At the village level, a communication committee was involved to facilitate implementation. These field activities were then used in the production of radio programmes that were broadcast by the local rural radio station between two field trips. Essential questions asked by community members were discussed in these programmes. Specialists would also comment on these questions and participate in a dialogue with community members.

These activities have opened up a space for dialogue in the communities about natural resource management problems, while promoting synergy between the different development actors working in the same locality. The decisions resulting from this dialogue and the exchanges of information have involved community members and engaged them in a process in which they actively search for solutions instead of passively waiting for external assistance (e.g. by getting rid of pest-infesting orange trees, by resuming a dialogue between farmers and pastoralists and by enabling women to have a voice at community meetings).

Nevertheless, this experience also showed the difficulties associated with a participatory approach – namely, the danger of raising expectations without the possibility of responding to identified needs. For example, after prioritizing the lack of access to drinking water in the locality, community members and the team did not have many solutions to offer because the communication initiative was not associated with any specific development action or equipped with a structure that had the technical and financial resources to answer those needs.

In 'And Our "Perk" Was a Crocodile: Radio Ada and Participatory Natural Resource Management in Obane, Ghana', Larweh describes a situation in which a community was confronted with a decision to either migrate or renew its waterway, which was now choked by weeds, trees and debris. In fact, it no longer existed for most of the year. The community radio was part of a process where the community discussed the situation and decided to clear 40 years of accumulated debris. Neighbouring communities joined in the collective work. Four years later, the river irrigates the fields and is navigable. Through participatory communication, the community was able to unite around a single goal and transform their situation using their own means.

Video, photography, posters and brochures

In other situations, especially those aiming at empowerment, community members will take the lead in using communication tools or in taking decisions regarding the design, production and use of communication materials. Community input is well documented in Odoi's chapter on introducing villagers to video production, photography and the making of posters and brochures (see 'Communication Tools in the Hands of Ugandan Farmers' in Part IV).

The chapter describes the experience of farmers who edited a video that had been produced by the research team to share the results of their activities with other farmers. In this case, the farmers rejected the video because they were convinced that they could do a better job in delivering their own messages and experiences. The farmers first had a meeting to decide who should show what and how, set a date for the new recording, and signalled to the researchers when they were ready. This would have never happened if the researchers had not undertaken a process of participatory communication with the farmers – a clear manifestation of the farmers' empowerment.

The same thing happened with the photographs. After the pictures were developed, the farmers rejected them and started anew. For some time, research team members were discouraged and wondered when the production process would end.

As for printed materials, Odoi's chapter explains that the farmers could easily produce a brochure, but that the production of the poster was more difficult because this was a new concept for them. On examining a poster depicting proper water and sanitation practices placed at the entrance of the community hall, farmers said that it was teaching someone how to write. Clearly, the tool was not adapted to this specific community.

Tools should also be considered from the viewpoint of their usage. In a case from Lebanon by Hamadeh et al (see 'Goats, Cherry Trees and Videotapes: Participatory Development Communication for Natural Resource Management in Semi-Arid Lebanon' in Part III), a video and a local users' network inspired by a traditional way of communicating and resolving issues were used to manage conflicts and to facilitate the expression of views by marginalized people.

This research focused on understanding changes in resource management systems in an isolated highland village that was in the process of moving from a traditional cereal livestock-based economy to a rain-fed stone-fruit production system, and of improving prospects for sustainable community development. Community members were involved at different stages, and capacity-building was sought by establishing a local users' network.

The network acted as a medium to bring together the different users (e.g. cherry growers, flock owners and women), researchers, development projects, government officials and representatives of traditional decision-makers. It also supported participation, communication and capacity-building efforts.

The project used a traditional way of communicating and resolving dilemmas called *majlis*, in which issues are brought up within the community. As the network grew, so did the researchers' understanding of communication principles and the need to develop specialized sub-networks. Three sub-networks were developed, two of them dealing with the main production sectors in the village (livestock and fruit growing), while a third one addressed women's needs.

Tools and practices were mainly interpersonal: round-table meetings, community outreach by students, joint field implementation of good NRM

practices, and workshops on different NRM themes. Short video documentaries were also produced and used during meetings.

Video was experimented with as part of an effort to involve the community
in dialogue and conflict resolution. Marginal groups could express their
points of view and the images helped to shed light on some aspects of conflict
and dissent. The videos were shown in the presence of all parties, and the
showings were followed by discussions that were also filmed and documented.
A revised video that included the earlier discussions was then shown to the
whole village until a positive dialogue started to emerge from the audience.
Video was also used to highlight the economic productivity of women and to
prompt discussions.

It was found that video helped marginal groups, who were usually shy in
formal meetings, to express themselves. Videos were also found to be useful
in generating discussions and awareness among and between different people
and factions.

Influencing or implementing policy

Promoting poverty alleviation, food security and environmental sustainability
also requires changes in the institutional and legislative environment. Local
and national authorities, policy-makers and service providers all contribute
to shaping and enforcing the regulatory environment in which the required
changes must take place. Therefore, it is important to facilitate dialogue
at that level in order to gain support for the initiatives developed by local
communities.

Facilitating dialogue at the level of local and national authorities, policy-
makers and service providers, or between communities and the policy
environment, or advocating changes in the policy environment, are other
roles for the NRM practitioner or researcher as a communication actor.

Two chapters from Cambodia in Part III ('Communication Across
Cultures and Languages in Cambodia' by Kimhy and Pinreak and
'Talking with Decision-Makers in North-Eastern Cambodia: Participatory
Development Communication as an Evaluation Tool' by Kimhy) provide
examples of how participatory communication can influence policy and help
with its implementation.

In 'Talking with Decision-Makers in North-Eastern Cambodia: Participatory Development Communication as an Evaluation Tool', Kimhy shares
the experiences of indigenous communities who evaluated a natural resource
management initiative implemented by the government and presented their
findings to government officials. The presentation also included recommendations to the government in a context where government representatives
usually tell communities what they should do. In this case, evaluation was used
both as an empowerment tool for community members and as an advocacy
tool for influencing the government.

In 'Communication Across Cultures and Languages in Cambodia', Kimhy
and Pinreak describe a situation in which a team was visiting villagers, in the

context of a new land rights extension law, in order to inform them of their rights and of existing laws. Transferring information or knowledge across cultural and language barriers is difficult; but it is much more difficult when some of the concepts do not even exist in the vocabulary of the people with whom we want to converse. This was the case in this initiative because concepts such as laws and land titles did not exist for the indigenous communities with whom the team was interacting. At the same time, communicating these concepts was an important task because powerful interests were taking away their lands and forests, and the communities did not know what to do.

At first, the team, who did not speak the indigenous languages and had prepared its information material without any community involvement, failed to reach the communities. It then experimented with a participatory communication approach, involving community members in preparing the sessions and communication material. It also included indigenous people as full members in the land rights extension work, which changed the team's whole approach to working with communities.

It is interesting to note that the team also used a tool called the 'Livelihood Framework' in the course of its discussions with the communities. It presented ideas expressed by the community in pictures that were painted and then revised by the community. The visuals, in this case, greatly assisted in the discussions and expressions of different viewpoints.

A chapter from the Philippines by Torres (see 'Paving the Way for Creating Space in Local Forest Management in the Philippines' in Part III) tells how participatory communication helped to implement community-based natural resource management with indigenous people. When community-based forest management was adopted as a national strategy in the Philippines, issues started to emerge with regard to the readiness and capacity of communities to handle the delegated tasks and functions.

On its part, an upland people's organization called the Bayagong Association for Community Development was able to assert, legitimize and sustain its control over a piece of forestland that it had been occupying *de facto* for years. To do so, community members underwent a process of participatory resource management planning. In the gender-sensitive PRA methodologies that were used during a year of action research, participatory communication was a core process.

This participatory experience helped participants to get a better grasp of their resource quality, to assess their own capacities and weaknesses, and to identify internal and external threats, as well as how these could be handled. It enabled them to gain the knowledge, attitudes and skills that were necessary to deal with the management of their forestland and to develop rational approaches to forest management. But they also learned to become more open and assertive about their rights, and soon became empowered in identifying and addressing their needs by using locally available resources before turning to outside sources for assistance.

Participatory development communication played a critical role in tempering the socio-political environment so that a climate favourable to the community's takeover of the forestland be created. However, success was not only due to communication. Other factors such as social capital, policy presence and assistance by external actors also played a role. What is unique is that participatory communication enabled the evolution of a 'participation-as-engagement' process veering away from the usual 'participation-as-involvement' process.

A chapter from Indonesia by Jahi (see 'From Resource-Poor Users to Natural Resource Managers: A Case from West Java' in Part III) tells of a research project that originated from a question that researchers asked themselves while they were doing a baseline study in a remote rural area. The researchers wondered whether poor farmers and landless farm labourers could participate in managing a strip of public land that stretched out along a river and thus be able to derive benefits from that activity.

By law, farming activities were prohibited on this land. Only grass cultivation was allowed on the riverbanks, which had been raised to prevent local flooding. However, regardless of the regulations, landless farmers continued their farming activities on the riverbanks. Meanwhile, officials of the Department of Public Works kept enforcing the regulations and eradicating the crops. A consensus then developed. The farmers could continue their activities provided that they grew grass at least 1m away from the river's edge. Sheep rearing was encouraged.

The researchers established links between university researchers, local government officials, extension services, village governments and local farm communities. Communication materials such as slide shows, posters and a comic book were developed and tested with farmers and extension workers. Different topics were developed for different audiences. For example, presentations on the potential of raising sheep were prepared for local policy-makers and decision-makers, and aspects of sheep production and rural family budgets were covered in productions for extension workers and farmers.

Capacity-building for livestock extension workers and farmer leaders was then provided. In-kind loans in the form of sheep were provided to the farmers, who agreed to return a certain number of the offspring to the project. Supervision and backstopping activities were also provided to farmer leaders, who agreed to share the information with other farmers after they had acquired enough experience.

Farmer-to-farmer communication was encouraged and supported. Indeed, it was found to be a more efficient way of raising farmers' interest than the methods that researchers or extension workers used to employ. The experience also raised public and private interest in supporting economic activities, such as sheep rearing, in the district. Fifteen years after the beginning of the project, livestock production in the district has developed significantly and small farmers can still earn their living in this way.

In the policy arena, there are also situations where participatory communication must coexist with inadequate policies and help to seek solutions. In a chapter presenting the case of the Kahuzi-Biega National Park in the Democratic Republic of Congo (see 'Conserving Biodiversity in the Democratic Republic of the Congo: The Challenge of Participation' in Part III), we find a situation in which a conservation measure (the creation of a park to protect a unique ecosystem and a population of mountain gorillas) was implemented in a top-down way without the involvement of the population living in, or on the fringes of, this newly protected territory. In such a conservation model, the population is excluded from managing natural resources. Consequently, local people do not participate in or support the new unpopular measure.

In this case, an alternative plan had to be developed. Using environmental communication and in collaboration with the population living in the area, community development activities compatible with the conservation of the park and its natural resources began to be planned and implemented. These activities quickly led to the development of mechanisms for participatory management. Soon, some 200 village parliaments were set up to facilitate the process. Not only have local opinions changed towards the park, but the communities have also started taking charge of its protection.

The promotion of policies goes hand in hand with collective action. One chapter in Part III (see 'Giving West African Women a Voice in Natural Resource Management and Policies' by Ouoba) depicts the daily life of a rural woman of the African Sahel and her difficulties with regard to natural resources: lack of access to water and fuelwood; problems of soil fertility; and lack of access to landownership. It also tells of the efforts of a rural women's association to find collective answers to these individual problems. The solutions to the NRM problems experienced by rural women must come from their own efforts, a process that can be facilitated by participatory communication. We can see that such initiatives are part of an empowerment process in which marginalized people, who are not used to expressing themselves on such issues, develop confidence and learn to voice their difficulties and needs, and to formulate specific actions to address these needs.

Capacity-building issues

Participatory development communication and, more broadly, the use of communication in the context of participatory development or participatory research have to be appropriated by NRM researchers and practitioners. This should also be the subject of exchanges and discussions with other stakeholders who participate in these activities, such as community members.

Five chapters in Part V ('Forging Links between Research and Development in the Sahel: The Missing Link'; 'Isang Bagsak South-East Asia: Towards Institutionalizing a Capacity-Building and Networking Programme in Participatory Development Communication for Natural Resource Management'; 'Implementing Isang Bagsak in East and Southern Africa'; 'Implementing Isang Bagsak: Community-Based Coastal Resource Management in Viet Nam'; and 'Implementing Isang Bagsak: A Window to the World for the Custodians of the Philippine Forest') discuss the implementation of Isang Bagsak, a learning and research programme in participatory development communication. The expression 'Isang Bagsak' comes from the

Philippines and means reaching a consensus, an agreement. Because it refers to communication as a participatory process, it has become the working title for this initiative.

The programme seeks to increase the capacity of development practitioners, researchers and associated stakeholders who are active in the field of environment and natural resource management in using participatory development communication to work more effectively with local communities and associated stakeholders. It aims at improving the capacity of practitioners and researchers to communicate with local communities and other stakeholders, and to enable them to plan, together with community members, communication strategies that support community development initiatives.

The programme combines face-to-face activities with a distance-learning strategy and web-based technology. With the distance component, the programme can answer the needs of researchers and practitioners who could not easily leave work for a campus-based programme. It is currently implemented in South-East Asia and Eastern and Southern Africa, and is in the start-up phase for the African Sahel.

In South-East Asia, Isang Bagsak is being implemented by the College of Development Communication of the University of the Philippines at Los Baños. It works in the Philippines, Cambodia and Viet Nam.

In the Philippines, the programme is implemented in partnership with the Legal Assistance Centre for Indigenous Filipinos (PANLIPI), an NGO devoted to legal assistance to indigenous Filipinos. The goal of the programme is to bring indigenous people into the mainstream of learning about NRM.

In Viet Nam, capacity-building in participatory development communication aims to improve approaches to coastal resource management, understand how to influence local policies in order to improve participatory management of coastal resources, and form a national network in community-based coastal resources management. Furthermore, a Vietnamese version of the programme, called *Vong Tay Lon*, is being prepared.

In Cambodia, participants come from the new forest administration department. This national body is responsible for formulating and implementing forest policies, which affect more than half the country's total land area. Its mandate includes the elaboration of a statement on National Forest Policy, which will be based on a consultative process inclusive of all stakeholders in national forestry policy formulation.

In Southern and Eastern Africa, the programme is implemented in Zimbabwe, Malawi and Uganda by the Southern Africa Development Community Centre of Communication for Development (SADC-CCD). By building capacity in participatory development communication, the programme aims to facilitate collaboration among decision-makers, planners, development agents and communities in order to improve the management of the environment and natural resources, as well as the research and development initiatives. The programme works in partnership with the National Agriculture Research Organization in Uganda, with a national resource management research initiative called the Desert Margins Initiative in Malawi, and with the Department of Agricultural Research and Extension in Zimbabwe.

Another programme is being initiated with an agroforestry network in Senegal, Burkina Faso and Mali, which will be led by the Sahel Programme of the International Centre for Research in Agroforestry (ICRAF-Sahel). In the Sahel, the starting point for implementing Isang Bagsak is the realization that new agroforestry technologies which should improve lives are not being widely adopted despite efforts to convince people to do so. The objective of the programme is to reinforce the capacities of the different actors so that they can co-produce and co-disseminate new knowledge in collaboration with all involved stakeholders.

El Hadidy addresses the issue of capacity-building in the context of the Arab region, but situates participatory development communication within the larger framework of participatory development (see 'Reflections on Participatory Development and Related Capacity-Building Needs in Egypt and the Arab Region', also in Part V). This chapter advocates that practitioners should engage in a critical reflection of their practices. It states that, in itself, the 'delivery of resources' mode of operation in the form of transfer of know-how and skills is not sufficient. It also indirectly implies that resources are transferred from those who have them to those who do not, instead of recognizing that every practitioner has skills and abilities that need to be brought to the surface. Unlike capacity-building that requires a 'how-to' approach, such as proposal writing or business planning, capacity-building in participatory development communication should focus on recognizing that communication is a natural process. It advocates an approach based on the facilitation of resourcefulness rather than providing resources. This process goes hand in hand with the documentation and discussion of local participatory practices.

The chapter by Thompson and Acunzo from the United Nations Food and Agriculture Organization (FAO) (see 'Building Communication Capacity for Natural Resource Management in Cambodia', Part V), presents a national capacity-building effort in Cambodia that was designed to help a communication team (which brought together the staff of two ministry communication units) design and implement targeted information and communication interventions to support plans and efforts made by local communities for natural resource management. The strategy was based on implementing information and communication strategies at the field level and providing in-service training in pilot sites. The learning process included participatory analysis, training of villagers, material design and production, as well as monitoring and evaluation for improving agricultural and fishing practices. The chapter describes the constraints and lessons learned in the course of this initiative. Among the challenges encountered, the authors mention that the lack of operational budgets makes it difficult for the newly trained communication team to apply their new skills. Similar trends have also been observed in other capacity-building initiatives. We need to address this situation as part of capacity-building efforts and to examine how the latter can be better integrated within the operational plans of targeted institutions.

Finally, capacity-building and co-learning efforts should also document and promote a systematic use of participatory development communication in natural resource management.

It is important to state that there is no single all-purpose recipe to start a participatory development communication process. Each time, we must look for the best way to establish the communication process among different community groups and stakeholders, and use it to facilitate and support participation in a concrete initiative or experimentation driven by a community to promote change.

However, participation in the planning process is important. We already stressed that using participatory development communication demands a change of attitude from researchers and development practitioners. Traditionally, the way in which many research teams and practitioners used to work was to identify a problem in a community and experiment with solutions, with the collaboration of local people. On the communication side, the trend was to inform and create awareness of the many dimensions of that problem and of the solution that community members should implement (from an expert viewpoint). This kind of practice has had little impact; but many researchers and development practitioners still work along those lines.

Working with participatory development communication means involving the local community in identifying the development problem (or a common goal), discovering its many dimensions, identifying potential solutions (or a set of actions) and making a decision on a concrete set of actions to experiment with or implement. It is no longer the sole responsibility of the researcher or development practitioner.

Participatory development communication supports a participatory development or research-for-development process. Such a process is usually represented through four main phases, which, of course, are not separated but flow into one another. These stages are problem identification, planning, implementation, and monitoring and evaluation. At the end of the process, a decision is made that either consists of going back to the beginning of the process (problem identification) and starting another cycle or revising the planning phase, or scaling up efforts and starting another planning, implementation and evaluation cycle.

The participatory development communication model supports this process. Bessette (2004) discusses the most common steps when planning and implementing participatory development communication in a natural resource management context:

- Step 1: establish a relationship with a local community and understand the local setting.
- Step 2: involve the community in identifying a problem and potential solutions, and in carrying out a concrete initiative.
- Step 3: identify the different community groups and other stakeholders concerned with the identified problem (or goal) and initiative.
- Step 4: identify communication needs, objectives and activities.
- Step 5: identify appropriate communication tools.
- Step 6: prepare and pre-test communication content and materials.
- Step 7: facilitate the building of partnerships.
- Step 8: produce an implementation plan.

- Step 9: monitor and evaluate the communication strategy, and document the development or research process.
- Step 10: plan the sharing and utilization of results.

The process, nevertheless, is not sequential. Some of these steps can be implemented in parallel or in a different order. They can also be defined differently depending upon the context. It is a continuous process, not a linear one. But the steps can guide the natural resource manager researcher or practitioner in supporting participatory development or research through the use of communication.

Institutional aspects

Implementing participatory development communication faces the same constraints as the participatory development process that it supports: it demands time, resources and practical modalities that can only result from negotiating with the donor organizations involved.

Initiating the process

In traditional development culture, financial support often comes after the revision and acceptance of a formal proposal, whether it is a research for development proposal or a development project proposal. In order to go through the different levels of revision and acceptance, the proposal must be clear and complete. The development problem or goal must be clearly identified and justified, the objectives must be outlined with precision and all the activities must be detailed. The full budget, of course, must figure in the proposal with all its budget notes.

Although some organizations are rethinking the process and promoting programme instead of project orientation, in most cases this is the situation we face. It is important to put this issue on the agenda of donor organizations and to demand a review of the procedure: if we want to develop a participatory development process and have community members and other stakeholders have their say during all phases of the process, starting with the identification and planning phases, this means that we need the time and resources to do so.

In the meantime, we can identify two modalities that can be proposed to the donor organization. The first consists of putting together a pre-proposal that will seek to identify and plan the project with all stakeholders. The second modality – which is really a second choice in case the first one is not possible – consists of building the proposal in a way that will permit its revision with community members and other stakeholders.

Changes during implementation

A participatory development or research process cannot be planned in the same way as the construction of a road: as participation is facilitated and more feedback is gathered, a wider consensus develops and decisions are made; things change. This is why it is always an iterative process and we must have the possibility of changing plans as we go along in order to attain the objectives that have been identified.

This must also be discussed with the donor organizations involved since traditionally, once a proposal has been accepted, nothing can be changed.

Time considerations

The length of activities is also a problem. Proposals often have to be developed within a two- or three-year time frame. But participation takes time, and in many cases, this span is barely enough to really start the process. Thus, even if the expected results have not yet happened, it is necessary to identify the progress made by the research and development activity and to build the case for the continuation of support. This also underlines the importance of a continuous evaluation mechanism when implementing the process.

Regional perspectives

Two chapters in Part II, from Africa and Asia, examine participatory development communication from a regional perspective.

In Asia, Quebral, who was the first to use the term 'development communication' more than 30 years ago, retraces the evolution of participatory approaches to development communication (see 'Participatory Development Communication: An Asian Perspective'). The chapter situates this evolution in the context of the communication units, departments and colleges in Asian universities and from the perspective of fighting poverty and hunger. Quebral notes that development communication does not identify itself with technology *per se*, but with people, particularly the disadvantaged in rural areas. Participatory development communication uses the tools and methods of communication to provide people with the information they need and to reinforce their capacity to make their own decisions.

The chapter insists on recognizing the beginnings of development communication and on expanding upon earlier achievements. Older models retain their validity in certain situations and can still be used when appropriate. The chapter also presents lessons and observations learned through this Asian experience.

In the context of natural resource management, Quebral insists on the importance of a balance between technology and people's empowerment, and on how participatory development communication can help people to zero in on their problems. It can also support people's choice of the technologies that they wish to experiment with.

Offering another regional perspective, in 'Participatory Development Communication: An African Perspective' Boafo describes and analyses the application of participatory development communication within the African context and stresses the linkages between communication and the different dimensions of development on the continent.

Since the 1960s and 1970s, many development communication strategies and approaches have been used in numerous development programmes by a large number of development organizations. However, greater efforts remain to be made to address the constraints facing the practice of participatory development communication, particularly in the context of rural and marginalized communities, where the majority of the populations in most African countries reside.

In this context, notes Boafo, community communication access points and traditional media are of particular importance. Effective applications of participatory development communication approaches and strategies at the grassroots and community level should necessarily involve the use and harnessing of these communication resources. With their horizontal and participatory approaches, they can contribute effectively to enhancing participation in cultural, social and political change, as well as in agricultural, economic, health and community development programmes.

Conclusions

In the field of natural resource management, participatory development communication is a tool that reinforces the processes of participatory research and participatory development. It aims to facilitate the participation of communities in their own development and to encourage the sharing of knowledge needed during these processes. It integrates communication, research and action within an integrated framework. Furthermore, it involves researchers, practitioners, community members and other stakeholders in the different phases of the development process. But most importantly, participatory development communication points out that natural resource management must be directly linked to the agenda of communities and must seek to reinforce their efforts to fight poverty and to improve their living conditions.

For communication to be effective in addressing the three interlinked development challenges of poverty alleviation, food security and environmental sustainability, it must fulfil the following functions: ensure true appropriation and ownership (not just the buy-in) by local communities of any NRM research or development initiative; support the learning needed to realize the initiative and facilitate the circulation of relevant knowledge; facilitate the building of partnerships, linkages and synergies, with the different development actors working with the same communities; and influence policy and decision-making processes at all levels (family, community, local and national).

To achieve these objectives, a major effort is required in capacity-building – more specifically, in participatory learning – for practitioners in the field of natural resource management. Development workers, NGOs, researchers, extension workers and government agents responsible for technical services need appropriate communication skills. The ability to work with local communities in a gender-sensitive and participatory way, to support learning processes, to develop partnerships with other development stakeholders and to affect the policy environment should be recognized as equally important as the knowledge needed to address technical issues in natural resource management.

At the same time, field practitioners, researchers and community members who are involved in natural resource management initiatives are experienced in the use of communication within participatory research and development initiatives. There is no recipe that can be used in all situations; but there is much to learn from sharing, discussing and reflecting on our own experiences. As advocated in El Hadidy's chapter 'Reflections on Participatory Development and Related Capacity-Building Needs in Egypt and the Arab Region' in Part V, we should use an approach that facilitates resourcefulness rather than provides resources.

Of course, such a process goes hand in hand with the documentation and discussion of our natural resource management and participatory development communication practices. This is why initiatives such as the Isang Bagsak programme and the FAO initiative in Cambodia should be developed, supported and multiplied in various contexts and situations. It is also why participatory learning in participatory development communication for both practitioners and stakeholders should be on the agenda of every organization supporting NRM research and development initiatives. It is only through such efforts that we can make participatory development happen, not only at the level of our discourses, but where natural resource management occurs in the field. It is also through such efforts that we can make sure that local actions can have a global impact by influencing the policy environment and making the knowledge available to those who really need it.

Finally, it is through such efforts that we can promote and cultivate the values which are at the core of our work, including the principle that people should be able to participate fully in their own development. In a recent report, Nora Quebral (2002) insisted that:

> *We now need to explicate those values more finely and cultivate them more rigorously in our actions. Our training procedures may have overly stressed skills at the expense of values. We need to make values more explicit, to deliberately pair them with the corresponding skills, if necessary. My first challenge, then, to development communicators is to make development communication values more pronounced in their practice.*

The same challenge can be extended to NRM practitioners and researchers: we need to make participatory development happen if we are to support communities and governments in their efforts to address the three interlinked development challenges of poverty alleviation, food security and environmental sustainability. Participatory development values, local and modern knowledge in natural resource management, and communication skills must be combined for this to happen.

References

Bessette, G. (2004) *Involving the Community: A Facilitator's Guide to Participatory Development Communication*, IDRC, Ottawa, Canada, and Southbound, Penang

Quebral, N. C. (2002) *Reflections on Development Communication (25 Years Later)*, College of Development Communication, University of the Philippines at Los Baños (UPLB), Los Baños, the Philippines

II

Regional Perspectives

Participatory Development Communication: An Asian Perspective

Nora Cruz Quebral

Asia is a region of many faces. This chapter speaks of the Asian experience with participatory development communication (PDC) from the perspective of one of its sub-regions – that grouping of nations known collectively as South-East Asia. More precisely, the chapter delimits itself to PDC as interpreted by communication units set up in South-East Asian colleges and universities as part of their agricultural extension or outreach function. The affinity of these units with the media offices of extension services in US land grant colleges[1] is obvious. Nonetheless, they have evolved – and continue to do so – into hybrid structures more appropriate to their cultures and to the state of knowledge in the field of development and communication.

There are other Asian viewpoints on PDC, notably in India and other parts of South Asia. They will be similar to the South-East Asian experience in some ways, different in others. All have lessons to offer in the continuing delineation of the relationship between communication and human development.

University communication units

The communication units referred to are found in Indonesia, Malaysia, the Philippines, Thailand and Viet Nam. They are at the incipient stage in transitional societies such as Cambodia, Laos and, perhaps, Myanmar. Regarded as adjuncts to the biological and physical science departments of their universities, the older units were initially tasked with extending the results of research generated by those departments, with some public relations and publicity jobs for administrators thrown in. This they were expected to accomplish through the media; hence, they were staffed with writers, editors, artists, audio and video specialists. Face-to-face interaction with farm families was considered something that extension workers do and, therefore, was outside the mandate of the communication staff.

An obsolete, ante-millennium model of PDC, you say? It is alive and kicking in South-East Asia in spite of globalization, state-of-the-art information and communication technology, participatory communication activism, terrorism and all other change-inducing phenomena now sweeping the world. Evidence of its endurance may well be mirrored, in greater or lesser degree, among the organizations usually represented in international events: the focus on proffered technology, the sidelining of communication practitioners within the organization, the forced merger of communication with other seemingly related units for reasons of efficiency, economy or whatever.

Evolution of participatory development communication in South-East Asia

There is another side to the picture, however. It was in this type of communication unit that PDC as study and practice first saw light in South-East Asia, was nurtured and then diffused to other developmental fields such as health and the environment, among others. At least seven of those university communication units have evolved into fully fledged teaching departments with their own research and outreach programmes. One has even achieved college status, although still under rather shaky circumstances at the moment.

Every forward step has meant greater latitude to break away from traditional characterizations and to chart their preferred direction while expanding their influence. In the College of Development Communication (CDC) at the University of the Philippines at Los Baños, for instance, the staff remains concerned with the agricultural content of PDC, but in the broader context of natural resource management (NRM) through their association with the United Nations Food and Agriculture Organization (FAO) and the

International Development Research Centre (IDRC), or of reproductive health through their projects with the Philippine Department of Health and the Johns Hopkins University Center for Communication Programs. Through formal and non-formal training programmes, CDC has produced hundreds of development communicators who have fanned out to other fields besides agriculture and to other countries outside South-East Asia. Through its various curricula and publications, then, reinforced by its links to research and action programmes such as Isang Bagsak,[2] CDC has become the nucleus of a major network engaged in the study and application of communication principles in or for development.

The participatory character of development communication has always been considered a given in most of South-East Asia, although the type and degree of participation may not always have been uniform. Until recently – for example, in Malaysia – 'participatory' did not always translate into direct critiques of government policies as in, say, the Philippines, where the political institutions are more Westernized – some would say too Westernized. On the other hand, even in an old democracy such as Thailand, participatory development communication as taught in the universities may still follow the top-down diffusion mode simply because of less exposure to ever-changing development communication thought as new insights are uncovered. As for a hierarchical society such as Cambodia's, particularly with its present form of government, participatory development is still uneven. There is less of it in formal communication encounters, but apparently a great deal more among peers in informal field settings. Clearly, PDC is a product of a society's culture, socio-political institutions and acceptance of current thinking in development and communication. It is also clear that PDC professionals everywhere have a great opportunity to enlarge the degree of citizen participation in their societies by always making visible in their work the principles of participatory development.

And what is the essence of PDC in South-East Asia today? Mindful of its beginnings, PDC aligns itself with those who would reduce, and possibly eliminate, hunger, poverty and sickness in the world. Yet, as a social science, it does not identify with technology *per se* but with the people who use or do not use it, particularly among the disadvantaged in rural areas. Thus, its ultimate goals are equality and social justice for all and freedom for everyone to develop their potential. It uses the tools and methods of communication chiefly to educate through non-formal ways so that people may have both the capacity and the information to make their own decisions.

Some observations and reflections

South-East Asian development communicators have had their problems and setbacks, to be sure. But they have also had their high moments and successes. Through it all they have learned from their own experience and that of others. A few of their more current observations and reflections on participatory development communication are shared below:

- The conceptual difference between communication as process and communication as media or channels seems to bear repeating every so often. As process, it is the exchange or interchange of all types and kinds of information within a society or social group, which is why communication is said to be the most basic of all social processes. It is also seen by many as communication through mass or community media. This traditional perception can be enlarged to include all the mechanical and personal avenues through which information flows between and among the members of a social group. Whether seen as process or as channel, communication can be consciously used for development. Development communication, then, is the process of multilevel exchange within a society of information whose intent is to advance human development and which is channelled through selected media.
- Communication media have long been dichotomized into mechanical and personal. Mechanical media, such as radio, television and now information and communication technologies (ICTs), have received much more attention. It is time to ferret out the nuances of interpersonal communication that promote development. Initiatives such as Isang Bagsak have made a good start. They can also explore the workable combinations of face-to-face and mediated communication that delineate process. In this way, they can take the concept of development communication process out of the generic stage that it is still in and give it more precision and specificity.
- It seems to have become standard in many disciplines for younger professionals to denigrate the work of their predecessors around the globe as being reactionary, perhaps forgetting that they do so from the vantage point of hindsight. And so they try to reinvent the wheel. Without the foundation laid down by those seminal thinkers around the world who have gone before, today's communicators, for instance, would not have had the concept of communication in, or for, development to begin with, and to which is now attached the 'participatory' label. A lesson worth sharing with other development professionals is this: do not turn your back on your beginnings. Acknowledge them, even as you build on them.
- In the practical realm, new models of communication do not necessarily replace older ones. They just co-exist. This is reflected today in the undiscriminating use of terms associated with both old and new models. As a case in point, 'target audiences' and 'beneficiaries' are spoken of alongside of 'stakeholders' and 'participants'. PDC professionals should set the example of being clear about the kind of communication they advocate and of adapting their terminology accordingly. At the same time, they should recognize that older models retain their validity in certain situations and can still be used where appropriate.
- It is now accepted that rural people and other disadvantaged groups have the right to participate in decisions affecting their lives. They need to be empowered – as the stock phrase goes – to realize their self-worth, and to have their opinions heard and factored into the development dialogue. The same can be said of another group in the development world

– the extension technicians, media practitioners and other rank-and-file fieldworkers. In the diffusion model of technology transfer, they are the faceless middlemen who connect the scientists to the local communities. In later communication models, they are hardly visible and are, perhaps, just as neglected. They need to be recognized, too, as valued participants in the development process and accorded equal rank with the other actors.

- As shorthand for innovation, technical content, improved practice or – in our case – natural resource management, technology is not a bad word. PDC professionals should make their peace with it. Development needs a balance between technology and people empowerment. Neither one by itself can go it alone. PDC can be a tool to help people zero in on their problems and apply the technology they wish, given an adequate array to choose from and the capability to make the choice.

- Still on the subject of balance, the trend seems to be for development communication as art and language to break new ground in areas where development communication as social science has only ventured peripherally. This is a welcome move for which some caveats may be offered. Unilateral answers have never worked before and there is no reason to believe that they will now. Development is a multifaceted pursuit and PDC practitioners must integrate within it as many facets as are feasible. On another note, anecdotal case studies without back-up systematic investigation could lead us back to equating PDC with its channels, whether mechanical or personal, rather than with it as process.

- Communicators have been accused of talking only to themselves. Should PDC professionals not also discuss overlapping concerns with researchers, practitioners and administrators involved in natural resource management? Many development professionals still operate within the old researcher–extension worker–farmer paradigm of technology transfer, perhaps because they have not been exposed to newer ones. PDC could facilitate that type of dialogue not only through mediated communication, but also at meetings in venues that NRM researchers, practitioners and administrators are familiar with.

- With the present state of world finances, many countries, including developed ones, can no longer support one-on-one intensive, but expensive, extension systems, potentially leaving the field to commercial companies. What alternatives does the richness of information and communication technology have to offer to small farmers with dissimilar needs? Isang Bagsak has piloted a possible community-based resource. More experiments like this are essential when done systematically and with an eye to their fiscal viability for poor countries.

- Is participatory development communication a means or an end, or both? Is Isang Bagsak meant to achieve better NRM in a community, or is it a way for researchers and community residents to internalize PDC? Or are both objectives valid? The answer will dictate what indicators should be used to gauge the success of projects such as Isang Bagsak.

- Finally, PDC can be institutionalized in two ways: in policy to ensure its adoption by field practitioners, and in theory to ensure its continuing viability and validity through research conducted by students and academics. Both will enrich communication for development as practice and as a field of study.

Conclusions

Participatory development communication, in its several variations across countries, is a young but dynamic field that is nurtured by many disciplines. At the same time, its unique window to human development allows it to pioneer new concepts and practices that other fields can emulate. It has come quite a way in the span of 30-odd years. Like science and art, it can contribute much more as long as its advocates, with their own kind of tools and expertise, hold fast to their vision of equality and social justice for all, and freedom for everyone to develop their potential.

Notes

1   In the US, land grant colleges are a set of state and territorial institutions of higher learning that receive federal support for integrated programmes of agriculture teaching, research and extension for agriculture, food and environmental systems.
2   Isang Bagsak is a learning and networking programme that aims to improve communication and participation among natural resource researchers, practitioners, communities and other stakeholders, and to provide communication support to development initiatives in helping communities overcome poverty.

Participatory Development Communication: An African Perspective

S. T. Kwame Boafo[1]

Since Nora Quebral (1971) of the University of the Philippines at Los Baños first used the term 'development communication', the concept has become entrenched in communication studies and practice in Africa. Several African communication scholars, researchers and practitioners have written about and attempted to contextualize the application of the concept in the African environment; a number of study materials have also been produced and virtually

all communication teaching and training institutions in the continent offer courses and programmes on the concept, often under such appellations as 'communication and development'; 'participatory development communication'; 'communication for development'; 'communication for social change'; and the older but more practice-oriented term 'development-support communication'.[2]

The concept and its different appellations have been very comprehensively defined in the communication literature and we will not attempt to do any in-depth definitional or operational analysis here.[3] A working definition that contains many of the tenets and assumptions of the concept, and which we will adopt in this chapter, refers to development communication as the planned and systematic application of communication resources, channels, approaches and strategies to support the goals of socio-economic, political and cultural development. Participatory development communication (PDC) puts accent on the process of planning and using communication resources, channels, approaches and strategies in programmes designed to bring about some progress, change or development, and on the involvement of the people or community in change efforts. As Ascroft and Masilela (1994) have aptly noted, in the African context just as elsewhere, participation translates into individuals being active in development programmes and processes; they contribute ideas, take initiative and articulate their needs and their problems, while asserting their autonomy.

This brief chapter describes and analyses the application of participatory development communication within the context of African countries. It attempts to situate the use of development communication within the social, economic, political and cultural development challenges and realities in the region and draws attention to a number of contextual factors that determine the effectiveness of participatory development communication in Africa.

Development challenges and development communication

The challenges in social, political, cultural and economic development and transformation in African countries are very well articulated in various documents, publications, conferences, plans of action and programmes, and are particularly well summarized in the recently launched New Partnership for Africa's Development (NEPAD).[4] Although it is not important here to enumerate the varied development challenges facing African countries, it is nevertheless relevant in the context of participatory development communication to stress that there are linkages between communication and the different dimensions of development in Africa, whether they are political, social, economic or cultural. Research studies and experience in diverse contexts and countries in Africa have clearly demonstrated that development communication approaches can be used to enhance participation in cultural, social and political change, as well as in agricultural, economic, health and community development programmes. In a word, regardless of the type of development challenges in African countries, there is some function for communication and information in the efforts made to address those development challenges.

Since the decades of the 1960s and 1970s, development communication strategies and approaches have been employed in numerous development programmes and projects across the length and breadth of Africa. A variety of development communication approaches and strategies have been used by international organizations, funding agencies, government departments, non-governmental organizations (NGOs) and civil society groups in development-oriented programmes and projects designed, *inter alia,* to improve agricultural production; tackle environment problems; prevent and manage health problems and pandemics such as malaria and HIV/AIDS; improve community welfare, the status of women and educational levels; promote or enhance democracy and good governance; and encourage local and endogenous cultural expressions and productions. But the practice of development communication in Africa has been done in the face of several major communication constraints in the region. These constraints are well documented in a number of publications and reports on communication in Africa[5] and include the following:

- weak and inadequate infrastructure and spread of communication and information systems, as well as limited financial resources to develop or strengthen them;
- dislocation and disparities in communication and information flows between urbanized areas and rural communities, as well as disadvantaged population groups, because of insufficient access for large segments of the national populations to modern communication and information means; and
- low priority given by policy- and decision-makers to communication and information as integral components of development programmes; this low priority often translates into the absence of effective policies and structures to guide, manage, coordinate and harmonize communication for development activities in virtually all African countries.

Considerable efforts are being made in African countries to address the above communication constraints and difficulties with financial and technical support from a number of United Nations agencies, international and regional organizations, multilateral and bilateral funding agencies, and professional bodies. These efforts have gone a long way towards enhancing communication and information infrastructure; strengthening communication capacities; nourishing the emerging independent and pluralistic media; increasing access to communication and information systems; and developing human resources in communication and information in Africa. The efforts have resulted in the rapid development of community radio stations in such countries as Cameroon, Ghana, Mali, Malawi, Mozambique, South Africa and Zambia; the development of community multimedia centres and tele-centres in Ethiopia, Mali, Mozambique, Senegal, Tanzania and Uganda; the growth of independent and pluralistic media structures in such countries as Botswana, Burkina Faso, Ghana, Kenya, South Africa and Tanzania; improved communication training programmes in a number of African countries leading to more professionally trained communication practitioners; and, along with the profound changes from monolithism to pluralism in the political landscape of several African countries, a communication milieu that facilitates the use of communication and information channels to express diverse views and opinions on national development concerns, particularly in South Africa, Ghana, Mozambique and Tanzania, among others. Qualitatively and quantitatively, much more effort remains to be done in Africa to address the constraints that confront the practice of development communication. However, given the correlation between communication development and development communication, efforts should contribute to an enhanced use of communication and information in socio-economic, political and cultural development processes in Africa.

The effectiveness of participatory development communication: Some contextual factors

A search through the literature on development communication in Africa indicates that a number of factors come into play in determining the effective application of development communication approaches in support of national programmes in the African context. Among these factors are:

- creating a participatory communication environment that not only gives room for the expression of diverse ideas on societal developmental concerns, but also facilitates grassroots-level interaction;
- strengthening the flow of public information and opportunities of public dialogue on development policies and programmes;
- informed popular participation based on enhanced access to pluralistic and independent communication media;
- producing and disseminating information content that reflects as well as responds to the local values and information needs of the people at the grassroots level;
- using culturally appropriate communication approaches and content;
- using community communication-access points, especially community radio and, more recently, community multimedia centres, as well as small-scale, localized and group media;
- ensuring access to information for women and young people and developing their competencies and skills in the use of communication and information technologies;
- harnessing the strengths of traditional media (drama, dance, songs, story-telling, etc.) and combining them with new information and communication technologies; and
- providing practitioners with appropriate training in the use of communication and information to support development programmes.

In the context of development programmes in rural and marginalized communities where the majority of the populations in most African countries reside, community communication-access points, traditional media and culturally appropriate communication approaches and content are of particular importance in participatory development communication. Alumuku and White (2004) have observed that the communicating capacity of the local community must be harnessed in the conception of development communication strategies in the region. In this regard, community media in African countries, especially community radio, provide the enabling space for local community members to make known their views and opinions on development problems and the possibility of participating in the resolution of those problems. Alumuku and White's (2004) study in Ghana, South Africa and Zambia reported that the communicating capacity of the local community in the form of community radio stations was harnessed to produce and disseminate programmes dealing with such issues as healthcare education; conflict resolution; gender equity; education for responsible democratic governance; defending local development interest; stimulating economic development; and promoting local culture. These are symptomatic of the development problems in many African communities, which the power of community communication resources (with their horizontal, participatory approaches) can help to resolve at the grassroots level.

Similarly, in the African communication environment, given the limited access that some national population groups, especially the marginalized segments living in remote villages and rural communities, have to mass communication media, the communicating capacity of the local community resides in the so-called traditional media resources and channels (traditional leaders, drama, concerts, songs, story-telling, puppetry, drumming, dancing, etc.). They serve as reliable channels of news and information gathering, processing and dissemination in many rural communities, and often address local interests and concerns in local languages and cultural contexts which the community members can easily understand and with which they can identify. Effective applications of participatory development communication approaches and strategies at the grassroots and community level should necessarily involve the use and harnessing of these pervasive traditional communication instruments and resources. Traditional media, especially story-telling, songs, drama and local street theatres, stem from local cultural norms and traditions; their content is usually couched in culturally appropriate ways and they often serve as effective means of channelling development issues. They have been used in communication interventions addressing issues related to improving agricultural productivity, natural resources and environmental management, HIV/AIDS and other development problems. Examples of such use abound in African countries.

Illustrative of the use of traditional media to address development challenges are:

- the Theatre for Community Action project in Zimbabwe, which uses theatre to support and involve rural community members in several

districts in Matabeleland in the combat against HIV/AIDS;
- the use of theatre and folk musical groups to disseminate agricultural information to farmers in rural communities in Nigeria;
- the transmission of messages about reproductive health and HIV/AIDS prevention through traditional dance and music in rural communities in Ghana, Kenya, Malawi, Mozambique, Tanzania, Uganda and Zambia, among others;
- the use of participatory drama and folk music to address issues of gender inequality and HIV/AIDS in Niger State, Nigeria; and
- the transmission of messages about natural resources management through dramatic performances in Uganda.

In sum, traditional media provide horizontal communication approaches to stimulating discussion and analysis of issues, as well as sensitizing and mobilizing communities for development. However, one must be cautious about romanticizing the abilities and impact of traditional media in development. Like other communication and information means, they have their weaknesses and limitations in time and space; they are particularly deficient in simultaneous dissemination of information about development issues across wide and geographically disperse populations. Research and experience in the use of traditional media indicate that they are most effective in participatory communication of development in rural communities when combined with mass communication resources, especially radio. The challenge facing practitioners of participatory development communication in African countries is to be sufficiently cognizant of the potentials and limitations of traditional media and knowledgeable about how to skilfully harness and combine them with other communication and information forms for development. The practical and technical guidelines for designing and implementing PDC interventions using traditional media approaches in combination with other communication forms (including community needs analysis; designing and pre-testing messages/content; training; costs analysis; raising of required funding; implementation; monitoring; and evaluation) lie outside the scope of this chapter.

Conclusions

Communication and information have significant functions to fulfil in supporting and fostering socio-economic, cultural and political development and transformation in African countries. These functions have been recognized by communication scholars, researchers, trainers and practitioners alike, and constitute the bulk of the literature on communication and development in Africa. They are equally stressed at different levels of communication educational and training programmes and provide the basis for communication practice across the continent. This chapter has attempted to situate the use of participatory development communication within the social, economic, political and cultural development challenges and realities in the region, and has drawn particular attention to a number of contextual factors that determine the effectiveness of PDC in Africa. The chapter has been based on the conceptual premise that the kernel of communication teaching, research and practice in Africa lies in their contribution to addressing the myriad development problems and challenges facing the continent.

Notes

1   The views expressed in this case study are those of the author and do not necessarily reflect those of the United Nations Educational, Scientific and Cultural Organization (UNESCO).
2   See, for example, Jefkins and Ugboajah (1986); Akinfeleye (1988); Boafo (1991); Moemeka (1994) and Kasoma (1994). On training, it is worth noting here, in particular, the introduction of the Masters programme in development communication in the Department of Mass Communication, University of Zambia; the programmes at the Southern Africa Development Community Centre of Communication for Development (SADC-CCD) in Harare, Zimbabwe; the Centre for Rural Radio Development (CIERRO) in Ouagadougou, Burkina Faso; and practical training provided by development institutions, NGOs and community groups. With regard to study materials, one can cite the training modules on development communication prepared by the African Council for Communication Education under a project funded by UNESCO in 1991.
3   For comprehensive discussions and analyses of participatory communication for development, see Servaes et al (1996); Dervin and Huesca (1997); Servaes (1999); Wilkins (2000); Melcote and Steeves (2001); and Huesca (2002).
4   For an analysis of current development problems and challenges facing African countries, see, for example, United Nations General Assembly (1994); OAU (1995); and Secretary-General to the United Nations Security Council (1998).
5   For a discussion of these constraints, see, for example, Boafo and George (1991); Moemeka (1994); and Agunga (1997).

References

Agunga, (1997) *Developing the Third World: A Communication Approach*, Nova Science, Commack, NY
Akinfeleye, R. (1988) *Contemporary Issues in Mass Media for Development and Security*, Unimedia, Lagos
Alumuku, P. and White, R. (2004) 'Community radio for development in Africa', Paper presented in the Participatory Communication Section, 24th General Conference of the International Association for Media and Communication Research, Porto Alegre, Brazil, July 2004
Ascroft, J. and Masilela, S. (1994) 'Participatory decision-making in Third World development', in White, S. A., Nair, K. S. and Ascroft, J. (eds) *Participatory Communication: Working for Change and Development*, Sage Publications, New Delhi
Boafo, S. T. K. (1991) 'Communication technology and dependent development in sub-Saharan Africa' in Suusman, G. and Lent, J. A. (eds) *Transnational Communications: Wiring in the Third World*, Sage Publications, Newbury Park, London and Delhi
Boafo, S. T. K. and George, N. (eds) (1991) *Communication Processes: Alternative Strategies for Development Support*, ACCE, Nairobi

Dervin, B. and Huesca, R.T. (1997) 'Reaching for the communicating in participatory communication: A meta-theoretical analysis', *The Journal of International Communication*, vol 4, no 2, pp46–74

Huesca, R. (2002) 'Participatory approaches to communication for development', in Mody, B. and Gudykunst, W. (eds) *Handbook of International and Intercultural Communication*, Sage Publications, Thousand Oaks, CA

Jefkins, F. and Ugboajah, F. (1986) *Communications in Industrializing Countries*, MacMillan, London

Kasoma, F. (ed) (1994) *Journalism Ethics in Africa*, ACCE, Nairobi

Melcote, S. and Steeves, H. L. (2001) *Communication for Development in the Third World: Theory and Practice for Empowerment*, Sage Publications, London

Moemeka, A. (ed) (1994) *Communicating for Development: A New Pan-Disciplinary Perspective*, SUNY Press, Albany, NY

OAU (Organization of African Unity) (1995) *Relaunching Africa's Economic and Social Development: The Cairo Agenda for Action*, OAU Council of Ministers 17th Ordinary Session, 25–28 March 1995, Document ECM/2 (XVII) Rev. 4

Quebral, N. C. (1971) 'Development communication in the agricultural context', Paper presented at the symposium In Search of Breakthroughs in Agricultural Development, College of Agriculture, University of the Philippines, Laguna, the Philippines

Secretary-General to the United Nations Security Council (1998) *The Causes of Conflict and the Promotion of Durable Peace and Sustainable Development in Africa*, Report of the Secretary-General to the United Nations Security Council, 16 April

Servaes, J. (1999) *Communication for Development: One World, Multiple Cultures*, Hampton Press, Cresskill, NJ

Servaes, J., Jacobson, T. and White, S. A. (eds) (1996) *Participatory Communication for Social Change*, Sage Publications, Thousand Oaks, CA

United Nations General Assembly (1994) *African Common Position on Human and Social Development in Africa*, United Nations General Assembly, Document A/Conf.166/PC/10/Add.1, January

Wilkins, K. (ed) (2000) *Redeveloping Communication for Social Change: Theory, Practice and Power*, Rowman and Littlefield, Boulder, CO

III

Participatory Development Communication in Action

The Old Woman and the Martins: Participatory Communication and Local Knowledge in Mali

N'Golo Diarra

If participatory communication is to bring about lasting change, it must give a prominent place to local knowledge. Yet, such knowledge is not always readily accessible, nor is it easy to deal with the ethical issues that can arise when we are looking for the best ways to put it to use. At that point we must ask ourselves whether the prerogatives associated with our role as communication facilitators give us the right to popularize knowledge that was hitherto the exclusive preserve of a few individuals for whom it may traditionally have been a source of advantage and even of power. The question becomes even

thornier when the knowledge in question could turn out to be very useful for the community as a whole.

During my field research for preparing a training kit[1] on ways to combat erosion, I came across a piece of indigenous knowledge that for decades has been improving agricultural output and human welfare in a village tucked away in Mali's cotton zone, more than 500km from the capital city of Bamako.

I was researching local farmers' knowledge as well as new, supplementary techniques that could be used for combating farmland erosion. My strategy was to hold meetings with various social and occupational groups in the village according to an agreed schedule that would fit with their activities.

In these discussions, farmers showed that they knew a lot about their environment and about the best ways to preserve it. They told me, moreover, that fighting erosion would mean curbing the wholesale cutting of trees on the plateaus surrounding the village. But they also pointed out that, because their fields are on hilly ground, they would have to align the furrows to counter the flow of rainfall runoff. Some of them even suggested that fallowing could be a solution; but most participants eventually discarded this idea because the extensive nature of local agriculture left little room for fallow fields.

After three days of investigation, I went to brief the village chief on my findings before reporting them to a public feedback session. I was very happy with the cooperation the villagers had shown and with the results I had obtained, and I was thinking only about getting on with developing my training kit. So I was surprised when the village chief put a question to me, seemingly out of the blue: 'Has anyone told you about the governess of the seasons?'

I had to confess that I knew nothing of her. But I had no hesitation in countering with another question by asking him who she was. Smiling at me in a kindly way, he tapped my shoulder and invited me to supper. The next day, very early in the morning, the chief woke me and without taking time for breakfast, he took me to the house of the old woman who greeted us as if she had been expecting our arrival. The chief made the introductions and explained briefly to the old lady what I was doing in the village. Then, excusing himself, he disappeared and left me alone with the 'governess of the seasons'.

She began by recounting to me something of her own life, including the tragic death of her husband and the loss of her six children. It was a lonely life, she said, but yet she felt it was a happy one. Suddenly, she went into a trance and began to speak in words that I could not understand. Somewhat disconcerted, I managed nevertheless to keep my cool. She soon got hold of herself, looked at me a moment and then asked me if the 'purple martins' had arrived. Surprised by this question, I had to say I had not noticed them. At that, she burst out laughing. She asked me to sit down beside her on the couch and began to tell me why she was known in the village as 'the governess of the seasons'.

'My father was the great traditional healer in the village', she said very matter of factly:

When he died, he would have no direct descendents he could pass on his knowledge to, except me. One day he called me to his hut. He said that he was at the twilight of his life, that there were things he knew and that he must pass on to his descendents. But he could not transmit everything to me because I was a woman married into another family. I told my father that whatever he could pass on to me would be of service to the whole village, for I no longer had any children in this world to whom I could transmit that knowledge. It was then that my father told me the story of the purple martins and how they govern the local seasons.

Looking me straight in the eye, she continued her story:

'Remember this well, daughter', said my father, 'the purple martins are fabulous birds. They arrive at the start of the rainy season and they make their nests in the forest along the great river that passes by the village. Their arrival, and that of the storks, heralds the beginning of the rainy season for people in our village and nearby.

When they build their nests down in the river valley, I have noticed that they take into account the volume of water expected in the river. Three weeks after their arrival, I find that their nests will be placed either high up in the trees or lower down. So I conclude that, since they are nesting when water levels are high, the position they choose for their nests will depend on how much water they expect the river to have, perhaps to prevent their eggs from being washed away. As soon as I have noted the positioning of their nests, I consult the spirits and my fetishes for permission that very evening before I tell the news to the village chief so he can order sacrifices. The announcement of this news is a big event for the entire village. To some extent, it determines what farmers will plant and where they will pasture their livestock during the coming season. All the families promise gifts if they get a good harvest and if the village workforce stays in good health.'

Continuing her story, the old woman confided to me that her father was now dead and that she continues to play this role of village seer. 'That has been the source of wealth for my family and for the village for generations, for after the harvest all the villagers keep their promises, and our village and the neighbouring area are assured of food self-sufficiency.' She concluded with a laugh: 'There you are, thanks be to the purple martins!'

With the old lady's concurrence, and my commitment not to reveal her name or the name of the village, I have been able to use this example as part of the scenario for the training package.

The community sees things differently

Two years after I heard the old woman's story, I had the chance to return to the same village for a farmer training session using the anti-erosion training package we had produced at the time. There were warm greetings all around and the occasion was marred for me only by learning that the village chief and the old woman had both died.

The training began well. As agreed, I used the example of the old woman and the martins, treating it as if the story had taken place somewhere else. The villagers were all for trying to protect these birds by taking better care of the riverbed. Yet, when I applied the example to the river that runs past the village where we were, there was a chorus of outcries from all sides and the training session was temporarily disrupted.

Some farmers went so far as to declare that this river was haunted by evil spirits, or djinns,[2] to whose wrath could be laid all the flooding that has afflicted the village and its surroundings. No one talked about how important the river was for the village's social and economic development and still less about the fact that after the rain all the water runs off the plateaus straight into the river. In the villagers' minds, the mystic dimension of the river far outweighed its role in terms of the use and conservation of local resources.

Local knowledge and ethical issues

Local knowledge is very difficult to get a handle on. It is usually held by individuals or families of great status and influence in the community, as was the case in this village. Because of this, we participatory communication practitioners must weigh the pros and cons before we disclose any such knowledge. I found myself in a very tricky situation that day and it left me hesitant for a long time about the wisdom of revealing the old lady's story. I finally decided to talk it over first with the new village chief. After some consultation, he gave me the go-ahead to tell the tale to participants in the training session.

When I started my story, there was dead quiet. Everyone was surprised at what they heard. Feelings of guilt, astonishment, curiosity and even disgust registered on their faces and in their statements. A few were against my revealing this secret of such importance to the community. Most of the participants, by contrast, were relieved to hear the story. Many of them felt guilty about their practices or about their behaviour towards me, as the trainer, but above all about their treatment of the river – this resource that was so vital to the community but that had been blamed by so many generations for the flooding it caused.

In this way, a training course in anti-erosion techniques to preserve farmland was transformed into a real discussion about changing the community's attitude towards the river. All the debate focused on how to save and protect the river. At the end of the session, we addressed the need to adopt measures to protect the river from erosion in order to avoid flooding in the villages. Then we went on to debate the importance of restoring farmland to ensure the community's economic development and its food security. Participants were unanimous on the need to take anti-erosion measures to protect the river and make the villages safe from flooding. To do this, they set the goal of creating a buffer zone along the river, which would also serve to protect the martins. They also undertook to use anti-erosion techniques on the plateaus in order to break the speed of the rivulets and the wind so that the fields could be restored and agricultural output improved.

All these commitments are very fine; but now that this secret is out, what will come of the vision and of people's attitude towards the possessors of local knowledge? What will become of places that used to be considered sacred? As development practitioners, how can we make judicious use of local knowledge, while respecting ethical, copyright and intellectual property issues?

Drawing lessons from the experience

One of the lessons I drew from this participatory communication experience was that the villagers were very happy to have me back and especially to see themselves in the film. They were greatly bolstered by the opportunity to express themselves and to be heard. They realized that they had something to say about everything that concerns their local development and they lost no time in letting me know that.

It is clear that, thanks to participatory communication, these farmers have been able to reach consensus on the key role that the river plays in their community and on the best ways of managing that resource. In the past, the old lady had been regarded in the village as merely an oracle who had to be consulted at the onset of the rainy season so that the good farmers could set their course. The martins were looked upon by many people as simple migratory birds whose arrival announced the beginning of the rainy season, while the river was considered a place haunted by evil spirits. Indeed, if the old lady's story had not been revealed, it would have been much harder to get people to protect the watercourse – they would have continued to treat it as the haunt of *djinns*, and it would have been left to its sorry fate. In that case, neither the farmers' fields nor the river would be safe today.

The achievements of participatory communication

In the experience recounted above, it is clear that new know-how for combating erosion, coupled with indigenous knowledge, including that of the old lady, led to a more rational and efficient approach to managing the community's natural resources – that is, the river and its surroundings, farmland, pastureland and the village itself, which was always subject to flooding from rain runoff.

Another important point is that this village has now become a pioneer in combating erosion through such techniques as planting hedges or 'green fences', building stone dykes and weirs or planting sodded strips, and it has even enlisted neighbouring villages in the cause. Farmers have put up stone-retaining walls all along the river and on the hillsides surrounding the village in order to break the speed of rain runoff. This has helped to protect the river and, at the same time, to make the hilly fields usable for dryland farming, while reserving the lower-lying lands for rice growing.

Villagers have also planted trees in a strip about 300m wide along each bank of the river. They have banned all woodcutting and fires in this zone, except with explicit permission from the village chief and his advisers, who make up the village's forest management committee.

Many activities are now flourishing along the river, including fishing, livestock raising, rice growing and, in particular, market gardening, to which the women devote themselves for a good part of the year. The village now has its own little forest along the embankments where the martins can come to lay their eggs every year without fear. The area surrounding the river has today been transformed into a tropical microclimate where people, animals and birds together share the bounty of the legacy left by the old woman and her martins.

Of course, there is still the question of sustainability for these achievements, particularly since the village chief and the old woman are dead. Yet, all signs suggest that farmers' involvement in efforts to protect the river will help to keep these actions going.

By way of conclusion, this experience has made clear to me that a community's local knowledge must have an important place in the entire process of participatory communication. It is up to us, communicators, to learn how to discover that knowledge, to be in phase with its holders and to introduce it into our training tools so that communities can put it to use in the interest of local development.

For the village chief and the 'governess of the seasons' who taught me to appreciate and to pass on this knowledge, my prayer is that they may rest in peace.

Notes

1   A training kit is a set of audiovisual and hardcopy training tools prepared in cooperation with the community. The process, as developed by the Centre de Services de Production Audiovisuelle (CESPA) in Mali, involves several stages during which we structure the contents in light of local people's knowledge about a topic, taking due account of their cultural codes and of the objectives for the training exercise. The outcome is, thus, the result of a joint effort among researchers, development technicians, communicators and rural people themselves.

2   In the cosmology of some Malian cultures, the *djinns* are evil spirits who usually dwell near watercourses. Natural disasters are often attributed to the wrath of the *djinns*, when they have been displeased by some human action.

Introducing Participatory Development Communication within Existing Initiatives: A Case from Egypt

Rawya El Dabi

Introducing participatory development communication (PDC) within existing initiatives is not an easy task. The necessary conditions must either be in place or put in place. Attitudes must also reflect the methodology that one is trying to introduce. When these are lacking, collaboration between the different stakeholders can be impaired. This is particularly true when the various partners work on different levels and weigh differently in the

decision-making process. This chapter presents the challenges faced when getting on board after an initiative has started and when trying to introduce a methodology that questions the way things have been done so far.

In Egypt, 95 per cent of the population is concentrated in 5 per cent of the entire country's surface area. Therefore, the government of Egypt has asserted that developing human settlements outside the Nile Valley in desert areas and frontier governorates is a national priority.

However, the expansion of human settlements outside the old valley faces several constraints, such as the inhospitable and vague institutional set-ups for managing development in desert areas, outside current jurisdictions. The poor coordination and cooperation between the existing governorates' local authorities and new communities is also a problem.

In that context, the government office in charge of urban planning has been assigned the responsibility of preparing regional, urban and structural plans for the new settlements. It is also responsible for monitoring their implementation. Hence, close working relationships with local authorities have to be in place, especially since the technical capacity of local governments does not yet allow for comprehensive decentralization.

With a view to responding to these difficulties, an initiative was put in place with funding from both the government office in charge of urban planning and the United Nations Development Programme (UNDP). Its main objective was to devise implementation mechanisms for strategic plans in order to create an enabling regulatory and procedural environment for sustainable development and growth. The initiative was to start with a pilot demonstration in two relatively new communities.

More specifically, the initiative aimed at improving the capacities of the government's urban planning office and of local partners in participatory planning and management, in which participatory development communication plays an essential role. It also aimed at helping the initiative's partners to institutionalize these new capacities, establish partnerships and create strategic alliances to attract investments and sustain implementation. These partners included the government's urban planning office, the local authorities of the two communities involved, small- and medium-sized entrepreneurs, business associations and the communities at large. The preparation and dissemination of guidelines for city and community consultations were also considered as part of this initiative, as well as holding consultations to prepare local development plans in light of the strategic plan.

Origins of the initiative

This initiative was born after three years of awareness building and sensitization efforts led by an institutional development expert on the importance and benefits of participatory approaches in development planning. Early diagnosis had shown that the participatory planning and management practices of the human settlements in the area under issue could greatly benefit from being expanded. It also showed that the initiative's partners needed assistance in adopting a more inclusive *modus operandi*, with a view to establishing partnerships and strategic alliances that would gear the strategic plan towards its implementation.

In the course of that preparation phase, it became clear that participatory development planning is much broader than the strategies strictly developed from a physical planning viewpoint, which have proved to be a complete failure. Indeed, their failure has been manifest in the fact that although new cities were designed and plans were handed down to local governments for implementation, the strategies were never fully implemented. Therefore, citizens were reluctant to move to these new cities. Those who did move faced huge difficulties as the services provided did not respond to their needs or were absent in the first place.

From theory to action

In its early phase, the initiative worked towards streamlining the inter-organizational interface within the existing institutional framework. It tried to detect the most serious barriers that needed to be reconsidered within that framework in the area under issue, and proposed realistic modifications in order to enhance the potential of public, private and civil society participation in local development.

Thus, through direct and constructive dialogue and consultations, the initiative tried to help stakeholders to define and rectify sources of friction and derailments, and proposed practical institutional changes. In addition, municipal policy options paved the ground for an improved business environment and participatory development.

In order to introduce participatory development communication within the initiative, several steps were envisioned. First, all the documents related to the initiative were to be reviewed in order to become aware, for example, of its nature, the main problems encountered, suggested solutions and the stakeholders involved.

The second step consisted of assessing and analysing the communication problems, as well as the communication channels and materials used by the immediate stakeholders.

Third, a training programme was to be developed and implemented in order to help the initiative's partners understand the concept of participatory development communication and acquire the capacity to apply it in their work. In order to do that, it was necessary to identify a unit within the existing organizational structure of the communities' local authorities that could become the key actor in maintaining the use of participatory approaches throughout the implementation of their communities' development strategies. Thus, the idea was for the staff of these units to be trained in participatory communication and development planning, together with staff members of the government planning office.

The training was envisioned to be of a practical nature and to emanate from the specific circumstances in which the participants live and work.

Hence, it was expected to facilitate the application of their newly acquired skills directly in the field. Initially, the idea was for the staff of these units to be trained in participatory communication and development planning so that they could implement their communities' development strategies. The training programme was to be held over a lengthy period of time in order to ensure that participants actually learn and practise the skills needed for a participatory development communication process.

The next step involved providing assistance to the immediate stakeholders in developing a participatory communication strategy for their communities' development plans. Finally, the plan included participation in general activities such as workshops that lead to the development of such elements as issue papers and city consultations.

Results fall short of expectations

Several obstacles hindered the implementation of the plan described above. First, the project team within the government's planning office continued to perceive participation as a process aimed at allowing all stakeholders to voice their problems, but without helping them to overcome these problems.

Furthermore, the team significantly interfered in the work of the communication and development planning specialists, with a disruptive effect. It did not allocate enough time to assess the training needs prior to organizing the training programme. It also continuously requested pre-prepared training material, as if distributing bulky materials to trainees would ensure the training's success.

Probably the greatest difficulty was that the team viewed training as a top-down process in which lecturing was to be the main training tool. In the eyes of team members, the training duration was not to exceed two to three days (for budgetary constraints). Moreover, these short training workshops included both communication and development planning. As a result, there was not enough time for participants to fully grasp what participatory development communication really means and where it fits within the development planning cycle. The same situation prevailed with regard to the institutionalization of participatory approaches in the government's planning office and the local authorities' structures: the resources allocated did not suffice in terms of time and money.

Finally, the staff of the special units had not yet been fully selected at the onset of the programme and there were many disagreements on the selection process (and too much politics was involved). Thus, the implementation of the training sessions was disrupted, with a significant impact on the participating structures' capacity to integrate participatory methodologies in a sustainable manner.

Looking back on the initiative

Despite its serious setbacks, this initiative provided lessons on introducing participatory development communication within existing initiatives, particularly if more than one funding partner is involved. In hindsight, it seems that several preparatory steps would have helped the initiative to better achieve its goals.

For one, the international organization supporting the initiative has to be aware of participatory development communication, its benefits and prerequisites. This awareness would facilitate the work of communication specialists in many ways. For example, the terms of reference would be formulated more adequately and the communication facilitator would find support when it comes to dealing with the governmental partner. Unfortunately, up to now, many donor agencies or international organizations still do not fully understand the principles of participatory development communication. Many of them seem to be more concerned with the rights-based approach and making efforts to introduce it within all their programmes. Participatory development communication ought to be treated in the same way.

Furthermore, participatory development communication should not be considered a 'complementary accessory' to the initiative to be introduced at a later stage (mainly the implementation stage), as was the case in the initiative described in this chapter. Participatory development communication should start with the idea of the initiative and this, of course, is only possible when the partner international organization is fully aware and convinced of its value.

Accordingly, the governmental partner should also be trained in participatory development communication's concepts and methodology. This training would enable the communication specialist to work hand in hand with the governmental partner to develop a participatory development communication strategy for the initiative that they are about to implement. If all these conditions are in place, one can expect that the governmental partner will support the PDC process throughout the project instead of hindering it.

Goats, Cherry Trees and Videotapes: Participatory Development Communication for Natural Resource Management in Semi-Arid Lebanon

Shadi Hamadeh, Mona Haidar, Rami Zurayk,
Michelle Obeid and Corinne Dick

When societies undergo in-depth changes, traditional structures, resource management systems and means of communication sometimes do not suffice in coping with the pace of transformation. Moreover, conflict may arise as a result of the increasing stress that people and their environment are faced

with. Enhancing existing communication and conflict resolution practices with novel tools and methodologies can help to build the social fabric anew and provide communities with a common sense of direction.

Traditional agro-pastoral Lebanese villages located on the marginal slopes of the semi-arid Lebanon mountains have been undergoing drastic changes in response to socio-economic pressures that have been developing over the last 25 years. During 1991 to 1993, a case study on changes in resource management systems conducted in Arsaal, a vast highland village, revealed a massive conversion from a traditional cereal/livestock-based economy to a rain-fed fruit production system.

Subsequently, a follow-up study was conducted with a view to analysing components of changes, trends and sustainability in the emerging production system and improving the prospects for sustainable community development. The land-use system in Arsaal, including socio-economic components, was then characterized and its resource base was assessed, with an emphasis on soil and water conservation strategies. Local beneficiaries were involved at different stages of the project and strengthening of local capacities was sought through the establishment of a local users' network. Avenues for non-agricultural income-generating activities for women were explored. A second phase was designed to test and evaluate technologies and management options developed by the users' network in phase 1 to assess progress towards sustainability in the major land-use systems, as well as to monitor, evaluate and strengthen the capacity of the local users' network towards sustainable natural resource management, with an emphasis on gender analysis.

The overall picture reflected a society in transition from an agro-pastoral system to a more diversified livelihood integrating rain-fed fruit production and off-farm jobs, as well as quarrying and related activities, in addition to the traditional sheep and goat and cereal/pulses production. Traditional management strategies based on community consensus had given way to conflict over land use among animal herders, fruit growers and quarry owners, with an increasing socio-political influence from the latter group. The conflict was highly entangled in the traditional family web of the village. In spite of the ramifications that these conflicts introduced within families and clans, traditional structures showed a good degree of resilience. In summary, the Arsaali society was in a prolonged state of disequilibrium, crisis and crisis management.

Furthermore, starting in the mid 1960s, the breakdown of traditional resource management practices had led to the dismantling and complete paralysis of the local municipality. This situation was perpetuated during the decades of civil unrest because the new emergent forces (political parties and militia) were more involved in national politics than in local resource management. This chaotic state of affairs continued following the return of hundreds of youth from Beirut, driven back by the demise of the leftist militia after the Israeli invasion of Lebanon and their hopes of prospective involvement in smuggling activities across the Syrian borders.

The year of 1998 saw the election of the first municipal council in 35 years, and the community acquired an administrative body that faced the challenge of managing several forms of land use and their conflicting requirements. The election, however, was largely thought of and conducted in terms of familial alignments, thus leading to the formation of a municipal council that knew little about local administration and lacked the perception required to develop local resources.

Creating a local users' network

Following discussions with our local facilitator, the Association for Rural Development in Arsaal (ARDA), it became crystal clear that some kind of medium was needed to facilitate interaction among the various local beneficiaries and other groups, such as researchers, development workers and non-governmental organizations (NGOs). The presence of such a medium would provide a platform for different stakeholders to assess and develop a common understanding of research and development needs and possible solutions to the lack of extension structures.

A local users' network was therefore conceived to bring together the different stakeholders and to fulfil the critical functions of participation, communication and capacity-building.

In the Arab world, the traditional way of communicating and resolving dilemmas is, largely, face-to-face interaction. This forms the basis of the tribal *majlis* during which issues are brought up in the community, usually at the house of the community leader.

The strategic role intended for the network was to form a participatory interactive platform based primarily on face-to-face interaction in informal group meetings as a variant of the traditional tribal *majlis*, this time extending beyond the community and involving all the development stakeholders: the community, researchers, development projects and government.

After consulting with the local NGOs and meeting with the local authorities, the *mukhtars* (heads of villages) and the acting municipality officials (the municipality had been dissolved since 1965), the mechanics for establishing the network were defined in order to ensure the representation of traditional decision-makers, as well as new emerging forces. ARDA played a facilitating role in contacting various groups of users (such as the cherry growers, flock owners and women).

The objectives of the project were discussed and evaluated, and network members agreed to actively participate in the activities.

It was hoped that the flexible structure of the network, the common interest of its members and the rewards to all involved in it would ensure a level of sustainability through its adaptive development. The rewards included benefits arising from farmer-to-farmer training, productivity improvement as a result of scientific research, improved training capacity of the NGOs involved, and reinforced links between farmers and local authorities.

As the network grew, our understanding of communication principles evolved with it and the need to define a workable, meaningful typology or a system of user categorization that considered their subjective nature became obvious. Specialized working groups were born and later developed into three sub-networks; two of these dealt with the main production sectors in the village: livestock and fruit growing. The third addressed women's needs. Local coordinators were designated to coordinate each sub-network. Specific on-farm trials were developed, discussed with the farmers and implemented. From its onset, the project made sure that network members were representative of the different resource user groups in the community. This helped to ensure that the needs of the community at large were voiced in the network, which meant that the solutions developed were relevant to the rest of the community. This, by itself, greatly enhanced a widespread knowledge-sharing.

Moreover, a specialized unit in the network called the environmental forum was specifically created to catalyse knowledge-sharing with the community at large. The forum was made up of Arsaali youth, mostly school teachers, who were trained by the project team in the good practices developed by the network and had as a mission their widespread distribution. In order to do this, they primarily used face-to-face interaction, especially during critical periods. In addition, they used complementary material, such as the 'best practices booklet' developed by the project to summarize and simplify project findings in a language accessible to farmers. The forum also served as a communication channel between the community and the local users' network, in which refinements and remedial measures were identified.

Special emphasis was placed on evaluating and analysing the observations and feedback of the network members. This was done with the purpose of assessing the response of the community to the new techniques, as well as adapting these techniques to community needs. Farmers' findings were fed into the research process by way of regular meetings and contacts with research team members. These farmers constituted a platform whose purpose was to spread research findings and exchange observations. The input and feedback of network members constituted the main elements for use in establishing intervention strategies to gain sustainable improvement.

Using a wide range of tools

The tools and practices used by the local users' network were mainly interpersonal. They included regular issue-centred, round-table meetings for members of the sub-networks; community outreach by students during their training programmes; 'live-in-the-village' and 'work-with-the-farmer' approaches; joint field implementations of good practices in natural resource management; short video documentaries on different issues, which were also used as powerful participatory tools; newsletters; a website; and, most importantly, a series of workshops on different themes related to natural resource management and community development.

The network functioned as a self-reinforcing interactive participatory communication platform that proved to be an effective and innovative experience by promoting economic development and socio-political empowerment, while exposing the community to other development interventions.

Art and visual communication tools were also part of the network's arsenal. In its second phase, the project developed a partnership with Zico House, an alternative cultural community house specialized in the use of art for community development. Video-making was experimented with as part of an effort to involve the community in dialogue and conflict resolution, the premise being that imagery has the power to shed light on aspects of conflict and dissent. More importantly, it constitutes a platform for freedom of expression for marginal groups and provides a visual reference of a specific development context over time.

Making videos to resolve conflicts

The network provided an environment in which conflict resolution could take place among different land users since the needs of conflicting parties could be voiced and compromises explored. Early meetings held to discuss the conflict issues among the parties revealed a reluctance to engage in dialogue. After a few stalling sessions, our communication unit suggested using visual images to facilitate the dialogue initiation process. Initially, the representatives of the conflicting parties who refused to discuss the issues at stake were interviewed and filmed separately. Then the video was projected in the presence of all parties, followed by a discussion that was also documented on film. The local actors who were videotaped were consulted during the video editing process. The final step was to project the new video to a larger audience, including more people from the whole village until a positive dialogue started to emerge from the audience. The moment of consensus was also filmed and formulated by a local facilitator into a set of specific recommendations for follow-up. For instance, the conflicting parties agreed to refer to the local authorities (the municipality and the *mukhtars*) and entrust them to develop and recommend different scenarios for land-use management in the village.

Videos to empower women

Another video was produced with the aim of highlighting the economic productivity of women in NGOs, in the co-operative for food provisions and in a pastoral society. This video particularly addressed the improved self-perception of women and the feeling of empowerment that accompanies production. The film was shown to a group of women and men, some of whom were in the film, over an *iftar* (breakfast) organized by ARDA. Those who were in the film felt quite empowered to see themselves on the screen, especially when they received compliments from others about their stated opinions in the film. Both men and women emphasized the importance of working extra-domestically. Women stressed that although money is essential and it elevates the status of women in their households, the mere act of exposure, learning and socialization that comes about from working is satisfying and, indeed, raises one's self-esteem.

There was also a discussion about women who are not involved in production and the importance of acknowledging their role in society. Some commented that the film should have also addressed other 'typologies' of women, such as school teachers and housewives.

The group agreed that it enjoyed watching the film and that film is a very appropriate means of documentation, especially in a context such as Arsaal where people enjoy 'watching' more than reading.

Conclusions

The use of a wide range of communication tools and methods – that is, from the traditional *majlis* setting, to workshops, to novel tools such as videos and video-making – in the local context was very revealing. Marginal groups usually shut out of the local power structure, suspicious of it and often shy in formalized *majlis* setting became very candid before the camera and expressed unvarnished opinions as if, for them, lenses were neutral objects and there was no need for the formal politeness of the *majlis* confrontation. The videos turned out to be valuable for generating discussions and awareness among and between different people and factions. The world of image was able to re-establish a communication platform to implicate local people, reflect their real needs, allow marginal groups free expression, shed light on the nature of conflicts amidst the village, and facilitate resolution of conflicts over natural resource management.

One of the most pressing issues we still face is how far the local communities will be able to use the product of participatory development communication to improve their livelihoods. Participatory development communication activities must be intimately linked to development activities – namely, the transfer of resources. Only when elements of development were injected into our community-based research process did changes in the behaviour and aspirations of the people start to emerge.

From Resource-Poor Users to Natural Resource Managers: A Case from West Java

Amri Jahi

Conservation efforts often fail to take into consideration that alternatives have to be available if local populations are to change their natural resource utilization patterns. Along the Cimanuk River in West Java, efforts to prohibit agriculture on the raised riverbanks to curb seasonal flooding irremediably failed until viable economic alternatives were introduced. Through participatory development communication, the dialogue between researchers, decision-makers and small and landless farmers made it possible for sheep raising to flourish as an alternative to the practice of agriculture on the riverbanks that was making villages more vulnerable to annual flooding.

Fifteen years ago, as we were preparing a baseline study to explore the development options that could improve the living conditions of small farmers and landless farm labourers living along the Cimanuk River in the district of Majalengka, West Java, we wondered whether those resource-poor farmers would be able to participate in managing a strip of public land stretched out along the river. We also wondered whether they would be able to derive benefits from such an undertaking.

At that time, the main concern of the local government and of the communities living along the river was flood control. Every rainy season, heavy downpours filled up the river and flooded the villages. To control this annual flooding, the Department of Public Works bought a strip of land along the river that goes through the district's villages and small towns, and raised the riverbanks to a certain height. Hence, the annual flooding was controlled and the communities were freed from the yearly threat of natural disaster.

However, this flood prevention technique created another problem for the small and landless farmers who had been farming on the riversides long before the banks were raised. According to the new government rules and regulations, farming activities were forbidden on the public land along the river due to the risks of destabilization of the land structure and hindrance of the waterways.

In spite of the government rules and regulations, the small and landless farmers pursued their farming activities as usual. On their part, officials of the Department of Public Works continued to enforce the regulations by eradicating the crops that could possibly weaken the raised riverbank's land structure.

Looking for alternatives

After several years of playing cat and mouse and tired of conflicting with the local communities, the Department of Public Works finally proposed a win–win solution to the farmers. The latter were finally allowed to farm along the riversides, provided that they grew grass at least 1m away from the river's brim and on both sides of the raised bank. However, they were not allowed to grow bananas or any other type of land crops that could destabilize the land structure.

To further motivate the farmers to grow grass, the government officials promised to supply them with sheep. However, this promise did not materialize very well. The farmers were disappointed with the few low-quality sheep delivered to them and the abundant grass was wasted.

At that critical time, a baseline study was undertaken in order to understand the availability of local natural resources such as land, crops, animals and water, as well as the limitations, problems and opportunities associated with their use and development. Moreover, the study also aimed at understanding the traditional systems and at assessing indigenous capacity.

The study concluded that there was potential for involving the small farmers in managing the public land located along the river and the irrigation channels, and for deriving benefits from it, provided that a funding body be put in place.

A follow-up pilot study was then undertaken. Essentially, the challenge consisted in helping farmers to solve their problems and benefit from the abundant grass, while ensuring the conservation of the riverbank. After discussing the best-suited approach, our team decided to use participatory development communication. Thus, linkages were established between university researchers, local government, the Department of Public Works, the local livestock extension service, the village authorities and local farm communities.

Using appropriate communication tools

Prior to the fieldwork, communication materials such as sound-slide shows, posters and even a comic book were prepared and tested with farmers and extension workers. Development communication graduate students were involved in producing and testing these materials, either through their message design course assignments or through their theses research.

Topics such as the social and economic potential of sheep raising in the district of Majalengka were specifically addressed with local decision-makers and livestock extension workers. Other topics concerning various technical aspects of sheep production, marketing and rural family budget management were developed for extension workers, farmers and their wives.

The fieldwork started once the communication materials were ready. In the first quarter of 1992, a memorandum of understanding was officially signed between the government of the district of Majalengka and the Faculty of Animal Science at Bogor Agricultural University.

During the ceremony, a sound-slide show regarding the social and economic potentials of sheep raising in the district of Majalengka was shown to the government officials and to the local House of Representatives, prior to signing the agreement. This created enthusiasm among the officials attending the ceremony. Moreover, the local media covered the event, which was broadcast on the evening news by the regional television station.

The next step consisted of promoting capacity-building for the local livestock extension personnel and farmer leaders. Once a month, researchers conducted an on-site training meeting for the livestock extension workers and other officials of the district livestock service. The technical aspects of sheep production, including housing, feeding, breeding and sheep healthcare, were addressed. Two sound-slide programmes about sheep production were shown and discussed with the extension workers.

A pre-test and a post-test on the training subjects were always administered to the extension workers prior to and after the slide show in order to build up the agents' interest and to focus their attention on the subjects. Moreover, the tests were also useful in indicating whether the sound-slide shows had been effective in improving the agents' knowledge on the topics that were being discussed.

Following the agents' training, a short meeting was conducted for farmer leaders of the four pilot sites. On that occasion, they were informed that they were entitled to receiving an in-kind loan in the form of sheep. If they agreed, they would have to return a certain number of the sheep offspring to the project for a period of four years.

After the meeting, participants were invited to visit a senior farmer leader in the village of Balida to witness how he collected forage. One month later, they joined an excursion to the district of Garut to visit a sheep-raising initiative. The Garut sheep were particularly well known for their fast growth and prolificacy.

Trying out new techniques

Upon their return, every farmer leader was asked to build a good sheep house capable of housing two adult rams and ten ewes, and to prepare a plot to grow improved varieties of grass and legumes to feed the sheep for demonstration purposes.

Intensive supervision and backstopping activities were provided to the farmer leaders in the following month to allow them to get first-hand experience in raising the new sheep breed in their own environment. After one month, the farmer leaders had enough experience in taking care of the sheep and were able to provide information about the Garut sheep raising to other farmers in the villages.

Upon completion of the trial run among the farmer leaders, monthly evening training meetings were conducted for the rest of the farmers in the four research sites. The first training meeting was about building a good low-cost and healthy sheep house using local materials, while the second one focused on the appropriate feeding for this specific breed of sheep. During the meetings, all farmers were given pre-tests and post-tests in order to measure their knowledge regarding the training subjects prior to and after the audio-visual presentations.

About two weeks after the training sessions, the farmer leaders were asked to check whether all the farmers had made or had improved their sheep houses for the incoming new sheep. Farmers who had not done so were encouraged to rebuild or improve their sheep houses.

Two weeks after completing the necessary preparations, a farmer leader with extensive experience in sheep breeding was asked to join the researchers to select rams and ewes for 100 farmers in the four research sites.

The training meetings continued after the sheep had been distributed to the farmers. The topics addressed ranged from technical aspects of sheep raising to rural family budget management.

Meanwhile, the research assistant, together with the farmer leaders, visited every farmer once a week to check on the sheep's condition and health and to provide further backstopping activities to the farmers.

About one year after the distribution of the sheep, all farmer leaders were invited to a meeting in the district capital to share their experience in Garut sheep raising with farmers from other villages. This farmer-to-farmer communication turned out to be an excellent medium for raising other farmers' interest in Garut sheep production, better than anything that researchers or extension workers could have done.

Promising results

The initiative had a first positive result when the head of the local livestock service, using the narrative progress report for the first year, was able to convince the local government to provide his office with extra funding

Towards the end of the second year, results began to be experienced at the grassroots level, when several farmers began to return the sheep offspring to the initiative. According to a previous agreement, the sheep offspring were to be split fifty–fifty. The returned offspring were to be revolved to other farmers interested in sheep raising. This was an indication that the initiative was having a positive impact on the small farmers involved. Indeed, between September and December 1993, the sheep population increased by 159 per cent. One year later, the increase reached 271 per cent, while at the end of 1995, it had risen to 306 per cent.

The value of the participating farmers' assets also increased accordingly. In less than three years, the value of the sheep almost tripled – a positive contribution to the local economy. Moreover, the welfare of most farmers involved in the initiative improved significantly since the income generated by selling the sheep was worth more than 1.5 times the capital loaned to them.

In addition to the economic benefits, such as having more sheep in barns and gaining additional income, the initiative also brought about some social benefits to the villagers involved, who gained respect from their families and communities. One of the farmer leaders even managed to send his son to the Bogor Agricultural University, where he obtained his degree in Animal Science last year.

In May 1996, after about five years of continued fieldwork, the initiative officially came to an end. However, since the ewes continued to give birth and farmers kept on returning the sheep offspring, the revolving activities continued on their own.

A new spur to the initiative

In early 1999, following the severe Asian financial crises that badly hurt the Indonesian economy, the initiative obtained additional funding to buy 55 new sheep. These sheep were distributed to two other villages in the district of Majalengka.

Meanwhile, around the same time, one former project group in the village of Kadipaten received a 300 million Indonesian rupiah loan through the World Bank's social security network funds. The purpose of this loan was to help the local communities cope with the economic crisis. The farmer group used it to further develop their sheep raising ventures.

Eight years after the initiative terminated, we are still serving the farmers, though at a slower pace and on a smaller scale. To cover the operational costs of supervision and backstopping activities, fieldworkers have been allowed to sell the culled sheep.

Today, livestock production, including sheep production, is growing at a good pace in the district of Majalengka. Both public and private investments have been increasing steadily, especially since the 1998 economic crisis. Between 2002 and 2004, the government of Majalengka invested over 9 million rupiahs in livestock production.

Public funds have been channelled to farmers as loans through local banks, mainly for beef and cattle production. In addition, a big livestock marketing infrastructure has recently been built in the region. Livestock traders from east and central Java bring and sell their sheep and cattle in the newly operated livestock market to local and other traders and buyers from Bandung, Bogor and Jakarta.

In other words, the situation of livestock economics in the district of Majalengka has greatly evolved compared to the situation that prevailed in 1989, when the research team first arrived. We believe that over the last 15 years, our work has contributed to this situation by raising a flag to both public and private audiences, informing them about the existing potential in a remote district of West Java that was waiting to be explored and developed.

During recent times, when driving along the river and the irrigation channels in the former research sites, one can see that the irrigation channels are well maintained and that farmers periodically harvest the grass on the riverbanks. From a distance, one can see numerous small white plastic flags stuck orderly in the ground every 10m, on both sides of the irrigation channel. What do these flags mean? Are they a border line? Yes, they are. No one is now allowed to harvest the grass except for the farmer who sticks the flags on the ground. The grass now means something to the rural communities as it has played a key role in local economic development, while contributing to solving environmental problems.

In summary, through participatory development communication, researchers, extension workers, government officials and cooperating small farmers were able to work hand in hand to solve certain community problems.

Through the introduction of sheep raising as an alternative economic activity, the small and landless farmers living in the farming communities along the sides of the Cimanuk River have been able to ensure the management of the meagre public natural resources in their environment.

Problems encountered

This story would not be complete without describing some of the problems that the researchers encountered in the course of the initiative.

The first problem occurred during the early phase of the fieldwork, when we were trying to gain acceptance and support from the village heads. Most of the village heads in the research sites had shown great interest in the benefits that they could obtain from the activities that were to occur in their areas, and they were keen on having a cut.

Denying this opportunity to the village heads could have meant troubles ahead or, worse, it could have led to the failure of the initiative altogether. In contrast, letting them have a share of the benefits could possibly spare the researchers and the farmer leaders many feuds and tensions with the village officials.

For those reasons, the wisest choice at that time seemed to be to give them a small slice of the benefits. With that in mind, we included the village heads' relatives on the list of sheep recipients, provided that they agreed to abide by the project's policies and rules.

It seemed to us that by doing this, we would save a lot of energy and avoid further conflict between farmer leaders and village heads. This compromise would also make it possible for us to use the village halls for training meetings.

As the activities unfolded, we found out that high mortality rates occurred among the project sheep raised by the village heads' relatives. Why? The answer was simple. They were not farmers and did not have the necessary skills to raise sheep. Moreover, they did not spend enough time on their tasks. Consequently, they failed in raising the sheep and also failed in returning the sheep offspring.

The second problem we encountered was the implementation of inappropriate feeding practices. Farmers were used to feeding natural grass to their animals and gave similar feeds to the new prolific sheep, without adding enough legumes in the rations. As a result, many animals suffered malnutrition.

Malnutrition brought about many miscarriages among the pregnant ewes and high mortality rates among newly born lambs. To solve these problems, the research team introduced urea molasses block as a protein and mineral supplement for the sheep.

The third problem was the lack of grass and forage supply during the dry season. This recurring problem was seasonal and continuously haunted the farmers year after year. At the beginning of the fieldwork, the research team, together with certain farmer leaders, initiated a demonstration plot to cultivate improved grass and legume varieties in the four research sites. However, very few farmers were willing to cultivate the grass and the legumes, arguing that the grass supplies had always been abundant. Why bother cultivating it?

Another difficulty encountered was that farmers did not respond well to the suggestion of silage-making as a way of preserving the grass, particularly during the rainy season, when there is a large supply of grass. They did not like its smell.

Lastly, after the project's end, when we had to reduce the frequency of our visits to farmers due to limited funds, many farmers did not keep their promise to return good-quality sheep offspring to the initiative. This situation badly affected the efforts to revolve the sheep offspring to other potential farmers. The sustainability question then came up. How much longer would the project last?

Despite these shortcomings, this initiative provided a number of learning opportunities. Looking back on the work accomplished, we can draw lessons and identify the most outstanding features that contributed to making it a success story.

First, conducting a comprehensive baseline study on the research sites was very useful in order to understand the situation, the problems and the opportunities related to local natural resource development, as well as the traditional systems and indigenous capacities of local communities. This also allowed for the development of a sound action plan to implement the initiative.

Second, the strong commitment of a funding partner to ensure adequate funding in order to implement the action plan and prepare the communication materials proved to be a key issue. Furthermore, support from local policy-makers, village authorities and farmer leaders also proved extremely important to ensure the successful implementation of the action plan.

The direct involvement of local extension workers and farmer leaders in disseminating the materials, in distributing the sheep and in providing backstopping activities also seems to have greatly contributed to building the necessary trust with local farmers. Moreover, holding monthly meetings proved to be a good way of getting feedback and monitoring activities, while providing further information to farmer groups about sheep raising and marketing their products.

Finally, holding dissemination seminars on the project's success, with the participation of farmer leaders, the academic community and policy-makers, was very important for the project's replication and expansion.

Participatory Research and Water Resource Management: Implementing the Communicative Catchment Approach in Malawi

Meya Kalindekafe

Water resources are so important in sustaining life and livelihoods that their management can be quite complex in light of the numerous and diversified stakeholders involved. However, the lack of a comprehensive and coordinated

management strategy often results in failure to resolve the water-related problems experienced by local communities. Researchers can greatly contribute to changing this situation, provided that they adopt participatory methodologies and do not limit the scope of their research to the biophysical environment at the expense of the social environment.

In Malawi, responsibility for water resources is considerably fragmented. For example, depending upon water use, the departments in charge may be either the Malawi Water Department, the city and town assemblies, the Malawi Fisheries Department or the Ministry of Agriculture and Irrigation. This situation impedes the promotion of integrated and sustainable water resource management schemes. Coupled with inadequate action research, this has led to a failure to resolve most of Malawi's water-related problems, which include the poor management of water resources at local levels; the lack of mainstreaming of gender issues in water resource management; pollution and associated water-borne diseases; catchment area degradation; and the lack of human capacity to assist local communities.

With these problems in mind, a research initiative has been undertaken by the Biology Department of the University of Malawi to investigate the extent of environmental degradation in relation to the use and availability of water in the Lisungwi, Mwanza and Mkulumadzi rivers area. The research initiative also aims at assessing and documenting indigenous knowledge on water resource management, including the coping strategies used locally.

The rivers under study are located in the Mwanza district of southern Malawi, which covers an area of 2239 square kilometres inhabited by 138,015 people. These rivers are important tributaries of the Shire River, an outlet for Lake Malawi and the main source of hydropower in Malawi. The interest in these rivers stems from the fact that they pass through areas with various degrees of environmental degradation, mainly as a result of human activities. Indeed, the rural location of the study area implies that the local population's livelihood relies heavily on the exploitation of natural resources, water being one of them.

In the long term, the study will be useful in developing sustainable and integrated water resource management strategies and fostering economic development in the area. In other words, the results of this study will provide a route map towards designing appropriate integrated water resource management strategies and plans for the area in order to lessen water-related problems and to help achieve local social, economic and environmental goals.

To do this, the study will more specifically try to determine the social and economic influences that should be taken into consideration in policy-making. It will also try to identify the indigenous knowledge that already exists at the local level and that can be used as a stepping stone to develop water resource management plans. It will also pay special attention to gender issues and possible ways of integrating them within the water management schemes. Furthermore, on a more technical level, the quality of water will be assessed in order to determine how safe it is for local communities to use these water supplies as drinking water. If need be, mitigation measures that can be put in place to avoid the degradation of catchment areas will be defined, together with ways of building human capacity at the local level.

Using participatory tools and approaches

In order to enhance participation, an approach called 'communicative catchment' has been used throughout the study. As described by Martin (1991), the communicative catchment approach is an action-based form of research (experts engage in theory-based participatory action with communities) based on systematic thinking (including interrelationship between social and natural environments). It is an approach that allows both experts and local residents to be involved in land-use decisions, as well as in evaluating the long-term effectiveness of their action. As full-fledged participants, communities manage the catchment, while resource managers play a facilitating and coordinating role for community involvement and action.

For a long time, most researchers in natural sciences have concentrated their research on the biophysical environment, while ignoring the social environment. However, experience has shown that the social environment plays a crucial role in the functioning of natural systems. Catchment ecology and management have also provided valuable information for achieving sustainable land-use practices. However, the input and cooperation of local residents is absolutely necessary (Brown and Kalindekafe, 1999). Although various participatory communication or data collection methods have been used over the years to involve local communities and gain their input, in most cases nothing returns to communities once the research project is completed.

In this case, the research team initially conducted a literature review in order to define general guidelines for the study. However, since the root causes of the environmental problems were not known, it was necessary to involve the local community. This was not easy because people have their own priorities when it comes to their livelihood. In order to overcome this difficulty, the study team had to come up with an effective participatory communication approach that could facilitate the relationship with local communities. The communicative catchment approach is being used in this study to ensure maximum involvement and feedback to communities.

The communicative catchment approach, like other like-minded methodologies such as community forestry, is based on concepts similar to those of participatory development communication (PDC), which can be defined as the effective exchange of ideas and information through the active involvement of communities and other stakeholders for improved welfare. In this particular study, the use of the communicative catchment approach is considered a proxy for participatory development communication.

In that context, a number of participatory tools have been used, including questionnaires, focus group discussions, resource mapping and interviews with key informants.

Training

As a first step, enumerators from the study area have been trained in a participatory fashion. The trainees contributed to the improvement of the questionnaires and the development of key questions for the focus group discussions. The training sessions have also allowed for capacity-building in terms of awareness of key issues and the roles that different groups play in natural resource management. Enumerators have helped the researchers by advising them on cultural norms – for example, how to approach women and elders to ensure their cooperation. During the training sessions, the sitting arrangement has been done in a manner that does not place the facilitator as a boss but as a co-participant, which makes the process very interactive. Postgraduate students at times accompany the principal researchers as a way of learning by observing and doing. The knowledge gained from the field experience is discussed further in class and incorporated within their own research projects.

Semi-structured questionnaires

The second step consists of questionnaires being administered for each river. The questions cover all the issues that are crucial to the objectives of the research. This approach requires a lot of time with local people. Enumerators administer the questionnaires under the supervision of the principal researchers. Apart from responding to the questions, people are also allowed to ask questions and make comments on the issues that they deem relevant, even those that are not covered by the questionnaire. This allows for the constant improvement of questionnaires by ensuring that the issues discussed with earlier respondents are not omitted.

Focus group discussions

Focus group discussions, which are open-ended and semi-structured conversations with smaller groups made up of men, women and traditional leaders, have also been conducted. The advantage of focus group discussions is that they encourage the participation and contribution of different interest groups who may not share freely their views and concerns in the presence of other groups' members because of customs and traditional beliefs. One full day is assigned for the focus group discussions for each river.

It is usually difficult to assemble people together in a context where their priority is to solve their immediate needs rather than discuss environmental issues. This is particularly true with women, whose work burden is often overwhelming. The approach adopted in this study consisted of holding discussions as the women continued with their normal activities (in this case, selling goods at a local primary school). The men were easier to assemble in groups.

Key informant interviews

Key informants for this study include retired officers, chiefs, village headmen, field assistants and community development assistants, depending upon availability. Most of these informants are very knowledgeable and cooperative. However, a few of them want their personal agenda to be met rather than that of the community.

Resource maps

As a follow-up to focus group discussions, resource maps are drawn by the same members of the focus group discussions on the following day to examine the different resources used by women, men and other gender groups and the personal and use value that women and men attach to such resources. The purpose of the exercise is twofold. First, it aims to map out the resources that are thought to be associated with dominant socio-cultural categories of 'women' and 'men'. Second, it seeks to map out the spaces used by individual gender groups. This enables the researchers to draw out contradictions between local ideology about gender roles and gender spaces (i.e. what should be) and daily gender practice (i.e. what is). In other words, this exercise reveals local social ideals regarding gender roles and the use of space and resources as it happens in everyday life.

Each gender group is then asked to map out the spaces, places and resources used by women and men using different colours, codes and symbols. Participants are asked to point out and comment upon the key places, features (such as their home or a nearby road), structures and resources that are important to them. They are then asked to identify and draw key places/ spaces that are essential (or peripheral) to their daily activities, as well as the places/spaces which they perceive to be important to men/women and to themselves personally. Participants are not to be interrupted unless they stop drawing, in which case questions are asked to prompt them.

This approach is challenging in that it takes up a lot of time. For this reason, participants are usually given drinks and local food such as *nsima*. Bringing drinks such as orange squash and food such as rice and meat is seen as a special event and attracts people. The meat is bought from the local people and the drinks from local shops. Both researchers and assistants help in the cooking. Although some consider it controversial, the team has discovered that through eating together, people feel that you are part of their group and open up more easily to discussions.

Benefit–analysis charts

The use of benefit–analysis charts, as a point of departure within the focus group discussions, allows for in-depth examination and analysis of who uses and benefits from particular resources. This gives data on who actually benefits from the different resources despite access, control and use. Flip charts are used in the focus group discussions to draw benefit–analysis charts. Researchers explore why women and men use natural resources (i.e. the benefits they receive from particular natural resources) by examining the attributes that women and men ascribe to different resources (nutrition, medicinal use and so on). We also explore who holds traditional knowledge and which resources are commonly sold to local/regional markets and by whom.

Transect walks

Together with the local people, the state of water and of biodiversity is assessed, as well as the general environmental degradation along each riverbank. Local people usually explain the different uses of various species, while the researcher explains the biological and ecological use and the importance of conserving those resources. Both parties learn from each other in the process.

Communication challenges and strategies

In this process, a number of communication challenges may arise. Possible strategies to address the most common challenges are outlined in Table 3.1.

The way forward

The researchers involved in this study have recently joined the Isang Bagsak forum, a capacity-building and networking programme in participatory development communication. Through the various themes posted on the programme's electronic forum, the researchers have been able to exchange ideas and to learn from others on the use of PDC in natural resource management. Ideas that can be applied to the local setting are then communicated to local communities for their comments.

Most people in the study area are poor. They have little formal education and very low levels of basic science, but plenty of traditional ecological knowledge. Therefore, the project will now focus on integrating basic science with sustainable traditional ecological knowledge.

A workshop for the community-based natural resource management (CBNRM) committee members and local extension workers who live and work in the vicinity of the three rivers is planned for the project's end in order to obtain final feedback. The results (including maps produced and photographs taken) will be disseminated to, and finalized at, the workshop. The material will also be used as a basis for future work.

The capacity of local communities is expected to improve through an increase in their knowledge base and the promotion of technological changes in water resource management. As a way of providing feedback to people, maps and photographs acquired during the study will be placed at strategic places throughout the area, such as schools, health centres, churches and

Table 3.1 *Communication challenges and strategies*

Communication challenges	Strategies
Low educational levels, coupled with poverty, result in less appreciation of the need for addressing long-term environmental problems than meeting people's immediate needs.	Before the actual work starts, the researchers spend more time discussing the concept of the study and the benefits of an integrated approach with community members.
Negative attitude towards researchers. In the past, other organizations have interviewed people; but the results have not been communicated back to the communities.	Research team members discussed with the water committees the possibility of identifying places where results could be displayed, but also how they would want the results to be communicated.
Sometimes local activities/events such as funerals affect the planning of activities. When working with a limited budget which does not provide for large contingencies, this can be a problem.	The researcher negotiates with local people an activity date that is not too distant. Researchers have learned to make alternative plans for activities that have to be postponed, such as engaging in further fieldwork or analysing data.
Some areas are difficult to access, so considerable time is spent walking.	Local people advise on viable footpaths to areas. In some cases, researchers start walking a few hours earlier than planned.
Questionnaires take up people's time. Some complain that their time was wasted.	Communication approaches such as focus group discussions are conducted with minimal disruption of daily activities (e.g. at a marketplace).
Rains began late and some scheduled meetings had to be cancelled due to people being engaged in gardening activities.	Interview times are agreed with the village headman.
Communication with the Youth Focus Group, especially with girls, was not possible due to early marriages.	This group is a social category that is no longer consulted and schoolchildren are interviewed, instead, at school.
Some individuals disrupt group meetings.	These individuals are usually disciplined by the local communities.
Few researchers are fluent in the local language.	Researchers make use of local research assistants and teachers.

other common places. It is believed that the visual impressions will constantly remind local people of existing environmental problems and possible solutions. With the sustainable use of natural resources as its objective, the study also aims to build trust in the communities in terms of the project's worth and the relevance of their involvement. As such, those involved in this study are accountable for setting the agenda, disseminating messages, approaching issues, etc. The communicative catchment approach allows for this accountability: once the study is completed, the existing community-based natural resource management committees will periodically monitor the natural resources of the area, especially the 'hotspots', and come up with better ways of managing the environment.

It is hoped that this approach will ensure the integrated management of resources and improve the welfare of local people.

References

Brown, R. and Kalindekafe, M. (1999) 'A landscape ecological approach to sustainability: Application of the communicative catchment approach to Lake Chilwa, Malawi', in FitzGibbon, J. E. (ed) *Advances in Planning and Management of Watersheds and Wetlands in Eastern and Southern Africa*, Weaver Press, Harare

Martin, P. (1991) 'Environmental care in agricultural catchments: Towards the communicative catchment', *Environmental Management*, vol 6, no 15, pp773–783

Communication Across Cultures and Languages in Cambodia

Lun Kimhy and Sours Pinreak

Sharing information or knowledge across cultural and language barriers is never easy. But it can be even more difficult when some of the concepts that one wants to share do not exist in the vocabulary of the other group, or when the latter does not even have a concept for what is being expressed. In such a case, images, pictures and photographs can act as bridges to convey ideas and concepts from one culture to the other. When used by minority groups to translate their worldview into words and ideas, pictures can help to describe the complex web of interrelations that make up their livelihood and improve cross-cultural understanding.

In the province of Ratanakiri in north-eastern Cambodia, the official language is Khmer. On their part, the indigenous people living in the area do not understand that language. Furthermore, they have been living in such

isolation that modern concepts such as laws and land titles do not exist in their vocabulary; hence, they do not have any concept of their meaning.

As part of its Land Rights Extension Programme, the provincial government had a team visit the villages to inform people of their rights and of the country's laws. This was deemed important because indigenous communities were in a desperate situation: outsiders and powerful people had been taking away their traditional land and forests. The communities did not know what to do; but they were worried that their children would be in big trouble. In some cases, newcomers showed them receipts of purchase pretending that these were official documents authorizing the sale of their land. Unaware of the country's laws and of land sale procedures, indigenous communities often believed what they were told and ceded their land. At other times, they were made to thumbprint papers for packets of salt or other goodies, only to realize that they had unknowingly agreed to sell their ancestral land.

Although it had very good intentions, the Land Rights Extension team had a slight problem: its members were all Khmer and did not understand the indigenous languages spoken in the area. Hence, their tools and materials were not suited to what they thought were indigenous people's needs, based on their own understanding of the situation. Most of the time, the indigenous communities did not understand what was being said. Occasionally, when someone was asked to translate the team's material into the local language, the translation was so literal that indigenous people still found it difficult to understand what was being said. Despite these setbacks, the team was enthusiastic and motivated.

Although it was already using participatory approaches in all its development activities, the Land Rights Extension team began looking for alternative mechanisms to increase the participation of communities. At this time, the team's adviser, Sous Pinreak, was invited to join a group which was being trained in participatory development communication (PDC) as part of the Isang Bagsak network. Mr Pinreak became interested in the ideas discussed at the Isang Bagsak meetings and presented them to the Land Rights Extension team, who showed great enthusiasm for this new approach. After discussing it further, team members agreed to try it out.

Deciding on how to go about doing it required further discussion. Team members felt that it would be useful to involve the Community Advocacy Network, which is comprised of community members across the province who have been participating in the formulation of the Forest Law and of the Community Forest Decree.

At that time, the Land Rights Extension team was using a model called the Livelihood Framework as a tool to talk with communities and to analyse their livelihood system. It included four aspects: socio-cultural, economic and institutional elements, as well as natural resources. All these aspects were deemed important for the community to be able to live together (see Figure 3.1). The framework was also found to be very useful in initiating discussions with the Department of Agriculture and for staff members to understand the livelihood of the indigenous communities with which they were working.

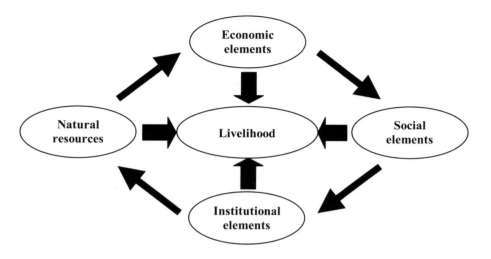

Source: Ratanakiri CBNRM project

Figure 3.1 Livelihood system analysis

The Livelihood Framework was presented to representatives of the net-work, who also thought it was a good tool to initiate discussions on land and resource tenure issues. However, they hinted that pictures, instead of words, would be a more appropriate tool to illustrate each aspect. Moreover, the selected pictures would have to be representative of indigenous people's reality. Thus, the pictures were drawn according to the advice of network members. Altogether, six sets of pictures were prepared. Each of them represented an aspect of the Livelihood Framework, while the last one was a combined picture representing all aspects. Network members were then asked how these pictures could be used to discuss tenure issues. It was agreed that the first picture would represent natural resources. The process unfolded as follows:

- Communities were first asked what each picture represented and how it related to their lives.
- One by one, the different aspects were presented and people were asked several questions in order to relate the pictures to their lives.
- Next, the picture depicting the five aspects and the interrelation between them was presented.
- As a second step, a picture of the same area, but where all the natural resources had been destroyed, was presented to people. They were asked what was in the picture and what would happen to their livelihood in such a situation.
- Once people had a better grasp of the framework, they were asked about the natural resource issues faced by their community.

- They were then asked if they could identify the reasons for the natural resource issues that they had identified.
- Finally, they were asked if they knew how to solve the problems, emphasizing traditional management systems. Most of the time, people replied that traditional systems could solve only the issues that involved indigenous people, but did not work with outsiders. This opening on cultural differences provided the opportunity to introduce new concepts, such as laws and rights.

Existing similarities between indigenous cultures and the Khmer culture made the process easier. Indeed, both cultures respect age and elders, both want harmony and dislike open conflicts, and in both cases decisions are primarily made by consensus. Conforming to existing social norms is also important in both cultures. Based on these similarities, the members of the Community Advocacy Network made suggestions on how to address the issues with the communities. They also emphasized the importance of using the local language during the sessions.

Once the materials had been developed, the Land Rights Extension team started planning its fieldwork. An idea then came up: why not include indigenous people in the team? Thus, a larger team consisting of indigenous people and government staff was formed. However, some difficulties soon arose: government team members, who were living in towns, had planned their activities without consulting the community members who, for their part, lived in the villages. As a result, the latter were forced to follow the timetable set by the government team, even if they had planned other activities. Other problems also came up with regard to payment and motorcycle use. All these predicaments had to be addressed tactfully so that nobody was blamed and a compromise could be found.

The team then started going to different villages with the aim of sharing ideas and concerns with the indigenous communities, and raising their awareness of new issues such as land rights. The results were very good. In a reflection workshop, community members said that 'facilitation as performed by this team was one of the best'. While in previous activities people would just sit and discuss among themselves during presentations, in this case they listened, asked a lot of questions and greatly contributed to the discussions.

It was interesting to see that community members responded very well to the pictures. Moreover, having the translation done by indigenous members of the team was definitely a strong asset since their translation was not as literal. This was possible because the translator and the trainer were both clear about the objectives of the workshops and about the use of the materials. Thus, they were able to explain the new concepts by primarily using locally applicable examples. This made the team understand the importance of using local language and locally applicable pictures.

Reflections on the use of participatory development communication

Following this experiment, the team had the opportunity to reflect on the process and on the possibility of mainstreaming the use of PDC in land rights extension activities. For one, team members concluded that participatory development communication sometimes involves getting community members to work with government staff. However, the organizers need to be clear about the agendas of the two sides. If they differ, they should be prepared to reduce or resolve these differences so that the two sides can work effectively together.

Furthermore, involving government staff in activities that successfully use PDC improves their familiarity with this approach and builds their confidence. It also increases their motivation to use PDC again.

Finally, using participatory development communication requires that the organizers have resources for transporting participants and paying *per diems* so that community members and government staff can participate efficiently in the process.

Talking with Decision-Makers in North-Eastern Cambodia: Participatory Development Communication as an Evaluation Tool

Lun Kimhy

Most of the time, external experts are responsible for evaluating development initiative, which they do with little or no participation of local communities. Although this kind of evaluation can undoubtedly respond to certain needs expressed by decision-makers or funding partners, it does very little in terms

of capacity-building at the grassroots level. Moreover, the viewpoints and opinions voiced by the communities can be significantly diluted in the process. In that sense, facilitating direct communication between communities and decision-makers as part of an evaluation process can greatly improve its transparency and effectiveness. Assisting communities to discuss and decide on the issues and concerns to be raised, on the best way to express them and on the most appropriate tools to be used can also contribute to building their sense of ownership of the development initiative, while improving local governance.

However, this is easier said than done, and making it a reality can be quite a challenge. In north-eastern Cambodia, the Partnership for Local Governance decided to take up that challenge and apply its newly acquired participatory development communication skills to an experimental evaluation process where indigenous communities communicated directly with government officials. Galvanized by the possibility of being heard at the highest level, participants did not even back up when faced with the difficulty of learning to use high-tech communication tools.

This story tells of indigenous communities who, although they speak very little Khmer (the national language of Cambodia), presented their project evaluation findings to high-level Khmer-speaking provincial government officials. The presentation included recommendations for the officials in charge of a large-scale development initiative, who are more used to telling communities what to do than listening to their concerns.

With a view to responding to the development needs of local communities, the provincial government of Ratanakiri has been implementing a comprehensive, multiyear initiative called the Community Natural Resource Management Project. This initiative aims at assisting highland indigenous comunities in their efforts to acquire greater control over the natural resources they have traditionally used and to improve their swidden farming systems. As such, it supports participatory development activities in community-based natural resource management (CBNRM); improvement of farming systems; local government planning; non-formal education; land titling and conflict resolution; public information; and gender issues. An evaluation is conducted at the end of each year by a team comprised of representatives from the different provincial line departments.

After six months of learning participatory development communication (PDC) through a distance-learning programme, the Partnership for Local Governance (PLG), which plays an advisory role to the project, approached the provincial government to discuss the possibility of a more participatory type of evaluation. Since they had been working together for several years,[1] a relationship based on mutual trust was already in place. Furthermore, these same people regularly socialized outside work on occasions, such as during picnics and dinners. Based on this strong relationship, both sides were used to working with the understanding that the activities undertaken were aimed at building the government's capacities, as well as those of communities. Therefore, the provincial government immediately showed great support for the idea and agreed to allow the communities to evaluate the project.

The various partners involved in the initiative were then asked to identify community members who had experience with workshops and meetings and were more confident compared to other community members. The selected community members were then asked if they were willing to participate in the evaluation process. At an introductory meeting, the director of the Provincial Department of Environment explained that the government had previously performed evaluations, but sensed that some issues raised by the communities had been diluted. This time, the government wanted to hear the voice of the communities directly. Thus, the objective of the evaluation was to empower communities by providing them with the opportunity to express their views directly to decision-makers. Most of the invited community members agreed to participate in the experiment.

After forming the team, government staff, project advisers and the selected community members discussed the indicators that were to be used and the information to be collected. This helped community members to get a better grasp of the project's rationale. They also developed checklists and questions, which they pre-tested. The next step aimed at discussing and deciding on how the information collected would be analysed (see Figure 3.2). At that point, three government staff members helped the community evaluation team, while three members of the Partnership for Local Governance staff ensured that government staff actually coordinated the process without dominating it.

The rationale behind the framework used to analyse the information is as follows: the facilitation of development activities is key to creating a feeling of ownership. Good facilitation, on its part, allows everyone to participate and decisions to be made by all or by a representative sample of the population in a way that is judged to be fair and equitably involves men and women. If these three factors are properly facilitated, the process results in good ownership by the community. Finally, it is believed that good ownership can bring about the changes that the project is trying to achieve.

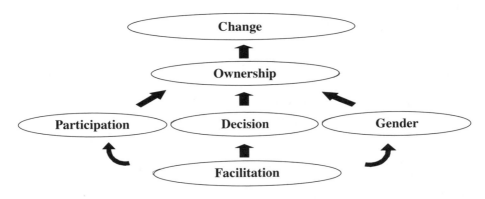

Figure 3.2 Framework for analysing the collected information

After the roles and responsibilities of the evaluation team were agreed upon, information was collected from 15 villages. The data was organized under the following headings: facilitation, decision-making, participation, gender, ownership and change. The team then summarized the data under each heading. The next step consisted in asking the team to reflect on what the information told them about the different headings.

The main findings indicated that, most of the time, government staff, officials and workers use the Khmer language, are impatient and don't listen to the opinion of communities. As a result, activities are mainly determined by government. The evaluation team requested that the project's goal and objectives be explained succinctly, using the local language as much as possible. Furthermore, government staff should consult and discuss clearly with community members before making any decision. They should also agree on a schedule of activities well ahead of time.

The communities then had to present these results to the government. At that point, it became important to consider the tools and methods that could possibly be used to convey the information in a manner that would be acceptable to the participants and within the capacity of the community. Based on a methodology typical of communication strategy design, the evaluation team was asked to consider who the audience was and what attitude it typically holds towards the communities. The issue of who should do the presentation was then addressed, as well as its design. Where did they think they could have problems? This showed that even though communities were able to produce good results, it was equally important to present them properly and to define the basic elements of any communication strategy.

It was decided that the indigenous members of the team would present the findings. Each team member was assigned a different part. Having decided that they would take turns in presenting their findings, people wanted to use simple, short and sharp messages with a clear sequence and flow. They also wanted pictures to be part of the presentation. Having recently bought a new liquid crystal display (LCD) projector and having experimented with Power-Point presentations, the Partnership for Local Governance representatives hinted that a PowerPoint presentation could be appropriate for this situation. The presentation was then developed jointly, and community members immediately tested it. The testing included questions by the advisers and by other team members.

During the workshop, the team made an effective and clear presentation. Government officials were very impressed since they never expected community members to be able to use this technology. Community members were able to answer all the questions raised by the government officials, who were very interested in finding out how the evaluation had been done. The two parties understood and listened to each other. This showed that the normal sender–receiver (government–community) relationship can be transformed when the receivers are in charge of the communication process.

The PowerPoint presentation was very new to Ratanakiri and had the audiences glued to their seats. As a result, everyone paid close attention to what was being presented. One of the lessons learned from this initiative is that, sometimes, new technology can be used very effectively to get messages across and to help people forget who the presenter is. But, more importantly, their participation throughout the evaluation process allowed community members to become more confident, capable of defending their findings and of clarifying what they had done. The practice sessions and going through the analysis step by step was very important in providing this confidence.

In general, the officials accepted the findings, with some changes. After the workshop, the project's information and education component began producing audio cassettes in the local language; translation into the local language became more frequent during discussions, and communities were allowed to discuss among themselves in their own language.

It is not completely sure that these changes were brought about only by the evaluation findings since a number of other factors were also influencing the provincial government at the time. Various organizations were, indeed, insisting on the use of local languages, on the importance of being patient and on the need to listen to community members.

To assist indigenous communities communicate with high-level provincial government officials requires time, resources, energy, patience and effort. Management can facilitate this process by developing a good working relationship with government partners, holding regular meetings with them and using project documents to get them interested in trying out better ways of implementing development initiatives and building a stronger relationship with community members. Management must also support community members as they learn new concepts and experiment with new technology.

Note

1   Advisers had worked with the government officials in the Cambodian Relief and Reconstruction Programme that had preceded PLG, as well as with two previous phases of an International Development Research Centre (IDRC) project.

From Rio to the Sahel: Combatting Desertification

Ahmadou Sankaré and Yacouba Konaté

It is estimated that 900 million people around the world are affected by the phenomenon of desertification, which is bringing with it drastic changes to the climate and to human activity. In 1994, in follow-up to the Earth Summit that was held in Rio de Janeiro two years earlier, an international Convention to Combat Desertification (the CCD) was signed. That event marked an important step in the worldwide effort against desertification. Of course, agreements negotiated in grand international forums do not always

have the hoped-for impact at the local level. Nevertheless, this convention represented an innovation in that it involved local people from the outset as the driving force in its implementation. It was in this context that the Permanent Interstate Committee on Drought Control in the Sahel (CILSS) set out a few years later to develop and try out a participatory communication approach. The idea underlying this approach is that those who see their food security threatened and the biodiversity of their lands impoverished are best placed to design and implement initiatives to protect their environment.

The idea emerged at a time when most countries of the Sahel were working to design their own national action programmes for implementing the CCD. Recognizing the importance of a participatory approach and of forging partnerships with all stakeholders and, most importantly, with the local people, the member countries of the CILSS attempted to find new ways of doing things. An analysis of the strategies used during the past showed clearly that their top-down approach severely limited their impact. By focusing on disseminating information and exhortations, those strategies missed the key point, which is to involve the people themselves.

The experiment with participatory communication as a methodology was conducted in two member countries of the CILSS: Burkina Faso and Chad. Seven sites were selected for the trial, four in Burkina and three in Chad. The experiments were designed to attack problems of great diversity. Yet, all were directly related to the phenomenon of desertification and its consequences on people's daily lives. This chapter discusses four of these experiments: two from Chad and two from Burkina Faso.

Doum-Doum: Halting the desert's advance

Located some 200km from N'Djamena, the capital of Chad, the sub-prefecture of Doum-Doum covers an area of 2647 square kilometres. Its population is estimated at 1000 inhabitants. The locality faces a problem with the sanding-up of its polders and wadis, the main zones suitable for farming. In fact, like most of Chad, Doum-Doum is a very arid zone, almost a desert. The polders (dyked lands) are the beds of dried-up lakes where the soil is highly fertile, containing a great deal of organic matter. The depressions between the dunes, known as wadis, have a silt-clay soil and a shallow water table, which makes them highly productive as well. The encroachment of sand into these lands thus constitutes a real threat to local food security.

It was in this context that the Doum-Doum rural development project undertook its experiment in cooperation with the CILSS participatory communication project.

The initiative began with a series of information sessions and discussions in order to identify the problem and to sound out local ideas for solutions.

To do this, the facilitators first organized information and discussion meetings with the key opinion leaders. They then visited the polders and the wadis in order to assess the scale of the problem. Next, they held chat sessions with local people to analyse the problem jointly and to propose solutions.

At the end of these discussions, people took upon themselves the task of reforesting their environment by growing 10,000 tree seedlings for each zone, to be transplanted when the rains came. Farmers were careful to choose species that were well suited to the zone and resistant to drought.

These chat sessions served to 'loosen the tongues' of local people and encourage them to express their needs – for example, the need for wide-diameter wells for themselves, for their livestock and for the plants that they would shortly have to start watering. Thanks to these sessions, the villagers also saw the advantage of organizing themselves in groups. Today there are 20 women's groups, whereas there were only 13 at the outset.

One activity involved visits to other zones and sites as a way of making local people familiar with solutions adopted by communities facing similar problems. Two delegates from each zone travelled to the sub-prefecture of Ngouri, where they were impressed with the successful use of hedges or 'live fences' and decided to replicate that experiment at home.

People also recognized the need to strengthen their capacities. The facilitators therefore arranged training sessions on sowing and watering techniques. At each session, the facilitators used flip charts to illustrate their points. This strategy served to trigger in people's minds an awareness of the role that each could play in halting desertification.

Moreover, during the sessions, the older people of the village were able to recall the environment as it was during the past when, although it was threatened, it was not so severely degraded. This provided a lesson for the younger generations who are gradually losing their farmland to the advancing sands.

The training session featured films on the environment, organized for each zone under the eye of the facilitators. Here, again, people were prompt to react: one 60-year-old man from Kouloudia exclaimed, after watching one of the films, that 'Sanding is not inevitable; we know that there can be solutions even here.' The solutions that these farmers are now investigating involve the planting, tending and maintenance of plants around their fields, whether in the wadis or in the polders.

Participatory communication thus helped the communities to identify their problems and analyse them, while bringing unspoken needs into the open and facilitating home-grown solutions.

The principal outcomes at Doum-Doum

Although the extreme heat and lack of water were hard on the seedlings, of the 30,000 planted in the nurseries, 10,341 survived and were transplanted to fix the dunes. The area ended up with 8km of plantings, four to five rows deep.

The experiment with participatory development communication (PDC) also highlighted the damage caused by cutting trees for firewood and the importance of using substitute fuel sources. Dialogue on these questions addressed the use of improved wood stoves and the burning of cornstalks as a way of saving wood.

The initiative also made farmers aware of their capacity to analyse problems and to find local solutions. This sparked in many of them the determination to take their fate into their own hands and to plan for the future. They also came to appreciate more thoroughly the value of natural resources and to draw up local regulations to combat desertification. In fact, the communities showed a real desire to be active rather than passive players.

Bol: The silent revolution

In the area around Lake Chad, there is no clear relation of ownership between the land and those who work it. The area is heavily Muslim and the land belongs theoretically to God. In fact, it belongs to the first village or clan that settled it. Villagers may have user rights to lands located on the dunes, in the polders or in the wadis. There are apparently no sales of land, which are, in fact, prohibited.

Outsiders have the right to cultivate vacant lands; but they have no permanent user rights. They can, however, acquire the land if they live long enough in the locality or if they marry a girl from the village. They then become full members of the community and this status gives them definitive rights to the land.

The situation is quite different for women, even those born in the region. Although they are actively involved in all stages of farm work alongside their husbands, they are seldom awarded title over a plot. Apart from working on their husband's land, their main input is confined to harvesting the crop, caring for the animals and selling livestock by-products. It is not for any shortage of land that this is so, and some of the polders developed by the Société de Développement du Lac, in particular those at Berin and Guini, are still underused and even unoccupied. Women could make use of these lands and contribute to family income and to the community's food security, while enhancing their own economic independence. Yet, tradition insists that women have no right to the land and this is holding back the development of the villages around these polders.

It was this very complex problem that the participatory communication initiative attempted to resolve. The results that the facilitators achieved, by sparking real dialogue on this issue in the villages, are nothing short of astounding. Some people even speak of a silent revolution.

Women as landowners

'Now we talk openly about things, whereas before we tended to keep quiet. The women's group that this project has helped to set up can only make us even more open.' Those were the words of a woman in Sawa, one of 30 small villages clustered around the polder, which is now being improved. The women of this village are making plans for the future: 'With the income from our plots we are going to buy a mill, a seed drill and some threshers.' The women of Bol now have full rights to the land and they are determined to become increasingly independent and take care of themselves. Were it not for the delay in building the irrigation ditches, they would already have their plots. They have been assured, however, that the plots will be distributed equitably once the works are finished.

The communication strategy

The communication strategy that brought the village to this point unfolded in several stages. First, the facilitators secured the support of the traditional authorities, particularly the Sultan of Bol, who is greatly revered by local people. He agreed to go up and down through the villages in an effort to get the men to change their minds about making room for women in their proceedings.

Enlisting this traditional personality was helpful in obtaining husbands' consent for their wives to take part in the initiative. In many parts of Africa, a woman cannot make any public moves without the consent of her husband. At a public meeting, women may accept a proposal; but until their husbands have given approval, they will not do anything. The Sultan's involvement helped to overcome the husbands' initial reluctance, and they were readier to accept their wives' participation.

The facilitators next held a series of meetings with the husbands, and then with the women. After a few chat sessions the husbands finally agreed that turning over the plots to their wives would be a plus for the family budget.

Yet, despite this consent from their husbands, women's right to speak out was not yet assured. This was something that developed over a series of stages. As always, at the initial discussions it was only the head of the women's association who spoke: what she said had an 'official' ring to it and was not particularly interesting. In such a setting, the spokeswoman says only what has been agreed, which is what her audience wants to hear. The same was true for the next three or four sessions.

Among the women of Bol, the floor at a meeting is awarded by seniority. The women's president is often the oldest woman of the village or someone who is close to the chief. The ones who follow her in speaking will be the older women. Their interventions cannot last very long, two or three minutes at most, filling a total time of 20 to 45 minutes.

The transition between this and the second stage is rather delicate. It depends upon several factors: the degree of trust that the facilitators have won, their tactfulness and their familiarity with the setting. The facilitator must be aware of local traditions and he must speak in proverbs and allegorical turns of phrase in order to put the crowd at ease and encourage them to speak.

It is during the third phase that the real give-and-take occurs, and this can happen without warning. It is noticeable when people begin to draw out their statements (from two or three minutes at the beginning to perhaps ten minutes) and when women no longer wait to be given the floor. A greater variety of topics are then addressed; people will speak about anything that comes into their minds. It is at this point that the women begin to discuss their problems and put forward their thoughts. At Sawa, it took more than an hour to reach this point. All those who spoke at the beginning paid reverence to the Sultan, betraying his considerable influence, before they got to the problem they wanted to raise. Then came a point when three women jumped up at the same time to speak, and the meeting had to be called to order so that everyone could have a turn.

One of the sessions ended with a spontaneous popular dance. This was something the husbands had not seen for a long time. A *marabout*, or holy man, confided to one of the facilitators: 'Your coming has liberated our women; we've never seen them that way before!'

Some valuable lessons

The village of Bol has decided to address a very sensitive problem: the marginalization of women in the distribution and farming of improved lands.

Our experience showed that when communication tools are properly used the community becomes a frontline player. In these cases, the participation of traditional authorities such as the Sultan, the district heads, the women leaders, the *marabouts* and the village chiefs was crucial. It was they who facilitated the meetings with the women and took part in them initially.

This approach did not involve any mass communication tools such as radio or television. The emphasis was on chat sessions, supported sometimes with photographs, videos or tape recordings dealing with experience in neighbouring communities.

One of the most effective communication tools turned out to be photographs from the discussion sessions. They were posted in the village square where everyone, and especially the women, could see them and could point to themselves taking part in debates about their own very real concerns.

The tape recorder was always at hand for recording the discussions, and listening to the cassettes afterwards sparked very positive reactions. The showing of films, followed by debate, encouraged people to speak. After viewing a video on how women in another region had secured rights to land, one woman in Sawa declared to the audience: 'We want to be like our sisters in the village of Matafo, who organized themselves and won access to land. Now they can use those harvests to meet family needs without turning to their husbands.'

Beli: Managing common pasturelands

The Beli transfrontier project is attempting to set up a suitable framework for the shared use and management of pasturelands. The idea was to facilitate communication as a means of creating this framework. In the end, the project led to the establishment of decentralized committees in the villages.

A fragile ecosystem

The district of Tin-Akoff lies along the border with Mali to the north and Niger to the east and has some of the best pasturelands in the province of Oudelan. The Beli River, which flows through the district from west to east, is the focal point for herders.

The river supports shrubby thickets all along its banks and affords rich pasture for the herds. Once the haunt of wild animals and visited only by the boldest herders, the land today draws herds from other districts of Oudelan and from other provinces, as well as from villages in the neighbouring countries.

The sparse, shrubby vegetation of this chronically rain-deficient region is at risk of rapid degradation from overexploitation. The river, which once ran deep, dries to a braiding of trickles by January because of the invasion of sand and the great numbers of livestock. Its banks are increasingly being taken over by fields and gardens. The lack of defined livestock paths to the water means that the whole ecosystem is heavily trampled.

Although hunting was once a favourite pursuit, the district is now losing its wildlife. One of the explanations proffered pointed to the widespread poaching that has decimated the population of wild animals or driven them to take refuge in neighbouring countries. Yet, there are still some flocks of gazelles to be seen and the occasional hyena or jackal, as well as many species of birds.

There are two social groups that pasture their animals in Tin-Akoff: the Kal-Tamachek and the Fulbe (or Peul). They have long shared the area and they both speak the same language, Tamachek. Herding is the main source of livelihood, but most people also do some farming. The area around the village is therefore divided into areas for pasture and for crops.

The fields were traditionally located upland among the dunes; but with recurrent droughts, they are now to be found increasingly along the banks of the river. As for the pasture areas, they shift from the dunes to the river, depending upon the season.

The communication strategy

The communication strategy has three main objectives: to share knowledge about natural resources, to introduce localized frameworks for cooperation and to make those frameworks more effective. The main partner responsible for carrying out the transboundary Beli project was the Waldé Ejef Association. It selected as facilitators men who were familiar with the locale and women who were socially respected (one of them came under the authority of the traditional chieftainship). The only drawback was that they were not based in the villages but came from the departmental capital.

To achieve the first objective, the facilitators and the villagers used chat sessions, posters and videos as their tools. Two films provided backup for the discussions. One dealt with the Oursi pond and the other with the management of hunting areas in eastern Burkina. Accompanied by local foresters (the facilitators are not natural resource experts), they engaged people in discussion about the status of their resources, the pressures acting on them and the best ways of preserving them.

The facilitators succeeded in getting the villagers to talk about what was happening to their natural resources, about their relations with migrant people and about the unbridled exploitation of vegetation and wildlife. In all the villages, people complained that wildlife was disappearing because of illegal hunting and they were disgusted with poachers who would arrogantly brandish a hunting permit issued by the authorities. They were unanimous on the need for committees to control hunting.

Taking the bull by the horns

The experiment at Beli did not produce any results of a physical kind. Yet, the facilitators' efforts encouraged a movement to social organization: in the eight villages there are now associations of farmers, herders, women, youth, and wildlife and fishery management units. Organizations of this kind can help to sponsor development activities. In some villages, steps have been taken to create tracks for livestock to reach the river. During the meeting with government officials and the project coordinators, people let the authorities know that they wanted to be involved in environmental management. Initiatives are now under way to obtain permits for wildlife and environmental management. If those efforts succeed, they could well lead to income-producing activities. In short, local people and government extension services are now working closely together.

Not everything has run smoothly ...

Despite these encouraging results, the Beli initiative ran up against obstacles that hobbled its activities. In addition, the distance of the facilitators from the field (at the start of the project only one facilitator resided in Tin-Akoff, while the others stayed in Gorom-Gorom, about 70km away) severely constrained contact with local people. Moreover, there were not enough facilitators (only three for eight widely scattered villages); even worse, the first group of trained facilitators had to be replaced, in time, by others who had received no instruction in the approach before they were dispatched to the field.

The lack of commitment on the part of the authorities also led to some problems. Although they were fairly familiar with the project and its objectives, they were still confused about the difference between the Beli project, Waldé Ejef and the participatory communication initiative. This confusion was not of great consequence because all players were working towards the same end. Government officials, however, complained of the lack of information and the shortage of financial and material support, which sharply curtailed their participation. In fact, when the authorities are completely empty handed (the provincial livestock office has no electricity or telephone, let alone vehicles and fuel), their field presence will be very limited.

Ouarkoye: Fighting bushfires

Every year fires consume 30 per cent of Burkina's territory, with devastating impacts on vegetation, soils, animals and people. The district of Ouarkoye, in the region of Boucle du Mouhoun, is no stranger to this scourge.

The most severe consequences of bushfires include the following:

- the decline in forested areas resulting from bushfires and from the expansion of the farming frontier;
- the inability of vegetation to regenerate itself because of persistent wild-fires;
- accelerated erosion along watercourses;
- reduced harvest yields;
- loss of soil fertility;
- the great frequency of fires (61 per cent of the surface area is burned every year); and
- shortage of pasturelands.

The communities in the district of Ouarkoye, like most of those in the western part of the country, are headless societies in which the chieftainship has little hold on the population. Decision-making power in all matters, including natural resources, lies in the hands of the land chief, who parcels out the land, particularly through indigenous families. Traditional customs remain strong and exert a firm hold on social life.

The traditional channels of communication operate through the markets that are held once a week or every three days, or through places of worship, events such as marriages and baptisms, and the many snack bars in the area.

The people of Ouarkoye, faced with the degradation of their natural resources, could never agree on how to plan and implement a proper fire management programme. Consequently, they seemed to have little interest in protecting their resources.

The communication strategy

The first step was to discuss with the villagers the causes and consequences of bushfires. After making initial contact with the customary leaders in the district, the facilitators turned to a theatre troupe to dramatize aspects of the issue by means of a play. This production toured the villages, sparking debate on the issue of fires and encouraging people to speak about it.

At the same time, the facilitators and the fire management committee heads met with different social groups such as women, village elders, leaders, young people, teachers and hunters to provide them with information and gather their opinions.

The local radio station in the province also helped to convey information and to feed discussion by broadcasting a series of programmes on the fire problem. These were transmitted in the local Dioula language and they struck a real chord with the villagers, some of whom went to the station for further information.

Gradually, the leaders of the village fire management committees, and then the general public, learned about the techniques of fire management.

Working with the traditional chiefs

The traditional chiefs played a key role in the project's success. Their support and involvement quickly won over all the stakeholders. The same could be said of the use of theatre – indeed, the village troupe that was forged during the first performances went on to carry the show to many other villages.

Results and constraints

At the end of the project, each of the 28 villages and farming settlements had established its own fire management committee. Subsequently, two training centres were set up and are now operational. They have trained 362 instructors, who in turn have provided training to the villagers.

Among the initiative's most important outcomes were the creation of a local theatre troupe (called Sininyasigi) and four community activity centres, as well as the production of eight programmes dealing with fire management, which were broadcast over the local radio station Fréquence Espoir.

According to one facilitator: 'With forum theatre and the video shows, farmers were able to see what it's like to live where there are no fires. They could compare that situation with their own burned lands.' Members of the forest management committee at Miena (a village surrounded by high grass that was thoroughly dry at this time, in March 2002, something that was unimaginable before the PDC project) explained: 'The soil has been enriched because where there's been fire, and where there hasn't, it's not the same thing at all!'

Apart from these concrete outcomes in terms of communication, the experiment also produced the following results.

A significant reduction in bushfires

There was a sharp decline in bushfires in most of the villages where the project was undertaken, thanks to action by the villagers themselves who built firebreaks several kilometres long to protect the sacred forests, the villages and the fields. As an indication of the improved environment, a number of wildlife species that had disappeared have now returned to the area. In August 2001, an elephant moved into a forest protected by the villagers, and this was quite a novelty.

A qualitative change in the relations between stakeholders

The project brought about a marked shift in the attitudes of the facilitators, the forestry officials and the local populace alike. 'In the past', admitted one forest officer, 'we acted like policemen over the peasants. We would call the farmers together and tell them things, and that was all. It was a completely unilateral approach. Today, it's the villagers who explain what has to be done or not done, thanks to the stage play.'

'Before, if a farmer saw a fire he would keep mum about it', said one of the facilitators. 'Today, the villagers are quick to sound the alarm and go put it out.' The facilitators also insist that, before the project, it was hard to find the people who set fires; but that is no longer the case today.

Farmers care for their land and help the forestry officers

Today, farmers feel responsible for their land and they look after it diligently. The facilitators found that farmers are fully committed to managing fires on their land. They settle minor legal disputes by themselves, with the help of the forestry officers. They set and enforce fines, a portion of which goes to the village. The traditional chiefs establish special areas for ceremonial fires, which are now kept under strict control. The village of Kebaba has even created a 2000ha forest reserve, and most of the other villages are hoping to emulate that initiative.

Renewed dialogue and trust between the forestry service and local people

Before the forest management committees were established, forestry officers were somewhat lax in performing their surveillance duties. This has now changed. 'No longer does a forestry officer simply draw up a programme alone in his office and then put it to a sparsely attended village meeting. Now, it's the fire management committees that organize the meetings. The people are more comfortable in talking about things and they take an active part in the debates', says one facilitator.

Stronger partnerships with the technical experts

Partnership with the main technical services (farming and livestock, the environment) has been strengthened and extension workers are now given training in the local language.

Communication tools at the service of local people

Throughout this PDC experiment, we used simple, inexpensive and locally adapted communication tools. We did not resort to mass media, such as radio or television, but relied on the local tools at hand. In all cases, the facilitators made local people the focus of their strategies, using home-grown modes and channels of communication. The emphasis was on maintaining contact with the local traditional authorities and on conducting chat sessions, sometimes supported with photographs, videos and tape recordings from experiments in neighbouring communities. Moreover, the experience showed that when tools are carefully selected in light of objectives and are placed in the hands of local people, it is they who become the principal players in development.

The first phase of this action research project wound up in May 2002. The approach taken and the successes achieved were analysed during a regional roundtable that brought together stakeholders from member countries of the CILSS with an interest in this approach. Convinced of its utility, all of them expressed their desire and their intention to extend this experiment to other member countries of the CILSS.

Engaging the Most Disadvantaged Groups in Local Development: A Case from Viet Nam

Le Van An

Policy changes aimed at protecting the environment can have a major impact on local communities when it implies radical changes to their way of life. This was the case in central Viet Nam, where the government established new regulations in order to better protect the forest. For the ethnic minorities whose livelihood was based on swidden farming, this meant that their ancestral knowledge was no longer useful in the new context. This could have signified further marginalization and impoverishment of the local population. However, the use of participatory methodologies and, later on, of participatory development communication (PDC) made it possible to engage local farmers

in a process aimed at discovering and trying out new ways of making a living, while safeguarding the natural resource base. Special attention was paid to those sectors of the population who usually do not benefit from development initiatives, with surprising results.

In Hong Ha, a commune located in the province of Thua Thien Hue in central Viet Nam, the livelihood of the upland minority people used to be largely based on slash-and-burn cultivation and on the natural resources from the forest. However, during recent years, a new government policy has been put in place to encourage these farmers to change their practices from swidden to sedentary farming in order to improve their access to social services and ensure the protection of the forest. As a result, they have had to change their agricultural systems. This is not easy for them as their indigenous knowledge and skills are no longer relevant in the new production system.

The main problems faced by the upland people in central Viet Nam are poverty and the degradation of natural resources. The lack of food causes serious hardships to families since many of them do not have enough rice to eat for three to five months every year. Cassava and upland rice used to be their traditional food crops; but the population growth, combined with the limits of swidden farming, has resulted in a sharp decrease in food production. Although wetland rice is being replaced with upland rice, the paddy field is small and there is no water supply for the cultivation of rice. As a result, during the food shortage season, poor farmers have to go to the forest to collect non-timber products and sell them in order to get enough income to buy their daily food.

Moreover, the area's natural forest was seriously damaged during the war due to fire and the use of the chemical defoliant known as Agent Orange. After the war, when people started to rebuild their lives, trees were cut down to be used as construction materials and to be sold for income. As a result, the quantity and quality of forested areas have greatly decreased over the last decades. Since the early 1990s, the government has been making great efforts to invest in forest management and has undertaken a number of replanting forest programmes. In order to support these forest and land management efforts, a forestry law was issued in 1991, followed by a land law in 1993. Thereafter, the government has made a number of decisions with the aim of increasing the rights of farmers with regard to land and forests. However, the necessary conditions to apply these new policies are not yet in place in the upland areas due to the poor management system of local organizations and the lack of understanding on the part of the local population.

As a matter of fact, most of the land in Hong Ha falls under a watershed management scheme. Local people only have access to about 1 per cent of the total land area for agricultural production. Given the population growth, the changes introduced in agricultural production techniques and their limited access to resources, one truly wonders how the upland people can possibly improve their livelihood and their resource management practices with a view to sustainable development.

Hence, given the existing needs in upland central Viet Nam, since 1998 the University of Hue has developed a research initiative entitled

Community-Based Upland Natural Resource Management. This initiative has been implemented by the research team of Hue University of Agriculture and Forestry. It aims to:

*   improve the livelihood of the upland poor;
*   advance human resource capacity; and
*   support policies that perform for the poor in the upland areas.

From the onset, the methodology of participatory action research has been used to work with farmers in order to identify problems, develop potential solutions and monitor the changes and the impact of the initiative on their livelihood, as well as on natural resources.

In 2002, participatory development communication was introduced to this initiative through a learning process that involved other teams in Cambodia and Uganda. The community-based natural resource management (CBNRM) initiative in Hong Ha was chosen as the study site in Viet Nam, where this new approach was to be experimented with. Henceforth, its researchers were trained to use its methodology and tools.

The research site

The Hong Ha commune belongs to the A Luoi district, which is located in the province of Thua Thien Hue in central Viet Nam. There are 21 communes in the district, of which Hong Ha is one of the poorest.

The commune has 230 households and approximately 1200 inhabitants, who pertain to different ethnic groups. The K'tu people are the most numerous, with 47 per cent of the total population. The Pa Co (including the Pa Hy) account for about 28 per cent of the population and the Ta Oi for 16 per cent, while 7 per cent are Kinh or lowland Vietnamese. The official land area that belongs to the commune covers 14,100ha, of which only 180ha are agricultural lands. Forestry lands cover about 11,000ha; but not all of it is considered 'good' forest. There are also bare hills, which cover about 2700ha. Most of the forestry land is under the management of the State Forestry Enterprise and the Watershed Protection Board. Previously, local people practised shifting cultivation in these forest lands; but they have started to turn to sedentary farming since swidden cultivation is no longer allowed by the local and national governments.

Due to these forced changes in their farming system and their limited access to assets and natural resources, production is low and the opportunities for income generation are very limited. Traditionally, the livelihood of these upland people was based on forest resources such as non-timber forest products for food and cash income. However, this is now a minor source of income to respond to their needs as the forests are becoming more and more degraded. Furthermore, the changes in policies and in the management system for these areas have also restricted the access to these assets.

Process, results and impact

The CBNRM initiative aims to develop a research methodology based on participatory approaches. Capacity-building of local people is also a major objective, with a view to increasing their participation and that of other stakeholders throughout the process. Participatory action research is a method aimed at developing a learning process for both farmers and researchers. Besides this new research methodology, the research team was trained in PDC through face-to-face and internet-based learning involving researchers from Viet Nam, Cambodia and Uganda. Nine learning themes were introduced, from approaching a community to developing and implementing a plan, monitoring the unfolding of activities, and evaluating and disseminating the results. In each team, researchers organized as a group to discuss each topic and to post their knowledge on a website. Thus, they learned from the experience of other teams and shared their own experiences in implementing participatory development communication. Indeed, for each theme that was addressed, researchers not only learned through theory, but also applied the knowledge to their research work in the field. Half way through the process, all members of the three teams gathered together in Hue to share their experiences. The final workshop was held in Uganda, where the whole process was evaluated.

In the context of the CBNRM initiative in Hue, the integration and application of PDC has increased the participation of women and the poor in order to improve food production and income generation. It also contributed to building the capacity of local people, especially that of women and disadvantaged people. Furthermore, it led to increased access to resources for local people, especially to land and forest resources. Finally, it put in place a learning and dissemination process to share the results of the initiative with other communities.

Increasing people's participation, especially among women and the poor

It is not easy to achieve 'real' participation by villagers and stakeholders in the whole process in a community-based approach. For many years, in Viet Nam and in other developing countries, technologies or solutions for agriculture and rural development were introduced by outsiders, such as scientists or development workers. The level of farmers' participation was very low. Recently, given that the 'top-down' approach has not yielded real results in terms of rural development and is increasingly being rejected by farmers, the participatory approach has been gaining ground. 'Real' participation can be induced by using different approaches and methods to meet the practical needs of farmers, as well as to enhance their confidence and their own human capital. Furthermore, the use of PDC in natural resource management helps researchers to better understand the situation of the people with whom they work.

Who participates in research processes?

Not all farmers have the same interest and conditions with regard to participating in an initiative. Usually, the advantaged people or farmers who are not poor and have better social capital in the community tend to participate more. Women and the poor, as 'marginal people', are rarely involved in these activities.

There are many ways to approach a community. Participatory development communication aims first and foremost at improving the understanding of researchers. Making a list of the farmers or having them classified according to their wealth levels by community leaders and by different groups of people can help in understanding the community. Each group of people has its own problems and interests. Communities should be approached in an appropriate way, compatible with the existing cultural, social and economic conditions. The more we understand a community, the better the participation will be.

Why do farmers participate?

There are various reasons why farmers participate in development initiatives. In many cases, farmers want to improve their production or income. However, in some instances, they simply want to be involved because development projects can bring them direct benefits.

At the beginning of a research initiative, it is very important to undertake a diagnosis of the situation and to define the research issues. The meetings with villagers should be organized separately for different groups, such as women, men, the poor and leaders. Since the results from these meetings will all be different, researchers and villagers will be able to better understand the situation and to develop the following steps.

How can participation be increased?

Participatory development communication aims to improve the quality of participation. Throughout their training, the researchers had the opportunity to acquire the necessary knowledge and skills to improve their capacity to encourage farmers' participation. They achieved this by developing communication strategies and using appropriate tools, such as video, cameras and leaflets. Learning by doing in the field is also useful in facilitating the involvement of farmers.

The type of questions and the way in which the dialogue is established are also important when talking with farmers. In-depth questions are used to help understand the situation better and to figure out what the farmers' ambitions are. Usually, open-ended questions are used to gather general information. If more information is needed, in-depth questions and discussions come in handy.

Many poor farmers and women benefited from the use of PDC in our study. Numerous farmers said that they had never participated in community development before as they thought that it was not an opportunity for them. Since PDC has been introduced, local people have had greater opportunities to participate in development initiatives.

Improving the local livelihood

Meetings were conducted to identify the needs and problems of local people, as well as the possible solutions to overcome these problems. Since people have different farming activities, meetings were organized with different groups: leaders, women, the poor and men. Each group had different opinions regarding priority activities that should be undertaken.

Therefore, based on the capacity and interests of farmers, different working groups were formed: a rice production group; a pig production group; a fish raising group; a home garden improving group; a cassava production group; and a forestry production group. The research team worked closely with these groups to help them understand that the research initiative was not a development agency, but that it could try to help them by improving the quality of participation.

Each group consisted of a number of farmers. In general, there were no more than 15 people per group to start with. Farmers discussed their problems and the possible solutions that they envisioned might solve them. With the facilitation of researchers, discussions also took place in groups so that an overall diagnosis of the situation could be suggested.

The next step consisted of developing possible solutions, based on the causes of the problems. For example, in the case of rice production, farmers decided to test three new varieties of rice and to try out the application of a fertilizer, as well as transplanting and direct-sowing methods.

In each experiment, about three to five farmers agreed to test the selected options. The group itself made the decision as to who would get to apply the different tests, based on their interests and on the group discussions. The other group members were invited to participate in the evaluation and learning meetings at least three times during the growing season: once at the beginning during the experimental design stage, the second time during the growth period and the last time at harvest time. At each meeting, farmers developed their own criteria to monitor and evaluate the results of the new technologies and made the decisions concerning which varieties were to be sown, which fertilizer was to be applied and which cultivation techniques were the most appropriate. The results from on-farm monitoring and evaluation were shared with other group members and with non-participants. Thus, the learning process also encouraged farmers to disseminate the results to other farmers in their community.

Similar steps were developed with the farmers interested in pig production and fish raising or in other types of activities. In the case of pig production, three experiments took place. The first experiment aimed at raising Mong

Cai as mother pigs or as sows, while the second one focused on raising cross-breeds and fattening pigs. Finally, an experiment was conducted on ensiling cassava root and leaves for feeding pigs.

However, some activities were not successful. For example, mother pigs of the Mong Cai breed were provided to ten farmers. They produced good piglets in the first year, thanks to the artificial insemination service from the university. However, without the support of researchers or students, their pigs could not get this service. Since no boar was kept in the village, the mother pig stopped producing piglets. Eventually, mother pigs stopped being kept in the village.

In general, the on-farm trials that used PDC yielded good results for the groups involved in pig production, fish raising, cassava production, and home garden and forestry production. For example, while only five farmers were using ensiled cassava leaves and roots to feed their pigs in 1998, 38 of them had started using that technique by 2003. Furthermore, the number of pigs raised in the community grew from 60 to 340. On its part, the fish-raising group, which grew from 12 farmers in 1998 to 54 in 2003, diversified its production from one type of fish to four, and increased productivity from about 4 tonnes per hectare at the beginning of the initiative to close to 15 at the end. Cassava production increased similarly, with more varieties now being grown for different uses, both domestic and commercial. Moreover, instead of growing cassava as a mono-crop, farmers now practise intercropping with beans and corn. They also apply contour lines to prevent soil erosion. However, one of the most drastic changes observed as part of this initiative occurred with home gardening. Indeed, while very few farmers used to work in their home gardens at the start, most of them now have home gardens linked to their production system (livestock, fishpond and home gardens). Animals, which used to wander around and destroy plants, are now under control in pens or have been fenced out, thus resulting in a reduction of conflicts in the village.

How can we work with the poor and the disadvantaged groups in a community?

This is a big challenge for any research or development group. Thus, it is important to find activities that really work for the poorer people. Usually, the poor and the disadvantaged do not participate in research or development programmes, and do not have access to the opportunities provided by outsiders or local leaders. For example, since extension workers or development officers want to achieve positive results from their demonstrations, they often work with the most advantaged people, who can demonstrate better results. Conversely, the capacity of the poor and the disadvantaged to apply production technologies that come from outside is limited. These challenges can be overcome if researchers or practitioners understand and follow the process of participatory development communication, while paying special attention to the poor. Second, farmers prefer to work in a group whose members experience similar conditions. In our case, similar-interest farmer groups proved to be a good way of working with the poor.

Furthermore, as part of this initiative, a PDC strategy was used to develop cooperation with the target people in the community. From that point on, action research was undertaken in order to meet the needs of farmers, and concentrated on identifying people, developing solutions and finding the best way to work with them.

Farmers also feel that participatory development communication has made them better understand the communication process between researchers and farmers. Because of their low education level, tools such as videos, cameras, leaflets, posters and role plays made them feel more confident when implementing initiatives aimed at improving their production.

Capacity-building with the poor

For most people in poor upland communities, the first need is food security, achieved by improving their production technology. However, in order to achieve sustainable development, capacity-building with local people is the next step after improving their immediate needs.

In any community, there are formal and informal organizations. Formal organizations are established to manage the community according to the current government system, such as the commune's people's committee, the farmers' association or the women's unionc. With regard to the development of upland communities, these organizations hold functions and responsibilities that have been assigned to them through formal regulations. In this case, the research team worked with the people's committee, the farmers' association and the women's union since these organizations have all established a good working relationship with the farmers. Meetings were organized with a view to empowering them and helping them improve their work.

Improving the capacity of leaders and commune organizations was the most important feature of building the communities' asset base. Participatory approaches were introduced and applied by these local organizations, thus encouraging people to define their own priorities and objectives, while making a contribution to the communities' plans and activities. Training and study visits were also organized to provide learning opportunities and to build people's confidence with regard to their relation to the outside world.

Training and cross-study visits for farmers were also organized, with an emphasis on the participation of the poor and women. The contents of training were developed to meet the practical problems of farmers. For example, before sowing their seeds, rice producers requested training. Given that they found it difficult to remember the material presented in a classroom setting, training sessions were organized in their fields. The fact that farmers learn better by doing became a communication challenge for researchers and practitioners alike since communication plans to be used in the training workshops had to be developed in order to better suit the farmers' conditions.

Given the existence of credit funds as part of this initiative, the women's union and the farmers' association were trained in managing small credit schemes and using credit money to invest in production. While 10 farmers in the commune initially participated in this scheme, after three years, 47 women had benefited from this credit fund.

Moreover, informal organizations such as groups of farmers were established to meet the demands of the poor. People usually have different interests and capacities in working. Each group included the farmers with the same set of interests. For example, groups were organized for farmers who preferred fish raising, pig production or home gardening. Within these groups, the farmers themselves established regulations on crucial issues such as membership and requests for financial or technical support.

Another form of asset-building consisted of working with the most disadvantaged in the community, particularly women. The fact is that the poor and women have a lot of potential for development. However, in most initiatives, the activities in which women and the poor are able to participate are usually quite limited. In this initiative, the team made special efforts to meet the poor in the community and give them various opportunities to become involved. There is no doubt that the design of a more appropriate communication strategy, coupled with the use of participatory approaches and tools, greatly facilitated the participation of the most disadvantaged groups within the community. Moreover, the fact that financial and technical support were made a priority also contributed to the successful participation of women and the poor, together with the support granted to their learning process.

A number of times during meetings held in the commune or in other districts, the leaders of the commune expressed the fact that the approach adopted by this initiative differed from previous ones. Indeed, giving farmers the opportunity to improve their understanding and working with them on what they are interested in has contributed to building their confidence. In the past, the ideas and priorities of local communities were largely ignored, while their local knowledge was simply not taken into consideration.

By way of conclusion, we can say that participatory development communication, as a new way of working with local people, improves the quality of the participation process. In order to use it properly, researchers must not just be open to acquiring new knowledge. They must also be willing to improve their skills in directions that may initially appear unclear to hard-core scientists. However, the benefits of their work can be greatly enhanced if these new forms of knowledge and skills are put to the service of the most disadvantaged.

Conserving Biodiversity in the Democratic Republic of the Congo: The Challenge of Participation

Pierre Mumbu

In places where biodiversity is under threat, governments often establish protected areas with a view to natural resource conservation. This intent is certainly laudable; but unless an effort is made to consult the people who have always lived from those resources, and unless they are involved in the decision-making process, conflicts are bound to arise. In this case, conservation efforts are likely to be in vain. The story of the Kahuzi-Biega National Park situated

in South Kivu in the eastern part of the Democratic Republic of the Congo (DRC) provides a prime illustration of the confrontational dynamics that are sparked when two contrasting approaches collide: that of the authorities seeking to protect a fragile ecosystem from over-exploitation, and that of indigenous people insisting on their ancestral rights over the territory. It also illustrates how participatory communication can help to find common ground between opposing views and generate cooperative action.

Located in the eastern part of the Democratic Republic of the Congo, the Kahuzi-Biega National Park (PNKB) is a major repository and sanctuary for biodiversity. It is home to 13 species of primates, 9 species of antelope and more than 400 bird species. Yet, since 1996 the park has figured on the list of World Heritage sites at risk because of intense human pressure on its resources. Until very recently, efforts to impose the prescribed ban on all exploitation of natural resources were met with stiff resistance from local people, especially the pygmies for whom the park's territory has been their natural homeland. The presence of indigenous people in the zone, who continue to pursue their traditional way of life, has tended to undermine conservation efforts, an effect exacerbated by uncontrolled poaching and the actions of armed militias.

The first efforts to protect the biological wealth of this zone were made a long time ago. As early as 1937, the Belgian colonial authorities established an integral reserve in a portion of this territory. In 1955, the protected zone was increased from 600 to 6000 square kilometres. It was officially made a national park in 1970 and was declared a United Nations Educational, Scientific and Cultural Organization (UNESCO) World Heritage Site in 1980.

The creation of the park was to serve two objectives: to preserve the typical mountain forest of Kivu and to protect a subspecies of gorilla that is found only in the forests of the eastern Congo. The park also constitutes the sole protected area in sub-Saharan Africa where there is continuity between lowland and highland forests, a circumstance that favours ecological exchanges between those two systems.

The park plays an important role, as well, in conserving the region's hydrography, rainfall and climatic equilibrium. Moreover, this natural forest has served to protect the water table upon which the inhabitants of the city of Bukavu and the villages located within the park depend for their water supply.

A complex set of issues

The portion of the park lying at higher altitudes is home to more than 0.5 million people. Only the original portion is uninhabited. But in the surrounding area, the density of human population is as high as 300 individuals per square kilometre. The natural forests in these areas have been virtually exterminated. The park has the appearance of a wooded island afloat in a sea of fields. Local people use the forest as a source of firewood and lumber, medicinal plants and foods of various kinds. The majority of the pygmies living in this section were expelled during the early 1970s and were resettled beyond the park boundaries, from where they continue to pursue their subsistence activities in the territory they previously occupied. It is from among the pygmies that the park recruits trackers for determining the location of animals and it is also the pygmies who provide the best guides for the poachers who are decimating the park's wildlife.

The 1975 expansion of the park brought 9000 villagers within its boundaries. At the same time, other people have been expelled. This is what happened to the Bananziga and the Mwanga-Isangi, who lost their traditional rights to these lands. In this area, the alternation of forest and village creates a mosaic within the park. Its inhabitants engage in extensive slash-and-burn agriculture, hunting, fishing and mineral digging, or small-scale mining, as well as many other activities necessary to their subsistence.

For the most part, it is the higher portion of the park that has been subject to demands for the return of lands. The traditional chiefs of this hierarchical society feel that they lost some of their power, which is essentially based on landholding, when a portion of their territory was incorporated within the park.

Tensions have been heightened by the fact that farmers have conspired with corrupt government officials to obtain and exploit fraudulent concessions within the park. When the farmers saw that these lands were to be taken away from them, they began to stir up opposition to the park's existence.

By the mid 1980s, it became clear that widespread local opposition to the very existence of the park was a serious obstacle to efforts to conserve its biodiversity. The authorities then recognized the need to hold discussions with the people concerned.

Problems raised by the local population

In the course of many meetings organized by the park authorities, local people described their view of the problem by saying that when the park was created, people were driven out of a territory that they had occupied for generations. Furthermore, when the park was expanded in 1975, people were neither consulted nor compensated. This fuelled resentment on the part of some village communities, who for this reason refused to recognize the park's existence. Previously, when the forest belonged to them, the indigenous pygmy population lived in harmony with its surroundings, and made rational use of the forest's natural resources.

Today the forest belongs to no one; it is government property. Consequently, the respect that ownership generates has vanished, and the traditional sound approach to the use of the forest has been replaced by a free-for-all to exploit its resources. There has also been a shift of mentality among some of the customary chiefs. The attentiveness that the chiefs traditionally bestowed upon the community as a whole has been replaced by a pronounced individualism. In some cases, the chiefs are now in the pay of rich and influential individuals who use them for commercial purposes. Great swaths of the territory have been ceded to rich farmers who are now growing monoculture commodities such as coffee, tea and quinine.

At the same time, the people who live around the park are poor and their numbers are growing. This exerts heavy pressure on the park since these people are in constant search for cultivable lands and staple necessities, such as wood for burning and for construction, wild game, mushrooms and medicinal plants.

There is also some confusion over the park boundaries, which have yet to be clearly defined. As a result, there have been some incursions of settlers back into the park, especially in its older portion. At this time, it would be difficult to force them out again without offering some form of compensation. Moreover, the migration of people with no 'forest smarts' into these wooded areas has caused considerable environmental degradation because they have no knowledge of traditional techniques of natural resource management.

In short, people see the park as a curse rather than a blessing. People struggling to eke out a living say that they are fed up with having their crops ravaged by the park's marauding elephants and they are tired of daily hassles with forest wardens.

Problems raised by the park authorities

For their part, the park's management felt that it was the lack of visible economic benefits that led people to challenge the park as an element of the region's economic development.

Although they do their best to respond to local people's demands, the park's managers must cope with a very heavy bureaucracy that prevents them from reacting promptly, and this is a further source of local discontent.

Finally, the villagers do not hold park staff in high esteem because of what they see as official harassment.

A development strategy for the area around the park

These discussions with local people revealed that, as with other national parks in the DRC, the Kahuzi-Biega National Park had always been run in a way that excluded local people from taking any part in its management. Not surprisingly, this exclusive model was found wanting. What was needed was an alternative model for managing and conserving the park's resources.

The authorities therefore undertook a number of development initiatives to help lift people out of their isolation and poverty. The intent was to foster activities that would, to some extent, compensate local people for the restrictions imposed on their use of the park's natural resources so that they would feel the park was making some tangible contribution to their development.

Beginning during the early 1990s, therefore, health and maternity centres were established, primary and secondary schools were built, and wells and water supply systems were installed, in collaboration with local people, while maintenance of infrastructure such as bridges and service roads was stepped up.

This new collaborative approach proved its worth during the two wars that tore South Kivu apart. In 1996 and in 1998, when the area was infested with armed militias, conservation staff dared not venture into the forest. Therefore, in places where there was good cooperation with the local people, such as in Kalonga and Nzovu, the people took it upon themselves to protect the park. On the other hand, where relations were less satisfactory, natural resources fell prey to massive and systematic destruction by the militias, by poachers and even by the local populace.

A 1996 survey found that around 36 per cent of people had a favourable opinion of the park's conservation, a surprisingly high percentage in light of the previous systematic opposition. However, an assessment conducted in 1999 found that the socio-economic activities supported by the development strategy, which was targeted at communities surrounding the park, were not really meeting the needs of local people, who were more interested in economic initiatives that would ensure their family livelihood.

In light of these findings and those of a more in-depth study of stakeholders in the park's natural resources, the authorities realized that they would have to strengthen collaboration with local people in order to achieve truly participatory management.

The participatory management strategy

Thus it was that, in October 2000, participatory management and communication mechanisms were established in certain villages on an experimental basis. These structures now offer a framework for cooperation, dialogue and decision-making. They are, in a sense, 'village parliaments', responsible for examining options and discussing and preparing a development plan for each of the villages bordering the park.

The members of these participatory management structures represent many viewpoints: public institutions, religious institutions, customary chiefs, leaders of local development initiatives, and other specific groups such as the pygmies, farmers and youth, as well as the forest wardens and women's associations. Each structure has a steering committee, a monitoring committee, a drafting committee and an anti-poaching committee. Since these participatory structures were set up, there has been a noticeable change in the kinds of initiatives put forward and greater attention is now paid to income-generating activities in order to meet locally expressed needs.

In addition to these participatory management structures, people have banded together in village social units to pursue collective self-help activities, rather than relying solely on individual initiatives. These units represent the backbone of management and support for locally inspired micro-projects.

Using the mass media

Along with the interpersonal communication on which the participatory structures are based, the park authorities have continued to resort to the mass media for the broader dissemination of information. Cooperative links have been established with a great many partners, who are provided with written and audiovisual materials that they can use in their activities. At the local level, use is also being made of the more traditional media, such as theatre, dance and popular songs.

Lessons learned from the process

The participatory management strategy has so far been confined to only about 10 per cent of the area surrounding the park because of security concerns generated by persistent armed confrontations. However, we can already draw some lessons from the process instituted to promote dialogue and cooperation between the park authorities and the neighbouring population.

The mode of communication that has proved itself most useful is that of interpersonal communication, where people address each other face to face at meetings or assemblies. The 'village parliaments' mentioned earlier are the prime example.

In the end, we may conclude that efforts to conserve biodiversity are more likely to succeed if policies and practices relating to management of the park's natural resources carry the stamp of consensus among stakeholders, rather than being imposed. In other words, free and open communication based on dialogue and mutual respect between the park authorities and the local people can overcome the resistance of those who initially felt excluded from decisions that were of vital concern to them.

To be sure, the communities concerned have yet to establish full 'ownership' over the development process, as they must if it is to be sustainable. In particular, once this experimental phase is over, it will be important to ensure that people are sufficiently equipped to continue with it and that they have the necessary means to carry on with the activities now begun. Otherwise, the crushing weight of poverty is likely to drive people back to helping themselves to the park's resources in the absence of viable options. This is a particular concern in the case of the pygmies, whose living conditions are still highly precarious. One of the greatest challenges, then, is to reinforce the existing income-generating activities and to expand their focus beyond sheer survival to embracing self-financing of local initiatives.

Yet, the long-term success of this strategy will depend in large measure upon strengthening dialogue between the authorities and the people, and maintaining attitudes of cooperation and collaboration on both sides. Here, again, the outlook is still uncertain, and it is not clear that the authorities are fully convinced that ongoing participatory communication is a key factor for future success.

Nevertheless, to date, the results are quite compelling. People in both places where this approach was tried out are playing a leading role in protecting the park's natural resources. Moreover, information now circulates more freely and an open climate of cooperation prevails among stakeholders. More concretely, since the participatory management strategy and the village economic units were established, goods taken from the park, such as game, bamboo and other forest products, are less in evidence in the village markets. Villagers are reporting poachers and have captured the occasional one themselves. They have even asked for portable telephones so that they can pursue poachers and smugglers more readily.

Yet, the sheer magnitude of the task is an obstacle to spreading the experiment to the entire local population. While there has been a considerable decline in poaching by villagers, there has been a sharp rise in the depredations of militias and other armed bands that have moved into the park. Against these illegal activities, the members of the participatory management structures are virtually powerless.

While awaiting a definitive end to the intermittent wars that have wracked South Kivu and that continue to prevent participatory communication and management from taking firm root in the villages that have tried it, another challenge is looming on the horizon: self-defence associations are springing up among the pygmies, who are claiming inalienable and unfettered aboriginal ownership rights over the park and its resources.

This situation, which would seem, at first glance, to contradict the participatory approach to managing natural resources, could be an important test for determining just how far participatory communication can contribute to preserving biodiversity without compromising the survival of cultural diversity.

The Word that Quenches Their Thirst: Rural Media and Participatory Development Communication in Burkina Faso

Souleymane Ouattara and Kadiatou Ouattara

Make water spring from the Earth and restore parched lands to greenery, and do this with empty hands? It can be done. The secret? Bring together people accustomed to working on their own, use the radio and share knowledge, including local know-how. This is a story that involves foresters, agronomists

and media people. But, above all, it is a story of the inhabitants of two villages (Nagreongo and Kriollo) in Burkina Faso and one (Kafela) in Mali. It tells of an experiment with action research that was marked by both successes and failures, but that provided plenty of lessons. The initiative was launched by the Journalists in Africa for Development (JADE) network, which was convinced that the rural media were missing a fine opportunity to support development because they lacked participatory resources and tools.

Barren soils, lack of water and accelerated deforestation are problems to which non-governmental organizations (NGOs) and government technical services are far from indifferent; but the solutions they try to implement are sometimes not very effective. In many cases, they pursue programmes already mapped out by their head offices, without any participation by the 'beneficiaries'. Moreover, even if they recognize the importance of community participation, they are not always well equipped to translate that concern into concrete action in the field.

It was to meet this gap that JADE undertook an experiment in participatory communication, an approach that involves local people in seeking local solutions to their own problems. The experiment was conducted in three villages, two of them in Burkina Faso and a third in southern Mali. Each of these villages faced natural resource problems that were holding back its development. Participatory communication served to rally stakeholders around solutions that they themselves had identified and agreed upon. This experiment produced lessons of three kinds, relating to the use of action research as a methodology, the conditions that must be present if interventions of this kind are to succeed and the choice of communication methods or tools.

Nagreongo: Too many women at the well!

In Mooré, one of the 60 languages spoken in Burkina Faso, the clearings where nothing grows are called *zippelé*. Exploited, abused, eroded by the rain and fierce wind, compacted by animals, the earth hardens into a sterile crust. The people of Nagreongo, a village in central Burkina, know all about the *zippelé*: for them it is a synonym for famine and thirst.

In March 2001, when the action research project was launched in Nagreongo, the village had 18,948 people and four wells. Not only were these water points insufficient, but two of the wells were out of service. With no management committee, maintenance work was random and haphazard. A famous healer lived in the village, and the sick would flock to consult him, at the same time imposing enormous demands for water and for medicinal plants. This village happened to be located within the broadcast range of a community radio station, Radio Yam Vénégré, which meant that people had access to information on farming techniques. They knew the history of their villages and they could listen to tales and legends.

Kriollo: When the water stretched as far as the eye could see

Further to the north in Burkina Faso is the village of Kriollo, an isolated settlement of some 2000 souls. When asked why people would establish a village in a place where there was no water for humans or animals, Dicko Issa Boureima, now around 50, recalled memories from his youth: 'I remember when I was still a teenager, perhaps 40 years ago, the water here was like a lake that extended all the way to Taaka, another village about 5km away.'

Today, farmland is in short supply, the quality of the soil is steadily collapsing and firewood is hard to find. In addition to these problems, the people are illiterate and women are not allowed to speak in public.

Kafela: Where a woman's worth is measured by the height of her woodpile

In Kafela, a Senoufo village of 510 people on the outskirts of the town of Sikasso in southern Mali, the forests are disappearing under heavy pressure from wood gatherers. The town of Sikasso generates tremendous demand for wood. To meet that demand, people are constantly hacking away at the forest, which is, in fact, their main source of income. Yet, the search for profits is not the only explanation. The community also carries some heavy social and cultural burdens. In these Senoufo lands, a wife's measure as a woman depends upon how much wood she can stack: a woman's social status can be gauged by the height of the woodpile in front of her house.

Action research as a methodology

Sharing the methodology

The first lesson from this experiment is that people have to agree upon the proposed initiative. But first, what is action research? Unfortunately, the term is not easily defined, especially when working with communities where the concept is foreign to the way in which they analyse problems and make decisions. This is why it is important, before undertaking any rural project of action research, to devote sufficient time to explaining the notion and to find the appropriate terms in the local language.

The weight of cultural heritage

Action research takes place against the backdrop of local traditions, which must be taken into account. For example, in many villages of West Africa, women are not allowed to speak in public. To get around this cultural prohibition, the project's fieldworkers set up homogeneous discussion groups consisting of women, men and youth. According to Awa Hamadou, a midwife in the village of Kriollo, 'Separating the groups meant that everyone could participate in the debate.' This was a first for a village that is heavily Muslim.

Today, the women are free to express themselves in public, in the presence of men, and on any topic they choose.

Voting and its limitations

The approach we used requires that the people identify and analyse their own problems, and set priorities among them on the basis of a direct vote. This is done by having each problem represented by a candidate. The approach helps to strengthen local democracy and the community's consensus-building capacity in a place where, by tradition, it was the elite who decided things for everyone.

This direct voting technique allows all the villagers to express themselves. With this rule, everyone can accept the majority choice. Yet, while the technique is certainly useful, it is far from perfect because people of the village are influenced in their vote by the views of the community leaders. One solution might be to establish private voting booths so that people cannot be swayed in their choice.

Putting things in context

We find that when communities have the chance to speak out, they will often raise matters that the project is not in a position to address. Faced with severe problems that threaten their very survival, people will often look for quick fixes, whereas a participatory approach places more emphasis on the search for lasting, consensus-based solutions. These two kinds of needs are not necessarily incompatible. The point is to help communities establish relations with other players who can provide the material support that they need. In the case of our three villages, the team was successful in this intermediation effort.

Choosing the players

Finally, the project's success will depend heavily upon the choice of players. They must not only have the required technical skills, but should be available and able to work in empathy with the communities, as well as with the technical partners.

Preconditions

Know your territory

The methodology requires a proper familiarity with the local setting as a precondition for subsequent phases. During a workshop held in Ziniaré in January 2001, extension workers in agriculture, animal husbandry and the environment explained their outreach techniques. Participants, including the local communicators, were then invited to comment on these presentations. They pointed out to the technicians the importance of the initial salutation phase, which is frequently cut short. They also recommended open discussions so that they too could propose solutions to the problems at hand. This exercise demonstrates the importance of 'knowing your territory', and here a contact person within the community can help. This person is usually known in Dioula, a language common to several ethnic groups in the region, as the *diatigitié* – roughly the 'host' – and it is he who will open your eyes and tell you what to do and what not to do while you are in the village. As the old proverb says: 'The stranger has big eyes but he can't see.'

Key players

During this initial phase, it is also important to identify the key players – that is, all those whose opinions count in making community decisions. In some cases, their influence will be such that they can stymie an initiative. The profile of the key player will vary from one region to another: he may be a healer, an imam, a customary chief or even an ordinary farmer.

Tools and methods for effective communication

The traditional means and vectors of communication

Each community will have its own means and vectors of communication. In most villages of West Africa, communication takes place at the church or mosque, at the well, in the market, in the village square, or at gathering points for young people such as kiosks, tea houses and popular celebrations. The project made good use of these traditional means of communication. In Nagreongo, the healer's representative, who was also the village information officer, had a megaphone that he used at meetings. In Kriollo, which is heavily Muslim, communication relied on the imams. In Kafela, the traditional chiefs transmitted information to the community. In all these areas, particularly in Burkina Faso, there is also a communication system based on the leaders of village groupings and on village administrative officers.

The project took advantage of these existing communication arrangements and was able to limit conflicts. But, at the same time, it ran the risk of becoming a prisoner of the system because discussions were heavily influenced by those in charge of circulating information, who thereby derived a certain power. Clearly, a proliferation of communication channels and tools is no guarantee that everyone (particularly the women and youth) will be involved in decisions that commit the community. The challenge, then, is to use existing means and vectors judiciously and to propose ways of improving them.

Facilitation

For the project coordinator, facilitation means encouraging people to pool their efforts and their skills in order to resolve local problems through natural resource management. The multidisciplinary nature of the group thus constituted appealed to people in the participating villages. One farmer explained:

> *In the past, visits to our village by members of this or that regional committee or agency were random and disorganized. They took up a lot of our time and people were tired of going to meetings, although we never said that to the experts. They did not even answer all our questions. Now that we have this group, we are finding answers to all our concerns and we are also saving time.*

As to the agencies, they are not only getting firsthand experience of local people's problems, but are now taking them into account in their follow-up activities. This new partnership goes well beyond the old working relationships that were based on institutional considerations and made no allowance for a common approach.

Using local languages

Using local languages is essential for sharing information, thinking and outcomes with all partners in the process. The local relay points, rooted in their home communities, are already using the national languages effectively to deal with communities in the course of their daily work. Radio stations, for example, are broadcasting in Mooré and Fulfuldé at Ziniaré and Dori, and in Bambara and Senoufo at Sikasso. A newsletter, called *Kuma*, has also been translated into Mooré and Dioula, to an enthusiastic community reception. Although people in these communities were literate, they were not always able to find publications in local languages, and were thus unable to keep up their reading skills. Kuma is now helping to fill this void.

Local languages offer many advantages; but translating concepts such as action research, participatory communication and research project into these tongues can be a complicated affair. A great deal of time was spent discussing translations within the team in order to test out new expressions before adopting them and disseminating them locally.

Finally, the diversity of languages is such that research team members will never be able to master all of them. During our meetings, we therefore had to arrange for translation from French into the local tongues and back, and this in effect doubled the length of time that would normally be devoted to such meetings.

Discussion groups

Discussion groups make it possible to compare different ways of addressing a problem. They can also be used to uncover conflicts of interest and to find solutions that will be fair to all parties involved. The advantage of these groups is that everyone has a chance to speak, including women and youth who are normally shy about expressing themselves in public.

The limitations of radio

The research team launched its activities by radio, which is a powerful information and communication tool, especially in rural settings where the oral tradition dominates. Editorial committees were established in each of the three zones (Sikasso, Ziniaré and Dori, where the project's partner radio stations were based) in order to produce regular information programmes. These committees operated in the conventional editorial mould, with the one difference that in addition to radio producers, they also included representatives from the communities and from development agencies.

The synergy that emerged among all these players enriched the content of the 'magazines' (reports, surveys and technical data) and represented a real experiment in partnership. Unfortunately, the development officials often steered the debate towards their own particular concerns. The village communities, the prime audience for the message, were not sufficiently involved in choosing the information topics. In order to overcome this constraint, we set up a network of local communicators in Dori and strengthened those already existing in Sikasso and Ziniaré. The local communicators were responsible for identifying topics in advance, preparing field trips for the radio team, taking part in the broadcasts themselves and providing feedback to the stations.

The workshops

The workshop meetings served a vital function throughout the participatory communication experiment. They allowed people to expand their normal circle of relationships and to replace an ethos of competition with cooperation.

Conclusions

Whenever communication problems existed within the communities, synergy among the stakeholders, through group discussions that identified and analysed problems, served to establish an open dialogue and to make decisions that would commit the entire community. In this way the community was able to find home-grown, consensus-based solutions to the water shortage, to the disappearance of the forest and to the collapse of soil fertility. This open approach, and particularly the participatory dimension of activities, also tended to temper the climate of competition that often exists among development workers in the field.

The communities who threw themselves most enthusiastically into the project believe that action research and participatory communication are appropriate and are a readily usable means of addressing their development problems, instead of waiting for unpredictable financing.

Growing Bananas in Uganda: Reaping the Fruit of Participatory Development Communication

Nora Naiboka Odoi[1]

Banana is one of the most important crops in Uganda, and in many homes, especially in the central part of the country, it is the staple food. Many small-scale farmers derive both income and food from the banana crop. But since the 1970s, many of them have been experiencing decreased farm yields as a

result of declining soil fertility, pests, diseases and socio-economic problems. In this changing context, traditional practices have proved insufficient to cope with the new challenges. On their part, researchers have come up with technologies that could be of benefit to farmers. But the latter have proved reluctant to integrate these new technologies within their practices, despite researchers' attempts to disseminate them.

This scenario has led some agricultural researchers to question their way of communicating with farmers and to experiment with more participatory approaches. This story is about farmers who have been so successful in solving their own problems that they have become teachers to other farmers and development communicators in their own right.

Despite the growing hardship and difficulties associated with their traditional agricultural activities, farmers tend to overlook the new resource management practices proposed by researchers. When they do use them, it is only during the period that researchers are with them. Therefore, agricultural research findings have not been integrated in a sustainable manner, even when the initiatives bear positive results.

One of the reasons for these failures points towards the fact that researchers and extension service providers have largely relied on top-down dissemination methodologies in which farmers are not involved in decision-making regarding which technology to implement in their gardens. They do not own the natural resource management initiatives being tried out in their own gardens. When researchers visit the villages, some farmers have been known to show the researchers two plots of gardens: their own and that of the researchers. 'This is our garden, and this one is yours', the farmers are often heard to say, the second garden being the one where farmers are putting into practice the technologies recommended by the researchers.

For the past few years, researchers have been on the lookout for a methodology that may result in farmers owning the natural resource management research initiatives so that they are able to implement them in their own banana plots in a sustainable manner. In this sense, participatory development communication (PDC), which is a relatively new concept in Uganda, promises positive results regarding information-sharing between farmers and researchers. During recent years, several organizations have started to implement participatory methods, albeit with varying degrees of local people's participation. Uganda's National Agricultural Research Organization (NARO) implemented one such initiative through its National Banana Research Programme. The two-year research initiative called Communication Among Banana Growers for Improvement of Soil and Water Management aimed at developing a two-way communication model suitable for a better flow of information between researchers and banana growers, at enhancing farmers' participation in experimenting with different banana improvement technologies, and at fostering farmer-to-farmer training. The study used PDC as a tool to foster active participation of the local community in identifying natural resource management problems in banana gardens, as well as their causes and solutions.

Initiating dialogue

After a series of in-house planning meetings, a team of researchers – comprised of a socio-economist, a communication specialist and counterpart, a soils expert and a specialist of the site – toured several banana-growing districts. They were in search of a suitable site that would host the natural resource management through PDC initiative. The team zeroed on Uganda's south-western district of Rakai, most particularly the sub-county of Ddwaniro. This choice was motivated by the fact that Ddwaniro is among the leading banana-growing areas in the district; it also has soil management problems as a leading farming constraint, a relatively good road network and is inhabited by hard-working farmers.

The research team first held a series of consultative meetings with district and sub-county officials, opinion leaders and extension agencies working in the sub-county before holding any meetings with farmers. This step facilitated the introduction of the initiative to the named groups, in addition to lobbying for their support. Farmers were encouraged to form farmers' groups, whose representatives later participated in the PDC initiative.

The researchers facilitated the identification and prioritization of natural resource management problems by the farmer representatives. In order to do this, they divided themselves into three groups, in line with problems pertaining to soil fertility, soil erosion and soil moisture retention. The farmers then defined their communication objectives and needs regarding the identified problems, the activities that could be undertaken to alleviate the problems and the communication tools that could assist them in sharing their new knowledge with their original farmer groups.

The researchers discovered that some farmers had extensive indigenous knowledge related to the three natural resource management concerns, but that it required validation. Moreover, farmers did not have a forum to share information with each other, hence the need for communication tools.

Horizontal communication processes

The researchers facilitated participating farmers in visiting other farmers who were already using more appropriate natural resource management practices. Following the visits and the discussions with other farmers, the participating farmers were more convinced than before of the new technologies' benefits. The researchers then facilitated the implementation, by the farmers themselves, of proper natural resource management in their own banana plots.

The farmers were amazed at the results of the new practices. Plots of land that they had previously abandoned, saying that they could never be productive, were now yielding good banana fruits. Their banana plots looked healthier, and they yielded better quality bananas in greater quantities than before. They were now able to sell some of the bananas for cash, while remaining at the same time with enough left over as food to eat in their homes. Banana farmers have a practice of selling bananas when the banana fruits are still on the parent plant. The bargaining process between the seller and the buyer often takes place within the banana plot while the plant is still intact with its fruit, a practice believed to be due to the perishable nature of the banana fruit. After implementing proper natural resource management, the participating farmers said that their sales fetched more money from buyers because the amount of invested labour was evident in the banana plot. A well-managed banana plot made people more inclined to buy the product, like a proper packaging of the banana fruit.

The farmers were now the proud owners of well looked after and high-yielding banana plots. Unlike before, farmers were confident to show their banana plots to other farmers and visiting dignitaries in their community. Local leaders started using the banana plots of the practising farmers as showpieces. Natural resource management through PDC raised the prestige of the practising farmers. Several of them became leaders in their communities. Some of them even managed to contest successfully for political leadership in their community.

But the practising farmers never forgot the fact that they were only representatives of other farmers in their local communities. After they had mastered the natural resource management technologies, they expressed the desire to share their new knowledge with other farmers and the farmers they had represented in the initiative. They recognized that they had to use communication tools that had the capacity to illustrate how to implement natural resource management technologies. They chose to use video, photographs, posters and brochures on soil fertility, soil erosion and soil moisture retention.

Looking back on the experience

The concept of participatory development communication was new to researchers. This resulted in the loss of valuable time since several activities had to be redone. The biggest challenge happened at the production stage, when both researchers and farmers were new to using the communication tools. Eventually, professionals (a graphic artist, an illustrator and a cameraman) had to be called upon in order to successfully produce the materials.

Although farmers wished to practise the new natural resource management techniques, they indicated constraints of money, labour and agricultural tools. Some of these problems were institutional in nature and could not be solved by communication alone. Thus, the researchers and the farmers themselves sought partnerships with other organizations that could help them to solve these problems.

The production process took longer than intended. This was because farmers first had to learn how to use the communication tools before focusing on natural resource management. The learning process was not straightforward: the farmers had neither handled still cameras nor video cameras before. Consequently, there was a period during which farmers were only building their confidence to handle the equipment, which they considered to be too expensive for them to replace in case of damage. Eventually, they became confident enough to handle the cameras. There followed a process of perfecting the capture of usable photographs. Many films returned half pictures: pictures of children with no heads or pictures of people with half faces.

Farmers discovered on their own what type of photographs could successfully communicate information regarding proper natural resource management. Thereafter, they set about trying to acquire those pictures. Unintentionally, the process of taking and retaking photographs deeply ingrained the subjects in the farmers' minds. Without realizing it, the farmers were learning more about proper natural resource management while their focus was on taking pictures of the technology used.

After the communication materials were produced, participating farmers showed them to other members of the community, who made comments about them. One such comment required a change to be made in how a good banana mat looks. The original picture showed banana mats with single banana plants; but it was pointed out that a recommended banana mat should have at least three banana plants: 'a mother, daughter and granddaughter'. A change was also made in the video. This followed a recommendation that a song should be included in the video. On the whole, the other members of the community appreciated and understood the communication materials.

Monitoring and partnership formulation proved to be another big challenge to the researchers. They only realized at the end of the phase that these two steps should commence at the beginning of the initiative and continue throughout the process.

By the close of the first phase of the Communication Among Banana Growers for Improvement of Soil and Water Management initiative, farmers had appreciated the power of belonging to a group. As a result, they founded an association called the Ddwaniro Integrated Farmers' Association, through which they hope to search for, access and share relevant information and services about common community problems. This is the origin of a two-way communication forum between farmers and other stakeholders, including service providers and researchers. Such communication will bridge the gap between researchers and farmers, and supplement the inadequate extension services.

The farmers can now tackle their own community problems instead of waiting for external assistance. They are proactive instead of passive observers of community problems. They do not fear approaching service providers regarding their community concerns. In addition, other people find it easier to assist farmers when they are in groups. But unlike other groups, the group that formed after farmers had undertaken a common activity is stronger and promises more sustainability and more focus. It is a secondary group, unlike other groups that are only primary groups. A secondary group has more chance of implementing farmer-to-farmer teaching because there is already another group of farmers waiting to get information from it. This illustrates a multiplier effect of the PDC methodology.

The researchers and other stakeholders who participated in the Ddwaniro initiative are now convinced of the power of participatory development communication in implementing natural resource management initiatives together with farmers. They have begun to incorporate PDC within their research initiatives. Farmers now master the three methods of natural resource management to the point where they are now confident enough to share the information with other farmers. The farmers of Ddwaniro have begun sharing their experiences with other farmer groups within Rakai and other districts. This is a result of other farmers recognizing the positive results of the Ddwaniro initiative, and asking for similar initiatives to be introduced in their own areas. On their part, women are no longer as shy as before. Men have seen the benefit of their wives taking part in the PDC ventures, and they now readily allow them to attend meetings.

The way ahead

Any initiative begins with a proposal. For a long time, this has been done by researchers alone. Recently, researchers have discovered that it is possible to write a research proposal together with other participating stakeholders, such as farmers. Once the research proposal has been approved, the researchers, together with other stakeholders, can agree on the actual implementation of activities, based on the participatory development communication methodology. After having started with problem diagnosis and prioritization, the process then moves on to identifying solutions and implementing them, followed by evaluation and report writing. Indeed, PDC is most successful when it is used throughout an initiative, and not only at certain stages.

When looking ahead, it appears seems that PDC is a skill that farmers can transfer to other areas of their lives. In this case, farmers have used it to solve natural resource management problems. They could also use it to solve medical, social, economic and political problems. Farmers have the potential to undertake many initiatives on their own; but this potential often needs to be re-awakened before they can make use of it. After an issue has been discussed, farmers gradually realize that they can undertake certain tasks that they initially did not think about, or that they initially thought they were not capable of implementing. Whereas some farmers may recognize on their own that PDC can be used in different contexts, there may be a need to facilitate other farmers in recognizing the possibility of using PDC in varying spheres of their lives.

For some, participatory development communication may initially appear too expensive, considering the time, staff, travel, money and other logistics required. In the case described here, it took some time before farmers and scientists began appreciating its power and benefits. But after implementing the first phase of this initiative, the benefits seem to have dwarfed the initial costs. The 60 farmers who participated in the first phase have multiplied the efforts of the one agricultural extension worker in their area about 30 times by teaching other farmers in their immediate neighbourhood and beyond.

The other farmers are extremely happy with the farmer teachers because they are easily accessible. Moreover, they understand their problems better than outside agricultural experts. The farmer teachers give hope to other farmers by showing them that the natural resource management issues they are teaching can be easily implemented. Other development agents are now planning to hire them to teach other farmers in agricultural-related initiatives. Indeed, these farmers have gained so much confidence that they have begun filing radio items related to their natural resources management teachings to local FM radio stations in order to extend their accomplishments and knowledge beyond their own communities.

Note

1   This chapter was written in consultation with Wilberforce Tushemereirwe, Drake N. Mubiru, Carol N. Nankinga, Dezi Ngambeki, Moses Buregyeya, Enoch Lwabulanga and Esther Lwanga.

Giving West African Women a Voice in Natural Resource Management and Policies

Rosalie Ouoba

In many West African cultures, education and cultural norms do not encourage women to speak out in public. Yet, women constitute around 53 per cent of the population and 80 per cent of them live in the countryside. They account for 90 per cent of agricultural labour and they do most of the transporting and processing of farm and forestry products. Prohibited from speaking out in their communities, women are thus denied the right to

participate in decisions about the management of natural resources, of which they are the primary users. The Union of Rural Women of West Africa and Chad (UFROAT) decided to tackle the situation in a new way by enlisting women from six countries[1] to participate in drawing up an action plan:

> *My name is Sali Fofana. I come from Dafinso, a little village in western Burkina Faso. My daily life revolves around the search for water, which takes longer and longer to find. Our village has only two pumps for 1800 people. One of the pumps has broken down, and so we have to spend half the day lining up in the heat of the sun to get our supply of water. For us, water means life, as you will hear people say in the Sahel. We have to find enough for our people and our animals. I am so overwhelmed that I have to send my daughter out for water. That means she does not attend school regularly, and she's not doing very well in her class work; but what can I do? Not only that, but the marsh has been dry since January. It's as if nature herself has abandoned us to our fate.*

Sali Fofana is not a real person; but what she describes is very real for millions of African women who feel directly the impact that environmental collapse is having on their living conditions. Her words could be those of any of these women. They sum up the experience that a great many rural women have related in the course of an initiative to make their voices heard in decisions about natural resource management policy in their region.

With the accelerating disappearance of vegetation cover from countries of the Sahel, not only is water becoming scarcer, but people have to travel ever greater distances for the firewood that is their chief source of domestic fuel. The situation falls heaviest upon women such as Sali Fofana, whose stories often go something like this:

> *In the past, when I needed wood for the kitchen, all I had to do was go into the field next to the house. Today, wood is scarce and I have to go a lot further to find it, or I have to buy it. Since this is now a lucrative business and the distances are great, it's the men, with their trucks and carts, who go out for the wood and then they sell it at ever steeper prices.*

It was no surprise that the 1992 Earth Summit in Rio de Janeiro emphasized the role of women in preserving a viable and stable environment, while stressing the need to reduce extreme poverty, which is closely linked to environmental degradation.

Given their key role, rural women should be a great asset in developing natural resource management policies. This was precisely the objective of an initiative to enlist rural women from six West African countries in drawing up an action plan for natural resource management in a process that was participatory from beginning to end. With the implementation of that plan, they would, in turn, become a force for pressure and a source for ideas concerning national and regional policies. Moreover, the process sought to demonstrate that participatory communication and, in particular, the exchange and sharing of knowledge can help rural women to analyse the problems they face and seek solutions by drawing upon the knowledge, experience and resources that they possess.

Methodology

The initiative has involved two stages to date. Initially, workshops were organized in each of the countries to help the women prepare the materials that would be used in the action plan. The second stage was to formulate the plan. In 2005, there was a third stage, which saw a regional workshop where women from the six participating countries validated the plan.

In methodological terms, the workshops were designed to let the women express themselves by defining and analysing their own situation and coming up with their own solutions that will take their specific circumstances into account. Talking is the main tool in these workshops. The initiative also exhibited the following characteristics.

Recognizing the diversity of rural women's concerns about ecology and farming

Although the degradation of natural resources is a generalized phenomenon, its severity and the constraints that it generates vary from one agro-ecological zone to the next. In order to reflect this diversity of problems, representatives were selected from every agro-ecological zone in each country, and were invited to express themselves in their own language.

Facilitating information on key concepts

These workshops employed two key concepts: natural resource management and action research. Experts from the environment and natural resource ministries and resource persons in different fields were asked to contribute to these workshops. After listening to all the viewpoints expressed by the women, they offered some explanations on these questions and on the national policies to which they related.

The organizers had to draw heavily on their experience in order to achieve the workshops' objectives, given the women's low level of education.

Participatory diagnosis of constraints in each zone

The women were divided into groups according to their home region. Focusing on the three components of natural resources (land, vegetation and water), each group described and analysed the status of natural resources in each agro-economic zone, as well as the role of women in resource management and the difficulties that they encountered in performing that role. Next, the groups compiled an inventory of possible solutions and the needs facing women under each of these components.

Interaction among the groups

The working groups reported their results to the plenary sessions, where the women debated both the difficulties they faced in common and those specific to certain zones. They then discussed and prioritized the activities to be included in the action plan.

Field visits

Visits were arranged to meet with women's associations working in the field of natural resource management. This brought the participants into contact with other women's associations engaged in activities similar to their own. The visits provided the opportunity for very useful exchanges between participants in the programme and the associations visited. They also laid the basis for an ongoing relationship among women from different agro-ecological zones.

Some eloquent stories

Throughout the process, participants discussed the impact of gender inequality on rural women's lives and on their efforts to preserve natural resources. The story recounted by Sali Fofana provided eloquent testimony of how working and living conditions for female farmers have deteriorated over the years:

> *My field is barely half a hectare. It's a long way from the village and from my husband's field. It's an old field and it's not very productive. I have to get up at the crack of dawn so I can get my own fieldwork done before I have to go and work on my husband's field. I have no one to help me and I don't have any modern tools or inputs, such as a plough or mineral fertilizers, and so I put a lot of hard work into my field for a very meagre return. This year, for example, I harvested only two bushels of peanuts. Of course, I also grew some okra, some sorrel and some eggplants to perk up the family diet.*
>
> *A few years ago, like all the women in my group, I learned how to produce organic fertilizer. But I don't make any use of it; if I enrich my field I'm afraid it will be taken away from me. It's the chief who gave me my field. In our village, women don't have the right to own land. This means they can't plant trees on the lands that are given to them, nor can they take any major steps to protect the soil.*
>
> *I like to gather shea nuts and seeds from the locust tree [néré] to use in the kitchen and to earn a little money on the side. Nobody grows these things. You find them in the fields and in the bush. Today the bush is steadily retreating and it's becoming harder to find these fruits. We can no longer get any soumbala [a kind of mustard from the néré, used for seasoning] or shea butter, and now we have to buy our shea nuts and locust seeds from the men. This is really serious because we're not*

allowed to go into the fields to gather them. Since these wild fruits are worth money, the men keep the harvesting of them to themselves. People are so greedy that they will even take fruits before they are mature and then they use chemicals to speed up the ripening, something that can be hazardous to your health.

The local organization as a talk forum

Yet, despite all these difficulties and setbacks, women are not giving up. In some cases, their resolve has led to the creation of groups that are now the main forums for sharing information and talking about their daily lives. These groups have produced leaders who are now invited to represent them nationally and internationally, as was the case with this initiative. As Sali Fofana puts it, things unfold as follows:

As you can see, I'm tied up with problems. But thank God I'm a fighter. So I got the women in my village to group together to try to resolve our problems. After all, we want a bit of happiness too. My village sisters asked me to head up the group. They chose me because I get along well with everyone, I lived abroad in Ivory Coast and I went to school as far as the intermediate level.

In the invitation letter, I was asked to organize a meeting with the women and the men to discuss the status of the environment in our village and to talk about the place and role of women. This was important because we have to consider all points of view if we are to take decisions that will help the whole village. If we, women, just got together alone to analyse things, that would be all very well; but the men might feel left out and they could put roadblocks in the way of our project. As it is now, they are our allies.

Kaya: Women and experts

Following this local initiative, a national workshop was held in each of the six countries, bringing together women from several regions. In Burkina, participants met at Kaya, north of the capital. This workshop was intended to pool the women's experiences and observations and wrap them up into a coherent action plan. For Sali Fofana, as for all the rural women who took part, the opportunity to express the viewpoint of her village and to compare notes with the other women was an invaluable experience. And when their knowledge and their observations were validated by experts in various disciplines, this served to reinforce the women's perception of the importance of their role within their communities. For all the Sali Fofanas who had invested time and effort in the process, this was a moment of great pride:

The great day had arrived; I was off to Kaya. There, I met women from the four corners of Burkina. Each of us spoke about her village, about rainfall patterns and their impact on the bush, on water, on people and on animals. Each of us talked, as well, about her role as a woman in exploiting and managing the bush, the water and the land. We realized that a woman's situation is just about the same all over the country. There were natural resource management experts present and they gave us some new information that we used to analyse the situation in depth. They admitted that what we were saying was right on the mark. Like other countries in the Sahel, Burkina has been suffering a prolonged drought since 1969 and rainfall has dropped sharply. This has had a severe impact on vegetation, on soils and on farming systems, leading to an explosion in the area under cultivation and a drop in fertility and in the availability of water. While national guidelines set a standard of one water point for every 500 people, the real figure in the countryside can be multiplied by five or six times, and around 57 per cent of households get their water from undrinkable sources.

We found some solutions and we proposed some actions. Take the fruits that we use for our condiments, for example. We said that men and women have to work together, within the home and in the village, to appreciate the usefulness of these fruits in feeding the family and to enforce village regulations that used to prohibit people from taking the fruits too early. The authorities used to decree a specific day when everyone could go out and gather the fruits. We waited until they were ripe and no one could take them before. That regulation has to be brought back into force in the village. Then everyone will have a little, and that will prevent the fruits from being gathered when they are still green and unusable.

We want to be involved in selecting the species for reforestation. Since the locust and the shea don't yield as much anymore, we need to plant more of them, as well as fruit trees that we women can use to feed and care for our family and to earn a little money. We hear that shea butter brings a lot of money today because the Europeans like it and they buy a great deal of it. We women know how to produce the butter, we are trained, and we can look after the trees and make a high-quality product that will bring even more money.

The locust tree produces soumbala, a spice that every woman knows how to prepare and that gives a good taste to our sauces. Since the locust no longer does well, soumbala is expensive. It is being displaced by the Maggi cube, which is widely advertised. People say that's why so many people are dragging along with high blood pressure. We want to bring back our native plants that allow us to eat and to earn a little money.

When it comes to water, we have asked to be involved in selecting well sites and to be given some decision-making positions on the water point management committees. We're also going to demand more water points in the villages so that water will be more readily accessible.

As to the women's fields, we want the ones they give us as our personal plots to be as close as possible to the family field, and we want to be able to use our husbands' tools and materials to cultivate our fields. We also want to see discussion in the village about the importance of the women's fields in terms of family food security. More and more, our families depend on the produce from our fields to tide them over until the next harvest. We have problems in producing the compost that the extension workers have taught us to make, and sometimes we have to steal it and put it on our fields. If we had more compost, we could cultivate our plots more intensively and produce more food. With all the solutions proposed, we discussed their feasibility and we selected the best ones.

Lessons learned from the experiment

Involve the women in data gathering

To promote community participation, the women were asked to do some advance preparation before the workshops whenever they could. In some cases, this preparation involved the entire community, while in others it was limited to the women's associations in their specific configurations. These discussions within the community not only served to inform the community about the important role that women play, but sensitized people to the problems women face in carrying out their tasks. The results formed the basis for activities in the workshop groups.

Have participants analyse the data

The approach used was shown to be very effective because it respected the ability of rural women to define their place and their role, as well as the constraints they face in managing natural resources. In their analysis, they took into account the pattern of distribution of household tasks while stressing the need for dialogue between men and women in order to re-establish a degree of balance and equity. They identified their needs very clearly in terms of natural resource management, and they put forward solutions that took account of the whole community.

The methodological approach

Allowing rural women to speak and appreciating the value of their knowledge can instil confidence in this social group, which suffers daily from injustice and inequalities, and can encourage them in constructing a vision of society.

The right to speak is very important in the life of every person, of every social group and of society as a whole. It lets individuals or groups express what they are, what they experience and what they feel. It also allows them to affirm what they know, what they can do, what they think and what they want. But, above all, speaking allows individuals or groups to organize themselves in order to improve the situation or achieve the goal by entering into communication with other people or other groups to enrich the social debate. Finally, speaking allows traditionally marginalized individuals or groups to participate in taking more relevant and equitable decisions for sustainable development.

In short, speaking is a fundamental human right. Yet, given the balance of forces within societies, this right is denied to some people, particularly women.

Despite the real progress that has been made during recent years, the overwhelming majority of rural women in West Africa still have no voice. Yet, when they are given the opportunity, they can define and analyse better than anyone the situations that affect them, provided that the facilitator respects their pace and gives free rein to their capacities for imagination and expression. The right to speak must mean the right to express oneself; but it also means listening and communicating, for it is through communication that new visions are constructed. Thus, if rural women express themselves and compare their views with those of other groups (through dialogue, exchange of knowledge and study visits), they will be able to decide the best actions to take as a group to secure sustainable development for their communities. At first, they will talk mainly about 'practical needs'; but they will very quickly move on to 'strategic interests' (transforming unequal relations between men and women, or between rich and poor). When they talk about the actions they want to take to meet their needs, women always demonstrate a concern to discuss them with the men and with their communities. In the end, most of the actions they select are things that they can do either by themselves or in conjunction with the partners working with them.

Problems encountered

The approach taken in the workshops in the six countries, while empowering, was beset by problems relating to the limitations, both intrinsic and extrinsic, that are imposed on women.

The diversity of languages

In some of the workshops, there were nearly as many languages as participants, and we had to find local facilitators who could serve as interpreters. Apart from Mali, where nearly all of the women in the workshop spoke Bambara, proceedings had to be conducted in at least three languages. It took a lot of work to prepare the interpreters to give translations that would respect all the subtleties. It was very frustrating to have a woman recount her experiences in her own tongue and with great emotion, and then to hear a translation that would render not half the story and nothing of the emotion. We can imagine that rural women have the same feeling when debates are conducted in French, and when they have to wait for the translation in order to understand proceedings and offer their own viewpoint.

Moreover, the time it takes to go through this exercise can discourage participants, who would rather engage directly with the audience.

Socio-cultural constraints

By tradition, women do not participate in the discussion of community problems. They do not express their views in public, unless they are mandated to do so by a group. Most of the women attending the workshops, even those representing associations, were unaccustomed to speaking publicly and their lack of experience made them uneasy about expressing themselves spontaneously.

In fact, given the social division of labour, the many domestic tasks that fall to women leave them little time to attend meetings. Thus, they never acquire the habit of participating or of speaking in front of men.

Religion (or its interpretation)

Because of their religion, some women are not allowed to frequent places where men are present. Their daily life is cloistered and they take little part in associations of the kind that allow women to open up and let down their guard. Those who get the chance to attend workshops of this kind often have trouble expressing their viewpoints clearly.

Women's illiteracy

The fact that most rural women are illiterate is an enormous constraint on their ability to interact with others. Among the group (from six countries), only perhaps one woman in four could read and write (either in French or in her native tongue). Despite the effort to provide illiterate adults with audiovisual tools to help them receive and exchange information, we must recognize that the rural world remains largely cut off from the exchanges of information that are essential to thinking about development.

Lack of education

Analysing a situation or problem requires capacities for assessment, reflection and reference to other experience or knowledge. Because they are confined in their environment and have no access to information that might deepen their understanding, rural women have trouble analysing a problem in depth, identifying all its causes and ranking them in order to find all the possible solutions. Their analysis, however relevant it may be, is limited. They jump too readily from a problem to the solution without appreciating all the underlying causes. Rural women need to broaden their horizons and interact with other people.

Some results from the workshops

During the national workshops, the women analysed the natural resource situation in their communities, listed their concerns and their needs relating to natural resources, and identified the main actions required to meet those needs.

The women's analysis of their situation showed that, to varying degrees, natural resources in all their countries are deteriorating at an unprecedented pace that threatens people's very survival. This sparks competition that throws into conflict the various components of a given community, or of two or more communities who share these resources. It is most often the weakest members of the community who are excluded from managing these resources, and women fall in this category.

These workshops also revealed the women's awareness that improving the conditions under which they work is a prerequisite to any solutions for preserving natural resources. They want to change their relationship with their communities through dialogue so that their family members and friends will appreciate their contribution differently. Despite differences from one agro-ecological zone to another, some common features stand out clearly, such as the need to overcome inequalities of access to land and property rights. Land, water and vegetation are at the very centre of women's lives, and they will be the first to try to preserve those assets if they are given the means.

Participants in this initiative wanted to produce an action plan that would be an instrument for reinforcing their capacities to analyse, to compare and to propose solutions so that they could constitute a critical mass for reversing current trends. In the end, they wanted to develop collective strength so that they could negotiate and influence decision-makers in order to change their own living conditions and those of their communities.

A first outline of the action plan became available in September 2004. To turn this into a real tool for members of the UFROAT rural women's associations, it will now have to be validated. A regional workshop with all the delegates from the six countries should serve to deepen the analysis and confirm or reject some of the proposed strategies.

By way of conclusion, we may say that overcoming the many obstacles that prevent rural women from speaking out means establishing a framework and conditions within which women can affirm themselves and their contribution to society, where they can broaden their horizons and encourage society's awareness of the importance of seeking and listening to their views and taking them into consideration at all levels of debate and decision-making bodies.

In a sense, facilitating dialogue among rural women, with other members of their communities and with development players, means unleashing at least 50 per cent of the region's potential energy and putting it to work for development.

Note

1   These six countries comprised Benin, Burkina Faso, Mali, Niger, Togo and Chad.

Water: A Source of Conflict, a Source of Social Cohesion in Burkina Faso

Karidia Sanon and Souleymane Ouattara

Women glaring daggers at each other, husbands ready to come to blows, and rivalries between former villagers who have moved to the city, between ethnic groups and between adherents of different religions. Can anyone imagine a more explosive mixture? Yet, this is the time bomb that the researchers of GUCRE (Management of Conflicting Uses of Water Resources project)[1] have set out to dismantle by facilitating peaceful access to water for everyone in the Nakambé Basin of Burkina Faso. They are succeeding in their

wager. They have made social cohesion their battle cry, and they are using participatory chat sessions, to which they have added other tools, such as forum theatre, video, radio and posters. Here is a story from Silmiougou, a village in the heart of Burkina.

This is not exactly the forested West, far from it. But neither is it the Sahel. Here at Silmiougou, acacias, ana trees and mangoes grow side by side with thorn bushes, the precursors of the sparse vegetation of the great North. Silmiougou, a village in the heart of Burkina Faso, still dreams of its verdant past. Listening to the village elders tell the story of this settlement, we can readily see how the environment has deteriorated. According to oral tradition, it was Peul herders who originally occupied this area. Under constant attack from rustlers stealing their flocks, the Peuls complained to the king of the Mossi, who sent them his stout warriors for their protection. Little by little, the Mossi protectors grew in number and took over the village.

Silmiougou, which originally had only a few Peul families living in huts and eking out an existence, today has around 1900 inhabitants, divided among 12 neighbourhoods. An initial consequence is that water is becoming scarce. People have started to supply themselves from a dammed-up pond fed by rain runoff. But during the dry season, the reservoir holds not a drop of water. Every woman who comes for water has to carry a basket. She uses this to remove the earth that she puts beside the reservoir to form a dike. The reservoir holds more water, but not all year. The village thus has to dig wells, to which have been added, over time, five boreholes hastily drilled by 'developers' without any involvement or participation of the communities. 'All we saw was some big machines that had come from Ouagadougou. They drilled some holes, and the kids went out to watch. Finally, people came to the chief of the village to have him tell us to drink that water', recalls one of the village inhabitants. Moreover, the boreholes are only a stopgap measure; some time later, two of them were rendered virtually unusable by polluted water.

The well: A battleground for women?

In September 1999, our research team visited Silmiougou as part of the GUCRE project. Consisting of university graduates (a hydraulic engineer, a sociologist and an economist) and a facilitator, our team looked into conflicts over water sources, particularly the way in which communities can participate in finding solutions.

In Silmiougou, the areas surrounding the wells and boreholes amount to boxing rings where women have a go at one another with unbelievable ferocity. Every day is replete with stories of broken water jugs, wounds of varying severity and family tensions. Cécile, the energetic president of the village women's association, describes the situation:

> *Every two or three days, the pump would break down. The men would
> search high and low for parts to repair it. They would accuse the women
> of causing the breakdowns through their daily disputes, and the women
> would point the finger at each other. That's when it would come to
> blows.*

Indeed, it was the law of the jungle that prevailed around the wells. Everyone
recalls the slogan of the time: 'Smash the jugs!' Old Cécile recalls those
moments with a good deal of emotion:

> *After the dispute, the woman at fault would usually go home and tell her
> husband her version of the events. He would then come back to the well,
> armed with a machete. After the explanations, he would see that his wife
> was in the wrong. He would go home really angry and take it out on his
> wife. Then the wife would come back to the well and ask who the liar
> was that told her husband stories and made him scold her. And the fight
> would start all over again.*

As if that were not enough, the children accompanying their mothers to the
well would relieve themselves there and no one would pick up the excrement.
The same thing happened with droppings from animals drinking at the well.

Tensions of many kinds

All over Nakambé, which is one of the four watershed basins of Burkina
where the GUCRE team has been active, water is the source of conflicts.
Those conflicts are exacerbated by power struggles. This is particularly
true in Silmiougou, where the researchers set out in particular to study
the local communication system. And what did they find? After the 1983
revolution, which brought Captain Sankara to power, the distribution of
official government information was left to the village administrator, who
represented the interface between the prefect and the local population. This
new centre of power created tension within the village. The administrator and
the village chief no longer even speak to each other. Two camps are forming,
and their confrontations paralyse any initiative for the village's development.
Elsewhere, as in Kora, a village located some 100km from Silmiougou, the
situation is hardly better. Former residents of the village now living in the
city quarrel with each other incessantly over anything to do with sponsoring
development activities back home, to the point that any new initiative that
the opposing camp might put forward is sure to fail. Assessing the overall
situation, GUCRE wrote in one of its reports:

> *There are many conflicts: there are too few water points and they are
> badly managed; there is a shortage of drinking water; the pumps have
> broken down; there is no hygiene; people fight with each other at the
> wells over their social status, their religion or their ethnic background.*

Tensions emerge whenever divergent interests, whether they are economic, political, social or cultural, meet around a water point without any civil, legal, administrative or customary mechanism for finding a compromise, a peaceful solution.

Facilitating social dialogue

This civil mechanism of which the researchers speak will involve instituting a culture of dialogue, an initial manifestation of which appeared in 2001 with the holding of a roundtable. This event brought together donors, government ministries (in particular, the ministries of the environment and of water, and their provincial offices), people involved in water projects and the local populations. The objective was to allow communities to present to policy-makers the results of their thinking in the wake of field surveys, and at the same time to learn about existing water management policies. During the roundtable, the experts pointed to a paradox: two-thirds of the well pumps are allowed to remain out of service, while communities face a drinking water shortage.

After the 2001 roundtable came the feedback phase, in which the conclusions were put to the communities and appropriate local solutions were sought. The team held discussions with local people about the causes of the pump breakdowns and possible solutions. It encouraged dialogue, joint efforts and interchange among stakeholders – communities, the research team, non-governmental organization (NGO) partners, project personnel and resource people – so that they could all help in solving conflicts relating to water use. To this end, the team resorted to various communication tools, such as discussion groups, a forum theatre and posters.

Local people took an active role in preparing and validating the messages during the discussion sessions. These meetings involved user groups that focused on issues of sharp concern to them, such as the cost of drinking water, the dysfunctional water management committees and the attitude of communities to the facilities. 'At the time', recalls Pascal Tandamba, the GUCRE facilitator, 'people did not feel particularly concerned about well management problems on the grounds that the wells belonged not to them, but to the government.'

The discussion sessions served not only to analyse the problems thoroughly and to review both local and imported solutions, but to select the most appropriate communication tools for resolving conflicts over water use. Thus, participants decided to produce a forum-theatre play with the village troupe based in Ziniaré, some 20km away. The content of the play reflected all the communication problems in the six villages involved in the project – namely, disputes over the well among women and between women and other users such as herders, who show up at peak hours to water their animals; hygiene; faulty water management (non-payment); and lack of infrastructure maintenance. Wherever the play was performed, the audience recognized themselves in the characters and in the plot. They were eager to join in the debate, which became very lively. After the performance in Bagré in eastern Burkina, another zone of GUCRE activity, a nun improvised a song that was soon on everyone's lips: 'GUCRE has come to untie our tongues. Let these practices go. Cohesion is the Royal Road to resolving our problems.' Apart from the theatre, communities also chose to use radio broadcasts, videos and posters. But a tool is not used for its own sake: it must meet several criteria, one of which is that it must be relevant to the objective at hand. It must also be properly adapted to the people who are going to use it. For example, the illustrations carried on the posters were drawn by the villagers themselves before they were redone by a commercial artist.

There are many accomplishments to point to: wells drilled or rehabilitated; water and health management committees established or reconstituted. The communities and the research team have drawn some lessons from all this.

The community viewpoint

People are pleased with the project team's approach, which is based on direct discussion with them and on developing an understanding of their own ways of communicating, and the nature of their social relationships. Every decision emerging from such an approach takes on greater credibility in the community's eyes.

Using discussion groups and pooling their results not only makes it possible to take account of the trepidation that women feel when speaking in front of men, but also facilitates the future sharing of ideas without gender distinction. People also feel that the project is not setting itself up in judgement over them, to award them good or bad scores. These are some of the conditions that facilitate the adoption of new practices, such as proper well maintenance.

The women are in favour of using this approach to decision-making, whatever the problem at hand: 'We could have a discussion like this about the mill to see how to manage it better and keep it running properly.'

Today, the issues under debate are rallying the entire community without any clan divisions: 'Amongst ourselves we often speak of Mr Pascal and of our two children, Judith and Alidobi,[2] and of the open spirit that he has brought us.'

The research team's viewpoint

The research team found that stepping up the pace of meetings among groups with divergent interests has helped to restore and strengthen social cohesion. The communities have begun to take a new view of the wells. People now feel themselves the owners of the wells and are therefore responsible for them. Disputes over the wells have declined considerably.

Social cohesion, then, constitutes the cornerstone for any process of participatory development, hence the importance of *Ned la to* ('man defines himself through his neighbour'). Nothing can be done without give and take, and give and take implies at least two individuals.

News of the project's success has been spreading. A number of villages along the river that participated in the project (for example, Pighin, Gweroundé, Ramitinga, Donsin and Voaga) have sent representatives to participate in the work as observers. The approach is winning converts even beyond the project zone, where people are calling upon the expertise of villages involved in the project to help them develop the same approach to resolving conflicts over water use, and in other matters as well.

Lessons learned

The experiment provided a number of lessons on how to improve the process. It is important to ensure that local people participate in defining the problem because there may be some differences in the way that researchers or facilitators see the problem and the way that the communities look at it. Researchers must not try to impose their own views of things or favour certain solutions over others that the community would like to try out. In other words, community participation must be effective right from the diagnostic stage and it must allow home-grown solutions to be sought, based on local experience and know-how.

In addition, it is important to create partnerships at all social levels, from the most important personality to the most marginalized. In effect, the most influential person may not necessarily possess the truth. We must also forge partnerships with individuals who, from close at hand or from far away, can influence activities undertaken with the local people. Those individuals must, of course, be identified in advance.

Problems and challenges

When experimenting with a new methodology, difficulties will frequently crop up, and it will be a challenge to address them.

Moreover, when conducting participatory development communication projects, we frequently find that people express tangible needs, such as women's credit, drilling a well or building a health centre. This raises an ethical question, for it will often be impossible to respond to all these material needs. The team is frequently asked how to deal with the situation, even though it has no immediate solutions to offer.

Another problem that the team faced was how to extend the impact of its intervention on a broader scale. No instrument had been established for capitalizing on the accomplishments or 'banking' them; therefore, it was hard to reproduce them in other villages of the Nakambé Basin or in other watersheds experiencing the same problem.

Finally, there is the question of the time and financing required to bring the process to a successful conclusion. Involving communities step by step throughout a participatory process takes a great deal of patience on the part of both the communities and of the other stakeholders, and it also requires a good deal of money. This point is not always fully appreciated. By means of this experiment, we can see that for a participatory communication endeavour to be effective requires the investment of enormous quantities of time and sufficient resources, in particular because of the frequent travel involved.

In conclusion, we may say that when communities participate in defining the problems holding back their development and in seeking solutions, they assert ownership over the outcomes, the benefits and the lessons. They will use this knowledge and these new ways of doing things to deal with development problems as they arise. This is the essence of the lessons that we can draw from the initiative, which used participatory communication to resolve conflicts over water resources in the Nakambé Basin.

Notes

1   GUCRE stands for Gestion des Usages Conflictuels des Ressources en Eau, or Management of Conflicting Uses of Water Resources, a project sponsored by the Centre d'Études pour le Développement Économique et Social (CEDRES) of the University of Ouagadougou in Burkina Faso.

2   In its work with rural communities, GUCRE collaborates with locally based facilitators. Judith and Alidobi were among these.

Experimenting with Participatory Development Communication in West Africa

Fatoumata Sow and Awa Adjibade

Long a neglected art, communication is today rightly treated as a key element in community development activities. Over the last decade, West Africa has seen numerous experiments for using the potential that communication offers people in finding solutions to their local problems. Many of these initiatives have sought to spark community dialogue about natural resource management. This chapter looks at two such projects, one for managing water use conflicts and the other focusing on rural communication and sustainable development.

Although the authors were not directly involved in these projects, they were associated with them in several ways, and from their observations they have drawn some lessons that could help to improve existing practices in participatory development communication (PDC).

Today, ten years after the first experiments with participatory communication got under way in West Africa, we can say that far from being cut off from local reality, this approach finds its wellspring in community life, in inter-group relations, in local knowledge and know-how, and in development practices and activities, as well as in existing systems and tools of communication.

In these communities, participatory communication has helped to identify the major concerns and conflicts related to natural resources. Disputes typically have to do with land occupancy or with crops trampled by livestock; disagreement over the location of a well or the boundaries of a property; or tensions sparked by deforestation through bushfires and illegal woodcutting. Power struggles between customary leaders and government authorities over the question of landownership also give rise to many problems.

These conflicts and tensions crop up so frequently that they can shatter the harmony of village life and pose an obstacle to local development. The lack of consensus about the underlying causes and on mechanisms for managing conflicts within the community or between neighbouring villages, and the lack of modern legislation governing the notion of property itself are just some of the elements that require good communication if local development is to be sustainable. The PDC approach then becomes a useful tool in identifying the sources of conflicts, and in seeking and implementing solutions and sustaining efforts for the sound management of natural resources.

By promoting dialogue and the horizontal exchange of views, PDC allows all social groups to contribute to resolving problems. With this approach, religious and customary chiefs, government experts and authorities, youth, the elderly, men and women can create the dynamics for sharing experiences and forging partnerships without distinction as to social class or caste, dynamics that can work not only within the community but between several communities or villages.

Another important outcome has been to raise the status of women, evident now in their readiness to speak in public, in the emergence of women's groups and in the greater openness of mind that characterizes women and men alike. Thanks to the knowledge and expertise acquired through these initiatives, women who were once excluded from outreach sessions have now themselves become agents of training and awareness raising for the community (and even for their husbands). They are quicker to speak out, they are putting on theatrical plays and they are creating new channels of access to credit, training and HIV/AIDS prevention.

A greater appreciation of local know-how is a further contribution that participatory communication is making to development. By enlisting people in diagnosing problems and seeking solutions, the project teams have offered them the opportunity to share their ancestral know-how and practices. This has helped to mobilize and involve local people who are no longer expected simply to accept and apply the advice they receive from researchers or technicians.

Conditions for success

If participatory communication is to be really successful, there are a number of conditions that should be met. Through their physical presence at the side of local people and participating partners, facilitators also play a key role in participatory communication while serving as relay points for field coordination. Their commitment and their mastery of the approach are important for achieving the expected results.

The process can be kept on track through an iterative approach that includes regular follow-up activities involving dialogue, clarification, training and capitalizing information. In other words, if all the stakeholders can agree at the outset on a possible solution and on ways of implementing it, there is no reason for it to fail.

If all stakeholders have a proper common understanding of the concept of participatory communication, this will avoid frustration and resistance that could hold up activities in the field. Stakeholders can be brought up to speed by organizing workshops where key players and partners are introduced to the concept of PDC, discuss it and seek clarifications, and through meetings to coordinate follow-up and support in the field. Regular feedback on results will provide further encouragement to participating communities and can be used to help spread the word about the activities.

Applying the PDC approach to concrete local concerns can facilitate an understanding of the initial context in which the activities are being pursued and an appreciation of the objectives to be achieved. It also helps to rally communities to resolve their problems and to take responsibility for implementing the adopted solutions, something that is essential in asserting community ownership over projects. Of course, keeping both the customary and administrative authorities informed and interested will contribute greatly to achieving the objectives and must not be overlooked.

Interpersonal communication is also very important in the PDC approach. This means that the facilitators must first be given proper training in interpersonal communication techniques.

An iterative process for evaluating progress and a mechanism for hands-on monitoring and coaching will help players to ask the right questions as they go along and make the necessary corrections, and thus take a qualitative leap forward in achieving project objectives.

Project governance can be enhanced by setting up a management committee composed of and led by community members. This will not only build mutual trust and maintain harmony within the project, but will also bring it transparency, which is essential if community members are to rally to the process and remain committed to it.

Participatory communication facilitators: Aptitudes, qualities and roles

Experience has shown that for the PDC approach to succeed in natural resource management or in any other development project, the facilitator or the person in charge of the process must be:

- attentive and willing to listen to others;
- available to pitch in at any time;
- considerate and respectful of other partners;
- sufficiently imbued with the subject and the problems to be addressed;
- able to put forth persuasive ideas;
- considerate of the local setting;
- open and readily reachable;
- motivated, confident and devoted to the task; and
- recognized and admired by the community and the authorities.

The facilitator, as the fulcrum in the PDC process, has the essential role of:

- reducing tensions (resolving social and cultural obstacles);
- putting specialists in contact with participants;
- fostering a good working relationship among all partners;
- engaging in advocacy work with donors and development partners;
- identifying the specific needs of partners and different local groups;
- coordinating all activities;
- negotiating agreements between groups, partners, authorities and donors;
- bringing success stories to the attention of all partners and local participants.

Problems encountered in the participatory communication approach

As a process for achieving social change and resolving communities' development problems, PDC is a long-term undertaking and a number of stumbling blocks may be encountered along the way.

The participatory communication approach can also run into other problems that communication alone cannot resolve. Moreover, the approach may well take longer than the planned life of a project and this may not fit with the time horizons of donors, partners or local people.

Budgetary constraints may affect the frequency and duration of the follow-up that such activities require. There are also bits of social and cultural baggage that make it hard to change mentalities and behaviour, something that takes a lot of time. Finally, managing power relationships within communities demands a great deal of skill and finesse.

Conclusions

The approach championed by participatory development communication can achieve results of great significance for communities, researchers and technical partners by fostering initiatives that involve dialogue about managing environmental problems. Experiments in West Africa over the last decade have provided clear indications that this methodology can be applied very widely to natural resource management projects and to any other development project. Yet, the facilitator must be careful not to arouse false expectations. Nor should the facilitator try to take the place of development agencies. Finally, the facilitator must refrain from imposing preconceived solutions and must try, instead, to promote the kind of synergy in which the community itself will take charge of its development.

The situation in Africa today, marked by poverty and recurrent environmental disasters, calls for a real break with the past in terms of community participation and the place and role of communication and information as key factors in reinforcing people's capacities for self-development.

The boom in community or co-operative-run radio stations in the countryside, and the steady spread of information and communication technologies are obviously important factors that must be taken into account in planning strategies for strengthening and expanding participatory development communication.

Strategic Communication in Community-Based Fisheries and Forestry: A Case from Cambodia

Jakob S. Thompson

Faced with tremendous communication needs, many natural resource management projects undertake to develop materials and implement activities without a spelled-out and systematic use of communication. When the need for a clearer and more effective strategy arises, reflecting on past and future challenges can help to select the most appropriate methodology to respond to the needs of communities.

This chapter looks at the numerous difficulties faced when dealing with a very large array of participants and communities where cultural, occupational

and geographical features seem to divide people rather than unite them. It is based on the reflection of a training and extension officer whose appreciation of development communication has been evolving along the way in view of the multiple challenges faced by a community-based fishery and forestry initiative in Cambodia. As a result, the approach has evolved from the production of materials to the systematic use of communication tools and, more recently, to the establishment of more participatory processes.

The Cambodian province of Siem Reap is located in the north end of the Tonle Sap Lake, which is the largest freshwater lake in South-East Asia. Every year, the flow of the river connecting the Tonle Sap Lake to the Mekong River is reversed because the flood level in the Mekong becomes higher than that of the Lake. This results in a fivefold increase of the area covered by the lake. During that time, the productive inundated areas become the nesting, feeding and spawning ground for hundreds of different fish, bird and animal species. Thus, this seasonal flooding is a condition for the existence of this rich ecosystem, which provides a source of food and income for a rapidly increasing population on and around the lake.

The overall development objective for the initiative, entitled Participatory Natural Resource Management in the Tonle Sap Region, now in its third phase, is to establish responsible, productive and sustainable management of forest and fishery resources by local communities in order to meet local needs and stimulate local development within the province of Siem Reap. More specifically, it aims at developing community fisheries and forestry, while promoting private and community-based development activities in support of natural resource management. It also pursues the objective of strengthening institutions and building local and regional capacity.

The initiative emphasizes the facilitation of productive and sustainable community-based natural resource management (CBNRM). In the seasonally flooded zone around the great lake, the initiative works with fishing communities to protect and manage thousands of hectares of inundated forest habitat for the production of both fish and forest products. In the upland areas, the concept of community forest management is spreading as communities seek to protect and conserve rapidly disappearing forest resources. Currently, the initiative is providing support to the provincial government departments of forestry, fishery and environment, which assist 44 community forestry organizations in the management of more than 20,000ha of forest lands and ten community fishery organizations in the management of close to 108,000ha of fishing grounds.

Besides community-based natural resource management, the initiative has identified and engaged in a number of supporting activities. Tree nursery development and the distribution of seedlings to villages and farmers, along with aquaculture development, has been introduced as a livelihood option to reduce people's dependency upon the natural resources base. Rural credit and training of savings groups is stimulating local development, and environmental education for children and adults has become a priority issue, with a view to strengthening the environmental awareness of current and future natural resource managers in Siem Reap. A geographical information system (GIS) unit is also providing maps and mapping support to the activities.

The communication challenge

The community fishery and forestry processes are participatory step-by-step approaches that start with a range of assessments. The first steps aim at setting up a management committee through village- and community-level elections; drafting, finalizing and gaining approval for bylaws and statutes; demarcating the community forestry or fishing grounds; and supporting the development, approval and implementation of a management plan regarding the use and protection of natural resources by the community in years to come. Open and effective communication in a broad sense is a key agent in making this community fishery or forestry process work: communication between project staff and stakeholders, between stakeholders and also among staff members. Communication also plays an important role in influencing and strengthening the external framework for community-based natural resource management, as well as in gaining the necessary recognition and support from authorities, decision-makers and other key actors.

As is the case in most countries, there is not a strong tradition for inter-departmental cooperation in the Cambodian governmental structures. The multidisciplinary approach of the initiative has, however, depended upon and created an opportunity, as well as a climate, for cooperation and exchange between the forestry and fishery departments, and with the Ministry of Environment. However, the existing power structures in Cambodian society are complex. In some cases, the lack of transparency and control functions makes it difficult for open and unambiguous communication to take place.

Variety of communication challenges

For a number of reasons, the communities involved in CBNRM in Cambodia represent a wide range of communication challenges. At one end of the spectrum, the initiative works to assist community forestry development in sites with one or a small number of villages located along a small road or within a limited geographic area. The people involved seem to have a sense of community belonging and they come from similar ethnic and religious backgrounds. To a large extent, people in these communities have corresponding priorities when it comes to the use and protection of forest resources. In this setting, a communication or extension worker can achieve a lot by arranging a village meeting with the help of the village chief or by posting information on the village information board. The success of a meeting or an information campaign in such a setting largely depends upon the quality of its planning and implementation, and not upon factors such as logistics, community structures or differences in interest.

At the other end of the spectrum, one can find community fishery sites on the lake, with multiple villages of different sizes and structures. Their inhabitants represent different ethnic groups with a variety of languages and religions. There is a wide range of resource users, from the fishermen who entirely rely on fishing for their income and livelihood to the rice farmers who see cleared agricultural land as more valuable than inundated forest. In addition, there is also a wide range of people influencing the priorities of fishermen: fish buyers and moneylenders have stakes in their catches; government officials influence the way in which they fish through the exercise of their power; offenders overexploit the resources with no long-term responsibility and often use fishing methods that are damaging to the environment; and migratory fishermen/farmers from outside the community settle by the lake during the best fishing season. Furthermore, tourism is becoming an increasingly important source of income for some fishing communities close to Siem Reap.

Besides the complex human aspects of a fishing community, the physical features also represent a challenge to the communication facilitator: a floating fishing village is in a constant state of moving. People relocate their floating houses about 15 times per year in order to keep up with the changing water level of the lake. After the rainy season, most houses are moored along a road with the morning market right on the doorstep and the main town of Siem Reap or another main market within easy reach. Later, in the dry season, the water level of the lake has dropped by 7m to 8m and the same houses are now located many kilometres further out, on the open lake. At that time of year, a trip to town involves a boat trip and a long and painful motorbike trip through the inundated zone, only to get you to where your house used to be some months before. When entering the inundated forest on foot at this time of year, one also has to look up to see the information signs about community fishery that were previously posted in the treetops during the wet season. In this setting, there are many obstacles to effective communication, and the skills and intentions of the communication or extension worker are sometimes not enough to inform and engage the people in the community. The communication process therefore has to rely on a much wider range of communication tools, and one has to build more effectively on local capacities in order to reach an audience or to encourage broad participation.

The communication approach

Preliminary assessments and experiences indicate that regardless of the initiatives under way, people rely on direct interpersonal communication for their information needs in the sense that people talk to each other to convey information. People see radio and meetings as the potentially best way of communicating within their communities; but so far, the mass media has mostly been used for entertainment and, to some extent, for news. Written materials or newspapers have little potential due to very low levels of literacy. Illustrated posters have a potential; but the lack of permanent structures and the harsh climatic conditions pose a challenge as to when they are to be produced and posted. Illustrated handouts or leaflets have been identified and used by the project as a means of communicating; but again, only limited information can be transferred in the form of pictures. However, when text is used, the understanding is generally quite low.

The initiative has used a number of different tools and approaches to communicate to, from and within communities and stakeholder groups in order to build awareness, inform, educate and encourage participation. These tools have primarily been developed by the initiative and participation has so far been a result of communication activities, rather than being used for building the communication itself. However, attempts have been made to include communities in the development of communication tools. Nevertheless, so far their participation has been limited to consultations on the content and layout of single tools, rather than full-fledged strategy development with participatory rural communication appraisal (PRCA) or similar tools.

Meetings, workshops, etc.

In this initiative, meetings are the main communication tool. At the community level, the management committees, assisted by the field facilitators, invite people to attend meetings to present and discuss different aspects of the community forestry and fishery processes. The drafting of rules, regulations and management plans depends upon the participation of community members and other stakeholders, and fair elections obviously depend upon all the groups participating. Issues of such importance that plenary discussions are required also arise independently of the community fishery or forestry process. Meetings are then called in order to inform and consult the views of the people. The project is also facilitating horizontal communication between communities in so-called community fishery and forestry network meetings, to which representatives of all the communities involved in CBNRM development are invited. These meetings give the communities a chance to communicate among themselves and to communicate directly with representatives of the provincial and centralized authorities. The monthly district-level planning meetings have also been another important channel of communication. At earlier stages, the project team leader would personally attend the meetings; but in the past years, this responsibility has been handed to field staff. The main goal is to inform community representatives about community-based natural resource management and to gain recognition amongst stakeholders. Requests for further expansion of community forestry to new areas are commonly put forward by communities in these meetings.

Extension and environmental education for adults

On the lake, assessments indicate that meetings are too time-consuming and difficult to attend for the working population. In order to raise general environmental awareness in a wider audience than the meetings normally attract, the environmental educators travel from village to village with an environmental educational programme that is conducted at a time of day when fishermen can attend (this time varies depending upon the season and the type of fishing gear they use). A video about the history and current state of the natural resources in their community is used as an entry point to a problem tree exercise that eventually leads to a discussion of the possible solutions to the problems they face. The environmental education is based on entertaining and highly interactive approaches, small group discussions and plenty of room for humour and flexibility.

Besides destructive fishing, the encroachment of forest for agriculture and the extensive use of pesticides, which is likely to affect the long-term fish productivity of the lake, also constitute environmental concerns. Pesticides are likely to influence the safety of eating fish from the lake. As a result, the project has used a farmer field school approach with the seasonal farmers who grow agricultural crops on a large scale within the inundation zone. Evaluation indicates that these initiatives have led to an increase in awareness, improved farming practices and a reduction in pesticide use.

There is also the case of migratory people who enter the community fishery areas to fish every year. Community fishery organizations are assisted in their work of communicating with these people. Typically, these people are farmers who spend the dry season in temporary shelters close to channels or ponds in the community fishery areas. They fish and process fish for a few months before going back to celebrate the Khmer New Year and resuming agriculture, when the rains eventually start again. This is considered a traditional practice that may prove difficult to curb. Consequently, there is a need for information about the local rules that apply in the community; fishing fees; the basic ecological reasons for banning some fishing gear; and about any management practices in the community that they could take part in, in order to contribute to the sustainable management of the resources. To achieve this, community members and project facilitators travel around the inundation zone by boat and on foot to localize and inform people as they go.

Environmental education for children

Cooperation with the provincial Ministry of Environment has primarily evolved around an extensive environmental education programme. Essentially, the programme includes the establishment and use of a floating environmental education centre on the lake called the Gecko Centre and the development and implementation of an environmental education manual for primary school teachers with three partner non-governmental organizations (NGOs).

The Gecko Centre is located in the floating community closest to the main town of Siem Reap. Twice a week, the Gecko and its staff host a day of environmental education activities for groups of schoolchildren in the area in order to raise their environmental awareness and knowledge about the lake ecosystem in particular. The Gecko Centre also works as an entry point for some of the tourist boats that are taking people on excursions to see the natural and cultural environment on the lake. Besides contributing to the general awareness about the lake, this is providing an opportunity for the Gecko to cover some of the operating expenses from voluntary contributions made by tourists or other visitors.

The environmental education programme also includes the design, production and use of an environmental education manual in Khmer. In collaboration with three other NGOs, a manual for teachers has been developed and 1000 copies have been distributed for use in primary schools. Some 100 teachers have already been trained in environmental education principles and how to use the manual. They are now implementing the book as part of their work. Today, monks are also being trained in how to implement environmental education activities in the pagodas for out-of-school children. The book will be updated and reprinted based on the results of this first phase in order to augment environmental education methods and the environmental curriculum in the Cambodian school system.

Supporting awareness building

Posters are developed and/or used by the project with general or site- or topic-specific content, and the community fishery/forestry organizations have also been supported financially to build signboards for information dissemination, general environmental awareness raising, etc. The posters are hung and the sign boards are fixed at central points in the communities or at places that fishermen visit, such as landing sites for fish or outside the pagoda.

Because it is located close to Angkor Wat, Siem Reap is a major tourist destination. Therefore, it was deemed important to contribute to the general awareness raising by disseminating information on the importance and unique features of the Tonle Sap Lake ecosystem and community-based natural resource management. In addition to the Gecko Centre, visitors will find posters, books, visiting cards and locally produced palm sugar candies on display or for sale at the airport, as well as in many hotels and restaurants in town. At the provincial Ministry of Agriculture, there is a documentation centre with project documentation and other relevant documents, books, videos and resources. This centre is being used by project staff for information searches; but it is also open to and frequently used by students, NGOs and other users. A richly illustrated coffee-table book with written contributions from local and international experts on the culture and nature of the lake is the latest contribution to increasing the international audience's awareness of the importance of the Tonle Sap Lake.

Communicating management activities

Another challenge for the community fishery and forestry organizations has been to make everyone aware of the existence of CBNRM and to build acceptance for the activities and the rights and responsibilities that community-based natural resource management implies for the intended members. This is being done through the registration of members and the collection of a small membership fee. Since people usually do not give money for something they know little about, they start asking questions about community fishery and forestry.

Other practical management activities, such as the development of fish sanctuaries, have also proved to engage people and to gain general acceptance for the role of community fishery. The large fenced-off areas are built to shelter a stock of brood fish throughout the most intensive fishing season. The fish can then migrate to the forest in the flood season to spawn. Coloured flags and explanatory signs on each corner mark where the sanctuaries are, and people don't pass one by without asking what it is for. In community forestry, the strips of cleared forest and vegetation to prevent fire from spreading have much the same effect, along with thinning activities and marked research plots.

Why a strategic approach to communication?

As this overview suggests, a wide range of tools and methodologies are applied in order to inform and educate people, as well as to promote participation. However, there has not been a clear strategy for communication work, and the different activities have not been applied as part of a larger communication plan. The general approach has, instead, been to apply available and appropriate tools to deal with single communication issues as they arise. As mentioned earlier, the increasing communication needs of the project have resulted in the establishment of a training, extension and communication unit that is supposed to deal with communication needs in a more systematic way, and in accordance with communication for development principles. This was also a response to a possible expansion of project activities to other provinces in Cambodia. In hindsight, one could argue that the project probably would benefit from a more systematic approach to communication for a number of reasons, as follows.

Improving effectiveness

Throughout the years, it has become clear that communication materials that do not involve people in the development or interpretation process have very limited impact. Consequently, the project involves community members in, for example, making specifically designed community fishery posters for each site. The management committees in the communities and the field facilitators seem to have great expectations regarding the effectiveness of such materials. However, the use of single materials outside a well-planned and properly scaled communication strategy seemingly works against its purpose. First, the huge needs of the communities to disseminate information cannot be covered by a single material such as a poster without seriously reducing its quality and effectiveness, even after a strict prioritization exercise is done with the community to limit the amount of information squeezed onto the poster. Second, the isolated use of a single tool such as a poster may be used to transfer the committees' responsibility of informing people to becoming informed, since the information is now theoretically accessible to the people in the community. The management committee therefore seems to think that a poster relieves them of the important task of talking with people and exploring other complementary means of communication in order to ensure that everyone has a chance to express their view.

Making better use of multidisciplinary aspects

The initiative strives to apply a multidisciplinary approach to deal with complex real-life issues. The project, however, works within governmental departments, and the development and application of cross-sectoral solutions in a sectoral system is one of the project's main achievements. However, an overarching communication strategy could probably enable the project to benefit even more from the multidisciplinary approach by assessing strengths and weaknesses in the different sectors as a basis for exchanging communication expertise and tools between sectors.

Building stronger project and community identity

Tonle Sap Lake is increasingly becoming the focus of small and large development initiatives. This has resulted in a multiplicity of actors with complementing, overlapping or even contradictory objectives and methods compared to those of the initiative. In this setting, it would be beneficial to clearly point out to people what is, and isn't, community-based natural resource management development. In the future, communities will probably have to make the distinction in order to avoid confusion and to decide what is worth spending time and effort on. If not, they risk spending too much time accommodating researchers, NGOs or other actors, with limited potential benefit for the development of the community or even with agendas that do not benefit the communities at all.

Learning lessons

Based on our impression and in-house assessments, the work to raise awareness, inform, educate and communicate with communities has clearly had an impact within the communities. People generally know that community fishery and forestry exist and their purposes. People are also ready to engage in the management of natural resources and the environment since they generally understand the importance of the resource base and their effect on it. It is, however, difficult to assign the successes to any part of the approach, or even to decide if the improvement is a result of this initiative or of other factors, since the lack of an overall strategy also means that there is no established system for monitoring and evaluating the communication. Of course, this also makes it more complicated to develop guidelines and recommendations for the future, or for others to learn from, based on the initiative's communication approach.

The initiative has worked closely with partners and within governmental structures to lobby and advocate for the importance of community-based natural resource management. It has also contributed to the formulation of policies on CBNRM. Some of this work has been obvious in speeches and written contributions and comments; but much of it is also achieved through less visible channels and media. This aspect of the initiative's impact might be one that is strengthened due to the somewhat random but, nonetheless, flexible approach to communication in the project. Again, the documentation, lessons learning and reproducibility of this aspect of the project will be difficult to ensure since its success is built, to a large extent, on personal devotion and communication skills. Thus, it may be difficult to make generalized recommendations or to propose specific methodologies.

Scope for participatory development communication

So far, there has been a strong demand in Siem Reap for expansion and for handing over new land to communities who will manage it. The framework for development has, by and large, been set by factors outside the communities, such as the formal requirements for community-based natural resource management. The further development and continuation of CBNRM activities, also beyond the lifetime of the initiative, will need to look at diversifying community-based natural resource management activities to accommodate a wider range of people in the communities. Strengthening CBNRM and achieving widened support will therefore depend more and more upon the participation of the people who benefit from, and take part in, the management of the resources. In other words, there is a need to allow people to interact, collaboratively learn and influence decision-making in order to make CBNRM suit their needs. Participatory development communication approaches can potentially identify the community interventions and management options that are best suited for each community in order to ensure the basic CBNRM ideals of participation, equity and sustainability. The sometimes complex nature of the communities will probably also call for strong communication efforts in order to fulfil the intentions of any intervention. In other words, PDC is meant to play an important role both as a tool and as an outcome of CBNRM in Siem Reap. However, in order for this to succeed, some important preconditions must be met:

- Skilled CBNRM facilitators with an understanding of participatory processes and an appreciation of communication as a tool for development must be supported in working with communities for the benefit of the latter.
- At all levels, there must be unconditional support for the communities' right to take the lead in managing, identifying and implementing management-related activities that sustain CBNRM.

Paving the Way for Creating Space in Local Forest Management in the Philippines

Cleofe S. Torres

When, after decades of dispossession and disempowerment, shifting government policies make it possible for a devolution process to take place, local people's organizations can only rejoice. But assuming full responsibility for the environment they live in requires that communities acquire new skills and new capacities, while building their social capital. Although not a panacea, participatory development communication (PDC) can greatly help

communities to take up the new challenges associated with decision-making and self-determination.

The Bayagong Association for Community Development Inc (BACDI) is an upland people's organization in Aritao, Nueva Vizcaya, the Philippines. Its members were victims of the involuntary relocation project brought about by the government's construction of a dam that submerged the people's ancestral lands. Instead of taking the government's offer of resettlement in a place not akin to their culture, they migrated to a nearby forested land. This experience precipitated their engagement in some kind of participatory development communication as they needed, then, to discuss their fate and how they could cope with it. Of course, these people did not know that they were involved in PDC!

Labelled as squatters and encroachers by government, they struggled for almost 28 years to hold on to the lands that they have been *de facto* occupying since they migrated in 1960. Periodically during the past, they had to resist and endure the efforts of government agents and threats from outside speculators to expel them from the area. To make their appeals known to the concerned authorities, these upland farmers engaged in dialogues with government officials and authorities. But when their voice seemed not to be heard, they resorted to street protests and lobbying. These strategies were outcomes of their frequent discussions and analysis of their situation.

A wind of change worked in BACDI's favour when the government's policy on natural resource management adopted community-based forest management as the national strategy for sustainable development of open forestlands. BACDI applied for and was awarded the Community Forest Stewardship Agreement, a tenurial instrument that formally legitimized their claim over the forestland they have been occupying. It granted them tenure of 25 years, renewable for another 25 years.

While the political and social intent of such a devolution strategy was highly appreciated, issues emerged regarding the community's readiness and absorptive capacity to handle the devolved tasks and functions. Devolution as used here refers to 'the process where the locus of power and control shifts from the state to the local communities' (Magno, 2003). Under community-based forest management and as part of their new set of responsibilities, BACDI members underwent the process of participatory resource management planning. Here, PDC took the form of a social preparation method, equipping the people with the knowledge and skills that they would need as forest resource managers. Through community mapping, resource identification and problem prioritization, they learned to draft their plan and called it their own.

The other gender-sensitive participatory rapid appraisal methodologies used for situational analysis included community resource profiling; gender-disaggregated household activity and decision-making; stakeholder identification and analysis; political-ecological mapping; historical-structural analysis; institutional analysis; participatory analysis of problems and options; and resource management action planning. In all these methods, PDC was a core process for learning and planning.

Supplementing this variety of PDC methods were communication tools such as Venn diagrams, maps, community billboards and posters, written documents of their organization, and policies agreed upon for managing their resources.

This exercise paved the way for the community members to have a better grasp of the quality of their resources, their cultural integrity, their social capital and their political capacity, as well as a sense of their prejudices, weaknesses and vulnerabilities – knowledge that they could have not unravelled had they not undergone PDC. This also enabled them to identify internal and external threats to their resource management and how these might be handled.

To a large extent, the participatory experience enabled them to implement their plan with more confidence and better direction. Whereas before, the community simply remained buried in their culture of silence and subservience, the knowledge acquired of their rights and responsibilities through community meetings and discussions enabled them to become more open and assertive. As a result, when they conducted social mobilization among their members and allies, they were already more aware of community organizing, government policies, their internal capacities, external possibilities and the given constraints in the field.

Participatory development communication paved the way for BACDI's link-up with other government and non-governmental organizations (NGOs). Members knew that they could not muster all the resources needed to manage their forest resources well. They eventually realized that they also needed to access external assistance, such as livelihood opportunities, *barangay* roads, markets, credit, schools, health centres and other social services. Partnerships became another imperative, and they were able to establish alliances by using PDC.

As they learned more about themselves and their resources from the various participatory rural appraisal methods used in the study, BACDI members were able to come up with more rational plans and approaches for managing their communal forest. In the past, they had left these matters to their leader and whoever was deemed influential in their group. PDC also enabled them to reorient their outlook and address their needs by first using locally available resources before turning to outside sources for assistance.

For monitoring and evaluation, they employed informal, unstructured discussions where they would ask each other and reflect upon where they were at a certain time relative to their plans and targets. The observations they made were then discussed in their regular community meetings. Here, members were allowed to clarify and validate their observations.

Used in the various facets of participatory resource planning, PDC has actually served as a mechanism and, at the same time, as the context of group learning. As they progressed in learning about their biophysical, sociocultural, economic and political environments, BACDI members became more enlightened and rational managers of their forest resource. Even though most of them had only attained a low level of education, and some were even unschooled, they felt that they had learned many things from their community natural resource management activities. As a result, they were able to delineate the bounds and limits of the physical and political space that they had claimed for themselves. This was reflected by the sample plan and set of policies formulated governing the sustainable use and protection of their resources.

Social dialogue became a frequent activity in which they engaged by virtue of the communal nature of their forest resource. Decisions had to be collective and inclusive. Although discussions were not always smooth sailing, the members of the community have gradually learned the techniques of negotiation and consensus-building. In the process, their dwindling social capital has been enhanced. Social capital refers to cooperative social relations and collective action processes. This is embedded in 'norms or reciprocity, networks of civic engagement, trust, and obligations that facilitate coordinated activities' (Asian Development Bank, 1994). Since social capital depreciates through time, constant dialogue and open communication prevented it from deteriorating. Strong social capital has enabled the BACDI community to protect their traditional resource system from outside encroachers and the centralizing tendency of the state.

As BACDI members learned to master the science and art of forest management, they were continuously bombarded with challenges. Dominant among these was the multi-stakeholder setting for decision-making and action planning. It was through PDC that community members were able to respect and manage the diversity of views on an issue. Constant dialogue opened them to many possibilities and made them realize that there could, indeed, be many possible solutions to a problem.

Likewise, PDC provided the venue for mainstreaming women in a men-dominated BACDI ethnic group. From mere cooks and servers during community meetings, women started assuming more substantive roles, such as being a liaison officer, secretary, treasurer and even *barangay* councillor.

It must be emphasized that it was not PDC alone that paved the way to local forest management by BACDI members. Other factors were also involved, such as forest culture, social capital, policy presence and assistance by external actors. It can be safely assumed, however, that participatory development communication played a critical role, tempering the socio-political environment, internal and external to the community, so that a climate favourable to the community's takeover of resource management was created.

Reflections

Just like any other communication methods, PDC is not a panacea. It also has certain caveats, nuances and limitations. As a tool, its users and practitioners have to understand its profound intricacies. It takes time and practice, and a lot of learning from failures, to appreciate what it can and cannot do. Hence, its participants have to understand the action–reflection–action dynamics that are built into it.

Participatory development communication, as a catalyst for change, should also be accompanied by the inputs necessary for development, such as credit or capital, roads, water and technical assistance. It can establish the link between and among the development players in the community. Information and consensus are necessary conditions; but they are not sufficient to bring about broad development as desired by the community.

Likewise, devolution does not necessarily ensure effective local forest management. Policy and political and technical support have to be provided. But the use of PDC as an ingredient in the process makes it more workable, socially acceptable and satisfying in meeting the democratization and efficiency agenda of natural resource management. Moreover, PDC enables the evolution of 'participation-as-engagement' process, veering away from the usual 'participation-as-involvement' process (Contreras, 2003). The latter simply implies that the 'subject' or actor is merely a participant in the process, in a position of powerlessness. The former, instead, transforms the actor into an 'active subject' and not just an object or a client. 'Participation as engagement' is a necessary condition for genuine empowerment and for creating space in local forest management, which can be best achieved through participatory development communication. The impact on the community, especially in terms of managing their own forest, would have been different had other forms of top-down communication been employed.

References

Asian Development Bank (1994) *Handbook for Incorporation of Social Dimensions in Projects*, Asian Development Bank, Manila, the Philippines

Contreras, A. P. (ed) (2003) *Creating Space for Local Forest Management*, La Salle Institute of Governance, Manila, the Philippines

Magno, F. A. (2003) 'Forest devolution and social capital', in Contreras, A. P. (ed) *Creating Space for Local Forest Management*, La Salle Institute of Governance, Manila, the Philippines

IV

Communication Tools and Participatory Approaches

Communication Tools in the Hands of Ugandan Farmers

Nora Naiboka Odoi[1]

Introduction

Participatory development communication (PDC) aims to establish two-way horizontal communication processes. However, in a situation where there is a need to introduce new information, communication processes can hardly be horizontal if the tools used to communicate with farmers remain in the hands of experts and professionals. This is what the research team of the National Banana Research Project in Uganda had in mind when it established a farmer-to-farmer training programme. Indeed, farmers were so empowered

by the process and enthusiastic about sharing their new knowledge with other farmers that they soon wanted to produce their own communication material. After pushing aside the material that had been initially produced for them, they assessed their communication needs, established objectives, defined activities, produced their own material and even selected indicators to measure their success. In other words, they took control of the communication process from beginning to end.

Farmer-to-farmer communication

Quiet reigned inside the sub-county headquarters building as farmers of Ddwaniro, within the south-western district of Rakai in Uganda, looked with apprehension at Nora and Moses, the research team's communication resource staff. Nora and Moses fidgeted with the television screen and video deck while Fred, the driver, fixed the power generator. As farmers peeked at the TV screen, they all seemed to be saying the same thing in their hearts: 'Let nothing go wrong now with this video. This is our chance to show other farmers what we have learned after all these months of learning and practising.'

The loud blast from the electric gadgets signalled that all was well and ready with the machines. Moses turned down the television volume as Nora began:

> *We can now settle down and watch what we recorded last time. Remember, this is the video we are going to use to share knowledge about proper banana management with other farmers. We are nearing the launching day of our farmer-to-farmer teaching programme when we shall invite other farmers, district leaders, researchers and all members of our community and neighbouring communities to show them the results of our endeavours.*

The farmers smiled at the thought of moving from the production stage onto the next stage when they would show other farmers what they had been doing during these past months.

The video began rolling. Farmers were visibly excited as they recognized themselves in the film:

> *That, indeed, is Mr Kubo talking. We saw him doing that... That is, indeed, his well-kept banana garden... But why is he talking from far away? It would be better if he were talking while standing near us [in the foreground]... Moreover, he is not looking at us from the screen... The pictures are not flowing well. We should not have used Mr Muganda to illustrate the mulching technology. We could have used Mrs Muganda instead... She is better at explaining issues. But where is the good banana bunch to show the result of a well-kept banana garden?*

The video continued to a stop just in time as numerous hands shot up simultaneously. Mr Kubo began the barrage of remarks that followed the viewing of the video clip: 'Madame, we have seen the video; but I for myself am not convinced that it will deliver our message.' The farmers unanimously rejected the video, agreeing with Mr Kubo and adding that they could produce a better video than that, but that this time they were going to make a more thorough plan of action. This marked the end of the farmers' meeting with Nora and Moses. The farmers immediately reconvened their own meeting and chose a chairman for the session. The agenda was how to produce a better video than the one they had just watched.

The farmers chose Mr Sebulime to be the overall presenter of their farmer-to-farmer teaching video. Mr Kubo was selected to demonstrate organic manure-making; Mr Lubwamira would demonstrate the digging of trenches in order to guard against soil erosion; and Mrs Muganda would explain the proper mulching of a banana garden. They fixed a date for the next video recording session and informed Nora and Moses about their decision.

On the day preceding the video recording, Nora and Moses arrived in Ddwaniro, together with a professional cameraman. They made contact with the farmers, who took them around the banana gardens that were going to be used for the video illustrations. A mini rehearsal was organized to determine what was going to take place the next day.

Farmers took the lead on recording day. They guided the team of researchers and the cameraman to the different banana gardens that were going to be used for demonstration purposes. They had prepared so well that they knew exactly what sequence the recording would follow. Consequently, the recording process took only a short time, and there was little footage to cut out during the editing of the video. Although this last phase took place in Kampala, away from the farmers, they unanimously accepted the video the next time that it was shown to them. They agreed that other farmers would understand and probably make use of the message contained in the video. This was confirmed later when they showed the video to other farmers of Ddwaniro.

Using photography

Slowly shaking his head with a sneer, Mr Sebulime examined the photographs that Moses had brought back from Kampala.

'These pictures cannot do. Look, the women appear as if they are going to a wedding feast. How can a farmer working in a garden dress up smartly like this?'

'That is exactly what was recorded last time', Moses replied.

'But it is not right,' Mr Sebulime retorted. 'Look at Mr Bazanya. He is looking directly into the camera; he is actually posing for the photograph. Look, this photograph is so crowded with people that its educative intention is completely lost. No, we cannot use this photograph to teach other farmers. A good photograph should have few people, at least not more than three, and the pictures should be big enough to be seen', Mr Sebulime confidently asserted.

From the side, Nora, who had only recently joined the participatory development communication research team, looked on in disbelief. The farmer was describing a good photograph as if he had attended a photography class. He was only missing the accompanying jargon of 'centre of visual interest' and 'foreground'.

'They always change what they last said', Moses explained. 'We have spent many months on these photographs and sometimes I wonder whether the production process will ever end. It seems as if we are only going round in circles. Farmers photograph what they wish to appear on their brochures; you take the film to Kampala for developing and printing and by the time you bring the photographs back, the farmers have changed their mind and they would like another type of shot.'

In another corner of the same room, Enock, the agricultural extension worker, looked on as the farmer described the picture he had wanted to capture in the photograph. The group of farmers also listened as Bazanya narrated how a good banana crop should be illustrated:

> *There should be a mother plant with a good banana bunch, a daughter who will take over from the mother, and a granddaughter. The picture should show good mulching practice where you do not put the grass right up to the banana plants. If you put the mulch very near the banana stem, the insects living in the grass will be able to attack the banana plants also.*

Listening to the dialogue, Nora realized that the farmer was doing the extension work instead of Enock, who was only listening. At the end of Bazanya's description, Enock asked him why he had not photographed exactly that. In fact, Bazanya had photographed what he had explained. But when the pictures were brought back for the farmers to choose from, the person who had appeared in that particular photograph had taken it for keeps! So it was no longer available for use in the brochure.

Making posters and brochures

Farmers turned up to make corrections on brochures and posters. It was a good gathering considering the long distances, the difficult terrain and the heavy dust on the roads due to the dry season. Although the farmers had previously been divided into three groups according to their different natural resource management problems, work on the brochures went smoothly, with farmers sharing photos when necessary. For example, if a group's photos did not depict what the farmers wanted to illustrate, the group approached other members for a more suitable photograph. This indicated that the farmers appreciated the fact that the exercise was not intended for competition amongst themselves, but for the purpose of sharing information with other farmers who had not yet taken up the new farming practices. It also suggested that the three initial working groups were slowly merging into one group.

The poster was not as easily constructed as the brochure, partly because farmers said that they did not know what posters look like. This was a challenge considering the fact that the community hall in which we were working had a poster on its wall. There were also several posters in the entrance of the community hall. Upon asking the farmers whether they had seen them, some said no, while others said yes – but that they had not examined them closely enough to know what they were about. When examining one poster that depicted proper water and sanitation practices, some farmers actually said that its aim was to teach people how to write. The concept of poster-making was difficult for the farmers to understand since they were expected to put several photos on one chart, which together would tell a story. They finally agreed to make posters with a limited number of pictures. Initially, the plan was to use still photography, with subjects that the farmers themselves photographed. But certain illustrations proved difficult to capture, such as a hole in the ground that was to be used in the making of organic manure and the trenches that capture soil erosion. In the end, farmers settled for artistic illustrations.

During the material production stage, despite time constraints, the production process appeared to have no end in sight. Facilitators then made it possible for farmers to work with a professional video cameraman and an illustrator in order to finalize the process. This arrangement also solved the problem of farmers taking the photographs for keeps and the limited expertise in the farmer-researcher team regarding the making of posters and brochures.

After the production stage, farmers devised a plan of how they were going to use the communication materials. In a workshop setting, farmers agreed on the geographical scope of their intended farmer-to-farmer information sharing. They confessed that, on their own, they could only manage to share information within their villages. This is because farmers can either walk on foot or ride bicycles to cover the relatively short distances in the villages. They also agreed that in order to be credible to other farmers, they had to make sure that their own banana gardens illustrated the recommended soil management techniques. They identified the different categories of people who needed to be informed about the three modes of soil and water management. These included fellow farmers like themselves, vulnerable groups such as women, as well as the disabled and orphans heading households. In order to get support for their farmer-to-farmer information sharing, farmers included local leaders, politicians and non-governmental organizations (NGOs). They identified institutions and channels through which they could share the information with other farmers. Farmers agreed to make use of radio, a farmers' newsletter and drama to supplement their information sharing through brochures and posters. In addition to one-to-one information sharing, farmers identified village meetings, churches and market days as possible fora. As a way of concretizing their plan of action, they made a time framework within which to accomplish specific activities related to the information-sharing objective. They also agreed on monitoring indicators and mechanisms.

Getting more specific

As we witnessed this participatory material-making process, we realized that there is another possibility regarding material production for the purpose of farmer-to-farmer information sharing. Three sets of communication materials may be produced. One set is intended for the 'teacher farmers' – the farmers who are going to share information with others. This set of communication materials should be illustrative and should enable interaction between the farmers on the identified subject. The teacher farmers should use this illustrative material to explain to the other farmer(s) the recommended soil management practices. This communication material should be produced in fewer numbers and should be durable since it is going to be used over and over again.

After sharing information with the other farmers, the teacher farmer should leave some material with the learner farmer for reference purposes. This communication material should be produced in larger numbers and need not be as durable as the material for the teacher farmers.

There could also be a third set of communication materials that would act as a back-up for the information-sharing activity by providing brief, general information on the subject at hand. For example, this could be a poster that would hang on the walls of farmers' houses, in shops or in churches; it could also be a radio programme.

In the end, it is important to remember that these materials will always be more effective and appropriate if farmers are closely associated with their production or, even better, if they end up taking charge of their production.

Note

1   This chapter was written in consultation with the other members of the research team: Wilberforce Tushemereirwe, Drake N. Mubiru, Dezi Ngambeki, Carol N. Nankinga, Moses Buregyeya, Enoch Lwabulanga and Esther Lwanga.

From Information to Communication in Burkina Faso: The Brave New World of Radio

Souleymane Ouattara and Kadiatou Ouattara

Introduction

In Burkina Faso, as everywhere in Africa, radio exerts an undeniable attraction. In countries where the government has given up its monopoly of the airwaves, nearly all the available frequencies are now occupied by radio stations run by community, commercial or religious broadcasters. Not only do people everywhere listen to the radio, but it has taken root throughout the land, even

in the most remote villages. The success of this tool, which some call 'Africa's internet', allows people to express their concerns in their own languages. Yet, this communication medium remains largely an unexploited resource when it comes to participatory development communication (PDC). An experiment conducted by the Journalists in Africa for Development (JADE) network at three places in Burkina Faso and in Mali demonstrates both the usefulness of such an approach and the importance of taking certain precautions to ensure its success.

From listeners to communicators

Somewhere in Burkina Faso a farmer sits down in a studio and belts out a few songs dedicated to his friends and relatives. The man is about 50 years old. He is happy to be able to make his voice heard in his home village. This is the first time that he has ever entertained a radio audience. Earlier, he had sent his sons with a present of some yams and a chicken for the station technicians.

Elsewhere, members of a rural listeners' club have just delivered their weekly programme sheet to the station. The form is filled out in the national language and shows the broadcasts that they have heard, together with listeners' observations and their requests. The station will take these into account in its coming broadcasts.

Examples of this kind represent amazing progress compared with the way things were once done, from the time the first radio stations were established in Africa until the 1990s, when many countries in West Africa began to deregulate the airwaves. Yet, despite this freedom of speech and the openness that radio maintains with its audience, considerable effort is still required if radio stations are to become true tools for participatory communication. It is all very well to free up the airwaves or to run radio games in the villages; to greet people throughout the day by their own names; to produce and broadcast programmes about farming, livestock and the environment; and to take account of listeners' expectations. All this allows better use to be made of radio; but it is not the same thing as making radio a tool for PDC, and still less for the management of natural resources.

There is no magic recipe for making radio a participatory communication tool. But the experiment that JADE conducted during 2000 to 2002 in three places (two of them in Burkina Faso and one in Mali) provides some food for thought. This chapter looks, in particular, at the most successful effort, launched in January 2001 at Ziniaré in central Burkina Faso.

Central Burkina Faso is a region of poor soils and one that faces severe water shortages. The 2001 project involved using participatory communication to share information among communities, development agents and radio producers in order to improve the management of natural resources. Besides JADE, the group includes technical experts from the agriculture and environment ministries, farmers organized as communication relay points in their communities (also called local communicators) and radio producers. The partner radio station belongs to a farmers' organization, Wend Yam.

Participatory communication: The key to natural resource management

Natural resources are, by their nature, common resources, and their exploitation and (above all) their preservation are in the hands of the community. These resources must be regarded as a fragile asset to be used wisely while being conserved for tomorrow, and protecting them depends in large part upon the community. It is the community who must decide whether or not to preserve forests, water sources and the entire natural environment in order to ensure its own subsistence and that of future generations. Yet, the general tendency is to exploit these resources with a view only to the short term, and there is little thought given to preserving them. We may safely say, then, that there is cut-throat competition between different users (farmers, herders, agro-businesses, etc.) that risks not only mortgaging the future of these common resources, but also creating conflicts over access to them. For all these reasons, natural resource management is a natural target for participatory communication as a means of creating dialogue among all stakeholders and helping them to make decisions that will serve the interests of everyone. Above all, PDC can help to organize joint activities based on a shared vision.

Promoting social groups through radio

It was not by chance that JADE chose to work with radio: it already had long experience in this field. But the first attempts, innovative though they were, did not meet with much success for two reasons. First, JADE was broadcasting with the help of technicians, radio producers and local communicators; but it was not working with the communities themselves, who were treated strictly as sources of information. Second, radio was not used to support participatory communication activities that might help to identify problems, analyse them and find solutions.

Moreover, despite its community radio designation, Radio Yam Vénégré, JADE's major partner in this project, remained a typical rural radio station. It was content to broadcast information to the rural public, the only difference being that it made much use of Mooré and Fulfuldé, the two most widely spoken languages in the region. The results produced by this top-down approach were disappointing and led JADE and its partner to try to use radio in another way.

The new approach recognizes and places value on community viewpoints. The public can express itself directly over the airwaves, instead of simply listening to broadcast advice from experts and radio producers. Discussion groups (for women, young people, adults and seniors), backed up by a team of development experts and radio producers, analyse problems with natural resource management, their causes, their consequences and their potential solutions. This process, which has been recorded from start to finish, involves four stages:

1   identifying problems in natural resource management;
2   programming produced by radio personnel and local communicators;
3   broadcasting the programmes;
4   gathering feedback.

Identifying problems in natural resource management

'What do you consider to be the priority problem with water, wood, fields and the land, and how could you help to resolve it?' This was the first question put to debate in the four discussion groups, and then in a plenary session. The idea was to highlight the community's capacity to find its own means for resolving its problems before turning to other sources for assistance.

At Nagreongo, an arid region of central Burkina Faso, three issues emerged from the discussions: water shortages, soil degradation and lack of wood. A poll was taken, and water shortages received the most votes. This issue was therefore ranked first in terms of priorities.

Among the reasons for the water shortage, people cited the inadequate wells, the great depth of the water table, the lack of erosion control and the absence of reservoirs. A more thorough analysis also revealed that the water was unhealthy for human consumption, and that simple techniques would have to be found to make it drinkable.

In terms of consequences and impacts, it was evident that the lack of clean drinking water led to public health problems, such as Guinea worm disease, childhood diarrhoea and other waterborne illnesses. Moreover, the shortage of water was destroying village communities as their able-bodied workforce gradually left, and was also harming livestock activities.

Programming produced by radio personnel and local communicators

In a participatory approach, producing a magazine-type programme involves not only radio producers, but also local communicators. These are farmers who are responsible for identifying village interests in advance, preparing field trips, taking part in producing a broadcast and gathering feedback from listeners. In this way, a series of 45-minute radio magazine programmes was produced, essentially based on ideas from members of the various discussion groups and interspersed with a few musical interludes. This was 'raw feed' that served to highlight participants' viewpoints.

Broadcasting the programmes

Public announcements were made, usually a week in advance, alerting people of the coming broadcast. This not only made it possible for people to tune in individually, but also, in some cases, to form discussion groups.

Gathering feedback

The local communicators were responsible for sounding out listeners' views on the broadcasts, with the assistance of the listeners' clubs (in Ziniaré and Sikasso). Typically, a local communicator would visit the main regional town on market days, and people would pass on their opinions to him or her. In some cases, listeners with literacy skills would send their viewpoints direct to the station. Generally, the station attempts to respond directly to listeners whenever possible. It also occasionally calls upon resource persons or institutions to deal with listeners' concerns.

Lessons from the experiment

Preliminary steps are needed to establish the sense of trust that is essential in getting the process up and running

Radio stations typically approach the communities with a preconceived idea of a broadcast. People rarely have the chance to make their expectations known in advance. Even today in rural areas, radio stations, despite their proliferation, are still inaccessible to most of the population. The fact is that when people are able to express themselves over the airwaves, this gives them some power, not only within their village but beyond. A good example of this arose in Nagreongo. Radio producers went to the village for a working session with resource personnel in order to establish a basis for collaboration and they found, to their surprise, that the entire village came out to meet them. Yet, this is a common practice, something communities do to stay in the good graces of development agencies. How could the situation be handled? It was here that local knowledge helped to avoid embarrassment. The working session originally scheduled was set aside, and the programme producers set about recording a broadcast on the history of the village, to the great satisfaction of everyone. They then held an interview with the local healer, an influential personality in the village and a great master of ceremony. These broadcasts were carried over Radio Yam Vénégré in the evening.

In this particular case, the radio team turned a dicey situation into a triumph by producing a radio broadcast on the spot. This ability to react quickly to situations allowed them to win the villagers' trust, and then to hold discussions with separate groups of men, women, youth and the elderly.

Collaboration with traditional authorities, technical experts and communities (in other words, teamwork) is an essential factor for success

The participatory approach demands real collaboration among all stakeholders – traditional authorities, research teams and communities – in order to minimize the risk that things will grind to a halt if certain parties are overlooked. Teamwork produces results that reflect the great diversity of the group. In Dori, as in Sikasso, field reporting was considerably enriched by the representatives of farmers' organizations and by rural development experts. The programme producers confined themselves to formulating pertinent questions and ensuring that proper use was made of the recording equipment. Yet, this approach requires a thorough knowledge of the local setting.

Radio gives added value to local folk knowledge

The importance of local folk knowledge has to do with the fact that it is slowly disappearing in the areas under study, without ever having been fully assessed and appreciated. Folk knowledge is not a simple store of information; rather, it has to do with the realm of understanding and of what is sacred. This is why it is important that communities are involved and that the appropriate approach, involving the following steps, is taken:

1   Pinpoint and prioritize the natural resource management problem.
2   Identify the resource persons who have local know-how relating to that problem.
3   Interview those resource persons.

The information collected must deal with the nature of local knowledge, where it was learned, its description, the way in which its holder uses it in practice, the results obtained and the constraints observed.

It is also useful to develop in advance a line of argument to support any information-gathering process in order to overcome the possible reluctance of local knowledge holders and to appreciate their knowledge, which is generally brushed aside by development officers.

That line of argument might point to the fact that as the processors of this folk knowledge move on in years, it becomes ever more urgent to record and preserve what they know and to make it available to the entire population. It may also be useful to point out the paradox inherent in looking to outside knowledge, which is often costly and inappropriate, when there are ready-made home-grown solutions at hand. Local knowledge, indeed, must be treated as a heritage to be preserved.

How can radio help to promote traditional local knowledge? Because it is immensely attractive to the public, radio can give to any information that it broadcasts an importance that it would not otherwise have, not only in terms of building an audience, but above all in terms of winning recognition for it. To be quoted on the radio leads to increased credibility.

The dual nature of radio

Radio is both a means of local communication and a way of spreading information widely. This dual characteristic also makes it a tool for participatory communication and mass communication. Not only does it reach the specific groups with whom the research team is working, but it is broadcast well beyond the bounds of the project's impact area. This concentric spread – the 'oil-spot effect' – gives radio broad scope. A leader of an 11,000-member women's organization in western Burkina Faso tells how the radio helped her in her work:

> *We used to send out flyers to our members to announce our meetings. Some of them were never received. The lucky ones who got the invitations would arrive two or three days after the meeting. Thanks to the radio, we can make an announcement and everybody knows about it at the same time. This has greatly strengthened our organization and its credibility.*

Radio is thus an especially useful medium, both for the mass dissemination of information and for participatory communication. In some cases, it can reach a large audience and amplify messages, while in other cases it allows for real popular participation via the airwaves in the search for solutions.

The question of ethics

Radio: A tool for changing mentalities

Through the new knowledge that it introduces within the community, radio can help to change mentalities. But if it is to do this, there are some preconditions that must be met. We frequently find that broadcasts reflect the viewpoint of the elite rather than that of the majority. The advantage of the participatory approach lies precisely in its capacity to give everyone a chance to express their expectations and their viewpoint, not only to the rest of the community but to people in other villages as well. Radio thus constitutes a powerful tool for changing mentalities, as many examples demonstrate. Women who doubted the effectiveness of vaccinations are now taking their children in for shots. Men who once scorned civil marriage are accepting it today. When it comes to managing natural resources, local knowledge broadcast by Radio Ziniaré is now being tried out in many local villages, where the mistrust that once prevailed has been replaced by a sense of recognition and gratitude.

The limitations of radio

There is a paradox about radio: it is highly seductive; yet it also has the capacity to misinform when it is improperly used. Moreover, it is impossible to interact with the radio host. 'Sometimes it can be hard to understand the message. For example, if someone doesn't get the meaning of a poster you can explain it to him; but with the radio that's not possible', explains one non-governmental organization (NGO) manager in Ouahigouya in northern Burkina.

Generally, subjects of concern to listeners are seldom addressed on the radio. When they are, the treatment may be only superficial. Other technical constraints relate to the station's broadcast range, the relevance of the topics, the cost of producing programmes, their nature, the timing of broadcasts, low audience interest, or the cost of powering listeners' receivers.

When it comes to natural resource management and the participatory approach, we find that radio producers are often poorly equipped professionally, particularly in the techniques of participatory communication. Moreover radio stations, while they may serve a community purpose, tend to become commercial enterprises more attuned to the needs of development agencies than those of the community. The PDC approach, particularly in the area of natural resource management, demands time and skills, which radio stations generally lack. They are more accustomed to entertaining their audience with music or carrying live round-table discussions because they do not have the means to produce programmes in partnership with the communities.

Tools and methods

Discussion groups

Our approach encourages members of the community to participate in debates. Thus, all the selected topics are discussed with the target groups (women, seniors, adults and youth) at the same time. According to Awa Hamadou, a midwife in the village of Kyollo in northern Burkina, 'Dividing people up into groups lets everyone participate in the debates.' A significant step forward has just been taken in this heavily Islamic village, where, in the past, it was only the men who were allowed to speak in public. Now, women are free to express themselves in front of the men on any topic.

In Nagreongo, the research team also relied on four working groups to identify natural resource management problems, analyse causes and consequences, and propose solutions. It is in these groups that the contrasting ways of approaching a problem within the village become clear. It is also in these groups that conflicts of interest are revealed and that solutions respectful of all parties can be found. The advantage of these groups is that everyone has a chance to speak, including women and youth who normally hesitate to express themselves in public.

Audio cassettes

The fleeting nature of oral messages remains a limitation inherent in radio broadcasting. When a programme is carried just once, it has little chance of reaching the entire potential audience. Many programmes are rebroadcast repeatedly to overcome this constraint. Yet, we still find that for many reasons the retention rate for radio information remains low. It is almost impossible to listen to a broadcast several times, as one can do with a cassette. The utility of the radio cassette is that a broadcast can be heard again and again, it can be stored and it can become a tool for sparking discussion about a given topic. Radio stations participating in the project are trading 'magazine' programmes produced and copied on cassettes. Since information on natural resource management issues is not event specific or particularly time sensitive, the audiocassette can be a more appropriate medium than direct broadcasting. Similarly, some broadcasts, particularly those dealing with local knowledge, have been recorded on cassettes to facilitate their circulation to listeners' clubs. Trading programmes among themselves allows individual communities to acquire new knowledge.

Traditional channels: Local communicators

The project on rural communication and sustainable development made some innovations in natural resource management by encouraging reliance on local skills, particularly local communicators. The local communicator was the cornerstone of the project and served as the interface between the relay points (i.e. the project representative in the area) and the community. The local communicator's many functions make him an indispensable player in any action-research effort.

In fact, local communicators perform many tasks: organizing and hosting chat sessions and supporting the research team in identifying natural resource management issues with the community. They also gather local knowledge to be dealt with by the broadcast, identify the resource persons (model farmers), organize recording sessions and gather feedback.

The importance of follow-up

The project's impacts are due, in large part, to follow-up by stakeholders in the field. Development technicians have taken it upon themselves to share experiences, experiments and lessons learned as a way of working with the communities. They can now foster synergy among themselves and, in particular, with the radio producers.

As to the radio producers, they can regularly be found collecting information in the field. They also work with the community in choosing issues, in handling them and in gathering feedback. For this purpose, they have retained the devices that were introduced (e.g. listeners' clubs and local communicators). Today, radio is using the participatory communication approach to go into the field, identify problems and seek solutions with local people. Moreover, thanks to local communicators who have their roots in the community, the issues of concern to local people can be inventoried and, in this way, can form the basis for radio programmes. Local people are now taking a much more active role in producing broadcasts.

In Sikasso, radio producers who were not involved in the project were, nonetheless, impressed by its achievements to the point where they wanted to use the approach in their own work. Within government technical services, as well, several managers recognized the usefulness of radio in dealing with bushfire prevention, sharing knowledge about production techniques, settling disputes over environmental management, preventing disease and reinforcing producers' groups.

Today, Radio Yam Vénégré has more direct and more regular contact with local people and is facilitating negotiation between them and the development agencies. With its experience in PDC, it is now playing an advisory role to development institutions. Moreover, it has helped a project that was designed to strengthen farmers' organizations to clearly explain the objectives of its surveys before beginning fieldwork. Programmes have been produced and broadcast for this purpose, and development officials are now working together. Even outside the project, broadcasts are being produced with other agencies attracted by the approach, something that also earns new revenues for the radio station.

By way of conclusion, we can take satisfaction from having used radio in a new way and we can congratulate ourselves on the resulting achievements. But the research team could have gone even further. It should have systematically produced audio cassettes in order to preserve the chain of communication, to reach those who missed the broadcasts and to allow certain points of debate to be revisited. Posters or booklets would also have helped to reinforce viewpoints. With respect to institutional information – that is, community questions about the project itself and the responses of the project coordinators – use of the radio could have cleared up a number of misunderstandings. This experiment provided a good opportunity for assessing the most efficient and effective types of programming. In addition, PDC could have been compared with the conventional communication format, which is simply based on broadcasting information. Recognizing that the broadcasts occurred both before and after the group discussions, a summary of the previous broadcast could have been replayed before the discussion sessions were held. The radio station, in fact, had the right equipment, including powerful loudspeakers, to perform this task properly.

Despite the importance of rural radio, it should not take the place of chat sessions. Radio should be seen as a supplement, one that relies on the results of these sessions for input. Chat sessions must also meet certain conditions if they are to be truly participatory. In the absence of all these conditions, people will, of course, continue to listen to the radio; but what they hear will be the voices of the producers, the dominant voices of the day – it will not be a tool at the service of the community that is truly participatory.

And Our 'Perk' Was a Crocodile: Radio Ada and Participatory Natural Resource Management in Obane, Ghana

Kofi Larweh

Introduction

Radio is usually seen as the ideal medium to reach people in remote areas who do not have access to other sources of information. But when committed to community participation and people's development aspirations, radio can

be much more than a source of information. It can also be a powerful tool to facilitate consensus-building and decision-making at the community level, thus becoming a catalyst for change. This is precisely what Radio Ada set out to do when it decided to accompany the efforts of the people of Obane in the eastern part of Ghana. After four decades of watching their environment deteriorate, the inhabitants of that community rolled up their sleeves to restore the waterway, which used to be at the heart of their lifestyle, and regain their lost prosperity. Day in, day out, Radio Ada was by their side, voicing their concerns and acting as a morale booster. In the hands of the community, the radio also became a tool for advocacy and mobilization. The results are quite astonishing.

A catalyst for collective action

Many towns and villages seem to have forgotten the reasons that they have come to be where they are. The people of Obane have not.

Obane is a rural community in Big Ada in the Dangme East district of Ghana, about 100km from the capital city of Accra. It is a brisk hour-long walk away on a hot, dusty road from Radio Ada, the community radio station of the Dangme-speaking people.

The people of Obane still remember that the waterway is the reason that their forefathers chose to live there. The waterway, the Luhue River, is a tributary of the mighty Volta River. It supplied fish, provided water for irrigation and served as a bustling transport course. The water was so abundant that Obane used to be the food basket for Big Ada. The women were, among other occupations, fishmongers, farmers, mat weavers and petty traders, while the men fished, farmed or hunted.

That was over 40 years ago.

The creation of the Volta Dam during the early 1960s reduced the flooding of the fields at Obane. Weeds, trees and debris choked the waterway. The surrounding lands became barren, leaving a fetish grove with isolated trees and some patches of green to the south. Not even the shadowy line of emaciated trees tracing the meander of the Luhue River in the background is thick enough to break the line of sight to the west.

Over time, Obane became one of the poorest communities in the district. For this reason, Radio Ada has always taken a special interest in Obane. Radio Ada is Ghana's first community radio station. On the air since February 1998, the station broadcasts 17 hours daily exclusively in Dangme, the language of its listening community.[1] It is staffed by volunteers drawn from the community and trained in its home-grown workshops. Radio Ada's identity is rooted in the culture of its listening community and inspired by their desire to improve their economic way of life while maintaining close community ties. The station actively pursues a participatory development philosophy. One of the main ways in which it tries to operationalize this philosophy is through 'narrowcast' programmes. These regular weekly programmes are recorded in different communities, with the main occupational groups in the listening community – fishmongers, fishermen, women farmers, men farmers and so on. The programmes are driven by the participating community members. They determine the content and it is their voices that predominate. Programmes are presented as a continuing dialogue, where producers simply act as facilitators. The station's holistic approach to community-initiated development as 'the voice of the voiceless' in content and the production process generates trust.

Thus, the programmes serve to affirm the knowledge and experience of group members, who share what they know, what they feel and what they believe. They fuel consultation and interaction between themselves and their counterparts, as well as between other listeners in the Dangme-speaking communities. They also serve to deepen the relationship with Radio Ada. As resources permit, the stations follow up the programmes with more extensive community consultation.

It was after a number of 'narrowcast' programmes had been produced with the people of Obane that one such consultation was initiated by Radio Ada. The consultation was held a few years ago, when the community was confronted with a decision to migrate or to renew their waterway.

The consultations generally involve the use of participatory rural appraisal (PRA) tools, such as the ranking-and-scoring matrix for needs or priorities. Despite the inherent friendliness of the tools, often when one asks communities about their priorities, the answer goes something like this: 'We need pipe-borne water, roads to transport our produce, electricity, employment, etc.' The litany never seems to end and tends to read like a shopping list.

In the case of Obane, possibly because of the ease and trust that had been built in their relationship with Radio Ada and almost certainly because they were at a decisive point in the life of their community, the responses were more detailed, reflective and textured. In every case, respondents had a story to tell connecting the present to the past.

The fishermen lamented that they had become goat and sheep rearers; one owns a cattle kraal. They were worried that the different fishing skills they had learned from their parents and friends were no longer of value to them and their children. The river did not exist during most of the year. It was choked, and even during the short rainy season it provided only limited access. The vast marshy lands that had boasted of crabs and reeds for weaving had all run dry.

The women could not bear sending their children off to school in Big Ada and beyond without enough food to support their stay. Formerly a little rise in the tide caused natural flooding of the fields for fresh crops even during the dry season. Although Obane lacks many basic amenities, the women identified 'unity' as their most important priority for development. They were sure that they, working with their husbands and joining other neighbouring communities, could revive the economic life of the community and restore it to its former status as a food basket.

During the plenary discussion, Divisional Chief Nene Okumo was invited to share his reflections. He narrated how during the past their fathers occasionally cleared the waterway to support ecological processes and to sustain flora and fauna. Then and there, the plenary decided to dredge the Luhue River using communal labour. Immediately, Madame Adjoyo Djangma, a mother of seven and redundant fishmonger, intoned a jubilant, melodious traditional chorus. The response was infectious as others took up the chant.

There were contour lines on some foreheads – more than 40 years of vegetable growth and silt to clear! The work extended over 10km, and would entail varying degrees of difficulty. In the meanwhile, the other neighbouring communities had their own priorities and there was the need for tools and other logistics. How could they possibly mobilize the resources needed?

'How can your community radio station, Radio Ada, help?' There was a measure of relief on some faces as the role of communication was put forward: 'Announce what we have decided to do … announce the day and time for the work … when others hear, they will join us … announce the names of those who have reported for work … put our needs on the radio … tell other people about the by-laws we shall make to prevent the river from choking.' The ideas kept coming.

Having facilitated consensus- and decision-making on the ground, Radio Ada fuelled the communication bonds within the Obane community and connected them to other communities and institutions to support the communal labour needs. Trusting the station to honour its offer of support, the groups actively turned it into a tool for advocacy and mobilization. They announced at their convenience and free of charge their work plans – whose turn it was to work – and raised issues on air about the environment, their occupations, their lives and the project's progress.

Both the women and the men of Obane conducted the clearing, and 60,000 seedlings of red and white mangrove have since been planted.

Before long, four groups from different communities – Obane, Gorm, Togbloku and Tekperkope – were working together to dredge the Luhue River. Other groups from other communities joined them to show solidarity and were duly acknowledged on air. They included communities from Dogo, Dorngwam and Atortorkope, as well as from Aminapa, Wasakuse and Midie.

District Wildlife Officer Dickson Yaw Agyeman stated: 'The radio broadcasts were a morale booster and a challenge. The Wildlife Department gave them Wellington boots, nylon ropes, cutlasses, hoses for weeding and food for work – kenkey[2] and fish and pepper.'

Their voices and active participation on air also won them other collaborators, allies and supporters. Organizations and institutes such as the Dangme East District Assembly, the Canadian High Commission, Green Earth and the Kudzragbe Clan of Ada Elders contributed to the initiative.

The men's group received a loan facility of 14 million Ghanaian cedis[3] to boost agriculture, while the women obtained a loan for 15 million cedis for food processing and marketing. In addition, a 12-seater KVIP toilet[4] was built. Funding was through the Wetlands Management and Ramsar Sites Project, which is supported by the World Bank. The magnitude of work accomplished over these four years might intimidate a stranger; but knowing what the river and the surrounding fields mean to their lives, the people of Obane persisted. With the support of other communities and of the Wildlife Department, as well as their community radio station, Radio Ada, they now have what matters most to them.

Day in, day out, as needed, Radio Ada followed the efforts of Obane on the air. Throughout the process, it took its cues from the leadership of the community, who were constantly at the station making requests for specific programmes and announcements. Radio Ada not only helped to mobilize concrete material support for the back-breaking work of the people of Obane, but also inspired the rest of the community with their enthusiasm, perseverance and sense of unified purpose. By consistently projecting a community endeavour into the public domain, as well as through the soft cheerleading style of the radio broadcasts, even disputes over land boundaries, customs and leadership were avoided. No one could afford to risk name or reputation by swimming against what was now literally a growing tide.

Today, one can travel on a boat from Big Ada through Luhuese to Obane and beyond. Year-round small-scale farming is possible by using irrigation methods, and the people now have access to freshwater. A replication is also in the offing at Totimekope near Ada Foah, the administrative centre that is the twin town of Big Ada. There, the Futue River is to be cleared in order to restore river transport, fishing and farming. The satisfaction of Radio Ada is in fulfilling its mission and sharing in the joy of the community's successes. Occasionally, however, there are more concrete rewards – perks, so to speak.

One sunny day, one of the dredging groups, led by Alfred Osifo-Doe (a professional mason), burst straight from the dredging site, sweating and jubilant, into Radio Ada, excitedly bearing a gift for the station – proof of the success of the common endeavour, they cried, fished out of the increasingly swelling waters of the Luhue.

Deeply touched, but mindful of its conservation role (as well as of the practical difficulties involved), the volunteers of Radio Ada managed to persuade the dredging team to return the gift to its natural habitat. So, be careful the next time you cross the Luhue River because by now it must have grown – the baby crocodile, that is!

Notes

1   The station has about 600,000 listeners, within a 100km radius. Although there are variations in the Dangme language, such as Klo, Gbugbla, Se, Ningo and Ada, the people are bound by a common history and culture. The station derives its name from the location in relation to the language. It is based at Tetsonya near the main town, Big Ada, a two-hour drive from the capital of Accra. The main towns in the radio's coverage area are Ada Foah, Big Ada, Kasseh, Sege, Goi, Akplabanya, Otekporlu, Agogo and Asesewa. The Atlantic Ocean and the Volta River form part of the boundaries of Dangmeland.

2   One of the local staples and delicacies, 'kenkey' is made from fermented maize dough rolled into balls and boiled in dried maize husks. It is usually eaten with

charcoal-grilled fresh fish and a relish of freshly ground hot pepper, onions and tomatoes.

3   At the time, US$1 was worth approximately 6500 Ghanaian cedis.

4   Kumasi ventilated improved pit (KVIP) toilets are an improved version of pit latrines.

Burkina Faso: When Farm Wives Take to the Stage

Diaboado Jacques Thiamobiga

Introduction

In some parts of Burkina Faso, women do not have the right to speak in public. Yet, they play a key role in the development of their communities. As farmers, they can observe the changes that are taking place in their environment, such as the impoverishment of the soil in the fields where they work. But because they are excluded from public debate, it is hard for them to help find solutions to these problems. In the case study related in this chapter, the women of two

villages in western Burkina Faso turned to their local traditions and culture to find a way of speaking out and launching discussion on problems such as soil erosion, soil productivity and even the property owning rights of women. This is a fine story to tell a cousin who will be glad to know that in his native region, participatory theatre has allowed women to address these questions publicly and, at the same time, to raise their status within their communities.

Letter to a cousin

Dear Cousin:

In this letter I want to tell you the story of the Burkina Faso farm wives who produced a first-class theatrical piece. As you know, people who have been to school think that those who haven't are unable to mount theatrical productions of the same quality that they can. Farm wives in the villages of Badara and Toukoro, in Burkina Faso, have now given the lie to this assumption. They never went to school, and they can't even read and write. Yet, they succeeded in putting on a truly high-quality 'debate theatre' production.

They were helped in this effort by a multidisciplinary team consisting of a sociologist, an agronomist, a theatre producer, a communicator and a video producer. With this support they were able to identify the problem that was to be the subject matter of the play, which they then created and presented in the villages of Badara, Toukoro, Tondogosso and Dou. After the show, the audience offered some criticisms, as well as some ideas for improving the play. Finally, the women drew some lessons from their experience.

The problem targeted by the play

The story I am going to tell you took place in two villages of western Burkina Faso, Badara and Toukoro. For your friends who are unfamiliar with Burkina Faso, you can tell them that the country is located in the heart of West Africa. It is one of the poorest countries in the world; indeed, the United Nations Development Programme ranks it 173 out of 175 countries in its development score.

Burkina has no oil or diamonds. Its only wealth is to be found in its men and women, who are dedicated to work, especially on the land. Its people depend upon farming and livestock for their livelihood. In the past, the harvests were abundant. Burkina villages hardly ever went hungry. This is not true anymore. The land no longer yields good harvests for it is worn out and impoverished. As if this were not bad enough, the rains are unevenly distributed over space and time. It's as if Mother Nature were angry with Burkina and its people.

That was the case this year. Many farmers have had no harvest from their fields. One of them said the other day on television that he had not harvested 'a single grain of millet'. The women in the two villages that participated in this project said the same thing on many occasions. The president of one of the women's groups in Badara told us: 'The village's poor harvests are due to a lack of soil fertility. The situation is so severe that the village is going to disappear.' She added that 'You don't need a degree from an agricultural school to know that the soil in the fields is no longer healthy. All you have to do is look at the stalks of millet or sorghum.' One of her friends said the same thing, adding that 'the land is worn out' and 'the fields are full of striga', a weed with violet flowers that springs up in fields that have lost their fertility. The words of these two women were backed up by other women in Badara and Toukoro. The team agronomist confirms that what the women say is true. The soils around these two villages are not as fertile as they were, although they have not deteriorated as far as those in other parts of the country.

The women point out that they do not often have the opportunity to talk about this problem with each other, and still less with their husbands. Most of the time, extension workers provide technical advice only to the husbands, who may not even talk about it with their wives. The women have no radio stations where they might get advice. Furthermore, they are not allowed to talk about the problem in public, for their husbands will say that's none of their business and may even get angry with them.

What we have here, then, is a big communication problem. We can say that through this theatrical experiment the women have sought to discover how they can address the problem of soil fertility so that their villages will understand that something has to be done about it.

The creative theatre process

When the women decided to take action, they asked themselves how they could put the problem to their villages without making their husbands angry. It was through discussion amongst themselves and with the team's theatre producer that the women of Badara were reminded of a traditional ceremony where they are allowed to dress up as men and speak a few plain truths, and the men have no right to be annoyed. It is village custom that gives the women this privilege, and it is binding on everyone.

The ceremony is held whenever the rains dry up in the middle of the rainy season, thereby threatening the annual harvest. During the ceremony the women appeal to the bounty of nature. They insist that when they come out in this disguise to make their supplications to heaven, the skies will open up and will send them home in the rain. That was the idea behind this debate theatre production. Members of the theatre troupe helped them to put together the production in six steps, which I am going to describe for you.

Step 1: Getting to know the villages

The women set out to see whether the debate theatre idea would be acceptable in their villages. As you know, nothing can be undertaken in the village without the approval of the customary, religious and administrative officials. That's why you have to know about local customs, what is allowed and what is not.

Moreover, mounting the production requires the consent of the husbands, without whom the women would have been unable to participate.

In the end, the women turned to their knowledge and their know-how concerning soil fertility, as well as communication. This step, which was assisted by the team sociologist, allowed everyone to understand the villages more thoroughly. It also helped to establish relations with everyone. The women were even allowed to move to another village with members of the theatre troupe for three weeks. Every Saturday evening they would go home to see their families, and on Monday morning they went back to their temporary quarters.

Step 2: Preparing the agronomic model

An agronomic model was then prepared with the help of the team agronomist. The women inventoried and analysed their knowledge and their know-how concerning soil fertility and they defined four key principles:

1   The land feeds the people, so the people must feed the land.
2   If you want to feed the land properly, you must understand it thoroughly.
3   If you use only inorganic fertilizer on the land, it's like putting water in a basket.
4   If you want the land to feed your children and your grandchildren, start taking care of it now.

The agronomic model includes techniques that allow producers to:

- combat soil erosion by building anti-erosion installations such as stone retaining walls and earthwork dikes;
- protect the soil against wind erosion and sunscald by mulching;
- prepare the seeding bed using the traditional zai technique, which involves digging small pits and putting organic manure in them, and then placing the seeds in them when the rains come;
- preserving the fields' existing tree cover, at the rate of 25 trees per hectare (the general standard);
- enriching the soil with organic manure at a rate of 2.5 tonnes per hectare (the general standard);
- combining organic with inorganic fertilizer; and
- combining different crops, such as cereals and legumes.

Step 3: Conceiving the play

Using the agronomic model and collected materials on patterns of communication in the villages, the theatre troupe representative created a play called Sétou's Challenge. This play tells the story of a village woman who was deeply concerned about the loss of soil fertility and about the tough living conditions of farmers, men and women alike. She took advantage of a big celebration in her village (a child's baptism) to talk with other women about the problems affecting them and their village.

It was during this celebration that a strange character named Doda, the village fool, appeared. To everyone's surprise, he began to dance with the women; but they shoved him away in fear. Seizing the occasion, Doda suggested that they do a play. At first, the women thought this was a silly idea; but then they listened to him more closely and accepted his challenge. They decided together to employ the device of debate theatre to address the problem of soil fertility that was impoverishing the women and their village. As Doda put it: 'Debate theatre is a game that not only amuses people, but makes them think about the problem of soil fertility.'

Step 4: Creating the play

This was the most interesting and, at the same time, the funniest step. It was, in a sense, a theatrical piece in itself. Creating the play involved multi-sided negotiation among:

- the women themselves within their associations, and between the women of the two villages (Badara and Toukoro), for selecting the actresses;
- the technicians, the husbands and the village authorities, on the one hand, and between the women and the technical members of the multidisciplinary team, on the other hand;
- the technicians of the multidisciplinary team.

In this way, 14 women (seven for each village) were selected to act in the play. In order to bring the women from the two villages together and to give them a couple of weeks to put together the production, it was decided, with the agreement of the husbands, to move the troupe to the rural activities centre of the village of Banakélédaga, midway between the two villages, where a theatre school was set up. It took some patience for the women to learn to be comic actresses, and they worked at it for four weeks with the help of the playwright and the two stage directors. Everyone approached the task with true professionalism; as a result, the women were ready to put on their play after four weeks.

Step 5: Presenting the play

When the play was ready, the women gave five performances (two in Badara, one in Toukoro, one in Tondogosso and one in Dou). Every performance was a big hit. It was especially fun for the women when they got to do the play in their own village; whenever they came on stage, the local audience was astonished to see their mothers, sisters or wives dressed up as men. Many husbands recognized them by their gestures and their manner of speaking. In fact, each performance sparked a big celebration in the villages.

Step 6: Evaluation

The debate theatre experiment was evaluated using several procedures. During the performances, and especially during the discussion that followed them, members of the audience offered some criticisms and proposals for improvement, and these were subsequently acted upon. In addition, the actresses held discussions after each step among themselves and with the multidisciplinary team about the accomplishments, the limitations and the difficulties of each phase. The multidisciplinary team, for its part, assessed each step in the process as it unfolded. Finally, a theatre expert conducted an external evaluation of the whole exercise.

Results of the theatre experiment

Through this debate theatre experiment, the women created a good atmosphere, that of a public celebration, within the villages where they gave their performances. Seeing the women dressed up as men made people laugh, of course; but it also made them think about the problem of soil fertility. One member of the audience told us: 'Seeing the women in costume set me to thinking more than laughing.' One woman expressed her thoughts to us in the following terms:

> By disguising themselves, the women were portraying a real-life situation; they are often called upon to fill the role of absent men. This is no longer a disguise but a reality that we have lived. Under these conditions, the women are no longer just wearing men's clothes; they are also taking charge of their families, something that used to fall to men, just as widows must do.

Next, the debate theatre served as a learning experience for everyone: men, women and children. People exchanged viewpoints about village life, they discussed male–female relations, and they shared technical information for improving and preserving soil fertility. Can you imagine? Farm wives giving technical advice on soil fertility to men! This was a first for our region.

Finally, there was a great turnout by people in all five villages where the women put on their play. In every case, there were at least 200 people – men, women, children and young people – in the audience.

There were, indeed, many other results, of which we may cite only a few:

- People responded enthusiastically to the play, making it a real tool of participatory communication for development.
- Villagers became more aware of the need to preserve and improve soil fertility.
- The social status of women was raised within their villages.
- Uneducated and illiterate women developed the capacity to speak in public through the device of the debate theatre.
- Members of the multidisciplinary team strengthened their skills.
- Tools were developed for reproducing this experiment in action research elsewhere and for capitalizing on its output (video, documentation and technical reports).

In the end, the play had some very positive impacts on families and on the villages. On the agricultural front, the people who took part in the debate theatre have put into practice some of the new techniques that they saw demonstrated there. For example, manure has become hard to find in the villages where the play was produced because farmers are now using it more frequently on their fields.

As well, the zai technique is now being used in their common fields by the women who took part in the play. People often ask the women to do the play again. A survey in the two villages (Badara and Toukoro) showed that the play had helped people to find solutions to soil fertility problems, and to overcome inequalities between men and women.

When we say that Burkina's lands are poor, this is both true and false. It is true in the north of the country, where most of the soils have become lateritic. But it is not true in the west, where the soils are still relatively fertile. Badara and Toukoro are located in this part of the country.

For that reason, the knowledge and know-how of southern women was limited to the use of chemical fertilizers. On the other hand, women from the north had experienced the acute problem of collapsing soil fertility in their native area. They were able to set an example, then, for using household wastes and tree branches to contain rain runoff, planting crops of niébé (cowpeas) to fix soil nitrogen, and using techniques such as mulching, zai and stone retaining walls. All these techniques were confirmed by the agronomist and by the theatre expert in the course of supplementary research in Kouni, a northern village where the soils have become highly degraded and infertile.

Yet, it is in terms of gender relations that the debate theatre experiment produced the most surprising results: by letting women speak out, it revealed some unsuspected aspects of the soil fertility problem. In fact, far from ignoring the importance of regular fertilization of the soil, through their stage characters the women explained that without property rights they had no interest in investing to improve yields from a piece of land that the men could simply take back once it had become productive again.

Lessons drawn from the theatre experiment

One of the lessons we can draw from this experiment has to do with the basic problem: we must undertake an in-depth study of villages in order to understand them properly. Research and development projects tend to downplay this phase – the local setting assessment – despite its importance. Financial partners often dispense with it as too costly. Yet, such a study improves our knowledge of the social and cultural habits of villages, their agronomic aspects and their communication practices.

The experiment also revealed the importance of involving local people from the outset in defining and analysing the problem because outside experts do not always see the problem in the same way as the inhabitants who have to live with it on a daily basis. On this point, the example of the women who worked the property ownership issue into the play is highly revealing.

In addition, the turnout in the villages where the play was performed shows that debate theatre can be an effective tool of participatory development communication. It can also facilitate dialogue between women and their communities on a given development problem because it is well suited to the realities of rural life in Burkina. It can thus allow women to play a greater role in developing their communities. This is a tool, then, that can rally different social categories (women, men and youth) and different occupational classes (farmers, herders, etc.) to the cause of developing their communities.

The experiment also showed that illiterate farm wives can accomplish great things if we will just believe in them. This is why it was essential to have them involved throughout the process. The play offered the chance to share the knowledge they had, and to acquire new knowledge, while releasing them from certain social and cultural constraints so that they could speak in public. They even ended up giving some advice to the men! This does not happen often in village communities, which are still in thrall to custom and tradition. If you should run across some of the women who took part in the play, you won't believe that they are simple farm wives. Some of them have had the chance to participate in workshops where they express themselves just as surely as the experts. In other words, debate theatre is an instrument for raising the status of farm wives and winning recognition for them as individuals, as much as for their knowledge and their know-how.

On the institutional front, debate theatre can create a setting conducive to effective partnership if each partner agrees to play his role openly and honestly. It provides the opportunity to strengthen institutional capacities through the effect of complementarity and synergy of efforts. It is essential, however, that institutions involved in the process have a proper understanding of their respective attributes from the outset.

Some criticisms of debate theatre

There is a village saying that the onlooker can dance better than the performer. In other words, when you are in the position of an observer, you are well placed to offer sound criticism. Similarly, when you have done something yourself and then look back on it some time later, you can also make sound criticism. By standing back a bit, we can offer some criticisms about this debate theatre experiment.

First, setting it up is a long and difficult process. That's mainly because of the approach taken for conceiving the play. Moreover, the play was first written in French by the theatre troupe manager, using material he had collected from the women. The play then had to be translated into Dioula, the national language. Yet, it could have been written directly in Dioula, using the women's own expressions, which could then have been more readily reflected in the text. Moreover, since the women were, for the most part, illiterate and had never performed a play, it was hard for them to learn their parts by heart. It would have been useful to work out ways of writing plays together with the women, and this would at the same time have helped to give greater recognition to their knowledge and their know-how.

Nor was the process facilitated by the decision to create a single theatre troupe involving women from the two villages, and to house them in another village half way between the two. This meant negotiating the husbands' consent and knocking off work every Saturday so that the women could go home to see their families. In the village, as you know, the woman is everything to her family. It's hard for her to leave her family to go and work in another village. That's why one of them quit the troupe at the beginning of the production, and she had to be replaced.

Next, the process was expensive. It took a lot of money, especially when several experts were brought in. It would have been a lot cheaper to use a village troupe that is used to putting on performances without great financial backing.

I must also point out that the team consisted of members from three different institutions,[1] and this was another source of difficulties, especially since they did not all live in the same city. Because they were located so far apart, it was not possible to stick to the work schedule, and things tended to drag out. While the play was supposed to be produced after six months, it actually took two years. It would have been simpler and more efficient to work with a multidisciplinary team from the same institution, with the time savings that would have implied.

Finally, the women and grassroots communities showed that they were ready and willing to commit themselves. But they were somewhat disappointed by the fact that not all the needs raised by the play could be met because of the lack of physical and financial resources. Moreover, the team agronomist has not been able to follow up with the farm wives who have been experimenting with the new techniques for soil fertility preservation that were addressed in the play.

Conclusions

In reading this story, you must have realized that this debate theatre experiment served to win new recognition and appreciation for the knowledge and know-how of Burkina Faso farm wives relating to soil fertility. Even better, they enhanced their own self-esteem by showing the men that they were quite able to discuss their communities' basic development problems in public through dialogue and participatory development communication. Debate theatre is, thus, a tool that people can use for dialogue about the development problems of their villages. Finally, it can be an instrument for rallying people around the villages' development efforts.

That's why I want you to recount this story to your friends. Above all, tell them that debate theatre can be the yeast that leavens the dough of participatory development communication. By doing so, you will be helping to put to use the wonderful work of the Burkina farm wives who experimented with that debate theatre production.

That's all for now. Goodbye and take care!

Your cousin from the village.

Note

1   The Centre d'Études Économiques et Sociales de l'Afrique de l'Ouest (CESAO), based in Bobo-Dioulasso, the Théâtre de la Fraternité, based in Ouagadougou, and Zama Publicité, with offices in both cities.

How the Parley Is Saving Villages in Burkina Faso

Diaboado Jacques Thiamobiga

Introduction

Because so many African villages have retained their oral traditions, ancestral forms of communication based on dialogue can be very useful tools in participatory development communication. Yet, the younger generation does not always keep up these traditions and practices, even though they are often better adapted to the local setting. In the process of rural development, they encourage people to speak their minds, they facilitate consensus and they promote joint endeavours. Here is a letter to a cousin, dealing with one of these forms of communication: the parley, or, as it is known in Burkina Faso, the palabre.

Letter to a cousin

Dear Cousin:
Do you remember the phrase that our grandfather liked so much and that
he kept repeating? He would say that 'the parley is what saves the village'.
When we were little, we could not grasp the wisdom of that saying. Today,
I'm going to try to explain it to you through two stories. One of them comes
from a journalist friend who describes how the parley allowed the village of
Silmiougou to put an end to its water wars. The other story has to do with ten
villages in eastern Burkina Faso that used the parley to control bushfires. To
help you understand the stories, let me tell you first about the parley and the
role that it has always played in our villages.

Talking eye to eye

When we were children, you'll remember, we often saw the elders seated
under the big tree in the village. We would ask our grandfather why the older
men got together and spent their time talking as if they had nothing else to
do. He would say: 'It's the parley that saves the village.' He would always
tell us the same thing, that the elders were looking for ways to solve the
village's problems through the parley, the African-style town hall meeting. He
would finish by telling us: 'Parleying, it means looking each other in the eye
or speaking a few plain truths.' In fact, grandpa was right. No problem can be
solved unless the people of the village are prepared to sit down and discuss it,
talk it over. That's what the parley is all about.

You see, parleying and discussing mean the same thing. But a village
parley doesn't happen just by chance. Remember how grandpa and the other
elders would bring no one but their grandsons to the meeting and let the kids
play while the elders parleyed? It was only the heads of the families and clans,
only the men with beards and white hair, who were allowed to participate in
the parley over village affairs. You understand: they couldn't just let everyone
into the parley when they were talking about witchcraft, death, the rape of
young girls and sensitive things like that.

The parley would not end until the village elders had arrived at solutions
that everyone could accept. There had to be consensus. And those solutions
also had to fit with custom. It's not always easy, as you know, to find solutions
that can command consensus and are consistent with custom. That's why
the parley often ran on for a very long time. Everyone had the chance to
hear what the others had to say, and then to have his own say. So the parley
was an opportunity for dialogue. Everyone could express himself freely; but
everyone was always respectful of the others and of the village's customs.

Today, we can say that the parley is a traditional way of establishing
communication, and one that is particularly well suited to our villages. This
approach fits very nicely with participatory development communication
(PDC), which is based on both participatory processes and on traditional or
modern media, as well as on interpersonal communication and facilitation

skills. The idea is to find a solution to development problems as identified and defined by the communities themselves. For those of us who did not go to school very long, this means that the whole village has to get together and discuss things in order to find solutions to problems that burden our village life.

With this approach, the village can:

- pinpoint the development problem that has sparked us to use participatory communication;
- identify the individuals or groups affected by the problem;
- define the needs, objectives and activities for participatory communication;
- select the channels, means and tools for participatory communication;
- test out those channels, means and tools;
- use those tools in the development process, once they have been tested;
- assess the outcomes at each significant stage.

Today, many development partners supporting our villages are using PDC. I'm going to tell you about two such cases, one of which had to do with managing conflicting water uses, and the other with the better handling of bushfires. These two initiatives both relied on the parley, although they modified it a bit. For example, the parley is now open to all social categories (men, women and young people). And it also uses new communication tools such as film, video, radio and cassettes. Let's see, then, how the parley put an end to the battle over water in Silmiougou.

Ending the water wars in Silmiougou

The elders of Silmiougou will tell you that the village was founded by the Peuls. They lived by raising livestock, which at that time was a flourishing activity. But it also attracted many rustlers, thieves who would come into the village to steal sheep, goats and even cattle. Fed up with this thieving, the original Peul settlers turned for help to the grand chief of the Mossi, Moogo Naba, who lived in Ouagadougou. He sent them a band of warriors who were used to dealing with rustlers, the Tapsobas, and they succeeded in driving the bandits out of the village. The Peuls could now live in peace.

However, the Mossi who came to protect the village decided to stay on to keep the village secure. Little by little their numbers grew, and they took over the local lands, including those on which the Peuls pastured their animals. This plunged them into permanent conflict with the Peuls, a conflict that the people of Silmiougou came to call 'the water wars'.

Not long ago, whenever the women went to draw water from the wells, they would get into a spat. The Peuls also had to fight to have their animals drink at the wells. Silmiougou was a daily battleground. To solve this problem, which could have ended in a real war within the village, people decided to turn to the team which was working with villages to stop water disputes. The project fielded Bila, who, instead of acting as a coach or extension worker, decided to use the traditional parley to bring the people of the village together to discuss ways of using the well without squabbling. In this way the village people were able to:

- think about how everyone could use the well water without fighting;
- exchange ideas about all the problems that could lead to water disputes;
- create groups to find solutions that would suit everyone;
- agree on solutions and how to apply them to everyone, even the village chief;
- apply those agreed solutions to everyone and enforce them on all water users; and
- hold frequent meetings to see what was working and what was not, and to find new solutions.

This is how the parley ended the water wars in Silmiougou. As you see, grandpa was right when he said: 'It's the parley that saves the village.' Thanks to the parley, the village of Silmiougou avoided a war over water. This story, told to me by my journalist friend, reminds me of another one about ten villages in the eastern part of the country where the parley made it possible to put out bushfires.

The bushfires are out

As you know, Burkina life depends upon the bush – that's where we find river water, wild animals, new farmland and trees. Women cut wood in the bush for cooking. They also collect the leaves and fruits of certain trees. And it is in the bush where livestock graze on grass, leaves and twigs. So anything that affects the bush in Burkina also affects development.

We must admit that people sometimes act as if they are unaware that the bush can be damaged and degraded. This is what happens with wildfires that destroy everything in their path. As you know, bushfires are frequent in Burkina. For various reasons, people make great use of fire during the dry season. Some claim that this is a traditional farming practice. But the problem is that some fires get out of control and lay waste to the whole area. This has become a real scourge that our country must address. That's why, in 1997, Burkina held a big meeting – a national forum – on fires. That meeting led to an initiative to promote sound fire management. Between 1999 and 2003, the initiative reached about 255 villages. It did so primarily through the parley.

In this case, the parley involved everyone in the village (men and women; young people and old; farmers, herders and merchants). It set up a permanent dialogue among these groups, and also with the technical support services in environment and agriculture. This is what happened with the ten villages that I'm going to tell you about. We discovered those villages during our research into the problem of bushfires – their causes, impacts and solutions – and the way in which people in these villages were handling fire management.

In this story, we will be talking about fire management committees, their work and their results. At the outset of the project, every village organized a fire management committee through a parley in which everyone participated. The committee has ten members: eight men (adults and youths) and two women. The committee's main role is to mobilize the entire village around fire management activities.

These activities involve selecting a fire management site, using controlled 'early burn' techniques,[1] protecting the site against wildfires and planting trees. All these activities are planned, implemented and assessed by the village population as a whole. Every social category has a role to play. The young people are responsible for surveillance and early burning. The women bring food and water to the young people working on the site, and the older folks provide advice to the youngsters and women.

As you can see, these activities could never happen without the parley, which takes place in the form of a general meeting of the village. During our research, we attended several general meetings. Let me tell you about one in the village of Kiparga, which we selected by chance. On that day, 70 people – men, women and young people – were in attendance. The participants expressed their viewpoints and exchanged knowledge about fire management, about the running of activities and about the problems encountered. Sometimes the parley can become very heated, and it is at this point that the elders intervene with words of advice for everyone.

Occasionally, to lighten things up, some cousins à plaisanterie[2] will jump in and get everyone laughing. They make jokes at the expense of those whom they are allowed to tease. They insult each other as if they were going to come to blows. Those who are unaccustomed to this practice don't understand it. They wonder how people can attack each other in this way and yet no one is allowed to get angry. This is what keeps the meetings interesting and fun, even if it means they may run on for a long time.

Things also get very lively at the meeting when the women use jokes to tell the men a few plain truths. Sometimes the very purpose of the meeting can be obscured by laughter. This gives an idea of the spirit and the philosophy that underlies the parley. It is a freewheeling and democratic debate. We could even call it a people's assembly. Thanks to the parley, people can plan their fire management activities: surveillance, protection of the bush and exploitation of its products. Working in this way, the villages have achieved real results and people are proud to point to them. People have rallied around the village efforts at fire management and they are committed to a process of permanent dialogue on this issue. The interests of the various social categories and occupational groups that make up the village can now be taken into account. Not only have the villagers managed to agree on the appropriate ways to manage bushfires, they have also launched other activities to protect natural resources, such as setting up and maintaining protected areas[3] or improving their herding methods.

Yet, there have also been some problems. Heavy pressure on natural resources, due to the demographic explosion, is one of these problems, as is the unconstrained exploitation of natural resources. Moreover, the villagers may be too busy with other activities to keep a permanent watch over the land. These problems, taken together, are causing the gradual disappearance of many useful plant species. Finally, because the village does not have a flour mill, it is sometimes hard for the women to play an active role on the committee.

Despite these difficulties, the people are emphatic about their determination to pursue bushfire control through surveillance, building gravelled firebreaks, site maintenance and planting trees at the site and in the village.

The parley: Advantages and demands

Each member of the village can see advantages in the parley. In fact, the parley affects a great many people through the general meetings and other assemblies. It is also a place for sharing ideas and experience, knowledge and know-how. It offers people a chance to think together about their common problems and to talk them over face to face. It is a means of dialogue and of negotiation that allows people to reach agreement on what they are going to do together. And, finally, if it is properly organized, it does not need much in the way of materials or money. All that's required is a meeting place and some mats or benches to sit on.

We must recognize, however, that there are some conditions that must be fulfilled. In the first place, someone has to take care of the town crier who calls people to the parley by buying him his cola[4] or his dolo.[5] Second, the facilitator has to be someone who takes the job seriously and enjoys everybody's respect, or he will never be able to guide the debate. And then, too, the elders have to be in attendance because they have a calming effect on people when the debate gets too heated.

On top of all this, there has to be a little money to buy food and refreshments for people if the parley is going to be a long one. Holding a good parley does take time, after all. Above all, people have to be patient. If you're planning to start the parley at 9.00 am, it will likely get rolling around 10.00 am or maybe even 11.00 am since everyone will have something to do at home before coming to the meeting. Then, when people arrive, it will take at least 30 minutes to greet each other properly. And each speaker will have to start with the customary string of salutations. Indeed, the parley takes a lot of time, and this can be a problem. The fact is that some people are now insisting on being paid before they will come to a parley, complaining that they have to leave their work to do so, and that 'parleying doesn't fill anyone's belly'.

It is also true that people may be called to many parleys. The extension worker will have his parley. The forester, the prefect, the deputy and the village chief will have theirs too. People may get fed up with all these meetings – after all, that's not the only thing they have to do with their time. For this reason, a parley will sometimes attract only a few people. These days, if you invite 100 people, you can be grateful if 30 actually show up at the meeting. Finally, people sometimes come to the parley thinking that the conveners are going to offer some tangible assistance to the village, whereas it is the process of dialogue itself that will eventually help the villagers find solutions to their own problems.

We must also recognize that the parley is one of those rare occasions when people can express themselves openly. Half of the available time may well be spent on issues other then those for which the parley was called. As you know, people do not all think the same thing. Often, each person will try to show the village that nothing can be done without him. The parley can become impossible if people from different districts are divided by long-standing rivalries and can agree on nothing. In one of the villages we visited, for example, there are two chiefs. You have to be a chameleon, or at least a true diplomat, to adapt to village realities. Otherwise you will just stir up conflicts and you will have everyone against you.

Under these conditions, it isn't easy to guide debate within a big group. You have to know the people and their natures. For example, when an elder speaks, he's going to start by retelling the village's history, before he comes to the point he wants to make. He may even speak in proverbs – and when someone speaks to you in proverbs, you have to respond in proverbs. Then there are the long customary salutations that I mentioned earlier. So, you see that the parley is a time-consuming process. And in the midst of it all, you have to know how to listen to people.

In the end, if you're going to have a successful parley you have to be patient, respectful, tolerant, intelligent and good at negotiating. Not everyone combines these qualities. There are some further requirements as well that I can't go into here, for they would take a whole story in themselves. To wrap up, let me offer some lessons from experience on how to use the parley.

The parley can teach us some lessons

Everything I have said above points to the fact that the parley, which is a real village institution in Burkina Faso, constitutes a space for dialogue or for participatory development communication. Yet, there are some requirements that must be respected if you are going to use it wisely. You must never forget that it will take a lot of time, and that using the techniques of group dynamics can help to make it run smoothly.

As you can see, the stories I have just recounted offer some real lessons. We saw how the parley allowed the village of Silmiougou to put an end to its water wars. In the eastern villages, it put out the bushfires. That's why we said in the title of this story that the parley can save villages. Now you understand why Grandpa kept talking about how 'the parley saves the village'. Now you can tell everyone that the parley – the African-style town hall meeting – is a practice that Burkina villages, like those elsewhere in Africa, use as a method of PDC. All we have to do is refine it so that it can become a real development tool. For all these reasons, I recommend its use in participatory development communication. That's the end of my story, and it's time to say so long and take care.

Your cousin from the village.

Notes

1   This is a conservation technique that consists of burning the bush immediately after the rainy season and before the vegetation, particularly the grass, has completely dried out in order to encourage regeneration.
2   Cousins à plaisanterie are a Burkina social innovation where different ethnic groups mock one another, which allows people to joke about sensitive subjects.
3   These protected areas are designated natural resource conservation zones, where management includes reforestation and fire prevention.
4   Cola are nuts that contain a stimulant substance and play an important symbolic role in social intercourse.
5   Dolo is millet beer.

V

Collaborative Learning in Participatory Development for Natural Resource Management

Forging Links between Research and Development in the Sahel: The Missing Link

Claude Adandedjan and Amadou Niang

Rural people in the Sahel, where the economy is based largely on farming and forestry, are still using inefficient production technologies. Moreover, the advance of the desert and the collapse of ecosystems that the region has witnessed for several decades are steadily impoverishing the population.

Researchers at the International Centre for Research in Agroforestry (ICRAF) have been working for several years to develop new technologies

that could improve output and thereby raise incomes and reduce poverty among rural people in the Sahel. Some of the most promising innovations have included:

- building fodder stockpiles so that livestock can be better fed, especially during the dry season;
- food banks to improve rural people's nutritional health and to save from extinction certain plant species, such as the baobab, that nourish and protect village fields;
- planting hedges to keep livestock from straying into fields and allowing more intensive planting of crops;
- improved and domesticated cultivars of forest fruit trees to raise orchard output and incomes;
- improved fallowing methods for restoring soil fertility and raising crop yields.

At the outset, ICRAF's strategy was to promote adoption of these innovations by working with development institutions and their agricultural extension services. ICRAF's contribution to this partnership was in the form of training, information and raw materials.

Yet, this strategy was not very successful. A number of reasons were put forward to explain its failure. First, the innovations were not based on the kind of local knowledge and expertise that the communities possessed, so they were little involved. Extension workers did not have the required technical expertise and were chronically short of resources. Moreover, no account was taken of the way in which farmers viewed these innovations. Finally, the extension services showed little interest in promoting the innovations, feeling that their tasks were simply designed to help the researchers.

Consortiums were created to address the situation by strengthening interaction between researchers and development workers, and to reinforce linkages by fostering an inter-agency team spirit.

While this new strategy improved the research–development relationship, rural communities still found themselves excluded from research and development discussions. Moreover, the strategy still failed to take into account farmers' perceptions and their resource and land management strategies.

In the end, despite all the effort that ICRAF has put into popularizing these technologies, it must be admitted that the adoption rate is still very low.

A different approach to research

Since 2000 a move has been under way among development stakeholders in the Sahel to exchange ideas and examine the reasons for these failures. Yet, while efforts initially focused on finding more efficient ways of disseminating the innovations, ICRAF has recently shifted its focus towards more participatory approaches. Field studies have shown that however effective the institutional partnership might be in getting research and development players to work together, when it comes to the farmers themselves the approach has remained very vertical. ICRAF has also recognized that most of the technologies proposed were based on a top-down model – that is, they were conceived and designed by researchers without any input from farmers. The researchers then had to try to 'sell' the innovations to the farmers. Participation was limited to ways of consulting farmers and rallying them around the activities proposed by the researchers.

Given the disappointing results from this approach, it became clear that the problem lay not only in the way in which the innovations were being disseminated, but in the very concept of the research process itself. The discussion sessions showed clearly that if people are not involved in the process from the outset, they will persist in seeing it as irrelevant to their concerns and their needs.

By way of example, a member of the ICRAF team recently related that some time after planting hedges in a community where stray livestock were ravaging crops, she went back to the village for a follow-up visit. To her great surprise, the people greeted her by asking whether she had come back to see 'her' trees. They felt so detached from the experiment that they still looked upon the trees as the property of the person who had planted them.

In another case, a participant in a training session reported that, after travelling several hundred kilometres to plant hedges in a village, he and his colleague were accorded a very chilly reception by the villagers. Some of them stayed away, refusing to come anywhere near the visitors. Taken by surprise, the two extension workers wondered why they were so unwelcome, particularly since they had obtained the consent of the village chief in advance. At this point the chief confronted them in person and angrily demanded to know what they were trying to do to his village. He insisted that they had abused his trust and had come solely to bring disaster on the villagers. Deeply disconcerted, the visitors finally realized that the type of shrubs they had brought with them were seen by the main local ethnic group as portents of evil spirits who would destroy the village.

Many examples of this type can be cited, where needs are analysed and solutions identified in a vacuum without taking into account local systems of understanding and the worldview that underlies them. Naturally enough, ICRAF has turned its thinking towards a new way of doing things. It was clear that it would have to think about more participatory approaches, in the course of which the following questions arose:

- How can we get farmers to participate more actively?
- Do these technologies really take account of people's priority needs?
- Among the target population, are all social groups considered, especially small farmers, women and youth?
- How can we help local people to participate in co-generation and co-dissemination of technologies?
- How can we ensure that our research results are actually put to use?

- Which tools of participatory communication must we use to promote change and innovation in strategies and methods?

Experimenting with a new approach

ICRAF-Sahel and its partners are currently undertaking a wide-ranging programme of capacity-building for stakeholders involved in farming and forestry in the Sahel in order to overcome the inadequacies of the current approach. To start with, three countries have been targeted: Mali, Senegal and Burkina Faso. There are 45 people participating from five different kinds of institutions: research facilities, training centres, development organizations, non-governmental organizations (NGOs) and women's associations.

The main challenge facing this consortium will be to involve rural people effectively in research and development efforts and in this way to launch a process of co-generation and co-dissemination of agroforestry innovations. The new approach will pay particular attention to farmers' knowledge and expertise, and this, in turn, will require a thorough familiarity and understanding of the local context.

Special attention will have to be given to strengthening consortium members' capacities in the use of participatory communication methods and tools, recognizing that the consortium consists for the most part of researchers who are trained in the 'hard sciences', and are consequently too rigid in their attitudes and practices. This effort will, in time, encourage them to change their practices and attitudes so that they can work more effectively with other stakeholders, particularly farmers.

The team is aware that it will have to reach out more effectively to farmers, with whom interaction is currently very weak. In fact, while the consortium has established a working relationship with farming organization representatives, their viewpoints do not necessarily reflect the aspirations of the different groups in rural communities.

Furthermore, the team hopes that, with the collaboration of these stakeholders, it will in time be able to develop and propose solutions and activities that will meet the needs of the various socio-economic groups in rural communities. The agroforestry innovations put forward in the past represented prototypes that took no account of the particular needs of certain groups. This was true, for example, in the case of small farmers, for whom the technology of 'living fences' remained inaccessible because they lacked the additional funds needed to adopt it. By revising their current approach to development communication to make it more participatory, ICRAF and its partners hope to strengthen mutual understanding among stakeholders, and to give new impetus and meaning to their research efforts, which, in the end, are devoted above all to improving living conditions for rural communities.

Isang Bagsak South-East Asia: Towards Institutionalizing a Capacity-Building and Networking Programme in Participatory Development Communication for Natural Resource Management

Maria Celeste H. Cadiz and
Lourdes Margarita A. Caballero

How are the facilitating factors enhanced and the challenges surmounted in institutionalizing an experience-based distance programme aimed at capacity-building and networking in participatory development communication (PDC) in natural resource management?

In essence, this question states the participatory study problem that the College of Development Communication (CDC) at the University of the Philippines, Los Baños, addresses in implementing its own pilot of Isang Bagsak in the South-East Asian region. This chapter reflects on PDC as something more than a community-based approach. It examines PDC at the project management level and partly at the learning programme level, where its principles also apply.

Isang Bagsak South-East Asia

'*Isang Bagsak!*' is a Tagalog expression signalling consensus, agreement or affirmation in a participatory meeting, and Isang Bagsak South-East Asia is this region's learning and networking programme emphasizing the participation of stakeholders in natural resource management (NRM) through PDC processes. The Isang Bagsak programme thus aims to improve communication and participation among researchers, practitioners, communities and other stakeholders in natural resource management and to reinforce the potential of development initiatives in helping communities overcome poverty. CDC is implementing its pilot of the programme, with the aim of institutionalizing it in the region after a 15-month pilot that included two participants from South-East Asia (Viet Nam and Cambodia) and one from Africa (Uganda).

The programme includes an introductory workshop on PDC and on the Isang Bagsak project; local discussions, practical study and the application of PDC in natural resource management; sharing and discussing syntheses of the previous themes at the regional level in a web-based forum; a face-to-face mid-term capacity-building workshop on specific PDC processes and techniques; and a final face-to-face evaluation and planning workshop.

Isang Bagsak South-East Asia began in August 2003. Three teams were carefully selected to participate in the first cycle of the programme: the Forestry Administration of the Ministry of Agriculture, Forestry and Fisheries in Cambodia (representing the government sector); the community-based coastal resource management programme (CBCRM) based in the Hue University of Agriculture and Forestry and the University of Fisheries in central Viet Nam (representing academe); and the Legal Assistance Centre for Indigenous Filipinos, better known as PANLIPI in the Philippines (representing the NGO and people's organization sectors). The introductory workshops have now been completed and the e-forum has been initiated.

Facilitating factors

The facilitating factors in implementing the programme include the following:

- high qualifications of the implementing and facilitating team, who are experts with advanced degrees and practical field experience in development communication, coupled with their institution's long history in development communication education, practice and research;

- collegial and congenial working relationships among team members reinforcing their high morale and making participatory project management possible;
- reasonable flexibility in administrative procedures of the partner agencies, the University of the Philippines at Los Baños Foundation, Inc and the International Development Research Centre (IDRC), respectively;
- high level of support from IDRC;
- sufficient documentation of the pilot phase of the programme, providing a template for the protocols of the current pilot in South-East Asia;
- prior training and experience of the implementing team members in handling distance education courses; and
- a culture of excellence and innovation in (including a high commitment of) the implementing agency.

Challenges

In spite of these facilitating factors, the implementing team has faced numerous challenges, as follows.

Challenge 1: Clearly presenting and reaching a common understanding of the programme cycle, rhythms and mandate among team members, as well as between implementers and learning participants

Part of the programme is an orientation workshop on the concept and processes of PDC and Isang Bagsak for the implementing team. At CDC, the decision was made to implement the programme as a team, rather than to assign one main facilitator employed by the project to run the programme. It was thus important that the different members of the implementing team, most of whom are already experts in development communication, come to an understanding of what the programme is about and how it is run.

The orientation workshop was also seen as an opportunity for CDC to share its own experiences in development communication with other implementers, and potential implementers, of the Isang Bagsak programme in other regions of the globe, with the aim of building a global Isang Bagsak network. Beyond being an in-house workshop, it thus became an international meeting where like-minded individuals from various disciplines and regions of the globe, all interested in uplifting the well-being of people in areas with critical states of natural resources, excitedly shared their experiences and insights as they discussed the PDC process.

Once the excitement brought about by the prospects of being part of a global network of PDC practitioners and scholars had died down, CDC came to grips with implementing the programme in its region. In an in-house forum requested by members of the CDC implementing team after the orientation workshop, the idea of incorporating PDC at the community level prevailed among team members. Yet, at the coordination level, the viewpoint was that Isang Bagsak, being essentially an experiential distance learning programme for PDC implementers, should primarily help to facilitate the learning process. Perhaps there was a strong desire by team members to directly undertake fieldwork, a valid sentiment coming from development communication teachers and researchers.

This led us again to validate the observation on the nature of communication as a process in project team management. Mutual understanding does not necessarily take place instantly; a single utterance of facts and principles does not bring about immediate understanding or acceptance of an idea. This sometimes made us wonder about our credibility as project coordinators among our own team members, a question that recurred whenever team members kept asking what the IDRC programme specialist would say when the need to settle issues related to implementing the programme – implying that our explanations were not sufficient. Arriving at a common understanding on implementing the programme as a team required plenty of meetings and discussion, as well as juggling schedules to find common time to do so in our multifaceted preoccupations as CDC faculty members.

IDRC has certainly assumed the role of a helpful adviser who does not categorically say 'yes' or 'no', but, in the true spirit of participatory management, refers back questions with his own question on why a certain direction was, or will be, taken. He has likewise taken the stance of openness that Isang Bagsak is a learning process and is in the process of continually evolving, as new and different partners implement it.

In itself, the participatory management style and its perspective challenge traditional ideas of project management and leadership, as well as the extent to which programme implementers may have internalized the participatory paradigm beyond conceptual understanding.

The same interpretation that the experiential learning dimension of the programme is itself the project surfaced within two learning teams. One prospective learning team thought that the programme would support new field initiatives; in another team, participation extended to the community level. It took a meeting with the IDRC programme specialist to clarify with members of the implementing team that Isang Bagsak is a learning programme, not a community outreach initiative in natural resource management – such an initiative being the intended participant of the learning programme.

The important lessons we offer in this experience are:

- the need to clarify the programme cycle, not just its content, by being careful in our use of terms, using credible channels and patiently developing mutual understanding on issues related to programme implementation; and
- incorporating the PDC concept and practices within the management of the programme and specific initiatives – an important dimension of participatory development communication.

Challenge 2: Exercising discernment and negotiation skills in selecting the right learning teams beyond clear-cut criteria set for participants

The process of selecting learning teams was in itself a learning process for the CDC. Misinterpretations of the nature of the learning teams' participation and support were largely brought about by the use of the term 'project' in referring to the learning programme. While the criteria for selecting the learning participants and what was expected from them were clearly spelled out, this was no guarantee that these criteria were clearly understood by the learning participants.

The criteria for selecting learning teams were as follows:

- having an ongoing project in community-based natural resource management funded for the following two years;
- sustained connectivity and access to the internet;
- willingness of the team members to learn PDC through experience;
- willingness of project managers to devote at least ten person hours per week to PDC activities, such as monitoring the use of PDC methods at the site; team meetings; studying the Isang Bagsak South-East Asia implementation manual and other resource materials; participation in the e-forum; and participation in workshops and evaluation meetings.

As the implementing team, the CDC team likewise considered the following factors in selecting learning team participants:

- willingness or interest to learn and adopt the PDC process;
- potential influence to advocate PDC for its institutionalization in community-based projects;
- participatory orientation;
- other practical considerations such as location and travel costs required for monitoring, and the prevalence of peace and order in the community site.

In fact, one of two prospective qualified teams was selected on the basis of need and the potential to institutionalize the PDC process within government. It turned out, however, that a month after finalizing negotiations, the selected team lost its internet connection, a requirement for participation in the programme, and sought support from the programme to restore it. Moreover, the supposed team leader was relocated to a provincial post and the new one, apparently not properly oriented by his colleague or supervisor about participation in the programme, initially gave the Isang Bagsak facilitators a cold reception in organizing the introductory workshop. Furthermore, their supervisor, who had endorsed and warmly received the project facilitators during the initial visit, became inaccessible during and after the introductory workshop, when the new team leader suddenly requested support for internet connections.

This particular experience was perhaps incidental. Yet, one lesson is perhaps that referrals should be an important input, in addition to the information gathered from application papers and visits, with face-to-face interviews when selecting learning participants. Just as students apply for admission to graduate studies, learning teams should apply to the programme and clearly understand the requirements and implications of such participation. Signing a commitment by learning teams and Isang Bagsak facilitators alike may be a necessary step in negotiating learning participation.

On the other hand, unexpected developments such as these continue to underscore that the programme cannot have a fixed system and procedure. Perhaps the more valuable lesson in this experience is that facilitators of such programmes should always be ready to engage in a balancing act between the need to be consistent with certain guidelines and procedures, and the need to be flexible where the unexpected arises.

Challenge 3: Negotiating a realistic programme schedule that allows for optimum participation of four different teams, including the implementing team, each with their own calendar of activities

Negotiating a workable timetable for the programme was in itself a complicated communication process. What made it doubly difficult was the mode by which this was negotiated, by email and/or websites. It seems that we Asians have not yet mastered communication via these channels. Often, a message sent is left hanging, with replies sent back by some recipients, while others neglect to answer even a simple 'Yes, the schedule is alright with us' or 'No, it is not workable; here is a counter-proposal.' Then again, the problem may be the recipients' limited or intermittent access to the internet.

The reality is that participants are busy themselves, and responses to questions about schedules do not necessarily entail a simple 'yes' or 'no'. Often, one's own team members need to be consulted first. Sometimes, due to the length and complicated nature of the process, the act of sending a reply is forgotten altogether.

This also has implications for the time it takes for exchanges on the web and among team members to complete loops in the e-forum. Based on the timetable drawn up, the whole programme from inception to the final evaluation workshop should take about 15 months, taking into account the slack expected during holidays and important occasions. We have received feedback that the time allotted for reacting to everyone's postings – one week – is too short for these series of exchanges within a team, and that two weeks would be more realistic. This would stretch the programme to about nine more weeks, or a total of approximately 17 months.

During the mid-term workshop when members of the three learning teams met together, participants suggested a more flexible schedule for the e-forum: two or three related themes to be simultaneously launched in separate conferences, and learning participants to be allowed to take as long as two to three months to complete the cycle of discussions, the posting of team discussions and debate. The programme is divided into four parts, comprising a total of nine themes. Each part includes two to three themes. The proposal, which has now been adopted for the rest of the e-forum and for the second programme cycle, is for each part with its two- to three-component themes to be launched together, instead of having a linear sequence of all nine themes. This was because the themes within the four parts were interrelated; the logic, therefore, was to discuss them together. However, postings of discussions will be grouped separately, based on each individual theme conference.

Challenge 4: Transcending and accommodating language and cultural barriers, and exercising cultural sensitivity in facilitating the learning process across cultures

In South-East Asia, the medium of exchange on PDC at the regional forum is English, a second language for all the learning teams. Proficiency in English varies across and within teams. Therefore, the process of translating postings and exchanges from the vernacular into English, and the other way around, requires additional time. However, it must be noted that, in hindsight, participants from the pilot phase found one unexpected gain from their participation in the programme: an enhanced proficiency and confidence in communicating in English.

Aside from the language barrier, facilitators found it a challenge to understand the participants' cultural differences and to exercise sensitivity in facilitating learning. For instance, during the introductory workshops, the Cambodian participants took a long while discussing and arriving at a consensus on their collective replies to questions in their own language before they translated them into English. The lengthy discussion is part of the Asian concern for 'saving face', interpreted as a form of social grace by some anthropologists. Asians would not want to belabour or burden the facilitators, who are visitors, with the details of their disagreements, uncertainties and tentative stances. They would just translate for the latter's consumption the resolutions agreed upon, already processed and deemed presentable.

In the e-forum, the lessons posted by participants looked too sanitized and 'correct', thereby lacking in richness precisely because the disagreements, uncertainties and questions were left out. The recently created private conferences among learning team members is an answer to this need for private conversations among members of a learning team before they post their own lessons, fit for 'public consumption'.

Yet, the question remains: how do we strike a balance between the Asian concern for saving face and the learning that comes from making mistakes and acknowledging them? Perhaps, in the Asian context, Isang Bagsak facilitators need to point out the reality and value of mistakes in the process of learning and capacity-building, rather than the view that making mistakes is a stigma. Should these 'mistakes' be kept concealed in private conferences? Likewise, will the habit of avoiding conflict and controversy as part of the Asian/Buddhist ethos of smooth interpersonal relations be a hindrance to participatory learning?

Another cultural challenge to learning PDC are the hierarchical structures and relationships found in many organizations in South-East Asia. One observation in capacity-building programmes for such organizations is that team leaders, who are supervisors, do not take part in the programme activities and do not mingle, but primarily relegate the learning to their team members. On the other hand, the leaders' participation in the capacity-building programme is crucial to ensuring the application of the approach in their national resource management research and activities. Thus, another challenge with regard to capacity-building in PDC for natural resource management is how an organization can learn to apply PDC without contradicting itself in its project management style. Will its credibility in applying PDC be compromised if it does not adopt a similarly participatory project management approach? Or will learning and applying PDC in communities bring about a shift in a hierarchical style of project management, albeit slowly and incrementally, towards a more participatory approach in project management?

On the other hand, inasmuch as our bias is for participation, to what extent does advocating PDC impose on the learning team's non-participatory, hierarchical culture an alien perspective and method that might be inappropriate?

The lawyer/executive director of PANLIPI, however, reminded us of the rights-based perspective that is compatible with the participatory perspective: a hierarchical leadership style may also be consultative and participatory if it upholds the basic human right of constituents to express themselves and communicate their honest views and insights. Upholding people's participation in development, therefore, should not be viewed as an imposition if it upholds basic human communication rights.

Challenge 5: Making the learning process participant driven and experience based, yet striking a balance with expert or theoretical knowledge in the discipline

Aside from its application in NRM communities and its implication in NRM project management, PDC is a philosophy that also applies to the capacity-building or learning process in the programme. Thus, the challenge that facilitators face deals with striking the perfect balance in combining participants' experiential learning with expert or theoretical knowledge in PDC. What is the best way of drawing out experiences and reflective thinking while identifying lessons and insights in participatory development communication? How should facilitators incorporate the wealth of knowledge and wisdom from experts and the existing body of knowledge within PDC when applied to natural resource management? Furthermore, how can a participatory learning process in PDC contribute to further building this body of knowledge?

Challenge 6: Learning as a community/team by alternating local face-to-face meetings in the local language with regional (international) virtual meetings in a foreign language using the internet

Aside from our previous experience, research and practical and theoretical knowledge on PDC, as well as our previous training and familiarity with distance learning mechanics, our only preparation in carrying out the Isang Bagsak programme in South-East Asia was, perhaps, our openness to learn the cycle and rhythms of the programme. We were learning along the way. The newer dimension of facilitating is the community or team mode of learning compared to individual-based distance learning.

The main lesson here is the need to painstakingly spell out this learning cycle and to explain it in detail, making such explanations and exercises part of the programme's orientation. Such exercises should not just focus on using the electronic forum's software, but on the whole process of discussions within a team, synthesizing the discussions, and then posting in and downloading from the electronic forum. This makes it imperative that facilitators are highly familiar with the Isang Bagsak process. Thus, facilitators should have gone through the programme themselves if they are to facilitate learning. A manual on Isang Bagsak facilitation may likewise prove helpful.

Challenge 7: Capturing the collective learning of participant teams while remaining concerned with the relevance of PDC and NRM efforts to enhance the well-being of grassroots communities with and for whom the teams are working

This final challenge spells out the need to strike a balance between the capacity-building process of NRM workers and the benefits of their efforts within the natural resource management community. This is primarily a reminder or caveat to Isang Bagsak facilitators that, in our concern with learning the PDC process, we should not lose sight of 'the big picture': the impacts on the natural resource management community. As the programme continues to unfold, we anticipate further insights and challenges related to the impact of the programme on the well-being of the communities with whom learning participants work. Participatory development communication is, beyond a body of knowledge and a practice, people's lives, as the PANLIPI executive director reminds us.

In all these challenges, the College of Development Communication takes comfort in the shared view with its regional partners of Isang Bagsak as an evolving programme and network that is continuously redefining itself. Its dream is for the Isang Bagsak programme to eventually evolve into a regular self-sustaining certificate distance-learning programme in PDC in natural resource management and other development concerns for various types of development workers in different contexts.

Implementing Isang Bagsak in East and Southern Africa

Chris Kamlongera and Jones Kaumba

This chapter seeks to show how the Southern Africa Development Community Centre of Communication for Development (SADC-CCD) is assisting national bodies working on environmental and natural resource management (NRM) in Malawi, Uganda and Zimbabwe. With support from the International Development Research Centre (IDRC), SADC-CCD is working on a project aimed at building the capacity of institutions

in participatory development communication (PDC). This work is part of ongoing activities at the centre, aimed at assisting governments of the region in their development efforts.

The problem with these efforts has been their failure to put ordinary people in the driving seat. The participation of rural communities (who constitute more than 70 per cent of the total population of the region) has never really been seriously considered as critical to bringing about positive change in the region. The voices of the people have not been seen as important in most development efforts. These voices are often, if not always, ignored by those who make decisions on development issues.

Those who agree that people's participation in rural development and poverty alleviation efforts is critical are still grappling with how best this can be brought about in a sustainable manner. There is still a need for a body of practical knowledge on how to involve rural communities fully in such work. SADC-CCD was set up to find ways of doing this.

To date, it has come up with participatory communication methodologies that are trying to answer the quest for such a body of knowledge, such as participatory needs assessment for development communication and rural development; participatory communication strategy development; participatory curriculum development for farmer field schools; participatory rural communication appraisal; including rural communities in the writing of proposals; and participatory evaluation of communication programmes and the use of folk media.

These participatory communication methodologies have been tested across several issues in rural development and have been seen to work. What remains as a challenge to the SADC-CCD is sharing the body of knowledge so far generated in a manner that is both cost-effective and sustainable.

Addressing the problems of rural development

Low participation of people in poverty reduction and rural development programmes often results in a low utilization or failure of these programmes. Explanations put forth to explain these shortcomings include poor planning with communities to be involved; a low sense of ownership by communities; inappropriate technical solutions; poorly packaged information and knowledge; ineffective training methodologies for rural/semi-literate clientele; and ineffective communication channels and wrong target groups. In many cases, these problems can be addressed by Communication for Development, a cross-cutting approach applicable to any area of development where lasting progress depends upon the informed choices and actions of the people involved. It applies equally to programmes for improved agriculture; nutrition; food security; health; water and sanitation; gender awareness; population and reproductive health; livestock; forestry; the environment; literacy; rural credit; income generation; and other key areas.

Communication for Development consists of the systematic use of communication to effectively involve people in development, most particularly in rural development. It is based on the principle of dialogue, using communication approaches, participatory activities, media and channels with all levels of people concerned as equal partners Instead of 'target groups' that characterize Western-style advertising and promotion, Communication for Development is mostly based on 'interaction groups', promoting a common understanding and exchange of knowledge and experience. It can be used at the interpersonal, group and mass levels.

Communication for Development can:

- ensure that planning takes into account the community's underlying concerns;
- build community ownership of projects and empowerment;
- ensure that the best technical solutions are considered, taking into account community perceptions and practices, as well as their indigenous knowledge;
- package information in ways that are useful and attractive to users, particularly rural people who may be semi-literate and/or traditional;
- enhance training and technology skills transfer with communication approaches and media materials suited to people with little or no formal education skills;
- identify both effective and preferred communication channels and media, influential sources of information and advice, and modern and traditional knowledge; and
- help to safeguard against biases and bring out the concerns of beneficiaries who may be marginalized.

Communication for Development can produce both qualitative and quantitative results, measurable as changes in awareness, knowledge, attitude and practices. Moreover, it is particularly geared for use in rural areas, with more traditional cultures, although its principles may be applied equally in the peri-urban context.

In order to address the problem of the lack of participation in rural development programmes, SADC-CCD has developed and used a methodology comprised of the following seven phases:

1 situation assessment;
2 participatory research with the community: problem and solution identification; baseline survey for awareness; knowledge, attitude and practice surrounding a specific development issue (a baseline survey whose instrument is built on the results of the participatory research also provides a benchmark for future evaluation of the communication programme);
3 communication strategy design;
4 participatory design of messages and discussion themes;
5 development of communication media materials and methods to be used;
6 field implementation (with training of field staff as necessary); and

7   monitoring and evaluation, including second survey (to measure results
    and plan the next intervention, if necessary).

This methodology has been adapted, tested, marketed and disseminated.
The solution has been based on methodologies that include participatory
approaches which aim at actively involving people, at all levels, in identifying
rural development problems and solutions, sharing knowledge, changing
attitudes and behaviours, making decisions and reaching consensus for
action.

It has been disseminated through experiential workshops, advisory and
consultancy services, and the production of communication materials using
participatory methods. Alongside the methodologies cited earlier, the centre is
producing manuals, case studies and handbooks to go with the workshops.

Dissemination of the methodology itself has been a key factor leading to its
adoption by rural development programmes and communities. A strategy for
inducting national staff teams from rural development projects into innovative
'learn-while-doing' and action-oriented workshops that include fieldwork and
follow-up implementation at the grassroots level has been developed.

This work has produced noticeable and positive changes in the awareness, knowledge, attitude and practice of those involved, including rural
communities, development field staff and their employers in government,
non-governmental organizations (NGOs), institutions and international
organizations.

In pursuit of its new strategy of partnerships, SADC-CCD has been
working with some organizations with which it shares common understanding
of the importance of Communication for Development. For example, it has
been working with the Centre for Rural Radio Development (CIERRO) in
Burkina Faso in an attempt to improve the status of rural radio in Africa.
It is also working with IDRC on PDC in environmental and natural
resource management. This latter programme operates in Uganda, Malawi
and Zimbabwe as the Isang Bagsak East and Southern Africa (IB-ESA)
programme. The programme provides one concrete example of participatory
development communication at work. Let us look at how this is taking place.

Isang Bagsak in East and Southern Africa

The introduction of the PDC programme for natural resource management
in East and Southern Africa has been welcomed as a major step in efforts
aimed at improving the management of the environment, natural resource
research and development initiatives in the region.

The programme recognizes that the level of communication skills of
those attempting to implement development programmes and projects is
vitally important to the success of any efforts whose objective is to ensure
sustainability. In the main, the programme is expected to help participants
design and apply new ways of reaching people more effectively, wherever
they may be, through interpersonal, group and mass communication. It

also aims at making the policy development process more transparent and open to all stakeholders and involving people in decision-making processes of designing and planning possible solutions to problems in their society. Furthermore, it seeks to help participants design multimedia development messages, materials strategies and campaigns to communicate new ideas and practices to those in need of them. Finally, it strives to assist governments through their various institutions and NGOs to formulate well-researched policies for reducing poverty and ensuring sustainable, gender-balanced and environmentally friendly rural development.

By improving the ability of researchers and development agents in working with communities, Isang Bagsak will open up whole new possibilities for communities to handle natural resource management and development issues in their own way and in line with their traditional and cultural requirements, thereby contributing to the sustainable use of available resources.

Recognizing that capacity-building in PDC should be an important programme area in the field of NRM, IB-ESA – through its eight projects spread out across the East and Southern African region – is striving to make a difference. Researchers see it as an important vehicle for working with communities.

Thus, the Isang Bagsak programme is providing face-to-face interaction and tuition between participants and facilitators, as well as interaction through an electronic forum. Participants can also communicate among themselves freely through a computer-based 'Village Square'. Moreover, they have access to the IDRC library resources, as well as SADC-CCD materials and books.

The programme has also ensured that these human and materials resources help participants to build their capacities to work with local communities in a participatory way, thereby making it easier to develop partnerships with other development stakeholders in their quest to influence effectively the policy environment at local and national levels. It has also offered participants an opportunity to quickly obtain up-to-date information on participatory methods in natural resource management without having to leave their place of work to receive such training. Most importantly, Isang Bagsak East and Southern Africa has promoted South–South cooperation to ensure that participants in the South draw on each other's expertise and experience in order to reduce dependence on the North. It is also noteworthy that participants in this initiative need not have the same level of experience since teams with little or no experience can share and learn from their more skilled counterparts.

The initiatives enrolled in the programme come from a variety of disciplines, which makes the implementation of PDC a multidisciplinary one. Today, three countries are participating in the programme: Malawi, Zimbabwe and Uganda. As described in the following sections, the supported initiatives greatly differ from one another. Yet, they all share the same views regarding the need to empower local communities and to increase their sense of ownership of research or development initiatives.

Malawi

Three initiatives are currently under way in Malawi. The first one, entitled the Macadamia Smallholder Development Project, is being implemented by the Department of Agricultural Extension Services (DAES), which is part of the Ministry of Agriculture, Irrigation and Food Security. Since 2000, DAES has a new policy that advocates pluralistic and demand-driven extension services. The challenge is to empower farmers so that they are able to demand services which address their needs and problems. At the same time, the department has to coordinate the activities of various extension service providers operating in the communities.

The Macadamia Smallholder Development Project covers two agricultural development divisions – namely, Kasungu and Mzuzu in northern Malawi. Both researchers and agricultural extension staff participate in implementing its activities. The initiative aims at improving the well-being of Malawians through poverty alleviation among rural people by promoting agricultural development. More specifically, it seeks to promote the production of macadamia nuts intercropped with other food crops, thereby ensuring food security and cash income for rural farmers. Conservation of the environment is an integral part of this initiative. According to the team leader, through their participation in Isang Bagsak, the team intends to acquire effective communication skills that will enable frontline extension staff to engage in dialogue and create a mutual learning environment within the communities with whom they work, thereby empowering farmers.

A second initiative, which focuses on indigenous fruit trees, is being undertaken by the University of Malawi's Department of Chemistry. Previous studies on the chemistry of indigenous edible wild fruits growing in Malawi have revealed their great nutritional value, including vitamins A, B, C and minerals. This initiative will focus on the development of those products and the promotion of processing and marketing at household levels in rural Malawi. More specifically, the research will address the utilization and commercialization of indigenous fruits of the Miombo eco-zone. Implementing this study requires working with communities, especially with women, who are the main processors of food in Malawi. Technologies that can remove the drudgery and labour load among the women processors have been developed. Further work in capacity-building of rural food processors to increase value and income for improved household welfare will also be undertaken.

The team also facilitated the identification of priority research and development activities among rural communities. These activities have necessitated working with NGOs and government departments, especially research and extension services.

Finally, a third initiative spearheaded by the Department of Biology of the University of Malawi is also under way, which has been involved, in the past, in various multidisciplinary projects where participatory development communication was lacking. In this case, the initiative will focus on water resource management in southern Malawi.

One of the most interesting aspects of this initiative is that it stems from natural scientists' interest to incorporate social issues within natural resource management research as a way of improving their work. So far, team members have carried out preliminary data collection. The five-people team will soon be conducting detailed surveys where questionnaires and focus groups discussions will be used to collect data. They believe that PDC can greatly help in this process. Among other things, it is believed that PDC will contribute to increasing the indigenous and modern community knowledge base by collating indigenous knowledge that already exists at the local level and using it as a stepping stone towards the development of water resource management plans. Another objective of this initiative consists of assessing the microbial and chemical quality of water to determine if it is suitable for human consumption. The relevant gender issues affecting water resource management in the communities in the study area will also be examined. Finally, this research will also identify the factors that contribute to the degradation of the catchment area and will propose mitigation measures that can be put in place.

This increase in knowledge, however, can only be shared and utilized effectively if the team members are equipped with appropriate skills through the PDC training programme.

Uganda

Three initiatives are also taking place in Uganda. The first, entitled Integration of Natural Resource Management in National Agricultural Advisory Services (NAADS), is located in Kabale, in the sub-county of Rubaya.

The area is mountainous and very steep. Farmers continually face a problem of soil erosion as they farm on steep slopes. Currently, the programme is encouraging them to integrate the planting of trees, such as apple trees and cariandra, within their farming in order to control soil erosion.

Interacting successfully with the communities involved in this initiative is vital to its success, and PDC is seen as the best methodology to do so. According to the programme's coordinator, this new methodology, which ensures the participation of the communities at every stage, is probably the best possible solution to tackling some of the communication issues inherent to the projects.

The second Ugandan initiative is being implemented under the responsibility of the Kawanda Agricultural Research Institute (KARI) of the National Agricultural Research Organization (NARO). It deals with communication among banana growers regarding soil and water management, post-harvest handling and the improvement of cropping systems.

The communication issues being addressed by the project include the need to share farmers' indigenous knowledge, the inadequacy of extension work due to poor facilitation and motivation, farmers' lack of access to adequate information, as well as the existence of communication gaps between farmers and researchers, farmers and farmers, and researchers and extension staff.

Farmer-to-farmer communication through posters, brochures and open days has been undertaken in order to alleviate the problem of inadequate extension staff. Joint planning between researchers, farmers and other participants has been contributing to the reduction of communication gaps among the different stakeholders.

Finally, the third initiative undertaken in Uganda deals with sustainable land use in banana production in central Uganda. The main partner for this initiative is an organization called Volunteer Efforts for Development Concerns. This initiative aims to conduct participatory rural appraisal sessions to develop food security calendars and action plans; train rural development extension workers in farm planning and layout, farming as a business, sustainable agriculture, post-harvest handling and communication; establish on-farm demonstrations; provide quality planting material (high yielding, pest- and disease-free); and conduct agricultural practical training in relation to spacing, organic manure preparation, integrated pest management, pruning/de-suckering and post-harvest handling.

Since this work involves substantial interaction with farmers, it is vital that a suitable methodology is used to ensure optimum results. Participatory development communication seems particularly appropriate in this case.

Zimbabwe

In Zimbabwe, an initiative called the Desert Margins Programme is being implemented with the aim of arresting land degradation in the desert margins through demonstration and capacity-building activities. The programme addresses issues of global environmental importance, national economic and environmental concerns and, in particular, the loss of biological diversity, reduced sequestration of carbon and increased soil erosion and sedimentation.

The initiative is part of a larger programme covering nine countries: Burkina Faso, Botswana, Kenya, Mali, Namibia, Niger, Senegal, South Africa and Zimbabwe.

Key sites harbouring globally significant ecosystems and threatened biodiversity have been selected in each of the nine countries. These sites are to serve as field labs for demonstration activities regarding the monitoring and evaluation of biodiversity status, the testing of the most promising natural resources options, as well as the development of sustainable alternative livelihoods and policy guidelines.

The approach is intended to be a holistic one, as the Desert Margins Programme takes an innovative participatory and integrated natural resource management approach that consists of the conservation of biological resources through restoration activities that reverse degradation processes, rather than the preservation of specific ecosystems or species in protected areas.

In Zimbabwe, the chosen sites are Matobo, Chivi and Tsholotsho, which are based on communal rule. This brings in issues of consensus-building before any work can be done. Ownership issues, policing systems, who controls resource utilization, who is going to implement the interventions and monitor them, who says there is a problem are all issues that must be discussed from the onset.

PDC has a role to play right from problem identification, including finding and choosing possible solutions. So far, work has involved surveys to gather baseline data on the areas. Some tools have already been used – for example, historical timelines and trend lines. The initiative will last six years and the intention is for the initiatives to be continued beyond that.

In this case, PDC comes in handy in empowering the community, thus facilitating the sustainable use of natural resources. Communities will be exposed to an open platform to discuss their natural resource management practices, problems, needs, opportunities and solutions. As asserted by the team leader, 'They will, in the process, gain knowledge to implement their initiatives through guidance from the facilitators.' PDC, therefore, becomes an important tool that ensures ownership of the project by the communities themselves, thereby giving it a much higher chance of success and resulting in improved livelihoods.

The Sedgwick Agricultural Development Project is located in the Tsholotsho district of Matabeleland North Province. It operates on 10,126ha of the estate, which is bound by a number of newly resettled farming households, where it works directly with 84 households and extends services to four other neighbouring villages from the Tsholotsho communal areas.

The project aims to ensure strategic crop production for national food security through irrigation development initiatives. It also seeks to develop capacity for the production of grain crops and other cash crops by the community and to enhance its livestock management practices, particularly animal nutrition and disease control. Finally, it aims to provide breeding services in order to preserve the indigenous Nkone herd for the community and to establish veld management systems, in collaboration with the Department of Livestock Development and Production.

Participation in the project was initially through the local authorities and leadership. The team had to arrange meetings with members of the community within the project area. Areas of intervention were identified by conducting problem tree analysis and ranking through participatory approaches.

This helped the team to ensure that the causes and effects of the problems, as well as their solutions, were identified from the community's point of view. Overall, people felt that the area is a drought-prone one, so there are bound to be problems with water and pastures. These problems, in turn, lead to animal deaths, ill health, low weight and, eventually, the loss of income from the sale of animals. Following the success of this new approach, the farmers have agreed to take part in the development of grazing management schemes and to coordinate the provision of water for dip tanks.

Challenges

Since the beginning of the Isang Bagsak programme, there have been some experiences or some questions that require addressing. These are:

- The lack of backstopping activities: these would ensure that researchers are assisted on the spot while in the field and would allow for effective documentation of activities by coordinating institutions at the field level.
- Accreditation: although the SADC-CCD gives certificates of completion to its training and workshop participants, the participants' organizations or institutions do not recognize these certificates. Recognition or accreditation coming from an academic institution would not only be recognized by other institutions, but would go a long way in rewarding genuine participation in the Isang Bagsak programme and could be used or added to in a degree programme.

Observations and reflections

The PDC programme has so far demonstrated a positive demand for its services to a diversity of stakeholders, as evidenced by the organizations that are currently participating. Bottlenecks must still be removed, however, and the programme must be offered to more research and development institutions in the region.

So far, the sharing of experiences through the e-forum is proving to be an innovative way of getting over the usual hurdles of embarrassment, fear and reluctance to open up that are often associated with adult learning. The facilitator seems to disappear from the scene once the 'theme for the day' has been introduced. The participants then comment on the facilitator's initial statements. They do this freely, digging deep into their past experiences and observations of life around them (including past mistakes) without any fear of embarrassment.

Participants also learn from each other as they post their own responses on the forum. Very often, they respond to their colleagues' entries, as well, seek advice or make comments on their own experiences or problems.

One very gratifying result of the Isang Bagsak experience is the readiness, by scientists, to consider issues of communication. This is a major breakthrough for the programme. Some of the academic participants are already changing their approach to hardcore science teaching, as the Isang Bagsak experience is asking them to consider how, in reality, they relate to the world around them. In doing this, hardcore science is being 'humanized'.

Reflections on Participatory Development and Related Capacity-Building Needs in Egypt and the Arab Region

Waad El Hadidy

Non-profit and other civil society organizations in the Arab region now face difficult challenges. On the legal front, the battle for redefining the boundaries of the non-profit sector is taking place. On the political front, the question of what participation means and how it can be controlled is intensely debated. On the social front, disruptions are affecting longstanding

strictly codified value systems and social structures. All these challenges are questioning current development practices. Although both government and non-governmental organizations (NGOs) realize the limitations of top-down, off-the-shelf interventions and sense the need for participation, participatory development has not become deeply embedded in practice.

What can we learn from a decade of grappling with the concept of participatory development and its application? This chapter presents the Centre for Development Services' thoughts on participatory development in the Arab region, with particular insights from Egypt. Participatory development here refers to both the paradigm and the participatory approaches that make it possible, such as participatory development communication (PDC), participatory rapid appraisal (PRA) and participatory learning and action (PLA). Related capacity-building needs will also be discussed.

Participation as a societal value

Many Arab societies were pioneers in ingraining the value of participation in their lives. Egypt, for example, was the first Arab, Muslim and African country to experience modern civil society organizations, as early as 1821. It was also the first to flirt with democratic governance, starting in 1866 (Ibrahim et al, 1991) The first modern NGO, the Egyptian Hellenic Philanthropic Association, was established in Alexandria in 1821. This association was qualitatively different from earlier traditional religious endowments, which were a function of single charitable individuals or families, a form known in Egypt for centuries before.

From a communal-cultural perspective, the diverse reality of civil society in the Arab world, which includes informal social networks and traditional organizations of kinship, tribe, village and religious community, is a reflection of a wide range of civic behaviour that honours participation (Saber, 2002). In Kuwait, the *diwaniyya* is a place where men and, during recent years, women meet informally and is widely acknowledged as the place where the current move towards democracy in Kuwait began (Al Sayyid, 1993). In parts of Lebanon, the tribal system creates a space for consultation among village members over issues that affect their community.

From a religious perspective, participation has always been a core pillar of Islam, especially in non-secular regimes, which have historically prevailed. As a concept and as a principle, Shura in Islam does not differ from democracy. Both Shura and democracy arise from the central consideration that collective deliberation is more likely to lead to a fair and sound result for the social good than individual preference. Both concepts also assume that majority judgement tends to be more comprehensive and accurate than minority judgement. As principles, Shura and democracy proceed from the core idea that all people are equal in rights and responsibilities. The Qur'an mentions Shura as a principle governing the public life of the society of the faithful, rather than a specifically ordained system of governance. As such, the more any system constitutionally, institutionally and practically fulfils the principle of Shura or, for that matter, the democratic principle, the more Islamic that system becomes (Sulaiman, 1998).

The paradigm of participatory development

Despite such deep-seated participatory principles in culture and religion, the paradigm and the institution of participatory development are not so well entrenched in the Arab world. Many international organizations have been keen on engaging local communities in decision-making regarding development initiatives in Egypt. Much of their earlier work during the 1970s began with working with committees of local citizens, as mobilized by these organizations. In order for the work of these committees to continue endogenously, they were advised to form NGOs. This was the formula pursued by most foreign organizations operating in Egypt during the 1970s and 1980s. Ironically, and despite their efforts, foreign organizations were met with theories of conspiracy and doubts about their real intentions.

For example, in a village of the Beni Souef area where an international organization had begun working, members of the committee were particularly doubtful. Throughout the initial meetings, the group of leaders requested a statement of expenses in order to better assess the situation. They found out that US$160,000 had been spent in meetings and trips between Brussels and Cairo. 'We didn't understand development at the time, and thought that every penny should be spent on improving conditions', stated Mr Badr. As members of the local council, they decided to expel the institute from Beni Souef. But the local leaders were pressured by the US embassy and the governor to let the organization stay. They were advised that 'This organization is going to stay, whether you like it or not. So it would be wise of you to make the best of them.'

While the essence of participatory development had been practised since the 1970s, it wasn't until the 1990s that the written rhetoric began to spread. Training courses and development projects all became prefixed with the term 'participatory'. Some organizations, including the Centre for Development Services and the Near East Foundation, took on the task of adapting the concept to the Arab region. Naturally, the efforts of a few organizations were not enough to reach thousands. Although many organizations are now using the language of 'participatory development', the conceptual framework is not well internalized since few genuine participatory development initiatives are taking place on the ground. Perhaps this is because 'participation in development' is often perceived as a Northern agenda, despite the fact that it first emerged in South America through the writings of Paulo Freire (1970).

There are limited known examples of successful participatory development initiatives. The documentation process of such initiatives usually targets the donor agency rather than local practitioners, decision-makers or community members. The critical reflection processes that precede documentation for the latter groups and result in organizational learning are frail in the Arab region. On the individual level, practitioners seldom engage in critical reflection because the predominant education systems do not encourage it. Organizational cultures are also non-conducive with regard to these aspects. Knowledge generation in our region usually revolves around an urgent need, whether it is a problem requiring resolution or an issue in need of addressing. The regular and consistent communication processes prevalent in rural communities, such as the tribal councils in Jordan or the Qat gatherings[1] in Yemen, create space for discussions, for local wisdom to emerge and for learning to take place. Such processes are not normalized within development organizations or among practitioners. This situation restrains contemplation on both an individual and group level and perpetuates the existing lack of knowledge about participatory development.

Moreover, the policy environment does not always enable civil society to apply participatory approaches. While it is true that in Egypt, Morocco, Tunisia and Algeria, governments encourage NGOs to complement public spending on social services, they fail to allow these organizations to freely empower people for fear that things may get out of hand (Kanawati, 1997).

Reflection on the Centre for Development Services' attempts to turn 'participatory development' into a meaningful concept in Egypt provides interesting insights. Indeed, using the PRA methodology adequately made a significant mark since many NGOs talked about participation as an integral part of their work. However, the emphasis on the NGO sector unnecessarily reinforced the separation between the government and NGO sectors. The overwhelming participatory discourse and its focus on NGOs implied that government cannot be participatory, and that the role of NGOs is to correct for the mistakes that government makes in planning development without involving people. The dichotomies created by the use of words such as 'putting the last first' and 'the marginalized', which dominated the discourse, tended to pit NGOs against government. In such a polarized situation, participation advocates were faced with what they wanted to avoid: a rift between NGOs and government. This only shows that even when efforts in promoting participatory approaches are expended, unwanted outcomes may result.

One region, many peculiarities

When speaking about the Arab region, it is important to realize the differing social, political and economic contexts that enable or stifle participatory development across different Arab countries. For instance, Egypt's social context is one of a patriarchal and authoritarian society reflected in communities' attitudes of dependence and ambivalence. For a long time, the government has taken on the role of the benevolent dictator invoking an attitude of 'government knows best'. The work of NGOs is largely controlled through legislation and scrutinized by state security. As a matter of fact, a study conducted to assess grassroots participation among Egyptian NGOs through indicators of membership, frequency of general assembly meetings and proportion of voluntary to salaried staff concluded that overall participation was very low, due in part to government inhibition of participation (Ibrahim,

1996). The situation in Palestine can be described as the reverse. In the absence of government, NGOs have taken on the role of provider of services, which in other countries would be categorized as public services. NGOs are dominant in Palestine, and the Palestinian Authority is now trying to position itself within the development realm. Such variations suggest that participatory development takes on different meanings, shaped by each country's context.

Implications for capacity-building

There is no doubt that more efforts are needed to contextualize participatory development so that it is more than a prefix to NGO programme titles. The question is how should this capacity-building take place?

The Centre for Development Services was originally established as a resource and service centre for NGOs. Training and technical assistance used to comprise most of its portfolio before it delved into long-term, self-conceived development programmes. Reflecting on the training-and-technical-assistance days led us to realize that this 'delivery of resources' mode of operation in the form of transfer of know-how and skills was essential but not sufficient. It implied that the beneficiaries of capacity-building require the transfer of resources from those who have to those who have not.

This has been and still continues to be the paradigm for capacity-building as perpetuated by NGOs and international agencies alike. As excerpted from a World Bank working paper:

> ... respect for independent civic action does not mean leaving community groups on their own to implement initiatives by trial and error. They generally lack specialized knowledge and the ability to apply it, and the success of their endeavours often hinges on receiving appropriate and sustained technical assistance in fields such as management informa-tion and project control, human resource development, and project formulation, monitoring and evaluation. (Siri, 2002)

It is rather paradoxical for organizations who aim to promote participatory development to be implementing capacity-building in such a way. Being participatory does not only mean engaging people in a transfer of knowledge, but rather shifting the outlook of capacity-building in order to recognize that people already have skills and abilities that only need to surface. This new outlook also recognizes that people internalize new knowledge not only through the transfer of knowledge, but also through experiencing this knowledge themselves.

More specifically, taking the example of participatory development communication and the Centre for Development Services' minimal, yet insightful, experience in training Arab organizations and providing technical assistance to a team of researchers on the concept, it would not be unfair to say that the essence of PDC was not conveyed. Discussions with practitioners who attended the training showed that their understanding of participatory development communication was limited to the production of materials. The research team perceived PDC as an extractive tool used to facilitate information gathering in research.

However, PDC, like other members of the participatory development family, is rather fluid and reinforces an alternative form of communication and partnership between communities and development practitioners. It is about communication, a process integral to our lives. It is also about being cognizant of communication needs, such as facilitating articulation, enabling collaboration and providing understanding, in addition to development needs that may focus on more concrete issues. Unlike capacity-building, which requires a 'how-to' approach, such as proposal writing or business planning, capacity-building in PDC should focus on recognizing that communication is a natural process. Participatory development communication supports the development project cycle. Hence, it addresses practitioners who are engaged in development. Rather than 'teaching' them about participation, it is important to recognize that these practitioners have rich experiences and what is needed is to facilitate their own learning. Rather than 'providing resources', an approach is required for 'facilitation of resourcefulness', a process of learning to learn. It is also critical to realize that participatory principles are not alien to the Arab world's cultural and religious heritage, but have deep-rooted traditions that need reviving.

For Arab practitioners, space is needed to reflect on experience and to develop insights, document stories and share with others. In a workshop held in Jordan in 1997, participants from nine countries in the Middle East reflected on the lack of documentation of local participatory practice, and indicated their enthusiasm for a forum that brings them together to discuss their experiences in the field (NEF and CDS, 1997). Before now, some attempts had been made; but none were sustained, probably because discussions were too formal and abstract and ended up as a series of sporadic and detached topics. Motivation factors were not integrally considered.

The suggested strategy for capacity-building assumes that self-reflection generates knowledge that is profound and better internalized than knowledge imposed from the outside. After all, how can we presume to intervene in others' development if we do not understand our own, or if we are not prepared to engage in our own (Kaplan, 1999)? This is not to say that this is a revolutionary strategy. In fact, the activities that it entails would not transcend the conventional workshops, networks and e-forums. The main difference would be in the approach, which would focus on facilitating reflection on experience and learning from it, adding an assets-based element to capacity-building. Such an approach aims to establish the case for mainstreaming project learning and learning review.

With this in mind, it is interesting to look at an existing capacity-building initiative in participatory development communication to see how it could be adapted to suit the specific needs of the Arab region.

Case in point: Isang Bagsak

Isang Bagsak, a capacity-building and networking programme in participatory development communication, provides a good basis for the learning-to-learn approach. The programme combines the use of internet technology with face-to-face meetings to discuss themes related to PDC. The following points highlight key aspects of the Isang Bagsak programme and discuss ways of tweaking these aspects so that they incorporate more reflection and self-learning:

- Overall strategy: Isang Bagsak is considered a capacity-building and networking programme that largely relies on distance education in its modalities. A resource person introduces each theme by sending it in a message to all participating teams for their comment and synthesis. Although not intended, the discussions have taken the form of questions and answers, where the resource person stimulates discussion on each theme through trigger questions and the teams gather to prepare their response to the questions.
- To create a reflection/learning programme, this attitude must be instilled from the start. The focus of the programme should be presented as one that seeks to enhance knowledge on participatory development communication and to strengthen critical reflection skills. Participants should feel that this is an opportunity to think critically about their development work and to draw out lessons learned.
- Content and discussion: the Isang Bagsak programme relies on retrospective reflection on previous projects, or on aspects of current projects that have already occurred. While this is essential in order to draw lessons learned from experience, it does not necessarily enable participants to apply lessons learned to new situations.

   In order to facilitate this, participants would be asked to pursue two tracks when discussing each theme. One track would be to consciously reflect on the theme in retrospect through discussions of previous experiences. This would reinforce the perception that organizations have some experience of participation, albeit unconscious. The other track would focus on the discussion of each theme in light of projects currently being pursued. The idea here is to provide a small fund for each participating organization in order to select a natural resource management initiative jointly with communities. This would start at theme 3: involving the community in the identification of a natural resource management problem and its solutions. The fund would be nominal, only enough to kick-start an initiative with the community. From theme 3 onwards, participants would be encouraged to share discussion themes with their communities as part of a joint learning and reflection process and an informal evaluation mechanism.
- Facilitation: in the Isang Bagsak programme, a resource person is responsible for introducing the theme and commenting on each team's synthesis. This may create a sense that the resource person is the expert in control.

While there should be a main resource person for each theme, participants should also take turns in co-facilitating themes. The resource person
should only intervene in the discussion when necessary.

- Design: the framework of the Isang Bagsak programme is largely
  predesigned. Participants take control over the discussions through
  synthesized responses.

In order to spur a learning and reflection environment, participants
should have space to participate in the design of the programme. For
instance, each participant can pose a critical question (for example, an
ethical dilemma or an issue worth pondering) to each theme that can
crosscut the discussions. Participants can also select for discussion the
participatory dynamic that is of relevance and importance to them.

- Participants: several country teams have already participated in the Isang
  Bagsak programme. A core team in each country was responsible for
  convening the country teams and synthesizing their discussions. There was
  concern that the richness of discussions may have been lost in too much
  synthesis, especially when there were several teams in each country.

Participating teams should be able to transfer their knowledge to other organizations. This can be achieved through two mechanisms. Each organization
participating in the programme should partner with another organization
of less experience. In this way, one organization 'mentors' the other in
implementing the joint initiative decided upon with the community. Each
country would be represented through two teams – hence avoiding dilution of
discussions. Another way to expand the programme beyond the participating
organizations would be to generate materials such as simple guidebooks,
training packages, frameworks and exercises that can be used by participants
to disseminate knowledge to others. Compilation of such material would be
the responsibility of a resource person other than the main facilitator. Such
material would include case stories provided by the participants and their
respective communities. Responsibility for using such material with other
NGOs would be the responsibility of the participants. This notion draws on
the experience of the Living University programme, initiated by Save the
Children US in Egypt. The programme is based on peer-to-peer learning
through NGOs. NGOs participate in an extensive training programme
and either graduate to become a 'Living University' or a 'learner'. Living
Universities then become ambassadors who transfer their knowledge to
other NGOs, and the cycle continues in a cascading effect. According to
the participating NGOs, learning from a peer NGO was a more positive
experience than learning from a contracted consultant, as is the case with
conventional training programmes.

Conclusions

To summarize, capacity-building in the Arab region should aim to bring out
in a systematic way knowledge which may exist at a tacit, subconscious level

within individuals or organizations. When we consider capacity-building in participatory development, we should also be considering capacity-building in learning. And consistent with the philosophy of participatory development of 'teaching a person how to catch the fish rather than providing the fish', capacity-building should also be about teaching how to learn, rather than providing ready-made learning.

Yet, it is not expected that Arab practitioners will jump at the idea of learning from experience. The issue of incentives and motivation needs to be carefully considered. Clear and tangible benefits must be demonstrated. Each practitioner will wonder about the personal gain from taking the time to participate in a learning initiative unless such time is factored into ongoing activities, perhaps as part of the 'dissemination' sub-item common to all development initiatives' budgets. Other incentives mentioned above include funding to start an initiative or the prospect of creating materials that are useful for training and other forms of knowledge dissemination.

Note

1   Qat is a sedative grown and consumed in Yemen and is considered an integral part of work and social life.

References

Al Sayyid (1993) 'A civil society in Egypt ?', *Middle East Journal*, vol 47, no 2, pp228–242

Freire, P. (1970) *Pedagogy of the Oppressed*, New York, Continuum

Ibrahim, S. E. (1996) *An Assessment of Grass Roots Participation in the Development of Egypt*, Cairo Papers in Social Science, vol 19, no 3, The American University of Cairo Press

Ibrahim, S. E. et al (1991) *Civil Society and Governance in Egypt*, Institute of Development Studies, University of Sussex, UK

Kanawati, M. (1997) *Consulting Egypt's Local Experts: Beyond Prince and Merchant*, The Institute of Cultural Affairs International, Pact Publications, New York

Kaplan, A. (1999) *The Development of Capacity*, United Nations Non-Governmental Liaison Service, Geneva

NEF (Near East Foundation) and CDS (Centre for Development Services) (1997) *Arabization for PRA Materials in the Documentation of the PRA Exchange Meeting: Challenging Practice Attitudes*, NEF and CDS, Amman

Saber, A. (2002) 'Towards an understanding of civil society in the Arab world', *Alliance Newsletter*

Siri, G. (2002) *The World Bank and Civil Society Development: Exploring Two Courses of Action for Capacity Building*, World Bank Institute, Washington, DC

Sulaiman, S. (1998) 'Democracy and Shura', in Kurzman, C. (ed) *Liberal Islam: A Reader*, Oxford University Press, New York

Implementing Isang Bagsak: Community-Based Coastal Resource Management in Central Viet Nam

Madeline Baguio Quiamco[1]

A young and fast-growing population and a robust tourism industry are providing the impetus for growth in the fisheries sector of central Viet Nam. As fish catch from the sea decreases in response to over-exploitation, aquaculture is responding to the demand for marine products. The culture of popular and valuable marine species such as lobster, shrimp, clams, crabs and preferred fish species in the lagoons and bays of central Viet Nam has become highly profitable, attracting investors and eventually creating the usual triumvirate

of problems that seem to follow demand-driven development: pressure on natural resources, displacement of local people from their livelihoods and inequity in the distribution of benefits.

Research aiming to understand the biological and social dimensions of fisheries resource management has been going on in central Viet Nam during the last decade. Between 1995 and 2001, as part of a research programme called the Lagoon Project, the Hue University of Agriculture and Forestry, the Hue University of Science and the Hue Department of Fisheries investigated the management of biological resources in the Tam Giang Lagoon. The research initiative also examined how global and national changes were affecting people's livelihoods in the province of Hue's coastal area.

The initiative, entitled Community-Based Coastal Resource Management (CBCRM) for Central Viet Nam, is a three-year project that seeks to utilize the results of the Lagoon Project and other research and development efforts to address coastal resource management problems in the region. It is being implemented by the three organizations involved in the Tam Giang Lagoon Project, as well as two others: the Research Institute for Aquaculture – Region 3 (RIA–3) and the Nha Trang University of Fisheries. Through the project, it is envisioned that approaches to coastal resource management throughout Viet Nam will be improved, that the complex livelihood connections will be better understood and that this new knowledge will influence local policy with a view to improving participatory management of coastal resources. Finally, a network of community-based resource management researchers will be initiated in Viet Nam to implement a capacity-building programme for all researchers.

Seeking grounds for partnership

The team's motivation to participate in Isang Bagsak[2] stems primarily from its desire to improve its scientists' capacity to use participatory development communication (PDC) to achieve coastal resource management goals. Team members saw in Isang Bagsak the answer to their need for 'social science' knowledge and skills. They expressed the belief that improved communication skills would enable them to conduct their people-related tasks more effectively and, ultimately, help the project to achieve its specific objectives, which are to:

- increase the capacity of the partner institutions for leadership in community-based coastal resource management in Viet Nam and to initiate a network of researchers as a means of providing capacity-building support for other Vietnamese institutions;
- understand the changes that have occurred in livelihood diversity, the coping strategies put in place by communities and local policy responses to those changes in lagoon resource use;
- identify and evaluate means and processes for scaling up fieldwork to include multiple communes using ecosystem-based management

principles while considering issues of larger socio-political organization and ecological aspects of living resources;

- assess and understand the impact of aquaculture on the livelihoods of traditional fishers; and
- explore options for improved participatory management of aquaculture and fisheries.

This rationale for participation convinced the Isang Bagsak South-East Asia coordinating group that its activities could, indeed, support the CBCRM's objectives. Isang Bagsak's other requisites for participation were also met: sustained connectivity, actual work going on in the community, and a willingness to apply PDC to their current work and to devote time to Isang Bagsak activities.

Initial results

Team members have been participating in cycle 1 of Isang Bangsak, together with three other teams in South-East Asia: the Forestry Administration in Cambodia, the Legal Assistance Centre for Indigenous Filipinos (PANLIPI) and the College of Development Communication (CDC) of the University of the Philippines at Los Baños, both based in the Philippines.

Collaborative implementation of introductory workshop

The introductory workshop for the Vietnamese team was held in Nha Trang, central Viet Nam on 20–22 February 2004. It brought together 16 fishery scientists from three implementing organizations: the Hue University of Agriculture and Forestry, the Nha Trang University of Fisheries and the Research Institute for Aquaculture – Region 3. During the workshop session aimed at levelling off expectations, it was understood that in participating in Isang Bagsak, participants expected to acquire the skills for working effectively with low literacy-level people, facilitating the activities of local people involved in development initiatives and helping them to solve their problems. They would also learn to plan PDC activities in community-based coastal resource management and acquire skills to involve more participants in community activities. Participants also expressed some concerns about some aspects of the programme and about PDC, which were discussed collectively.

E-forum

As far as connectivity and aptitude for the technical side of the e-forum are concerned, all three organizations have internet capability and the team members have access to it, although one team seemed to have less access than the others. At the e-forum briefing session during the introductory workshop, everyone demonstrated working knowledge of the internet. An hour after the briefing, the participants were finding their way around with the forum software.

Their problem was in responding to the forum questions, in constructing their messages and in posting them 'for all to see' since this needed to be done in English, the only language they had in common with the other teams from the Philippines and Cambodia. There was a general hesitation to post their thoughts because 'their English is not very good'. It was difficult for them to get beyond the salutation and greeting phase.

While some commented that the procedure was complex because it involved many steps, they also thought that the e-forum was a helpful and convenient way of communicating, and that it would be useful to them. To solve their language problem, they suggested a facility apart from the e-forum (and seen only by them) where they could construct their e-forum posting. They even suggested e-mailing their posting first to the country facilitator so that she or he could edit it before posting. A number of reasons explain this attitude:

- English is not spoken every day in Viet Nam, even by people such as scientists;
- posting on the e-forum can feel like one is publishing something (and therefore should be 'correct');
- offering one's thoughts and reacting to those of others in public is not easy for most Asians; these have to be done with care.

Learning participatory development communication through online sharing

Through their participation in the Isang Bagsak e-forum as the country team for Viet Nam, team members are increasing their understanding of the PDC concept and its application. For a team that began only recently with the major concern that it did not have sufficient knowledge of communication tools or a mastery of the English language, the Vietnamese team is doing well.

The team has maintained its participation in the e-forum, contributing richly to the knowledge exchange by posting its experiences in working with the coastal communities on the e-forum. At the same time, online exchanges of PDC experiences, ideas and analyses with the participating teams in the Philippines and Cambodia have broadened the Vietnamese team's understanding of the communication facet of its work with coastal communities. Isang Bagsak has introduced the perspective of PDC as a cycle of ten interrelated participatory steps. Using this new perspective as a framework, the Vietnamese team, along with the other participating country teams, is finding new meaning in the people-related work that it does in natural resource management. Through the e-forum, the Vietnamese team has shared the following experiences in PDC with the other teams.

Using communication to facilitate participation

The team has used a wide range of communication tools to facilitate participation, ranging from informal and formal conversations to training, and from participatory rural appraisal methodologies to mass media, specifically radio broadcast. The team explained that communication has helped to achieve natural resource management goals by motivating people to participate, clarifying cooperation among stakeholders, and convincing people and gaining their support, as well as through the enactment of laws and the issuance of policies, through the implementation of plans or projects, by mitigating boundary or resource-use conflicts, and by providing more choices or options. Communication cannot help, the team opined, when information is not updated to suit the changing context; when there is only one-way communication and no system for feedback, evaluation or monitoring; and when there is no transparency, causing people to lose confidence.

To be effective, researchers and development workers have to have certain skills and abilities. The team identified these as communication and presentation skills, the ability to motivate and work with people, and knowledge of local issues, concerns and socio-political conditions. Research and development workers, they said, must have planning skills, an open mind, fairness and foresight. They must also have cultural sensitivity, resourcefulness and an approachable personality.

Approaching the community

For the Vietnamese team, approaching a community involved first contacting the local authority or leadership, rather than going directly to the people. It required understanding the community's customs, beliefs and culture in advance. It also required understanding the local people and listening to them, including those whose ideas or opinions were in conflict with those of the team, those whose trust had been betrayed by other people from outside the community, and those with whom the team had limited contact. Approaching a community also meant clearly explaining the goals and objectives of the activity, sharing information with them, establishing relationships and becoming like local people themselves (i.e. eating, living with them and behaving as they did) to gain people's acceptance and pave the way towards working with them.

In doing this, the team encountered some initial difficulties. One such difficulty involved the community leaders, who would not allow the team to go into the village. The team solved this by spending time asking for help from higher-level officials – that is, from the commune or district authorities. Another difficulty the team encountered while initiating the project involved working time, which differed between the local people and the team. Team members worked with farmers and fisher folk. Fisher folk worked at night and rested during the daytime. Their solution was to adapt to the fisher folk's working schedule, highlighting the importance of flexibility and adaptation rather than bureaucratic efficiency among implementers.

In order to approach the community successfully and manage initial difficulties, the team needed more than what official documents could provide. The team felt that it required general information and secondary data, socio-demographic information on the people, and information on development projects and activities currently being implemented, as well as those planned for the next couple of years. The team gathered all this information by using participatory rural appraisal methods such as focus group discussions and timelining, and conducting household interviews.

The team described the people in the community as either fisher folk or farmers. Fisher folk are either fixed-gear fishers or mobile-gear fishers. In addition, the group of village leaders, women folk and the youth may also be identified. In general, the people in the community are very poor and have little education. They are friendly, but very conservative. The majority of them practise both fishing (fish capture) and aquaculture. Most farmers are rice farmers.

People in the community generally hold a very simple perspective about natural resources. They believe that the rice field is private, but that the fishing ground is a common resource. Thus, while they should not plant rice in someone else's rice field, they can fish and put fish cages anywhere in the lagoon. The team observed that traditional and modern media are used by the community. Activity organizers inform each household about important events, write news or conduct meetings; at the same time, people have access to mass media such as radio, television and the press (newspapers), as well as to posters and people's forums or exchanges of information. Since beginning to work in the community in 2003, the CBCRM team has shared with the people its knowledge and experience in community management and credit management; information about environmental and natural resource protection; and know-how in planning for aquaculture development.

Entering the community

The team approaches a community with an outline of an initiative's proposal. This document is the outcome of the team's discussions and states the overall objective of the initiative, the target area and the target group. However, specific objectives are still to be finalized based on the comments of the local people after the outline is presented to them. For example, one priority was for community activities that enhance the living standards of local people in environmentally friendly ways. Thus, research activities and interventions should follow this priority objective. But the team took time to build consensus in the community on what the priority activities are, and where/when these should be implemented.

People were asked to participate in identifying problems and their solutions. Through individual interviews and subsequent group discussions, they were asked what they had done to improve their livelihood, the problems they had encountered and how they thought these could be solved. In the group discussion, the team helped local people to share their experiences and ideas with each other, to hear one another's viewpoints, and to form a consensus on their problems, solutions and priority actions.

While the process had the potential to enable local people to understand the issues, form a consensus and decide on actions, in practice it still was constrained by several problems. Participation, it was observed, was influenced, first, by each participant's status and the culture of the group, and, second, by people's understanding of the project's activities and benefits. The team described the situation as follows:

> *A person's position in the community will either make him confident or not confident enough to participate in community activities. That is, a village leader is more confident than a poor person, even if he/she is normal. If a poor and not very educated person expresses his/her thoughts, it may not be treated as important as that of a well-educated villager's idea, even if the two ideas were similar. In a community we worked with, culture prevented women from participating. Women there were not used to talking at village meetings. According to an old concept: 'Women are inward, men are outward.' This prevents women from participating in such a meeting. Differentiation through social stratification (that is, small mobile versus large fixed fishing gear groups) also prevents the people from participating in the common action.*
>
> *Another problem that prevented participation was that our ideas to help people were sometimes not clear to them, so they did not realize the benefits from the actions to be undertaken. This limited people's participation.*

Testimonials about the initiative from people in the same circumstances and clear explanations of its goals, objectives and benefits helped mitigate these factors, which initially hindered people's participation.

Conclusions

Thus far, Isang Bagsak and PDC are helping CBCRM to enhance its effectiveness among its many stakeholder groups: fisher folk, fishery entrepreneurs and investors, policy-makers and implementers, other scientists and students of fishery. Through their interest in PDC, the team has demonstrated its capability to mobilize itself, collaborate and share resources. The team has maintained its active participation in the e-forum by sharing its experiences and ideas on the different participatory development communication themes discussed, although it still has to demonstrate its skill in reacting to the experiences of other teams and posting its insights.

Interaction with the other Isang Bagsak teams has also helped the Vietnamese team to identify its communication skills needs. Indeed, before the mid-term workshop was conducted in Viet Nam last August, team members were able to spell out the communication skills that they needed in order to improve their effectiveness in the field. On their request, the Isang Bagsak project team from CDC conducted skills development sessions on radio script writing and video production.

Through exposure to PDC concepts and its applications in the field in South-East Asia, the team is improving its understanding and knowledge of how it can utilize PDC for more effective research and development work in natural resource management. Isang Bagsak implementers are helping the team to nurture this understanding and knowledge by providing an atmosphere for participatory learning – that is, by demonstrating resolve, clarity of purpose, respect for differences, transparency and a genuine intent to empower through capacity-building. Consequently, this should build a sense of responsibility for Isang Bagsak activities, provide mutual trust and enhance the capacity for participatory development communication all around.

Note

1   This chapter was written with inputs from the community-based coastal resource management team at the Hue University of Agriculture and Forestry in Viet Nam.
2   Isang Bagsak is a learning and networking programme that aims to improve communication and participation among natural resource management researchers, practitioners, communities and other stakeholders, and to provide communication support to development initiatives aimed at helping communities overcome poverty.

Building Communication Capacity for Natural Resource Management in Cambodia

Jakob S. Thompson and Mario Acunzo

The government of Cambodia has established as a priority achieving and maintaining food security for its rapidly growing population. Existing national policies promote the sustainable multiple-use management of natural resources at the community level. However, this cannot be accomplished without information and communication interventions aimed at altering the negative attitudes and reducing the unsustainable and damaging practices of natural resource users. This, in turn, requires a strengthened national institutional capability to design and implement targeted information and

communication interventions in support of local community natural resource management plans and efforts. Failure to establish this capacity will result in diminished institutional capacity to prevent and mitigate environmental degradation, with resulting negative impacts on agricultural productivity and national food security.

Objective and activities

The overall objective of the United Nations Food and Agriculture Organization (FAO) initiative entitled Information and Communication for Sustainable Natural Resource Management in Agriculture is to contribute to the improvement of natural resource management in Cambodia through building national capacity in the systematic design and use of information and communication strategies, methods and materials. The primary activity undertaken by the initiative is the training of central- and provincial-level staff from the Ministry of Agriculture, Fisheries and Forestry and the Ministry of Environment in the theory, design and use of participatory development communication (PDC). During the 36-day training programme, the 19 participants who have been trained so far carried out the design and preliminary practical implementation of a strategy for PDC with villagers in two pilot sites. This learning-by-doing process included participatory analysis, training of villagers, and material design and production, as well as monitoring and evaluation for the improvement of agricultural and fishing practices. The training programme was carried out by a team of trainers from the College of Development Communication (CDC) of the University of the Philippines at Los Baños.

Training strategy

The project's capacity-building strategy is based on building field experience through in-service training and learning-by-doing strategic planning, implementation, monitoring and evaluation in local-level pilot sites. The step-by-step training approach used in this initiative can be summarized as follows:

- establishment of a project coordination unit and a communication team, comprising 16 communication unit staff from the two ministries;
- training of eight central-level communication team members in the design, implementation and evaluation of information and communication strategies;
- selection of pilot sites;
- study tour for two selected team members to the University of the Philippines at Los Baños;
- theoretical and practical training of team members through participatory situation analysis; planning, monitoring and evaluation; material development; and implementation of the plan, in collaboration with 73 natural

resource users, with a focus on identified natural resource management issues;
- training of team members in equipment handling, operation and care;
- training of team members in cost recovery.

So far, the communication team has undergone five training workshops for a total of 36 days, and another three workshops are planned to be held before the end of the project.

Achievements

The impressions and feedback gathered at all levels strongly suggest that the approach has led to a notable improvement of knowledge, skills and behaviour of both government staff and villagers. This is most evident in the way in which the work is being carried out by the communication team and the way in which people have changed their lives in the pilot sites. However, the institutional impact of the project is equally important.

Project impact

Team members have expressed the fact that the project has helped to improve their technical skills and the way in which they work with people in the communities. They say that they now handle desktop publishing and video editing effectively. At the same time, their appreciation of peoples' involvement in the development and implementation of information and communication initiatives has improved.

In the communities, people say that the activities have led to improved practices and increased awareness, such as improved production in pig raising and less cutting of flooded forests. People also raise new issues, such as chicken breeding and precautions relating to avian flu, as communication challenges for the community. This illustrates their appreciation of PDC as a way of dealing with the problems that they face.

A project coordination unit has been operational since the onset, based on the will and ability of two ministries at the provincial and central levels to join forces. For the future, this cooperation has demonstrated the benefits of working together to tackle PDC challenges as the strong links between staff in the two ministries now form a base for continued cooperation. The project has also made a valuable contribution in the form of the Cambodian government's acknowledgement of the importance and potential benefits of planned and targeted information and communication interventions when dealing with complex natural resource issues. To go with it, this initiative can provide examples in the form of training and field-level methodologies, as well as communication tools and materials. These aspects are also likely to have improved the conditions under which community-based natural resource management will develop into an integral part of the Cambodian government's work to increase food security in the future.

Equipment

The project has equipped the ministries with a digital video production and editing system, a digital still camera and a desktop publishing system that is in constant use, and trained staff members are becoming more and more confident and effective in the use of high-tech tools for media production. At the provincial level, the communication team has also received the same equipment. However, the opportunities to practise are not as frequent and the equipment is not as much in use as is the case at the central level, where the equipment is kept and used in an easily accessible area of the office, also under constant supervision and maintenance.

Additional capacity

The training capacities attained by the communication team were demonstrated and further strengthened when the project provided strategic communication training to a FAO community-based natural resource management initiative in Siem Reap Province. The communication team trained staff in designing and producing multimedia tools as part of a one-week training course. The outcome of the training and feedback from participants supports the impression that this is a task that the communication team can fulfil in times to come. Currently, the communication team is also providing training and technical support to the Special Programme for Food Security and to the Community Fishery Development Office in order to implement a communication strategy design and implementation process. As a result, the communication team is able to strengthen its skills and further develop its capacity as trainers, which is likely to constitute a major part of the support that it is intended to provide to programmes or projects in the future.

Current situation

Today, an important step has been made for PDC to contribute to sustainable natural resource management in Cambodia. As mentioned earlier, the communication team is currently providing support to two natural resource management programmes to build and implement information and communication components. Besides providing team members with an opportunity to practise and perfect their newly acquired skills, these large and highly visible natural resource management programmes are able to display the quality of materials and products of the communication team as a service provider. This, again, can form the basis for the team to become a responsive and capable service provider in mainstreaming and scaling up PDC in the natural resource management sector and in the evolving development scene in Cambodia.

Constraints faced by the initiative as a result of ...

... relying on foreign training capacities

So far, the main constraints faced by the project have been that project activities have more or less come to a halt when the international consultants were not in the country to provide training or to push for activities. This has been dealt with by hiring designate national staff to ensure the continuity of activities between missions. A general lack of operational budgets, for example, to travel and pay for materials is making it difficult for the communication team to work and practise the new skills. Relying on out-of-country training capacities is also making continued backstopping and follow-up difficult – which is something the team members clearly express a need for, especially in relation to the use and maintenance of computer-based tools.

... language

With regard to the actual training, Cambodia, as many other countries, poses a challenge when it comes to language. The use of foreign training capacities hampers the effectiveness of the training, and there is a constant risk that important points are lost in translation.

... skills and attitude of trainees

The prior attitude of fieldworkers is also something that one needs to take into consideration when designing and implementing training schemes. The trainees often have a more direct approach to teaching, which often contradicts participatory approaches to learning. The relatively low level of fieldworker's natural resource management knowledge also influences the will to assume an important underlying principle of participation – namely, flexibility. An underlying reason for this is probably that handing the leading role to people sometimes amounts to stepping into unpredicted fields of knowledge that are alien to the field worker. This again poses a threat to their authority, which is probably why trainees tended to focus more on the underlying subject than on the communication process, at least during the early stages of the training programme.

Lessons learned

Selection and levelling of content

The complexity and wide-ranging content of the project's training programme has proven to be a challenge. The trained team was comprised of staff from different levels and different backgrounds. Some had prior video production training and computer skills, some had experience working with communities, but few came to the training with a mix of the two. These two groups expressed different training needs some time after the training: 'practise-oriented' staff wanted to further strengthen their community organization skills and general ecological knowledge, whereas technology-oriented staff clearly expressed the need to learn more about the use of new software or about the technological part of development communication. Training needs appeared to be related to what team members already knew. There was a tendency to want to strengthen existing knowledge and to specialize, rather than to adopt the wide range of skills that a participatory development approach requires of the facilitator.

There is a need to adapt the curriculum to local conditions and to adjust the amount of information to be provided to intended learners accordingly. It is also necessary to assess the usefulness of high-tech solutions. As an example, team members now wish to learn how to use different video editing software since a local TV station is relying on them to provide material according to a certain format. A thorough assessment of the prior knowledge of the trainees and the framework for their work must be used to shape a training programme.

Allowing for practice

The high-tech equipment provided by the project is expensive, and people tend to treat it with care – or don't use it at all. If people are to become effective users of tools such as digital cameras or desktop publishing devices, they have to practise these techniques on a regular basis. This is particularly true in the provincial office, where the communication team is waiting for additional training before they put the tools to use. It is therefore important to emphasize that one should not sit and wait for opportunities or actual tasks to put these important tools to use, but to continually practise. In this way, skills are strengthened and can be ready when the need arises. It is also better to have a worn-out camera that many people can use, than to have an unused camera that no one knows how to use effectively.

Sustainability

In order for new knowledge and skills to become part of local capacity, there must be practical opportunities, as well as personal and cooperative adaptation. In this case, most targeted trainees were from central-level government units, whose aim was to create central-level capacity and to ensure the institutionalization and visibility of project activities. However, the staff had few opportunities to gain field-level experience unless with external support since operational budgets for travel are commonly lacking in government institutions. Consequently, central-level staff members engaged in TV programme production and computer-based work, while the necessary conditions for the field aspects of PDC to develop into practical skills were lacking. At a provincial level, there is never a lack of opportunities to work with communities and to practise the field aspects of the training programme; but the trainees have expressed the need for greater post-training support at the implementation stage, a lack of which is preventing them from further using the high-tech skills developed by the project, such as digital media production. One solution is to train staff at different levels on complementary skills and to have them work on a participatory, technical or cooperative basis depending upon their level of government. A second option is to create opportunities for field staff to practise technical skills and *vice versa* in order to lay the basis for cost-recovery schemes, thus facilitating independent field travel or backstopping in the future.

Implementing Isang Bagsak: A Window to the World for the Custodians of the Philippine Forests

Theresa H. Velasco, Luningning A. Matulac and Vicenta P. de Guzman

Isang Bagsak South-East Asia is an enabling programme in more ways than one. The implementation of this networking programme, through its basic strategy of participatory development communication (PDC), is giving rise to experiences and insights on involving the community in development work through communication. More significantly, this involvement hinges on the use of information and communication technology – a 21st-century

tool whose applications for development are currently gaining ground in developing countries.

The programme is being implemented in the Philippines by the College of Development Communication (CDC) of the University of the Philippines at Los Baños and its partner in the first cycle, Tanggapang Panligal Para sa Katutubong Pilipino, or the Legal Assistance Centre for Indigenous Filipinos (PANLIPI). The decision regarding the Philippine partner for the Isang Bagsak learning network was not an easy one to make. There were many mainstream organizations with vast experience in natural resource management. Resource constraints also figured prominently on the choice of partner.

In the end, the CDC posed one question that made all the difference: who would benefit most from the Isang Bagsak experience? The following rhetorical question also tipped the balance in favour of PANLIPI: why not share the learning from Isang Bagsak directly with the indigenous peoples, the custodians of Philippine forests?

According to the Indigenous Peoples' Rights Act of the Philippines, indigenous peoples refer to a group of people or homogeneous societies identified by self-ascription or ascription by others who have continuously lived as organized communities in a communally bounded and defined territory and who have, under claims of ownership, since time immemorial, occupied, possessed and utilized such territories sharing common bonds of language, customs, tradition and other distinctive cultural traits. Indigenous peoples have, through political, social and cultural inroads of non-indigenous religions and cultures, become culturally differentiated from the majority of Filipinos. Indigenous peoples will include people who are regarded as indigenous on account of their descent from populations who occupied the country before colonization and who retain some or all of their own socio-cultural or political institutions.

A showcase of synergy

PANLIPI is an organization of lawyers and advocates of indigenous peoples' concerns. Established in 1985, PANLIPI primarily aims to assist the indigenous peoples in their struggle for the recognition of their rights to their ancestral domains, culture and traditions, and other basic rights. The end goal is to empower the indigenous peoples to the fullest so that they can actively participate in every aspect of Philippine society. PANLIPI's mandate includes the provision of development legal assistance; legal education and outreach; institutional capacity-building; ancestral domains delineation; and resource management planning.

The networking among Isang Bagsak, CDC and PANLIPI may be viewed as a synergistic one right from the start. From PANLIPI's viewpoint, the promotion of the rights of indigenous peoples to their ancestral domains and the issue of natural resource management go hand in hand. The indigenous peoples are, after all, the best caretakers or stewards of the natural resources mainly because their socio-economic system is sustainable and not destructive.

For its part, CDC found in PANLIPI a partner who could best make use of the very tenets of development communication – that is, the use of communication towards improved socio-economic growth of a community that makes for social equity and the larger unfolding of individual potential. Through Isang Bagsak, the indigenous peoples of the Philippines are given their window to the world of learning and are sharing their own knowledge, insights and experiences with members of the network from various parts of the globe. Likewise, the other members of the network are allowed access to the rich heritage of peoples hitherto unwired to the Isang Bagsak network.

Lessons learned

Working with PANLIPI is, indeed, a work in progress that is bringing forth a lot of valuable insights for researchers and development workers in general. The introductory stage of the partnership was already fraught with lessons, foremost of which were the ethical considerations. For one, is it ethical to bring in the indigenous peoples to a programme that is quite alien to them in terms of the tools (information technology) to be used? Second, will they agree to become part of the undertaking, and how do we engage their participation?

Mutual respect and trust

There was conscious effort on the part of the CDC to show respect and to gain the indigenous peoples' trust at the start of the project. Before any agreement was inked, the project team, with PANLIPI's help, made a point of observing two things:

- understanding the cultural context; and
- securing the indigenous peoples' free and prior informed consent.

Henceforth, a series of consultations with the indigenous peoples who would be involved in the learning network was undertaken. The CDC team, together with the PANLIPI lawyers, literally sailed, crossed rivers and hiked through mountains to dialogue with the elders of the cultural communities. This was part of securing their free and prior informed consent, in keeping with the PANLIPI protocol and the nature of participatory development communication. These are two critically important values that many development workers seem to be taking for granted, but which the Isang Bagsak experience is bringing to the fore in this undertaking.

Genuine dialogue between and among the CDC, PANLIPI and the indigenous peoples paved the way for the participation of people from four areas: Mangyans from Mindoro; Tagbanuas from Palawan; Kankanaeys from the Cordilleras; and Aetas from Zambales. Iterative consultations leading to consensus-building were held. First, the CDC team conferred with the PANLIPI staff and lawyers assigned to the areas. Once PANLIPI's cooperation was ensured, dialogues were held with indigenous elders and leaders, and eventually with the members of the community.

The series of dialogues highlighted the following points:

- Knowledge sharing: Isang Bagsak offered an opportunity for the PANLIPI network and the indigenous peoples in the four areas to not only learn from others and from one another about natural resource management and PDC, but to contribute their own knowledge to the network.
- Assurance that the indigenous knowledge systems (IKS) would be respected: the indigenous peoples voiced a strong concern about the possible piracy of their traditional knowledge, as was their experience in the past. They were wary of outsiders 'stealing' their indigenous knowledge in the guise of development work. Prior experience of being exploited by researchers from outside (foreign and local) was a big stumbling block to securing the indigenous peoples' free and prior informed consent. Team members took pains to explain that nothing that did not come from the people, through their representatives in the programme, would be posted on the website. This was, in essence, a manifestation of respect for intellectual property rights.

Relevance to indigenous peoples' concerns

The indigenous peoples' decision to participate in Isang Bagsak rested largely on the realization that the skills they would develop would be useful to the concerns that they were advocating, such as the Ancestral Domains Sustainable Development and Protection Plan and other important provisions of the newly passed Indigenous Peoples' Rights Act.

PANLIPI participants saw the Isang Bagsak experience as an excellent opportunity to interact with professionals from other countries. Through the sharing of experiences, one important insight surfaced: the affirmation that all along, PANLIPI and the indigenous peoples have been practising the principles of development communication, in general, and those of PDC, in particular. The training that the PANLIPI participants have undergone in the course of their participation in Isang Bagsak has also afforded them the opportunity to write and communicate their ideas clearly.

Information technology for indigenous peoples

Technology could be an awesome development for indigenous peoples, only very few of whom have been initially exposed to computers, much less to the internet. Part of their initial reluctance to be part of Isang Bagsak was apprehension about the technology itself. Connectivity/accessibility also loomed as a potential problem.

To address these concerns, PANLIPI, for its part, has begun sponsoring a series of training sessions on information technology for the indigenous leaders involved in Isang Bagsak. CDC's information technology team trained

Iraya Mangyan leaders in Mindoro during February 2004. A user-friendly sourcebook on using the computer and the internet was developed by the CDC especially for this training. The sourcebook, in turn, would be useful in future information technology training for indigenous peoples. In addition, PANLIPI has been very supportive in enabling the indigenous peoples to have access to computers and the internet.

CDC and PANLIPI team members agree that one of Isang Bagsak's 'by-products' is the gradual mainstreaming of the indigenous peoples into the digital age.

PDC through the net: Teamwork at its best

By the time the project was in full swing, the PANLIPI–Isang Bagsak network had 45 members, 9 each from the 5 indigenous groups. A network within a network was created by the partners. As the saying goes: *together, everyone achieves more.*

The members were selected by the indigenous communities themselves on the basis of their knowledge and skills on natural resource management issues. Other criteria were also included in the list of qualifications. Prospective team members had to be knowledgeable about natural resource management and indigenous knowledge systems and practices. They had to be capable of facilitating community participation. Finally, they were also expected to understand development work and to be knowledgeable and skilled in promoting a rights-based approach to development.

The resulting team was dubbed *Limang Tagupak*, a Mangyan term representing victory. The name also signifies a five-point star, representing the configuration of the five-team network at the PANLIPI level. The star is also symbolic of a guide.

The sharing and learning process on PDC in natural resource management was carried out in the following manner:

- Local teams undergo orientation and training on Isang Bagsak.
- Local teams participate in the forum:
  - The CDC posts theme questions.
  - PANLIPI National Capital Region translates theme questions and sends them to local teams.
  - Local teams discuss the theme questions, formulate answers and send them to PANLIPI National Capital Region.
  - PANLIPI National Capital Region translates answers and posts reply.
  - PANLIPI National Capital Region downloads and translates country comments and sends to local teams.
- Local teams discuss country comments and make their own reflections, which they send to PANLIPI National Capital Region.
- PANLIPI National Capital Region translates and posts reflections.
- The CDC resource person synthesizes.
- PANLIPI National Capital Region translates the synthesis and sends it to the local teams.
- Local teams discuss the synthesis and extract the learning.

The process appears to be a tedious one. True enough, a number of problems hampered the implementation of Isang Bagsak at the PANLIPI level. For one, translations into English took up quite some time; hence, delays in posting were inevitable. Delays were also due to the team members' tight work schedules or their preoccupation with economic activities, leaving them less time to participate in the forum. Access to the internet (lack of equipment, telephone lines, internet terminals/connection) constituted the hardware part of the limitations. Financial costs of consultations and regular meetings, as well as internet rental, posed very real problems as well. So did forces of nature, such as bad weather preventing team members from meeting.

PANLIPI's response to the resource-related problems mirrored its commitment to the partnership. PANLIPI put up counterpart funds for training indigenous peoples on basic computer use and internet navigation, as well as hardware for internet accessibility. It forged cooperative undertakings with non-governmental organizations (NGOs) and other friends to boost its investment in equipment and internet access. To facilitate meetings among the team members, PANLIPI made arrangements for Isang Bagsak activities to coincide with regular meetings of indigenous peoples' organizations.

The foregoing are just four of the insights arising from several months' implementation of Isang Bagsak in the Philippines. Many more could be drawn from the researchers' and the indigenous peoples' experiences as they go through the cycle in the remaining months. Isang Bagsak South-East Asia is a work in progress.

VI

Conclusion

Facilitating Participatory Group Processes: Reflections on the Participatory Development Communication Experiments

Chin Saik Yoon

The case studies in this book are a special collection of attempts aimed at putting people first in the development process. They are challenging but very worthwhile efforts intended to find ways of using development communication

tools and methods in decidedly participatory ways that advance the collective priorities of communities as determined by themselves.

These experiments are grounded in issues related to natural resource management. This is a sector for which participatory approaches seem most apt. Natural resource management is unlikely to be effective without the involvement and support of the communities. Natural resources such as land, water, air and forests are vested in the 'commons' – spaces and locations that are open to all people and often protected by nobody – belonging to everyone, yet cared for by none.

Successfully facilitated participatory approaches enable communities to assume collective ownership of the commons and to manage them in a way that safeguards the people's long-term interests. In the absence of such participation, the depletion of natural resources is hastened as people rush to extract maximum use and benefit of these resources before others do the same. Participatory approaches may help to replace destructive competition with sensible cooperation to preserve rather than plunder resources.

The triggering of such cooperation, in turn, rests on our abilities to strengthen and activate our 'internal commons'. These invisible commons comprise our cultures, values and sense of community. They are the elements that ultimately inspire people to set aside desires for selfish short-term benefits in return for the long-term security of the community. Degradation of our internal commons will lead to the destruction of the physical commons and *vice versa*.

Focusing on facilitation

Natural resource management is a challenge to most existing development-promoting strategies because they have largely been designed to advance short-term interests. Such promotional and marketing techniques understandably fail when the ultimate objective is for people to consume less and to set aside their self-interests and preferences for the greater good of the community, which in turn ensures the sustainability of future generations.

The participatory method attempted in most case studies reviewed in this book is participatory development communication (PDC). The opening chapter of this book by Guy Bessette, 'Facilitating Dialogue, Learning and Participation in Natural Resource Management' details the multifaceted work of communicators supporting community participation. It also provides a wide-ranging review of the tools, methods and strategies that the case studies in this book applied. The overarching role of the PDC practitioner is of a facilitator of participatory group processes that serve communities in building awareness and reaching consensus. This concluding chapter sets out to discuss and identify the group dynamics that the PDC practitioner facilitates.

The focus on facilitating participatory group processes rather than implementing one-way communication strategies is what sets PDC apart from classical communication for development (ComDev). In the latter, practitioners apply an integrated set of strategies and tools to mobilize people to meet a set of predefined development goals.

In PDC, facilitators align themselves with a community and help to organize activities and facilitate processes that bring people together to strengthen their sense of community, sharpen their awareness of shared aspirations and problems, and undertake collective action to realize what they aspire to be. The collective approach builds confidence in people, while at the same time making their action more effective through pooling resources and sharing of risks.

Because social communication is one of the principal elements that bind communities, PDC practitioners are able to enhance the usefulness of their facilitating role with their expertise in communication. In helping communities to build their capacity to communicate, they not only strengthen the ability of communities to organize themselves, but also to reach out to others within and outside their communities.

Facilitating key processes

Some of the key communication and group processes that PDC practitioners focus on facilitating are summarized in Box 6.1.

The processes that PDC practitioners concentrate on facilitating may be clustered into four categories:

1   communicating effectively;
2   creating knowledge;
3   building communities; and
4   enabling action.

Conventional ComDev efforts usually concentrate on the first and last clusters (communicating effectively and enabling action). PDC covers two additional clusters (creating knowledge and building communities) that aim to self-empower people through augmenting and validating their knowledge of critical issues and subjects that affect their lives, and through forging strong alliances among people, groups and communities so that they can consult and act effectively together in order to address problems and realize aspirations.

It is the facilitation of these two additional sets of group processes that provides PDC with its participatory bias and its potential to support initiatives that are sustainable in the long term. Many development efforts have been undertaken in the past by focusing on effective communication and enabling of action; however, this narrow focus often leaves such efforts vulnerable to eventual failure. They fail because people lack ownership and relevant in-depth knowledge to assume control of activities in the long term and, more importantly, because they lack the sense of a community.

Box 6.1 *Key participatory development communication processes*

Key participatory development communication (PDC) processes include the following:

- Effective communication:
  - self-expression;
  - listening;
  - understanding.
- Creating knowledge:
  - sourcing information;
  - tapping indigenous knowledge;
  - processing and validating information;
  - sharing knowledge.
- Building communities:
  - building trust;
  - managing conflict and competition;
  - forging partnerships;
  - reinforcing self-identity;
  - reflecting on the past and present;
  - visioning the future;
  - affirming values;
  - adjusting values;
  - enabling transparency in decision-making;
  - sharing benefits.
- Enabling action:
  - identifying problems;
  - evolving solutions;
  - nurturing a sense of guardianship of the commons;
  - managing expectations;
  - taking stock and pooling resources;
  - sourcing complementary resources;
  - advocating to stakeholders;
  - mobilizing for action;
  - evaluating action;
  - iterating and refining action.

Creating knowledge

Although ComDev efforts often devote quality attention to the dissemination of information, knowledge is sometimes not created among the people who are targeted with such information. Dissemination efforts also rarely take into account the rich knowledge base resident within communities. This is especially significant in natural resource management where indigenous knowledge is just as important as scientific information. Indigenous knowledge is distilled over generations through people's close observation of their environment and is therefore particularly valid for the long time spans involved in managing natural resources.

N'Golo Diarra from Mali retells his unforgettable encounter with the 'governess of the seasons' of a distant village in his chapter 'The Old Woman and the Martins: Participatory Communication and Local Knowledge in Mali' in Part III. This woman had inherited an important piece of indigenous knowledge from her father, the great traditional healer of the village. He had observed that the purple martins build their nests on tree branches that were always above the floodwaters of the river that ran by the village. He shared this critical piece of knowledge with his daughter just before he died. Armed with this precious element of indigenous knowledge, she became the much-revered governess of the seasons, helping farmers in her village to decide where and when to plant their crops, as well as where to pasture their livestock during the coming season.

PDC attempts to facilitate the fusing of indigenous knowledge with scientific information. So, instead of starting with scientific information, PDC approaches set out to empower people by validating indigenous knowledge, which they know intimately, before progressing to the introduction of scientific information. Participatory approaches adopted in filtering, processing and applying new and existing information help to create knowledge that allows people to take charge and make decisions.

The importance of processes that create knowledge is best appreciated in the chaos–wisdom continuum:

$$\text{Chaos} \rightarrow \text{Data} \rightarrow \text{Information} \rightarrow \text{Knowledge} \rightarrow \text{Wisdom}$$

The continuum begins with *chaos,* fragmented and disorganized data that is not of use to people. *Data* may be clusters of numbers and visual observations that have been processed and made ready for use. *Information* refers to data that has been organized into meaningful chunks that provide meaning to people. Information becomes *knowledge* when it has been successfully communicated to and understood by the people. 'Knowledge is the product of information plus thought and ideas. It implies a value judgement because knowledge marks the processing by a human of useful and relevant information' (Green et al, 2005). *Wisdom* occurs when knowledge is used to make sound judgements.

The strategies often adopted by conventional communication approaches focus just on the dissemination of data and information to people. Frequently absent are the participatory group processes that permit people to convert these raw inputs into useful knowledge that they can use as a community.

Building communities

The sense of community varies across communities. It may be very strong among rural communities which are bound by cultural, religious and social ties forged over numerous generations. Or it may be weak, as among landless communities comprising a high percentage of transient members.

PDC approaches try to tap into existing community processes whenever possible in order to strengthen rather than undermine communities. Facilitators will therefore adopt traditional and folk media, and communicate via existing channels and networks to which they have access.

In the case of groups with weak ties, quality efforts need to be devoted to bringing people together to forge a sense of community. This is often attempted in conjunction with efforts to create knowledge. Here, information and data are shared, and people are helped to process the information and data in groups so that they not only create knowledge for themselves but discover, at the same time, the many similar experiences and problems that they share, building a sense of community in the process (see the following section on 'Evolving participation in West Java' for a case study of this approach).

In other cases, stress and conflict created by competition for depleted supplies of natural resources can seriously test social and community ties. Karidia Sanon and Souleymane Ouattara from Burkina Faso provided us with such a case from the Nakambé Basin of Burkina (see 'Water: A Source of Conflict, a Source of Cohesion in Burkina Faso' in Part III). The areas surrounding the wells in Silmiougou, a village in the heart of Burkina, had degenerated into 'boxing rings' where women would compete each day for limited access to water pumps and water. The competition had become so fierce that fights broke out among the women every day, when they would smash each other's water jars. The authors reflect on how the PDC team was able to help the community to find resolution to the conflict via carefully facilitated chat sessions, which were primed with PDC tools, such as forum theatres, participatory video and radio, and posters.

Tools and methods

ComDev's previous emphasis on effective communication and enabling action has led to more tools and methods being developed in these two areas. This is apparent in the cases presented in this book where we see well-tested communication strategies being competently applied.

The potential of PDC becomes apparent when practitioners go beyond these strategies and begin to innovate tools and methods that support and facilitate people in the other two areas: creating knowledge and building communities. PDC practitioners have considered various group processes in their search for appropriate tools. Participatory media methods, participatory rural appraisal (PRA) (Chambers, 1994) and Visualization in Participatory Programmes (UNICEF Bangladesh, 1993) have all contributed effective techniques and methods for PDC.

The experimentation with novel tools for facilitating participatory communication processes undertaken within the cases reviewed in this book is significant not only for the efforts of the PDC practitioners, but more importantly for the active participation of the thousands of people from the villages and communities where these cases took place. The 'new' tools, techniques and methods reported in this book are therefore as much the innovations of the villagers and development fieldworkers as they are of the PDC practitioners who wrote up the cases.

Evolving participation in West Java

The case reported by Amri Jahi in Indonesia (see 'From Resource-Poor Users to Natural Resource Managers: A Case from West Java', in Part III) is important partly because it is the longest running development project studied in this book. Stretching over more than 15 years, the initiative was sustained during much of this time by the communities and the PDC team on their own. More significantly, it began in Indonesia during an authoritative regime when participation was not encouraged and the development model was dominated by top-down, expert-to-farmer initiatives. Although the case reads very much like a classical ComDev project, starting with a traditional agricultural extension intervention, it evolved gradually and subtly over a decade and a half into a vibrant PDC initiative managed entirely by the landless goat herders who succeeded in engaging constructively with stakeholders in various government agencies. The natural resource management problem of eroding riverbanks that had previously set landless communities in conflict with the authorities has been amicably resolved to the interest of everyone. The future of the river valley is secure and so is the livelihood of the goat herding communities.

The switch over to PDC began with the involvement of farmer leaders and farmer cooperators in monthly meetings to share information on how to care for the sheep. These meetings aimed not just to disseminate information, but also to build a strong knowledge base within the community. The meetings also served a second but more important goal of building a sense and purpose of community. An immediate task before the community was the distribution of lambs that were returned to the community in fulfilment of earlier agreements by farmers to repay their loans of breeding stock in the form of the offspring of their animals.

The project team discovered one of the most effective PDC methods early on in its work when it brought together farmer leaders to share their experiences with farmers from other villages. The farmer-to-farmer communication proved to be more effective than earlier methods attempted by the researchers and extension workers. The participatory approach was proved to be sound.

Challenges continue to face the community 15 years on. The PDC facilitators have noticed that the quality of the lambs returned to the community has declined recently. This threatens the sustainability of the revolving stock of young animals that enable other landless villagers to take part in the scheme.

Participatory videos and media in Lebanon and Uganda

Participatory video is one of the earliest PDC tools. It was developed on the Fogo Islands of Canada around 1967 by Don Snowden, then director of the Extension Department at Memorial University in Newfoundland. This method came to be known as the Fogo Process (Williamson, 1991) and has since been adapted and used around the world. In the case contributed by Shadi Hamadeh and his colleagues, we see the Lebanese application of this method to manage conflict and to provide women with a medium for empowerment (see 'Goats, Cherry Trees and Videotapes: Participatory Development Communication for Natural Resource Management in Semi-Arid Lebanon', in Part III).

The project in Lebanon relies mainly on interpersonal methods in its work. Additional PDC tools were used when interpersonal approaches turned out to be unsuitable. This was the case when people who were in conflict refused face-to-face meetings as a means of resolving their differences. The facilitators decided to interview the people involved in the conflict separately in order to obtain their perspectives on the issues. A video comprising the individual interviews was then screened to a meeting of all parties involved in the conflict. The video succeeded in restarting stalled face-to-face discussions on solutions to the people's differences. The meeting itself was videotaped. The new footage of the meeting, together with the individual interviews, was made into a new tape and was screened at another meeting involving others from the community in order to expand discussions on resolving the conflict.

The Lebanese facilitators then produced a video that featured interviews with women highlighting their economic productivity in a pastoral society. The video was shown to both men and women. The women who were featured in the video felt empowered after the viewing. This is the 'mirroring' effect discovered in the Fogo Process. The video had served as a mirror of their community demonstrating and recognizing the women's role within it.

Although the Fogo Process started as a 16mm documentary production before evolving into video, the process can also be applied to a whole range of modern media, ranging from radio to the internet. This is a promising area for research and development among PDC practitioners in the years ahead.

The case from Uganda contributed by Nora Naiboka Odoi features multiple adaptations of the Fogo Process, but using still cameras, brochures and posters in addition to video (see 'Growing Bananas in Uganda: Reaping the Fruit of Participatory Development Communication', in Part III and 'Communication Tools in the Hands of Ugandan Farmers', in Part IV). The iterative processes followed in developing the video and print material may look confused at first reading, but they served the very important purpose of processing information to create knowledge about banana cultivation, in addition to PDC materials production.

Community radio in Ghana

Kofi Larweh's case about Radio Ada, Ghana's first community radio station, describes how the medium taps into the oral traditions of the community served by the station (see 'And Our "Perk" Was a Crocodile: Radio Ada and Participatory Natural Resource Management in Obane, Ghana', in Part IV).

In community broadcasting, the listeners determine the content that is put on air and take turns in producing the broadcasts. This is an adaptation of the Fogo Process to the medium. Radio Ada staff see themselves as facilitators: they record many of the programmes in the villages, rather than in the studios. The villagers decide which topics they want to put on tape and the way in which they are presented to the listeners.

This case discusses the role played by the station in people's efforts to dredge a clogged 10km long river. It was a daunting task. Radio Ada first provided people with a medium to mobilize others living along the river to join in the huge efforts of dredging the waterway. The station then became a source of encouragement as the back-breaking work progressed. The names of the villagers who took part in the dredging were announced over the air, thereby recognizing their contributions, while at the same time encouraging others to pitch in.

The community succeeded after four years of hard work and triumphantly reopened the waterway to boats and fishing after nearly 40 years of neglect. Radio Ada had stuck with the people through these years. The unclogged river could now also channel water to the irrigation canals of riverside farms, and crops could once again thrive. The people had by now discovered the power of radio and began using it as an advocacy tool, raising issues about the environment and their livelihood over the air. They had not only rediscovered a waterway, but a communication channel as well.

Debate theatre

Diaboado Jacques Thiamobiga's chapter reports on one PDC tool developed to overcome a challenge found in some parts of Burkina Faso where women tend not to speak in public due to cultural restrictions (see 'Burkina Faso: When Farm Wives Take to the Stage', in Part IV). This difficult aspect of the communities' tradition became a major obstacle when actions to curb farmland erosion and desertification were initiated in the villages of Toukoro and Badara. Given the key role that women play in agriculture, the project was primarily aimed at working with peasant women. However, it soon became clear that another tradition that had not yet been discussed openly was hindering the full participation of women in these environmental efforts. Indeed, in this part of Burkina Faso, women are not entitled to landownership and are only granted the temporary right to grow food crops in some fields, which are usually the least productive. This makes it pointless for them to invest time and resources in upgrading a piece of land that will most likely be taken away from them once they become more productive, thus affecting the whole community. Since they were not allowed to speak up, they faced a big communication problem on top of all of the other problems.

The women of Badara were then reminded of a traditional ceremony where women are allowed to dress up as men and publicly talk about issues in a forthright manner. The women decided to exercise this right and to speak up in the form of debate-theatre performances.

The PDC team worked closely with the women in researching the messages to be presented, and helped them to prepare for their performances. When the play was ready, the women gave five performances: two in Badara, and one each in Toukoro, Tondogosso and Dou. Every performance was a big hit. It turned out to be fun for the women when they got to perform in their own village. They astonished the audience when they made their entrance on stage. Members of the audience were pleasantly surprised to see their mothers, sisters or wives dressed up as men. Many husbands recognized their wives through their gestures and their manner of speaking. In fact, each performance sparked a big celebration in the villages. The debate-theatre performances served, more importantly, as learning experiences for everyone: men, women and children. People exchanged viewpoints about village life; they discussed male–female relations and landownership; and they shared technical information for improving and preserving soil fertility. The unthinkable happened: the women succeeded in giving technical advice on soil fertility to the men. At the same time, the women found it an empowering experience during which they enhanced their own self-esteem and proved to the men that they were able to discuss their communities' development problems effectively in public and to propose useful solutions and courses of action.

Multiple methods along the Lisungwi, Mwanza and Mkulumadzi rivers

The case contributed by Meya Kalindekafe in Malawi made use of an interesting combination of participatory methods (see 'Participatory Research and Water Resource Management: Implementing the Communicative Catchment Approach in Malawi', in Part III). Meya's team began with the familiar research and development steps of literature review, training, surveys using semi-structured questionnaires, focus group discussions, key informant interviews and the Harvard Analytical Framework. It then proceeded to use more participatory methods involving members of the riverside communities in preparing resource maps and benefit–analysis charts. The team also went on transect walks with members of the communities, a method used frequently by PRA practitioners.

The work in Malawi is an example of the direction that many PDC teams will take, in the future, of carefully selecting tools and methods from across a number of disciplines for application within one initiative. Communication researchers have adopted multidisciplinary approaches after recognizing long ago that the field of communication is inherently multidisciplinary and not just narrowly confined within the boundaries of media tools and methods.

Efforts in the future

PDC is probably the youngest component in the relatively new field of communications. The experiments reported in this book, together with work attempted elsewhere, have proven that the assumptions behind PDC are sound. They also show that PDC approaches do help people to facilitate long-term development efforts on their own terms.

The recent advent of PDC has meant that many tools and methods remain to be fully developed. The complex group dynamics that PDC attempts to facilitate will require PDC practitioners to innovate these new tools and methods. The experiences in this book show that media-based methods are suitable for a good number of the processes that practitioners need to engage with. These experiences also demonstrate that many of the processes are interpersonal in nature or involve participatory group processes that seem best served by interpersonal or group communication techniques. This indicates the need for a range of interpersonal communication techniques and group facilitation methods to be developed or acquired by PDC practitioners. People-embodied methods are some of the most challenging to develop and to share; an effective facilitator of participatory group processes is easy to appreciate, but very difficult to emulate. Table 6.1 summarizes the availability or lack of various tools and methods for the different processes that PDC needs to be equipped with in order to support or facilitate.

Table 6.1 shows that many of the missing or emerging tools and methods are clustered around the two areas of creating knowledge and building communities. This reflects the novelty of PDC within these two areas, not previously highlighted by the communication field.

PDC probably has much to learn from the field of education in the area of creating knowledge. The possibilities of interdisciplinary efforts in this area are numerous and are potentially of mutual benefit to both the education and communications sectors.

The area of building community also stands to benefit from the experiences of work conducted in organizational communication; community organizing; cultural studies; ethnography; indigenous knowledge; studies of power structures; negotiation and mediation; and psychology.

Conclusions

Nora Cruz Quebral pointed out in her chapter at the opening of this book ('Participatory Development Communication: An Asian Perspective', Part II) that communications theories and models have not replaced each other, but instead coexist productively. This coexistence is due to the different roles that communication plays in human and community interactions. Each theory and model serves a particular set of interactions and sets out to accomplish a different goal. PDC's role in natural resource management and sustainable development seems most promising from the set of case studies published in this book. But it is not an easy role to play. It carries with it all the trials and tribulations of the performing arts, where actors must spend long years of arduous apprenticeship before they are ready to perform on stage. Even then, the challenge changes from one performance to another as the mood and composition of the people in the theatre alter. It is not cold science: people rest at the core of what is attempted.

Table 6.1 *Availability of tools and methods for participatory development communication processes*

Processes	Availability of tools and methods		
	Exists	Being developed	Largely absent
Effective communication			
Self-expression	√		
Listening and understanding		√	
Creating knowledge			
Sourcing information	√		
Tapping indigenous knowledge		√	
Processing and validating information		√	
Sharing knowledge			√
Building communities			
Building trust		√	
Managing conflict and competition			√
Forging partnerships		√	
Reinforcing self-identity			√
Reflecting on the past and present	√		
Visioning the future	√		
Affirming values			√
Adjusting values			√
Enabling transparency in decision-making			√
Sharing benefits		√	
Enabling action			
Identifying problems	√		
Evolving solutions	√		
Nurturing a sense of guardianship of the commons		√	
Managing expectations		√	
Taking stock and pooling resources	√		
Sourcing complementary resources	√		
Advocating to stakeholders	√		
Mobilizing for action	√		
Evaluating action	√		
Iterating and refining action	√		

PDC, however, is very different from the performing arts in many other aspects. The effort is not over between two predetermined curtain calls. PDC takes time, stretching more than 15 years in one case reported here. It also needs to be underpinned by a set of very clear principles and values.

While PDC is very much like other areas of communication in terms of the media and tools with which it works, it is radically different in its philosophy. So, while most areas of communication require their practitioners to articulate messages lucidly, PDC often requires us to be silent and to listen carefully.

References

Chambers, R. (1994) 'The origins and practice of participatory rural appraisal', *World Development*, vol 22, no 7, pp953–969

Green, L., Lallana, E. C., Shafiee, M. and Nain, Z. (2005) 'Social, political and cultural aspects of ICT: E-governance, popular participation and international politics', in Chin, S. Y. (ed) *Digital Review of Asia Pacific*, Southbound, Penang

Stonier, T. (1990) *Information and the Internal Structure of the Universe: An Exploration into Information Physics*, Springer, London

UNICEF (United Nations Children's Fund) Bangladesh (1993) *Visualization in Participatory Programmes: A Manual for Facilitators and Trainers Involved in Participatory Group Events*, UNICEF, Dhaka

Williamson, H. A. (1991) 'The Fogo Process: Development support communications in Canada and the developing world', in Casmir, F. L. (ed) *Communication in Development*, Ablex Publishing Corporation, Norwood

Selected Readings

C. V. Rajasunderam

This bibliography is divided into four parts:

1 introducing communication as a tool to improve community participation in participatory natural resource management (NRM) research;
2 communication for community participation in planning participatory NRM research;
3 communication for community participation in implementing and monitoring participatory NRM research;
4 communication and sharing knowledge.

I: Introducing communication as a tool to improve community participation in participatory natural resource management (NRM) research

Arnst, R. (1996) 'Participatory approaches to the research process', in Servaes, J. et al (eds) *Participatory Communication for Social Change*, Sage Publications, New Delhi, and Thousand Oaks, London, pp109–126
- *Content*: two major approaches to participatory communication are discussed in this contribution.
Belbase, S. (1994) 'Participatory communication in development: How can we achieve it', in White, A. et al (eds) *Participatory Communication: Working for Change and Development*, Sage Publications, New Delhi, and Thousand Oaks, London, pp446–461
- *Content:* describes the components of a participatory communication project with particular reference to a participatory development communication (PDC) project in Asia.
Cohen, S. I. (1996) 'Mobilizing communities for participation and empowerment', in Servaes, J. et al (eds) *Participatory Communication for Social Change*, Sage Publications, New Delhi, and Thousand Oaks, London, pp223–248
- *Content*: the focus is on participatory communication for community empowerment.
Diegues, S. A. C. (1992) 'Sustainable development and people's participation in wetland ecosystem conservation in Brazil: Two comparative case studies', in Ghai, D. and Vivian, M. (eds) *Grassroots Environmental Action – People's Participation in Sustainable Development*, Routledge, London and New York, pp141–158

- *Content*: this chapter documents the traditional resource management practices of two Brazilian communities and explores the ecological and social impacts of the disruption of those practices by state-supported schemes for increased levels of resource exploitation.

Egger, P. and Majeres, J. (1992) 'Local resource management and development: Strategic dimensions of people's participation', in Ghai, D. and Vivian, M. (eds) *Grassroots Environmental Action – People's Participation in Sustainable Development*, Routledge, London and New York, pp304–324

- *Content*: this chapter draws on a wide range of experiences to provide general lessons on participation in community-level sustainable development projects that are supported and/or initiated by external agencies. The example points to a number of strategic dimensions regarding the interaction between people, resource management and development.

Fraser, C. and Villet, J. (1994) 'Communication in practice', in Fraser, C. and Villet, J. (eds) *Communication: A Key to Human Development*, United Nations Food and Agriculture Organization, Rome, pp8–23

- *Content*: covers better planning and programme formulation; people's participation and communication mobilizing; changing lifestyles; improved training; rapid spread of information; effective management and coordination; generating the support of decision-makers.

Ingles, A. et al (1999) *The Participatory Process for Supporting Collaborative Management of Natural Resources: An Overview*, United Nations Food and Agriculture Organization, Rome

- *Content*: the overview to this book describes the extent and nature of participation in the collaborative management of natural resources. The processes and practical aspects of promoting collaborative management of natural resources are also discussed. This book is part of a new set of materials on participatory processes currently being developed by the Community Forestry Unit of the United Nations Food and Agriculture Organization (FAO).

Kennedy, T. (1989) 'Community animation – An open-ended process', *Communication for Community, WACC Media Development* 3/1989, World Association of Christian Communication (WACC), London, pp5–7

- *Content*: focuses on the variety of applications of the community animation approach as an alternative to the public hearing process, participatory research, organizational development, conflict resolution, urban renewal development and development communication.

Narayan, D. (1995) 'The contribution of people's participation: Evidence from 121 rural water supply projects', *Environmentally Sustainable Development Occasional Paper Series No 1*, World Bank, Washington, DC

- *Content*: this study examines efforts to induce participation as a means of creating effective rural water systems and building the local capacity to manage them.

Rahnema, M. (1992) 'Participation', in Sachs, W. (ed) *The Development Dictionary: A Guide to Knowledge as Power*, Zed Books, London and New Jersey, pp116–129

- *Content*: a useful contribution on the many dimensions of the participatory process in development with a focus on the philosophic premises underlying the concept.

Saik Yoon, C. (1996) 'Participatory communication for development', in Bessette, G. and Rajasunderam, C. V. (eds) *Participatory Communication for Development: A West African Agenda*, International Development Research Centre, Ottawa, Canada, and Southbound, Penang, pp37–61

- *Content:* a good introduction to the conceptual bases of PDC and the practical aspects of using this approach in field projects.

Servaes, J. and Arnst, R. (1993) 'First things first: Participatory communication for change', *Media Development Journal of the World Association of Christian Communication*, 2/1993, vol XL, pp44–47

- *Content:* focuses on a comparison of the 'mechanistic' and 'organic' models of development communication work.

Slocum, R. and Thomas-Slayter, B. (1995) 'Participation, empowerment and sustainable development', in Slocum, R. et al (eds) *Power, Process and Participation: Tools for Change*, Intermediate Technology Publications, London, pp3–8

- *Content:* provides the conceptual background for participatory research methodologies.

Thomas, P. (1994) 'Participatory development communication: Philosophical premises', in White, A. et al (eds) *Participatory Communication: Working for Change and Development*, Sage Publications, New Delhi, and Thousand Oaks, London, pp49–59

- *Content:* the focus of this contribution is on the potential and limitations of PDC.

Thomas-Slayter, B. (1995) 'A brief history of participatory methodologies', in Slocum, R. et al (eds) *Power, Process and Participation: Tools for Change*, Intermediate Technology Publications, London, pp9–16

- *Content:* a comparison of traditional research approaches with participatory research methods.

Vivian, J. M. (1992) 'Foundations for sustainable development: Participation, empowerment and local resource management', in Ghai, D. and Vivian, M. (eds) *Grassroots Environmental Action – People's Participation in Sustainable Development*, Routledge, London and New York, pp50–77

- *Content:* this contribution demonstrates in specific terms the importance of traditional resource management systems and locally based popular environmental initiatives.

White, S. A. (1994) 'The concept of participation: Transforming rhetoric to reality', in White, A. et al (eds) *Participatory Communication: Working for Change and Development*, Sage Publications, New Delhi, and Thousand Oaks, London, pp15–32

- *Content:* a good introduction to the basic principles of PDC.

II: Communication for community participation in planning participatory NRM research

The focus of this section is on the following themes:

- approaching a local community;
- collecting and sharing information;
- involving the community in identifying NRM problems and solutions.

Allen, W. and Kilvington, M. (2001) *ISKM (Integrated Systems for Knowledge Management)*, Landcare Research, Manaaki Whenua, www.landresearch.co.nz/research/social/iskm.asp

- *Content*: an outline of a participatory approach to environmental research and development initiatives. Managing the constructive involvement of stakeholders in NRM research is a skill. The Integrated Systems for Knowledge Management (ISKM) approach is designed to support such an ongoing process of constructive community dialogue. It provides clear communication pathways to support dialogue and collective action.

Allen, W. et al (2001) 'Using participatory and learning-based approaches for environmental management to help achieve constructive behaviour change, section 5: Concepts for managing participation in practice', Landcare Research, Manaaki Whenua, www.landcareresearch.co.nz/research/social/par_rep5.asp
- *Content:* this article examines specific mechanisms that collectively support an overall framework designed to facilitate behaviour change for environmental management. These mechanisms are social capital; empowering people and communities; levels of participation; managing a participatory process; stakeholder analysis; and participatory monitory and evaluation.

Bergdall, D. (1993a) *Methods for Active Participation – Experiences in Rural Development from East and Central Africa*, Oxford University Press, Nairobi, Kenya
- *Content:* this book comprises the first two parts of the final report on *The Methods for Active Participation Research and Development Project (MAP)*. The MAP project was implemented in Kenya, Tanzania and Zambia from 1988 to 1991.

Bergdall, D. (1993b) 'The MAP facilitator's handbook', in *Methods for Active Participation – Experiences in Rural Development from East and Central Africa*, Oxford University Press, Nairobi, Kenya, pp146–200
- *Content:* contains simple guidelines on the role of facilitators in catalysing community participation in development; techniques for enabling broad participation; factors for creating a participatory environment; and ensuring quality in the planning process.

Bessette, G. (2004) *Involving the Community: A Guide to Participatory Development Communication*, International Development Research Centre, Ottawa, Canada, and Southbound, Penang, Malaysia
- *Content:* this guide is intended for people working in research and development. It introduces PDC concepts, discusses the use of effective two-way communication approaches and presents a methodology to plan, develop and evaluate PDC projects.

Bidol, P. et al (eds) (1986a) 'Working definition of alternative environmental conflict management', in *Alternative Environmental Conflict Management Approaches: A Citizen's Manual*, School of Natural Resources, University of Michigan, Ann Arbor, Michigan, pp17–19
- *Content:* provides practical guidelines on conflict management approaches and related group problem-solving skills.

Bidol, P. et al (eds) (1986b) 'Overview of conflict management', in *Alternative Environmental Conflict Management Approaches: A Citizen's Manual*, School of Natural Resources, University of Michigan, Ann Arbor, Michigan, pp23–60
- *Content:* provides practical guidelines on conflict management approaches and related group problem-solving skills.

Bidol, P. et al (eds) (1986c) 'Effective citizen teams', in *Alternative Environmental Conflict Management Approaches: A Citizen's Manual*, School of Natural Resources, University of Michigan, Ann Arbor, Michigan, pp194–204
- *Content:* provides practical guidelines on conflict management approaches and related group problem-solving skills.

Chandrasekharan, D. (2000a) *Proceedings: Electronic Conference on Addressing Natural Resource Conflicts through Community Forestry*, United Nations Food and Agriculture Organization, Rome

- *Content:* this serves as a good introduction to the different types of conflicts in natural resource management.

Chandrasekharan, D. (2000b) 'Categorisation of conflicts – Annex F', in *Proceedings: Electronic Conference on Addressing Natural Resource Conflicts through Community Forestry*, United Nations Food and Agriculture Organization, Rome, pp137–143

Chandrasekharan, D. (2000c) 'Participatory approaches – Annex G', in *Proceedings: Electronic Conference on Addressing Natural Resource Conflicts through Community Forestry*, United Nations Food and Agriculture Organization, Rome, pp145–150

Chandrasekharan, D. (2000d) 'Definitions – Annex H', in *Proceedings: Electronic Conference on Addressing Natural Resource Conflicts through Community Forestry*, United Nations Food and Agriculture Organization, Rome, pp151–155

Cornwall, A. et al (1993) *Acknowledging Process: Challenges for Agricultural Research and Extension Methodology, Discussion Paper 3,* Institute of Development Studies, Brighton, UK, and International Institute for Environment and Development, London

- *Content:* this discussion paper locates agricultural research and extension practices within wider social processes. Six participatory approaches to research are discussed.

Feldstein, S. and Jiggins, J. (eds) (1994a) 'Participatory methodologies for analysing household activities, resources and benefits', in *Tools For The Field: Methodologies Handbook for Gender Analysis in Agriculture*, Kumarian Press and Intermediate Technology Publications, London, pp36–44

- *Content*: discusses tools for gender analysis in agriculture.

Feldstein, S. and Jiggins, J. (eds) (1994b) 'Workshops for gathering information', in *Tools For The Field: Methodologies Handbook for Gender Analysis in Agriculture,* Kumarian Press and Intermediate Technology Publications, London, pp55–61

- *Content*: strategies for collecting information on the role of women in agricultural initiatives at the community level.

Feldstein, S. and Jiggins, J. (eds) (1994c) 'Using focus groups with rural women', in *Tools For The Field: Methodologies Handbook for Gender Analysis in Agriculture,* Kumarian Press and Intermediate Technology Publications, London, pp62–65

- *Content:* discusses the focus group as a communication tool for catalysing community action.

Feldstein, S. and Jiggins, J. (eds) (1994d) 'Practical considerations for improving gender-based research', in *Tools For The Field: Methodologies Handbook for Gender Analysis in Agriculture,* Kumarian Press and Intermediate Technology Publications, London, pp236–238

- *Content*: useful guidelines for enhancing the effectiveness of research on gender issues in agriculture.

Feldstein, S. and Jiggins, J. (eds) (1994e) 'Women's agricultural production committees and the participative-research action approach', in *Tools For The Field: Methodologies Handbook for Gender Analysis in Agriculture,* Kumarian Press and Intermediate Technology Publications, London, pp239–243

- *Content*: an example of the learning process involved in the participative-research action approach.

Kaner, S. et al (1996a) 'Introduction', in *Facilitator's Guide to Participatory Decision-Making,* New Society Publishers, Philadelphia, pp XIII–XVI

- *Content*: an overview of the participatory decision-making process.

Kaner, S. et al (1996b) 'Participatory values', in *Facilitator's Guide to Participatory Decision-Making,* New Society Publishers, Philadelphia, pp23–29

- *Content*: discusses full participation; mutual understanding; inclusive solutions; and shared responsibility.

Kaner, S. et al (1996c) 'Introduction to the role of the facilitator', in *Facilitator's Guide to Participatory Decision-Making*, New Society Publishers, Philadelphia, pp31–37
- *Content*: outlines the expertise that supports a group to do its best thinking.

Kaner, S. et al (1996d) 'Facilitative listening skills', in *Facilitator's Guide to Participatory Decision-Making*, New Society Publishers, Philadelphia, pp41–53
- *Content*: presents techniques for honouring all points of view.

Kaner, S. et al (1996e) 'Facilitating open discussion', in *Facilitator's Guide to Participatory Decision-Making*, New Society Publishers, Philadelphia, pp55–67
- *Content*: promotes the free-flowing exchange of ideas.

Kaner, S. et al (1996f) 'Dealing with difficult dynamics', in *Facilitator's Guide to Participatory Decision-Making*, New Society Publishers, Philadelphia, pp113–167
- *Content*: discusses how a facilitator can intervene in a context of difficult group dynamics.

Kaner, S. et al (1996g) 'Building a shared framework of understanding', in *Facilitator's Guide to Participatory Decision-Making*, New Society Publishers, Philadelphia, pp169–182
- *Content*: describes the principles and tools that support groups struggling in the service of integration.

Mikkelsen, B. (1995) 'Participation: Concepts and methods', in *Methods for Development Work and Research, A Guide for Practitioners*, Sage Publications, New Delhi, and Thousand Oaks, London, pp61–82
- *Content*: presents useful guidelines on collecting development information by using participatory approaches.

Narayan, D. (1996a) 'Introduction', in *Towards Participatory Research*, World Bank Technical Paper, no 307, World Bank, Washington, DC, pp1–15
- *Content*: discusses why participatory processes in data collection are important.

Narayan, D. (1996b) 'What is Participatory Research?', in *Towards Participatory Research*, World Bank Technical Paper, no 307, World Bank, Washington, DC, pp17–31
- *Content*: discusses the roles of a participatory researcher, and how conventional and participatory research/data collection differ.

Narayan, D. (1996c) 'Defining the purpose of the study', in *Towards Participatory Research*, World Bank Technical Paper, no 307, World Bank, Washington, DC, pp33–45
- *Content*: presents methods for clarifying the research purpose and defining the scope of the study.

Narayan, D. (1996d) 'Choosing data collection methods', in *Towards Participatory Research*, World Bank Technical Paper, no 307, World Bank, Washington, DC, pp59–79
- *Content*: discusses matching methods with information needs.

Narayan, D. (1996e) 'Data analysis dissemination and use', in *Towards Participatory Research*, World Bank Technical Paper, no 307, World Bank, Washington, DC, pp129–138
- *Content*: methods for analysing data and sharing information are discussed in detail.

Norris, J. (1995) *An Introduction to Participatory Action Research Excerpted from the Guide to the Film* From the Field, PAR Trust, Calgary, Alberta
- *Content*: a lucid description of the participatory action research process.

Oltheten, T. M. P. (1995a) 'Major lessons learned from the case studies', in *Participatory Approaches to Planning for Community Forestry: Results and Lessons from Case Studies Conducted in Asia, Africa and Latin America – A Synthesis Report,* United Nations Food and Agriculture Organization, Rome, pp25–36

- *Content*: the focus of this publication is on participatory planning for community forestry as a learning process.

Oltheten, T.M.P. (1995b) 'Conclusions and recommendations', in *Participatory Approaches to Planning for Community Forestry: Results and Lessons from Case Studies Conducted in Asia, Africa and Latin America – A Synthesis Report,* United Nations Food and Agriculture Organization, Rome, pp37–43

Riano, P. (1991) 'Myths of the silenced: Women and grassroots communication', *Media Development,* 2/1991 vol XXXVIII, World Association of Christian Communication (WACC), London, pp20–22

- *Content*: this chapter focuses on the social functions and perceived roles of women in development.

Riano, P. (ed) (1994a) 'Women's participation in communication: Elements for a framework', in *Women in Grassroots Communication: Furthering Social Change,* Thousand Oaks, London, and Sage Publications, New Delhi, pp3–29

Riano, P. (ed) (1994b) 'Process video: Self-reference and social change', in *Women in Grassroots Communication: Furthering Social Change,* Thousand Oaks, London, and Sage Publications, New Delhi, pp131–148

Riano, P. (ed) (1994c) 'The WEDNET Initiative: A sharing experience between researchers and rural women', in *Women in Grassroots Communication: Furthering Social Change,* Thousand Oaks, London, and Sage Publications, New Delhi, pp221–234

- *Content*: focuses on the communicative roles of women at the grassroots level and on the use of participatory approaches for research on environmental issues. WEDNET stands for Women's Environment and Development Network.

Spring, A. (1996) 'Gender and environment: Some methods for extension specialists', in *Training for Agriculture,* United Nations Food and Agriculture Organization, Rome, pp104–122

- *Content*: the focus of this chapter is on the methods and techniques for analysing gender and natural resource management. The examples are taken from Africa, Asia and Latin America.

Srinivasan, L. (1990a) 'Community participation for development', in *Tools for Community Participation: A Manual for Training Trainers in Participatory Techniques,* Prowwess/UNDP, New York, pp15–19

- *Content*: a concise introduction to the concept of community participation.

Srinivasan, L. (1990b) 'Planning a participatory training programme', in *Tools for Community Participation: A Manual for Training Trainers in Participatory Techniques,* Prowwess/UNDP, New York, pp21–30

- *Content*: useful guidelines on planning a training programme in participatory techniques.

Srinivasan, L. (1990c) 'Designing the participatory workshop', in *Tools for Community Participation: A Manual for Training Trainers in Participatory Techniques,* Prowwess/ UNDP, New York, pp35–43

- *Content*: describes the sequence of steps involved in designing a workshop on tools for community participation.

Srinivasan, L. (1990d) 'Simple daily evaluation activities and techniques', in *Tools for Community Participation: A Manual for Training Trainers in Participatory Techniques,* Prowwess/UNDP, New York, pp45–50

- *Content*: outlines participatory methods of evaluating daily training activities.

III: Communication for community participation in implementing and monitoring participatory NRM research

This section focuses on the following themes:

- developing communication strategies;
- using communication tools;
- evaluating and documenting communication activities.

Allen, W. et al (2001) 'Building group capacity for environmental change', in *Using Participatory and Learning-Based Approaches for Environmental Management to Help Achieve Constructive Behaviour Change*, Landcare Research, Manaaki Whenua, www.landcareresearch.co.nz/research/social/par_rep6.asp
- *Content:* the focus of this article is on participatory techniques for working with groups and teams.
Berrigan, F. J. (1979a) 'The practice of community communication', in *Community Communication: The Role of Community Media in Development*, UNESCO, Paris, pp18–27
- *Content:* a discussion of the challenges involved in the practice of participatory communication at the community level.
Berrigan, F. J. (1979b) 'The community media methodology', in *Community Communication: The Role of Community Media in Development*, UNESCO, Paris, pp28–41
- *Content:* community media methodology is described in detail, with three examples of successful projects.
Braden, S. and Huong, T.T.T. (1998a) 'The role of video in participatory development', in *Video for Development*, Oxfam, Oxford, pp13–26
- *Content:* this casebook examines an experiment in the Ky Nam village, Viet Nam, where a team of non-governmental organization (NGO) workers from four countries were trained in the participatory use of video for community development. The team then worked with the villagers in Ky Nam to make videotapes about issues identified and researched by the community. The casebook gives an objective account of the workshop, showing how community-made video can be used locally for purposes of conflict resolution and advocacy. This particular chapter outlines some of the main concerns that lay behind the Oxfam initiative in Ky Nam and the key principles underlying the uses of participatory video.
Braden, S. and Huong, T. T. T. (1998b) 'Participatory research and analysis in Ky Nam', in *Video For Development*, Oxfam, Oxford, pp41–50
- *Content:* this chapter describes the ways in which the villagers used video to research and retrieve information about their needs and problems. It also focuses on the learning processes involving the villagers and the visiting team of NGO workers.
Carey, H. A. (1999a) 'Visual communication', in *Communication in Extension: A Teaching and Learning Guide*, United Nations Food and Agriculture Organization, Rome, pp53–66
- *Content:* guidelines for conducting a training/teaching session on designing visual aids.
Carey, H. A. (1999b) 'Social action processes', in *Communication in Extension: A Teaching and Learning Guide*, United Nations Food and Agriculture Organization, Rome, pp69–73

- *Content*: this chapter outlines a plan for conducting a training/teaching session on applying social action processes.

Carey, H. A. (1999c) 'Your communication/interaction traits', in *Communication in Extension: A Teaching and Learning Guide*, United Nations Food and Agriculture Organization, Rome, p116

- *Content*: this chapter presents questions that enable you to see yourself as others see you.

Carey, H. A. (1999d) 'Communication self-evaluation form', in *Communication in Extension: A Teaching and Learning Guide*, United Nations Food and Agriculture Organization, Rome, p117

- *Content*: a discussion of self-evaluation of communication skills.

Chambers, R. (2002) *Participatory Workshops: A Sourcebook of 21 Sets of Ideas and Activities*, Earthscan, London

- *Content:* this sourcebook of practical approaches and methods for participatory workshops draws on a rich variety of experiences and is a very useful book for facilitators and trainers.

CMN (Community Media Network) (1998a) 'What is community photography?', in *Tracking: Community Photography Report*, CMN, Dublin, p1

- *Content*: community photography as a tool for empowerment.

CMN (1998b) 'Some poor cousin?', in *Tracking: Community Photography Report*, CMN, Dublin, p2

- *Content*: current approaches to community photography.

CMN (1998c) 'The business of images', in *Tracking: Community Photography Report*, CMN, Dublin, p5

- *Content*: community photography as a tool for reflecting a community's life experiences.

Estrella, M. et al (eds) (2000a) 'Methodological issues in participatory monitoring and evaluation', in *Learning from Change: Issues and Experiences in Participatory Monitoring and Evaluation*, Intermediate Technology Publications, London, and International Development Research Centre, Ottawa, Canada, pp201–216

- *Content*: practical aspects and challenges of participatory monitoring and evaluation.

Estrella, M. et al (eds) (2000b) 'Laying the foundation: Capacity building for participatory monitoring and evaluation', in *Learning from Change: Issues and Experiences in Participatory Monitoring and Evaluation*, Intermediate Technology Publications, London, and International Development Research Centre, Ottawa, Canada, pp217–228

- *Content*: key elements of capacity-building for participatory monitoring and evaluation.

Estrella, M. et al (eds) (2000c) 'Learning to change by learning from change: Going to scale with participatory monitoring and evaluation', in *Learning from Change: Issues and Experiences in Participatory Monitoring and Evaluation*, Intermediate Technology Publications, London, and International Development Research Centre, Ottawa, Canada, pp229–243

- *Content*: focuses on participatory monitoring and evaluation as a learning process involving many different stakeholders. This chapter also explores some of the social and political dimensions of participatory monitoring and evaluation, especially in relation to scaling up.

FAO (United Nations Food and Agriculture Organization) (1990) 'The communication system for Proderith II', in *Towards Putting Farmers in Control: A Second Case Study of the Rural Communication System for Development in Mexico's Tropical Wetlands*, FAO, Rome, pp11–15

- *Content*: focuses on the network approach and the various types of media used in this approach to rural development.

FAO (2000a) 'An alternative to literacy: Is it possible for community video and radio to play this role? A small experiment by the Deccan Development Society in Hyderabad, India', *Forests, Trees and People Newsletter,* no 40/41, December 1999/January 2000, FAO, Rome, pp9–13

- *Content*: these articles are devoted to the use of popular communication tools within a participatory process.

FAO (2000b) 'The development of participatory media in southern Tanzania', *Forests, Trees and People Newsletter,* no 40/41, December 1999/January 2000, FAO, Rome, pp14–18

FAO (2000c) 'Participatory video and PRA: Acknowledging the politics of empowerment', *Forests, Trees and People Newsletter,* no 40/41, December 1999/January 2000, FAO, Rome, pp21–23

FAO (2000d) 'What is all this song and dance about?', *Forests, Trees and People Newsletter,* no 40/41, December 1999/January 2000, FAO, Rome, pp24–25

FAO (2000e) 'TV Favela, a voice of the people: Experience with local TV in environmental work in Brazil', *Forests, Trees and People Newsletter* no 40/41, December 1999/January 2000, FAO, Rome, pp26–29

FAO (2000f) 'On becoming visible: The role of video in community strategies to take part in the discussion about the future of the forest', *Forests, Trees and People Newsletter,* no 40/41, December 1999/January 2000, FAO, Rome, pp30–34

FAO (2000g) 'Questions and answers about participatory video', *Forests, Trees and People Newsletter,* no 40/41, December 1999/January 2000, FAO, Rome, pp35–44

FAO (2000h) 'Community forestry radio over Nepal', *Forests, Trees and People Newsletter,* no 40/41, December 1999/January 2000, FAO, Rome, pp45–47

FAO (2000i) 'Networking for dialogue and action', *Forests, Trees and People Newsletter,* no 40/41, December 1999/January 2000, FAO, Rome, pp54–58

Feek, W. and Morry, C. (2003) *Communication and Natural Resource Management: Theory-Experience*, FAO, Rome, and The Communication Initiative, Victoria, British Columbia, www.fao.org/documents/show_cdr.asp?url_file=DOCREP/005/Y4737E/Y4737E00.HTM

- *Content*: this book has been written as a tool for people involved or interested in communication and natural resource management (CNRM). It presents short case studies, reflections and exercises that guide the reader through a self-learning process about CNRM practice and theory.

de Fossard, E. and Kulakow, A. (n.d. a) 'Overview of the steps of the planning process for development communication', in *A Planning Guide for Development Communication,* Foundation For International Training, Markham, Ontario, and Academy of Educational Development, Washington, DC, pp9–23

- *Content*: useful guidelines on the communication planning process. This chapter includes brief descriptions of 17 major steps involved in planning a development communication project.

de Fossard, E. and Kulakow, A. (n.d. b) 'Media glossary', in *A Planning Guide for Development Communication,* Foundation For International Training, Markham, Ontario, and Academy of Educational Development, Washington, DC, pp86–93

- *Content*: summary of the potential and limitations of different media used in development communication.

de Fossard, E. and Kulakow, A. (n.d. c) 'Guidelines for sample testing', in *A Planning Guide for Development Communication,* Foundation For International Training, Markham, Ontario, and Academy of Educational Development, Washington, DC, pp139–149

- *Content*: outline of a plan for pre-testing development communication materials.
de Fossard, E. and Kulakow, A. (n.d. d) 'Evaluation guide', in *A Planning Guide for Development Communication,* Foundation For International Training, Markham, Ontario, and Academy of Educational Development, Washington, DC, pp149–158
- *Content*: describes the instruments for formative and summative evaluation of development communication projects; it also provides useful guidelines on devising the evaluation questionnaire.
Guijt, I. (1998a) *Participatory Monitoring and Impact Assessment of Sustainable Agricultural Initiatives,* SARL Discussion Paper no 1, International Institute for Environment and Development, London
- *Content*: this discussion paper is a practical methodological introduction to setting up a participatory monitoring process for sustainable agricultural initiatives.
Guijt, I. (1998b) 'Definitions', in *Participatory Monitoring and Impact Assessment of Sustainable Agricultural Initiatives,* SARL Discussion Paper no 1, International Institute for Environment and Development, London, pp12–21
- *Content*: clarifies key concepts.
Guijt, I. (1998c) 'The key steps', in *Participatory Monitoring and Impact Assessment of Sustainable Agricultural Initiatives,* SARL Discussion Paper no 1, International Institute for Environment and Development, London, pp22–26
Guijt, I. (1998d) 'Indicators', in *Participatory Monitoring and Impact Assessment of Sustainable Agricultural Initiatives,* SARL Discussion Paper no 1, International Institute for Environment and Development, London, pp27–29
Guijt, I. (1998e) 'Annex I', in *Participatory Monitoring and Impact Assessment of Sustainable Agricultural Initiatives,* SARL Discussion Paper no 1, International Institute for Environment and Development, London, pp53–110
Harford, N. and Baird, N. (1997a) 'Introduction: Guidelines for making visual aids', in *How to Make and Use Visual Aids,* Volunteer Services Overseas, Heinemann Educational Publishers, Oxford, pp5–19
- *Content*: this chapter covers planning for making visual aids according to the type of activity; it also provides simple guidelines on pre-testing and evaluating audiovisual materials.
Harford, N. and Baird, N. (1997b) 'Re-usable visual aids which carry information', in *How to Make and Use Visual Aids,* Volunteer Services Overseas, Heinemann Educational Publishers, Oxford, pp20–43
- *Content*: this chapter discusses objects that carry information; these objects can be used over and over again, conveying different information each time.
Harford, N. and Baird, N. (1997c) 'Visual aids which display information', in *How to Make and Use Visual Aids,* Volunteer Services Overseas, Heinemann Educational Publishers, Oxford, pp44–66
- *Content*: in this chapter, the reader is introduced to displays of information, which are made more effective by using simple graphic design techniques.
IIED (International Institute for Environment and Development) (2000a) 'Keys to unleash mapping's good magic', in *PLA Notes, Participatory Learning and Action 39 – Special Issue on Popular Communications,* IIED, London, pp10–13
- *Content*: this article describes the eight steps involved in community-based mapping and the key questions that should be answered at each stage.
IIED (2000b) 'A participatory GIS for community forestry user groups in Nepal: Putting people before the technology', in *PLA Notes, Participatory Learning and Action 39 – Special Issue on Popular Communications,* IIED, London, pp14–18
- *Content*: this article explores some of the benefits and concerns of using geographical information systems (GIS) as a participatory tool.

IIED (2000c) 'Flying to reach the sun', in *PLA Notes, Participatory Learning and Action 39 – Special Issue on Popular Communications*, IIED, London, pp59–60
- *Content*: an interview with Anna Blackman of PhotoVoice, a UK-based non-profit organization that works on participatory documentary and photography projects around the world.

Kennedy, T. W. (1982) 'Beyond advocacy: A facilitative approach to public participation', *Journal of the University Film andVideo Association*, University Film and Video Association, vol XXXIV, 3/summer 1982, pp??[Q206]
- *Content*: the focus of this article is on the use of film and video as a catalyst for community discussions of development issues.

Linney, B. (1995a) 'Approaches to communication', in *Pictures, People and Power: People-centred Visual Aids for Development*, Macmillan, London, pp5–8
- *Content*: reflections on the authoritarian approach (one-way communication) and the people-centred approach (two-way communication).

Linney, B. (1995b) 'Making and using people-centred pictures', in *Pictures, People and Power: People-centred Visual Aids for Development*, Macmillan, London, pp87–131
- *Content*: guidelines on making and using a range of visual aids for people-centred communication.

Linney, B. (1995c) 'Pictures and empowerment', in *Pictures, People and Power: People-centred Visual Aids for Development*, Macmillan, London, pp173–188
- *Content*: the role of people-centred pictures in empowering individuals, groups and communities.

Narayan, D. (1993a) 'Introduction', in *Participatory Evaluation Tools for Managing Change in Water and Sanitation*, World Bank Technical Paper no 207, World Bank, Washington, DC, pp1–7
- *Content*: the role of evaluation in community-managed projects.

Narayan, D. (1993b) 'What is participatory evaluation', in *Participatory Evaluation Tools for Managing Change in Water and Sanitation*, World Bank Technical Paper no 207, World Bank, Washington, DC, pp9–19
- *Content*: focuses on problem-solving orientation, generating knowledge and involving users in analytical data.

Narayan, D. (1993c) 'Measuring sustainability', in *Participatory Evaluation Tools for Managing Change in Water and Sanitation*, World Bank Technical Paper no 207, World Bank, Washington, DC, pp27–30; pp43–61
- *Content*: issues covered in this article include human capacity development; managing abilities; local institutional capacity; knowledge and skills; systems for learning and problem-solving; and supportive leadership.

Nelson, N. and Wright, S. (eds) (1995) 'Theatre for development: Listening to the community', in *Power and Participatory Development, Theory and Practice*, Intermediate Technology Publications, London, pp61–71
- *Content*: an interesting case study of a community environment project in east Mali, which used a drama unit to tune into the mood and views expressed by the community.

Oakley, P. et al (1991a) 'Emerging methodologies of participation', in *Projects With People: The Practice of Participation in Rural Development*, International Labour Organization, Geneva, pp205–238
- *Content*: discusses the methodological tools used to promote participation in development.

Oakley, P. et al (1991b) 'Evaluating participation', in *Projects With People: The Practice of Participation in Rural Development*, International Labour Organization, Geneva, pp239–268

- *Content*: describes the conceptual challenges of evaluating the process of community participation.

Pretty, J. N. et al (1995a) 'You the trainer and facilitator', in *A Training Guide for Participatory Learning and Action*, International Institute for Environment and Development, London, pp13–38

- *Content*: focuses on the principal roles, skills and techniques that the trainer-facilitator should consider before undertaking participatory training activities.

Pretty, J. N. et al (1995b) 'Group dynamics and team-building', in *A Training Guide for Participatory Learning and Action*, International Institute for Environment and Development, London, pp39–53

- *Content*: the main features of managing group dynamics and building inter-disciplinary teams that are essential in practising participatory research and development.

Pretty, J. N. et al (1995c) 'Principles of participatory learning and action', in *A Training Guide for Participatory Learning and Action*, International Institute for Environment and Development, London, pp54–71

- *Content*: a summary of the core principles of participatory learning and action.

Pretty, J. N. et al (1995d) 'Training in participatory methods in the workshop', in *A Training Guide for Participatory Learning and Action*, International Institute for Environment and Development, London, pp72–89

- *Content*: describes the process of workshop training in three groups studying participatory methods.

Pretty, J. N. et al (1995e) 'The challenges of training in the field', in *A Training Guide for Participatory Learning and Action*, International Institute for Environment and Development, London, pp90–110

- *Content*: the complexities of training in a real world setting and how to deal with them.

Pretty, J. N. et al (1995f) 'Organising workshops for training, orientation and exposure', in *A Training Guide for Participatory Learning and Action*, International Institute for Environment and Development, London, pp11–129

- *Content*: the conditions necessary for preparing a training course on participatory methods.

Ramirez, R. and Quarry, W. (2004) *Communication for Development: A Medium for Innovation in Natural Resource Management*, IDRC, Ottawa, Canada, and FAO, Rome

- *Content*: this report presents, through stories and examples, the experiences of many people and projects worldwide where communication methods and approaches have been applied to address natural resource management.

Ray, H. E. (1984) 'Guidelines for planning and implementation of communication strategies', in *Incorporating Communication Strategies into Technology Transfer Programmes for Agricultural Development*, Academy of Educational Development, Washington, DC, pp85–133

- *Content*: a set of useful guidelines for planning and implementing effective communication programmes as integral components of agricultural projects.

Richardson, D. and Rajasunderam, C. V. (2000) 'Training community animators as participatory communication for development practitioners', in Richardson, D. and Paisley, L. (eds) *The First Mile of Connectivity*, *www.fao.org/docrep/x0295e/x0295e17.htm*

- *Content*: this contribution explores key questions and issues related to the training of animators/facilitators in PDC processes.

Servaes, J. et al (eds) (1996) 'Fitting projects to people or people to projects', in *Participatory Communication for Social Change*, Sage Publications, New Delhi, and Thousand Oaks, London, pp249–265
- *Content*: describes nine steps to develop an effective communication strategy for projects.
Shaw, J. and Robertson C. (1997a) 'Background, approaches and benefits', in *Participatory Video: A Practical Guide to Using Video Creatively in Group Development Work*, Routledge, London and New York, pp7–27
- *Content*: the benefits of participatory video (community-building; critical awareness; consciousness raising; empowerment; and self-reliance) are discussed in terms of group purpose, community purpose and individual purpose.
Shaw, J. and Robertson C. (1997b) 'Applications and project outcomes', in *Participatory Video: A Practical Guide to Using Video Creatively in Group Development Work*, Routledge, London and New York, pp166–191
- *Content*: a comprehensive guide to using video in a participatory way in group development work.
Slocum, R. et al (1996a) 'Community drama', in *Power, Process and Participation: Tools for Change*, Intermediate Technology Publications, London, pp72–74
- *Content*: this selection offers a description of field tested communication and participatory research tools for securing people's participation in development projects.
Slocum, R. et al (1996b) 'Conflict resolution', in *Power, Process and Participation: Tools for Change*, Intermediate Technology Publications, London, pp75–87
Slocum, R. et al (1996c) 'Focus groups', in *Power, Process and Participation: Tools for Change*, Intermediate Technology Publications, London, pp95–99
Slocum, R. et al (1996d) 'Network formation', in *Power, Process and Participation: Tools for Change*, Intermediate Technology Publications, London, pp155–158
Slocum, R. et al (1996e) 'Photography', in *Power, Process and Participation: Tools for Change*, Intermediate Technology Publications, London, pp167–171
Slocum, R. et al (1996f) 'Social network mapping', in *Power, Process and Participation: Tools for Change*, Intermediate Technology Publications, London, pp186–190
Slocum, R. et al (1996g) 'Study trips', in *Power, Process and Participation: Tools for Change*, Intermediate Technology Publications, London, pp191–193
Slocum, R. et al (1996h) 'Transects', in *Power, Process and Participation: Tools for Change*, Intermediate Technology Publications, London, pp198–204
Slocum, R. et al (1996i) 'Video', in *Power, Process and Participation: Tools for Change*, Intermediate Technology Publications, London, pp205–213
White, S. A. et al (1994a) 'Facilitating communication within rural and marginal communities: A model for development support', in *Participatory Communication: Working for Change and Development*, Sage Publications, London and New Delhi, pp329–341
- *Content*: focuses on strategies to mobilize community groups for development action.
White, S. A. et al (1994b) 'Participatory message making with video: Revelations from studies in India and the USA', in *Participatory Communication: Working for Change and Development*, Sage Publications, London and New Delhi, pp359–383
- *Content*: the two studies described in this contribution focus on the process of using the participatory message development model.
White, S. A. (ed) (1999) *Facilitating Participation: Releasing the Power of Grassroots Communication*, Sage Publications, New Delhi, and Thousand Oaks, London

- *Content*: this book is divided into three broad sections: 'The art of activation', 'The art of technique' and 'The art of community building'. Part 1 includes a presentation of the important concept of the catalyst communicator. Part 2 explores significant issues related to the competencies of facilitators. Part 3 contains three case studies that focus on the link between participatory communication and community-building.

Windahl, S. et al (1992a) 'The nature of communication planning', in *Using Communication Theory: An Introduction to Planned Communication,* Sage Publications, London and New Delhi, pp19–29

- *Content*: a broad conceptualization of the communication planning process.

Windahl, S. et al (1992b) 'Categorisation of basic strategies', in *Using Communication Theory: An Introduction to Planned Communication,* Sage Publications, London and New Delhi, pp39–49

- *Content*: this chapter discusses basic types of communication strategies.

IV: Communication and sharing knowledge

This section focuses on the following themes:

- facilitating the identification and sharing of local knowledge;
- planning dissemination of research results to different stakeholders;
- facilitating extension to other communities.

Allen, W. K. and Harmsworth, M. G. (2001) *The Role of Social Capital in Collaborative Learning,* Landcare Research, Manaaki Whenua, www.landcareresearch.co.nz/research/social/social_capital.asp

- *Content*: 'social capital' (a term used in development and organizational learning literature) can be thought of as the networks of communication and cooperation that facilitate collaborative learning. The role of social capital in fostering the social networks and information exchange required for collective action is the focus of this contribution.

Bolliger, E. et al (1992a) 'Extension methods', in *Agricultural Extension: Guidelines for Extension Workers in Rural Areas – Part B. List of Questions,* Swiss Centre for Development Cooperation in Technology and Management (SKAT), ??[Q239], Chapter 4

- *Content*: guidelines on agricultural extension methods – for example, demonstrations, field trips and competitions.

Bolliger, E. et al (1992b) 'Extension accessories', in *Agricultural Extension: Guidelines for Extension Workers in Rural Areas – Part B. List of Questions,* Swiss Centre for Development Cooperation in Technology and Management (SKAT), ??[Q241], Chapter 5

- *Content*: guidelines on the use of extension accessories – for example, technical leaflets, plots, displays and rural newspapers.

ETG Netherlands (1997) 'Spreading and consolidating the PTD process', in *Developing Technology With Farmers: A Trainer's Guide for Participatory Learning,* Zed Books, London and New York, and ETG, The Netherlands, pp191–213

- *Content*: participatory technology development (PTD) in agriculture is a process of purposeful and creative interaction between local people and outside facilitators in order to develop more sustainable farming systems. The focus of this chapter is on the last phase of the PTD process, in which the outcome of experimental activities by and with farmers is spread to other farmers.

FAO (United Nations Food and Agriculture Organization) (1996a) 'Integrating science and traditional knowledge to achieve sustainable development in Morocco', in *Training for Agriculture and Rural Development 1995–1996*, FAO, Rome, pp79–85

- *Content*: describes two Moroccan projects in order to demonstrate the importance of integrating traditional and scientific knowledge.

FAO (1996b) 'Group-based extension programmes in Java to strengthen natural resource conservation activities', in *Training for Agriculture and Rural Development 1995–1996*, FAO, Rome, pp123–133

- *Content*: describes how agricultural extension programmes in Indonesia are carried out through a group-based participatory approach. This chapter also records Indonesia's experience with field schools – an effective method for promoting farmers' learning and participation in agriculture.

Grenier, L. (1998a) 'What about indigenous knowledge?', in *Working with Indigenous Knowledge: A Guide for Researchers*, International Development Research Centre, Ottawa, Canada, pp1–11

- *Content*: describes some characteristics of indigenous knowledge and its contribution to sustainable development.

Grenier, L. (1998b) 'Developing a research framework', in *Working with Indigenous Knowledge: A Guide for Researchers*, International Development Research Centre, Ottawa, Canada, pp31–55

- *Content*: presents considerations for those who are developing frameworks for indigenous knowledge research.

Grenier, L. (1998c) 'Data collection', in *Working with Indigenous Knowledge: A Guide for Researchers*, International Development Research Centre, Ottawa, Canada, pp57–62

- *Content*: presents details of 31 field techniques for data collection.

Grenier, L. (1998d) 'Assessing, validating and experimenting with IK', in *Working with Indigenous Knowledge: A Guide for Researchers*, International Development Research Centre, Ottawa, Canada, pp71–86

- *Content*: assesses the products of IK research in terms of sustainability and looks at developing IK through validation and experimentation.

Grenier, L. (1998e) 'Sample guidelines', in *Working with Indigenous Knowledge: A Guide for Researchers*, International Development Research Centre, Ottawa, Canada, pp87–99

- *Content*: presents three sets of formal procedural guidelines for conducting IK research. The guidelines can be adapted for other situations.

Haverkort, B. et al (eds) (1991) 'Farmers' experiments and participatory technology development', in *Joining Farmers' Experiments: Experiences in Participatory Technology Development*, Intermediate Technology Publications, London, pp3–16

- *Content*: outlines the six steps in participatory technology development:
  1 how to get started;
  2 looking for things to do;
  3 designing the experiment;
  4 trying out;
  5 sharing results with others; and
  6 sustaining and consolidating the process of participatory technology development.

Inglis, J. T. (ed) (1993a) 'Traditional ecological knowledge in perspective', in *Traditional Ecological Knowledge: Concepts and Cases*, International Programme on Traditional Ecological Knowledge and International Development Research Centre, Ottawa, Canada, pp1–6
- *Content*: outlines the reasons for preserving traditional ecological knowledge (TEK), with a focus on the practical uses of TEK.
Inglis, J. T. (ed) (1993b) 'Transmission of traditional ecological knowledge', in *Traditional Ecological Knowledge: Concepts and Cases*, International Programme on Traditional Ecological Knowledge and International Development Research Centre, Ottawa, Canada, pp17–30
- *Content*: addresses the key issue of how ecological knowledge is transmitted from one generation to the next.
Inglis, J. T. (ed) (1993c) 'Integrating traditional ecological knowledge and management with environmental impact assessment', in *Traditional Ecological Knowledge: Concepts and Cases*, International Programme on Traditional Ecological Knowledge and International Development Research Centre, Ottawa, Canada, pp33–39
- *Content*: provides perspectives on the use of traditional ecological knowledge in environmental assessment.
Inglis, J. T. (ed) (1993d) 'African indigenous knowledge and its relevance to sustainable development', in *Traditional Ecological Knowledge: Concepts and Cases*, International Programme on Traditional Ecological Knowledge and International Development Research Centre, Ottawa, Canada, pp55–62
- *Content*: discusses the relevance of African indigenous knowledge to environment and development issues.
Johnson, M. (ed) (1992a) 'Research on traditional environmental knowledge: Its development and its role', in *LORE: Capturing Traditional Environmental Knowledge*, Dene Cultural Institute and International Development Research Centre, Ottawa, Canada, pp3–22
- *Content*: focuses on the problems of integrating traditional environmental knowledge and Western science.
Johnson, M. (ed) (1992b) 'Documenting and applying traditional environmental knowledge in northern Thailand', in *LORE: Capturing Traditional Environmental Knowledge*, Dene Cultural Institute and International Development Research Centre, Ottawa, Canada, pp164–173
- *Content*: describes the efforts of the Mountain People's Culture and Development Education Programme to document and apply the TEK of the highlanders of northern Thailand.
Narayan, D. (1996) 'Data analysis, dissemination and use', in *Towards Participatory Research*, World Bank Technical Paper no 307, World Bank, Washington, DC, pp129–138
- *Content*: explores techniques for data analysis; dissemination and utilization of research results; media options for communicating results; and follow-up activities.
Starkey, P. (1999a) 'Network types and network benefits', in *Networking for Development*, International Forum for Rural Transport and Development (IFRTD), London, pp14–20
- *Content*: examines the benefits of networking in development work; some network models are also discussed.
Starkey, P. (1999b) 'General guidelines for networking', in *Networking for Development*, International Forum for Rural Transport and Development (IFRTD), London, pp31–46

- *Content*: sets out concise, practical guidelines for an effective network operation, with examples from a range of development networks.

Van Veldhuizen, L. et al (ed) (1997) 'Kuturaya: Participatory research, innovation and extension', in *Farmers' Research in Practice: Lessons From The Field*, Intermediate Technology Publications, London, pp153–173

- *Content:* a case study of a research project on conservation tillage initiated by the Zimbabwe Agricultural Extension Service, AGRITEX Kuturaya, which emphasized learning and improving through experimentation. Farmers found a renewed confidence in their own knowledge and abilities as a result of their experimentation.

Warner, K. (1991) 'Local technical knowledge, shifting cultivation and natural resource management', in *Shifting Cultivators: Local Technical Knowledge and Natural Resource Management in the Humid Tropics*, United Nations Food and Agriculture Organization, Rome, pp1–10

- *Content:* examines shifting cultivation as a natural resource management strategy for the tropics, with a focus on utilizing local technical knowledge.

Warren, M. et al (eds) (1995a) 'Indigenous communication and indigenous knowledge', in *The Cultural Dimension of Development: Indigenous Knowledge Systems*, Intermediate Technology Publications, London, pp112–123

- *Content*: describes the interface between indigenous knowledge and indigenous communication. A framework for studying this interface is also outlined.

Warren, M. et al (eds) (1995b) 'Farmer know-how and communication for technology transfer: CTTA in Niger', in *The Cultural Dimension of Development: Indigenous Knowledge Systems*, Intermediate Technology Publications, London, pp323–332

- *Content*: describes the model of communication elaborated and applied by the Communication for Technology Transfer in Agriculture (CTTA) programme in its research on farmer innovation and communication in Niger.

Warren, M. et al (eds) (1995c) 'Indigenous soil and water conservation in Djenne, Mali', in *The Cultural Dimension of Development: Indigenous Knowledge Systems*, Intermediate Technology Publications, London, pp371–384

- *Content*: describes indigenous soil and water conservation practices and techniques in sub-Saharan Africa in the circle of Djenne, central Mali.

Index

Selected Studies in
MARRIAGE AND THE FAMILY

Selected Studies in
MARRIAGE AND THE FAMILY

REVISED EDITION

Edited by

ROBERT F. WINCH, Northwestern University
ROBERT McGINNIS, Cornell University
HERBERT R. BARRINGER, Northwestern University

HOLT, RINEHART and WINSTON
New York

PREFACE

The second edition of *Selected Studies in Marriage and the Family* builds—we like to think—on the shoulders of the first. As we began to think about putting the first edition together, we were mindful of the difficulty that besets many anthologies—a lack of organization and, all too frequently, a lack of coherence. Accordingly, we began with the organization and, after agreeing on this, we searched widely for articles which would fit our outline.

The first step in the contemplation of a revision was to review and revise the outline; the second step was to select the sixty best articles in the literature which would fit our outline, as revised. About one third of the selections are held over from the first edition. Virtually all of the other two thirds have been published since May 1953, the month in which the first edition was published.

In the first chapter we summarize some of the elements of scientific procedure and relate them to problems involved in studying the family. We are convinced that, as with other classes of phenomena, durable knowledge about the family can be achieved only through the use of scientific method.* We hope that the first chapter will provide the reader with a set of criteria for judging the contributions in the rest of the book.

After chapter 1, the organization of this book is rather similar to, although not identical with, Winch's *The Modern Family*. Chapters 2–6 present some general formulations concerning, and specific descriptions of, familial structures and functions. Familial structure is viewed as kinship and as roles differentiated on the bases of age, sex, and occupation. The traditional Chinese family is presented as the embodiment of the full complement of functions, to which both the family of Communist China and of urban America stand in contrast. The suggestion is made that the suburban American family may be a functionally oriented reaction to familial functionlessness.

* One of the editors (H.R.B.) would substitute "best" for "only."

Viewed in the large, the family cycle is endless. In chapter 7 we break it arbitrarily by considering patterns of fertility and social influences on the infant, prenatally and postnatally. The first paper in chapter 8 considers the literature on socialization in infancy and early childhood with a view to pointing out the major variables; the impact, if any, of family-life education on parental practices is the topic of the other paper in this chapter. Various aspects of social structures as determinants of children's behavior occupy chapters 10–12, and in chapters 13 and 14 we are concerned with the relations and strains in relations between adult offspring and their parents.

Dating, mate-selection, marriage, marital dissolution, and remarriage take up the remainder of the book. In chapter 15 the cross-cultural perspective leads us to see love as a rather unusual basis for mate-selection. Viewing our own society in chapter 16 we see a set of determinants of the choice of mates that have nothing to do with romance except in the limiting sense of determining our associates, that is, the field of eligibles from which we are likely to choose spouses. Dating as a selective procedure, the nature of love, and an experiment at scientific matchmaking are the topics of chapter 17. In chapter 18 we take a summary look at the findings of studies of marital adjustment and at the criteria they have used.

The stability and instability of marriages concern the last two chapters. Social class and the relative power of the two spouses are related to marital stability and marital satisfaction in chapter 19. In chapter 20 articles on such correlates of marital stability as socioeconomic status, mixed and nonmixed religions, and premarital versus post marital pregnancies are followed by a paper which assesses the relative stability of first marriages and remarriages.

To edit a volume of this sort provides an opportunity—indeed makes it obligatory—to assess the state of the discipline. To edit a second edition provides an opportunity to see how the discipline is developing. On the negative side candor requires the explicit admission that our predictions are still not very good; both the proportion of the total variance that is "explained," as measured by the squared coefficient of correlation, and the small number of findings that stand up through successive replication lead to this admission of weakness.

On the positive side is the strengthening and increasing awareness of theory and research design. As compared with the first edition, the second contains papers with more formally stated theory and more sophisticated analyses.

An expression of special appreciation is due Dr. John Clausen, who was very helpful in facilitating access to the papers by Eleanor Maccoby and William H. Sewell.

R. F. W.   R. McG.   H. R. B.

May 1962

CONTRIBUTORS

Conrad M. Arensberg, Department of Anthropology, Columbia University

Margaret K. Bacon, Department of Psychology, Yale University

Herbert Barry III, Department of Psychology, University of Connecticut

Howard W. Beers, Council on Economic and Cultural Affairs, Bogar, Indonesia

Hugo G. Beigel, Department of Psychology, Long Island University

Ivan Belknap, Department of Sociology, University of Texas

Orville G. Brim, Jr., Russell Sage Foundation

Paul J. Campisi, Department of Sociology, Southern Illinois University

Ruth Shonle Cavan, Department of Sociology, Rockford College

Irvin L. Child, Department of Psychology, Yale University

Harold T. Christensen, Department of Sociology, Purdue University

Allison Davis, Department of Education and Committee on Human Development, University of Chicago

Kingsley Davis, Department of Sociology, University of California at Berkeley

Antonio J. Ferreira, Psychiatrist, San Jose, California

Linton C. Freeman, Department of Sociology and Anthropology, Syracuse University

Paul C. Glick, U. S. Bureau of the Census

Joseph Golden, Atlanta University School of Social Work

William J. Goode, Department of Sociology, Columbia University

William E. Henry, Department of Psychology and Committee on Human Development, University of Chicago

Reuben Hill, Minnesota Family Study Center and Department of Sociology, University of Minnesota

Karen G. Hillman, Board of Education, Evanston, Illinois

August B. Hollingshead, Department of Sociology, Yale University

Morton M. Hunt, free-lance writer, contributor of "Profiles" to The New Yorker magazine

E. Gartly Jaco, Department of Sociology, Western Reserve University

Clifford Kirkpatrick, Department of Sociology, Indiana University

Melvin L. Kohn, Laboratory of Socioenvironmental Studies, National Institute of Mental Health

Marvin R. Koller, Department of Sociology and Anthropology, Kent State University

Mirra Komarovsky, Department of Sociology, Barnard College, Columbia University

Thomas Ktsanes, Department of Sociology, Tulane University

Virginia Ktsanes, Department of Sociology, Tulane University

Judson T. Landis, Institute of Human Development, University of California at Berkeley

Shu-Ching Lee, Department of Sociology, Southeast Missouri State College

Harvey J. Locke, Department of Sociology, University of Southern California

Eleanor E. Maccoby, Department of Psychology, Stanford University

George Herbert Mead (deceased), Department of Philosophy, University of Chicago

Hannah H. Meissner, Department of Sociology, Purdue University

Thomas P. Monahan, The County Court of Philadelphia

George P. Murdock, Department of Anthropology, University of Pittsburgh

William F. Ogburn (deceased), Department of Sociology, University of Chicago

Talcott Parsons, Department of Social Relations, Harvard University

Robert F. Peck, Department of Educational Psychology, University of Texas

A. R. Radcliffe-Brown (deceased), Department of Social Anthropology, Rhodes University

Julius A. Roth, New York School of Social Work, Columbia University

Georges Sabagh, Department of Sociology, University of Southern California

Alvin L. Schorr, Social Security Administration, U. S. Department of Health, Education and Welfare

William H. Sewell, Department of Sociology, University of Wisconsin

René A. Spitz, Department of Psychiatry, University of Colorado

Fred L. Strodtbeck, Departments of Sociology and Psychology, University of Chicago

Etsu Inagaki Sugimoto (deceased), lecturer and writer

Marvin B. Sussman, Department of Sociology, Western Reserve University

Dorrian Apple Sweetser, School of Nursing and Department of Sociology and Anthropology, Boston University

Mary Margaret Thomes, School of Social Welfare, University of California at Los Angeles

Sheila S. Tillotson, student, Northwestern University

Jackson Toby, Department of Sociology, Rutgers University

Karl M. Wallace, Department of Sociology, Los Angeles State College

William H. Whyte, Jr., writer

Peter Willmott, Institute of Community Studies, London

Donald M. Wolfe, Institute for Social Research, University of Michigan

C. K. Yang, Department of Sociology, University of Pittsburgh

Michael Young, Institute of Community Studies, London

CONTENTS

Selected Studies in
MARRIAGE AND THE FAMILY

1. EDITORS' INTRODUCTION: SCIENTIFIC METHOD AND THE STUDY OF THE FAMILY

Reports in this volume have been prepared by members of a curious breed—the social scientist. The breed is curious in that it attempts to focus a powerful mechanism, the scientific method, on the most mundane of matters, man's routine behavior. Why, one may wonder, should the social scientist study who marries whom when we all know that a man marries the first woman with whom he falls in love, whom he seeks to marry, and who accepts his offer.

The simplicity of this "truth" is, of course, deceiving. It fails to provide valid answers to questions the researcher is studying: What general principle will predict the kind of woman a specific man is likely to meet? What general principle will predict the one woman among those he meets with whom the man is most likely to fall in love? In mate-selection, as in other topics, it is the scientist's task to determine relevant variables, to weave the relations among the relevant variables into a theory, and then to assess the utility of that theory as an explanation of systematic observations as to what actually occurs.

The term "science" will be discussed in some detail on the following pages. The important point to grasp at the outset is that science consists not of orderly sets of facts, not of established truth. Rather science is a *method,* a procedure for arriving at reasonably adequate explanations of various aspects of the state of nature; it does not arrive at unalterable statements of final truth. Science is method, not fact. Social science, then, is the process of applying the method of science to the analysis of social behavior. In the present collection we consider the scientific analysis of marital and familial structures and functions.

Scientific study of marriage and the family has a single purpose: to acquire stable, systematic knowledge about these phenomena. Although such knowledge can be used to reduce human grief, the main purpose of such studies, and of this collection, is not to promote human happiness and welfare.

This introductory chapter provides the reader with some bench marks by means of which each article in the book can be judged as to its scientific merit. Some of the articles may seem eloquent and others heavy with jargon, such as "homogamy" and "complementariness." While treasuring literary style, the editors give top priority to scientific merit.[1] For this reason the editors suggest to the reader that questions of the following order should guide his evaluation of each report: What is asserted? Can the truth of the assertion be tested by the scientific method? Does the author actually bring evidence to bear on the question so that a negative outcome is possible, and thus does he test the assertion? If a test is offered, is it sufficiently conclusive that the reader is justified in accepting the assertion as tentative truth and using it in his life as a guidepost? If the author has provided what appears to be satisfactory empirical support for the assertion, then still is it an important assertion? Let us consider two criteria of importance. Can the assertion be generalized so that it leads to predictions about a class of events? Is the assertion linked to other assertions so as to form a theoretical web, or "nomological net," as it is sometimes called?

Science

In this section we deal with two major topics: the nature of science and how social researchers utilize scientific method in studying marriage and the family. The purpose of this discussion is to enable the reader to answer for himself the questions raised in the preceding paragraph, especially: Is this or that conclusion accurate enough that I may accept it as tentative truth?

Although facts are highly useful, the goal of science is not the collecting and classifying of facts. Rather the objective is the development of a completely general and systematic set of theories from which hypotheses are deduced and verified and in terms of which the structures and changes of animate and inanimate phenomena may be explained.

The objective of the scientist is to obtain satisfactory answers to *why* questions: Why do structures appear as they do, and why do they undergo

[1] While deploring turgid and opaque writing in any context the editors believe with Bertrand Russell that the language of daily life does not suffice for technical discourse. Russell remarks that what he calls "the cult of 'common usage' " excuses ignorance of mathematics and other fields "in those who have had only a classical education." Bertrand Russell, *Portraits from Memory and Other Essays,* New York, Simon and Schuster, 1956, p. 166. Relevant here is the discussion under "Utilizing the Scientific Outlook" near the end of this chapter.

the processes of change which characterize them?[2] Before the scientist can even learn how to ask the *why* questions properly, however, he must first ask and answer another set—the *how* questions. Before one can begin to explain why phenomena exist and change as they do, one must know how they are structured and how slowly or rapidly they change. To phrase it differently, before a scientist tries to explain a phenomenon, he should have an accurate description of that phenomenon. To put it more systematically, we can consider two distinct types of investigation. The first focuses on a *property,* such as income distribution in the United States or incidence of a particular disease. The second focuses on the *relations* that obtain among two or more properties in a specified population.[3] The papers by Radcliffe-Brown on kinship and by Glick on the life cycle of the family illustrate the focus on properties. Those by Brim on the association between sibling arrangement and sex-role socialization and by Wolfe on the correlates of marital power exemplify studies of relations.

There is a second important distinction among scientific investigations. This is the distinction between studies—whether of properties or of relations—that are not guided by theory or hypotheses and studies that are undertaken for the purpose of evaluating a theory or testing hypotheses. Logically both property and relation studies may be either theory-free or theory-bound; in practice, however, property studies are usually theory-free, or at least free of explicitly stated theory, whereas studies of relations are usually bound up in theory and the hypotheses are predictions based upon the theory. Studies of properties and of relations, studies that are theory-free and those that are theory-bound are all essential to science; they all contribute to the construction of theories. Studies of relations and theory-bound investigations do, in general, stand closer to the goals of science and thereby represent a more highly developed state of a discipline, but it is important to recognize that the study of properties and theory-free investigations lay the necessary base for the more advanced undertakings.

Theory

A scientific theory is a statement of the way in which abstract variables are related and from which verifiable hypotheses are deducible. The key word here is "verifiable." Although a theory may be suggested by an insight, an intuition, a dream, or any other bit of experience, to satisfy the

2 Needless to say, our use of the word "why" does not imply questions of ultimate causation which are answerable only on metaphysical grounds.

3 The distinction between these two types of investigation sometimes is referred to as that between descriptive and analytical studies. These terms, however, have created such dispute in their interpretation that we prefer to avoid them.

requisites of science it cannot end there. A theory must culminate in a statement from which testable hypotheses are deduced, tested, and verified or rejected before we can accept or reject the theory. Thus, the term "verifiable" in our definition of a scientific theory frees us from any obligation to regard as scientific knowledge any conclusion based exclusively on insight, intuition, or revelation.

Let us consider an example of the level of scientific explanation in the field of marriage and the family. On the basis of census data it has been concluded that, for white males at least, there is a rough inverse correlation between socioeconomic status and tendency to marital instability.[4] It has also been established that the incidence of divorce—at least in one county in Indiana—is positively correlated with premarital pregnancy.[5] In the present state of our knowledge it seems certain that an unknown number of other factors are also related to the incidence of marital instability. We may summarize these observations by asserting that in recent times (say, over the past fifteen years) the incidence of marital instability in the United States has been a function of factors a (socioeconomic status), b (premarital pregnancy), c (unknown), . . . n (unknown).[6] The statement is incomplete because the factors from c through n have not been identified. What remains is to discover what these factors are and how they are related to marital instability.[7]

Single-Factor and Multifactor Theories. In the last paragraph we said in effect that the theory accounting for the incidence of marital instability in the United States is incomplete. With reasonable security we may generalize to all of social science the observation that no complete theory exists. Instead, we find theories that are in various stages of development, that is, theories that contain statements of relations of varying degrees of complexity and abstraction. The least complex type of theory may be labeled the "single-factor" theory. Such theories attempt to explain variation in a class of phenomena in terms of the operation of a single variable. In other words, a single-factor study is a relational study, but one in which

[4] Karen Hillman, "Marital Instability and Its Relation to Education, Income, and Occupation: An Analysis Based on Census Data."

[5] Harold T. Christensen and Bette B. Rubinstein, "Premarital Pregnancy and Divorce: A Follow-up Study by the Interview Method."

[6] And possibly of interaction among these factors. In the above sentence n is some positive integer which is itself unknown.

[7] Of course the mere fact that there is a significant correlation between X and Y does not tell us that X causes Y or that Y causes X. All it tells us is that for every increment of change in X there is on the average some degree of change in Y. A correlation is said to be "spurious" when the covariation between the two variables under consideration is caused by some third variable that is not being considered explicitly.

the investigation is confined to one relation. Koller's analysis of mate-selection in terms of the factor of propinquity is illustrative of this type.[8]

From the work of Koller and others who have studied the relation between mate-selection and propinquity, then, we can predict that men tend to marry women who live near them rather than those whose residences are remote. But from studies based upon New Haven data we can also predict that men are more likely to marry women similar to themselves with respect to race, religion, ethnic grouping, social class, and age group than to marry women who differ from themselves in these social characteristics.[9] When we consider all of these factors together, we have a "multifactor" theory, and we are able to go considerably farther toward the goal of predicting who marries whom than we could have gone by considering any one of these factors by itself. In slightly different words, the greater the number of relevant factors in a theoretical system (that is, the more complex the system), the more complete will be the explanation. Moreover, the fusing of these several correlates of mate-selection into the single concept, the "field of eligibles," [10] illustrates the shift to a higher level of abstraction and how such a shift depends upon multifactor theories.

Since the goal of science is to explain as much about a given phenomenon as is practically possible, one may wonder why all theories are not considerably more complex. One important reason is that the more complex the theory, the more difficulty the researcher has in subjecting it to empirical test and in using it for predictive purposes. With single-factor hypotheses one may use simple techniques of analysis,[11] but with every added factor the methods of testing hypotheses become more difficult and tedious.[12]

While the articles in this volume contain conclusions, it must be realized that these conclusions are far from complete. And, like the conclusions in all scientific fields, they are certainly not final. The present status of the sociology of the family is indicated by the fact that some of these papers are (theory-free) studies of properties and others are single-factor theories. But a scientific discipline cannot reach maturity without first acquiring a fund of knowledge about properties and about simple relations between pairs of properties.

[8] Marvin R. Koller, "Residential and Occupational Propinquity."

[9] August B. Hollingshead, "Cultural Factors in the Selection of Marriage Mates"; Ruby Jo Reeves Kennedy, "Single or Triple Melting Pot? Intermarriage Trends in New Haven, 1870-1950," *American Journal of Sociology,* 58, 1952, 56-59, 201.

[10] Thomas Ktsanes and Virginia Ktsanes, "The Theory of Complementary Needs in Mate-Selection."

[11] Such as the t test and the zero-order r.

[12] The tedium of multivariate analysis has been markedly reduced by the development and availability of high-speed computers.

Making hypotheses researchable.   Theories consist of statements of relation between *abstract* categories. Abstraction is a process of reducing heterogeneous data to a homogeneous core. The "average man," for example, is an abstraction. It is quite probable that no human being is statistically average in every respect. Yet the idea of an average man is useful in that it gives us a statistical norm from which we may measure deviations in specific cases. An abstraction is an abbreviation of nature created in order that man's limited mind may be able to deal with the great variation of phenomena found in the universe.

Since the variables in scientific theories are abstract, it is impossible to find data that will directly verify or reject a theory. Before verification is possible, the theory must be related to something concrete which we can measure or count. This step is accomplished by means of the logical processes called "index construction" and "deduction." [13]

The essence of deduction is "the derivation of conclusions *necessarily* involved in the premises." [14] To illustrate, if we set forth the premises (1) that all men are mortal and (2) that William Jones is a man, then we must conclude that Mr. Jones is a mortal because this conclusion is a necessary consequence of the premises. Deductive logic is important to scientists because it is by means of deduction that they proceed from abstract theory through index construction to concrete data.

The researcher uses deduction in the following manner. The first premise is a statement of a relationship among abstract variables. The second premise contains indexes of the abstract variables in the first premise.[15] From these premises he deduces hypotheses that can be tested empirically. For example, if we wish to test Hollingshead's idea that family stability is a function of class position,[16] we may state our premises in this manner: (1) Family stability is positively correlated with position in the class structure. (2a) Divorce is an index of family instability (or is a nega-

[13] For a discussion of deduction, see Morris R. Cohen and Ernest Nagel, *An Introduction to Logic and Scientific Method,* New York, Harcourt Brace, 1934, esp. chs. 7 and 14. For a discussion of deductive reasoning and theory verification in sociology, see Hans L. Zetterberg, *On Theory and Verification in Sociology,* New York, Tressler, 1954.

[14] Cohen and Nagel, *op. cit.,* p. 273. (Italics in original.)

[15] A strict operationist would label the second premise an "operational definition." For a discussion of operationism, see Gustav Bergmann and Kenneth W. Spence, "Operationism and Theory Construction," in Melvin H. Marx (ed.), *Psychological Theory,* New York, Macmillan, 1951, pp. 54-66.

[16] August B. Hollingshead, "Class Differences in Family Stability." The above statement is an oversimplification of Hollingshead's actual position since he states that the "new upper-class family" is more unstable than the "nuclear upper-middle-class family." We may view Karen Hillman's paper as an effort to make such a test.

tive index of family stability). (2b) Income is an index of social class position. From these premises we hypothesize that divorce is negatively correlated with income. Thus by using incidence of divorce and income as indexes of the abstract variables (family stability and class position, respectively) in the first premise, we are able to test the statement of relationship between abstract variables. The adequacy of our test depends in part upon how good the indexes are and in part upon the adequacy of the research design.

Counting and measurement. Hypotheses concern characteristics of members of classes of phenomena. We speak of these classes of phenomena as populations (or universes) whether or not the units of observation are human beings. Verification involves counting or measuring. In the crudest case the scientist counts individuals to determine the number who do and who do not have a given characteristic. If the characteristic can be measured, the scientist will ordinarily prefer to measure the degree to which each individual possesses it.

Scientific measurement involves considerably more than Geiger counters and oscilloscopes. It may include any technique of recording the phenomenon in question with a minimum of fluctuation resulting from human idiosyncrasies. It may depend upon a multimillion dollar cyclotron or a man standing on a street-corner counting passers-by. Following are criteria for assessing the worth of a scientific instrument (that is, for telling how good an index is): relevance, validity, and reliability. By relevance we mean an affirmative answer to the question: Does it measure what the scientist happens to be interested in? By validity we mean the degree to which it measures what it is supposed to measure. Reliability refers to the degree to which the instrument gives constant measurements of an unchanging entity on different occasions.

In a broad sense, measurement consists of transforming physical characteristics into numbers or, more properly, into numerals. In this sense, even the assignment of serial numbers to soldiers or to equipment constitutes measurement. Similarly, the transformation of academic classifications, freshman, sophomore, junior, senior, into the numerals 1, 2, 3, 4 respectively is a form of measurement. The important point here is that such numerals should not be confused with real numbers for they do not necessarily have the algebraic properties of numbers. Addition, for example, is a property of numbers, but if we try to apply the arithmetic truth $1 + 1 = 2$ to the above illustrative measurement, we obtain the ridiculous proposition: "Two freshmen equal one sophomore." All too frequently social scientists fall into the trap of confusing numerals with numbers and incorrectly investing the former with mathematical properties of the latter. Whenever an investigator reports that he has added or multiplied scores of some sort, the reader whould ask himself whether or not the operation is justified by the measurement that was used. [17]

Sampling in studies of properties. Studies of properties and studies of relations call for different approaches to the problem of sampling. Let us discuss first what is involved in the study of properties. We notice first that the end toward which such a study is aimed is to provide the number which completes such a statement as: The crude birth rate of the United States in 1961 was ——— per 1000 population, or the median age at first marriage for white males in the United States in 1961 was ——— years. Ideally the procedure for arriving at the answer to a question about some property would be to make an observation on every member of the referent population and then to compute the desired parameter by means of simple arithmetic.

Because of the expense involved it is usually impossible to count all the individuals in an entire population who do or do not have a characteristic, or to measure a characteristic for the entire population. The compromise of the scientist (both social and natural) is to draw a sample of the population which he tries to make as representative of the whole population as he can, that is, a sample in which the proportion of each of the constituent segments, strata, etc. is the same as in the population. Since "even a large sample confined to a portion of the population is devoid of information about the excluded portions," [18] and since the purpose of this kind of investigation is to arrive at a statement about the entire population, representativeness of the sample is very important. With a representative sample the scientist is able to make an estimate of the value in the population of the property in which he is interested, or more technically, he is able to estimate the parameter of that property in the population. If, as is usually the case, the scientist also wishes to calculate the probability that his estimate deviates by any given amount from the true value, he must also sample randomly. In its most elementary form, labeled *simple* random sampling, this process consists of selecting a sample of size n (where n is some positive integer) in such a fashion that any such sample of size n is just as likely—but no likelier—to be chosen than any other sample of size n.[19] Random sampling also assures a degree of

[17] For a further discussion of measurement or, as it is sometimes called, scaling, see Warren S. Torgerson, *Theory and Methods of Scaling*, New York, Wiley, 1958, chs. 1-3.

[18] G. W. Snedecor, *Statistical Methods* (4th ed.), Ames, Collegiate Press, 1946, p. 460.

[19] The emphasis above on the representativeness of the sample should not be construed as implying approval of the so-called "quota" sample. For a detailed but ele-

representativeness of the sample, although this method does not maximize representativeness.

Sampling in studies of relations. The purpose of studies of relations is to find the nature of the relation, if any, between two or more variables. (In the next section we shall see that the conclusion of such a study may assert the probability that an observed state of affairs might have occurred by chance in a population in which no relation exists between the variables.) It frequently happens that the scrutiny of a relation or the test of a hypothesis is best carried out with something other than a representative sample. In the study by Strodtbeck, for example, there was interest in the relation between orientation to achievement in adolescent boys on the one hand and ethnicity and socioeconomic status on the other. If Strodtbeck had sampled representatively, the Jews would have been largely in the middle class whereas there would have been many lower-class Italians. For this reason ethnicity and socioeconomic status would have been highly confounded. Accordingly, it would not have been possible for Strodtbeck to conclude that status differences seemed to account for the differences in achievement and that "ethnic differences in family interaction are not of great relevance in explaining [differential achievement]." Because he had a problem in the study of relations, Strodtbeck drew samples of equal numbers of middle-class Jews and Italians and of lower-class Jews and Italians. With this procedure the effects of ethnicity and socioeconomic status could be segregated.

Again it is important to add that randomness of sampling is essential if the investigator expects to rely upon statistical analysis for the purpose of making inferences about the truth or falsity of the relationships in the population sampled, that is, about the probability that his results show a chance phenomenon.

Design of research: the "classical" experiment. From the discussion to follow we shall see that the conclusions which we may draw from a piece of research are determined in part by the way in which the study is set up or, to use the more technical term, are dependent in part on the "design" of the research. Although it is frequently impossible to use such a rigorous and ideal design in social research, we shall first consider a classical design of experiment. Then we shall note the ways in which certain practicable designs diverge from the ideal and with what logical consequences.

The simplest design that contains the essentials of the experimental method is as follows: A random sample is divided into two groups (or

mentary discussion of sampling procedures in social research, see Leslie Kish, "Selection of the Sample," in Leon Festinger and Daniel Katz (eds.), *Research Methods in the Behavioral Sciences*, New York, Holt, Rinehart and Winston, 1953, ch. 5.

subsamples), one called the "experimental" group, and the other, the "control" group. At this stage the two groups should be as much alike as possible in all relevant respects. If the assignment of individuals from the original sample to the two subsamples is done in a random manner, then the experimental and control groups should be sufficiently similar, and both groups should be representative samples of the population.[20]

After the two groups have been formed, every individual in each group is measured with respect to the index of the variable under study, or a count is made in each group to determine how many individuals possess or lack the attribute in question.

Next the experimental group—but not the control group—is subjected to a certain controlled condition (sometimes called "treatment"). After the experimental group has undergone the treatment, measures are obtained from both groups with respect to the variable about which the experimenter is concerned.

We now have two measurements (before and after treatment) on each of the two groups (experimental and control). Concerning these four measurements the researcher sets up what he calls "null hypotheses." A null hypothesis asserts that the obtained difference is *no greater than might reasonably have occurred by chance alone.*[21]

The first of these null hypotheses is employed in a comparison of the experimental-before and control-before measurements. The researcher hopes that here the test will sustain the null hypothesis because such an outcome is evidence that the two groups were initially similar. When the difference between the experimental-after and the control-after cells is greater than might reasonably be imputed to chance, this difference is interpreted to mean that the experimental treatment has resulted in a measurable effect.

Now let us summarize in schematic form this discussion of what we shall call the "fourfold" or "four-cell" experimental design:

1. A random sample of some specified population is drawn.
2. By random assignment or some other procedure the sample is divided into two subsamples, called "experimental" group and "control" group.

---

[20] While random assignment to the two groups is probably the usual procedure, if conditions are favorable, the matched-pair method is more efficient. For a discussion of the effects of random assignment of subjects, see Robert McGinnis, "Randomization and Inference in Sociological Research," *American Sociological Review*, 23, 1958, 408-414.

[21] Strictly speaking, a null hypothesis asserts that the two subsamples being considered, only one of which has been subjected to an experimental treatment, differ so little with respect to the measured variable that the probability is acceptably high that they could have been drawn from a single population; that is, a null hypothesis implies that the experimental treatment has had no measurable effect.

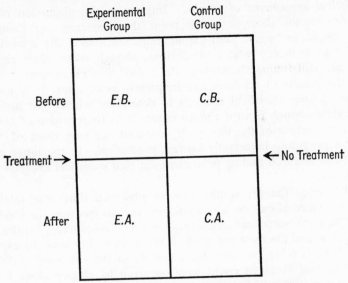

Figure 1.   Schematic diagram of four-cell experimental design.

3. Every individual in each group is measured with respect to the index of the variable under study. Or if the variable is an attribute, frequencies of "present" and "absent" are counted. The counts or measurements are entered in cells "E.B." and "C.B." of the fourfold table in Figure 1.
4. The two groups are compared with respect to the distribution of the variable measured or counted. Since the two groups are intended to be alike at this stage, it is expected that the difference between them as revealed by the measure will be no greater than might reasonably be expected by chance; that is, it is expected that the null hypothesis will be accepted.
5. The experimental group is subjected to some controlled condition, stimulus, or treatment. The control group does not have this experience.
6. Every individual in both groups is again measured with respect to the index of the variable under study. The counts or measurements are entered in cells "E.A." and "C.A." of the fourfold table in Figure 1.
7. Again the two groups are compared with respect to the distribution of the variable measured or counted. Since this comparison is made after the treatment and since the two groups were alike before treatment, a greater than chance difference at this point (a rejection of the null hypothesis) is interpreted as an effect of the treatment.[22]

22 The term "experiment" has been used by various writers with a variety of denotations. The editors of this volume regard the following as necessary conditions for a study to be regarded as an "experiment": experimental and control groups, controlled treatment or stimulus, and observations made both before and after treatment. Accordingly, the fourfold design is the simplest prototype of what we call the "experimental method."

Statistical assessment of results. Implied in the discussion of single-versus-multifactor theories is the point that, because our theories are mostly simple, or "weak," our explanations are limited. By a weak theory we refer to a theory whose predictions, though more often right than wrong, are still frequently wrong.[23] In a discipline characterized by weak theory the results of research are frequently ambiguous. It is necessary, therefore, to use statistical analysis to determine whether or not the results deviate enough from a chance outcome to be regarded as consonant with the prediction of the theory. By contrast, we may think of "strong" theory as providing invariantly correct predictions. In the latter case the trend of the research finding is so clear cut that statistical analysis is quite unnecessary.

To determine the probability that an observed difference might have arisen by chance alone, or, as statisticians say, to measure the significance of the difference, statistical techniques such as chi-square (χ^2), the critical ratio $(C.R.)$, and the t test are applied. When an author uses the expression "$P<.05$," he is telling us that, for a sample of the size used, a difference as great as that obtained could have occurred by chance alone less than 5 times out of 100.[24] This knowledge makes us reasonably certain that the difference is not a chance occurrence, but the evidence would be more compelling if $P<.01$ or if $P<.001$. These symbols tell us that if the null hypothesis is correct, a difference as large as that observed would not be expected to occur as often as once in 100 or 1000 trials respectively. Such a result, which is said to be statistically significant at the .01 or .001 level, is fairly convincing evidence that the null hypothesis is false and that instead there is a real difference between the population characteristics under investigation.

Interpretation of findings. If the results of a research are statistically significant[25] and in the direction hypothesized, they are interpreted as

[23] We may phrase this point in the language of correlation analysis. Theories in social science frequently lead to correlations no greater than .4. This may be phrased that such a theory "explains" 16 per cent [$(.4)^2 \times 100$] of the variance in the dependent variable and leaves 84 per cent [$100 - 16$] "unexplained."

[24] A critical ratio of 1.96 is always significant at the .05 level, and a critical ratio of 2.58 is always significant at the .01 level. The significance of an obtained t or chi-square, however, varies with the degrees of freedom involved. Tables of areas of the normal curve (for interpretation of the critical ratio), of t, and of chi-square are available in Herbert Arkin and Raymond R. Colton, *Tables for Statisticians,* New York, Barnes and Noble, 1950. Discussion of the application of such tests to social data is presented in such standard statistical texts as Margaret J. Hagood and Daniel O. Price, *Statistics for Sociologists* (rev. ed.), New York, Holt, Rinehart and Winston, Inc., 1952. The symbol "P" above stands for probability.

[25] What constitutes statistical significance is largely a matter of convention in the various disciplines. In fields where it is difficult to produce significant results, scientists

supporting the hypothesis. Then the researcher tentatively concludes (draws the generalization) that whenever a representative sample of the population studied is subjected to the same treatment under the same conditions, the result will be approximately the same. This conclusion is of course qualified: (*a*) by the fact that, no matter how small the *probability,* the *possibility* always exists that the result might have occurred by chance; (*b*) by the assumption that the sample was representative of the population to which the result is to be generalized; and (*c*) by the assumption that all relevant conditions remain constant or else vary with predictable consequences.

Such a generalization is founded on one of the assumptions basic to science: that *there is uniformity in nature,* or in other words, that what results today from a particular combination of elements under specified circumstances will result tomorrow from an identical combination of elements under duplicated circumstances. This assumption is not demonstrable with any finality, either logically or empirically. Since scientists do seem capable of making better than chance predictions of some kinds of future events, however, the assumption is pragmatically tenable. We must emphasize that an act of faith is involved in the scientist's assumption of uniformity in nature.

Approximations to the "classical" experiment. In a sense the experimental method described above is an ideal design of research. As we shall see, there are many problems in social science (as there are in natural science) wherein the utilization of this design is impossible. Accordingly research designs have been established which constitute approximations of the ideal. In the following example we can readily note the difficulty of applying the fourfold experimental design in sociological problems. Let us assume that we wish to test the hypothesis that nonmobile couples are less likely to be divorced than couples whose socioeconomic status is changed. In this case the treatment would consist in manipulating the family's social mobility. Since it is at least conceivable that the direction of mobility would make a difference, we would need two experimental groups—one which we would move upward and the other downward. Our control

---

are frequently satisfied if the probability that the observed result might have arisen by chance is no more than 1 in 20; in other disciplines 1 in 100 or 1 in 1000 may be the minimum chance probability for a result to be regarded as significant. What the statistician refers to as type II error (or β error) enters into the choice of significance level. Type II error is the acceptance of a hypothesis when it is false; type I error (or α error) is the rejection of a hypothesis when it is true. A .01 significance level implies that for each 100 samples drawn from a population in which the null hypothesis is true, one (type I) error will be made, and the null hypothesis will be rejected. Because of the reciprocal relationship between type I error and type II error it is not necessarily sound scientific judgment to minimize type I error.

group would be permitted no mobility in either direction. We would have to keep our families under these conditions for a "reasonable" period, in this kind of problem probably no less than ten years. And for the same period we would have to maintain very strict control over the activities of all families in our sample—both experimental and control. Naturally it is difficult to obtain random samples willing to undergo such experimental treatment over such a prolonged period. This difficulty has led sociologists to seek other designs of research that are less onerous to their subjects and still yield scientifically useful information.

Such an approximation occurs in the study by Roth and Peck, where couples are classified by intergenerational mobility (although these authors were not studying the incidence of divorce). Logically, this design is roughly comparable to the use of only the two bottom cells in the fourfold table, that is, the experimental-after and the control-after. The similarity is only rough because, instead of having one group that has been subjected to the treatment of mobility and a control group, there are two experimental groups—the upwardly and downwardly mobile—and a control group—the nonmobile. But the important logical difference is that both "before" cells are missing. A research in which the "before" cells are missing but in which there are two or more "after" cells is sometimes spoken of as a "correlational" study; the coordinate adjective for a study using the full fourfold design is "experimental."

With designs in which the "before" cells are missing we can never be certain whether the obtained differences result from the supposedly differentiating factor (for example, mobility) or are more meaningfully associated with some other factor or factors for which the sample was not controlled (for example, level of adjustment in engagement). The fourfold method with its random assignment of subjects to control and experimental groups gives us this assurance, telling us that the groups were alike before the treatment was administered.[26]

Because of the frequency of its appearance in the literature of social science we should also mention what we may call the "clinical" design or

[26] This set of considerations has led Selvin to take the extreme position that tests of significance are therefore useless and meaningless in correlational studies. (See Hanan C. Selvin, "A Critique of Tests of Significance in Survey Research," *American Sociological Review*, 22, 1957, 519-527.) It is the editors' position (*a*) that in the absence of the complete fourfold design it is frequently scientifically useful and statistically legitimate to determine whether or not a "real" difference exists (that is, to determine whether or not it is "significant" at some specified level), (*b*) that to turn up a significant difference is not equivalent to showing causation because without the fourfold design the probability rises sharply that some uncontrolled variable is influencing the result, and (*c*) that therefore the researcher must be more cautious in interpreting results from a correlational study than from an experimental study.

method. This method involves taking from the bottom two cells individuals who show the effect and then trying to ascertain the nature of the treatment to which they have been subjected. To illustrate this method let us assume that a psychiatrist examines a number of juvenile delinquents in a child guidance clinic and observes what appears to be a high incidence of broken homes among them. From these observations he concludes that juvenile delinquency is a consequence of broken homes.

Now let us examine what he has done. He has seen an aggregate of subjects who are probably not a random sample of any population. These subjects have in common a characteristic, delinquency, which the psychiatrist regards as the effect of the treatment of broken homes. Note that under the conditions specified he has no nondelinquent subjects. For this reason he has no way to determine whether the proportion having been subjected to the treatment (broken homes) is greater among those who do or among those who do not show the effect (delinquency). He is assuming the proportion of broken homes to be greater among the delinquents; his data cannot certify this to be true.

In other words, the clinical method has its own logical limitations in addition to those associated with correlational studies. We have already noted that when a study involves only the two "after" cells, it is never possible to conclude that the observed difference between those two cells is a consequence of the variable (or treatment) under study and of that variable only. The additional limitation resulting from the clinical method is that the researcher has no way of knowing how great a difference he should try to explain, or, indeed, whether or not the difference which he is trying to explain exists.

Before concluding our remarks about the clinical method, let us note one further way in which it is frequently used. This is the single case. In some fields it is not unusual to find articles in which some general proposition is advanced, the evidence for which is from a single case. To continue with our example of the foregoing paragraphs, we would have an illustration of this procedure if the psychiatrist had seen just one delinquent child, had discovered that the child came from a broken home, and then had presented this finding as a generally valid proposition.

What may we say about the logical status of this kind of evidence? We may note that all the logical limitations of the clinical method are of course still operative and that one more has been added. This limitation results from the fact that with only one case the investigator does not know if the treatment which he has identified as causative is common to, or occurs with "high" incidence among, individuals showing the experimental effect. To refer to our example again, simply because he had seen one delinquent who came from a broken home, the psychiatrist would be completely unjustified in concluding that all or most delinquents came from broken homes.

It is commonplace that it is extremely hazardous to generalize from a single case. From this analysis we can see that the hazard consists in all the limitations of the two-cell design, plus the additional limitations of the clinical design, plus the lack of evidence that the treatment observed occurs with any appreciable frequency among individuals manifesting the experimental effect.

Assessment of research designs. Our consideration of designs of research has been from the standpoint of studying relations among variables. When we view as our objective the testing of hypotheses, it is clear that of the designs that we have discussed the four-cell experimental method is best. If, however, our objective is the generation of new hypotheses, there is much to be said for the other methods, especially the clinical method. For example, when it is at all empirical, the Freudian literature consists almost entirely of clinical designs, frequently with but a single case. While the foregoing logical analysis shows that such studies can never be conclusive in establishing hypotheses, the Freudian literature has been remarkably fruitful as a source of new hypotheses. The clinical type of study has great value in the generation of hypotheses because, although it may involve only one or a very few cases, the cases are generally studied in great detail and thus they enable the reader to view the processes of change through time. An ethnographic description of a single culture, moreover, is logically a clinical design based on a single case.[27] Although such studies can never establish propositions that are generally valid in any crosscultural sense, they have provided social scientists with many ideas for hypotheses. Finally, we may note that the single case may be presented to facilitate communication, that is, to *illustrate* (but of course not to prove) a proposition. This is the purpose of the case presented by Ktsanes and Ktsanes.

Utilizing the scientific outlook. Before proceeding further, the reader should be cautioned about a possible difficulty which he may experience in undertaking to apply the findings of scientific research. Because the language of science and the language of everyday life do not stand in a one-to-one relationship, one can be misled by a direct translation. Human thought, including that of the scientist, is no more profound nor precise than the language used for expressing it. To facilitate communication and to refine his abstractions the scientist develops a language which, while borrowing terminology from the language of common sense, strips this

---

[27] When describing a culture, the anthropologist is concerned with a single case (the culture in question) despite the possibility that thousands or even millions of individuals may participate in that culture.

terminology of ambiguity, or "surplus meaning." The concepts of the scientist are, so to speak, skeletons of the concepts of common sense.

The conversations of common sense, on the other hand, usually proceed on the assumption that A knows what B means by a particular term, though neither may be quite sure. For example, the concept "love" can take on a variety of meanings. One cannot always be certain whether it refers to sexual desire, infatuation, brotherly love, "mature" love, or any other of a number of distinctions commonly made. In "The Theory of Complementary Needs in Mate-Selection," on p. 521, Winch offers a scientific definition of this same concept. Is this what we mean in everyday conversation? Probably not, for Winch has sought to abstract the concept to the point where surplus meaning and the ambiguity of the above terms is minimized. Since in interpersonal interaction a less abstract and more connotative concept is more likely to be used, one should not attempt to apply scientific propositions to one's personal problems or affairs without carefully examining the definitions involved in these propositions. The layman tends to think of his problems in the language of common sense, and this form of discourse is somewhat remote from the abstractions of science. Certainly these remarks do not imply that everyday problems are not "real" problems, nor that scientific propositions are of no practical use. Rather it *is* implied that care should be exercised to avoid confusing the two realms of discourse.

We have just concluded a discussion on the use of scientific method in social research. Our purpose has been to provide the reader with some means of evaluating the papers to follow. We, the editors, regard these articles as being among the best studies of marriage and the family. In view of the scientific immaturity of social science generally, however, the reader need not be surprised in discovering that none of these studies employs the fourfold experimental design and that many of them will prove vulnerable to criticism from the standpoint of ideal scientific procedure. As a consequence of the foregoing discussion, we hope that the reader will examine the following papers with such questions as the following in mind: How sound are the premises? How representative is the sample? Of what population is it alleged to be representative? How rigorous is the design of research? Do the conclusions necessarily follow from the design and the data? Over what range may the conclusions be generalized? What are the probable consequences of the conclusions? If the article is not an empirical study, does it offer hypotheses which are amenable to test? Can we move from abstract variables to concrete data without too much logical violence? Such questions are in order with respect to each article in the book. For that matter, such questions are in order whenever someone asserts: "The truth is that . . ."

2. FAMILIAL STRUCTURE IN GENERAL AND IN THE UNITED STATES

Let us think of a family as a group of two or more persons joined by ties of marriage, blood, or adoption, who interact with each other in their respective familial roles. (For some purposes it may be useful to add other conditions such as common residence.) In this chapter and the next two we shall examine the structure of the family; familial functions will be considered in chapters 5 and 6. Structure will be analyzed largely in terms of roles, but also in terms of kinship and size.

The sexual division of labor and the resulting differentiation of marital roles is taken up by Murdock, who also draws the convenient distinction between the nuclear and extended families. The relation of the nuclear to the extended family and of both of these to nonkin is illuminated by Radcliffe-Brown's analysis of kinship. The subculture of the American middle class values the small family as a residential unit or household. Arensberg argues that industrialization cannot be a sufficient cause for this value, as some believe, since there are industrialized societies which do not value the small-family system.

A generation ago Louis Wirth characterized "Urbanism as a Way of Life" as an impersonal interaction among anonymous actors.[1] In recent years there has been a reaction against Wirth's view of the city. In particular, studies have reported a considerable amount of visiting and mutual aid among relatives in different households.[2] One such study is by Sussman, who has investigated mu-

[1] *American Journal of Sociology*, 44, 1938, 1-24.

[2] See, for example, Detroit Area Study, *A Social Profile of Detroit: 1955*, Ann Arbor, Survey Research Center, 1956. The most frequent category of relative in the area was sibling and sibling-in-law, reported by three-quarters of the respondents. Two-thirds of the respondents saw one or more relatives outside the nuclear family as often as once a

tual aid among kin in Cleveland and has concluded that the view of the nuclear family as isolated is more fiction than fact. To the editors it is not clear exactly where "truth" lies in this dispute. If Wirth is interpreted literally, it seems clear that his view must be rejected for these studies show that urban families are not necessarily isolated from all extended kin. If Wirth is interpreted as meaning that urban life is more anonymous than nonurban life, these studies have not refuted him because they have been single-variable (or property), rather than relational, studies. One example of the consequence of taking other variables into account comes from a finding that Catholic and Jewish middle-class families interact more intensively with extended kin than do Protestant families of the same class.[3] The studies referred to above have been done in large cities, and the proportion of Jews and Catholics among the white population within the limits of large cities is greater than in surburbs, small cities, and rural areas. It may be that,—with religious affiliation controlled,—urbanism will turn out to be correlated with the isolation of the nuclear family, or perhaps there will be no correlation at all.

STRUCTURES AND FUNCTIONS OF THE FAMILY

George Peter Murdock

Adapted from *Social Structure* by George P. Murdock. Copyright 1949 by The Macmillan Company and used with their permission.

The family is a social group characterized by common residence, economic cooperation, and reproduction. It includes adults of both sexes, at least two of whom maintain a socially approved sexual relationship, and one or more children, own or adopted, of the sexually cohabiting adults. The family is to be distinguished from marriage, which is a complex of customs centering upon the relationship between a sexually associating pair of adults within the family. Marriage defines the manner of establishing and terminating such a relationship, the normative behavior and reciprocal obligations within it, and the locally accepted restrictions upon its personnel.

Used alone, the term "family" is ambiguous. The layman and even the

---

week. See also "Family, Class, and Generation in London" by Peter Willmott and Michael Young.

[3] Robert F. Winch, *Identification and Its Familial Determinants*, Indianapolis, Bobbs-Merrill, in press, table 8. Highly relevant is a study of "family circles" and "cousins' clubs" among New York City Jews. William E. Mitchell and Hope J. Leichter have some as yet unpublished papers on these "urban ambilineages."

social scientist often apply it undiscriminatingly to several social groups which, despite functional similarities, exhibit important points of difference. These must be laid bare by analysis before the term can be used in rigorous scientific discourse.

. . . The most basic type of family, called herewith the *nuclear family,* consists typically of a married man and woman with their offspring, although in individual cases one or more additional persons may reside with them. The nuclear family will be familiar to the reader as the type of family recognized to the exclusion of all others by our own society. Among the majority of the peoples of the earth, however, nuclear families are combined, like atoms in a molecule, into larger aggregates. These composite forms of the family fall into two types, which differ in the principles by which the constituent nuclear families are affiliated. A *polygamous*[1] *family* consists of two or more nuclear families affiliated by plural marriages, *i.e.,* by having one married parent in common.[2] Under polygyny, for instance, one man plays the role of husband and father in several nuclear families and thereby unites them into a larger familial group. An *extended family* consists of two or more nuclear families affiliated through an extension of the parent-child relationship rather than of the husband-wife relationship, *i.e.,* by joining the nuclear family of a married adult to that of his parents. The patrilocal extended family, often called the patriarchal family, furnishes an excellent example. It embraces, typically, an older man, his wife or wives, his unmarried children, his married sons, and the wives and children of the latter. Three generations, including the nuclear families of father and sons, live under a single roof or in a cluster of adjacent dwellings.

. . . The nuclear family is a universal human social grouping. The reasons for its universality do not become fully apparent when the nuclear family is viewed merely as a social group. Only when it is analyzed into its constituent relationships, and these are examined individually as well as collectively, does one gain an adequate conception of the family's many-sided utility and thus of its inevitability. A social group arises when a series of interpersonal relationships, which may be defined as sets of reciprocally adjusted habitual responses, binds a number of participant individuals collectively to one another. In the nuclear family, for example, the clustered relationships are eight in number: husband-wife, father-son,

[1] The terms "polygamy" and "polygamous" will be used throughout this work in their recognized technical sense as referring to any form of plural marriage; "polygyny" will be employed for the marriage of one man to two or more women, and "polyandry" for the marriage of one woman to two or more men.

[2] Cf. M. K. Opler, "Woman's Social Status and the Forms of Marriage," *American Journal of Sociology,* XLIX, 1943, 144; A. R. Radcliffe-Brown, "The Study of Kinship Systems," *Journal of the Royal Anthropological Institute,* LXXI, 1941, 2.

father-daughter, mother-son, mother-daughter, brother-brother, sister-sister, and brother-sister. The members of each interacting pair are linked to one another both directly through reciprocally reinforcing behavior and indirectly through the relationships of each to every other member of the family. Any factor which strengthens the tie between one member and a second, also operates indirectly to bind the former to a third member with whom the second maintains a close relationship. An explanation of the social utility of the nuclear family, and thus of its universality, must consequently be sought not alone in its functions as a collectivity but also in the services and satisfactions of the relationships between its constituent members.

The relationship between father and mother in the nuclear family is solidified by the sexual privilege which all societies accord to married spouses. As a powerful impulse, often pressing individuals to behavior disruptive of the cooperative relationships upon which human social life rests, sex cannot safely be left without restraints. All known societies, consequently, have sought to bring its expression under control by surrounding it with restrictions of various kinds. On the other hand, regulation must not be carried to excess or the society will suffer through resultant personality maladjustments or through insufficient reproduction to maintain its population. All peoples have faced the problem of reconciling the need of control with the opposing need of expression, and all have solved it by culturally defining a series of sexual taboos and permissions. These checks and balances differ widely from culture to culture, but without exception a large measure of sexual liberty is everywhere granted to the married parents in the nuclear family. Husband and wife must adhere to sexual etiquette and must, as a rule, observe certain periodic restrictions such as taboos upon intercourse during menstruation, pregnancy, and lactation, but normal sex gratification is never permanently denied to them. . . .

As a means of expressing and reducing a powerful basic drive, as well as of gratifying various acquired or cultural appetites, sexual intercourse strongly reinforces the responses which precede it. These by their very nature are largely social, and include cooperative acts which must, like courtship, be regarded as instrumental responses. Sex thus tends to strengthen all the reciprocal habits which characterize the interaction of married parents, and indirectly to bind each into the mesh of family relationship in which the other is involved.

To regard sex as the sole factor, or even as the most important one, that brings a man and a woman together in marriage and binds them into the family structure would, however, be a serious error. If all cultures, like our own, prohibited and penalized sexual intercourse except in the marital relationship, such an assumption might seem reasonable. But this is emphatically not the case. Among those of our 250 societies for which information is available, 65 allow unmarried and unrelated persons complete freedom in sexual matters, and 20 others give qualified consent, while only 54 forbid or disapprove premarital liaisons between non-relatives, and many of these allow sex relations between specified relatives such as cross-cousins.[3] Where premarital license prevails, sex certainly cannot be alleged as the primary force driving people into matrimony.

Nor can it be maintained that, even after marriage, sex operates exclusively to reinforce the matrimonial relationship. To be sure, sexual intercourse between a married man and an unrelated woman married to another is forbidden in 126 of our sample societies, and is freely or conditionally allowed in only 24. These figures, however, give an exaggerated impression of the prevalence of cultural restraints against extramarital sexuality, for affairs are often permitted between particular relatives though forbidden with non-relatives. Thus in a majority of the societies in our sample for which information is available a married man may legitimately carry on an affair with one or more of his female relatives, including a sister-in-law in 41 instances. Such evidence demonstrates conclusively that sexual gratification is by no means always confined to the marital relationship, even in theory. If it can reinforce other relationships as well, as it commonly does, it cannot be regarded as peculiarly conducive to marriage or as alone accountable for the stability of the most crucial relationship in the omnipresent family institution. . . .

In view of the frequency with which sexual relations are permitted outside of marriage, it would seem the part of scientific caution to assume merely that sex is an important but not the exclusive factor in maintaining the marital relationship within the nuclear family, and to look elsewhere for auxiliary support. One such source is found in economic cooperation, based upon a division of labor by sex.[4] Since cooperation, like sexual association, is most readily and satisfactorily achieved by persons who habitually reside together, the two activities, each deriving from a basic biological need, are quite compatible. Indeed, the gratifications from each serve admirably to reinforce the other.

By virtue of their primary sex differences, a man and a woman make an exceptionally efficient cooperating unit.[5] Man, with his superior physical

[3] A cross-cousin is the child of a father's sister or of a mother's brother. The children of a father's brother and of a mother's sister are technically known as "parallel cousins."

[4] See W. G. Sumner and A. G. Keller, *The Science of Society,* New Haven, 1927, III, 1505-18.

[5] *Ibid.,* I, 111-40.

strength, can better undertake the more strenuous tasks, such as lumbering, mining, quarrying, land clearance, and housebuilding. Not handicapped, as is woman, by the physiological burdens of pregnancy and nursing, he can range farther afield to hunt, to fish, to herd, and to trade. Woman is at no disadvantage, however, in lighter tasks which can be performed in or near the home, *e.g.,* the gathering of vegetable products, the fetching of water, the preparation of food, and the manufacture of clothing and utensils. All known human societies have developed specialization and cooperation between the sexes roughly along this biologically determined line of cleavage.[6] It is unnecessary to invoke innate psychological differences to account for the division of labor by sex; the indisputable differences in reproductive functions suffice to lay out the broad lines of cleavage. New tasks, as they arise, are assigned to one sphere of activities or to the other, in accordance with convenience and precedent. Habituation to different occupations in adulthood and early sex typing in childhood may well explain the observable differences in sex temperament, instead of *vice versa.*[7]

The advantages inherent in a division of labor by sex presumably account for its universality. Through concentration and practice each partner acquires special skill at his particular tasks. Complementary parts can be learned for an activity requiring joint effort. If two tasks must be performed at the same time but in different places, both may be undertaken and the products shared. The labors of each partner provide insurance to the other. The man, perhaps, returns from a day of hunting, chilled, unsuccessful, and with his clothing soiled and torn, to find warmth before a fire which he could not have maintained, to eat food gathered and cooked by the woman instead of going hungry, and to receive fresh garments for the morrow, prepared, mended, or laundered by her hands. Or perhaps the woman has found no vegetable food, or lacks clay for pottery or skins for making clothes, obtainable only at a distance from the dwelling, which she cannot leave because her children require care; the man in his ramblings after game can readily supply her wants. Moreover, if either is injured or ill, the other can nurse him back to health. These and similar rewarding experiences, repeated daily, would suffice of themselves to cement the union. When the powerful reinforcement of sex is added, the partnership of man and woman becomes inevitable.

Sexual unions without economic cooperation are common, and there

---

[6] See G. P. Murdock, "Comparative Data on the Division of Labor by Sex," *Social Forces,* XV, 1937, 551-3, for an analysis of the distribution of economic activities by sex in 224 societies.

[7] Cf. M. Mead, *Sex and Temperament in Three Primitive Societies,* New York, Morrow, 1935.

are relationships between men and women involving a division of labor without sexual gratification, *e.g.,* between brother and sister, master and maidservant, or employer and secretary, but marriage exists only when the economic and the sexual are united into one relationship, and this combination occurs only in marriage. Marriage, thus defined, is found in every known human society. In all of them, moreover, it involves residential cohabitation, and in all of them it forms the basis of the nuclear family. Genuine cultural universals are exceedingly rare. It is all the more striking, therefore, that we here find several of them not only omnipresent but everywhere linked to one another in the same fashion.

Economic cooperation not only binds husband to wife; it also strengthens the various relationships between parents and children within the nuclear family. Here, of course, a division of labor according to age, rather than sex, comes into play. What the child receives in these relationships is obvious; nearly his every gratification depends upon his parents. But the gains are by no means one-sided. In most societies, children by the age of six or seven are able to perform chores which afford their parents considerable relief and help, and long before they attain adulthood and marriageability they become economic assets of definite importance. One need only think here of the utility of boys to their fathers and of girls to their mothers on the typical European or American farm. Moreover, children represent, as it were, a sort of investment or insurance policy; dividends, though deferred for a few years, are eventually paid generously in the form of economic aid, of support in old age, and even, sometimes, of cash returns, as where a bride-price is received for a daughter when she marries.

Siblings[8] are similarly bound to one another through the care and help given by an elder to a younger, through cooperation in childhood games which imitate the activities of adults, and through mutual economic assistance as they grow older. Thus, through reciprocal material services sons and daughters are bound to fathers and mothers and to one another, and the entire family group is given firm economic support.

Sexual cohabitation leads inevitably to the birth of offspring. These must be nursed, tended, and reared to physical and social maturity if the parents are to reap the afore-mentioned advantages. Even if the burdens of reproduction and child care outweigh the selfish gains to the parents, the society as a whole has so heavy a stake in the maintenance of its numbers, as a source of strength and security, that it will insist that parents fulfill these obligations. Abortion, infanticide, and neglect, unless confined within safe limits, threaten the entire community and arouse its

[8] The term "sibling" will be employed throughout this work in its technical sense as designating either a brother or a sister irrespective of sex.

members to apply severe social sanctions to the recalcitrant parents. Fear is thus added to self-interest as a motive for the rearing of children. Parental love, based on various derivative satisfactions, cannot be ignored as a further motive; it is certainly no more mysterious than the affection lavished by many people on burdensome animal pets, which are able to give far less in return. Individual and social advantages thus operate in a variety of ways to strengthen the reproductive aspects of the parent-child relationships within the nuclear family.

The most basic of these relationships, of course, is that between mother and child, since this is grounded in the physiological facts of pregnancy and lactation and is apparently supported by a special innate reinforcing mechanism, the mother's pleasure or tension release in suckling her infant. The father becomes involved in the care of the child less directly, through the sharing of tasks with the mother. Older children, too, frequently assume partial charge of their younger siblings, as a chore suited to their age. The entire family thus comes to participate in child care, and is further unified through this cooperation.

No less important than the physical care of offspring, and probably more difficult, is their social rearing. The young human animal must acquire an immense amount of traditional knowledge and skill, and must learn to subject his inborn impulses to the many disciplines prescribed by his culture, before he can assume his place as an adult member of his society. The burden of education and socialization everywhere falls primarily upon the nuclear family, and the task is, in general, more equally distributed than is that of physical care. The father must participate as fully as the mother because, owing to the division of labor by sex, he alone is capable of training the sons in the activities and disciplines of adult males.[9] Older siblings, too, play an important role, imparting knowledge and discipline through daily interaction in work and play. Perhaps more than any other single factor, collective responsibility for education and socialization welds the various relationships of the family firmly together. . . .

Agencies or relationships outside of the family may, to be sure, share in the fulfillment of any of these functions, but they never supplant the family. There are, as we have seen, societies which permit sexual gratification in other relationships, but none which deny it to married spouses. There may be extraordinary expansion in economic specialization, as in modern industrial civilization, but the division of labor between man and wife still persists. There may, in exceptional cases, be little social disapproval of childbirth out of wedlock, and relatives, servants, nurses, or pediatricians may assist in child care, but the primary responsibility for bearing and rearing children ever remains with the family. Finally, grandparents,

[9] Cf. R. Linton, *The Study of Man,* New York, 1936, p. 155.

schools, or secret initiatory societies may assist in the educational process, but parents universally retain the principal role in teaching and discipline. No society, in short, has succeeded in finding an adequate substitute for the nuclear family, to which it might transfer these functions. It is highly doubtful whether any society ever will succeed in such an attempt, utopian proposals for the abolition of the family to the contrary notwithstanding. . . .

TABLE 1.

Relative (of man)	Premarital Intercourse		Postmarital Intercourse		Marriage	
	Forbidden	Permitted	Forbidden	Permitted	Forbidden	Permitted
Mother	76	0	74	0	184	0
Sister	109	0	106	0	237	0
Daughter			81	0	198	0

Perhaps the most striking effect of family structure upon individual behavior is to be observed in the phenomenon of incest taboos. . . . Despite an extraordinary variability and seeming arbitrariness in the incidence of incest taboos in different societies, they invariably apply to every cross-sex relationship within the nuclear family save that between married spouses. In no known society is it conventional or even permissible for father and daughter, mother and son, or brother and sister to have sexual intercourse or to marry. Despite the tendency of ethnographers to report marriage rules far more fully than regulations governing premarital and postmarital incest, the evidence from our 250 societies, presented in Table 1, is conclusive.

The few apparent exceptions, in each instance too partial to appear in the table, are nevertheless illuminating, and all those encountered will therefore be mentioned. Certain high Azande nobles are permitted to wed their own daughters, and brother-sister marriages were preferred in the old Hawaiian aristocracy and in the Inca royal family. In none of these instances, however, could the general population contract incestuous unions, for these were a symbol and prerogative of exalted status. Among the Dobuans, intercourse with the mother is not seriously regarded if the father is dead; it is considered a private sin rather than a public offense. The Balinese of Indonesia permit twin brothers and sisters to marry on the ground that they have already been unduly intimate in their mother's womb. Among the Thonga of Africa an important hunter, preparatory to a great hunt, may have sex relations with his daughter—a heinous act under other circumstances. By their special circumstances or exceptional character these cases serve rather to emphasize than to disprove the universality of intra-family incest taboos.

A major consequence of these taboos is that they make the nuclear family discontinuous over time and confine it to two generations. If brother-sister marriages were usual, for example, a family would normally consist of married grandparents, their sons and daughters married to one another, the children of the latter, and even the progeny of incestuous unions among these. The family, like the community, the clan, and many other social groups, would be permanent, new births ever filling the gaps caused by deaths. Incest taboos completely alter this situation. They compel each child to seek in another family for a spouse with whom to establish a marital relationship. In consequence thereof, every normal adult in every human society belongs to at least two nuclear families—a *family of orientation* in which he was born and reared, and which includes his father, mother, brothers, and sisters, and a *family of procreation*[10] which he establishes by his marriage and which includes his husband or wife, his sons, and his daughters.

ANALYSIS OF SYSTEMS OF KINSHIP

A. R. Radcliffe-Brown

Adapted from "Introduction" to *African Systems of Kinship and Marriage* by A. R. Radcliffe-Brown and Darryl Forde (eds.) New York, Oxford University Press, 1950, 1-43, and the International African Institute.

[Analysis] of kinship systems all over the world . . . reveals that while there is a very wide range of variation in their superficial features there can be discovered a certain small number of general structural principles which are applied and combined in various ways. It is one of the first tasks of a theoretical study of kinship to discover these principles by a process of abstractive generalization based on analysis and comparison. . . .

A system of kinship and marriage can be looked at as an arrangement which enables persons to live together and co-operate with one another in an orderly social life. For any particular system as it exists at a certain time we can make a study of how it works. To do this we have to consider how it links persons together by convergence of interest and sentiment and how it controls and limits those conflicts that are always possible as the result of divergence of sentiment or interest. In reference to any feature of a system we can ask how it contributes to the working of the system. This is what is meant by speaking of its social function. When we

[10] For these very useful terms we are indebted to W. L. Warner.

succeed in discovering the function of a particular custom, i.e. the part it plays in the working of the system to which it belongs, we reach an understanding or explanation of it which is different from and independent of any historical explanation of how it came into existence. . . .

We have first of all to try to get a clear idea of what is a kinship system or system of kinship and marriage. Two persons are kin when one is descended from the other, as, for example, a grandchild is descended from a grandparent, or when they are both descended from a common ancestor. Persons are cognatic kin or cognates when they are descended from a common ancestor or ancestress counting descent through males and females.

The term "consanguinity" is sometimes used as an equivalent of "kinship" as above defined, but the word has certain dangerous implications which must be avoided. Consanguinity refers properly to a physical relationship. The difference is clear if we consider an illegitimate child in our own society. Such a child has a "genitor" (physical father) but has no "pater" (social father). Our own word "father" is ambiguous because it is assumed that normally the social relationship and the physical relationship will coincide. . . .

The complete social relationship between parent and child may be established not by birth but by adoption as it was practised in ancient Rome and is practised in many parts of the world to-day.

In several regions of Africa there is a custom whereby a woman may go through a rite of marriage with another woman and thereby she stands in the place of a father (pater) to the offspring of the wife, whose physical father (genitor) is an assigned lover. . . .

The closest of all cognatic relationships is that between children of the same father and mother. Anthropologists have adopted the term "sibling" to refer to this relationship; a male sibling is a brother, a female sibling is a sister.[1] The group consisting of a father and a mother and their children is an important one for which it is desirable to have a name. The term "elementary family" will be used in this sense in this essay.* (The term "biological family" refers to something different, namely, to genetic, relationship such as that of a mated sire and dam and their offspring, and is the concern of the biologist making a study of heredity. But it seems inappropriate to use the word "family" in this connexion.) We may regard the elementary family as the basic unit of kinship structure. What is meant by this is that the relationships, of kinship or affinity, of any person are all connexions that are traced through his parents, his siblings, his spouse, or his children. . . .

The elementary family usually provides the basis for the formation of

[1] In Anglo-Saxon "sibling" meant "kinsman."

* Cf. Murdock's concept of the "nuclear family," p. 20 above—*Eds.*

domestic groups of persons living together in intimate daily life. Of such groups there is a great variety. One common type is what may be called the "parental" family in which the "household" consists of the parents and their young or unmarried children. We are familiar with this type of family amongst ourselves, but it is also a characteristic feature of many primitive peoples. The group comes into existence with the birth of a first child in marriage; it continues to grow by the birth of other children; it undergoes partial dissolution as the children leave it, and comes to an end with the death of the parents. In a polygynous parental family there are two or more mothers but only one father, and a mother with her children constitute a separate unit of the group. . . .

The first determining factor of a kinship system is provided by the range over which these relationships are effectively recognized for social purposes of all kinds. The differences between wide-range and narrow-range systems are so important that it would be well to take this matter of range as the basis for any attempt at a systematic classification of kinship systems. The English system of the present day is a narrow-range system, though a wider range of relationship, to second, third, or more distant cousins, is recognized in rural districts than in towns. China, on the contrary, has a wide-range system. Some primitive societies have narrow-range systems, others have wide-range. In some of the latter a man may have several hundred recognized relatives by kinship and by marriage whom he must treat as relatives in his behaviour. In societies of a kind of which the Australian aborigines afford examples every person with whom a man has any social contact during the course of his life is a relative and is treated in the way appropriate to the relationship in which he or she stands. . . .

A part of any kinship system is some system of terms by which relatives of different kinds are spoken of or by which they are addressed as relatives. The first step in the study of a kinship system is to discover what terms are used and how they are used. . . .

There is one type of terminology that is usually referred to as "descriptive." In systems of this type there are a few specific terms for relatives of the first or second order and other relatives are indicated by compounds of these specific terms in such a way as to show the intermediate steps in the relation. It is necessary in any scientific discussion of kinship to use a system of this kind. Instead of ambiguous terms such as "uncle" or "cousin" we have to use more exact compound terms such as "mother's brother," "father's sister's son," and so on. When we have to deal with a relationship of the fifth order, such as "mother's mother's brother's daughter's daughter," and still more in dealing with more distant relations, the system presents difficulties to those who are not accustomed to it. . . .

In the eighteenth century Lafitau[2] reported the existence amongst American Indians of a system of terminology very unlike our own.

"Among the Iroquois and Hurons all the children of a cabin regard all their mother's sisters as their mothers, and all their mother's brothers as their uncles, and for the same reason they give the name of fathers to all their father's brothers, and aunts to all their father's sisters. All the children on the side of the mother and her sisters, and of the father and his brothers, regard each other mutually as brothers and sisters, but as regards the children of their uncles and aunts, that is, of their mother's brothers and father's sisters, they only treat them on the footing of cousins." . . .

In the nineteenth century Lewis Morgan . . . found systems of terminology similar to that of the Iroquois in many parts of the world, and such systems he called "classificatory." The distinguishing feature of a classificatory system of kinship terminology in Morgan's usage is that terms which apply to lineal relatives are also applied to certain collateral relatives. Thus a father's brother is "father" and mother's sister is "mother," while, as in the type described by Lafitau, there are separate terms for mother's brother and father's sister. Consequently in the next generation the children of father's brothers and mother's sisters are called "brother" and "sister" and there are separate terms for the children of mother's brothers and father's sisters. A distinction is thus made between two kinds of cousins, "parallel cousins" (children of father's brothers and mother's sisters), who although "collateral" in our sense are classified as "brothers" and "sisters," and "cross-cousins" (children of mother's brothers and father's sisters). There is a similar distinction amongst nephews and nieces. A man classifies the children of his brothers with his own children, but uses a separate term for the children of his sisters. Inversely a woman classifies with her own children the children of her sisters but not those of her brothers. Classificatory terminologies of this kind are found in a great many African peoples. . . .

In classificatory systems the principle of classification may be applied over a wide range of relationship. Thus a first cousin of the father, being his father's brother's son, whom he therefore calls "brother," is classified with the father and the same term "father" is applied to him. His son in turn, a second cousin, is called "brother." By this process of extension of the principle of classification nearer and more distant collateral relatives are arranged into a few categories and a person has many relatives to whom he applies the term "father" or "mother" or "brother" or "sister."

The most important feature of these classificatory terminologies was pointed out long ago by Sir Henry Maine. "The effect of the system," he wrote, "is in general to bring within your mental grasp a much greater

---

[2] Lafitau, *Mœurs des Sauvages Ameriquains,* Paris, 1724, vol. i, p. 552.

number of your kindred than is possible under the system to which we are accustomed." [3] In other words, the classificatory terminology is primarily a mechanism which facilitates the establishment of wide-range systems of kinship.

There is more to it than this, however. Research in many parts of the world has shown that the classificatory terminology, like our own and other non-classificatory systems, is used as a method of dividing relatives into categories which determine or influence social relations as exhibited in conduct. The general rule is that the inclusion of two relatives in the same terminological category implies that there is some significant similarity in the customary behavior due to both of them, or in the social relation in which one stands to each of them, while inversely the placing of two relatives in different categories implies some significant difference in customary behaviour or social relations. . . .

The reality of a kinship system as a part of a social structure consists of the actual social relations of person to person as exhibited in their interactions and their behaviour in respect of one another. But the actual behaviour of two persons in a certain relationship (father and son, husband and wife, or mother's brother and sister's son) varies from one particular instance to another. What we have to seek in the study of a kinship system are the norms. From members of the society we can obtain statements as to how two persons in a certain relationship ought to behave towards one another. A sufficient number of such statements will enable us to define the ideal or expected conduct. Actual observations of the way persons do behave will enable us to discover the extent to which they conform to the rules and the kinds and amount of deviation. Further, we can and should observe the reactions of other persons to the conduct of a particular person or their expressions of approval or disapproval. The reaction or judgement may be that of a person who is directly or personally affected by the conduct in question or it may be the reaction or judgement of what may be called public opinion or public sentiment. The members of a community are all concerned with the observance of social usage or rules of conduct and judge with approval or disapproval the behaviour of a fellow member even when it does not affect them personally.

A kinship system thus presents to us a complex set of norms, of usages, of patterns of behaviour between kindred. Deviations from the norm have their importance. For one thing they provide a rough measure of the relative condition of equilibrium or disequilibrium in the system. Where there is a marked divergence between ideal or expected behaviour and the actual conduct of many individuals this is an indication of disequilibrium; for example, when the rule is that a son should obey his father but there are

[3] *The Early History of Institutions,* 1874, p. 214.

notably frequent instances of disobedience. But there may also be a lack of equilibrium when there is marked disagreement amongst members of the society in formulating the rules of conduct or in judgements passed on the behaviour of particular persons.

In attempting to define the norms of behaviour for a particular kind of relation in a given system it is necessary to distinguish different elements or aspects. As one element in a relation we may recognize the existence of a personal sentiment, what may be called the affective element. Thus we may say that in most human societies a strong mutual affection is a normal feature of the relation of mother and child, or there may be in a particular society a typical or normal emotional attitude of a son to his father. It is very important to remember that this affective element in the relation between relatives by kinship or marriage is different in different societies.

We may distinguish also an element that it is convenient to refer to by the term "etiquette," if we may be permitted to give a wide extension of meaning to that word. It refers to conventional rules as to outward behaviour. What these rules do is to define certain symbolic actions or avoidances which express some important aspect of the relation between two persons. Differences of rank are given recognition in this way. In some tribes of South Africa it would be an extreme, and in fact unheard of, breach of the rules of propriety for a woman to utter the name of her husband's father.

An important element in the relations of kin is what will here be called the jural element, meaning by that relationships that can be defined in terms of rights and duties. Where there is a duty there is a rule that a person should behave in a certain way. A duty may be positive, prescribing actions to be performed, or negative, imposing the avoidance of certain acts. We may speak of the "performance" of a positive duty and the "observance" of a negative duty. The duties of A to B are frequently spoken of in terms of the "rights" of B. Reference to duties or rights are simply different ways of referring to a social relation and the rules of behaviour connected therewith. . . .

One principle that may be adopted [in tracing kinship] is the simple cognatic principle. To define the kin of a given person his descent is traced back a certain number of generations, to his four grandparents, his eight great-grandparents, or still farther, and all descendants of his recognized ancestors, through both females and males, are his cognates. At each generation that we go backwards the number of ancestors is double that of the preceding generation, so that in the eighth generation a person will have sixty-four pairs of ancestors (the great-grandparents of his great-great-grandparents). It is therefore obvious that there must be some limit to tracing kinship in this way. The limit may simply be a practical one depending on the inability to trace the genealogical connexions, or there may be a theoretically fixed limit beyond which the genealogical connexion does not count for social purposes.

Another way of ordering the kindred may be illustrated by the system of ancient Rome. Within the body of a person's recognized cognates certain are distinguished as agnates. Cognates are agnates if they are descendants by male links from the same male ancestor.[4] In the Roman system there was the strongest possible emphasis on agnatic kinship, i.e. on unilineal descent through males.

In some other societies there is a similar emphasis on unilineal descent through females. With such a system a person distinguishes from the rest of his cognates those persons who are descended by female links only from the same female ancestress as himself. We can speak of these as his matrilineal kin. . . .

One important way in which the unilineal principle may be used is in the formation of recognized lineage groups as part of the social structure. An agnatic lineage consists of an original male ancestor and all his descendants through males of three, four, five, or *n* generations. The lineage group consists of all the members of a lineage alive at a given time. A woman belongs to the lineage of her father, but her children do not. With matrilineal reckoning the lineage consists of a progenetrix and all her descendants through females. A man belongs to his mother's lineage, but his children do not. Lineage groups, agnatic or matrilineal, are of great importance in the social organization of many African peoples. . . .

It is desirable to illustrate by examples the differences in the ordering of kindred as the result of relative emphasis on the cognatic principle or on the unilineal principle. As an example of a cognatic system we may take the kinship system of the Teutonic peoples as it was at the beginning of history. This was based on a widely extended recognition of kinship traced through females as well as males. The Anglo-Saxon word for kinsfolk was *mae*g (*magas*). A man owed loyalty to his "kith and kin." Kith were one's friends by vicinage, one's neighbours; kin were persons descended from a common ancestor. So, for "kith and kin" Anglo-Saxon could say "his magas and his frŷnd," which is translated in Latin as *cognati atque amici*.

The arrangement of kin by degrees of nearness or distance was based on sib-ship (English *sib,* German *Sippe*). A man's sib were all his cognates within a certain degree. One method of arranging the sib was by reference to the human body and its "joints" (*glied*). . . . "Sib" may be defined as meaning computable cognatic relationship for definite social purposes . . . [This cognatic relationship] was used for fixing the de-

---

[4] "Sunt autem agnati per virilis sexus personas cognatione juncti" (Gaius); "Agnati sunt a patre cognati" (Ulpian).

grees within which marriage was forbidden. After the introduction of Christianity the relation between godparents and godchildren was included under sib. The godfather and godmother were "god-sib" (modern "gossip") to their godchildren and marriage between them was forbidden. Sib-ship also regulated the inheritance of property. Persons who were not related to a deceased person within a certain degree had no claim to inherit.

Where the functioning of the Teutonic sib can be best studied is in the customs relating to wergild, which was the indemnity that was required when one person killed another. It was paid by the person who had killed and his sib, and was received by the sib of the deceased . . . The nearest kin paid and received most, the most distant paid and received least.[5]

The payment of wergild was an indemnity for homicide paid to those persons who had possessive rights (rights *in rem*) over the person who was killed. In the Teutonic system these rights were held by the cognatic relatives of the slain man by what was essentially a system similar to partnership. Each relative held as it were a share in the possession, and the consequent claim for indemnity and the share of any relative depended on the nearness of the relation so that, for example, the share claimed by the second cousins was twice that belonging to third cousins.

We have now to ask what use was made of the unilineal principle in the Teutonic systems. A man's kin were divided into those of the spear side (his paternal kin) and those of the spindle side (his maternal kin). . . .* The patrilineal principle also appears in the preference given to sons over daughters in the inheritance of land. But in default of sons, daughters might inherit land in some Teutonic societies. It appears that the principle of unilineal descent was only used to a limited extent in the Teutonic system. . . . An example of an arrangement of relations of kinship on the basis of unilineal descent may be taken from the Masai of East Africa. . . . The terminology is classificatory. . . . The Masai system arranges a man's kin into a few large categories. (1) His agnatic relatives belonging to his own lineage (or sub-clan); (2) his *apu* kin, if we may call them so, the descendants of those women of his lineage who are his father's sisters or his sisters, and on the other hand, the members of the mother's lineage: the relationship is established through a female and is then continued in the male line only; (3) his *aputani* relatives, i.e. relatives by marriage, either the kin of his wife or persons who have married one of his kinswomen; (4) those relatives to whom he applies the

---

[5] The most readily accessible account in English of wergild payments is Bertha S. Phillpotts, *Kindred and Clan,* Cambridge, 1913.

* Also known as the distaff side—*Eds.*

classificatory terms "grandfather," "grandmother," "grandson," "grand-daughter," some of whom belong to his lineage while others do not; (5) his *sotwa* relatives, a sort of fringe of persons not belonging to the first four classes, with each of whom some indirect connexion can be traced: they are vaguely his "relatives," persons with whom he should be at peace. . . .

We are here concerned with the ways in which different societies provide an ordering of the kin of an individual within a certain range, wide or narrow. One way is by tracing kinship equally and similarly through males and through females. There is a close approximation to this in modern European societies and in some primitive societies. In various societies we find some greater or less emphasis placed on unilineal descent, but there are many different ways in which this principle can be applied. In the Masai system there is marked emphasis on the male line, so that the most important of a man's social connexions are with the members of his own agnatic lineage and with those of his mother's agnatic lineage and those of the agnatic lineage of his wife. In other societies there may be a similar emphasis on the female line, so that a man's chief relations through his father are with the latter's matrilineal lineage. . . . There are a great variety of ways in which the unilineal principle may be used. It is therefore only misleading to talk about matrilineal and patrilineal societies as was formerly the custom of anthropologists. Some more complex and systematic classification is needed to represent the facts as they are. . . .

To some Europeans this use of the terms "female father," "male mother" may seem the height of absurdity. The reason for this is simply a confusion of thought resulting from the ambiguity of our own words for father and mother. There is the purely physical relation between a child and a woman who gives birth to it or the man who begets it. The same relation exists between a colt and its dam and sire. But the colt does not have a father and a mother. For there is the social (and legal) relation between parents and children which is something other than the physical relation. In this sense an illegitimate child in England is a child without a father. In the African tribes with which we are dealing it is the social and legal relationship that is connoted by the words which we have to translate "father" and "mother." To call the father's sister "female father" indicates that a woman stands in a social relation to her brother's son that is similar in some significant way to that of a father with his son. It is more exact, however, to say that a father's sister is regarded as a relative of the same kind as a father's brother, with such necessary qualifications as result from the difference of sex.

The principle of social structure with which we are here concerned is therefore one by which the solidarity and unity of the family (elementary or compound) is utilized to order and define a more extended system of relationships. A relationship to a particular person becomes a relation to that person's sibling group as a social unit. This shows itself in two ways. First, in some similarity in behaviour, as when the kind of behaviour that is required towards a father's brother is in some respects similar to that towards a father. Second, in the provision that in certain circumstances one relative may take the place of another, the two being siblings. Thus in some African societies the place of a father, a husband, or a grandfather may be taken by his brother. In the custom known as the sororate the place of a deceased wife is taken by her sister. In one form of the levirate the brother of a deceased man becomes the husband of the widow and the father of her children. . . . Miss Earthy[6] says that amongst the Lenge the father's sister (*hahane*, female father) "ranks as a feminine counterpart of the father, and sometimes acts as such, in conjunction with or in the absence of the father's brothers." She may offer a sacrifice on behalf of her brother's child, in case of illness, in order that the child may recover. Such sacrifice would, of course, normally be made by the father or his elder brother. . . .

Within the elementary family there is a division of generations; the parents form one generation, the children another. As a result, all the kin of a given person fall into generations in relation to him, and there are certain general principles that can be discovered in his different behaviour towards persons of different generations.

The normal relation between parents and children can be described as one of superordination and subordination. This results from the fact that children, at least during the early part of life, are dependent on their parents, who provide and care for them and exercise control and authority over them. Any relationship of subordination, if it is to work, requires that the person in the subordinate position should maintain an attitude of respect towards the other. The rule that children should not only love but should honour and obey their parents is, if not universal, at least very general in human societies.

There is therefore a relation of social inequality between proximate generations, and this is commonly generalized so that a person is subordinate and owes respect to his relatives of the first ascending generation—that of his parents. To this rule there may be specific exceptions, for the mother's brother in some African societies, for example, or for the father's sister's husband in some societies in other parts of the world,

[6] E. Dora Earthy, "The Role of the Father's Sister among the Valenge of Gazaland," *South African Journal of Science*, xxii, 1925, pp. 526-9. Also in *Valenge Women*, 1933, pp. 14 et seq.

whereby these relatives may be treated disrespectfully or with privileged familiarity. Such exceptions call for explanation.

The relation between the two generations is usually generalized to extend beyond the range of kinship. Some measure of respect for persons of the generation or age of one's parents is required in most if not all societies. In some East African societies this relation is part of the organization of the society into age-sets. Thus among the Masai sexual intercourse with the wife of a man belonging to one's father's age-set is regarded as a very serious offence amounting to something resembling incest. Inversely, so is sexual connexion with the daughter of a man of one's own age-set.

The social function of this relation between persons of two proximate generations is easily seen. An essential of an orderly social life is some considerable measure of conformity to established usage, and conformity can only be maintained if the rules have some sort and measure of authority behind them. The continuity of the social order depends upon the passing on of tradition, of knowledge and skill, of manners and morals, religion and taste, from one generation to the next. In simple societies the largest share in the control and education of the young falls to the parents and other relatives of the parents' generation. It is their authority that is or ought to be effective. All this is obvious, and it is unnecessary to dwell upon it. . . .

If the exercise of authority on the one side and respect and obedience on the other were simply, or even primarily, a matter of relative age, we should expect to find these features markedly characteristic of the relations between grandparents and grandchildren. Actually we find most commonly something almost the opposite of this, a relation of friendly familiarity and almost of social equality.

In Africa generally there is a marked condition of restraint on the behaviour of children in the presence of their parents. They must not indulge in levity or speak of matters connected with sex. There is very much less restraint on the behaviour of grandchildren in the presence of their grandparents. In general also, in Africa as elsewhere, grandparents are much more indulgent towards their grandchildren than are parents to their children. A child who feels that he is being treated with severity by his father may appeal to his father's father. The grandparents are the persons above all others who can interfere in the relations between parents and children. . . .

In what might perhaps be called the normal use of the generation principle in kinship structure, the generations provide basic categories. . . . In order to deal with this subject we have to consider the question of interpersonal rank or status in relationships. In a social relation two persons may meet as equal or approximately equal in rank, or one may be of superior rank to the other. Differences in rank may show themselves in many different forms. They are perhaps most easily seen in the rules of etiquette, or in an attitude of deference that the inferior is expected to show to his superior. . . . A relationship of unequal rank is necessarily asymmetrical. A symmetrical relationship is one in which each of the two persons observes the same, or approximately the same, pattern of behaviour towards the other. In an asymmetrical relationship there is one way of behaviour for one of the persons and a different, complementary behaviour for the other; as when a father exercises authority over a son, and the son is deferential and obedient. Terminology can also be symmetrical or asymmetrical. In the former case each applies the same term of relationship to the other; in the latter there are two different, reciprocal terms, as uncle—nephew. Where terminology in a relationship is symmetrical it is frequently an indication that the relationship is thought of as being approximately symmetrical in respect of behaviour. . . .

It is usual to apply the term "clan" to both patrilineal and matrilineal groups, but some American ethnographers use the term "clan" only for matrilineal groups and "gens" for patrilineal. . . . There are, of course, many different kinds of clan systems, but the term should be used only for a group having unilineal descent in which all the members regard one another as in some specific sense kinsfolk. One way of giving recognition to the kinship is by the extensive use of the classificatory terminology, so that in a system of patrilineal clans a man regards all the men of his clan as being his classificatory "fathers," "brothers," "sons," "grandfathers," or "grandsons." Frequently, but not universally, the recognition of the kinship bond uniting the members of the clan takes the form of a rule of exogamy which forbids marriage between two members of the same clan. Where clans are divided into sub-clans it may be only to the smaller group that the rule of exogamy applies.

Membership of a clan is normally determined by birth: where clans are matrilineal the children of a woman belong to her clan; where they are patrilineal the children belong to the father's clan. But in some tribes there is a custom of adoption. Where a man is adopted into a patrilineal clan, thereby abandoning his membership of the clan into which he was born, his children belong to the clan of his adoption, not that of his birth. In some African tribes the position of a child in the social structure depends on the source of the marriage payment for his mother. Thus, among the Lango children belong to the clan that has provided the cattle for the marriage payment for their mother. The father might not be a clansman, but might be a war captive or the sister's son of a clansman who was provided with a wife through cattle belonging to the clan, or the children might have been born outside marriage and the mother later married into the clan.[7]

If we look at a structure of clans or lineages from the point of view of an individual it appears as a grouping of his relatives. In a patrilineal system the members of his own clan are his agnatic kinsfolk, and the nearest of these to him are the members of his own lineage. The members of his mother's clan or lineage are also his kin, through his mother. He may apply to them the appropriate classificatory terms, and in some systems he may be forbidden to marry any woman of his mother's patrilineal clan. The members of his father's mother's clan and his mother's mother's clan may also be recognized as relatives, and those of his wife's clan or lineage may all have to be treated as relatives by marriage.

A clan system, however, also provides a division of the tribe into a number of distinct separate groups, each having its own identity. The clans may then, as groups, play an important part in the social, political, or religious life of the tribe. The extent to which they do this depends on the degree to which they are corporate groups. A group may be spoken of as "corporate" when it possesses any one of a certain number of characters: if its members, or its adult male members, or a considerable proportion of them, come together occasionally to carry out some collective action—for example, the performance of rites; if it has a chief or council who are regarded as acting as the representatives of the group as a whole; if it possesses or controls property which is collective, as when a clan or lineage is a land-owning group. In parts of Africa it is very common to find that land is held or owned by lineage groups, which are thus corporate groups. . . .

It should be noted that as a rule it is the adult men who really constitute the corporate kin group, and this is so for those systems that have descent through females. A good example is provided by the tribes of the lower Congo, described by Dr. Richards. . . . Villages or hamlets are formed of matrilineal lineages; all the men of a single lineage live together with their wives and young children, boys when they reach a certain age leaving their parents to join their mother's brother and his village. It is therefore the men of the lineage who form the corporate group, holding rights over land and acting collectively in various ways. . . .

The unilineal principle of reckoning relationship in one line (male or female) is utilized in a great variety of ways in different kinship systems. Where it is used to create a system of clans it facilitates that wide-range recognition of relations of kinship to which there is a tendency in many societies. A person will thereby find himself connected by specific social ties, subject to established institutional modes of behaviour, with a large

[7] T. T. S. Hayley, *The Anatomy of Lango Religion and Groups,* 1947, p. 40.

number of other persons. In the absence or weak development of political structure this gives an effective system of social integration. It is not possible to provide such very wide range in a system based on cognation, since that implies the tracing of genealogical relationships through all lines. But even more important is that unilineal reckoning makes it possible to create corporate kin groups having continuity in time extending beyond the life of an individual or a family. There are innumerable social activities that can only be efficiently carried out by means of corporate groups, so that where, as in so many non-literate societies, the chief source of social cohesion is the recognition of kinship, corporate kin groups tend to become the most important feature of social structure.

Thus it is the corporate kin group, whether clan, sub-clan, or lineage, that controls the use of land, whether for hunting, for pastoral life, or for cultivation; that exacts vengeance for the killing of a member, or demands and receives an indemnity. In the sphere of religion the kin group usually has its own cult, whether of its ancestors or connected with some sacred shrine. A continuing social structure requires the aggregation of individuals into distinct separated groups, each with its own solidarity, every person belonging to one group of any set. The obvious instance is the present division of the world into nations. In kinship systems cognatic kinship cannot provide this; it is only made possible by the use of the principle of unilineal descent. This is, indeed, obvious, but there have been writers who have used much misplaced ingenuity in trying to conjecture the origin of clans. . . .

THE AMERICAN FAMILY IN THE PERSPECTIVE OF OTHER CULTURES

Conrad M. Arensberg

Adapted from Eli Ginzberg (ed.) *The Nation's Children*, vol. 1, New York, Columbia University Press, 1960, 50-75, prepared for the Golden Anniversary White House Conference on Children and Youth. Reprinted by permission of the National Committee for Children and Youth, copyright holders.

That the family is part of the universal experience of mankind we know to be true. It is also true, however, that the family experience of the modern United States has very special features. In considering American

families and their effect upon children at home and in society, it is necessary to be clear as to universal characteristics of the American family and
as to its special or unique features. . . . Families, marital unions, kinship
systems are present in every human society and culture. But they are
shaped differently; they interconnect in many various ways; they assume
different relative importances in the functions of support of every kind,
from livelihood to affection, they perform for human beings, both the
grown-up ones and the children. Let us see more closely where American
family, marriage, and kinship, with their special American interconnec-
tions, fit in.

The Middle-Class Ideal

. . . the American family is distinguished by the great importance, emphasis upon, and independence of the small, immediate or "biological"
family of father and mother and minor children. American custom attempts to generalize this small unit, free it, trains most persons for roles
heading it in adult life, delegates societal and legal authority over and
responsibility for children almost exclusively to immediate parents in it.
In spite of some recent increases in the birth rate this unit is small; on an
average households are four and five persons at most; they begin with a
marriage of two potential parents, the spouses, who are urged to take up
residence, ideally, by themselves and away from others, "undoubling"
the larger households of larger, three-generation families still common in
many of our recent European immigrant and even our Southern populations; they swell for some years while minor children appear and grow to
young adulthood; they contract thereafter as children leave for an existence and a family life of their own.

The unit is not only small, so that households are small and mobile, the
family following the husband as he moves from job to job, position to
position, or town to town, increasing its isolation not only from kindred
but from neighbors and fellows of the community, in the great fluidity of
American occupational and residential life, but it is often very short-
lived. Not only are divorces common, contributing the major cause of
family dissolution (rather than war deaths or famine or emigration of
husbands, as in less fortunate countries) but the termination of family life
in a period of "the empty nest," with the spouses returned to a life together without children, is a standard, approved, and even planned-for
regularity of American social life. Just as the children are trained for the
day when they will "leave home" and "have a family of their own," so
old people are (ideally) expected to live apart and alone, visited perhaps
by adult children but not sharing a household with them, an eventuality
perfectly natural in most parts of the world, where gaffers and dowagers

may even rule the roost and certainly more often continue in it than leave it as here. But here even the small family endures, in an American's life time, only twenty years or so, especially when the parents ideally have all their children in their younger married years.

All this custom, most of it ideal middle-class American family life whose real prevalence in our mixed and varied population we can only guess at, reflects, obviously, the individual and equalitarian ideals of our country's social and political life, the spread of a wage-earning and money-and-credit consuming way of economic life among most of our people as well as the already mentioned traditional cultural emphasis upon the small family, with its connections to the free choice of mate and residence and occupation and to the open mobility between places and statuses of our society. All those things, together with the reduction of extended family relationships of kinship, inherent in the freeing of individuals from fixed and hereditary placements and categorizations, have marked our civilization since the overthrow of the "ancien régime." We have already cited the historical influences. But the special traditional cultural descent of this kind of family custom which present American conditions continue to deepen and generalize should be noted as well.

The Joint-Family

The best anthropological classification of the families of mankind treats them first as they vary in progressive size of the family unit, particularly as that unit forms the usual households of a society. Largest are the joint-families of India, the patriarchal families of the Chinese gentry of yesterday, the large households of the Middle Eastern countries, of much of Africa where they may be also polygynous, the *zadrugas* and other patriarchal households of the peasant lands which in the remote Balkans still today practice a household economy like that of ancient Rome. Here a founder, his sons, his sons' sons and all their wives, children, grandchildren, dependents and servants or slaves live their lives out in a house or compound of many rooms with common fields, gardens, and larder under central authority and in common defense for a lifetime. Eventually such a family usually splits to make more like it; the common lands or joint economy make greater size of household equivalent to strength and security; and the continually splitting households often retain ties of common defense, including even blood vengeance, to form far-flung clans of common unilineal descent.

We tend to forget how widespread even today, especially in the underdeveloped countries, are such great families and how common such clans, with the security and the trammels they bring, still are in the world. Because we have forgotten them, or belong to traditions which never knew them, does not mean they are any the less viable alternate ways of organizing individual and community life, imposing imperatives, and requiring virtues of their own kind, in many parts of the world where the national state is still new and where kinsmen and patrons rather than the national police protect individuals. In such lands they are still to be found, still opposing or braking the individualizing forces of modern pecuniary economics and of modern civil law. Some of our American ethnic groups, both immigrant and native, have strong and recent memories of joint households and clan ties, so different from the individuation of the small family of our majority tradition. When their households, for example, give over child care to grandmothers or take in nephews and cousins on the same basis as immediate children, sometimes in direct clash with our family law and our welfare procedures based on our small-family custom in which such relatives have no claim or right of care and protection, the difference in custom and family organization goes unacknowledged and the clash between public procedures and private interests and capabilities unresolved.

The Stem-Family

Our small-family tradition is based, of course, on quite other cultural antecedents than the joint-family and the clan uniting forever all the sons and grandsons of so-and-so. The next classification of families and households common in many parts of the world bases them on a size intermediate between the great households of the joint-families and our own small ones. American experience, indeed American social science, does not recognize this classification and fails to note that it is very widespread in the world, particularly in Europe, but also in Asian peasant lands, especially where small proprietorship has fostered the growth and transmission of inherited family farms. In the European countries, especially in those of small peasant holdings, France, Germany, Ireland, northern Italy and northern Spain, etc., but also in Japan, the Philippines (Ifugao, for instance) and in parts of peasant India and China both, an intermediate size of family and household, living for generation after generation on a family holding, has often become standard and customary. This counted in the homestead in each generation the peasant holder, his wife, his minor children, his unmarried brothers or sisters, living as unpaid farm laborers and helping him until they should move away or marry off, his father and mother, perhaps retired from active work but still influential and assisting. If one of the standard disasters of peasant subsistence agriculture was to be avoided, namely the endless equal subdivision of the plot among children until no one child inherited land enough for subsistence, then in each generation the family homestead and plot should be kept intact and undivided. One child, or at most two, should be heirs of the whole, becoming the new holder in turn, and his noninheriting brothers and sisters should have to find for themselves some other provision in life than a bit of the family lands or else remain at home forever in a minor, farm-helper status.

Through matchmaking and other mechanisms such restriction of inheritance to a single heir in each generation often became standard, acceptable, even ideal. The household and lands remained a stem or source of new heirs and new emigrants in each successive generation; a long line of holders kept the homestead in the line or stem; it even, usually, carried the name of the farm as a family name. Each generation knew a three-generation household of retired parents, heir and his spouse (either a son or a daughter might get the land as heir of the intact holding and Norman-French primogeniture and estate entail was merely one version of such custom). Each generation knew new waves of brothers and sisters, noninheriting children who must go out into the world to "make their fortunes" elsewhere, on new farms, in marriages outside, in the apprenticeships leading to artisan or other work in the cities.

This kind of family organization became and is still standard in most of the European countries, whence its name coined by the great French family sociologist LePlay comes: the *famille-souche* or the stem-family (*Stammfamilie*, in German). It seems historically in Europe to have grown up with the medieval transition between tribal landholding and peasant tenancy and proprietorship.

So deeply is it ingrained in European tradition, whether peasant or of higher class, that many discussions between Anglo-Saxons and Europeans founder on the unrecognized adherence of Anglo-Saxon tradition to the small-family and the usual European to the stem-family. Where an American, and an Englishman, in the small-family tradition, may be enjoined by his own desires, his wife, and his columnist of manners and personal problems, such as a Mary Hayworth, to set his old mother up to live alone and think it a hardship to have her under the same roof with his wife and children, a Frenchman may define the *foyer* (intimate family) to include her and regard it as unthinkable that grand'mère live anywhere else. Much of the "Americanization" of modern Americans involves undoubling of such stem-family households today, the dissolution of family kitties which pool the incomes and the salaries of even adult children, a usual and expectable European practice in many countries—indeed even necessary where "family allowances" and state pensions do not even presume individual wage equalities or reckon a living wage to include a family livelihood as with us. Countless thousands of Americans of second-generation or third-generation immigrant origin or even of American

Southern and Southern Hill background are new and transitional to the small-family, individualizing family tradition, moving toward it from the other moralities of the stem-family tradition, in a way analogous to that in which we can today find Yugoslavia moving from joint-family (*zadruga*) to stem-family (European peasant) organization with attendant difficulties of social change and adjustment. Our social sciences are still too young to let us know and recognize the many modern cultural and social transitions of this kind and to let us deal adequately with their personal and psychological costs. Only in recent years have the social work and welfare professions begun to recognize such transitions and to learn that caseworkers must be prepared to face them. Public and legal recognition of private customary differences of family interests and definitions is, indeed, not yet in sight.

Family Transformations

The general European movement of family organization during the Middle Ages seems to have been much that of Yugoslavia in recent decades, a movement from joint-family and clan protection for individuals and great-household economy, even for peasants, to smaller peasant subsistence holdings, of stem-family kind, with proprietorship passed down the line of family heirs. There is reason to believe that some parts of Europe, like some parts of the non-European world, never took part in this transition, chiefly because, as we shall see, small families and weak kinship units were aboriginal, part of another way of life than peasant subsistence. Deferring that suggestion for a moment, let us see what kind of kinship evolution took place as stem-families, if not conjugal small ones like ours, succeeded, at least in Europe, joint-families and great-households.

One change was certainly the spread of bilateral as opposed to unilateral kinship units, a shift from exclusive clans . . . to diffuse and general kindreds of the sort we know today, in which all the blood descendants of the same grandparents and great-grandparents as our own, are our cousins, regardless of whether they come through the male or through the female lines. We still reckon as relatives upon whom we have some claim, if only a bed in emergencies, the whole diffuse circle of such natural kin; no longer can the world be divided into the sons of my fathers, whom I must defend to the death, and the sons of my mother's clan, who may have to shoot me on sight. Only the family name still, with us as with other Europeans, descends down the paternal line, as a vague identifier. We can trace through European history, as we can trace it still in the spread of the national state today, the shift over to such stem-names, giving each man a family name. The custom reached Turkey only with Kemal Atatürk's reforms, and has yet to reach either Indonesia or (oddly enough) Iceland. We can likewise trace the dissolution of clans and phratries, still alive in Arabia or Pakistan, with the shift to the kind of bilateral, diffuse, cousin-counting kinship we ourselves know. In this shift to diffuse, relative reckoning of significant kinship, from a former counting instead of exclusive and corporate groups of special legal and moral force, we can still see a background to the individuation and the liberation from status and adherence prescribed at birth that has gone so far, as we have pointed out, in our own American treatment of kinship.

Let us at last return to that part of the European tradition in which, as with our own Anglo-Saxon heritage, neither the stem-family of the peasantries nor the fixities of joint-family and clan figured. Other parts of the world, as we said, have been found by social anthropologists to possess small-family organization. Notably these are some of the hunting peoples organized for a subsistence requiring great movement and fluidity among small bands of persons and, oddly enough, many of the civilized peoples of South East Asia; Malays, Thai, Burmese, etc. There is some evidence, too, that in periods of rapid urbanization, as in ancient Roman days, great movement and migration of persons and extreme fluidity of occupational life and easy social mobility have tended more than once to dissolve kinship rigidities, to isolate and free individuals and generalize small families, just as in recent British and American history.

A great argument of social science can be waged today whether pecuniary civilization, industrialization, the factory system in themselves do not force a generalization of small families, and indeed the European practice is to treat the small family, which we call the "democratic" type of family organization, as the "proletarian" or the "disorganized" one. But the argument is better left to one side, the more so as Japan, India, the Middle East, and even such countries as Belgium and Germany seem to be able to undergo industrialization without a wholesale or even a widespread adoption of American and British small-family social patterns. The only causative argument or association we can advance for the distribution in the world of small families as the standard family system of a culture is that any pattern of economic subsistence requiring fluid movement of persons and alternate sources of hands for impermanent productive units, whether bands of gatherers or hunters, or crews of fishing boats, or short-lived reindeer herds, or new factories recruiting temporary labor forces, seems to favor small-family generalization.

The historical dominance of small-family organization in Great Britain and the nearer parts of North Europe is another problem, not at all to be solved by reference to the Industrial Revolution. The villages of North Europe and of Britain, under the manorial system, seem to have known both stem-families and small-families, and the precarious conditions of medieval farming, with their requirements that a fluid, quickly formed plowteam be formed from any and all neighbors, whether kinsmen or no, at the end of the winters in which only a few oxen or plow-cattle survive to plow the next spring, may well be thought of as favoring the small-family, unobligated peasant, ready to turn to a chance neighbor in the village as quickly as to a cousin or clansman. Certainly by the time the Enclosure Acts had cleared the English villages, destroyed the yeomen who might have duplicated the Grossbauer (big, homesteading peasant) of the continent, and sent out the Puritans and other Dissenters into town life and overseas colonization, the English tradition of small-family life, the generalization of independent starts for children, and the whole apparatus of our modern family system seem already to have been well established.

Our own frontier seems to have spread the Scots-Irish, Southern-Appalachian stem-family tradition. But it also served to spread the Anglo-Saxon, post-medieval, and Puritan small-family way as it spread the English tongue. The Middle West combined and generalized the regional-sectional traditions of our earlier colonial times. Sociologists could note, as late as 1925, that homesteading in Iowa in the sense that farms went to heirs and stayed on with the family line—a definition of the term that stresses holding on to a farm rather than originally "nesting" it—was confined to German and Polish and Czech Americans. The Old Americans, "Anglo-Saxons," in that state as elsewhere preferred to start all the children alike and "independently," setting a boy upon the "agricultural ladder," helping him "start on his own," eventually selling out and dividing the money equally among all the children, moving on to the West or to an old age in which the family was dissolved and the retired farmer and his wife lived apart. The average period of a farm's stay in one owner's hands became twenty-five years, the exact duration indeed of a small-family's life, from the time the young tenant managed (or was helped) to buy his own farm (if only from father) to the time that he in his turn retired, sold, and divided the money equally among his heirs.

The Iowan procedure we are describing here is an excellent example of the American small-family way, with its independent, self-reliant children, all equally on their own, in contrast to the European *famille-souche:* the peasant household forever in the family line, in which a grown man is still a boy, still under the family council headed by his mother and his father, pooling the family's resources, arranging match and dowry for him and sister to the end of the old people's lives. The attendant dissolution of kinship, in which a neighbor or fellow-community member is oftener to be relied on than a cousin, let alone a now nonexistent clansman, is just another step in the reduction of household and family size, in the concentration of roles on small-family personnel, and in the sweeping away of intermediate supports or obligations between the small-family on the one hand and the community and state on the other.

The particular North European and British, even village and Puritan English, descent of the American small-family system is thus quite fateful in the especial evolution of our family system and its values. The special features of American family experience we noted earlier have legitimate origins in the cultural history of the country as well as in the special economic, legal, and political historical conditions of the country's growth. These special features pose special problems, psychological or other, for Americans. They pose such problems for Americans both in their own persons as sharers and movers of the American customs of family life and in their special difficulties of child welfare and child care. Many of these latter problems we have already cited: the isolation of the small family; the brittle dependence for physical and emotional security, as well as home training and discipline, upon the competence, cooperation, and adjustment of the spouses; the great and growing age separation segregating old people and their experience out of family and even occupational life; the unacknowledged transitions and exceptions from the ideal small-family morality of the majority, middle-class, and institutionally-official traditions, with their conflicts for individuals torn between values taught at home and values taught in school and community. Most recent indeed is the continuing weakening of what parental authority still remains in the parent-spouses, in the spread of permissive and "democratic" doctrines of family consultation and enlistment of child interests and prejudices. The father who is not so much a man, a model of adult manhood for his son, as a "pal" and another boy, absent and out of sight in the important, non-familial roles of his work existence, has already worried psychiatrists, especially in our newer, dormitory metropolitan suburbs, with their enforced segregation of women and children of limited like age and interests.

Most of these problems, social, legal, and psychological, seem to flow from the continuing evolution of our particular traditions, with the attendant individuation and dissolution of stabilizing and assisting personal contacts in our lives and their replacement by professional and community services. The trend is one that our long evolution of small-family independence and diffusion of kinship and other fixed-status ties long ago began. It is certainly irreversible, even if we wished to reverse it, which our people do not seem to wish to do. But if some information about its special historical character, its special place in the alternate ways of family and community organization in the history of mankind, and its special demands upon ourselves can help us manage better the trends and currents of social change in which we are caught, then perhaps this brief summary of the place of American family life in the perspective of other cultures will have served a purpose.

THE ISOLATED NUCLEAR FAMILY: FACT OR FICTION

Marvin B. Sussman

Adapted from *Social Problems*, 6:4 (1959), 333-340.

Current family theory postulates that the family in American Society is a relatively isolated social unit. This view of the family stems largely from theories of social differentiation in more complex societies.

A neolocal nuclear family system, one in which nuclear families live by themselves independent from their families of orientation, is thought to be particularly well adapted to the needs of the American economy for a fluid and mobile labor market. It is also suggested that differences in occupational status of family members can best be accepted if such individuals live some distance from each other [6, 7]: Support for these theories is found in the high residential mobility of Americans: one in five families makes a move during a given year and presumably these families are nuclear ones. The existing patterns of occupational and social mobility underlie this movement. It may be said that these mobility patterns "demand" a type of flexible and independent family unit.

The extended American family system, for the married person, consists of three interlocking nuclear families; the family of procreation, the family of orientation, and the one of affinal relations (in-laws) whose interrelationships are determined by choice and residential proximity and not by culturally binding or legally enforced norms. The isolation of these nuclear related families from one another is given further support in current conceptualization of family socialization patterns. Freudian analysis has stressed the difficulties confronting the individual as he seeks emancipation from the family of orientation and has interpreted many emotional problems in terms of in-law and parent-child conflict. Child rearing specialists have emphasized that the warmth and affection of the parent-child relationship should not be chilled as a consequence of competitive activity from the grandparent, aunt, or uncle. There is, it is said, sufficient threat to an already fragile nuclear family structure through sibling rivalry and parent-child differences. Thus having parents or collateral relatives living in the home or even close by adds . . . difficulties to the complicated problem of child rearing. Parents and their young offspring can therefore presumably attain a high level of functioning as a family if they are un-encumbered by the presence of relatives.

Some students of ethnic relations have interpreted the breaking away of first generation members from their immigrant families as a necessary prelude to growth and assimilation into American society. In still another field, many social class theorists have emphasized the fluidity of our class system and the necessity of the individual to be shorn of family and other ties which appear to hinder his upward movement within the class system. In our values we maintain that the son is better than the father, or that he should be, and in the process of becoming somebody, or achieving a higher status, more frequently than not it is necessary to discard former identifications (particularly those with parents and kin) for newer and more appropriate ones.

Despite these basic positions there are some empirical indications that many neolocal nuclear families are closely related within a matrix of mutual assistance and activity which results in an interdependent kin related family system rather than the currently described model of the isolated nuclear family. This development, while not superseding in importance the primacy of the nuclear family, may provide a new perspective with regard to its position. This does not mean that nuclear families of procreation are increasingly living with their kin in the same households: they still live in separate residences but frequently they reside within a community with their kin and engage in activities with them that have significant mutual assistance, recreational, economic, and ceremonial functions. Nor does this mean that a new kin structure and concomitant activities have emerged. It is probable that a functioning system of kin related nuclear families has always been in existence.

The suggestion that a re-examination of the position of social isolation of the nuclear family of procreation from members of the extended family is necessary has come from many sources. Professors Sharp, Axelrod, and Blood in Detroit, Edmundson and Breed in New Orleans, Deutscher in Kansas City, Dotson in New Haven have studied family relations in the urban setting.[1] From these and the author's 1951 study of intergenerational family relationships [10, 11] it became evident that the concept of the atomized and isolated nuclear family is not being substantiated by empirical research.

The Cleveland Studies

Data on kin and family relationships in Cleveland were obtained in 1956. Area probability samples were drawn from two census tracts classi-

[1] See, for example (2; 3; 9). British sociologists find similar evidence of kin in East London families (14, 15).

fied by the Bell-Shevky Social Area Analysis method [1] as lower middle class and working class respectively. An adult member of 27 working class and 53 middle class households representing 3.5 per cent of the total households in each area was interviewed.[2] The general analysis of kin and family relationships is based upon the 80 cases. Specific analyses by social class are based upon a comparison of 25 matched working and middle class family systems.

Information on kin relationships among these Cleveland families was obtained from questions about help and service exchanges, the functions of ceremonial occasions, and inter-family visitation. Help items included caring for children, help during illness, financial aid, housekeeping, advice, valuable gifts, etc. Ceremonial occasions included holidays, birthdays, and anniversaries. Questions of visits between kin included the preparing of get-togethers with relatives who lived in or out of town.

Concerning help and service exchanges, practically all families (100 per cent of the middle class and 92.5 per cent of the working class) were considered to be actively involved in a network of inter-familial help by virtue of giving or receiving one or more items of assistance listed above within a one month period preceding the interview.

This help pattern appears to exist along with high residential propinquity of related kin. Seventy per cent of the working class and 45 per cent of the middle class have relatives living in the neighborhood.[3]

The research findings suggest that modern means of transportation permit relatives to live in scattered areas and still operate within a network of mutual aid and service. In Table 1 are enumerated specific types of help and service found within kin networks.

Help during illness is the major form of assistance provided by members of kin related families. Such assistance was given in 92 per cent of the reported illnesses which occurred among kin related families living in the neighborhood in the twelve-month period preceding the interview. Respondent-parents and respondent-sibling reciprocal patterns do not differ significantly. However, service during illness for members of kin related families living some distance from each other show a different pattern. Quantitative data are lacking on this point but case data indicate that help between distant families is given when a family member is critically or believed to be critically ill or is suffering from a long term

[2] The choice of census tracts was based on data reported in (4).

[3] A note of caution is called for on propinquity of kin. Respondents were permitted to define their own neighborhood without a spatial limitation. It was deliberately left vague because of the difficulties in arriving at an acceptable definition of neighborhood. Thus, respondents might construe the "neighborhood" to be the city of Cleveland, or a census tract, or even the street in which the house was located.

illness. In this situation disruption of family routines is expected and a member of the well family, most frequently the middle aged parent, volunteers or is asked to come and help. There are no expectations for help in routine illnesses as found among kin related families living in the same neighborhood.

The amount of financial aid, care of children (babysitting), advice, and valuable gifts exchanged between members of kin related families is higher in this sample than found by Sharp and Axelrod in Detroit [9]. The trend toward mutual aid is in the same direction. Differences in magnitude are probably due to differences in sampling and characteristics of the two populations.

Twenty-five middle were matched with 25 working class family systems, using the number of nuclear related families (parent and child) as the matching variable. That is, if the respondent middle class family was composed of a middle aged couple (parental family) who had two married children, each of whom lived in separate households, then a working class unit of similar composition was selected to complete the matched set. An effort to match family systems on a second variable, namely, sex of the children in the family of orientation, resulted in too few cases for comparison. Matching on the sex factor as well as age of family members would be very important in the study of specific patterns of help and service between kin related families. These would include the quantity and type of aid exchanged in connection with movement of the immediate family through the life cycle. This problem will be investigated in the next phase of a longitudinal study on urban family networks. For the purpose of establishing the existence of aid networks, the matching technique used in this study is adequate.

TABLE 1. Direction of Service Network of Respondent's Family and Related Kin by Major Forms of Help

Major Forms of Help and Service	Direction of Service Network				
	Between Respondent's Family and Related Kin	From Respondents to Parents	From Respondents to Siblings	From Parents to Respondents	From Siblings to Respondents
	Percent*	Percent*	Percent*	Percent*	Percent*
Any form of help	93.3	56.3	47.6	79.6	44.8
Help during illness	76.0	47.0	42.0	46.4	39.0
Financial aid	53.0	14.6	10.3	46.8	6.4
Care of children	46.8	4.0	29.5	20.5	10.8
Advice (personal and business)	31.0	2.0	3.0	26.5	4.5
Valuable gifts	22.0	3.4	2.3	17.6	3.4

* Totals do not add up to 100 per cent because many families received more than one form of help or service.

The desideratum underlying this matching technique is that similar family systems of two social classes have equal opportunities to develop help and service patterns. The significance of differences between social classes was determined by the differences between matched pairs and not by differences between the samples. The test for differences between samples is insensitive to specific differences between paired sets and does not take into account the opportunity variable. If the structure of the family system is important in determining help and service patterns then it must be established as a control in the analysis of these patterns by social class.

In Table 2 are found statistically significant differences in four items of help and service by social class.[4] So, for example, middle class more than working class parental and child families give and receive financial aid: the P value of .004 indicates that differences more extreme than that observed would occur in only four out of one thousand trials if the hypothesis of no difference was true. On the other hand, the P value of .80 indicates that in eight out of ten trials the flow of aid from parent to child would be the same for both social classes.

There is no significant difference between classes on the amount of help given or received during an illness of a family member.

TABLE 2.   Differences in Help and Service Exchanged by Social Class

Help and Service Items	Middle Class	Working Class	P Values
Help during illness	same	same	.80
Financial aid			
Amount exchanged	more	less	.004
Flow of aid:			
Parent to child	same	same	.81
Care of children	more	less	.03
Advice (personal and business)	more	less	.04
Valuable gifts	more	less	.007

Controlling for distance between parents and married child's households, middle more than working class grandmothers are called upon to "take care of grandchildren." The latter are more often than the former gainfully employed. Among working class couples there is a tendency to use the available married brother or sister rather than parents for this service.

Middle more than working class family systems exchange advice and give valuable gifts to one another. The network of giving is between parents and children rather than between young married couples and their married sibling families. These differences probably reflect the eco-

[4] A discussion of the statistical procedure used may be found in (13, pp. v-x).

nomic condition and educational attainment of members of these two social classes.

Middle class more than working class parental and child families give and receive financial help. The network of giving is between parents and children rather than between young married couples and their married sibling families. The flow of financial aid is from parents to children; with respect to such flow, differences by social class are not statistically significant.

Regular social visits between parent and married child families as well as those between siblings occur mainly between those in the same neighborhood. Difference between social classes are not significant. The median frequency for visits between young married couples and parental families during the 12 month period preceding the interview was two to three times weekly and once a week between sibling families.

Seventy-four per cent of working and 81 per cent of middle class families have large family gatherings at least once a year. Ceremonial occasions such as Christmas, anniversaries, and other holidays are used largely for the gathering of kin who live outside of the neighborhood. Approximately half of the families in both classes have large family gatherings at one of the major holidays such as Christmas, New Year's, Easter, Thanksgiving or the Fourth of July. Birthdays and anniversaries, while not as popular for family reunions, are more frequently used by the middle class than the working class family as an occasion for a family gathering.

Data from another Cleveland study on population change in a given geographical area [12] supports the idea that many neolocal nuclear families are closely related within a matrix of mutual assistance and activity which results in a kin related family system.

Recently, large numbers of Southern mountain whites and Negroes moved into an area of Cleveland called Hough, containing over 21,000 households. Based on a random sample of 401 households in 1957, it was found that the non-white population has risen from less than 5 per cent in 1950 to 59.3 per cent in 1957. Ninety per cent of the non-white population have lived in their present houses five years or less, while 59.7 per cent of the whites have lived in their present houses for the same period of time. This demonstrates rapid mobility and changeover from white to Negro occupancy. In the midst of this invasion-succession process we find a fascinating network of kin ties. Sixty-seven per cent of the whites and 86 per cent of the non-whites have relatives living in the Cleveland Metropolitan Area. Of these, whites have 34 per cent and non-whites 60 per cent of their kin living in the two square mile Hough area. Eleven per cent of the whites and 19.5 per cent of the non-whites have three or more families of relatives living in Hough.

Moreover, in Hough relatives are sources for financial aid, second only to banks, and for assistance in times of personal trouble second only to clergymen. When first, second, and third sources of assistance are considered, relatives are most frequently sought. Additional data on inter-family visitation and exchange of services such as child care indicates an intricate matrix of inter-family activities in Hough. Thus, even in high transitional areas of the central city social interaction revolves around kin related nuclear families. It is suggested that the evidence of propinquity of kin related nuclear families has been overlooked in researches on the urban family and in such areas as use of leisure time, social participation and population mobility.

Discussion

Data have been presented which indicate that many kin related families live close to one another and are incorporated within a matrix of mutual assistance and activity. It has been suggested in the literature that individuals tend to locate wherever there is the best economic opportunity. This purports to explain, in part at least, why individuals usually establish neolocal residence some distance from their kin. Kin ties are considered *least* important in the individual's choice regarding a job, the choice being made in terms of a "best for me" ideology. The fact is that the rate of economic mobility may be influenced by kin ties. Moreover, persons seeking a new job may now locate where their kin are already established. In such cases relatives can help new arrivals adjust to the new community and provide many of the services already described. Even new arrivals into a community may soon be followed by their relations. This has certainly been the case of ethnic and racial migrations across the United States, the Puerto Rican movement is a good example. The high residential mobility suggested by Peter Rossi [8] and others may actually be a large scale movement of families into communities where their kin are already established. The notion of economic opportunity as related to mobility is still well founded but it needs modification in view of the findings that kin ties have far more significance in the life processes of families today than we have been led to believe. Parental support to newly married couples may provide the necessary anchorages for reducing mobility. It can be hypothesized that population mobility is not higher than it is because of parental support and other kin dependencies.

Class differences in the type of help and service exchanged reflect more the differences in life styles than willingness to participate in the mutual aid network. The middle more than the working class are in a better position to give financial help and expensive gifts; and non-working middle class grandmothers are more likely to be "free" to take care of grandchildren. The offering of advice, most likely to be given by the parent to the child, reflects middle more than working class occupational status and child rearing patterns.

A question can be asked as to why there is today in sociological writings so much emphasis upon the social isolation of the nuclear family. Scott Greer and Ella Kube ask a similar question concerning the emphasis upon the isolation and anomic state of the urban dweller [5]. The urbanite is said to be dependent upon secondary rather than primary group relationships. This view may exist because of a time lag between urban and family theory and research. It may also reflect a cultural lag between what was believed to be a generation ago (or may actually have been) and what exists today. The writings of such men as Durkheim, Simmel, Toennies, and Mannheim contain early 20th century views of family and social life in a growing urban industrial society. Durkheim's research on suicide indicated weaknesses in family structure and the effects of isolation upon the individual. In no way did he indicate the basic features of family structure which did, does today, and will tomorrow sustain its continuity on through time. In other words, a theoretical view tinted towards the ills of social and family life was implanted and subsequent research sought to ferret out the disorganizing features of social life.

The consequences of this process, using the family as an illustration, have been in abundant researches on "what is wrong with the American family," followed by a series of proposals on what should be done, and a dearth of studies on "what is right with the family." The "non-problem" functioning family, representing the majority of any society, carrying on the many daily tasks necessary for survival has not been the subject of much study. Yet examination of the non-problem family evolves an empirical base upon which there can be established the means for accurate diagnosis, evaluation, and treatment of the problem family.

Implied, therefore, is a revision of the social problem orientation and approach to the study of social systems. The approach suggested is that empirical studies of the "normal" or "functioning" units of a social system accompany or precede those made on the problem units of the system. Once the base of what is functioning is established then it becomes possible to evaluate a social problem and to propose adequate alternative solutions.

Conclusion

The answer to the question "The Isolated Nuclear Family, 1959: Fact or Fiction?" is, mostly fiction. It is suggested that kin ties, particularly intergenerational ones, have far more significance than we have been led to believe in the life processes of the urban family. While these kin ties are by no means replicate the 1890 model, the 1959 neolocal nuclear family is not completely atomistic but closely integrated within a network of mutual assistance and activity which can be described as an interdependent kin family system.

REFERENCES

1. Bell, Wendell, "Social Area Analysis," in Marvin B. Sussman, ed., *Community Structure and Analysis* (New York: Crowell, 1959).
2. Deutscher, Irwin, "Husband Wife Relations in Middle Age: An Analysis of Sequential Roles Among the Urban Middle Classes," unpublished report, 1954.
3. Dotson, Floyd, "Patterns of Voluntary Association Among Urban Working Class Families," *American Sociological Review*, 16 (October, 1951), 689-693.
4. Green, Howard W., *Census Facts and Trends by Tracts*, Cleveland Real Property Inventory, Special 1954 Report.
5. Greer, Scott and Ella Kube, "Urbanism and Social Structure: Los Angeles Study," in Marvin B. Sussman, ed., *Community Structure and Analysis* (New York: Crowell, 1959).
6. Parsons, Talcott, "The Kinship System of the Contemporary United States," *American Anthropologist*, 45 (January-March, 1943), 22-38.
7. Parsons, Talcott and Robert F. Bales, *Family, Socialization and Interaction Process* (Glencoe, Ill.: The Free Press, 1955).
8. Rossi, Peter H., *Why Families Move* (Glencoe, Ill.: The Free Press, 1955).
9. Sharp, Harry and Morris Axelrod, "Mutual Aid Among Relatives in an Urban Population," in Ronald Freedman, et al., eds., *Principles of Sociology* (New York: Holt, 1956), 433-439.
10. Sussman, Marvin B., "Family Continuity: Selective Factors which Affect Relationship Between Families at Generation Levels," *Marriage and Family Living*, 16 (May, 1954), 112-120.
11. ———, "The Help Pattern in the Middle Class Family," *American Sociological Review*, 18 (February, 1953), 22-28.
12. Sussman, Marvin B., and R. Clyde White, *Hough Area: A Study of Social Life and Change* (Cleveland: Western Reserve University Press, 1959).
13. *Tables of the Binomial Probability Distribution* (Washington: U.S. Government Printing Office, National Bureau of Standards, June 6, 1949).
14. Townsend, Peter, *The Family Life of Old People: An Inquiry in East London* (London: Routledge and Kegan Paul, 1957).
15. Young, Michael and Peter Willmott, *Kinship and Family in East London* (Glencoe, Ill.: The Free Press, 1957).

3. FAMILIAL STRUCTURE VIEWED AS ROLES

This chapter proceeds from the more or less particular—roles in the American family and society—to the general—an analysis of roles and role strain.

Demographic trends affect familial roles. Glick shows that, whereas in 1890 the average American marriage was broken by the death of a spouse before the couple's last child was married, more recently the parents can expect upwards of a dozen years together (on the average) after their last child has married and left home.

In his article Parsons asserts that the two social categories which are primary determinants of social roles are age and sex but that they do not operate with equal importance at each level. Adolescence he characterizes as a period of the "glamour girl" and the "swell guy," a period typified by a youth culture that is different from the patterns characteristic of both childhood and adulthood. It is a period dedicated to irresponsibility and to glamour. Transition to adult roles requires the sloughing off of many of the elements of glamour that typify the youth culture and the acceptance of status based on occupational specialization for males and the reflected status of the mate for females.

The article asserts that there is a particular period of development toward which earlier roles are geared. Parsons believes that in the Western world it is the period of youth that is idealized. Parsons suggests that as social roles become more anticipatory of a particular age period, preparation for later periods becomes more inadequate. From this we may infer that in American society there is a decided lack of preparation for age categories which succeed youth.

Viewing societal structures, including the family, as made up of roles, Goode argues that the individual's role obligations tend to be overdemanding. Goode's analysis is perfectly general; its applicability to the family is clear. For example, a man's boss may be expecting him to demonstrate his commitment to the company by working nights and weekends whereas his wife believes he "owes" this time to the development of his children. Goode indicates a number of ways in which an actor can manipulate his role structure in order to reduce his role strain and uses an economic model to represent how an actor selects a course of action to effect this end.

THE LIFE CYCLE OF THE FAMILY

Paul C. Glick

Adapted from *Marriage and Family Living,* 17:1 (1955), 3-9.

Within the life cycle of a given family, a host of demographic and economic changes take place that require continuous readjustments of the habits and values of the family members. Moreover, the secular and cyclical changes in age at marriage, size of completed family, and length of life have greatly affected the patterns of family formation, development, and dissolution. In this paper, some of the implications of these several types of change are analyzed with respect to the family in the United States.[1]

In the discussion which follows, frequent reference is made to both "married couples" and "families." The former is defined by the U.S. Bureau of the Census as a married man and his wife who are living together. A "family" is defined as two or more persons related to each other who are living together; thus, if two married couples live together and are mutually related, they are counted as one family. The source materials include numerous reports of the decennial censuses of population, the Current Population Surveys (sample surveys covering about 25,000 households), and annual vital statistics.[2] In some instances, the available

---

[1] A similar analysis, based largely on data for 1940 and earlier dates, was published by the author as "The Family Cycle" in the *American Sociological Review,* XII: 164-174, April, 1947.

[2] A selected bibliography of census and vital statistics reports used is as follows: U.S. Bureau of the Census, *U.S. Census of Population: 1950,* Vol. II, *Characteristics of the Population,* Part 1, U.S. Summary (tables on marital status and relationship to head of household); Vol. IV, *Special Reports,* Part 2, Chapter D, Marital Status; and Vol. IV, *Special Reports,* Part 1, Chapter A, Employment and Personal Characteristics. U.S. Bureau of the Census, *U.S. Census of Housing: 1950,* Vol. II, *Nonfarm Housing Characteristics,* Chapter 1. U.S. Bureau of the Census, *Current Population Reports,* "Marital Status, Number of Times Married, and Duration of Present Marital Status: April 1948," Series P-20, No. 23; "Marital Status and Household Characteristics: (date)," Series P-20, Nos. 33 (March 1950), 38 (April 1951), and 44 (April 1952); "Fertility of the Population: April 1952," Series P-20, No. 46; "Marital Status, Year of Marriage, and Household Relationship: April 1953," Series P-20, No. 50;

data are fragmentary; in other instances, they provide only approximations to the information desired. The shortcomings are not sufficiently
serious, however, to invalidate the central point of the discussion, namely,
that married couples now have many more years of family life remaining
after their children have married than did couples of earlier generations.

Stages of the Life Cycle of the Family

Marriage. The average young man in the United States in 1950 entered marriage for the first time at about the age of 23 years and his wife
at about the age of 20. (See table 1 and figure 1.) Both the groom and
the bride in 1950 were more than a year younger, on the average, than
the corresponding young persons who were entering their first marriage a
decade earlier. This decline stands in contrast with the fact that during
the entire 50-year period from 1890 to 1940, the average (median) age
at first marriage for grooms had declined only about two years and that
for brides only about one-half year, according to the best estimates available.[3]

Earlier marriages have become more common during recent years when

---

"Household and Family Characteristics: April 1953," Series P-20, No. 53; "Marital
Status of Workers: April 1953," Series P-50, No. 50; and "Family Income in the
United States: 1952," Series P-60, No. 15. U. S. Bureau of the Census, *Differential
Fertility, 1940 and 1910—Fertility for States and Large Cities.* National Office of
Vital Statistics, U. S. Public Health Service, Department of Health, Education, and
Welfare, *Vital Statistics of the United States,* reports for selected years from 1917 to
1950.

A closely related report was published by the National Office of Vital Statistics
shortly after the meeting in 1954 of the World Population Conference; this report is
entitled "Demographic Characteristics of Recently Married Persons: United States,
April 1953," *Vital Statistics—Special Reports,* Vol. 39, No. 3. It contains the results
of a sample survey conducted for the National Office of Vital Statistics by the Bureau
of the Census.

[3] Averages are in terms of medians and represent estimates based on marital status
and age distributions from decennial census reports for the respective dates. For an
explanation of the methodology, see *Current Population Reports,* Series P-20, No. 38,
p. 7. The report, "Demographic Characteristics of Recently Married Persons" (see
footnote 2), shows median ages at first marriage for men (23.2) and women (20.4)
who married in 1950. These figures are quite close to those shown in table 1 of this
paper; both sets of figures are based on census data by single years of age. In Vol. I
of the 1950 *Vital Statistics of the United States,* the median ages at first marriage for
grooms (23.9) and brides (21.5) in 19 reporting states were computed from distributions of marriages by 5-year age groups and are therefore somewhat higher. The differences between the various sets of medians may be attributed not only to the use of
different age groupings but also in part at least to the use of different types of basic
data and in part to the over-representation of highly urbanized states among the 19
reporting.

married women have found it easier to gain employment outside the home. More and more women now work for a period before marriage,

TABLE 1. Median Age of Husband and Wife at Selected Stages of the Life Cycle of the Family, for the United States: 1950, 1949, and 1890

Stage of the life cycle of the family	Median age of husband			Median age of wife		
	1950	1940	1890	1950	1940	1890
A. First marriage	22.8	24.3	26.1	20.1	21.5	22.0
B. Birth of last child	28.8	29.9	36.0	26.1	27.1	31.9
C. Marriage of last child	50.3	52.8	59.4	47.6	50.0	55.3
D. Death of one spouse*	64.1	63.6	57.4	61.4	60.9	53.3
E. Death of other spouse**	71.6	69.7	66.4	77.2	73.5	67.7

* Husband and wife survive jointly from marriage to specified age.
** Husband (wife) survives separately from marriage to specified age.

continue their employment after marriage until they start the childbearing period, then in a few years return to work outside the home. Women who are in the labor force have received more education, on the average, than those not in the labor force. In the marriage boom of the last decade, greater gains in the proportion married in the United States were made by the more-educated than by the less-educated sections of the population.[4]

An interesting sidelight on the changing age at marriage is an apparent decline in the gap between the median ages of husbands and wives at first marriage. The average husband of recent years is his wife's senior by about three years, whereas his grandfather was likely to have been senior by about four years.[5]

To simplify the treatment of our subject, we have limited our discussion to first marriages. It is recognized, of course, that many of these marriages become broken within a relatively short time;[6] in such cases, most

[4] John Hajnal, "Differential Changes in Marriage Patterns," *American Sociological Review*, 19: 148-154, April, 1954. Similar findings were reported by Calvin L. Beale in "Some Marriage Trends and Patterns Since 1940," an unpublished paper which was presented at a special meeting of the District of Columbia Sociological Society, May 3, 1952.

[5] The median difference between the ages of husbands and wives in their first marriages in 1948 was 2.8 years. (See Paul C. Glick and Emanuel Landau, "Age as a Factor in Marriage," *American Sociological Review*, XV: 517-529, August, 1950.) This figure is of approximately the same order of magnitude as the difference (3.3 years) between the median age of husband at first marriage and the median age of wife at first marriage derived from the same census data. In the absence of a direct measure of the difference between the ages of spouses in 1890, it is assumed that the difference between the median ages of spouses provides a usable approximation for that date.

[6] Paul H. Jacobson, "Differentials in Divorce by Duration of Marriage and Size of Family," *American Sociological Review*, XV: 236-244, April, 1950.

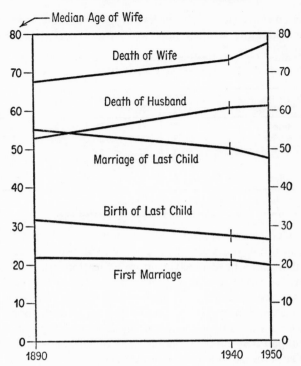

Figure 1. Stages of the life cycle of the family in the United States: 1890, 1940, and 1950.

of the marriage partners are at ages when remarriage rates are relatively high. In 1948, about 13 per cent of the married women living with their husbands had remarried after the dissolution of an earlier marriage; in 1910, the corresponding proportion was probably about seven or eight per cent.[7] Among persons who married since the end of World War II, about one fifth were entering a second or subsequent marriage.[8]

Childbearing. The average mother who was having her first child in 1950 was 22.5 years old, according to vital statitsics data on order of birth by single years of age of mother. The difference between the median age at first marriage based on census data and the median age at birth of first child based on vital statistics data, however, provides an unsatisfactory measure of the average interval between marriage and the birth of the first child. Similar data are used here to approximate the average in-

[7] Paul C. Glick, "First Marriages and Remarriages," *American Sociological Review,* XIV: 726-734, December, 1949.

[8] *Op. cit.,* and *Current Population Reports,* Series P-20, No. 53.

terval between marriage and the birth of the last child only because the relative error is much less in this case. More precise measurement of child-spacing intervals is now being undertaken by the Bureau of the Census in co-operation with the National Office of Vital Statistics, on the basis of data from the 1950 Census of Population and from the Current Population Survey.[9]

For women who had married and had reached the end of their reproductive period (45 to 49 years old) by 1952, the average number of children born per woman was about 2.35. By making use of this fact in conjunction with 1950 statistics on order of birth, it is estimated that approximately half of the women have borne their last child by the time they are 26 years old. Thus, the median length of time between marriage and the birth of the last child is probably close to six years.[10]

Because families have declined so sharply in size, the usual span of the childbearing years has become only about half as long as it was two generations ago. The average mother whose family reached completion in 1890 had borne 5.4 children, with an estimated interval of ten years between marriage and the birth of the last child. She had not given birth to her last child until she was about 32 years old. For 1940, the last of three (3.0) children was born when the mother was about 27 years old.

Women who had never borne a child constituted only about eight per cent of all women who had married and completed their period of fertility by 1890. This percentage approximately doubled by 1940 (15 per cent); it continued to rise by 1952 to 19 per cent for women 45 to 49 years old

[9] Several studies have provided significant information about child spacing for selected areas within the United States. For example, see P. K. Whelpton and Clyde V. Kiser, "Social and Psychological Factors Affecting Fertility. VI. The Planning of Fertility," *The Milbank Memorial Fund Quarterly,* XXV: 63-111, January, 1947. See also Harold T. Christensen and Hanna H. Meissner, "Studies in Child Spacing: III—Premarital Pregnancy as a Factor in Divorce," *American Sociological Review,* 18: 641-644, December, 1953.

[10] For each stage of the life cycle of the family, the ages given in table 1 for a given year are based on experience relating to marriages, births, or deaths as near to that year as available nation-wide data permit. The average number of children per ever-married woman 45 to 49 years old in 1952 (2.35) was used as an estimate of the average number of children that women who married in 1950 will have. Women who married in 1950, however, will not be at the end of their childbearing period until about 1975. If they were to bear, say, 2.8 children, on the average, instead of 2.35 as assumed, the estimated median age of mother at birth of the last child (and at the time the last child leaves home) would be increased about one year. As the practice of family limitation becomes more nearly universal, the number and spacing of children may vary more widely with changes in economic conditions. For a valuable analysis of past and probable future changes in size of completed families, see P. K. Whelpton, *Cohort Fertility: Native White Women in the United States* (Princeton, N.J.: Princeton University Press, 1954).

but there was evidence that it would fall sharply for younger women.

During the next decade or two the average number of children per completed family is likely to rise moderately and the proportion of women who remain childless throughout their reproductive years is certain to decline. Changes in patterns of marriage and childbearing which have developed since about the beginning of World War II will apparently have the effect of reversing, at least temporarily, the 150-year decline in the average size of completed family.

Children Leaving Home. From the time the last child is born until the first child leaves home, the size of family usually remains stable.

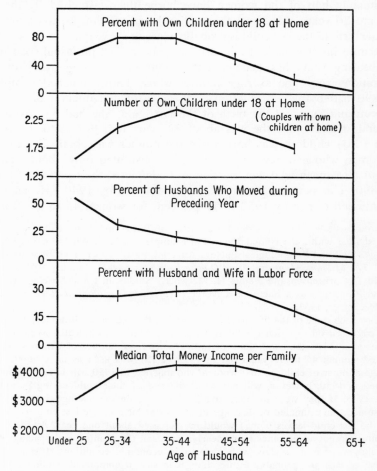

Figure 2. Characteristics of married couples by age of husband, for the United States: 1952.

Changes in family living during this period are those related to the growth and maturation of the children and the changing economic status of the parents.

If we make some allowance for mortality among the children, and if we assume that the children will marry and leave home at the same age that their parents married, we find that the average woman who married in 1950 will be about 48 years of age when her last child leaves home. By comparison, the average woman of her grandmother's day was about 55, if she lived that long, when her last child got married.

Dissolution of the Family. The combined effects of earlier marriages, smaller families, and longer average length of life have produced a remarkable change in the length of time that married couples live together after their children have set up homes of their own. Under conditions existing in 1950, a couple could expect to have about 41 years of married life before either the husband or the wife died; during their last 14 years together, the couple would ordinarily have no unmarried children remaining with them. Thus, the couple would have half as many years of married life with no young children at home as they would have with children at home.

By contrast, conditions existing in 1890 assured only about 31 years of joint survival for the husband and wife; in fact, the chances were 50-50 that one spouse or the other would die at least two years before their youngest child married.

Because men are usually older than their wives at marriage and have higher mortality rates, age for age, wives generally outlive their husbands. The wife can expect to live much longer after her husband's death if she is the survivor than the husband can expect to live after his wife's death if he is the survivor.[11] In this final stage of the family life cycle, the length of time that the remaining marriage partner survives has not changed greatly in the last two generations, but the expected ages during which the lone survivor is in this phase have been advanced several years.

[11] All estimates of age at death were based on chances of survival from age at marriage. Age at "death of one spouse" is the age to which half of the married couples are expected to survive jointly. Age at "death of other spouse" is the age to which half of the husbands (wives) are expected to live, without regard to the age to which their spouse lives. The difference between age at "death of one spouse" and age at "death of other spouse" is somewhat less than the expected period of widowerhood (widowhood) for the surviving marriage partner who does not remarry; but if age at "death of other spouse" had been calculated with age at "death of one spouse" as the point of departure, the overall length of the family cycle would have been somewhat elongated. United States life tables for all races in 1890 and 1940, and abridged life tables for whites in 1950, were used.

TABLE 2.   Characteristics of Married Couples by Age of Husband, for the
United States: 1952

Age of Husband	Per Cent of Couples with No Own Children under 18 at Home	Own Children under 18 at Home Per Couple with Children	Per Cent of Husbands Who Moved during the Preceding Year	Per Cent of Couples with Husband and Wife in Labor Force*
All ages	44.8	2.15	20.2	24.5
Under 25 years	42.4	1.51	55.8	24.9
25 to 34 years	20.9	2.08	32.9	24.9
35 to 44 years	20.9	2.46	18.2	29.5
45 to 54 years	49.4	2.08	⎰ 10.3	30.9
55 to 64 years	79.6	1.71	⎱	20.5
65 years and over	96.2	**	6.6	6.5

* Based on 1953 data.
** Fewer than 100 cases in sample.

Changes in Composition

As the family proceeds through its life cycle, it expands in size with
the addition of each child and eventually contracts as the children marry
and depart from their parental home. Many variations in this pattern
exist, of course. Some families have no children; in others, one or more
children remain at home longer than usual, often after marriage; and in
still other cases, one or more of the parents or parents-in-law spend their
later years with their children. Adjustments in living accommodations are
commonly made in order to meet the needs of the family and changes in
the economic activity of the family members generally occur as conditions
make such changes feasible and advantageous.

These dynamic aspects of family living are traced in this section by
studying changes in characteristics of married couples as the age of the
husband advances. (See table 2 and figure 2.)

Family Composition.   On the basis of data for a recent year (1952),
about 45 per cent of the married couples of all ages combined have no
sons or daughters under 18 years of age in their homes. About four tenths
of the husbands below the age of 25 years have no dependent children,
but only 21 per cent of those between 25 and 44 have none in the home.
Above the age of 45, the proportion of husbands without young children
of their own at home rises sharply and continuously until, among those
above the age of 65, nearly all have none living with them.

The number of children in homes with children rises until it reaches
two or three, on the average, by the time the husband is 35 to 44 years
old, then declines. Thus, the average family group, comprising the husband, wife, and young children, grows from two persons to four or five
and then diminishes gradually to the original two parents.

There are seldom any additional relatives living with the couple while the husband is under 40 years of age. From that time until old age approaches, however, there are likely to be one or more adult relatives (usually grown children of their own) in about one half of the homes and one or more other young relatives (usually grandchildren) in about one home out of ten. When a young married couple lives with the husband's or wife's parents, the chances are nearly two out of three that the couple will stay with the wife's parents. This arrangement is most common perhaps largely because the wife is likely to spend more time in the home than her husband, and because close daily contacts between a mother and her daughter are less likely to create tensions than similar contacts between a mother-in-law and her daughter-in-law.

Residential Location. About four fifths of the persons who marry change residences at the time of marriage or within the ensuing year. Thereafter, the mobility rate decreases sharply as the number of years married increases. By the time couples have been married 10 to 15 years, only about 20 per cent move to another home in the course of a year's time. By this time, most of the changes of residence required to provide room and a measure of privacy for the various family members have been made. Moreover, the difficulty of moving all of the belongings of the family when it is at its maximum size probably serves as a deterrent to residential changes during this stage of family life. The continued decline, rather than an increase, in mobility during the later years of life perhaps suggests, among other things, that families do not ordinarily move into smaller quarters after their children have left home. Data from the 1950 Census of Housing also suggest that the shifts to smaller homes are relatively few in number during this period of life and that most of them take place after the husband reaches 65 years of age. . . .

Concluding Statement. From the foregoing analysis, it is evident that the average family in the United States undergoes many significant changes in the course of its life cycle. At the same time, the pattern of these changes is different in important respects from that which prevailed a generation or two ago.

The effects of most of the changes since 1890 have been cumulative. Trends toward earlier marriage, smaller families, and longer length of life have culminated in the fact that couples now spend one third of their married life with no unmarried children of their own in the home. In 1890, the average married couple did not survive jointly to see their last child get married.

The recent upsurge in the number of births has resulted from increases in the number of women of childbearing age, in the proportion of persons in this age group who have married, and in the average number of children per married couple. The rise in the birth rate since 1940 is less significant for its effect on the current pattern of family development, however, than it is for its implications that more of the young adults are marrying and having a moderate-sized family.

AGE AND SEX IN THE SOCIAL STRUCTURE OF THE UNITED STATES

Talcott Parsons

Adapted and reprinted from the *American Sociological Review, 7* (1942), 604-16. . . . The present paper will not embody the results of systematic research but constitutes rather a tentative statement of certain major aspects of the role of age and sex in our society and of their bearing on a variety of problems. It will not attempt to treat adequately the important variations according to social class, rural-urban differences, and so on, but will concentrate particularly on the urban middle and upper-middle classes.

In our society age grading does not to any great extent, except for the educational system, involve formal age categorization, but is interwoven with other structural elements. In relation to these, however, it constitutes an important connecting link and an organizing point of reference in many respects. The most important of these for present purposes are kinship structure, formal education, occupation, and community participation. In most cases the age lines are not rigidly specific, but approximate; this does not, however, necessarily lessen their structural significance.[1]

In all societies the initial status of every normal individual is that of

[1] The problem of organization of this material for systematic presentation is, in view of this fact, particularly difficult. It would be possible to discuss the subject in terms of the above four principal structures with which age and sex are most closely interwoven, but there are serious disadvantages involved in this procedure. Age and sex categories constitute one of the main links of structural continuity in terms of which structures which are differentiated in other respects are articulated with each other; and in isolating the treatment of these categories there is danger that this extremely important aspect of the problem will be lost sight of. The least objectionable method, at least within the limits of space of such a paper, seems to be to follow the sequence of the life cycle.

child in a given kinship unit. In our society, however, this universal starting point is used in distinctive ways. Although in early childhood the
sexes are not usually sharply differentiated, in many kinship systems a
relatively sharp segregation of children begins very early. Our own society is conspicuous for the extent to which children of both sexes are in
many fundamental respects treated alike. This is particularly true of both
privileges and responsibilities. The primary distinctions within the group
of dependent siblings are those of age. Birth order as such is notably
neglected as a basis of discrimination; a child of eight and a child of five
have essentially the privileges and responsibilities appropriate to their
respective age levels without regard to what older, intermediate, or
younger siblings there may be. The preferential treatment of an older
child is not to any significant extent differentiated if and because he happens to be the first born.

There are, of course, important sex differences in dress and in approved
play interest and the like, but if anything, it may be surmised that in the
urban upper-middle classes these are tending to diminish. Thus, for instance, play overalls are essentially similar for both sexes. What is perhaps
the most important sex discrimination is more than anything else a reflection of the differentiation of adult sex roles. It seems to be a definite
fact that girls are more apt to be relatively docile, to conform in general
according to adult expectations, to be "good," whereas boys are more apt
to be recalcitrant to discipline and defiant of adult authority and expectations. There is really no feminine equivalent of the expression "bad boy."
It may be suggested that this is at least partially explained by the fact
that it is possible from an early age to initiate girls directly into many important aspects of the adult feminine role. Their mothers are continually
about the house and the meaning of many of the things they are doing is
relatively tangible and easily understandable to a child. It is also possible
for the daughter to participate actively and usefully in many of these activities. Especially in the urban middle classes, however, the father does
not work in the home and his son is not able to observe his work or to
participate in it from an early age. Furthermore, many of the masculine
functions are of a relatively abstract and intangible character, such that
their meaning must remain almost wholly inaccessible to a child. This
leaves the boy without a tangible meaningful model to emulate and without the possibility of a gradual initiation into the activities of the adult
male role. An important verification of this analysis could be provided
through the study in our own society of the rural situation. It is my impression that farm boys tend to be "good" in a sense which is not typical
of their urban brothers.

The equality of privileges and responsibilities, graded only by age but not by birth order, is extended to a certain degree throughout the whole range of the life cycle. In full adult status, however, it is seriously modified by the asymmetrical relation of the sexes to the occupational structure. One of the most conspicuous expressions and symbols of the underlying equality, however, is the lack of sex differentiation in the process of formal education, so far, at least, as it is not explicitly vocational. Up through college, differentiation seems to be primarily a matter on the one hand of individual ability, on the other hand of class status, and only to a secondary degree of sex differentiation. One can certainly speak of a strongly established pattern that all children of the family have a "right" to a good education, rights which are graduated according to the class status of the family but also to individual ability. It is only in post-graduate professional education, with its direct connection with future occupational careers, that sex discrimination becomes conspicuous. It is particularly important that this equality of treatment exists in the sphere of liberal education since throughout the social structure of our society there is a strong tendency to segregate the occupational sphere from one in which certain more generally human patterns and values are dominant, particularly in informal social life and the realm of what will here be called community participation.

Although this pattern of equality of treatment is present in certain fundamental respects at all age levels, at the transition from childhood to adolescence new features appear which disturb the symmetry of sex roles while still a second set of factors appears with marriage and the acquisition of full adult status and responsibilities.

An indication of the change is the practice of chaperonage, through which girls are given a kind of protection and supervision by adults to which boys of the same age group are not subjected. Boys, that is, are chaperoned only in their relations with girls of their own class. This modification of equality of treatment has been extended to the control of the private lives of women students in boarding schools and colleges. Of undoubted significance is the fact that it has been rapidly declining not only in actual effectiveness but as an ideal pattern. Its prominence in our recent past, however, is an important manifestation of the importance of sex role differentiation. Important light might be thrown upon its functions by systematic comparison with the related phenomena in Latin countries where this type of asymmetry has been far more sharply accentuated than in this country in the more modern period.

It is at the point of emergence into adolescence that a set of patterns and behavior phenomena which involve a highly complex combination of age grading and sex role elements begins to develop. These may be referred to together as the phenomena of the "youth culture." Certain of its elements are present in pre-adolescence and others in the adult culture. But the peculiar combination in connection with this particular age level is unique and highly distinctive of American society.

Perhaps the best single point of reference for characterizing the youth culture lies in its contrast with the dominant pattern of the adult male role. By contrast with the emphasis on responsibility in this role, the orientation of the youth culture is more or less specifically irresponsible. One of its dominant notes is "having a good time" in relation to which there is a particularly strong emphasis on social activities in company with the opposite sex. A second predominant characteristic on the male side lies in the prominence of athletics, which is an avenue of achievement and competition which stands in sharp contrast to the primary standards of adult achievement in professional and executive capacities. Negatively, there is a strong tendency to repudiate interest in adult things and to feel at least a certain recalcitrance to the pressure of adult expectations and discipline. In addition to, but including, athletic prowess, the typical pattern of the male youth culture seems to emphasize the value of certain qualities of attractiveness, especially in relation to the opposite sex. It is very definitely a rounded humanistic pattern rather than one of competence in the performance of specified functions. Such stereotypes as the "swell guy" are significant of this. On the feminine side there is correspondingly a strong tendency to accentuate sexual attractiveness in terms of various versions of what may be called the "glamour girl" pattern.[2] Although these patterns defining roles tend to polarize sexually—for instance, as between star athlete and socially popular girl—yet on a certain level they are com-

[2] Perhaps the most dramatic manifestation of this tendency lies in the prominence of the patterns of "dating," for instance among college women. As shown by an unpublished participant-observer study made at one of the eastern women's colleges, perhaps the most important single basis of informal prestige rating among the residents of a dormitory lies in their relative dating success—though this is by no means the only basis. One of the most striking features of the pattern is the high publicity given to the "achievements" of the individual in a sphere where traditionally in the culture a rather high level of privacy is sanctioned—it is interesting that once an engagement has occurred a far greater amount of privacy is granted. The standards of rating cannot be said to be well integrated, though there is an underlying consistency in that being in demand by what the group regards as desirable men is perhaps the main standard.

It is true that the "dating" complex need not be exclusively bound up with the "glamour girl" stereotype of ideal feminine personality—the "good companion" type may also have a place. Precisely, however, where the competitive aspect of dating is most prominent the glamour pattern seems heavily to predominate, as does, on the masculine side, a somewhat comparable glamorous type. On each side at the same time there is room for considerable difference as to just where the emphasis is placed—for example as between "voluptuous" sexuality and more decorous "charm."

plementary, both emphasizing certain features of a total personality in terms of the direct expression of certain values rather than of instrumental significance.

One further feature of this situation is the extent to which it is crystallized about the system of formal education.[3] One might say that the principal centers of prestige dissemination are the colleges, but that many of the most distinctive phenomena are to be found in high schools throughout the country. It is of course of great importance that liberal education is not primarily a matter of vocational training in the United States. The individual status on the curricular side of formal education is, however, in fundamental ways linked up with adult expectations, and doing "good work" is one of the most important sources of parental approval. Because of secondary institutionalization this approval is extended into various spheres distinctive of the youth culture. But it is notable that the youth culture has a strong tendency to develop in directions which are either on the borderline of parental approval or beyond the pale, in such matters as sex behavior, drinking, and various forms of frivolous and irresponsible behavior. The fact that adults have attitudes to these things which are often deeply ambivalent and that on such occasions as college reunions they may outdo the younger generation, as, for instance, in drinking, is of great significance, but probably structurally secondary to the youth-versus-adult differential aspect. Thus, the youth culture is not only, as is true of the curricular aspect of formal education, a matter of age status as such but also shows strong signs of being a product of tensions in the relationship of younger people and adults.

From the point of view of age grading, perhaps the most notable fact about this situation is the existence of definite pattern distinctions from the periods coming both before and after. At the line between childhood and adolescence "growing up" consists precisely in ability to participate in youth culture patterns, which are not for either sex the same as the adult patterns practiced by the parental generation. In both sexes the transition to full adulthood means loss of a certain "glamorous" element. From being the athletic hero or the lion of college dances, the young man becomes a prosaic business executive or lawyer. The more successful adults participate in an important order of prestige symbols but these are

---

[3] A central aspect of this focus of crystallization lies in the element of tension, sometimes of direct conflict, between the youth culture patterns of college and school life, and the "serious" interests in and obligations toward curricular work. It is of course the latter which defines some at least of the most important foci of adult expectations of doing "good" work and justifying the privileges granted. It is not possible here to attempt to analyze the interesting, ambivalent attitudes of youth toward curricular work and achievement.

of a very different order from those of the youth culture. The contrast in the case of the feminine role is perhaps equally sharp, with at least a strong tendency to take on a "domestic" pattern with marriage and the arrival of young children.

The symmetry in this respect must, however, not be exaggerated. It is of fundamental significance to the sex role structure of the adult age levels that the normal man has a "job" which is fundamental to his social status in general. It is perhaps not too much to say that only in very exceptional cases can an adult man be genuinely self-respecting and enjoy a respected status in the eyes of others if he does not "earn a living" in an approved occupational role. Not only is this a matter of his own economic support but, generally speaking, his occupational status is the primary source of the income and class status of his wife and children.

In the case of the feminine role the situation is radically different. The majority of married women, of course, are not employed, but even of those that are a very large proportion do not have jobs which are in basic competition for status with those of their husbands.[4] The majority of "career" women whose occupational status is comparable with that of men in their own class, at least in the upper-middle and upper classes, are unmarried, and in the small proportion of cases where they are married the result is a profound alteration in family structure.

This pattern, which is central to the urban middle classes, should not be misunderstood. In rural society, for instance, the operation of the farm and the attendant status in the community may be said to be a matter of the joint status of both parties to a marriage. Whereas a farm is operated by a family, an urban job is held by an individual and does not involve other members of the family in a comparable sense. One convenient expression of the difference lies in the question of what would happen in case of death. In the case of a farm it would at least be not at all unusual for the widow to continue operating the farm with the help of a son or even of hired men. In the urban situation the widow would cease to have

---

[4] The above statement, even more than most in the present paper, needs to be qualified in relation to the problem of class. It is above all to the upper-middle class that it applies. Here probably the great majority of "working wives" are engaged in some form of secretarial work which would, on an independent basis, generally be classed as a lower-middle-class occupation. The situation at lower levels of the class structure is quite different since the prestige of the jobs of husband and wife is then much more likely to be nearly equivalent. It is quite possible that this fact is closely related to the relative instability of marriage which Davis and Gardner (*Deep South*) find, at least for the community they studied, to be typical of lower-class groups. The relation is one which deserves careful study.

See papers by Hollingshead and Hillman, pp. 573-582 and 603-608 respectively for evidence on this point—*Eds.*

any connection with the organization which had employed her husband and he would be replaced by another man without reference to family affiliations.

In this urban situation the primary status-carrying role is in a sense that of housewife. The woman's fundamental status is that of her husband's wife, the mother of his children, and traditionally the person responsible for a complex of activities in connection with the management of the household, care of children, etc.

For the structuring of sex roles in the adult phase the most fundamental considerations seem to be those involved in the interrelations of the occupational system and the conjugal family. In a certain sense the most fundamental basis of the family's status is the occupational status of the husband and father. As has been pointed out, this is a status occupied by an individual by virtue of his individual qualities and achievements. But both directly and indirectly, more than any other single factor, it determines the status of the family in the social structure, directly because of the symbolic significance of the office or occupation as a symbol of prestige, indirectly because as the principal source of family income it determines the standard of living of the family. From one point of view the emergence of occupational status into this primary position can be regarded as the principal source of strain in the sex role structure of our society since it deprives the wife of her role as a partner in a common enterprise. The common enterprise is reduced to the life of the family itself and to the informal social activities in which husband and wife participate together. This leaves the wife a set of utilitarian functions in the management of the household which may be considered a kind of "pseudo-" occupation. Since the present interest is primarily in the middle classes, the relatively unstable character of the role of housewife as the principal content of the feminine role is strongly illustrated by the tendency to employ domestic servants wherever financially possible. It is true that there is an American tendency to accept tasks of drudgery with relative willingness, but it is notable that in middle-class families there tends to be a dissociation of the essential personality from the performance of these tasks. Thus, advertising continually appeals to such desires as to have hands which one could never tell had washed dishes or scrubbed floors.[5] Organization about the

---

[5] This type of advertising appeal undoubtedly contains an element of "snob appeal" in the sense of an invitation to the individual by her appearance and ways to identify herself with a higher social class than that of her actual status. But it is almost certainly not wholly explained by this element. A glamorously feminine appearance which is specifically dissociated from physical work is undoubtedly a genuine part of an authentic personality ideal of the middle class, and not only evidence of a desire to belong to the upper class.

function of housewife, however, with the addition of strong affectional devotion to husband and children, is the primary focus of one of the principal patterns governing the adult feminine role—what may be called the "domestic" pattern. It is, however, a conspicuous fact that strict adherence to this pattern has become progressively less common and has a strong tendency to a residual status—that is, to be followed most closely by those who are unsuccessful in competition for prestige in other directions.

It is, of course, possible for the adult woman to follow the masculine pattern and seek a career in fields of occupational achievement in direct competition with men of her own class. It is, however, notable that in spite of the very great progress of the emancipation of women from the traditional domestic pattern only a very small fraction have gone very far in this direction. It is also clear that its generalization would only be possible with profound alterations in the structure of the family.

Hence, it seems that concomitant with the alteration in the basic masculine role in the direction of occupation there have appeared two important tendencies in the feminine role which are alternative to that of simple domesticity on the one hand, and to a full-fledged career on the other. In the older situation there tended to be a very rigid distinction between respectable married women and those who were "no better than they should be." The rigidity of this line has progressively broken down through the infiltration into the respectable sphere of elements of what may be called again the glamour pattern, with the emphasis on a specifically feminine form of attractiveness which on occasion involves directly sexual patterns of appeal. One important expression of this trend lies in the fact that many of the symbols of feminine attractiveness have been taken over directly from the practices of social types previously beyond the pale of respectable society. This would seem to be substantially true of the practice of women smoking and of at least the modern version of the use of cosmetics. The same would seem to be true of many of the modern versions of women's dress. "Emancipation" in this connection means primarily emancipation from traditional and conventional restrictions on the free expression of sexual attraction and impulses, but in a direction which tends to segregate the element of sexual interest and attraction from the total personality and in so doing tends to emphasize the segregation of sex roles. It is particularly notable that there has been no corresponding tendency to emphasize masculine attraction in terms of dress and other such aids. One might perhaps say that in a situation which strongly inhibits competition between the sexes on the same plane the feminine glamour pattern has appeared as an offset to masculine occupational status and to its attendant symbols of prestige. It is perhaps significant that there is a common stereotype of the association of physically beautiful, expensively and elaborately dressed women with physically unattractive but rich and powerful men.

The other principal direction of emancipation from domesticity seems to lie in emphasis on what has been called the common humanistic element. This takes a wide variety of forms. One of them lies in a relatively mature appreciation and systematic cultivation of cultural interests and educated tastes, extending all the way from the intellectual sphere to matters of art, music, and house furnishings. A second consists in cultivation of serious interests and humanitarian obligations in community welfare situations and the like. It is understandable that many of these orientations are most conspicuous in fields where through some kind of tradition there is an element of particular suitability for feminine participation. Thus, a woman who takes obligations to social welfare particularly seriously will find opportunities in various forms of activity which traditionally tie up with women's relation to children, to sickness and so on. But this may be regarded as secondary to the underlying orientation which would seek an outlet in work useful to the community following the most favorable opportunities which happen to be available.

This pattern, which with reference to the character of relationship to men may be called that of the "good companion," is distinguished from the others in that it lays far less stress on the exploitation of sex role as such and more on that which is essentially common to both sexes. There are reasons, however, why cultural interests and interest in social welfare and community activities are particularly prominent in the activities of women in our urban communities. On the one side, the masculine occupational role tends to absorb a very large proportion of the man's time and energy and to leave him relatively little for other interests. Furthermore, unless his position is such as to make him particularly prominent his primary orientation is to those elements of the social structure which divide the community into occupational groups rather than those which unite it in common interests and activities. The utilitarian aspect of the role of housewife, on the other hand, has declined in importance to the point where it scarcely approaches a fulltime occupation for a vigorous person. Hence the resort to other interests to fill up the gap. In addition, women being more closely tied to the local residential community are more apt to be involved in matters of common concern to the members of that community. This peculiar role of women becomes particularly conspicuous in middle age. The younger married woman is apt to be relatively highly absorbed in the care of young children. With their growing up, however, her absorption in the household is greatly lessened, often just at the time when the husband is approaching the apex of his career and is most heavily involved in its obligations. Since to a high degree this humanistic aspect of the feminine role is only partially institutionalized it is not surprising that its patterns often bear the marks of strain and in- security, as perhaps has been classically depicted by Helen Hokinson's cartoons of women's clubs.

The adult roles of both sexes involve important elements of strain which are involved in certain dynamic relationships, especially to the youth culture. In the case of the feminine role marriage is the single event toward which a selective process, in which personal qualities and effort can play a decisive role, has pointed up. This determines a woman's fun- damental status, and after that her role patterning is not so much status- determining as a matter of living up to expectations and finding satisfying interests and activities. In a society where such strong emphasis is placed upon individual achievement it is not surprising that there should be a certain romantic nostalgia for the time when the fundamental choices were still open. This element of strain is added to by the lack of clear- cut definition of the adult feminine role. Once the possibility of a career has been eliminated there still tends to be a rather unstable oscillation between emphasis in the direction of domesticity or glamour or good com- panionship. According to situational pressures and individual character the tendency will be to emphasize one or another of these more strongly. But it is a situation likely to produce a rather high level of insecurity. In this state the pattern of domesticity must be ranked lowest in terms of prestige but also, because of the strong emphasis in community sentiment on the virtues of fidelity and devotion to husband and children, it offers perhaps the highest level of a certain kind of security. It is no wonder that such an important symbol as Whistler's mother concentrates primarily on this pattern.

The glamour pattern has certain obvious attractions since to the woman who is excluded from the struggle for power and prestige in the occupa- tional sphere it is the most direct path to a sense of superiority and im- portance. It has, however, two obvious limitations. In the first place, many of its manifestations encounter the resistance of patterns of moral conduct and engender conflicts not only with community opinion but also with the individual's own moral standards. In the second place, the highest mani- festations of its pattern are inevitably associated with a rather early age level—in fact, overwhelmingly with the courtship period. Hence, if strongly entered upon, serious strains result from the problem of adapta- tion to increasing age.

The one pattern which would seem to offer the greatest possibilities for able, intelligent, and emotionally mature women is the third—the good companion pattern. This, however, suffers from a lack of fully institu- tionalized status and from the multiplicity of choices of channels of expression. It is only those with the strongest initiative and intelligence who achieve fully satisfactory adaptations in this direction. It is quite clear that in the adult feminine role there is quite sufficient strain and insecurity so that wide-spread manifestations are to be expected in the form of neurotic behavior.

The masculine role at the same time is itself by no means devoid of corresponding elements of strain. It carries with it to be sure the primary prestige of achievement, responsibility, and authority. By comparison with the role of the youth culture, however, there are at least two important types of limitations. In the first place, the modern occupational system has led to increasing specialization of role. The job absorbs an extraordinarily large proportion of the individual's energy and emotional interests in a role which often has relatively narrow content. This in particular restricts the area within which he can share common interests and experiences with others not in the same occupational specialty. It is perhaps of considerable significance that so many of the highest prestige statuses of our society are of this specialized character. There is in the definition of roles little to bind the individual to others in his community on a comparable status level. By contrast with this situation, it is notable that in the youth culture common human elements are far more strongly emphasized. Leadership and eminence are more in the role of total individuals and less of competent specialists. This perhaps has something to do with the significant tendency in our society for all age levels to idealize youth and for the older age groups to attempt to imitate the patterns of youth behavior.

It is perhaps as one phase of this situation that the relation of the adult man to persons of the opposite sex should be treated. The effect of the specialization of occupational role is to narrow the range in which the sharing of common human interests can play a large part. In relation to his wife the tendency of this narrowness would seem to be to encourage on her part either the domestic or the glamorous role, or community participation somewhat unrelated to the marriage relationship. This relationship between sex roles presumably introduces a certain amount of strain into the marriage relationship itself since this is of such overwhelming importance to the family and hence to a woman's status and yet so relatively difficult to maintain on a level of human companionship. Outside the marriage relationship, however, there seems to be a notable inhibition against easy social intercourse, particularly in mixed company.[6]

---

[6] In the informal social life of academic circles with which the writer is familiar there seems to be a strong tendency in mixed gatherings—as after dinner, for the sexes to segregate. In such groups the men are apt to talk either shop subjects or politics whereas the women are apt to talk about domestic affairs, schools, their children, etc.,

The man's close personal intimacy with other women is checked by the danger of the situation being defined as one of rivalry with the wife, and easy friendship without sexual-emotional involvement seems to be inhibited by the specialization of interests in the occupational sphere. It is notable that brilliance of conversation of the "salon" type seems to be associated with aristocratic society and is not prominent in ours.

Along with all this goes a certain tendency for middle-aged men, as symbolized by the "bald-headed row," to be interested in the physical aspect of sex—that is, in women precisely as dissociated from those personal considerations which are important to relationships of companionship or friendship, to say nothing of marriage. In so far as it does not take this physical form, however, there seems to be a strong tendency for middle-aged men to idealize youth patterns—that is, to think of the ideal inter-sex friendship as that of their pre-marital period.[7]

In so far as the idealization of the youth culture by adults is an expression of elements of strain and insecurity in the adult roles it would be expected that the patterns thus idealized would contain an element of romantic unrealism. The patterns of youthful behavior thus idealized are not those of actual youth so much as those which older people wish their own youth might have been. This romantic element seems to coalesce with a similar element derived from certain strains in the situation of young people themselves.

The period of youth in our society is one of considerable strain and insecurity. Above all, it means turning one's back on the security both of status and of emotional attachment which is engaged in the family of orientation. It is structurally essential to transfer one's primary emotional attachment to a marriage partner who is entirely unrelated to the previous family situation. In a system of free marriage choice this applies to women as well as men. For the man there is in addition the necessity to face the hazards of occupational competition in the determination of a career. There is reason to believe that the youth culture has important positive functions in easing the transition from the security of childhood in the family of orientation to that of full adult in marriage and occupational status. But precisely because the transition is a period of strain it is to be expected that it involves elements of unrealistic romanticism. Thus, significant features in the status of youth patterns in our society would seem to derive from the coincidence of the emotional needs of adolescents with those derived from the strains of the situation of adults.

---

or personalities. It is perhaps on personalities that mixed conversation is apt to flow most freely.

[7] This, to be sure, often contains an element of romantization. It is more nearly what he wishes these relations had been than what they actually were.

A tendency to the romantic idealization of youth patterns seems in different ways to be characteristic of modern western society as a whole.[8] . . . The German "youth movement," starting before World War I, has occasioned a great deal of comment and has in various respects been treated as the most notable instance of the revolt of youth. It is generally believed that the youth movement [had] an important relation to the background of National Socialism, and this fact as much as any suggests the important difference. While in Germany as everywhere there has been a generalized revolt against convention and restrictions on individual freedom as embodied in the traditional adult culture, in Germany particular emphasis . . . appeared on the community of male youth. "Comradeship," in a sense which strongly suggests that of soldiers in the field, has . . . been strongly emphasized as the ideal social relationship. By contrast with this, in the American youth culture and its adult romantization a much stronger emphasis has been placed on the cross-sex relationship. It would seem that this fact, with the structural factors which underlie it, has much to do with the failure of the youth culture to develop any considerable political significance in this country. Its predominant pattern has been that of the idealization of the isolated couple in romantic love. There have, to be sure, been certain tendencies among radical youth to a political orientation but in this case there has been a notable absence of emphasis on the solidarity of the members of one sex. The tendency has been rather to ignore the relevance of sex difference in the interest of common ideals.

The importance of youth patterns in contemporary American culture throws into particularly strong relief the status in our social structure of the most advanced age groups. By comparison with other societies the United States assumes an extreme position in the isolation of old age from participation in the most important social structures and interests. Structurally speaking, there seem to be two primary bases of this situation. In the first place, the most important single distinctive feature of our family structure is the isolation of the individual conjugal family. It is impossible to say that with us it is "natural" for any other group than husband and wife and their dependent children to maintain a common household. Hence, when the children of a couple have become independent through marriage and occupational status the parental couple is left without attachment to any continuous kinship group. It is, of course, common for other relatives to share a household with the conjugal family but this scarcely ever occurs without some important elements of strain. For in-

[8] Cf. E. Y. Hartshorne, "German Youth and the Nazi Dream of Victory," *America in a World at War*, Pamphlet, No. 12, New York, 1941.

dependence is certainly the preferred pattern for an elderly couple, particularly from the point of view of the children.

The second basis of the situation lies in the occupational structure. In such fields as farming and the maintenance of small independent enterprises there is frequently no such thing as abrupt "retirement," rather a gradual relinquishment of the main responsibilities and functions with advancing age. So far, however, as an individual's occupational status centers in a specific "job," he either holds the job or does not, and the tendency is to maintain the full level of functions up to a given point and then abruptly to retire. In view of the very great significance of occupational status and its psychological correlates, retirement leaves the older man in a peculiarly functionless situation, cut off from participation in the most important interests and activities of the society. There is a further important aspect of this situation. Not only status in the community but actual place of residence is to a very high degree a function of the specific job held. Retirement not only cuts the ties to the job itself but also greatly loosens those to the community of residence. Perhaps in no other society is there observable a phenomenon corresponding to the accumulation of retired elderly people in such areas as Florida and Southern California in the winter. It may be surmised that this structural isolation from kinship, occupational, and community ties is the fundamental basis of the recent political agitation for help to the old. It is suggested that it is far less the financial hardship[9] of the position of elderly people than their social isolation which makes old age a "problem." As in other connections we are here very prone to rationalize generalized insecurity in financial and economic terms. The problem is obviously of particularly great significance in view of the changing age distribution of the population with the prospect of a far greater proportion in the older age groups than in previous generations. It may also be suggested, that through well-known psychosomatic mechanisms, the increased incidence of the dis-

[9] That the financial difficulties of older people are in a very large proportion of cases real is not to be doubted. This, however, is at least to a very large extent a consequence rather than a determinant of the structural situation. Except where it is fully taken care of by pension schemes, the income of older people is apt to be seriously reduced, but, even more important, the younger conjugal family usually does not feel an obligation to contribute to the support of aged parents. Where as a matter of course both generations shared a common household, this problem did not exist.

Since the original publication of this paper in 1942 evidence has appeared which mitigates the belief that the old are as isolated as Parsons suggests. See the editors' introduction to ch. 2 and the papers by Sussman and Schorr, pp. 18-19, 49-57, and 417-435, respectively—*Eds.*

abilities of older people, such as heart disease, cancer, etc., may be at least in part attributed to this structural situation.

A THEORY OF ROLE STRAIN

William J. Goode

Adapted from the *American Sociological Review*, 25:4 (1960) 483-496.

The present paper is based on the general view that institutions are made up of role relationships, and approaches both social action and social structure through the notion of "role strain," the felt difficulty in fulfilling role obligations. Role relations are seen as a sequence of "role bargains," and as a continuing process of selection among alternative role behaviors, in which each individual seeks to reduce his role strain. These choices determine the allocations of role performances to all institutions of the society. Within the limited compass of this paper, only a few of the possible implications of role strain as a theoretical approach can be explored.

The widespread notion that institutions are made up of roles is fruitful because it links a somewhat more easily observable phenomenon, social behavior, to an important but less easily observable abstraction, social structure. In functionalist terms, this notion also links the observed acts and inferred values of the individual with the institutional imperatives or requisites of the society. At the same time, by focusing on the elements in the individual's action decision, it avoids the pitfall of supposing that people carry out their obligations because these are "functional" for the society.

Approaching role interaction in terms of role strain offers the possibility of buttressing more adequately the empirical weaknesses of the most widely accepted theoretical view of society,[1] according to which the continuity of social roles, and thus the maintenance of the society, is mainly a function of two major variables: the normative, consensual commitment of the individuals of the society; and the integration among the norms

---

[1] I prefer to call this the "Lintonian model" (Ralph Linton, *The Study of Man*, New York: Appleton-Century, 1936), although Linton is not, of course, the creator of this model. Rather, he summed up a generation of thought about social structure in a clear and illuminating fashion, so that for many years the definitions and statements in this book were widely cited by both anthropologists and sociologists. Of course, "everyone knows" these weaknesses, but our basic model is not thereby changed to account for them.

held by those individuals. Although this view is superior to earlier ones,[2] it fails to explain how a complex urban society keeps going[3] because it does not account for the following awkward empirical facts:[4]

1. Some individuals do not accept even supposedly central values of the society.
2. Individuals vary in their emotional commitment to both important and less important values.
3. This value commitment varies by class strata, and by other characteristics of social position, for example, age, sex, occupation, geographic region, and religion.
4. Even when individuals accept a given value, some of them also have a strong or weak "latent commitment" to very different or contradictory values.[5]
5. Conformity with normative prescriptions is not a simple function of value commitment; there may be value commitment without conformity or conformity without commitment.
6. When individuals' social positions change, they may change both their behavior and their value orientations.
7. The values, ideals, and role obligations of every individual are at times in conflict.

Under the current conception of roles as units of social structures, presumably we should observe the role decisions of individuals in order to see how the society continues. On first view, as can be seen from the above list of points, the basis of social stability or integration seems precarious, and the decisions of the individual puzzling. For even when "the norms of the society" are fully accepted by the individual, they are not adequate guides for individual action. Order cannot be imposed by any *general* solution for all role decisions, since the total set of role obligations is probably unique for every individual. On the other hand, the individual may face different types of role demands and conflicts, which he feels as "role strains" when he wishes to carry out specific obligations.

In the immediately following sections, the major sources and types of

[2] For a systematic statement of several earlier models, see Talcott Parsons, *The Structure of Social Action*, New York: McGraw-Hill, 1937, Chapter 2.

[3] For an earlier discussion relevant to this paper, see William J. Goode, "Contemporary Thinking about Primitive Religion," *Soziologus*, 5 (1955), pp. 122-131; also in Morton Fried, editor *Readings in Anthropology*, New York: Crowell, 1959, Vol. II, pp. 450-460.

[4] For a good exposition of certain aspects of dissensus as they apply to American society, see Robin W. Williams, *American Society*, New York: Knopf, 1956, esp. pp. 352 ff.

[5] Charles H. Page has reminded me that role diversity is not confined to modern societies, as the work of functionalist anthropologists (e.g., Malinowski's *Crime and Custom* and Benedict's *Patterns of Culture*) has shown. This empirical fact is of considerable theoretical consequence, especially for the relations between adjacent social strata or castes, or between conquerors and the conquered.

role strain are specified, and thereafter the two main sets of mechanisms which the individual may use to reduce role strain are analyzed.

Types of Role Strain

It is an axiom, rarely expressed, of social theory that the individuals who face common role obligations *can* generally fulfill them. Indeed, most theories of stratification and criminality require such an assumption, and common opinion uses it is as a basis for its moral demands on the individual. We may suppose, as a corollary, that there are theoretical limits to the specific demands which societies may make of men. In addition, the "theorem of institutional integration"[6] is roughly correct as an orienting idea: people generally *want* to do what they are supposed to do, and this is what the society needs to have done in order to continue.

Yet, with respect to any given norm or role obligation, there are always some persons who cannot conform, by reason of individuality or situation: they do not have sufficient resources, energy, and so on. A wider view of all such obligations discloses the following types or sources of role strain:

First, even when role demands are not onerous, difficult, or displeasing, they are required at particular times and places. Consequently, virtually no role demand is such a spontaneous pleasure that conformity with it is always automatic.

Second, all individuals take part in many *different* role relationships, for each of which there will be somewhat different obligations.[7] Among these, there may be either contradictory performances required (the bigamous husband; the infantry lieutenant who must order his close friend to risk his life in battle) or conflicts of time, place, or resources. These are conflicts of allocation (civic as against home obligations).

Third, each role relationship typically demands *several* activities or responses. Again, there may be inconsistencies (what the husband does to balance his family budget may impair his emotional relations with the members of his houshold). There may be different but not quite contradictory norms which may be applied to the various behavioral demands of the same role (the clergyman as the emotionally neutral counselor, but as a praising or condemnatory spiritual guide). Perhaps most jobs

---

[6] This is the label which Talcott Parsons has suggested for the view, generally accepted since Durkheim's *Division of Labor,* that the maintenance of the society rests on desires of individuals to do things which must be done if the society is to survive. *The Social System,* Glencoe, Ill.: Free Press, 1949, pp. 36-43.

[7] In this paper, I distinguish role and status on the basis of only "degree of institutionalization": all role relations are somewhat institutionalized, but statutes are more fully institutionalized.

fall into this category, in that their various demands create some strain as between the norms of quantity and quality, technical excellence and human relations skills, and universalism and particularism.

Finally, many role relationships are "role sets," that is, the individual engages, by virtue of *one* of his positions, in several role relationships with different individuals.[8]

The individual is thus likely to face a wide, distracting, and sometimes conflicting array of role obligations. If he conforms fully or adequately in one direction, fulfillment will be difficult in another. Even if he feels lonely, and would like to engage in additional role relationships, it is likely that he cannot fully discharge all the obligations he already faces. He cannot meet all these demands to the satisfaction of all the persons who are part of his total role network. Role strain—difficulty in meeting given role demands—is therefore normal. In general, *the individual's total role obligations are overdemanding.*

Consequently, although the theorem of institutional integration, or the assumption of norm commitment, offers an explanation for the fulfillment of the duties imposed by a single norm, it does not account for the integration of an individual's total role system, or the integration among the role systems of various individuals, which presumably make up the social structure. The individual's problem is how to make his whole role system manageable, that is, how to allocate his energies and skills so as to reduce role strain to some bearable proportions. For the larger social structure, the problem is one of integrating such role systems—by allocating the flow or role performances so that various institutional activities are accomplished.

The Reduction of Role Strain: Ego's Choice

A sensitizing or orienting notion in functionalist as well as system theory —perhaps more properly called one element in the definition of a "system" —is that a strain is likely to be associated with some mechanisms for reducing it.[9] The individual can utilize two main sets of techniques for

[8] Cf. Robert K. Merton, *Social Theory and Social Structure,* Glencoe, Ill.: Free Press, 1957, pp. 369 ff. For its use in an empirical study, see Mary Jean Huntington, "The Development of a Professional Self Image," in R. K. Merton *et al.,* editors, *The Student-Physician,* Cambridge: Harvard University Press, 1957, pp. 180 ff.

[9] Merton's "mechanisms" operate to *articulate* role sets; see Robert K. Merton, "The Role Set: Problems in Sociological Theory," *British Journal of Sociology,* 8 (June, 1957), pp. 113 ff. Here, we are concerned with a more general problem, which includes role sets as a special source of role strain. Moreover, Merton is concerned with only one of our problems, integrating the total role systems of all individuals in a demarcated social system; while we are, in addition, concerned with the problem of the individual in integrating his own role system. Several of our mechanisms, then, are

reducing his role strain: those which determine whether or when he will enter or leave a role relationship; and those which have to do with the actual role bargain which the individual makes or carries out with another.

Ego's Manipulation of His Role Structure. Ego has at his disposal several ways of determining whether or when he will accept a role relationship:

1. *Compartmentalization:* This may be defined on the psychological level as the ability to ignore the problem of consistency. Socially, role relations tend toward compartmentalization because the individual makes his demands on another and feels them to be legitimate, in specific situations where he can avoid taking much account of the claims on that person. There seems to be no over-all set of societal values which explicitly requires consistency or integration from the individual. The process of compartmentalization works mainly by (a) location and context and (b) situational urgency or crisis. The latter process permits the individual to meet the crisis on its own terms, setting aside for the moment the role demands which he was meeting prior to the crisis.

2. *Delegation:* This may be seen, at least in part, as one way of achieving compartmentalization. If, for example, secular counseling is inconsistent with the clergyman's moral leadership role, he may be able to delegate it. If secular manipulation by a church is inconsistent with its sacredness, it may delegate some secular acts to lay leaders or to specialized religious orders. A wife may delegate housekeeping, and some of the socialization and nursing of the child. Note, however, that the societal hierarchy of values is indicated by what may *not* be delegated: for example, the professor may not hire a ghost writer to produce his monographs, and the student may not delegate examinations.

3. *Elimination of role relationships:*[10] Curtailment may be difficult, since many of our role obligations flow from our status positions, such as those in the job or family, which are not easily eliminated. Of course, we can stop associating with a kinsman because of the demands he makes on us, and if our work-group sets norms which are too high for us to meet we can seek another job. Aside from social and even legal limits on role

---

paralleled by Merton's. Compartmentalization partly corresponds, for example, to two of his—observability of the individual's role activities and observability of conflicting demands by members of the role set. Our mechanism of hierarchy or stratification, assigning higher or lower values to particular role demands, corresponds to two of Merton's—the relative importance of statuses and differences of power among members of the role set.

[10] This, again, is a general case of which Merton's "abridging the role-set" is a special example (*ibid.,* p. 117).

curtailment, however, some continuing role interaction is necessary to maintain the individual's self-image and possibly his personality structure: for example, many people feel "lost" upon retirement—their social existence is no longer validated.

4. *Extension:* The individual may expand his role relations in order to plead these commitments as an excuse for not fulfilling certain obligations. A departmental chairman, for example, may become active in university affairs so that he can meet his colleague's demands for time with the plea that other duties (known to his colleagues) are pressing. In addition, the individual may expand his role system so as to *facilitate* other role demands, for instance, joining an exclusive club so as to meet people to whom he can sell stocks and bonds.

5. *Obstacles against the indefinite expansion of ego's role system:* Although the individual may reduce his felt strain by expanding his role system and thereby diminishing the level of required performance for any one of his obligations, this process is also limited: After a possible initial reduction, *Role strain begins to increase more rapidly with a larger number of roles than do the corresponding role rewards or counter-payments from alter.* This differential is based on the limited role resources at individual's commands. The rewards cannot increase at the same rate as the expansion even if at first he increases his skill in role manipulation, because eventually he must begin to fail in some of his obligations, as he adds more relationships; consequently, his alters will not carry out the counter-performances which are expected for that role relationship. Consequently, he cannot indefinitely expand his role system.

6. *Barriers against intrusion:* The individual may use several techniques for preventing others from initiating, or even continuing, role relationships —the executive hires a secretary through whom appointments must be made, the professor goes on a sabbatical leave. The administrator uses such devices consciously, and one of the most common complaints of high level professionals and executives is that they have no time. This feeling is closely connected with the fact that they *do* have time, that is, they may dispose of their time as they see fit. Precisely because such men face and accept a wider array of role opportunities, demands, and even temptations than do others, they must make more choices and feel greater role strain. At the same time, being in demand offers some satisfaction, as does the freedom to choose. At lower occupational ranks, as well as in less open social systems where duties are more narrowly prescribed, fewer choices can or need be made.

Settling or Carrying Out the Terms of the Role Relationship. The total role structure functions so as to reduce role strain. The techniques outlined above determine whether an individual will have a role relationship with another, but they do not specify what performances the individual will carry out for another. A common decision process underlies the individual's sequence of role performance as well as their total pattern.

1. *The role relationship viewed as a transaction or "bargain":* In his personal role system, the individual faces the same problem he faces in his economic life: he has limited resources to be allocated among alternative ends. The larger social system, too, is like the economic system, for the problem in both is one of integration, of motivating people to stop doing X and start doing Y, whether this is economic production or religious behavior.

Because economic structures are also social structures, and economic decisions are also role decisions, it might be argued that economic propositions are simply "special cases" of sociological propositions.[11] In a more rigorous methodological sense, however, this claim may be viewed skeptically, since at present the former body of propositions cannot be deduced from the latter.[12] Rather, economic theory may be a fruitful source of sociological ideas, because its theoretical structure is more advanced than that of sociology. Since the precise relation between economic and sociological propositions is not yet fully ascertained, economic vocabulary and ideas are mainly used in the succeeding analysis for clarity of presentation, and the correctness of propositions which are developed here is independent of their possible homologs in economics.[13] In this view, economic performance is one type of role performance, a restricted case in which economists attempt to express role performance, reward, and punishment

[11] Some structural differences between the two cases, however, should be noted: (1) There are specialized economic *producers,* for example, wheat farmers who offer only one product on the market, but no corresponding sociological positions in which the individual offers only one type of role performance. Some political, religious, military, or occupational leaders do "produce" their services for a large number of people, but they must all carry out many other roles as well in the "role market." Every adult must take part as producer in a minimum number of such role markets. (2) Correspondingly, in the economic sphere all participate in several markets as *buyers;* in the role sphere they act in several markets as *both* sellers and buyers. (3) Correlatively, our entrance into the economic *producer* or *seller* activities may be long delayed, and we may retire from them early if we have enough money, but as long as we live we must remain in the role market: we need other people, and they demand us. (4) We may accumulate enough money so as to be able to purchase more than we can use, or produce more than we can sell, in the economic sphere; but in the role system we probably always ask more on the whole than our alters can give, and are unable to give as much as they demand.

[12] The most elaborate recent attempt to state the relations between the two is Talcott Parsons and Neil J. Smelser, *Economy and Society,* Glencoe, Ill.: Free Press, 1956.

[13] Again, however, correctness is independent of their origin. It is equally clear that they parallel certain conceptions of psychodynamics, but again their sociological value is independent of their usefulness in that field.

in monetary terms.[14] In both, the individual must respond to legitimate demands made on him (role expectations, services, goods, or demands for money) by carrying out his role obligations (performances, goods, or money payments). Through the perception of alternative role strains or goods-services-money costs, the individual adjusts the various demands made upon him, by moving from one role action to another. Both types of transactions, of course, express *evaluations* of goods, performances, and money.

In his role decisions, as in his economic decisions, the individual seeks to keep his felt strain, role cost, or monetary and performance cost at a minimum, and may even apply some rationality to the problem. At the same time, a variety of pressures will force him to accept some solutions which are not pleasant. His decisions are also frequently habitual rather than calculated, and even when calculated may not achieve his goal. Rather, they are the most promising, or the best choice he sees. Since the analysis of such behavior would focus on the act and its accompanying or preceding decision, the research approach of "decision analysis" seems appropriate to ascertain the course of events which led to the act.[15]

In role behavior, we begin to experience strain, worry, anxiety, or the pressures of others if we devote more time and attention to one role obligation than we feel we should, or than others feel we should. This strain may be felt because, given a finite sum of role resources, too much has already been expended; or because the individual feels that relative to a given value the cost is too high. The relative strength of such pressures from different obligations determines, then, the individual's role allocation pattern within his total role system. This system is the resultant of all such strains. Analysis of role allocation requires, of course, that we know the individual's *internal* demands, that is, the demands which he makes on himself, and which thus contribute to his willingness to perform well or not.[16]

[14] Anthropologists have noted for over a generation that economic theory needs a more general framework to take account of the non-monetary aspects of economic action in non-Western societies. Cf. Bronislaw Malinowski, "Primitive Economics of the Trobriand Islanders," *Economic Journal* 31 (March, 1921), pp. 1-16. See also Malinowski's earlier article, "The Economic Aspects of the Intichiuma Ceremonies," *Festskrift Tillagnad Edward Westermarck,* Helsingfors: 1912; and *Argonauts of the Western Pacific,* London: Routledge, 1922. Also Raymond Firth, *Primitive Economics of the New Zealand Moari,* New York: Dutton, 1929. For a discussion of the interaction of economic roles and religious roles, see William J. Goode, *Religion Among the Primitives,* Glencoe, Ill.: Free Press, 1951, Chapters 5, 6.

[15] See the several discussions in Paul F. Lazarsfeld and Morris Rosenberg, editors, *The Language of Social Research,* Glencoe, Ill.: Free Press, 1955, pp. 387-448.

[16] Price in an elementary economics textbook, determined by the intersection of supply and demand, requires no such datum (i.e., why or whence the demand is not

The process of strain allocation is facilitated somewhat, as noted above, by ego's ability to manipulate his role structure. On the other hand, the structure is kept in existence by, and is based on, the process of allocation. For example, with reference to the norms of adequate role performance, to be considered later, the White child in the South may gradually learn that his parents will disapprove a close friendship with a Negro boy, but (especially in rural regions) may not disapprove a casual friendship. The "caste" role definitions state that in the former relationship he is over-performing, that is, "paying too much." Such social pressures, expressed in both individual and social mechanisms, are homologous to those in the economic market, where commodities also have a "going price," based on accepted relative evaluations. Correspondingly, the individual expresses moral disapproval when his role partner performs much less well than usual, or demands far more than usual.

In all societies, the child is taught the "value of things," whether they are material objects or role performances, by impressing upon him *that* he must allocate his role performances, and *how* he should allocate them.[17] These structural elements are considered in a later section.

2. *Setting the role price in the role bargain:* The level of role per-formance which the individual finally decides upon, the "role price," is the resultant of the interaction between three supply-demand factors: (a) his pre-existing or autonomous norm commitment—his desire to carry out the performance; (b) his judgment as to how much his role partner will punish or reward him for his performance; and (c) the esteem or dis-esteem with which the peripheral social networks or important reference groups ("third parties") will respond to *both* ego's performance and to al-ter's attempts to make ego perform adequately.

The individual will perform well ("pay high") if he *wants* very much to carry out this role obligation as against others. He will devote much more time and energy to his job if he really enjoys his work, or is deeply committed normatively to its aims. The individual's willingness to carry out the role performance varies, being a function of the intrinsic gratifica-tions in the activity, the prospective gain from having carried out the ac-tivity, and the internal self-reward or self-punishment from conscience pangs or shame or a sense of virtue, or the like.

With reference to what ego expects alter to do in turn, he is more

---

relevant), and thus the model is simpler than the role model. However, more sophisti-cated economics, as well as the economic practitioner, must distinguish various com-ponents or sources of demand.

[17] Doubtless, however, the lessons can be made more explicit and conclusive when the "value" can be expressed in dollars rather than in the equally intangible (but more difficult to measure) moral or esthetic considerations.

likely to over-perform, or perform well if alter can and will (relative to others) reward ego well or pay him well for a good or poor execution of his role obligations. Thus, my predictions as to what will make my beloved smile or frown will affect my performance greatly; but if she loves me while I love her only little, then the same smiles or frowns will have less effect on my role performances for her.[18] Similarly, as will be noted in more detail, alter's power, esteem, and resources affect ego's performance because they allow alter to punish or reward ego more fully. The individual perceives these consequences cognitively and responds to them emotionally. If, then, the individual aspires to be accepted by a higher ranking individual or group, he may have to perform more adequately than for one of his own rank. This last proposition requires a further distinction. Alter may be able, because of his position, to reward ego more than alter rewards others in a similar status, but this additional reward may be no more than a socially accepted premium for extra performance. On the other hand, the additional reward may sometimes be viewed by those others as beyond the appropriate amount. Or the individual may over-perform in one activity of his role relationship to compensate for a poor performance in another—say, the poor breadwinner who tries to be a good companion to his children. Such further consequences of higher performance and higher reward may at times be taken into account by both role partners in making their role bargain.

If alter asks that ego perform consistently better than he is able or willing to do, then he may ease his allocation strain by severing his relationship with that individual or group and by seeking new role relationships in which the allocation strain is less.[19]

3. *Limitations on a "free role bargain":* The third component in ego's decision to perform his role bargain is the network of role relationships— "the third party" or parties—with which ego and alter are in interaction. If either individual is able to exploit the other by driving an especially hard role bargain, such third parties may try to influence either or both to change the relationship back toward the "going role price." Not only do they feel this to be their duty, but they have an interest in the matter as well, since (a) the exploiting individual may begin to demand that much, or pay that little, in his role relations with them; and (b) because the exploited individual may thereby perform less well in his role relations with

---

[18] Cf. Willard Waller's "Principle of Least Interest" in Willard Waller and Reuben Hill, *The Family,* New York: Dryden, 1952, pp. 191-192.

[19] Cf. Leon Festinger's Derivation C, in "A Theory of Social Comparison Processes," *Reprint Services No. 22,* Laboratory for Research in Social Relations, University of Minnesota, 1955, p. 123. Also see No. 24, "Self-Evaluation as a Function of Attraction to the Group," by Leon Festinger, Jane Torrey, and Ben Willerman.

them. These pressures from third parties include the demand that either ego or alter punish or reward the other for his performances or failures.[20]

It is not theoretically or empirically clear whether such third parties must always be a limited reference group, or can at times be the entire society. Many of the norms of reference groups appear to be special definitions or applications of similar norms of the larger society. Certain groups, such as criminal gangs or power cliques in a revolutionary political party, may give radical twists of meaning to the norms of the larger society. Under such circumstances, the interaction of pressures on ego and alter from various third parties can be complex. We suppose that *which* third party is most important in a given role transaction between ego and alter is a function of the degree of concern felt by various third parties and of the amount of pressure that any of them can bring to bear on either ego or alter.

Structural Limits and Determinants

Strain-Reducing Mechanisms. The individual can thus reduce his role strain somewhat: first, by selecting a set of roles which are singly less onerous, as mutually supportive as he can manage, and minimally conflicting; and, second, by obtaining as gratifying or value-productive a bargain as he can with each alter in his total role pattern.

As the existence of third parties attests, however, both sets of ego's techniques are limited and determined by a larger structural context within which such decisions are made. Not all such structural elements reduce ego's role strain; indeed, they may increase it, since they may enforce actions which are required for the society rather than the individual. Essentially, whether they increase or reduce the individual's role strain, they determine which of the first set of mechanisms ego may use, and on what terms. Similarly, they determine whether ego and alter may or must bargain freely, to either's disadvantage, or to what extent either can or must remain in an advantageous or costly bargaining position. The most important of such elements are perhaps the following:

1. *Hierarchy of evaluations:* Social evaluations are the source of the individual's evaluations, but even if only the frequently occurring types of choices are considered, such evaluations reveal complex patterns.[21] Some sort of overall value hierarchy seems to be accepted in every society, but

[20] Note, for example, the potential "seduction" of the mother by the child; the mother wishes to please the child to make him happy, and may have to be reminded by others that she is "spoiling" him.

[21] Norman Miller has used data from the Cornell Values Study to show how various combinations of social positions affect expressions of value in *Social Class Differences among American College Students,* Ph.D. thesis, Columbia University, 1958.

aside from individual idiosyncrasies, both situational and role characteristics may change the evaluation of given acts. Indeed, all individuals may accept contradictory values in some areas of action, which are expressed under different circumstances. Here, the most important of these qualifying factors are: (a) the social position of ego (one should pay some respect to elders, but if one is, say, 30 years of age, one may pay less); (b) the social position of alter (the power, prestige, or resources of alter may affect ego's decision); (c) the content of the performance by ego or alter (mother-nurse obligations are more important than housekeeper-laundress duties); and (d) situational urgency or crisis. When ego gives the excuse of an urgent situation, alter usually retaliates less severely. However, when several crises occur simultaneously, the allocation of role performances is likely to be decided instead by reference to more general rankings of value.[22]

These illustrations suggest how structural factors help to determine ego's willingness to perform or his performance, in an existing role relationship (Type 1b), and though they reduce his uncertainty as to what he should or must do, they also may increase his obligations. At the same time, these same factors also determine in part whether or when ego will include the relationship at all (Type 1a) in his total role system (for example, a noted physicist should engage in a technical correspondence with a fellow physicist, but may refuse to appear on a popular television show; even a passing stranger is expected to give needed aid in a rescue operation).

This set of interacting factors is complex, but gives some guidance in role interaction. Since there is a loose, society-wide hierarchy of evaluations, and both individuals and their reference groups or "third parties" may be committed to a somewhat different hierarchy of values, at least the following combinations of evaluations may occur:

	Evaluation of:		
Evaluations by:	Task content	Rank of alter	Situational urgency
Society			
Reference groups or third parties			
Alter			
Ego			

2. *Third parties:* Though third parties figure most prominently in the bargaining within an existing role relationship, especially those which are more fully institutionalized (statuses), they also take part in ego's manip-

[22] In an unpublished paper on "doubling" (the living together of relatives who are not members of the same nuclear family), Morris Zelditch has used approximately these categories to analyze the conditions under which the claim to such a right is likely to be respected.

ulation of his role structure, since they may be concerned with his total social position. For example, families are criticized by kinsmen, neighbors,
and friends if they do not press their children in the direction of assuming
a wider range of roles and more demanding roles as they grow older.

3. *Norms of adequacy:* These define what is an acceptable role performance.[23] Norms of adequacy are observable even in jobs which set
nearly limitless ideals of performance, such as the higher levels of art and
science, for they are gauged to the experience, age, rank, and esteem of
the individual. For example, a young instructor need not perform as well
as a full professor; but to achieve that rank he must perform as well as his
seniors believe they performed at his rank. Such norms also apply to the
total system of roles assumed by the individual. The individual may criti-
cize another not only when the latter's specific performance fails to meet
such criteria, but also when the latter's range of roles is too narrow (the
wife complains, "We never go out and meet people") or too wide (the
husband complains, "You take care of everything in the community except me").

The individual must assume more roles in an urban society than in
primitive or peasant society, and the norm of functional specificity applies
to a higher proportion of them. This norm permits individuals to bargain
within a narrower range, but also, by limiting the mutual obligations of
individuals (and thus tending to reduce role strain), it permits them to
assume a larger number of roles than would otherwise be possible. This,
then, is a role system basis for the generally observed phenomenon of
Gesellschaft or secondary relations in urban society.

4. *Linkage or dissociation of role obligations in different institutional
orders:* The fulfillment of role obligations in one institutional order either
rests on or requires a performance in another.[24] Thus, to carry out the
obligations of father requires the fulfillment of job obligations. Such doubled obligations are among the strongest in the society, in the sense that
ego may insist on rather advantageous terms if he is asked to neglect them
in favor of some other obligation. Linking two institutional orders in this
fashion limits ego's freedom to manipulate his role system.

At the same time, there are barriers against combining various roles,
even when the individual might find such a linkage congenial. (For example: a military rule against officers fraternizing intimately with enlisted
men; a regulation forbidding one to be both a lawyer and a partner of a

---

[23] This mechanism is akin to, though not identical with, Merton's "mutual social
support among status occupants." Merton, "The Role Set . . . ," *op. cit.,* p. 116.

[24] I have described this mechanism of institutional integration in some detail in
Religion Among the Primitives, op. cit., Chapters 5-10.

certified public accountant.) In an open society, some role combinations are permitted which would be viewed as incongruous or prohibited in a feudal or caste society.

Such pressures are expressed in part by the punishments which the individual may have to face if he insists on entering disapproved combinations of roles. The barriers against some combinations also apply to a special case of role expansion: entrance into certain very demanding statuses, those which require nearly continuous performance, are subject to frequent crises or urgencies, and are highly evaluated.[25] Even in our own society there are not many such statuses. The combination of two or more sets of potential crises and responsibilities would make for considerable role strain, so that few individuals would care to enter them; but in addition organizational rules sometimes, and common social attitudes usually, oppose such combinations.—The priest may not be a mother; the head of a hospital may not be a high political leader.

5. *Ascriptive statuses:* All statuses, but especially ascriptive statuses, limit somewhat ego's ability to bargain, since social pressures to conform to their norms are stronger than for less institutionalized roles. Some of these require exchanges of performances between specific individuals (I cannot search for the mother who will serve my needs best, as she cannot look for a more filial child) while others (female, Negro, "native American") embody expectations between status segments of the population. The former are more restrictive than the latter, but both types narrow considerably the area in which individuals can work out a set of performances based on their own desires and bargaining power. Because individuals do not usually leave most ascriptive statuses, some may have to pay a higher role price than they would in an entirely free role market, or may be able (if their ascription status is high in prestige and power) to exact from others a higher role performance.[26] The psychological dimensions of these limitations are not relevant for our discussion. It should be noted, however, that at least one important element in the persistence of personality patterns is to be found in these limitations: the role structure remains fairly stable because the individual cannot make many free role bargains and thus change his role system or the demands made on him,

---

[25] Perhaps the third characteristic is merely a corollary of the first two.

[26] Although the matter cannot be pursued here, it seems likely that in economic terms we are dealing here with the phenomena of the "differentiated product"—ego cannot accept a given role performance from just anyone, but from the specific people with whom he is in interaction—and of oligopoly—ego can patronize only a limited number of suppliers or sellers. Moreover, with respect to certain roles, both supply and demand are relatively inelastic.

and consequently the individual personality structure is also maintained by the same structural elements.[27]

6. *Lack of profit in mutual role deviation:* Since two role partners depend in part on each other's mutual performance for their own continuing interaction with other persons, mutual role deviation will only rarely reduce their role strain. It might be advantageous to me if my superior permits me to loaf on the job, but only infrequently can he also profit from my loafing. Consequently, both ego and alter have a smaller range of choices, and the demands of the institutional order or organization are more likely to be met. When, moreover, in spite of these interlocking controls, ego and alter do find a mode of deviation which is mutually profitable—the bribed policeman and the professional criminal, the smothering mother and the son who wants to be dependent—concerned outsiders, third parties, or even a larger segment of the society are likely to disapprove and retaliate more strongly than when either ego or alter deviates one-sidedly.

On the other hand, there is the special case in which ego and alter share the same status—as colleagues or adolescent peers, for example. They are then under similar pressures from others, and may seek similar deviant solutions; they may gang together and profit collectively in certain ways from their deviation.[28]

Less Desirable Statuses: Efforts to Change the Role Bargain. The preceding analysis of how ego and alter decide whether, when, or how well they will carry out their role obligations permits the deduction of the proposition that when an individual's norm commitment or desire to perform is *low* with respect to a given status—in our society, many women, Negroes, and adolescents reject one or more of the obligations imposed on them; perhaps slaves in all societies do—then alter must bring greater pressure to bear on him in order to ensure what alter judges to be an adequate performance. If the individual does under-perform, he is less likely to have strong feelings of self-failure or disesteem; he may feel no more than some recognition of, and perhaps anxiety about, possible sanctions from alter.

Individuals are especially conscious that they are "training" others in both child and adult socialization, if those others are suspected of being

---

[27] Cf. C. Addison Hickman and Manford H. Kuhn, *Individuals, Groups, and Economic Behavior,* New York: Dryden, 1956, p. 38.

[28] Albert K. Cohen has discussed one example of this special case at length in *Delinquent Boys,* Glencoe, Ill.: Free Press, 1955. It requires, among other factors, special ecological conditions and the possibility of communication among those in the same situation.

weakly committed to their role obligations. Thus, most Whites in the South have for generations held that all Whites have an obligation to remind the Negro by punishment and reward that he "should keep in his place," and that punishment of the Negro is especially called for when he shows evidence that he does not accept that place. It is the heretic, not the sinner, who is the more dangerous. It is particularly when the members of a subordinate status begin to deny normatively their usual obligations that third parties become aroused and more sensitive to evidence of deviation in either performance or norm commitment. On the other hand, individuals in a formerly subordinate status may, over time, acquire further bargaining power, while those in a superior status may gradually come to feel less committed to the maintenance of the former role pattern.

The Family as a Role Budget Center

For adult or child, the family is the main center of role allocation, and thus assumes a key position in solutions of role strain. Most individuals must account to their families for what they spend in time, energy, and money outside the family. And ascriptive status obligations of high evaluation or primacy are found in the family. More important, however, is the fact that family members are often the only persons who are likely to know how an individual is allocating his *total* role energies, managing his whole role system; or that he is spending "too much" time in one role obligation and retiring from others. Consequently, family relations form the most immediate and persistent set of interactions which are of importance in social control. Formal withdrawal from these relationships is difficult, and informal withdrawal arouses both individual guilt feelings and pressures from others.

Moreover, since the family is a role allocation center, where one's alters know about one's total role obligations and fulfillments, it also becomes a vantage point from which to view one's total role system in perspective. Because it is a set of status obligations which change little from day to day and from which escape is difficult, role alternatives can be evaluated against a fairly stable background. Consequently, other family members can and do give advice as to how to allocate energies from a "secure center." Thus it is from this center that one learns the basic procedures of balancing role strains.

Finally, family roles are "old shoe" roles in which expectations and performances have become well meshed so that individuals can relax in them. In Western society, it is mainly the occupational statuses which are graded by fine levels of prestige, just as achievement within occupations is rewarded by fine degrees of esteem. It is not that within jobs one is held to standards, while within families one is not.[29] It is rather that, first, socialization on the basis of status ascription within the family fits individual expectations to habitual performances and, second, rankings of family performances are made in only very rough categories of esteem. The intense sentiments within the family cushion individual strain by inducing each person to make concessions, to give sympathy, to the others. Of course, greater strain is experienced when they do not. These status rights and obligations become, then, "role retreats" or "role escapes," with demands which are felt to be less stringent, or in which somewhat more acceptable private bargains have been made among the various members of the group. One's performance is not graded by the whole society, and one's family compares one's family performance to only a limited extent with that of other people in other families.

The existence of unranked or grossly ranked performance statuses or roles may permit the individual to give a higher proportion of his energy to the ranked performance statuses. The institutions which contain such statuses vary. For example, in contemporary Western society, the layman's religious performances, like his familial performances, are ranked only roughly, but at one time evaluation of the former was more differentiated. However, familial performance apparently is never ranked in fine gradation in any society.[30] Here, an implicit structural proposition may be made explicit: the greater the degree of achievement orientation in a system of roles, the finer the gradation of prestige rankings within that system or organization.[31]

Role Strain and the Larger Social Structure

Social structures are made up of role relationships, which in turn are made up of role transactions. Ego's efforts to reduce his role strain determine the allocation of his energies to various role obligations, and thus de-

[29] See Melvin Tumin's discussion of incentives in various non-occupational statuses in "Rewards and Task Orientations," *American Sociological Review,* 20 (August, 1955), pp. 419-422.

[30] Partly because of the difficulty of outsiders observing crucial performances within it; partly, also, because of the difficulty of measuring relative achievement except in universalistic terms, as against the particularistic-ascriptive character of familial roles. Note, however, the creation by both Nazi Germany and Soviet Russia of a family title for very fertile mothers (an observable behavior).

[31] Note in this connection the case of China, the most family-oriented civilization. In comparison with other major civilizations, the Chinese developed a more complex ranking of kinship positions—and a more explicit ranking of familial performances. (See Marion J. Levy, Jr., *The Family Revolution in Modern China,* Cambridge: Harvard University Press, 1949, esp. Chapter 3.) Various individuals have figured in Chinese history as "family heroes," that is, those who performed their family duties exceedingly well.

termine the flow of performances to the institutions of the society. Consequently, the sum of role decisions determines what *degree* of integration exists among various elements of the social structure. While these role performances accomplish whatever is done to meet the needs of the society, nevertheless the latter may not be adequately served. It is quite possible that what gets done is not enough, or that it will be ineffectively done. As already noted, the role demands made by one institutional order often conflict with those made by another—at a minimum, because the "ideal" fulfillment in each is not qualified by other institutional demands and would require much of any person's available resources. Many such conflicting strains frequently result in changes in the social structure. Within smaller sub-systems, such as churches, corporations, schools, and political parties, the total flow of available personal resources may be so disintegrative or ineffective that the system fails to survive. In addition, the total role performances in some societies have failed to maintain the social structure as a whole.

Thus, though the sum of role performances ordinarily maintains a society, it may also change the society or fail to keep it going. There is no necessary harmony among all role performances, even though these are based ultimately on the values of the society which are at least to some extent harmonious with one another. Role theory does not, even in the general form propounded here, explain why some activities are ranked higher than others, why some activities which help to maintain the society are ranked higher, or why there is some "fit" between the role decisions of individuals and what a society needs for survival.

The total efforts of individuals to reduce their role strain within structural limitations directly determines the profile, structure, or pattern of the social sysem. But whether the resulting societal pattern is "harmonious" or integrated, or whether it is even effective in maintaining that society, are separate empirical questions.

Summary and Conclusions

The present paper attempts to develop role theory by exploiting the well-known notion that societal structures are made up of roles. The analysis takes as its point of departure the manifest empirical inadequacies, noted in the first section, of a widely current view of social stability, namely, that the continuity of a social system is mainly a function of two major variables: (a) the normative, consensual commitment of the individuals of the society; and (b) the integration among the norms held by those individuals. Accepting dissensus, nonconformity, and conflicts among norms and roles as the usual state of affairs, the paper develops the idea that the total role system of the individual is unique and overdemanding. The individual cannot satisfy fully all demands, and must move through a continuous sequence of role decisions and bargains, by which he attempts to adjust these demands. These choices and the execution of the decisions are made somewhat easier by the existence of mechanisms which the individual may use to organize his role system, or to obtain a better bargain in a given role. In addition, the social structure determines how much freedom in manipulation he possesses.

The individual utilizes such mechanisms and carries out his sequences of role behaviors through an underlying decision process, in which he seeks to reduce his role strain, his felt difficulty in carrying out his obligations. The form or pattern of his process may be compared to that of the economic decision: the allocation of scarce resources—role energies, time, emotions, goods—among alternative ends, which are the role obligations owed by the individual. The role performances which the individual can exact from others are what he gets in exchange.

It is to the individual's interest in attempting to reduce his role strain to demand as much as he can and perform as little, but since this is also true for others, there are limits on how advantageous a role bargain he can make. He requires some role performance from particular people. His own social rank or the importance of the task he is to perform may put him in a disadvantageous position from which to make a bargain. Beyond the immediate role relationship of two role partners stands a network of roles with which one or both are in interaction, and these third parties have both a direct and an indirect interest in their role transactions. The more institutionalized roles are statuses, which are backed more strongly by third parties. The latter sanction ego and alter when these two have made a free role bargain which is far from the going role price. The demands of the third parties may include the requirement that ego or alter punish the other for his failure to perform adequately.

Under his conception of role interaction, the bargains which some individuals make will be consistently disadvantageous to them: the best role price which they can make will be a poor one, even by their own standards. However, no one can ever escape the role market. The continuity of the individual's total role pattern, then, may be great even when he does not have a strong normative commitment to some of his less desirable roles. Like any structure or organized pattern, the role pattern is held in place by both internal and external forces—in this case, the role pressures from other individuals. Therefore, not only is role strain a normal experience for the individual, but since the individual processes of reducing role strain determine the total allocation of role performances to the social institutions, the total balances and imbalances of role strains create whatever stability the social structure possesses. On the other hand, precisely because each individual is under some strain and would prefer to be under less, and in particular would prefer to get more for his role performances than he now receives, various changes external to his own role system may alter the kind of role bargains he can and will make. Each individual system is partly held in place by the systems of other people, their demands, and their counter-performances—which ego needs as a basis for his own activities. Consequently, in a society such as ours, where each individual has a very complex role system and in which numerous individuals have a relatively low intensity of norm commitment to many of their role obligations, changes in these external demands and performances may permit considerable change in the individual's system.

The cumulative pattern of all such role bargains determines the flow of performances to all social institutions and thus to the needs of the society for survival. Nevertheless, the factors here considered may not in fact insure the survival of a society, or of an organization within it. The quantity or quality of individual performances may undermine or fail to maintain the system. These larger consequences of individual role bargains can be traced out, but they figure only rarely in the individual role decision.

With respect to its utility in empirical research, this conception permits a more adequate delineation of social structures by focusing on their more observable elements, the role transactions. This permits such questions as: Would you increase the time and energy you now give to role relationship X? Or, granted that these are the ideal obligations of this relationship, how little can you get away with performing? Or, by probing the decision, it is possible to ascertain why the individual has moved from one role transaction to another, or from one role organization to another. Finally, this conception is especially useful in tracing out the articulation between one institution or organization and another, by following the sequence of an individual's role performance and their effects on the role performances of other individuals with relation to different institutional orders.

4. FAMILIAL, OCCUPATIONAL, AND SEX ROLES

The interplay between occupational and familial roles is a central theme in the first two articles of this chapter. After examining a number of business executives, Henry concludes that being emotionally "weaned" from parents is a necessary condition of success.

The article by Whyte, which appeared in *Life,* is an abridged version of two articles that were originally published in *Fortune.* Having conducted a number of interviews with executives, Whyte finds that for the upwardly mobile man in a large corporation the acceptability of his wife to his superiors (and their wives) is a consideration of some importance. Disadvantage, according to Whyte, falls to the man whose wife's social behavior is not appropriate to the social class which he is seeking to enter. The approval of the higher echelons is bestowed upon the wife who is not troublesome, who stays out of the way, and who accepts the proposition that her husband belongs to the corporation. Whyte asks whether or not the ambitious junior executive would be well advised to marry a girl of social standing above his own. A later article will report that marital adjustment is not especially good when the man "marries upward." [1]

An overdrawn, if telling, analogy may be drawn between the obligations of the traditional Chinese husband and those of the corporation husband. In neither case is the prime obligation of the man to his wife. In traditional China a man's parents came first. The point was dramatized by the fact that if his parents demanded it, he would be expected to divorce his wife.[2] Similarly, Whyte points out that the demands of the corporation are great, as are its powers to dispense rewards.

[1] J. Roth and R. F. Peck, Social Class and Social Mobility Factors Related to Marital Adjustment."

[2] Shu-Ching Lee, "China's Traditional Family, Its Characteristics and Disintegration."

The article by Henry notes certain characteristics of personality that are rewarded in business: acquisitiveness and achievement, self-directedness and independence of thought. These are not characteristics that, according to Whyte, are approved in the "company wife," nor do they seem to constitute a good description of the maternal role. For women, therefore, marriage and career seem to require and to reward different—indeed conflicting—patterns of response. Many girls of the American middle class train for a career while looking for a husband. The discrepancy between the prescription for the occupational role and for the familial role results in confusion concerning the prescription for the feminine sex role, as reported in the study by Komarovsky. (Wallin, who repeated Komarovsky's study in a different setting, confirmed her finding that a substantial proportion of college women feel called upon from time to time to pretend inferiority to men, but he concluded that the respondents in the second study were not agitated by the necessity for deception.[3]) Some reactions to conflicts in the feminine sex role are portrayed in the last two selections of the chapter.

THE BUSINESS EXECUTIVE: A STUDY IN THE PSYCHODYNAMICS OF A SOCIAL ROLE

William E. Henry

Adapted from *The American Journal of Sociology* 54 (1949), 286-91, by permission of The University of Chicago Press.

The business executive is a central figure in the economic and social life of the United States. His direction of business enterprise and his participation in informal social groupings give him a place of significance in community life. In both its economic and social aspects, the role of the business executive is a highly visible one sociologically. It has clearly definable limits and characteristics known to the general public. These characteristics indicate the function of the business executive in the social structure, define the behavior expected of the individual executive, and serve as a guide to the selection of the novice.

Social pressure, plus the constant demands of the business organization of which he is a part, direct the behavior of the executive into the mold appropriate to the defined role. "Success" is the name applied to the whole-hearted adoption of this role. It assumes that the individual behaves in the manner dictated by the society and society rewards the individual

3 Paul Wallin, "Cultural Contradictions and Sex Roles: A Repeat Study," *American Sociological Review*, 15, 1950, 288-293.

with "success" if his behavior conforms to the role. It punishes him with "failure" should he deviate from it.

The participation in this role, however, is not a thing apart from the personality of the individual participant. It is not a game that the person is playing, it is the way of behaving and thinking that he knows best, that he finds rewarding, and in which he believes.

Thus, the role as socially defined has its counterpart in the personality structure of the individuals who participate in it. To some extent the personality structure is reshaped to be in harmony with the social role. The extent to which such reshaping of the adult personality is possible, however, seems limited. An initial selection process occurs in order to reduce the amount of time involved in teaching the appropriate behavior. Those persons whose personality structure is most readily adaptable to this particular role tend to be selected to take this role, whereas those whose personality is not already partially akin to this role are rejected.

This paper describes the personality communalities of a group of successful business executives. The research upon which it is based was undertaken to explore the general importance of personality structure in the selection of executive personnel. Many aptitude tests have been employed in the industry to decrease the risk involved in the hiring of untried personnel and to assist in their placement. These tests have been far less effective in the selection of high level executive personnel than in the selection of clerical and other non-administrating persons. Many business executives have found persons of unquestioned high intelligence often turn out to be ineffective when placed in positions of increased responsibility. The reasons for their failure lie in their social relationships. No really effective means has yet been found to clarify and predict this area of executive functioning. It is to this problem that our research [1] was directed.

From the research it became clear that the "successful" business executives studied had many personality characteristics in common. It was equally clear that an absence of these characteristics was coincident with "failure" [2] within the organization. This personality constellation might be

[1] The research undertaken will be described in its entirety in a subsequent report. In summary it involved the study of over 100 business executives in various types of business houses. The techniques employed were the Thematic Apperception test, a short undirected interview and a projective analysis of a number of traditional personality tests. The validity of our analyses, which was done "blind," rested upon the coincidence of identical conclusions from separately analyzed instruments, upon surveys of past job performance, and the ancedotal summary of present job behavior by the executive's superiors and associates. . . .

[2] Success and failure as here used refer to the combined societal and business definition. All of our "successful" executives have a history of continuous promotion, are thought to be still "promotable" within the organization, are now in positions of major

thought of as the minimal requirement for "success" within our present business system and as the psychodynamic motivation of persons in this occupation. Individual uniqueness in personality was clearly present but despite these unique aspects, each executive had in common this personality pattern.

Achievement Desires

All show high drive and achievement desire. They conceive of themselves as hard working and achieving people who must accomplish in order to be happy. The areas in which they do their work are clearly different, but each feels this drive for accomplishment. This should be distinguished from a type of pseudo-achievement drive in which the glory of the end product alone is stressed. The person with this latter type of drive, seldom found in the successful executives, looks to the future in terms of glory it provides him and the projects that he will have completed—as opposed to the achievement drive of the successful executive, which looks more to the sheer accomplishment of the work itself. The successful business leader gets much satisfaction from doing rather than merely from contemplating the completed product. To some extent this is the difference between the dreamer and the doer. It is not that the successful executives do not have an over-all goal in mind, nor that they do not derive satisfaction from the contemplation of future ease, nor that they do not gain pleasure from prestige. Far more real to them, however, is the continual stimulation that derives from the pleasure of immediate accomplishment.

Mobility Drive

All the successful executives have strong mobility drives. They feel the necessity to move continually upward and to accumulate the rewards of increased accomplishment. For some the sense of successful mobility comes

---

administrative responsibility, and are earning salaries within the upper ranges of current business salaries. Men in lower level supervisory positions, men who are considered "failures" in executive positions, and men in clerical and laboring jobs show clear deviations from this pattern. This suggests, of course, that this pattern is specific for the successful business executive and that it serves to differentiate him from other groupings in industry. The majority of these executives come from distributive (rather than manufacturing) businesses of moderately loose organizational structure where cooperation and teamwork are valued and where relative independence of action is stressed within the framework of a clearly defined overall company policy. In organizations in which far greater rigidity of structure is present or where outstanding independence of action is required, it is possible that there will be significant variations from the personality pattern presented here. We are currently extending our data in these directions.

through the achievement of competence on the job. These men struggle for increased responsibility and derive a strong feeling of satisfaction from the completion of a task. Finished work and newly gained competence provide them with their sense of continued mobility.

A second group rely more upon the social prestige of increased status in their home communities or within the organizational hierarchy. To them the real objective is increased status. Competence in work is of value and at times crucial. But the satisfactions of the second group come from the social reputation, not from the personal feeling that necessary work has been well done. Both types of mobility drive are highly motivating. The zeal and energy put into the job is equal in both instances. The distinction appears in the kinds of work which the men find interesting. For the first group the primary factor is the nature of the work itself— is it challenging, is it necessary, is it interesting? For the second group, the crucial factor is its relation to their goals of status mobility—is it a step in the direction of increased prestige, is it appropriate to my present problem, what would other people think of me if I did it?

The Idea of Authority

The successful executive posits authority as a controlling but helpful relationship to superiors. He looks to his superiors as persons of more advanced training and experience whom he can consult on special problems and who issue to him certain guiding directives. He does not see the authority figures in his environment as destructive or prohibiting forces.

Those executives who view authority as a prohibiting and destructive force have difficulty relating themselves to superiors and resent their authority over them. They are either unable to work smoothly with their superiors, or indirectly and unconsciously do things to obstruct the work of their bosses or to assert their independence unnecessarily.

It is of interest that the dominant crystallization of attitudes about authority of these men is toward superior and toward subordinates, rather than toward Self. This implies that most crucial in their concept of authority is the view of being a part of a wider and more final authority-system. In contrast, a few executives of the "Self-made," driving type characteristic of the past of business enterprise maintain a specific concept of authority with regard to Self. They are the men who almost always forge their own frontiers, who are unable to operate within anyone else's framework, to whom cooperation and teamwork are foreign concepts. To these men, the ultimate authority is in themselves and their image does not include the surrounding area of shared or delegated power.

Organization and Its Implications

While executives who are successful vary considerably in their intelligence test ratings, all of them have a high degree of ability to organize unstructured situations and to see the implications of their organization. This implies that they have the ability to take several seemingly isolated events or facts and to see relationships that exist between them. Further, they are interested in looking into the future and are concerned with predicting the outcome of their decisions and actions.

This ability to organize often results in a forced organization, however. Even though some situations arise with which they feel unfamiliar and are unable to cope, they still force an organization upon it. Thus, they bring it into the sphere of familiarity. This tendency operates partially as a mold, as a pattern into which new or unfamiliar experiences are fit. This means, of course, that there is a strong tendency to rely upon techniques that they know will work and to resist situations which do not readily fit this mold.

Decisiveness

Decisiveness is a further trait of this group. This does not imply the popular idea of the executive making quick and final decisions in rapid fire succession, although this seems to be true of some of the executives. More crucial, however, is an ability to come to a decision among several alternative courses of action—whether it be done on the spot or whether after detailed consideration. Very seldom does this ability break down. While less competent and well organized individuals may become flustered and operate inefficiently in certain spots, most of these men force their way to a conclusion. Nothing is too difficult for them to tackle at least and try to solve. When poorly directed and not modified by proper judgment, this attitude may be more of a handicap than a help. That is to say, this trait remains in operation and results in decision-making action regardless of the reasonableness of the decision or its reality in terms of related facts. The breakdown of this trait (usually found only in cases where some more profound personality change has also occurred) is one of the most disastrous for the executive. As soon as this trait shows disturbance, the executive's superiors become apprehensive about him. This suggests an interesting relationship to the total executive constellation. Whenever a junior executive loses this quality of decisiveness, he seems to pass out of the socially defined role. It is almost as though the role demanded conviction and certainty as an integral aspect of it. The weakening of other aspects of the ideal executive constellation can be readily reintegrated into the total constellation. The questioning of the individual's certainty and decisiveness, however, results in a weakening of the entire constellation and tends to be punished by superiors.

Strong Self-Structure

One way of differentiating between people is in the relative strength or weakness of their notions of self-identity, their self-structure. Some persons lack definiteness and are easily influenced by outside pressures. Some, such as these executives, are firm and well-defined in their sense of self-identity. They know what they are and what they want and have well developed techniques for getting what they want. The things they want and the techniques for getting them are of course quite different for each individual, but this strength and firmness is a common and necessary characteristic. It is, of course, true that too great firmness of sense of self-identity leads to rigidity and to inflexibility. And while some of these executives could genuinely be accused of this, in general they maintain considerable flexibility and adaptability *within the framework of their desires and within the often rather narrow possibilities of their own business organization.*

Activity and Aggression

The executive is essentially an active, striving, aggressive person. His underlying personality motivations are active and aggressive—although he is not necessarily aggressive and hostile overtly in his dealings with other people. This activity and aggressiveness are always well channelized into work or struggles for status and prestige. This implies a constant need to keep moving, to do something, to be active. This does not mean that they are always in bodily movement and moving physically from place to place (though this is often true), but rather that they are mentally and emotionally alert and active.

This constant motivator unfortunately cannot be shut off. It may be part of the reason why so many executives find themselves unable to take vacations leisurely or to stop worrying about already solved problems.

Apprehension and the Fear of Failure

If one is continually active and always trying to solve problems and arrive at decisions, any inability to do so successfully may well result in feelings of frustration. This seems to be true of the executives. In spite of their firmness of character and their drive to activity, they also harbor a rather pervasive feeling that they may not really succeed and be able to do the things they want. It is not implied that this sense of frustration comes only from their immediate business experience. It seems far more likely to be a feeling of long standing within the executives and to be only accentuated and reinforced by their present business experience.

This sense of the perpetually unattained is an integral part of this constellation and is part of its dilemma. It means that there is always some place to go, but no defined point at which to stop. It emphasizes the "self-propelled" nature of the dynamics of this role and highlights the inherent need to always keep moving and to see another goal always ahead. This also suggests that cessation of mobility and of struggling for new achievements will be accompanied by an inversion of this constant energy. The person whose mobility is blocked, either by his own limitations or by those of the social system, finds this energy diverted into other channels. Psychosomatic symptoms, the enlargement of interpersonal dissatisfactions, and the development of rationalized compulsive and/or paranoid-like defenses may reflect the re-direction of this potent energy demand.

Strong Reality Orientation

The successful executives are strongly oriented to immediate realities and their implications. They are directly interested in the practical and the immediate and the direct. This is of course generally good for the immediate business situation, though an overdeveloped sense of reality may have secondary complications. If a man's sense of reality is too highly developed, he ceases to be a man of vision. For a man of vision must get above reality to plan and even dream about future possibilities. In addition, a too strong reality sense that does not find the realities in tune with individual's ambitions may well leave a further sense of frustration and unpleasantness of reality. This happens to many executives who find progress and promotion too slow for their drives. The result is often a restlessness rather than activity, a fidgetiness rather than a well channelized aggression, and a lack of ease that may well disrupt many of their usual interpersonal relations.

The Nature of Their Interpersonal Relations

In general, the mobile and successful executive looks to his superiors with a feeling of personal attachment and tends to identify with them. His superior represents for him a symbol of his own achievement and activity desires and he tends to identify himself with these traits in those who have achieved more. He is very responsive to his superiors—the nature of this responsiveness of course depending on his other feelings, his idea of authority, and the extent to which his sense of frustration is present.

On the other hand, he looks to his subordinates in a detached and impersonal way seeing them as "doers of work" rather than as people. He treats them impersonally, with no real feeling of being akin to them or having deep interest in them as persons. It is almost as though he viewed his subordinates as representatives of things he has left behind, both factually and emotionally. Still uncertain of his next forward step, he cannot afford to become personally identified or emotionally involved with the past he has left behind. The only direction of his emotional energy that is real to him is upward and toward the symbols of that upward interest, his superiors.

This does not mean that he is cold and treats all subordinates casually. In fact he tends to be generally sympathetic with many of them. This element of sympathy with subordinates is most apparent when the subordinate shows personality traits that are most like those of the superior. Thus, the superior is able to take pride in certain successful young persons without at the same time feeling an equal interest in all subordinates.

The Attitude Toward His Own Parents

In a sense the successful executive is a "man who has left home." He feels and acts as though he were on his own, as though his emotional ties and obligations to his parents were severed. It seems to be most crucial that he has not retained resentment of his parents, but has rather simply broken their emotional hold on him and been left psychologically free to make his own decisions. We have found those who have not broken this tie to be either too dependent upon their superiors in the work situation, or to be resentful of their supervision (depending of course upon whether they have retained their dependent parental ties or rather they are still actively fighting against them).

In general, we find the relationship to the mother to have been the most clearly broken tie. The tie to the father remains positive in the sense that he views the father as a helpful but not restraining figure. Those men who still feel a strong emotional tie to the mother have systematically had difficulty in the business situation. The residual emotional tie seems contradictory to the necessary attitude of activity, progress, and channelized aggression. The tie to the father, however, must remain positive—as the emotional counterpart of the admired and more successful male figure. Without this image, struggle for success seems difficult.

The Nature of Dependency-Feelings and Concentration Upon Self

A special problem in differentiating the type of generally successful executive is the nature of his dependency-feelings. It was pointed out above that the dependency upon the mother-image must be eliminated. For those executives who work within the framework of a large organization where cooperation and group-and-company loyalty are necessities, there must remain feelings of dependency upon the father-image and a need to operate within an established framework. This does not mean that the activity-aggression need cannot operate or that the individual is not decisive and self-directional. It means only that he is so within the framework of an already established set of over-all goals. For most executives this over-all framework provides a needed guidance and allows them to concentrate upon their achievement and work demands with only minimal concern for policy making of the entire organization. For those executives who prefer complete independence and who are unable to work within a framework established by somebody else, the element of narcissism is much higher and their feelings of loyalty are only to themselves rather than to a father-image or its impersonal counterpart in company policy. These feelings differentiate the executives who can cooperate with others and who can promote the over-all policy of a company from those who must be the whole show themselves. Clearly there are situations in which the person of high concentration upon self and low dependency-loyalty feelings is of great value. But he should be distinguished in advance and placed only in such situations where these traits are of value.

The successful executive represents a crystallization of many of the attitudes and values generally accepted by middle-class American society. The value of accumulation and achievement, of self-directedness and independent thought and their rewards in prestige and status and property are found in this group. But they also pay the price of holding these values and by profiting from them. Uncertainty, constant activity, the continual fear of losing ground, the inability to be introspectively leisurely, the ever-present fear of failure, and the artificial limitations put upon their emotionalized interpersonal relations—these are some of the costs of this role.

THE WIFE PROBLEM

William H. Whyte, Jr. | Adapted from *Life,* January 7, 1952, 32-48.

Over the last few decades American corporations have been evolving a pattern of social community able to provide their members with more and more of their basic social wants. Yet, the corporation now concedes, one of the principal members of its community remains officially almost unnoticed—to wit, the Wife. For the good of the corporation, many executives believe, it is time the matter was remedied. "We control a man's environment in business and we lose it entirely when he crosses the threshold of his home," one executive says mournfully. "Management, therefore, has a challenge and an obligation to deliberately plan and create a favorable, constructive attitude on the part of the wife that will liberate her husband's total energies for the job." Others, though they might not put it quite so baldly, agree that the step is logical.

Just how to do this is a problem that has many a management understandably baffled. On one very basic matter, however, management is not in the slightest baffled. It knows exactly what kind of wife it wants. With a remarkable uniformity of phrasing, corporation officials all over the country sketch the ideal. In her simplest terms she is a wife who (1) is highly adaptable, (2) is highly gregarious, (3) realizes her husband belongs to the corporation.

Are the corporation specifications presumptuous? It would appear not. The fact is that this kind of wife is precisely what our schools and colleges —and U.S. society in general—seem to be giving the corporation.

Let us define terms: we are discussing the wives of the coming generation of management, whose husbands are between 25 and 40, and in junior or middle echelons of management or with logical aspirations of getting there. There is, of course, no sharp dividing line between age groups, but among older executives there is a strong feeling that this younger generation of wives is the most cooperative the corporation has ever enlisted. "Somehow," says one executive, "they seem to give us so much less trouble than the older ones." "Either the girls are better or the men are marrying better," says another. "But whatever it is with these people, *they get along.*"

The Negative Role

Perhaps it is merely that this generation of wives has not yet grown older and more cantankerous. Perhaps. But there is evidence that this group-mindedness is the result of a shift in values more profound than one might suppose. The change is by no means peculiar to the corporation wife but by the nature of her job she may be the outstanding manifestation of it. And a preview, perhaps, of what is to come.

First, how do the wives conceive their own role? Critical literature has been answering the question rather forcefully, with the result that many Americans (and practically all Europeans) assume that the wife of the American businessman not only is the power behind the scenes but wants to become more so. The picture needs considerable revision. For the striking thing that emerges from wives' comments is the negativeness of the role they sketch. As they explain it, the good wife is good by *not* doing things—by *not* complaining when her husband works late; by *not* fussing when a transfer is coming up; by *not* engaging in any controversial activity. Moreover, they agree heartily that a good wife can't help a husband as much as a bad wife can hurt one. And the bad wife, clearly, is one who obtrudes too much, whether as a "meddler," a "climber," a "fixer" or, simply, someone who "pushes" her man around.

Resolutely antifeminist, the executive wife conceives her role to be that of a "stabilizer"—the keeper of the retreat, the one who rests and rejuvenates the man for the next day's battle.

This stabilizing calls for more than good homemaking and training the kids not to bother daddy before dinner. Above all, wives emphasize, they have to be good listeners. They describe the job somewhat wryly. They must be "sounding boards," "refueling stations," "wailing walls." But they speak without resentment. Nurturing the male ego, they seem to feel, is not only a pretty good fulfillment of their own ego but a form of therapy made increasingly necessary by the corporation way of life. Management psychologists couldn't agree more. "Most top executives are very lonely people," as one puts it. "The greatest thing a man's wife can do is to let him unburden the worries he can't confess to in the office."

A Social Operator

In addition to listening she can do some judicious talking. If she is careful about it she can be a valuable publicity agent for the husband. "In a subtle way," says one executive, "they put in a plug for the husband. They tell things he wouldn't dare tell for fear of seeming immodest." In similar fashion they can humanize him if he's a boss. "About the time I get fed up with the bastard," says a junior executive, "here I am, going over to dinner at his house. And she's so nice. She jokes about him, kids him to his face. I figure he can't be so bad after all."

Low-key "stabilizing," then, the wife sees as her main task. There is another aspect to her role, however, and it is considerably less passive. For the good corporation wife must also be a social operator, and when husbands and wives sketch out the personal characteristics of the ideal wife it is the equipment for this role that comes first to their minds. What they ask for, more than any other quality, is gregariousness, or a reasonable facsimile. Here are some of the ways in which they spell it out.

EXECUTIVE: "She should do enough reading to be a good conversationalist. . . . Even if she doesn't like opera she should know something about it so if the conversation goes that way she can hold her own. She has to be able to go with you if you're going to make a speech or get an award, and not be ill at ease."

EXECUTIVE: "The hallmark of the good wife is the ability to put people at their ease."

WIFE: "The most important thing for an executive's wife is to know everybody's name and something about their family so you can talk to them—also, you've got to be able to put people at their ease."

EXECUTIVE: "Keeping herself so she is comfortable with people on the boss's level is important. I don't think reading and music and that kind of stuff are vital."

EXECUTIVE: "The kind you want is the kind that can have people drop in any time and make a good show of it even if the baby's diapers are lying around."

WIFE: "It's a very worthwhile bunch we have here. Edith Sampson down on Follansbee Road is sort of the intellectual type, but most of the gang are real people."

For the corporation wife, in short, being "sociable" is as important as stabilizing. Like the army wife, an analogy she detests, she must be a highly adaptable "mixer." In fact, she needs to be even more adaptable than the army wife, for the social conditions she meets are more varied. One year she may be a member of a company community, another year a branch manager's wife, expected to integrate with the local community— or, in some cases, to become a civic leader, and frequently, as the wife of the company respresentative, to provide a way station on the route of touring company brass.

"It Makes Me Laugh"

As a rule, she is inextricably bound up in the corporation "family," often so much so that her entire behavior—including what and where she drinks—is subtly conditioned by the corporation. "It makes me laugh," says one wife in an eastern city dominated by one corporation. "If we were the kind to follow the Pattern, I'll tell you just what we would do. First, in a couple of years, we'd move out of Ferncrest Village (it's really pretty tacky there, you know). We wouldn't go straight to Eastmere Hills —that would look pushy at this stage of the game; we'd go to the hilly section off Scrubbs Mill Pike. About that time, we'd change from Christ Church to St. Edwards, and we'd start going to the Fortnightlys—it would be a different group entirely. Then, about 10 years later, we'd finally build in Eastmere Hills." It makes her laugh, she says, because that would be the signal to everybody that she had become a wife of the top-brass bracket. Which she probably will.

Few wives are as articulate as that on the social role, but intuitively they are generally superb at it; their antennae are sensitive, and they know the rules of the game by heart. Second nature to the seasoned wife, for example, are the following:

> Don't talk shop gossip with the Girls, particularly those who have husbands in the same department.

Don't invite superiors in rank; let them make the first bid.

Don't turn up at the office unless you absolutely have to.

Don't get too chummy with the wives of associates your husband might soon pass on the way up.

Don't be disagreeable to any company people you meet. You never know . . .

Be attractive. There is a strong correlation between executive success and the wife's appearance. Particularly so in the case of the sales wife.

Be a phone pal of your husband's secretary.

Never—repeat, never—get tight at a company party (it may go down in a dossier).

One rule transcends all others: *Don't be too good.* Keeping up with the Joneses is still important. But where in pushier and more primitive times it implied going substantially ahead of the Joneses, today keeping up means just that: keeping up. One can move ahead, yes—but slightly, and the timing must be exquisite. Whatever the move, it must never be openly invidious.

Perhaps it is for this reason that, when it comes to buying an auto, the Buick is so much preferred: it envelops the whole executive spectrum and the jump from a Special to a Super, and from a Super to a Roadmaster, can be handled with tact.* Not always, though. In one eastern steel town, where cars have always been the accepted symbol of rank, the chairman of the board has a Cadillac—certainly a high enough ceiling. The president, however, has taken to buying Buick Supers, with the result that people in the upper brackets are chafing because it would be unseemly to go higher. Except for the chairman, accordingly, only the local tradespeople drive Cadillacs and Roadmasters.

The good corporation wife, the rules continue, does not make friends uncomfortable by clothes too blatantly chic, by references to illustrious forebears or by excessive good breeding. And she avoids intellectual pretensions like the plague.

Are these rules of the game merely the old fact of conformity? In part, yes. But something new has been added. What was once a fact has now become a philosophy. Today's young couples not only concede their group-mindedness; they are outspokenly in favor of it. They blend with the group not because they fear to do otherwise but because they approve of it.

While few young wives are aware of the sacrifice involved, the role of the boss's wife is one that they very much covet. In talking about the qualities of the ideal wife—a subject they evidently had thought over long and often—they were at no loss. In one third of the cases the American woman's favorite cliché "gracious" came instantly to them, and in nearly

---

* The present equivalents of yesteryear's ranks shows a reach for the exotic. Special has become Le Sabre; Super is now Electra; and Roadmaster is Electra 225—*Eds.*

all the others the descriptions spelled out the same thing. Theirs is a sort of First Lady ideal, a woman who takes things as they come with grace and poise, and a measure of *noblesse oblige*; in short, the perfect boss's wife. But how near do they come to the ideal?

What a Wife Faces

What, for example, of the listening job that wives take such pride in? How well can they listen? Consensus of a cross section of U.S. executives: not very well. ("And for God's sake, don't quote me.") There are excuses aplenty. "If he has had a rough day," says one wife, "I don't want to hear about it. He'd only get mad and say things the children shouldn't hear." The husband, however, may be the one chiefly to blame. He asks for active, intelligent listening, yet seldom wants advice ("Women just don't understand").

And how well does she handle the special social problem? In advancing the husband in the office, the corporation is quite likely to advance him socially as well. There is no easy out for the couple in such cases, and for the wife the inward tug of war between the social *status quo* and the prospect of advancement can be extremely poignant. "I must have made some terrible mistakes," laments one wife now in mid-passage. "I love people and I've made many intimate friends in the company, but since Charlie got his new job it's just been hell on us. He has so much control over their lives, and it's all gotten so complicated."

The fact that the office can spell sanctuary for the husband does not go unresented. Perhaps this is why the Christmas office party provokes such surprisingly bitter, if concealed, feeling from many wives. It dramatizes the wife's exclusion. Here, on this appointed day, is the world she can never share, and for all her brave little chuckles at the standing jokes of the office gang, she comes face to face with the fact. That is, if she's allowed to attend.

Burning though this exclusion may be to the wives, it is a topic they dislike intensely to talk about or to think about. And for them, indeed, the waters may well be better left muddy: to peer too deeply is to uncover an underlying point even more provoking. Where, the awful question comes up, does the man find his major satisfactions?

A common feminine observation is that a man's major satisfactions come from the home. If he's happy there, he can be happy in his work, and vice versa. The belief is probably necessary. Is it correct as well?

Item: As management psychologists note, the average executive shows a remarkable ability to repress his home worries while on the job; rarely, however, can he shut out office worries at home.

Item: The reaction to this Hobson's-choice question: "If you had to make the choice, which would you take: an increasingly satisfying work life and a proportionately souring home life—or the opposite?" The answers would surprise wives. "This business of doing it all for the family," as one husband confesses, "it's just a rationalization. If I got a windfall today I'd still knock myself out."

"Man's love is of man's life a thing apart," Byron once observed. " 'Tis woman's whole existence." So, for all the group integration and communication skills she can muster, it will probably remain.

The schism between Home and Office has been even more accentuated recently. Thanks, in part, to the way the tax structure has accumulated, the corporation now provides the man with a higher standard of living in his work than in his home—and, it might be added, a higher one than his wife enjoys. From 9 to 5 he may be a minor satrap, guiding the destiny of thousands, waited on by secretaries and subordinates; back in his servantless home he washes the dishes. Nor is it merely the fact of his satrapy; the corporation virtually rigs it so that he can have more fun away from home.

The expense account has become a way of life. There is not only travel. There are also luncheon clubs, company retreats, special conventions, parties and perquisites, and, though the wife may be thrown an occasional convention as a crumb, the expense-account world rarely encompasses her. It is primarily a man's word—and if the man is at a low salary, he is likely to find the pattern of life at 7118 Crestmere Road dull in comparison.

"The company has spoiled Jim terribly, " says one wife. "Even when he was only earning $7,500 a year he used to be sent to Washington all the time. He'd go down in a Pullman drawing room and as J. R. Robinson of the General Company, take a two-room suite. Then he used to be asked by some of the company officers to a hunting and fishing lodge that the company kept in the north woods. When he went to New York, he'd entertain at 21, the Barberry Room and the Chambord. Me, meanwhile, I'd be eating a 30¢ hamburger and, when we went away together on vacation, we would have to go in our beat-up old car or borrow my sister's husband's. This taste of high life gives some of these characters delusions of grandeur. Small wonder that they get to fidgeting after they have been home a couple of weeks."

"What the hell can you say?" says one executive. "Here I am eating high off the hog, meeting interesting people, while Jo is slaving back home. I get a big bang out of all this, but I also have a sort of guilty feeling, so I say to her, 'Gee, honey, I hate all this traveling, but I just have to do it.' " Of the wives *Fortune* interviewed, many mentioned, commiseratingly, how their husbands looked forward to coming home, how awful it was sleeping in hotel beds, rattling around on trains and eating bum food.

There are some things, however, that cannot be explained away. For more than sirloins and drawing rooms are at issue; over the long pull this disparity aggravates perhaps the most subtle problem of marriage: equality of growth. If marriage, as Sociologist Everett Hughes puts it, is a "mutual mobility bet," for whom are the cards stacked?

Growth can mean many things. To the younger generation of executives it seems to mean an increasing ability to handle and mix with people. And the terms are the same for the wife. "The wife who is not very sociable," goes a highly typical male observation, "might not affect the husband directly, but she can hurt him just the same. A lot of business is done weekends. If she doesn't go for this, her lack of growth can hold the man back." "I have seen it happen so many times," says another executive sadly. "He marries the kid sweetheart, the girl next door or a girl from the jerkwater college he went to. They start off with a lot in common. Then he starts going up. Fifteen years later he is a different guy entirely. But she's stayed home, literally and figuratively." Even the old idea of a wife as a sort of culture carrier is virtually dead; she is still expected to read and things like that, but for functional reasons. "Sure I want her to read good books and magazines," as one executive puts it. "I don't want her to make a fool of herself in conversation."

Fundamentally, of course, the problem goes back to whom the executive chooses in the first place. Is the moral that he should marry a girl "superior" to him? Thanks to the commonly accepted saw that a woman can pull a man up, but not vice versa, there are many who think he should. ("My best executives," remarks one boss, "are the ones who 'outmarried' themselves.") But the pitfalls are many. Her qualities may drive the man to preoccupation with office prestige in order to prove himself to her; furthermore, unless she is excellent at hiding her superiority— or lets it rest fallow—she can hurt his chances in a close "family" community. The Bryn Mawr accent can be absolute death for a career in some Midwestern corporations.

What kind of background for the woman, then, is the optimum? A serious career can be dismissed easily; there is almost universal agreement among wives, husbands and corporations on this score. Work before marriage, however, is generally approved. "I feel the fact that I worked before marriage," says one wife, "is a help. I know what goes on in an office and can understand what Charles is up against."

College? Here is the *summum bonum*. There are some obvious reasons; because virtually all executives now go to college, the couple in such cases starts off with shared values. But corporation people mention a reverse factor almost as much. It is not so important for the wife, they say, to have gone to college, but it is very important not to have *not* gone to college. If she hasn't, corporation people warn, she is prey to an inferiority complex that makes it difficult for her to achieve real poise. Some corporations, accordingly, make it their business to find out whether or not the wife has a degree.

More and more corporations these days are interviewing the wife before hiring an executive, and some are not uninterested in fiancées. There are many holdouts ("This railroad picks its executives and lets its executives pick their wives and so far it's been okay"), but roughly half of the companies on which *Fortune* has data have made wife-screening a regular practice and many of the others seem about ready to do so. And the look-see is not academic. About 20 per cent of its otherwise acceptable trainee applicants, one large company estimates, are turned down because of their wives.

Ordinarily, the screening is accomplished via "informal" social visits. Many executives, for example, make it a point to call on the wife in her own home. Louis Ruthenburg, board chairman of Servel (which never hires an executive without a look at the wife), likes to recall how one college president used to insist on eating breakfast with a candidate's family; the wife who didn't fix her husband a good breakfast, he used to say, wasn't a good risk. To help them spot such key indicators many executives rely heavily on their own wives. "My wife is very, very keen on this," says one president. "She can spot things I might miss. And if the gal isn't up to par with her, it's no go."

How to Screen a Wife

But the initial screening is only the beginning of the corporation's interest. In one way or another the corporation manages to keep an eye on the wife, and more and more the surveillance is deliberately planned. At the Container Corp. of America, for example, it is the duty of all vice presidents to get acquainted with their subordinates' wives, and on their travels they are expected to meet the wives of executives in the field. Thus, when a man's name comes up for promotion the company has the answers to these questions: What is the health of the family? What is their attitude toward parenthood? How does the wife run her home? Does she dress with taste?

The effect of all this surveillance on the husband's career is substantial. In the home office of an insurance company, to cite one not untypical example, the president is now sidetracking one of his top men in favor of a less able one; the former's wife "has absolutely no sense of public relations." In another company a very promising executive's career is being similarly checked; his wife, the boss explains, is "negative in her attitude toward the company. She feels that business is her husband's life and no part of hers." Wives who have donated income of their own to raise the family living standard may also call down sanctions on the husband. Says one president, "When a man buys a home he can't afford on his salary alone, we either question his judgment or feel that the wife wears the pants." In either case his career is not likely to profit.

So with alcohol. The little woman who gets tipsy in front of the boss is not quite the joke her celebration in cartoon and anecdote would indicate; indeed, it is almost frightening to find out to what degree executive futures have been irretrievably influenced by that fourth Martini. And it need happen only once. Recently, the president of a large utility felt it necessary to revise his former estimate of two executives. At the last company dinner their wives drank too many glasses of champagne. "They disported themselves," he says regretfully, "with utter lack of propriety."

Interestingly, divorce rarely disqualifies a man. Because of the phenomenon of the outgrown wife, the regret of most companies is tempered by the thought that the executive's next and, presumably more mobile, wife will be better for all concerned; one company, as a matter of fact, has a policy of sending executives away on extended trips if they need separating from nagging or retrograde wives.

One company has arranged for the team of consulting psychologists it retains to help out in delicate situations (currently they are making progress with an alcoholic wife). In most cases, however, the salvage task is up to the top man himself. "A lot of the 'company business' that presidents do," says one of them, "covers this sort of work. Take a situation I've got to wrestle with now. In one of our branch plants the wives of two vice presidents have started a feud. The men get along fine, but one of the wives is a real troublemaker. So I guess it's up to me to take a trip halfway across the continent—for other reasons, of course—and try and see what I can do about it."

Important as the wife-screening process may be, most executives realize that it is, at best, only a negative measure. For even with the most cooperative wives there can be much misunderstanding over such topics as travel and long hours. Therefore it is the company's duty, they argue, to *sell* the wife on the corporation's point of view. The result is an increasing use of such media as films, brochures and special mailings to drive home, in effect, the idea that the corporation isn't stealing her husband from her.

But something far more important is being brewed for the wife. It is not enough, in the view of many companies, that she merely be "sold" on the company; she should, they believe, now be integrated *into* it. "When a man comes to work for us," says William Given, chairman, American Brake

Shoe Co., "we think of the company as employing the family, for it will be supporting the entire family, not merely the breadwinner." "The days of the strictly home wife," says a bank president, "are gone. She has become indispensable to our entire scheme of business." Among U.S. corporations, easily the most conspicuous and successful example of this kind of integration has been Thomas J. Watson's International Business Machines Corp. "Our wives," Watson explains, "are all part of the business. We started with just a few hundred people in 1914 and decided that no matter how large we grew we would carry it on in the family spirit. We always refer to our people as the 'I.B.M. Family' and we mean the wives and children as well as the men."

As a result the company can correctly claim that it makes available "complete social satisfactions." For $1 a year I.B.M. people enjoy a country club with swimming pool, bowling, 18-hole golf course, softball, tennis, picnics, and parties of all kinds. Even the children are integrated. At the age of 3 they may be enrolled in a special children's club, and at 8 go on to become junior members of the big club.

In keeping with the family spirit Mrs. Watson, a very gracious, modest woman, sets an example for other wives. "She's made my work play," her husband explains. "She has a great gift for human relations. I confer with her about personnel because she knows all the people. She has met them at luncheons where we hold a regular receiving line, and every year she goes to the 100-Percenter Club meetings." In addition to this, Mrs. Watson travels with her husband all over the world and keeps in touch with I.B.M. people; last year she traveled 38,046 miles and met 11,845 I.B.M. men and their wives.

Social integration, however, does not mean that the corporation necessarily *likes* the wife. A great many, as we have seen, do. But in some cases the corporation welcomes her largely as a means of defending itself against her. Amiable as it may be about it, the corporation is aware that the relationship is still triangular—or, to put it another way, if you can't beat 'em, join 'em. "Successes here," says one official, "are guys who eat and sleep the company. If a man's first interest is his wife and family, more power to him—but we don't want him." "We've got quite an equity in the man," another explains, "and it's only prudence to protect it by bringing the wife into the picture."

In fairness to the corporation wife, she must be recompensed somehow for the amount of time the company demands from her husband. Companies recognize the fact and are consequently more and more providing social facilities—from ladies' nights to special clubs—to hypo the sense of identification.

One corporation has gone considerably further. Via the wife of the heir apparent to the presidency, there has been set up, in effect, a finishing school so that the wives can be brought up to the same high standards. As soon as the husband reaches the $8,000-to-$10,000 bracket the wife becomes eligible for the grooming. It is all done very subtly: the group leader drops helpful advice on which are the preferred shops, where to dine, what to wear when doing it and, somewhat like a good cruise director, has a way of introducing newcomers to congenial people. "Her supervision is so clever and indirect," says one wife, "that the other wives appreciate it probably."

When the corporation turns to the Sales Wife, its attention becomes even more intense. As an economic lever on the salesman, companies have learned, there is no stimulus quite so effective as the wife, if properly handled. Some sales executives make a habit of writing provocative letters to the wife, reminding her of the sales-contest prizes her husband could win for her and how he is doing at the moment (not so well as he should be).

As an extra employee, the wife's potential is so great that with some concerns the "husband and wife team" is not only desirable but mandatory. And the wife is not always merely the junior member. "Wives can do a lot on their own," explains the president of a large paper-box company. "A lot of important business connections have grown from friendships between our wives and wives of executives of other companies. One of our executives' wives recently was down at Miami for two weeks, and a friendship she struck up with a woman there resulted in a big order from an account we hadn't been able to crack in 15 years."

Insurance companies, among the first to exploit this "team" potential, bear down heavily on the theme through a constant stream of literature addressed to wives. Through magazine articles penned by veteran wives they are told of the psychological requirements ("Earl Made a Believer of Me," Mrs. Earl Benton explains to wives in a typical article).

The question of integration is by no means simple. What we have been talking about so far is the kind of integration deliberately planned by companies. But there is another kind. Quite beyond the immediate control of the corporation there are forces at work to draw the bonds between wife and corporation even tighter.

Paradoxically, perhaps the greatest of these is the very decentralization of industry. Thanks to this growing trend, it is now a commonplace that the road to advancement is through transfer to the different seats of the corporation empire.

With their talent for adaptability, the younger generation of wives is in most respects well prepared for this new way of life. Most accept it philosophically, and a good many actually prefer it to staying put in one place.

"Any time the curtains get dirty," says one wife, "I'm ready to move. I enjoy meeting new people and seeing new places. And it's kind of a vacation sometimes."

There are, nevertheless, some very real tensions produced. And for no one more than the wife. It is she, who has only one life in contrast to her husband's two, who is called upon to do most of the adjusting. The move at once breaks up most of her community friendships, severs her local business relationships with the bank and the stores, takes her from the house and the garden on which she worked so long, and if the move takes her to a large city it probably drops her living standards also.

But it is the effect on the children that concerns wives most. While the children are very young, most wives agree, the effect is not harmful; they make and forget friends easily. As they reach junior-high age, however, a transfer can become a crisis. Recalls one wife: "Every time my daughter made a place for herself at school with the other kids, we'd move, and she'd spend the next year trying to break in at another school. Last year, when she was a senior in high school, she had a nervous breakdown. She was sure she was an outsider." The effect is not often this drastic but, while most children sweat out their adjustment without overt pain, the process is one parents find vicariously wrenching. One executive who recently changed to a nontransferring company has no trouble recalling the exact moment of his decision. One night at dinner his little boy turned to him. "Daddy," he said, "where do you really live?"

While constant transfer exposes the couple to many environments, it is, nevertheless, one of the most powerful of all the forces for integration. Because moving makes their other roots so shallow and transitory, the couple instinctively clings all the harder to the corporation.

What are the wife's basic unadjusted feelings about all this? The answer is clear: she likes the way of life. To picture her as a helpless sort of being pushed around by the corporation would be to attribute to her a sense of plight she does not feel; she must be considered not only an object of the integration but a force for it in her own right. She has become such an ally of the corporation, in fact, that on several matters it would almost appear that she and the corporation are ganging up on the husband.

Whatever else she may think of the corporation, on three main points she and her sisters agree:

The corporation means opportunity. The big company, wives explain, plays fair. "We went over all the pros and cons of bigness before Jim joined Du Pont," says one wife, "and we've never regretted joining. The bigness holds out a challenge for you."

The corporation means benefits. "Eastman Kodak has wonderful goodwill policies," a wife explains. "I used to have to attend to all the home details like insurance and bills. Now the company has someone who does those things for you—they even plan vacations for you."

The corporation means security.   "Some companies may pay more at the start, but employment is not so secure. Here they never fire anybody, they just transfer you to another department."

Few wives go on to articulate their image of "the Company." But there is an image, nonetheless, that of a beneficent "system," at once impersonal and warm—in a nice kind of way, Big Brother.

There is, of course, another side to the picture. Many companies that have extensive wife programs do not attempt social integration, and some not only look on the wife—to borrow one executive's explanation—as none of their damn business, but take active steps to see that she *doesn't* get close to them. A sampling of executive views—oil company: "We are just as happy if we never see her at all." Tool company: "If wives get too close to management they always get too status-minded. That means trouble." Motor company: "Wives' activities are their own business. What do some of these companies want for their $10,000? Slavery too?"

In Praise of the Ornery Wife*

Having concluded the report, we find that some second thoughts are in order—one being a fleeting wish that we had never brought the subject up; too many wives read it.

But the news is out, and if it is even a half-way accurate representation of what is in store for the coming generation of management, it gives us the heebie jeebies. The picture that emerges, in brief, is that of a society in which the individualist, rugged or otherwise, seems to be out, definitely. What the modern corporation wants is "group integration"; to them, the "good" wife is the wife who subordinates her own character and her aspirations to the smooth functioning of the system; the wife, in short, who "adapts." And a small, but growing number of companies are taking active measures to bring the personal lives of their members within the corporation's domain.

This kind of thing is disturbing enough. But what is much more disturbing is the wife's reaction to it. For the fact is that the group-life is precisely what she seems to *like*. Getting along with people, she indicates —in a hundred different ways—has become more than mere expediency; it has become a dedicated purpose.

This devotion to group values is by no means peculiar to the corporation way of life; it shows there more, perhaps, but it does not stem from it. Some corporations, to be sure, have been giving it a powerful assist,

* Reprinted from *Fortune,* November, 1951.

but the basic forces behind "social integration" are far more universal. The wives who join the corporation come already equipped with an amenable philosophy. And in this, they are not only the reflector of the values of a whole generation, but the tutors of another, a preview perhaps, of what is to come.

By all indications, the philosophy they embrace is growing steadily in appeal. And if one word could define a philosophy, "adapt" would be it. Already, thanks in part to the growing impact of "social engineering" (a phenomenon *Fortune* will examine more closely in a future issue), "adapt" has been so well articulated and so intricately rationalized that its value as a guide to conduct is no longer questioned.

Which is just the trouble.

The many virtues of adaptability, certainly, need no defense. But how much more are we to adapt? There is no ceiling in sight for the word. Indeed, it is becoming, in itself and without any qualification, an almost obsessive watchword. Whatever the circumstances, the good scout *adapts*. As The First Rule of "The Mental Hygiene Creed" (from the pamphlet, "The Doctor's Wife") typically enjoins: "I shall adapt to life, immediately, completely, and gracefully."

This is fatuous advice. "Adapt" is a meaningless word unless it is considered in relation to what one is supposed to be adapting to—and that tacit "life" needs lots of defining. To illustrate: there are a good many case histories at hand in which the husband has given up a job at a new post because his wife did not take to the community. Was she wrong? In the new lexicon of values, yes; as the obeisance paid "adapt" indicates, it is the environment that should be the constant; the individual, the variable. But might not she have been right after all? Some towns *are* stifling and backward, and one can adapt to them only by demeaning oneself. Should she, then, adapt? And if so, why?

Not only is the individual demeaned, but society as well. The status quo is institutionalized; the person who adapts can adapt only to something already existent. By extension, therefore, the advice implies that our creative capital is complete, and that we may live happily on the interest by merely refining and perfecting what we already have.

True, the very process of adapting may itself be dynamic. Many of our old notions of individual creativity are more sentimental than accurate; in "the well-adjusted group," as the social scientists have demonstrated, there is a total far above the sum of the individuals, and we have yet to exploit its full potential. With considerable persuasiveness, some go on to argue that the sheer working together by an increasingly group-minded people will furnish all the creative power society needs.

Just possibly, however, they may be wrong. We will go part way with their thesis; it is undeniably true that you don't have to be an s.o.b. to be creative. It is equally true, however, that a real advance in any field inevitably involves a conflict with the environment. And unless people temper their worship of environment, they may well evolve a society so well adjusted that no one would be able—or willing—to give it the sort of hotfoot it regularly needs. We would all be too busy participating and lubricating and integrating and communicating.

Several months ago a top official of one of the most group-integrating of corporations fell to musing over the death of a fellow official and his wife. It made him think a bit, he told one of his associates, of the drift of the company's personnel policy. "You know, they were terrifically stimulating people," he said. "They were the last *characters* I ever knew."

"I wonder," he added, thoughtfully, "whether we'll ever get any more."

It's not a trivial question.

CULTURAL CONTRADICTIONS AND SEX ROLES

Mirra Komarovsky

Adapted from *The American Journal of Sociology*, 52 (1946), 184-89, by permission of The University of Chicago Press.

Profound changes in the role of women during the past century have been accompanied by innumerable contradictions and inconsistencies. With our rapidly changing and highly differentiated culture, with migrations and multiplied social contacts, the stage is set for myriads of combinations of incongruous elements. Cultural norms are often functionally unsuited to the social situations to which they apply. Thus, they may deter an individual from a course of action which would serve his own, and society's, interests best. Or, if behavior contrary to the norm is engaged in, the individual may suffer from guilt over violating mores which no longer serve any socially useful end. Sometimes culturally defined roles are adhered to in the face of new conditions without a conscious realization of the discrepanices involved. The reciprocal actions dictated by the roles may be at variance with those demanded by the actual situation. This may result in an imbalance of privileges and obligations[1] or in some frustration of basic interests.

Again, problems arise because changes in the mode of life have

[1] Clifford Kirkpatrick, "The Measurement of Ethical Inconsistency in Marriage," *International Journal of Ethics*, XLVI (1936), 444-60.

created new situations which have not as yet been defined by culture. Individuals left thus without social guidance tend to act in terms of egotistic or "short-run hedonistic" motives which at times defeat their own long-term interests or create conflict with others. The precise obligation of a gainfully employed wife toward the support of the family is one such undefined situation.

Finally, a third mode of discrepancy arises in the existence of incompatible cultural definitions of the same social situation, such as the clash of "old-fashioned" and "radical" mores, of religion and law, of norms of economic and familial institutions.

The problems raised by these discrepancies are social problems in the sense that they engender mental conflict or social conflict or otherwise frustrate some basic interest of large segments of the population.

This article sets forth in detail the nature of certain incompatible sex roles imposed by our society upon the college woman. It is based on data collected in 1942 and 1943. Members of an undergraduate course on the family were asked for two successive years to submit autobiographical documents focused on the topic; 73 were collected. In addition, 80 interviews, lasting about an hour each, were conducted with every member of a course in social psychology of the same institution—making a total of 154 documents ranging from a minimum of five to a maximum of thirty typewritten pages.

The generalization emerging from these documents is the existence of serious contradictions between two roles present in the social environment of the college woman. The goals set by each role are mutually exclusive, and the fundamental personality traits each evokes are at points diametrically opposed, so that what are assets for one become liabilities for the other, and the full realization of one role threatens defeat in the other.

One of these roles may be termed the "feminine" role. While there are a number of permissive variants of the feminine role for women of college age (the "good sport," the "glamour girl," the "young lady," the domestic "home girl," etc.), they have a common core of attributes defining the proper attitudes to men, family, work, love, etc., and a set of personality traits often described with reference to the male sex role as "not as dominant, or aggressive as men" or "more emotional, sympathetic."

The other and more recent role is, in a sense, no *sex* role at all, because it partly obliterates the differentiation in sex. It demands of the woman much the same virtues, patterns of behavior, and attitude that it does of the men of a corresponding age. We shall refer to this as the "modern" role.

Both roles are present in the social environment of these women throughout their lives, though, as the precise content of each sex role varies with age, so does the nature of their clashes change from one stage to another. In the period under discussion the conflict between the two roles apparently centers about academic work, social life, vocational plans, excellence in specific fields of endeavor, and a number of personality traits.

One manifestation of the problem is in the inconsistency of the goals set for the girl by her family.

Forty, or 26 per cent, of the respondents expressed some grievance against their families for failure to confront them with clear-cut and consistent goals. The majority, 74 per cent, denied having had such expe-· riences. One student writes:

How am I to pursue any course single-mindedly when some way along the line a person I respect is sure to say, "You are on the wrong track and are wasting your time." Uncle John telephones every Sunday morning. His first question is: "Did you go out last night?" He would think me a "grind" if I were to stay home Saturday night to finish a term paper. My father expects me to get an "A" in every subject and is disappointed by a "B." He says I have plenty of time for social life. Mother says, "That 'A' in Philosophy is very nice dear. But please don't become so deep that no man will be good enough for you." And, finally, Aunt Mary's line is careers for women. "Prepare yourself for some profession. This is the only way to insure yourself independence and an interesting life. You have plenty of time to marry." . . .

A student reminisces:

All through high school my family urged me to work hard because they wished me to enter a first-rate college. At the same time they were always raving about a girl schoolmate who lived next door to us. How pretty and sweet she was, how popular, and what taste in clothes! Couldn't I also pay more attention to my appearance and to social life? They were overlooking the fact that this carefree friend of mine had little time left for school work and had failed several subjects. It seemed that my family had expected me to become Eve Curie and Hedy Lamar wrapped up in one.

Another comments:

My mother thinks that it is very nice to be smart in college but only if it doesn't take too much effort. She always tells me not to be too intellectual on dates, to be clever in a light sort of way. My father, on the other hand, wants me to study law. He thinks that if I applied myself I could make an excellent lawyer and keeps telling me that I am better fitted for this profession than my brother.

Another writes:

One of my two brothers writes: "Cover up that high forehead and act a little dumb once in a while"; while the other always urges upon me the importance of rigorous scholarship.

The students testified to a certain bewilderment and confusion caused by the failure on the part of the family to smooth the passage from one role to another, especially when the roles involved were contradictory. It seemed to some of them that they had awakened one morning to find their world upside down: what had hitherto evoked praise and rewards from relatives, now suddenly aroused censure. A student recollects:

I could match my older brother in skating, sledding, riflery, ball, and many of the other games we played. He enjoyed teaching me and took great pride in my accomplishments. Then one day it all changed. He must have suddenly become conscious of the fact that girls ought to be feminine. I was walking with him, proud to be able to make long strides and keep up with his long-legged steps when he turned to me in annoyance, "Can't you walk like a lady?" I still remember feeling hurt and bewildered by his scorn, when I had been led to expect approval. . . .

The final excerpt illustrates both the sudden transition of roles and the ambiguity of standards:

I major in English composition. This is not a completely "approved" field for girls so I usually just say "English." An English Literature major is quite liked and approved by boys. Somehow it is lumped with all the other arts and even has a little glamour. But a composition major is a girl to beware of because she supposedly will notice all your grammar mistakes, look at your letters too critically, and consider your ordinary speech and conversation as too crude.

I also work for a big metropolitan daily as a correspondent in the city room. I am well liked there and may possibly stay as a reporter after graduation in February. I have had several spreads [stories running to more than eight or ten inches of space], and this is considered pretty good for a college correspondent. Naturally, I was elated and pleased at such breaks, and as far as the city room is concerned I'm off to a very good start on a career that is hard for a man to achieve and even harder for a woman. General reporting is still a man's work in the opinion of most people. I have a lot of acclaim but also criticism, and I find it confusing and difficult to be praised for being clever and working hard and then, when my efforts promise to be successful, to be condemned and criticized for being unfeminine and ambitious.

Here are a few of these reactions:

My father: "I don't like this newspaper setup at all. The people you meet are making you less interested in marriage than ever. You're getting too educated and intellectual to be attractive to men."

My mother: "I don't like your attitude toward people. The paper is making you too analytical and calculating. Above all, you shouldn't sacrifice your education and career for marriage."

A lieutenant with two years of college: "It pleased me greatly to hear about your news assignment—good girl."

A Navy pilot with one year of college: "Undoubtedly, I'm old-fashioned, but I could never expect or feel right about a girl giving up a very promising or interesting future to hang around waiting for me to finish college. Nevertheless, congratulations on your job on the paper. Where in the world do you get that wonderful energy? Anyway I know you were thrilled at getting it and feel very glad for you. I've an idea that it means the same to you as that letter saying 'report for active duty' meant to me."

A graduate metallurgist now a private in the Army: "It was good to hear that you got that break with the paper. I am sure that talent will prove itself and that you will go far. But not too far, as I don't think you should become a career woman. You'll get repressed and not be interested enough in having fun if you keep after that career."

A lieutenant with a year and a half of college: "All this career business is nonsense. A woman belongs in the home and absolutely no place else. My wife will have to stay home. That should keep her happy. Men are just superior in everything, and women have no right to expect to compete with them. They should do just what will keep their husbands happy."

A graduate engineer—my fiancé: "Go right ahead and get as far as you can in your field. I am glad you are ambitious and clever, and I'm as anxious to see you happily successful as I am myself. It is a shame to let all those brains go to waste over just dusting and washing dishes. I think the usual home life and children are small sacrifices to make if a career will keep you happy. But I'd rather see you in radio because I am a bit wary of the effect upon our marriage of the way of life you will have around the newspaper."

Sixty-one, or 40 per cent, of the students indicated that they have occasionally "played dumb" on dates, that is, concealed some academic honor, pretended ignorance of some subject, or allowed the man the last word in an intellectual discussion. Among these were women who "threw games" and in general played down certain skills in obedience to the unwritten law that men must possess these skills to a superior degree. At the same time, in other areas of life, social pressures were being exerted upon these women to "play to win," to compete to the utmost of their abilities for intellectual distinction and academic honors. One student writes:

I was glad to transfer to a women's college. The two years at the co-ed university produced a constant strain. I am a good student; my family expects me to get good marks. At the same time I am normal enough to want to be invited to the Saturday night dance. Well, everyone knew that on that campus a reputation of a "brain" killed a girl socially. I was always fearful lest I say too much in class or answer a question which the boys I dated couldn't answer.

Here are some significant remarks made from the interviews:

When a girl asks me what marks I got last semester I answer, "Not so good— only one 'A'." When a boy asks the same question, I say very brightly with a note of surprise, "Imagine, I got an 'A'!"

I am engaged to a southern boy who doesn't think too much of the woman's intellect. In spite of myself, I play up to his theories because the less one knows and does, the more he does for you and thinks you "cute" into the bargain. . . . I allow him to explain things to me in great detail and to treat me as a child in financial matters.

One of the nicest techniques is to spell long words incorrectly once in a while. My boyfriend seems to get a great kick out of it and writes back, "Honey, you certainly don't know how to spell."

When my date said that he considers Ravel's *Bolero* the greatest piece of music ever written, I changed the subject because I knew I would talk down to him.

A boy advised me not to tell of my proficiency in math and not to talk of my plans to study medicine unless I knew my date well.

My fiancé didn't go to college. I intend to finish college and work hard at it, but in talking to him I make college appear a kind of a game. . . .

It embarrassed me that my "steady" in high school got worse marks than I. A boy should naturally do better in school. I would never tell him my marks and would often ask him to help me with my homework.

Mother used to tell me to lay off the brains on dates because glasses make me look too intellectual anyhow. . . .

How to do the job and remain popular was a tough task. If you worked your best, the boys resented the competition; if you acted feminine, they complained that you were clumsy. . . .

On dates I always go through the "I-don't-care-anything-you-want-to-do" routine. It gets monotonous but boys fear girls who make decisions. They think such girls would make nagging wives.

I am a natural leader and, when in the company of girls, usually take the lead. That is why I am so active in college activities. But I know that men fear bossy women, and I always have to watch myself on dates not to assume the "executive" role. Once a boy walking to the theater with me took the wrong street. I knew a short cut but kept quiet.

I let my fiancé make most of the decisions when we are out. It annoys me, but he prefers it.

I sometimes "play dumb" on dates, but it leaves a bad taste. The emotions are complicated. Part of me enjoys "putting something over" on the unsuspecting male. But this sense of superiority over him is mixed with feeling of guilt for my hypocrisy. Toward the "date" I feel some contempt because he is "taken in" by my technique, or if I like the boy, a kind of a maternal condescension. At times I resent him! Why isn't he my superior in all ways in which a man should excel so that I could be my natural self? What am I doing here with him, anyhow? Slumming?

And the funny part of it is that the man, I think, is not always so unsuspecting. He may sense the truth and become uneasy in the relation. "Where do I stand? Is she laughing up her sleeve or did she mean this praise? Was she really impressed with that little speech of mine or did she only pretend to know nothing about politics?" And once or twice I felt that the joke was on me: the boy saw through my wiles and felt contempt for me for stooping to such tricks.

Another aspect of the problem is the conflict between the psychogenetic personality of the girl and the cultural role foisted upon her by the milieu.[2] At times it is the girl with "masculine" interests and per-

[2] Margaret Mead, *Sex and Temperament in Three Primitive Societies,* New York, Morrow & Co., 1935.

sonality traits who chafes under the pressure to conform to the "feminine" pattern. At other times it is the family and the college who thrusts upon the reluctant girl the "modern" role.

While, historically, the "modern" role is the most recent one, ontogenetically it is the one emphasized earlier in the education of the college girl, if these 153 documents are representative. Society confronts the girl with powerful challenges and strong pressure to excel in certain competitive lines of endeavor and to develop certain techniques of adaptations very similar to those expected of her brothers. But, then, quite suddenly as it appears to these girls, the very success in meeting these challenges begins to cause anxiety. It is precisely those most successful in the earlier role who are now penalized.

It is not only the passage from age to age but the moving to another region or type of campus which may create for the girl similar problems. The precise content of sex roles, or, to put it in another way, the degree of their differentiation, varies with regional class, nativity, and other subcultures.

Whenever individuals show differences in response to some social situation, as have our 153 respondents, the question naturally arises as to the causes. It will be remembered that 40 per cent admitted some difficulties in personal relations with men due to conflicting sex roles but that 60 per cent said that they had no such problems. Inconsistency of parental expectations troubled 26 per cent of the students.

To account for individual differences would require another study, involving a classification of personalities in relation to the peculiar social environments of each. Generally speaking, it would seem that it is the girl with a "middle-of-the-road personality" who is most happily adjusted to the present historical moment. She is not a perfect incarnation of either role but is flexible enough to play both. She is a girl who is intelligent enough to do well in school but not so brilliant as to "get all 'A's' "; informed and alert but not consumed by an intellectual passion; capable but not talented in areas relatively new to women; able to stand on her own feet and to earn a living but not so good a living as to compete with men; capable of doing some job well (in case she does not marry or, otherwise, has to work) but not so identified with a profession as to need it for her happiness.

A search for less immediate causes of individual reactions would lead us further back to the study of genesis of the personality differences found relevant to the problem. One of the clues will certainly be provided by the relation of the child to the parent of the same and of the opposite sex. This relation affects the conception of self and the inclination for a particular sex role.

The problems set forth in this article will persist, in the opinion of the writer, until the adult sex roles of women are redefined in greater harmony with the socioeconomic and ideological character of modern society.[3] Until then neither the formal education nor the unverbalized sex roles of the adolescent woman can be cleared of intrinsic contradictions.

THE PROBLEM OF PANTS

Adapted by permission of the editors from *The Daily Northwestern*, Exchange Section, May 9, 1952. Taken from the *Syracuse Daily Orange*.

A challenging letter answered an editorial attack on coeds for wearing slacks. The female of the species had this to say:

"Before entering college we had great expectations concerning the college man. Needless to say we were gravely disappointed.

"What has happened to the rugged, outdoor man? He is no longer rugged. He eats soft food, sleeps too much, and considers the slightest physical exertion too much for him. He is never outdoors, his social life being centered around the parlor.

"An energetic game of chess or a snappy bull session is all the exercise he gets. . . . One glance at his apparel would make you doubt whether he is even a man.

"Masculine individuality has become a mirage. It seems that everything he does is inspired by the group to which he belongs. His aim in life is determined by what others have decided to be worthwhile goals. He no longer has the power to think and decide for himself.

"As an example, when he is on a date, the girl must be prepared to decide what movie they will see, what they will do afterwards, and she must even plan to spend the evening entertaining the man who has lost the power to take an active part in conversation.

". . . You denounced us girls for wearing slacks and jeans, etc. We would be only too glad to give them back to you, if you'd begin to earn your pants."

[3] See excellent discussions in Talcott Parsons, "Age and Sex in the Social Structure of the United States," *American Sociological Review,* VII (1942), 604-16, and in the same issue, Ralph Linton, "Age and Sex Categories," pp. 589-603, and Leonard S. Cottrell, Jr., "The Adjustment of the Individual to His Age and Sex Roles," pp. 617-20.

GOLDEN GIRL: EXAMINING A CERTAIN KIND OF COED

Sheila Tillotson

Adapted from *The Daily North-western Dimensions*, April 18, 1960, 2-5.

co' ed', *or* co'-ed' (ko' ed'), *n. U.S.* A female student in a coeducational institution, esp. a college or university.

Co' ed', (ko' ed'). The same, spelled with capital "C." Delete the word *student*.

Angela is a Coed. One of delightful breed that is sadly diminishing before the onslaught of Sputnik-scared educators who are screaming "College is for learning!" and who, alas, are being successful in their campaign to shift campus emphasis from the bar stool to the library chair.

Luckily, Northwestern is still a stronghold, albeit one of the last, where the Coed flourishes, frolicking gaily back and forth across Sheridan rd. during the day and up and down Sheridan rd. at night, bringing cheer to the lives of students and faculty who would otherwise stagnate in the atmosphere of lectures, concerts, and intellectual discussions.

What is this Coed; this colorful animal, who skips through one, two, three, or even—rare occurrence—four years of higher education and then sheds its camel-colored coat and bright neck covering to assume the drabber shades of Pillar-of-the-Community, Officer-of-PTA, and Adoring-Wife-of-Rising-Young-Executive? What goes on beneath the silken page-boy hairdo before mating and molting season when the page-boy is exchanged for a sleek coiffure under a chic hat?

Under Angela's golden hair is all the essence of the Coed, a romantic being full of youthful sophistication; the darling of fashion magazines and short story writers.

Angela lavishly showers her personal gaiety on all the campus. "What are you studying for? You get the same diploma for an A or a C." With this clarion call she releases her roommate from the bondage of a desk and welcomes her to the joy of the bridge table. "Who *was* Senator Neuberger? Did someone get shot in Africa? I would vote for what's-his-name, he's cute!" With these soothing words Angela calms fear-torn news-paper readers, shows them one can exist quite pleasantly unaware, and that there will be no terrified neurotic marking the ballot in years to come.

The cheerful tinkle of her charm bracelet heralds her entrance into the library and the cunning pad-pad of her sneaker-clad feet beats a happy rhythm in the brains of those studying as she goes for cigarette after cigarette. What would Deering [Library] be like without Angela's stage whisper, "How ARE you? I didn't even SEE you!"

Ah yes, there is the excitement of watching Angela subtly alter her plumage so it is always . . . Right. She listens when Paris decrees UP! or DOWN! about her hemlines. She owns many crewneck sweaters but dumps them willingly for shaggy cardigans or long sleeved pullovers when progress demands it. She has her dress clothes, carefully graded— simple wool sheath, lower-cut black taffeta, full-skirted Triad dress; all slightly, but only slightly, different from those of her mates. Angela will never be the object of the whispers, "Did YOU see . . ." She "frankly" tells her roommate that she prefers "quality to quantity," but due to a generous (or hen-pecked, as it were) daddy, she has both.

Her talons are long and smoothly oval, and discreetly rosy. But it is her face, ah, her face, that shows Angela's true Coedness. Lined brows, carefully pruned, arch over round eyes with delightfully curled lashes (who minds if the curl is sometimes a decided right-angle bend?). The innocent blush of youth and health is on her cheeks (what matter if today youth and health come out of a bottle?). The pleasantly bowed lips part to reveal even white teeth (the nasty name "metal-mouth" is long forgotten). The gay redness of her lips has led some disgruntled males to complain of dry cleaning bills, but what price beauty?

Yes, Angela is a Coed. She treads the labyrinth in the ivory-tower, as confusing as the mazes in palaces of old with their stone towers, secure behind the talisman on her left breast. She knows her friends and foes by the simple expedient of checking the insignia they wear. This enables her to pick her associates without the bother of discovering their personal worth for herself. In inter-personal relationships it saves her from the worrisome process of decision. Everyone in the ivory tower is nicely catalogued: Good, Medium, Poor, and Independent.

Her moral armor has chinks and cracks that show only at the appropriate time and place, and are smoothly welded before the eager eyes of dormmates and worried parents. But Angela isn't concerned about the cracks and chinks in her armor. She understands the American code of behavior—anything is all right as long as you don't get caught. And Angela won't get caught, Angela is a True Coed.

She progressed from the innocent giggle of her high school days to the knowing giggle of the Coed by diligent perusal of many books with titles like *Love Without Fear* and *Understanding Yourself*. Angela knows, and understands, and she is not afraid. In fact, she has been heard to remark, "I KNOW it's a line, but he's so GOOD at it!"

Never will Angela's brain be fogged by the pink clouds of pure pleasure. As she is preserved from the depths of despair by the steady tick-ticking of her well-ordered brain, so she is barred from the heights of ecstasy by the prosaic belief that everyone is essentially like herself, and most of them are an inferior edition. The bland life is for Angela and she feeds the ulcer of her insecurity with the milk of cautiousness.

As Angela swirls across the polished floor of the M & M Club through the blue mist of smoke that weaves the music and the people into the throbbing fabric of pleasure, in that once-only time, too soon swallowed by advancing years, is she blissfully caught up in the joy of the moment? Tick-tick goes the brain, "I wonder if the pin in my bra strap is showing?"

By accident Angela is at a table where over foamy glasses words have begun to flow. Searching words, original only because they have been found again by young minds. Does she wonder? Is she spurred by the example of curious introspection to dig inside her own head? Tick-tick goes the Coed brain, "All that talk about Being and Non-Being, Beauty and Truth they are just stringing a lot of words together, and nobody is saying anything new."

The professor has just told a very funny joke. About Jews. He is attacking discrimination with the rapier of ridicule. Angela laughs (everyone else is). Does the not-so-subtle message reach her real life—the next time her roommate has a date with Bernstein will she repress the recoil? Tick-tick, and the point of the joke is neatly pigeon-holed in the corner of her brain marked "facts to remember for exam."

Every once in a while Angela goes to church on Sunday morning. It is interesting to note that these excursions usually come after a particularly wild Saturday night with an anatomically-curious young man in a parked car . . . but in any case, she goes, and the words from the pulpit don't have the punch they used to have. That nice cushiony foundation she sat on all through high school is wearing thin in spots. "Oh well," comments the brain, "I am going through a period of doubt. All intelligent people doubt. I'll doubt for a while and then everything will be all right."

And then one sad day Tragedy visits Angela. A telegram comes, "Mother died this morning." And life is smashing and splashing all around her in ugly little dribbles of nothingness. Why? Why? Why? To whom can she go? Does anybody care? Terror is seeping into the upset

Coed brain. A quick peek inside her head shows an appalling barrenness. Run! Run! There must be someone, somewhere. . . . And there is. Good old Bill. He is only a "Medium" but this is an emergency. Bill is all too willing to bask for a while in the reflected glory of Angela's insignia. He knows how to sooth away the hurt . . . "Relax, just relax. I'm not going to hurt you . . ."

Slowly the clock resumes its steady ticking. Whoops! What happened? There is a Medium insignia hanging from the talisman. Oh well, Bill is nice and solid, and after all, I *was* saving *that* for my husband . . . we can have pretty yellow kitchen curtains . . . and I'm sure he will make a lot of money . . . and he wants to tour Europe for a month sometime, after he's seen the "good ole USA" . . . and he only wants three children, too . . . and it is so much fun to pass the candle . . . tick tick tick.

And so Love comes to the Coed. Not in any bright shower of sunlight and flowers and moonbeams, not with any soft mystery or dim yearnings not, in other words, with any foolishness. It chugs in smoothly, covering the gross accident of its birth with deft strokes of the mental paint brush.

Thus the Coed mates and molts, assumes the dignity of age without the foundation of maturity. Doubly secure now, behind the dual shield of two talismans, Angela has rediscovered the all-encompassing guidepost for her life. "What will people say?" With this most solid of all foundations, she can build a life, Good Citizen, Good Wife, Good Mother. In the sheltered retreat of a $35,000 home, Angela can rear little Coeds, giving them their camel colored coats and plaid neck scarves in grammar school. Train them from the very beginning for their role as joy-bearers for the campus.

5. IMPACT OF SOCIAL CHANGE ON FAMILIAL STRUCTURE AND FUNCTIONS: THE CHINESE CASE

Traditional Chinese society was family centered. The first article in this Chapter describes the structure and functions of the family in traditional China. As Yang notes in the opening lines of the second selection, the forces that were to bring about a revolution in the Chinese family had been gathering for a half century before the Communists came to power. The Chinese "family revolution" is evidenced in the disappearance of family-owned land and in the decline in influence of the clan, of the aged, and of males.

CHINA'S TRADITIONAL FAMILY, ITS CHARACTERISTICS AND DISINTEGRATION

Shu-Ching Lee

Adapted from the *American Sociological Review*, 18 (1953), 272-280.

This paper attempts to define the traditional Chinese family system which persisted through two thousand years without any substantial change. In the pages that follow the model of this system is described to represent the family of the gentry and officialdom, although the deviation of a peasant household would be slight, for in China the peasant is an earnest imitator of the life patterns established by the gentry-scholar class. However lasting as was China's traditional family, the early half of the twentieth century witnessed a rapid disintegration of this age-old institution among the modern-educated intellectuals in the large cities and trade ports, and the emergence of the conjugal family as a sharp competitor. Under the impetus of the present social revolution the destruction of this system has penetrated to all groups of society and to the vast interior.

Chinese Familism

. . . For practical purposes, it seems advisable in describing the term "familism" to include the following five essential features. (1) *Emphasis on the father-son relationship.* Although the family as a social organization always involves parent-child or husband-wife relationships, unless the relationship between the father and son is stressed there would never be familism, in some cases not even a stable family. Accompanying the family system arranged on this basis are patriarchy, patrilocal residence after marriage, patrilineal descent, agnate clan organization, and the inheritance of property by one or all legal (male) heirs. (2) *Family pride.* The family's standing is important in that any member's glory is taken to be the glory of the family, and his or her disgrace is the disgrace of the household. In China's dynastic days, if a man gained fame in civil examinations or in officialdom, not only his parents became distinguished but his wife and sons were also honored. (3) *Encouragement of the large family.* Traditional values persistently and invariably encourage the living together of blood relatives and regard with shame any division of the household. It is unimportant what the actual size of an average family in China may be; whenever economic circumstances permit, the family membership will continue to grow. The enormously extended family of Chang Kung-i, which comprised nine generations (including of course many collateral descendants) living together under the same roof, was personally visited and decorated by an emperor of the Tang Empire. A large organization such as this, though rare, did exist and was commemorated in history. (4) *The cult of ancestor worship.* The father-and-son relationship, being a link of an unending chain between generations, leads upward to ancestor worship and downward to the "sin" of no posterity. A natural result of this emphasis is demonstrated in the keeping of elaborate pedigrees and the building of splendid ancestral halls.[1] (5) *Common ownership of property by the family.* Beside the ideological reinforcement, the economic basis of familism is its common ownership of property. Perhaps no family could achieve unity unless the destiny of its members were bound together by a mutual sharing of prosperity and disaster. As a common practice, disintegration of the household generally begins when the patriarch has become too

[1] Cf. Hsien Chin Hu, *The Common Descent Group in China and Its Functions,* New York: Viking Fund Publications in Anthropology, No. 10, 1948.

weak to enforce the family's rules, and collectively owned funds are used for private interests.

These five attributes taken tentatively as the content of familism bear, of course, varying weights in the support of the Chinese large family system. They serve, nevertheless, as indispensable parts of an integral organization. It must be pointed out, however, that with the strong backing of these elements and through some twenty centuries of development the large family in China was found largely in the wealthy class, especially the gentry. In an ordinary village there might be some big households with thirty or forty members, but the remaining families, although similar to them in basic structure, each comprised no more than five or six persons. To put it another way, the large family is the universally exalted ideal, but only those who possess ample wealth are in a position to attain it. The family generally begins to disintegrate at the moment when there are no longer enough resources for all, and different branches are forced to take what can still be shared from the commonly owned property and go elsewhere. This is why the popular notion in China always takes the division of a house as signalizing a misfortune, and the maintenance of a fairly large household as an indication of prosperity and prestige.

The Family as an Institution

As a rule, life in the large family centers on consanguinity, whereas life in the small family centers on matrimony. The Chinese family definitely constitutes an institutionalized type of the former. Its fundamental characteristics are stability, continuity, and perpetuation through generations. A long period of living together by a number of persons in the same household naturally develops some peculiar ways of life which are called the family's tradition. In order to preserve this tradition, each family generally sets up its own rules, written or unwritten, to discipline its members. Both the tradition and the rules are important means of maintaining the family as an institution.

The family tradition, composed of a variety of minute manners unique to each family, gives its members a strong feeling of *esprit de corps*. This tradition is conscientiously kept intact and proudly transmitted through generations. When strengthened by the cult of ancestor worship, it becomes the code of conduct and permeates the minds of family members. It determines the proper relations between members of the family, inheritance of the family's property, and succession to a title, an occupation, or even a craft. In traditional China, the complex of family etiquette, which distinguished members of a good family from those of a poor one, was generally taken by society as a yardstick of the prosperity and the status of a family.

In intellectual families, the spirit of tradition is codified into a written constitution, or the family laws,[2] which enable the patriarch to regulate behavior of the family members. His power and authority extend far into the realm of what the West would consider to be personal matters of an individual. The question arises as to how this power and authority can be maintained, especially after members have reached maturity. Aside from moral sanction and the social pressure exerted by the community, a crucial answer to this question is the inheritance of or succession to the patrimony or craft. In traditional China when education was not socialized, the means of securing a livelihood was primarily transmitted within the family. Not only was the tangible property such as land, house, and livestock largely inherited, but in addition such crafts or skills as brewing, weaving, drug-making, or painting were also acquired through the father-son relationship. If a son wanted to be a legal heir for any of these, the first duty he learned at childhood was to obey the family's rules, or better still, to live up to the family's traditional spirit. It must be remembered further that severe punishment could be inflicted on a wayward son, including hauling to the ancestral hall, flogging, ostracism, or even execution upon the consent of the elders of the agnate clan.

Since family property or craft symbolizes the achievements of many forefathers, sentiments and emotions are naturally attached to it. It is held to be a duty for the heirs to expand or improve heritage and not to lose or permit it to deteriorate. If the latter should unfortunately happen, the degree of disgrace to the incumbent, a typical prodigal in the eyes of his neighbors, usually became so unbearable that the only way out was to leave the native community. It may be stated further that since the property inherited in the family bears sentimental values, any division of it would hurt the feelings of a son. The most desirable solution is naturally to keep the house from being partitioned and to encourage as many generations as possible to live together. If circumstances in the household have become such that a division is inevitable, the manner of apportioning the family property is strictly arranged along the line of the father-son relationship. Tangible property is allotted equally among the sons, except that some special privileges are possessed by the eldest in descent, though not to the extent of primogeniture, and the sons of a concubine are recognized as having little claim. As regards the craft, however, in order to safeguard it within the family as a guaranteed livelihood, the general rule of succession is to transmit the skill only to male heirs, and not to daughters and daughters-in-law.

[2] The book which is used to record these codified rules is called the *chia-li-p'u.* Cf. Y. K. Leong and L. K. Tao, *Village and Town Life in China,* London: George Allen & Unwin, 1915, pp. 7, 24 and 70-71.

From this kind of family-centric situation derives a popular concept—the emphasis on the family's descent. It is this concept that makes an insult to one's mother the greatest offense, and the remarriage of a widowed wife a sign of disgrace to the family. Another effect of this notion is the esteem placed on genealogy. Among the distinguished, large families, an elaborate pedigree must be kept in good order in the ancestral hall. It is a glory to trace the family tree back to historically renowned figures who may or may not be blood kin of the family. On the other hand, dignity and grace require descendants to delete the names of disreputable persons from the genealogical records. To revise or to add new biographies to the pedigree is regarded as so solemn an undertaking that not only members of the whole clan participate in celebration, but well-known scholars are invited to take charge of the editorship. All this, coupled with the discipline of family members, has fostered an attitude in society that due respect is always paid to members of an old but decaying family, whereas acrimonious remarks and scorn are generally directed to the misdemeanors committed by members of the *nouveau riche* households.

Functions, Roles, and Status

The way in which the members of a Chinese family are trained or disciplined can best be understood by an examination of the functions and roles assigned by Confucian traditions to each category of different relations within the family. As a primary institution, the family takes no account of any individual, but places all its emphasis on the identification of individual members with the established roles according to consanguineous or matrimonial principles.

The basic relatives of a family are, of course, parent and child, husband and wife, and brothers and sisters; the rest are only extended forms of these three relationships. The father is a patriarch (or a "stern sovereign" as the Chinese refer to him) whose solemn duty, beside representing the family in financial and other matters in the community, is to make sure that no member violates the family's established traditions and rules and that nothing happens to lower the family status. Naturally, he loves his sons and daughters just as much as his wife does, but socially he, as a patriarch, is traditionally entrusted with the responsibility for the rightful and proper conduct of other members in the family and also for their success in society. The proverb, "spare the rod, spoil the child," illustrates well the situation. In fulfilling his duty, the father has to exercise his authority to punish a misbehaving member in order not only to correct him (or her) but to maintain harmony and order in the household. Because of this, he is generally *respected*, but rarely *loved*.

Confucian doctrine, especially as amended by the scholars of the former Han Dynasty, inculcates two principles with regard to the roles of the father and the mother; that of *esteem* and that of *affection*. The father commands his descendants' esteem and the mother, their love. The chief responsibility of the former is to discipline and provide or the education of his children, while to his wife is entrusted their care and rearing. It is quite beyond the mother's role to punish her sons, except to warn the stubborn ones that "If you continue to do such and such things I am going to report to your father." Difficulties arise, however, when the father dies early and leaves no brother in the family to take care of disciplinary measures. Many a distinguished family has been ruined by such unfortunate circumstances.

According to the established convention in traditional China, the husband as a husband has no clearly defined role, except to support his wife; his position is found in other connections such as a son or as a father. This does not mean that he cannot assert authority over his wife—very often he does. The wife's place in the family deserves lengthy discussion. As one who has been brought up in a different family and different circumstances, she is more or less a stranger. She becomes the wife of a husband whom she may have never met before marriage and assumes the role of an active member of his family. In contrast to the situation in a conjugal family, her affiliation with her husband's family is considered to be far more important than the simple fact of her being a spouse. Because of this, she is rigidly required to comply with the traditions and rules of the family into which she has married, no matter how disagreeable and unreasonable they may be to her. Should she have difficulty in getting along with the family, not to mention any misconduct on her part, her husband, regardless of what his feelings may be, is forced to repudiate her to maintain the integrity of the family institution.

The basic function of a wife is in brief to perpetuate the family's name, in other words, to procreate male children. Unless this is fulfilled, her position in the family is precarious. As is well known, the first of the seven conditions under which a wife may be repudiated is infecundity.[3] To quote from Mencius again, "Three things are considered to be unfilial, and to have no posterity is the greatest of them." Even if a barren wife

---

[3] The seven conditions upon which a wife can be repudiated are: (1) No posterity, (2) licentiousness, (3) disobedience to parents-in-law, (4) quarreling, (5) theft, (6) jealousy, and (7) diseases regarded as vicious. On the woman's side, however, there are three conditions under which she may refuse to go: (1) Her close relatives through whom the marriage was arranged are all deceased, and, therefore, she has no place to go to; (2) she participated for three years in the mourning for her husband's parent; and (3) the family's present posperity and fame was attained through years of struggle in which she shared the hardships. Cf. Daniel H. Kulp, *Country Life in South China*, New York, Columbia University Press, 1925, p. 184.

is tolerated, her husband is obliged to take a concubine who is young, attractive, and often fascinating.[4] Although when she has borne her husband several sons her status is raised, the mother never enjoys a very lofty position. According to Chinese tradition, she can never have a life of her own. Should her husband die while her sons are still boys, the rule of three obediences[5] requires her to "obey" or rather to depend upon her son. Perhaps the most miserable and unbearable life is that of the young, childless widow (even worse than a spinster); no matter what virtue a widow may possess, she is taken by legend as the most unlucky sign to any male. No bachelor would venture to marry her, even if she, herself, had the courage to face the stigma of unchastity and disgrace of remarrying.

Under a patriarchal system, the roles played by sons and daughters (or brothers and sisters) are inevitably different. It is obvious that sons are the bearers of the family name, legal heirs of the family property, to whom are entrusted the family reputation and tradition. Naturally they have a bigger voice in family affairs. On the other hand, daughters are reared with an eye to the marriage that must take them out of the family. The only claim a daughter can make is a dowry which amounts to a fraction of the family property.[6] If forced by economic circumstances, however, the poor patriarch simply sells his young daughters as concubines or even prostitutes.

Marriage and the Selection of a Mate

Since marriage in traditional China means taking a new member into the family rather than simply getting a wife for a husband, marriage is defined in the Confucian Classics as "to make a union between two persons of different families, the object of which is to serve, on the one hand, the ancestors in the temple, and to perpetuate, on the other hand, the coming generation." It must be arranged through "the orders of the parents and the words of the go-between." Among young boys and girls, dating is unknown and romantic love nonexistent—both seem to be the devices for a conjugal union, not for an institutional family.

[4] A popular saying in China with regard to this reads: "To marry a wife is for her virtue; to take a concubine is for her beauty."

[5] A woman should obey her father before marriage, obey her husband after marriage, and obey her son when her husband has deceased.

[6] It is a customary practice that a certain amount of jewels and garments is demanded from her husband's family before marriage known as *tsai-li,* which, in turn, is generally bestowed on the bride as a part of her dowry. The actual worth of the *tsai-li* varies according to the economic circumstances of the two families concerned. If the bride's family is poor and the bridegroom's rich, it is not uncommon for the *tsai-li* to be converted into cash which exceeded several times the value of the dowry.

In the Western world the choice of a mate, based on mutual compatibility, is made in the light of life-long companionship. In the traditional society of China, however, a wedding was not considered as a matter between two individuals, but rather as a conjugation between two families, and, therefore, personal adjustment between husband and wife constitutes only a minor factor in the selection of a spouse. With regard to a wife, consideration was chiefly centered on the following three qualifications: (1) capability of bearing children, (2) compliance with the family's traditions and rules, and (3) ability to endure household drudgery. It is readily seen that these qualities can be more or less objectively estimated, or at least, they are not so subtle as "personal compatibility" which the Western husband and wife may not be able to determine at the moment of marriage.

Among the three qualities for which a wife is chosen, the ability to bear children is of course the most difficult one to know ahead of time. Beside what the fortune tellers may have to say on this, a traditional way to estimate it is to note the number of children her mother has borne. A daughter who is an only child is not taken as a good risk and generally has some difficulty in getting married. Physical strength is also considered, although it may have more to do with her laboring ability. Perhaps the most important factor in selecting a wife is the likelihood of the girl's getting along with other members, especially male members, of the family. It is on this consideration that the traditional marriage is arranged on the ground of the two families being equal in status. The family status as mentioned above, is an indication of its good traditions and promising prospects, although this account is generally balanced by the amount of property the family actually has in its immediate possession. Traditions are based on the common ideology of Confucianism, and the role played by each category of relatives in the family is universally defined by it. Living in a homogeneous culture such as China's, the equal status of their families provided a good chance for the young couple to get along well. Should any difficulties arise between them, the institutional bonds of wedlock are so tight that both of them, but in particular the wife, would have to be resigned to the situation.[7]

Under the marriage traditions of China, the boy's family always takes the initiative, while the girl's may either accept or reject the proposal. In the process of making such an arrangement, few decent families would be willing to expose their young girls to inspection by members of any family other than close relatives. Shyness on the part of the girl counts for only a part of pre-marital avoidance; the chief reason seems to be the family's abhorrence to any suggestion that it is engaged in selling its daughter.

[7] Cf. Leong and Tao, *op. cit.*, pp. 100-101.

Thus any reliable information on the girl must be obtained through a person, usually a woman, who is acquainted with the girl and possesses intimate knowledge of her. Yet, she (or he), an amateur or a professional match-maker, must be one whose words can be trusted by both parties concerned. It is not uncommon, especially in the towns and cities, for a professional go-between to coax or cheat two completely incompatible families or persons into marriage, with matrimonial tragedy as the result. To this it may be added that if the girl in the Chinese traditional family actually lives in seclusion, the service of match-making is not only to carry on negotiations between the two families, but also to transmit messages between the boy and girl who are engaged, thus serving to stimulate their romantic feelings toward each other and to give each some knowledge of the other.

Perhaps a word or two may be said relative to the life of a young couple after marriage. Demonstration of love and affection between them in public is strongly discouraged, and the ideal of matrimonial life is epigrammatized as "husband and wife in bed, but friends out of it." That this inhibition may be advisable is apparently on account of the mother-in-law's feelings. Naturally she does not like her son to be taken away by another woman, and disagreements and quarrels between the mother-in-law and daughter-in-law are rather unavoidable. Humiliation and submission on the part of the latter, who is the loser on all issues, lead only to an intensification of the daughter-in-law's desire for revenge against her future daughter-in-law. One way out of this vicious circle is of course a cross-cousin marriage which, quite prevalent in traditional China, is generally made between a married woman's son and her brother's daughter, but not *vice versa*.[8] The advantages of this arrangement are that identity of the family background of the mother-in-law and the daughter-in-law would at least reduce a part of their differences, and that since they are doubly related, any misunderstanding that might arise could be resolved through the daughter-in-law's parents.

[8] Although joking privileges are commonly found among all cross-cousins related by marriage, and such privileges are generally taken as a sign of potential mating relationship, traditions in China somehow discourage and, in some regions, prohibit the "return-home-marriage" on the mysterious ground that it is *ku-hsieh-tao-liu* (the reverse flow of a married aunt's blood). This may very well be due to the fact that by marrying a married woman's son, it was felt that there would be greater harmony between the mother-in-law and the daughter-in-law, for the former would be constrained to act less harshly toward her brother's daughter. While, in the opposite case, there would be a tendency for the brother's wife to take out her pent-up frustrations on this girl who is the granddaughter of her own mother-in-law. Cf. also Francis L. K. Hsu, "Observations on Cross-Cousin Marriage in China," *American Anthropologist*, XL (January-March, 1945), pp. 83-103.

The Maintenance of a Large Household

If the family based on the father-son relationship may be called the consanguine family, and that based on the husband-wife relationship the conjugal family, one of the great differences between the two is that a conjugal family ends in itself, whereas a consanguine family can be extended to include a great number of relatives in the same household. Furthermore, in the former case, as stated by Margaret Mead, the society has to rely upon other types of social groups for permanent form.[9] On the other hand, the family system based on consanguinity can and generally does color the society with many of its attributes. In the case of China, it seems to represent the familistic type of social organization *par excellence*.

The father-son relationship within the family is the extensible link of an endless chain which vertically leads to the cult of ancestor-worship and horizontally to the extended or large family, the agnate clan, and finally the nation. Since brothers and cousins are bound together by patrilineal descent, encouragement of their living together in the same household is only natural. However, when the size of the family has grown so large as to comprise four or five generations, the head of the family is inevitably confronted with, among other things, the difficult problem of disciplining and educating the ever increasing numbers of children and training the wives of the younger generations.

The training of the daughter-in-law and grand-daughters-in-law, is entrusted to their mothers-in-law, especially to the patriarch's wife. After the first day of the wedding ceremony, the bride is carefully advised as to the behavior to be expected of her in conformance with the established traditions and rules of her new family. Should a breach of these conventions arise, the husband concerned is in difficulty. Unless he punishes his wife by word or act in public before the family it is most likely that he together with wife and children will be ostracized from the household. A son expelled on this ground is commonly regarded as unfilial and his conduct, therefore, is viewed as extremely disgraceful.[10]

The way in which the children are disciplined and educated is different. Under the large family system, a child is born into a group, a primary institution, and brought up in a milieu where, through personal contact and intimate reactions to each other, young and old, identification

[9] Cf. her article on "Family" in *Encyclopedia of the Social Sciences*, VI (1931 edition), p. 57.

[10] In the Chinese society there are many comic stories and jokes told at the expense of the hen-pecked husband. Under the family system a wife could hardly domineer over her husband, and yet, when the rare event occurs it is sensational.

with and conformity to the family and its rules become more or less a natural process of integration. Members of a lower generation address members of the one immediately above reverently as fathers and mothers, and address each other as brothers and sisters. It is the feeling of "we" (not of "I" as in the family of the Western world), which is cherished, cultivated, and finally incorporated in the personality of the grown-up adults. "A Chinese," wrote Leong and Tao, "does not live for himself and for himself alone. He is the son of his parents, the descendant of his ancestors, the potential father of his children, and the pillar of the family."[11] Observation of the cardinal virtue, filial piety and obedience, is exhorted in the name of the ancestors, rewarded by the inheritance of the patrimony, and reinforced by the family traditions and rules which sanction punishment of any offender. It is the binding force of this virtue and of its far reaching effects that has reduced juvenile delinquency to a minimum in the Chinese family; and it is also this aspect of life which leads an able American observer to remark: "Of almost no moral law in any civilization can it be more confidently said that it is not honored in the breach. In fact it has become so deeply ingrained, so firmly interwoven in the unconscious, as to constitute almost a biological principle."[12]

The cultivation of a strong "we" feeling in family life naturally engenders an equally strong pride in the family. This pride is very persistent and popular among the high standing families, and has led not only to friendly relations and marriage arrangements, but also to family feuds and even bloody conflicts between families or clans in traditional China. The implications of this concept go far beyond the family's locality. When any member of the household has succeeded in gaining an official position, it is both his duty and his honor to get jobs for other members, capable or incapable, through his association and influence. A brilliant son may be supported not only by the resources of his immediate family, but many a large and prosperous clan generally makes practice of raising enough funds to support a gifted member to gain fame and position. This is done mainly on the ground that members of the entire family or clan may all be benefited some day.

The Family in Transition

. . . Since the advent of Western influence in China a century ago, . . . this traditional and time-honored system has begun to decay with ever-increasing speed. Modern education and the importation of Western

11 Leong and Tao, *op. cit.*, p. 68.
12 Nathaniel Peffer, *China: The Collapse of a Civilization*, New York: John Day Co., 1930, p. 39.

ideas have gradually weakened the hold of Confucian teaching upon which most of the family ethics are based. Urbanization, together with the creation of job opportunities and of better living quarters for middle and upper classes, have affected a great number of people through a loss of traditional values.

This transition which has taken place in the Chinese family represents but a phase of a worldwide trend. It is a shifting of the center of family life from consanguinity to matrimony, or to adopt Dr. E. W. Burgess' terminology, from institution to companionship. The social change in China which has thoroughly shaken the traditional structure of society has shifted the emphasis of the family to comradeship. The whole process may be described in the following three stages:

(1) *The Decline of the Large Family System.* The changes of socio-economic environment since the turn of the century have brought about many effects fatal to the traditional family system in China. The intellectuals who have the responsibility of upholding the traditional model of the family are no longer interested in traditional ways of life. The rise in the cities and trade ports of the new group of "upstarts" in the form of compradors, big merchants, and warlords, rendered the situation in the family still worse. One common characteristic of this group of people is a weakness of conviction and virtue, traditional or Western, and yet an exploitation of all the privileges that work in their own favor under the old system. For instance, many of these parvenus, irrespective of their occupations, have not only bought domestic maids but also taken one or more attractive concubines of obscure and often disreputable origin, even though their formally married wives may still be alive and have borne to them many sons to bear the family names. They, as family heads, continue to exercise power and authority over their offspring in accordance with the prerogatives of the old patriarchy, but it now devolves upon the public schools and colleges to provide their education. In the household, because of heterogeneity of cultural background, enforcement of family rule is very difficult, and bitter dissensions and even licentious relations between family members and others are frequent.

In the villages, the breakdown of the large households is not so much the work of outside cultural impact as it is of the profound effects of economic forces. The ever-growing decay of the rural economy has forced many of the peasants' sons or daughters to leave home and to seek jobs elsewhere, and thus has reduced the family size to a minimum. Although the family of the landlord class is still relatively big and continues to be managed more or less along the traditional pattern, its educated sons or daughters, after completing their schooling, are most likely to stay away from home and set up their own small families in the cities.

(2) *The Emergence of the Conjugal Family.* The real conjugal family in which the members live together both in spirit and in form is largely a phenomenon of the large cities and is found among Westernized intelligentsia. Failure to make distinctions between form and spirit has caused many writers, Western as well as Chinese, to exaggerate the prevalence of the conjugal type of family.[13] The poor peasants in the villages may live in nominally conjugal families, but the way in which their wives and children are treated and disciplined is fundamentally traditional. The modernized professional is the only one whose economic independence from the support of the old household gives him freedom to marry the girl he loves, and whose college-educated wife cannot tolerate even a day of the mother-in-law's meddling in her business. Under the influence of modern education, the young talk about the out-datedness of the traditional family, and vigorously demand the right to choose their own mates, to set up their own conjugal families in the cities, and to live in their own way without interference.

(3) *The Emancipation of Women.* The destruction of the traditional family structure was accelerated by the revolutionary tide which has swept over China since 1946. The family system has been the target for heavy attack by the Chinese Communists. The spirit of family collectivism in terms of "we" feeling and group solidarity seems to have been shifted to apply to the Party and the nation, while the low status of women, together with the frequent maltreatment of wives, concubines, daughters-in-law, and domestic maids, has been used to point out the evils of the traditional family system. In mass meetings and public trials, mistreated women are encouraged to voice their grievances against their husbands, parents-in-law, and others. These latter persons are severely condemned, and can scarcely escape some punishment. Enslaved maids begin to settle accounts with their former owners by demanding wages covering all past years.

The attempt of the Peking regime to bring about a fundamental revolution in the family system is being implemented by inculcating in the minds of people, and especially of the youth, new ideas with regard to marriage and the family. All the traditional family ethics and Confucian virtues are discarded as vestiges of feudalism. Families are urged to sign "a pact of patriotism" under which all individual members pledge their allegiance to the nation, and not to any particular persons within the family. Hence, newspapers often report open trials, in which a wife accuses her husband, or a son or daughter brings charges against the father. Marriage is no longer taken as a matter involving two families as in

[13] Cf., for instance, the statistics gathered by Olga Lang in her *Chinese Family and Society,* New Haven: Yale University Press, 1946, pp. 134-54.

traditional China, nor between two individuals as in the Western world, but a spiritual union of two comrades of different sexes; and the first task of the couple is to strengthen and cherish their commonly shared belief of communism, and then to engage in production to build a new society. The prospect of success of this indoctrination campaign cannot at present be estimated, but one thing seems certain: familism and the large family system as known historically are irretrievably gone from the land of the Middle Kingdom.

ALTERATIONS IN THE KINSHIP SYSTEM, IN THE EARLY COMMUNIST TRANSITION

C. K. Yang

Adapted from *A Chinese Village in Early Communist Transition,* by C. K. Yang, Cambridge, Mass., The Technology Press, 176-180; copyright 1959 by the Massachusetts Institute of Technology.

Alteration of the traditional Chinese kinship system was one of the significant objectives of the "family revolution" which had been gathering force for some fifty years before the Communists came to power; but in the pre-Communist period it was confined to the upper and middle classes in the urban centers. The national success of the Communist revolution was followed by official action which not only quickened the tempo of change of the kinship system but also caused a rapid extension of that change from the upper and middle classes to the urban working class and from the urban centers to the countryside.

The first concrete evidence of such action in the neighborhood of Nanching was the introduction of the new marriage law. Toward the end of 1950 a notice was posted in front of the subdistrict government office in Pingan Chen announcing that henceforth marriages dictated by parents would be illegal, that a marriage must be based on the free will and free choice of the marriage partners, and that all new marriages must now be registered with the subdistrict government in order to be legally recognized. Later a young woman officer came to Nanching, held a meeting with a group of the villagers, and explained these same points. The principles of the government notice and the young woman's speech were not entirely new to the villagers, who lived too close to the big city not to have heard of marriage as a result of free choice of partners; but they had regarded such a marriage as another strange fashion in the strange modern urban living, especially among the "foreignized" rich city folks without any thought that it could one day apply to their own intimate life.

We observed two marriages in Nanching in 1951 which, in compliance with the new law, were registered at the subdistrict government. In each case, a young cadre member asked the engaged young man and woman whether they had been forced into the marriage by parents or other parties, or were being married of their own free will and choice. Each said it was by free will. The man paid the small registration fee and received a printed marriage certificate, a totally strange object to the villagers, who had never heard of the government taking a hand in the private marital affairs of the people. But appearances were deceiving. In each case the parents in the families concerned had arranged the marriage and prompted the young couple in what to say in going through the registration. Before the marriage, the traditional practice of the boy's family offering the girl's family a ceremonial price was followed—although the new law prohibited paying a "body price" for the bride; and after the registration the full traditional wedding ceremony was enacted in the village, including the clan feast in the ancestral hall.

It was obvious that these Nanching marriages typified the conservative peasants' desire to stick to the old system, a deep-rooted conviction that could not be changed merely by the passage of a new law. In Nanching the individual still depended upon the family for economic security and development, and the parents will and authority over the children's marriage carried the weight of economic power. A village girl's predominant notion about marriage was still that it provided her a home where she could eat and have children to support her in her old age. Arranged marriage as a means of carrying on family production had been a proven formula of economic security; and until a new system of economic security became equally proven of its worth, it was not easy to induce the peasants to give up economic security for romance. Furthermore, the village clung to the strong tradition of sex segregation, and opportunity for normal social life between the sexes was still too limited for the development of romantic associations. Lee Ying, the old Confucian scholar, at first refused to be interviewed by our girl student investigator. After we convinced him that it was perfectly normal to talk to a young girl, he consented to answering questions, but he talked to her with his face averted, stealing frequent sidelong glances at her.

After the Communists were established in the village, they insisted that both men and women should come to the increasing number of political meetings, and an amateur theatrical group was organized in which boys and girls worked together. The New Democratic Youth League established a branch in the village, and among the small membership in 1951 three or four were girls. These were the small beginnings of a social life between the sexes.

Whereas the effect of Communism on the kinship system as measured by its impact on the traditional marriage procedure was bound to be a gradual one, its destructive influence on the internal kinship status system was immediate. We have already noted the diminishing value of age as a serious factor in the growing difficulty of the older generation in trying to continue the subordination of the young and maintain the social order built on age and generational level. The Communist accession to power greatly accelerated this development. The peasants' association, the new center of power in the village, was mostly staffed by young men under thirty, and Wong Ping, the head of the association, was only in his mid-forties. The New Democratic Youth League branch was a political organization of considerable importance in the public affairs of the village. Older villagers were frightened of its members because of the new power they wielded and fear that they would betray family secrets to the police or cadres. One of the notable features of Communism was that the young, who as a group had never been taken seriously by traditional society or given responsible duties, now came to the fore in social prestige and political importance.

Communist policy aimed explicitly at changing women's status. The new marriage law granted women a higher status in the family and in society than before; and the new legal freedom of divorce gave her an outlet from extreme mistreatment by her husband and mother-in-law. At least one woman in Nanching learned about and used this new right in our time there. Women began to emerge from family confinement to attend meetings and take part in public activities sponsored by the cadres and the peasants' association. The new amateur theatrical group in the village included women, something previously unheard of; and there was a Women's Association in Pingan Chen to which the village sent several women representatives. This association was doing little more than participate in political campaigns under the guidance of local cadres, but the very presence of an independent organization of women dealing with public affairs was in itself a new phenomenon of considerable importance, something entirely out of context with the traditional social order based on sex segregation and the exclusion of women from public affairs.

Land reform had a direct effect on the internal structure of the family. Since land was redistributed not to the family as a whole but to each member on an equal-share basis regardless of age and sex, land reform gave the young and the women in unprecedented sense of importance in contrast to the traditional system of family property ownership, in which the head of the family had sole right to dispose of the family property and female descendants enjoyed no inheritance rights. Moreover, the land reform regulations stipulated that each member of the family might take his or her share of the family land out of the family, for instance in case of a divorce, an egalitarian arrangement of roles of the members and the economic leverage that clearly strengthened the position of the young and the women.

The drastic alterations in the traditional kinship system were most dramatically symbolized by the disintegration of the formal structure of clan authority, the inevitable result not of specific Communist government orders but of the operation of the whole devastating revolutionary process of remaking the village's political and economic life.

Local public affairs were now so much more intimately related to regional politics and formal government than in the pre-Communist period that the elders' council of the clan, ignorant of the new political situation and faced by the rising importance of the modern educated younger generation in this field, were helpless. After the fall of 1950, we did not hear of a single meeting of the elders' council, although more problems involving the whole village had occurred since then than in any previous year that we knew of. The position of business manager as an agency of the clan's formal organization was practically abolished, there being now no clan property to manage as the clan land and other income-yielding properties were redistributed during land reform. Moreover, the Communists arrested many clan managers in other villages on the accusation of embezzlement of clan funds; and official directives on land reform pointed to the office of managing clan finance as a system of "feudalistic exploitation" of the clansmen, a system ridden with corruption and embezzlement.

The "class struggle," especially during the land reform, seriously weakened the solidarity of the clan by introducing a non-kinship criterion of group interest. The peasants' struggle against the landlords, the disclosing of hidden ownership of land to the Communist cadres, the tenants' accusation of cruelty and exploitation against rent collectors, the victims' charges of injustice and oppression by "local bullies," the activists' efforts in ferreting out counterrevolutionaries—all these turned many close kin against each other. The process of struggle had difficulty in getting started, especially when the case involved the young bringing charges against the old, but after the inhibition of kinship was broken down, the struggle became bitter.

We noted that during the Chinese New Year in early 1951 the ancestral halls were no longer cleaned and decorated as before; and no community operas were staged by the clans after land reform. Instead, there were the new recreational groups such as the village theatrical group which were not organized on a clan basis. Ceremonial occasions such as spring and autumn sacrifices to the ancestors were reduced to simple affairs performed only by a small number of older clansmen who also privately contributed toward the sacrificial expense which was formerly financed by the income from the now confiscated clan property. To the Communist cadres, ancestor worship was a superstition like other forms of supernatural worship, and they would do nothing to remedy the situation. Most of the ancestral halls in the village were transformed into classrooms for literacy and political training and offices for the new organizations; they no longer served as places of religious inspiration for the perpetuation and development of the clan.

Our observations in Nanching confirmed the impression that the clan was unlikely ever to recover its traditional importance in the operation of village life, and that the whole kinship framework of collective action, supported by its traditional status system, would diminish in strength while the universalistic type of political and social order would increase its influence.

6. IMPACT OF SOCIAL CHANGE ON FAMILIAL FUNCTIONS: THE AMERICAN CASE

Postulating seven functions—economic, status giving, educational, religious, recreational, protective, and affectional—Ogburn asserts that six—all but the affectional function—have been reduced as family activities in recent times. The following two papers present transitional families. The second paper concerns the farm family of the middle 1930s in central New York, whose forebears had been in the country for several generations at least. The third selection concerns the Sicilian peasant family in America as seen in the middle 1940s. Evidence of what Ogburn calls loss of function is seen in both of these settings in the impinging of urban values, the reduction of the authority and influence of family elders, and the increasing importance of contacts outside the family and the local community.

Some take Ogburn's analysis and other formulations of similar import and through extrapolation foresee the demise of the family. Others profess to believe that, figuratively speaking, the family is not beaten but is merely regrouping its forces to defend itself against the ravages of industrialization, urbanization, and secularization. A development more consistent with the latter view, but of course not proving it, is the flight of the masses to the suburbs, where there is a burgeoning of family life and an astonishing rise in the birth rate. Jaco and Belknap conclude the chapter with some observations about family life outside the city limits.

THE CHANGING FUNCTIONS OF THE FAMILY

William F. Ogburn

Adapted from "The Changing Family," *The Family*, 19, 1938, 139-43.

The dilemma of the modern family is due to its loss of function. Throughout the period of written history the family has been the major social institution. Indeed, in the long period of prehistory, as well as in historical times, the family has been a larger social institution than it is in the Twentieth Century in the United States and western Europe.

Prior to modern times the power and prestige of the family was due to seven functions it performed:

Foremost was the economic function. The family was the factory of the time. It was a self-sufficient unit, or nearly so. The members of the family consumed only what they produced. Hence money, banks, stores, factories were not needed. A wife was a business partner, a good foreman, or competent worker.

As a result of this economic function the family became a center of prestige and gave status to its members, its second function. A member of a family was less an individual and more a member of a family. It was the family name that was important, rather than the first name. Most families stayed for generations on the same pieces of land in or near a small community and hence had an opportunity to establish reputations. It was important to marry into the right family, as well as to marry the individual. The family name was a badge and had to be guarded at all cost and at all times.

The nature of the household economy was such as to make the home the center for education, not only of the infant and child of pre-school age, but also the youth for his vocational education, physical education, domestic science, and so on. The higher education was often obtained by employing a tutor who lived with the family.

A fourth function was that of protecting the members. The husband protected the wife by virtue of his physical prowess, a protection now furnished by the police. The elders found a place readily in the household of the child to spend the twilight of their lives. Children were an old age insurance.

The family exercised a religious function, also, as evidenced by grace at meals, family prayers, and the reading together of passages in the Bible. Husbands and wives were supposed to be members of the same church.

Recreation in those days was not a function of industry; that is, it was not commercialized. There was some community recreation but it was often at the homestead of some family. Recreation centers outside the home were few.

A final function was that of providing affection between mates and the procreation of children.

These seven functions—economic, status giving, educational, religious, recreational, protective, and affectional—may be thought of as bonds that tied the members of a family together. If one asks why do the various members of the family stay together instead of each going his way, the answer is that they are tied together by these functions. If they didn't exist, it is not easy to see that there would be any family.

The dilemma of the modern family is caused by the loss of many of these functions in recent times. The economic function has gone to the factory, store, office, and restaurant, leaving little of economic activity to the family of the city apartment. About half of education has been transferred to the schools, where the teacher is a part-time or substitute parent. Recreation is found in moving pictures, parks, city streets, clubs, with bridge and radio at home. Religion doesn't seem to make as much difference in family matters as formerly, grace at meals and family prayers are rare. As to protection, the child is protected at home, but the state helps also with its child labor laws and reform schools. The police and social legislation indicate how the protective function has been transferred to the state, as has the educational function. Family status has been lost in marked degree along with these other functions in an age of mobility and large cities. It is the individual that has become more important and the family less so. On the other hand the family still remains the center of the affectional life and is the only recognized place for producing children.

From this survey it may be seen that at least six of the seven family functions have been reduced as family activities in recent times, and it may be claimed that only one remains as vigorous and extensive as in prior eras.*

The loss of these functions from the family institution does not mean

* ". . . it may be said that the affectional function is still centered in the family circle and that no evidence is recorded of any extensive transfer elsewhere. The evidence of increased separations and divorces does not prove that husbands and wives now find marriage less agreeable than their ancestors did. It may mean only that certain functions and traditions which once operated to hold even an inharmonious family together have now weakened or disappeared. . . . The future stability of the family will depend . . . [largely] . . . on the strength of the affectional bonds." William F. Ogburn, "The Family and Its Functions" in *Recent Social Trends*, McGraw-Hill, 1933, pp. 663, 708— Eds.

that they have been lost to society. They have not disappeared from society as they have from the family. Rather they have been transferred from the family to other institutions, schools, factories, stores, clubs, commissions, and so on. What is the family's loss is the gain of the state and of industry.

One other point may be noted as to these changes in the family. Their causes can be traced largely to the inventions using steam as power. The old family existed with the handicrafts in the city and with subsistence farming in the country. Steam power made possible cities, factories, modern transportation, mass production, and specialization, which are part of the process of the transference of functions away from the family.

There are a number of consequences of the uses of this power. One is the increase in separation and divorce. A sample of the census of 1930, weighted slightly in favor of cities, showed about one in ten families broken by separation, annulment, or divorce. It is well known that one in every five or six marriages contracted will end in a divorce court.* The reason is clear. The bonds that hold married couples together are weaker and fewer. Hence husbands and wives fall apart. Women can get jobs outside the home, and men can get meals and mending done elsewhere. The one function remaining more or less as strong as formerly, the affectional tie, however, is not as strong alone as the seven ties together. The affectional bond snaps and there follows separation and divorce.

The situation is affected not only by steam but by one other invention, the contraceptive. It seems that this invention increases the amount of marriage and promotes early marriage, rather than the contrary as is sometimes claimed. But it would also tend to result in more families without children. Divorce is many times as frequent among couples without children as with them.

Another consequence of the transfer of functions from the family is the decline of the authority of the family. There are no longer families that dominate societies as was once the case. Much greater authority rests with state and industry. So also authority in the family declines. The husband's authority over the wife is not what it used to be. The state challenges the authority of the parent over the child, for instance, as to its education and as to its labor. The child grows up accustomed to authorities, many of them elsewhere than at home. The respect which a child has for its parents rests more upon their personalities than upon authorities they possess. So the respect one member has for another is not bolstered up by powers and sanctions. So if the respect is not based upon personality it is not likely to exist.

* For more recent figures concerning divorce, see p. 95 and pp. 633-635 below—*Eds.*

Another result of the shift of functions from the family to other institutions is the change in the nature of marriage. Marriage was at one time a semi-business proposition, which parents and elders realized fully and the young people realized in part. The young man looked for a good homemaker, who was diligent, thrifty, and capable. It was worth while for a young woman to have a reputation among the neighbors in this regard. The young man was certainly expected to be a good provider, to come from a good family, and to have status. If either had property, that was an item of consideration. Under this framework there was a chance for some romance, unless the marriage was arranged by the parents and unless dowries were of overshadowing importance. Marriage was viewed as an institution, a business. On the other hand, romantic love alone was another thing. It came and went. You were in love today but not tomorrow. It was not considered a phenomenon stable enough upon which to erect a business, to raise and rear a family. There must be something else, efficiency and ability.

With the shift of functions away from the family, romantic love has taken over marriage, aided by moving pictures and the pulp magazines. Whether the wife is a good cook is a secondary consideration. It is not necessary that she be a good seamstress any more than she needs to know how to spin and weave. Hence, there are more hasty marriages. It has become necessary for states to pass laws requiring a certain amount of time to elapse between the purchase of a license and marriage.

Another result of the decline of the family functions is the conflict between the new conditions of family life and the old attitudes surviving from an earlier type of family life. Thus the older philosophy said that woman's place was the home. True enough it was when she made soap, wove cloth, and prepared medicine from herbs. But the maxim is not so clear for women with no children living at home. . . . Often these women live in small apartments quite unsuitable for economic activity. Besides, one in every 8 or 9 married women helps out the family income by drawing wages for work done outside the home. Many men feel it reflects on them to have their wives work for wages. Others feel that they are head of the house, a position that had more significance under the household economy. The conflict is apparent in the case of girls, who do not know whether to prepare for marriage or for jobs. It is not only difficult to do both, but there is also a psychological conflict between the new economic freedom of self-support and the lifelong devotion to husband, children, and home.

One effect of the invention of the contraceptive has been a loss of a family function rather than a transfer of that function to another social institution. I refer to childbearing. No other social institution produces children and illegitimacy is probably on the decline. Thus, at the time when the American colonies won their independence from Great Britain, 10 wives bore 78 children; one hundred and fifty years later 10 wives bear only 23 children. The cost of rearing a child, especially in the city, is great today. Not many fathers could provide opportunities for education and health to seven or eight children, especially when the law forces the family to care for them until they are almost twenty years of age. Fewer children mean, then, more advantages and opportunities, and no doubt better food, less illness, and superior physical well-being.

But the gain on the psychological side is not so clear. The only child and the oldest child are a much larger percentage of all children now than formerly. They receive relatively more attention than middle children. They are with adults more. These conditions cause more geniuses but also more failures. They are said to be more narcissistic and exhibitionistic. It is claimed by psychoanalysts that neurotics are drawn proportionately more from the only, oldest, and youngest children. Indulgent parents are more likely to "spoil" an only child than those of a large brood. Anxious mothers are more likely to inculcate anxiety in an oldest child than in a middle child. Such problem children in youth are a responsibility of the family and school in wealthy neighborhoods. But in poor neighborhoods, where mothers work away from home and where the streets are the playground, such problem children become a responsibility of the state, for their gang life leads to delinquencies of various kinds that may bring them up against the law.

The problem of the family rearing of children in a modern city is due in part to the survival of the older attitudes which are in a practical way incompatible with modern urban conditions, and to the absence of a definite pattern of guidance in a changing society for parents whose intelligence quotients may not be very high. No ethic has as yet risen to take the place of the one followed in the Victorian era.

Not all the difficulty is due to conditions within the family. The conditions outside the family make successful family life difficult. For the family does not exist in a vacuum, as the saying goes, but is a part of society. The inventions which have so changed the institution of the family have also changed society. These changes in society that impinge on the family, often with disastrous effects, may be summarized by the word heterogeneity. There are in a modern city many groups to which the members of a family belong. Formerly they were members of the church, and of perhaps two or three clubs. Now the men of the family belong to a business group, to a church, to a union or trade association, and to some clubs. The wife may belong to a business group, a card club, a church, a social club. The children belong to a school group, perhaps a play group, and perhaps a club.

The meaning of these various memberships lies in the important role the group plays in shaping one's conduct. We conform to the folkways and mores that are set by the group. We become like the group within which we live. We cannot long resist the pressure of Main Street. A man does not rise much above the level of the group in which he lives, nor does he fall much below this level. Our self is really what the group influences make it. Hence, personality is a social product.

Now in modern times the group influences that determine the character of the members of the family do not flow from the family alone, or from just one group. The members of the family belong to many groups, each one having its own folkways and social evaluations. The boys' gang has a set of standards different from those of the school children, or of the family, or the church. Often these values conflict. Pavlov is said to have produced a neurosis in a dog. He did it by conditioning the dog to respond positively to a great circle of light thrown on a screen. The same dog was conditioned to respond negatively to a great ellipse of light thrown on the same screen at another time. Then when the great experimenter changed the light from a circle and an ellipse to a midway type of figure, the dog did not know what to do, and broke down in a fit of trembling. The conflicting values and standards of the different societies, business associations, churches, athletic groups, and pleasure groups may produce somewhat similar conflicts.

There is a competition with the family by other groups for the control over its members. The family no longer holds sway. The nature of family life is such that it is necessarily important, but it often breaks down under the competition with other groups. Parents thus lose control over their children, husbands and wives lose influence with each other.

This situation is affected by one other condition, namely, city life. When the village is small, consisting of a few hundred persons, and there is little travel and communication with the outside, everyone knows everything about everybody else. Life is in a goldfish bowl. The result is a homogeneity. The set of values for a particular club or association perforce conforms to a general village pattern.

But under city life, the members of the different groups do not know one another. They often come from different neighborhoods, one may never see another member of a club except when the club meets. Hence, city life presents isolation and heterogeneity. In addition communication from the outside world brings—via magazine, radio, and moving picture

—the folkways of other places and other lands. The result is an individualization of the members of the family. The individual no longer has his moral problems solved for him by the family group. The heterogeneity of society and the rapidity of social change make impossible specific formulae which tell one what to do in different situations. Right and wrong have to be figured out by the individual, which calls for a high I.Q. and some ability to think in an emotional situation.

Such are some of the consequences following from the loss of functions by the family due to modern invention.

A PORTRAIT OF THE FARM FAMILY IN CENTRAL NEW YORK STATE

Howard W. Beers

Adapted from the *American Sociological Review*, 2 (1937), 591-600.

Pictures of the grandparents or great-grandparents hang today beside needlework samplers and faded hair flowers in heavy frames on the walls of an occasional farmhouse parlor. Heirlooms now, they typify a cultural period that had also its characteristic patterns of social arrangement. The early designs for living are heirlooms, too, but not yet as completely relegated to the walls of memory as are the paintings and the handwork in those parlor frames. Sometimes, indeed, we have nostalgic urges to recall them entirely from the past to play again the old roles of certain status in the present period of greater social confusion. But cultural systems change and if there were to be any permanence of role and status it would be too often a kind of social *rigor mortis*. Specific patterns of family life are therefore neither universal nor permanent. Our pictures of the farm family, as of every other social grouping, must be adjusted at intervals to cultural change in each locale and in each social stratum.

A classic picture of the early farm family in New York State has been worded by James Mickel Williams.[1] The figure in his portrait is an English puritan family reaching New York via New England, gradually reshaped by the conditions of pioneering, but with basic patterns enduring throughout the period of subsistence farming and continuing even into the recent periods of commercial agriculture and contemporary metropolitan dominance. The pioneer American family was large, biologically vital, and of

[1] James Williams, *Our Rural Heritage*, New York, A. A. Knopf, 1925, pp. 46-80.

strong social texture. It was "the beginning and the end of rural social organization."[2] Family groups were geographically isolated, economically self-sufficient and socially self-contained. Parents were often "the school, the church, in extreme cases the state."[3] Fathers were austerely dominant. Wives were obedient, faithful, subordinate in person and in law. Strict obedience was required of children. Actually, the subjection of wife and children to the father exemplified their common submission to natural processes, never completely understood, always uncertain. The common and paramount interest of family members in the outcome of the farm enterprise necessitated agreement on all matters. Farming and living were synonymous. There was need for an executive in each family who could give direction to the process of living. The natural executive was the father, hence our usual judgment that the farm family was patriarchal. There followed from these conditions a strong family pride and exclusiveness, rigid adherence to custom. Self-restraint, thrift and industry were predominant in attitude and in action. There were strong standards of modesty and morality. There was respect for authority, whether parental, religious or legal.

The matrix of rural custom in which these family-forms were set has been only stiffly flexible, yielding slowly to urban encroachment. The very strength of its original position has not only retarded change but it has added to the discomforts of change. Social confusion is in proportion to the rate of change of the mores. Life is well-ordered and relatively easy when standards and rules are fixed, commonly known and commonly unchallenged. The psychological strains and tensions accompanying rapid social change are most acute where there has been greatest reliance on precept and formula. The potentialities for intragroup conflict, therefore, have recently been great in the farm family. City folk were earlier inured to the presence in one family group of widely variant interests and activities than were farm people. Perhaps this is why Burgess found evidence that adolescent-parent relationships are less well adjusted in the rural than in the city home.[4] Alterations of farm family life may be occurring today in a maximum-discomfort stage of cultural change.

The family type described by Williams was at one time common in many parts of the northeastern hay and dairy sections. It was the bio-

[2] P. A. Sorokin, C. C. Zimmerman, and C. J. Galpin, *A Systematic Source Book in Rural Sociology,* University of Minnesota Press, 1931, vol. II, p. 4.

[3] J. F. Brown, *Psychology and the Social Order,* New York, McGraw-Hill Book Co., 1936, p. 224.

[4] White House Conference on Child Health and Protection, *The Adolescent in the Family.* Report of the sub-committee on the Function of Home Activities in the Education of the Child, E. W. Burgess, chairman, New York, D. Appleton-Century Co., 1934.

logical and social ancestor of present day farm families in Central New York State. These families are living today on family-size farms which they own or rent and from which they derive their chief income. In the sample studied,[5] all members are native born.

. . . The husbands average . . . 43.5 years in age, two years older than the wives. The parents were married an average of 18 years ago. An average of 3.6 children have been born per family. An average of 2.7 of these children per family are living with their parents at the time of observation. The families have been settled on their present farms an average of 10.8 years. Three-fourths of the wives and four-fifths of the husbands were born on farms. The families are similar, then, with respect to race, general culture, and occupation. They differ among themselves with respect to age, stage of family development, economic status and other factors. There are many interfamily variations like the individual differences among persons. The family patterns in this group are only once or twice removed from those of native white families in town, for they all stem from the early patriarchal type. The country cousin, however, remains biologically more vital than its urban kin. Even though farm birth rates have declined, perhaps more rapidly than city birth rates, the farm family still is formed earlier and is larger.[6] The age of marriage in the United States has been increasing among classes of higher economic status, but it has been declining among farm people, as among other low income classes.[7]

This portrait of contemporary farm families in a particular area will emerge more distinctly if one model sits against the background of descriptive data for all the families studied. Excerpts from a case narrative will help to clarify the outlines of our discussion.[8]

The X family lives eight miles from a village of 2500 people in a rugged south central dairy section of New York State. To reach their home, one drives out through the broad valley and up a winding black-top road through the Gully. Turning right into an uphill lane, one stops between the frame house (straw yellow with white trimmings) and the unpainted barns. A flashlight beam points the way through a dark, rainy night to the back entrance.

This has been their home for 17 years. Mr. X is 43 years old, and his wife is 41. Both were born and reared on nearby farms. He finished common school. She attended the village high school for two years. Neither has had any occupa-

[5] Howard W. Beers, *Measurements of Family Relationships in Farm Families of Central New York,* Ithaca, N.Y., Cornell Univ. Agr. Exp. Station Memoir 183, 1934, p. 38.

[6] Dwight Sanderson, "The Rural Family," *Journal of Home Economics,* 29, April 1937, 223.

[7] Dwight Sanderson, *ibid.,* p. 223.

[8] The case excerpts included with the text of this article are all drawn from the narrative of one family, sufficiently representative to illustrate general statements.

tional experience other than farming, and the farm is now their sole source of income.

Four children have been born to the mother, but the first boy died of pneumonia in his second year. The second boy, now eighteen, is a sophomore at college. The next child is a boy of eleven, in seventh grade. The youngest, a daughter of six, is in school for the first time—four born, three living, and two at home.

When married 20 years ago, Mr. and Mrs. X lived for three years with Mr. X's parents, working the homestead on shares. Then they bought their present farm of 96 acres, going in debt for the full cost. Their small cash reserve was invested in repairs to buildings.

An average season's work on this farm involves handling 15 acres of hay, 12 of spring grain, five of ensilage corn, seven of buckwheat, three of potatoes, one acre of field beans, two acres of wheat, four of alfalfa, the care of nine milk cows, some young stock and 100 hens. Mr. X is now rated by local leaders as a careful and successful farmer.

The basic differences between early farm and city families were due both to rural isolation and to occupation. Today we find the kind of work that families do is becoming relatively more important than their place of residence. The social length of a physical mile varies from moment to moment. Farmers in central New York State undoubtedly knew of the Hindenburg Zeppelin disaster before the flames were extinguished. Certainly, many of them while at breakfast took vicarious part in the coronation of a British King. Furthermore, as the influence of distance declines and the influence of occupation becomes more marked there comes increasingly into the picture a third differentiating factor, namely, that of economic status. Socio-economic differentiation within the rural community is becoming more pronounced. It is less and less possible to portray sharp contrasts between "the rural" and "the urban" family because of greater social heterogeneity within both rural and urban groups. Hitherto rural people in America have belonged largely to the middle classes. Recently, increasing numbers have moved into social positions of lessened status. As this happens it is important to note that "the climbing of the ladder of gentility" [successfully accomplished by Mr. X] "has suddenly become a much more difficult task than has heretofore been the case."[9]

Underlying or accompanying the vital and economic changes of the farm family, there are changes in psychosocial relationships, changes in status and role. One factor basic to the definition of status is the division of labor among family members. Hence, any change in work pattern is significant. . . . In the farm family of today, specialization is more marked than it was during an earlier period. Formerly, it was not unusual for girls to help with the outdoor work and for boys to help in household

[9] Roy H. Holmes, *Rural Sociology*, New York, McGraw-Hill Book Co., 1932, p. 73.

routines. Today, however, the processes of economic production have largely left the house for field and barn. . . .

Child labor is traditional on farms and it has been reduced in quantity more by compulsory school attendance than by any shifts in rural attitude. This is well evidenced by nation-wide rural negation of the proposed constitutional amendment regulating the labor of children. In an earlier time, custom allowed rural fathers the privilege of getting all the work they could from their sons while the latter were legal minors. Social maturity was recognized at 21, but not before. This father-son work relationship is still almost unique to the farm family. It constitutes a type of vocational training inherent in family farming. . . .

Although there is greater division of labor in today's farm home, many activities are still shared, and propinquity still fosters family solidarity. Farm families vary in these customs, but the X family is representative for central New York.

The family members are all home together an average of six evenings per week. When asked what the family usually does in the evening to pass time pleasantly, Mrs. X said, "Well, when the boys are home they like to have Mr. X play checkers with them or something like that. I read a great deal. Mr. X reads as much as he can with his poor eyes. Then we have music, too. Lots of times we get around the organ and sing." The family always gathers at meal time with the exception of luncheon on school days when the children are not at home. But when the family is at home, each waits for the others to assemble before starting to eat. Reading aloud is customary, as it was in both of the parental homes. The Bible is read aloud once each day. As a rule, shopping in town is a family activity. There is family observance of the usual holidays. On Christmas the family goes to the home of Mrs. X. On Thanksgiving they go to the home of Mr. X. On New Year's Day they observe a holiday at home. On Decoration Day they go to the cemetery to decorate the graves of their first-born and their dead kin. Birthdays are always celebrated with at least a cake . . .

Daily Bible readings and strong social dependence upon church activity appear in the case of the X family. Neither they nor their neighbors follow the old custom of "saying grace" at meals. Yet every one of the marriages among these farm families were performed by a minister of religion rather than by a civil officer. Wives in particular declared that they "would not feel married unless a minister had performed the ceremony." The marriage mores of rural life are still intimate with institutional religion, but family activities are affected less directly by religion than they were in an earlier period.

The shared activities, propinquity and group rituals in these families are not ordinarily accompanied by overt demonstrations of affection. Here, as elsewhere, traditions of restraint and habits of emotional control are vestiges of the pioneer attitude. It is likely, however, that practices with respect to shared activity, demonstration of affection or family ritual vary more from family to family now than they would have varied three generations ago in the same area.

Along with propinquity and the patterns of work and leisure, division of executive authority in the home is equally basic to family structure. As Sims has written, "Although the rural family inclines to the patriarchal type, it often manifests noteworthy democratic traits." [10] It might now be argued that the farm family inclines more and more to the democratic type of organization. The old conditions of risk and uncertainty in agriculture have not been entirely supplanted but there is now less mystery about the processes of production. Uncertainties of biology and weather, although replaced by uncertainties of the market, are no longer so insistent that each family group have a patriarchal head.

The present importance of markets, and the consequent emphasis upon intense, specialized production is introducing certain new influences on role and status within the family, giving some impetus to the democratic trend.

On a check list, which Mr. and Mrs. X completed independently, each gives the other credit for helping earn the family income. Each of them reports it to be earned by "father and mother together." Mrs. X is responsible for buying food. Purchasing children's clothing is a shared responsibility. Borrowing money is a matter that rests largely with Mr. X, although both parents discuss any problem of this sort before action is taken. If a problem directly concerns the children, they are called into council. Buying machinery is a matter for Mr. X's decision. He decides what crops to plant, when and where to plant them. If there is any remodeling to be done in the house, a joint decision is made. Contributing to the church is a matter for consensus. Mr. and Mrs. X select together the papers to which they will subscribe. Writing checks is done only by the husband. He buys the insurance, although whether or not it shall be taken is first agreed upon. Training the children is shared; seeing that children study lessons is also shared. Giving the children permission to leave the home or to go away is joint; punishing children is done by both. Both parents give the children spending money. Both of them help in planning the children's education, although Mrs. X said, "Now some of these things, like choosing the children's vocation, neither one of us ever thought that was our place."

There is little evidence here of uncompromising paternal dominance. The husbands in these families rarely take upon themselves the sole responsibility for making decisions, even when problems of business are up for solution. A discussion involving at least the husband and wife precedes the reaching of a decision in seven homes out of eight. In nearly half of the homes, children also are consulted. However, the importance of family discussion varies according to the kind of problem awaiting

[10] N. L. Sims, *Elements of Rural Sociology*, Thomas Y. Crowell Co., 1934, p. 434.

solution. Questions relating primarily to the farm business are more likely than others to be decided by the husband without consulting the wife. Decisions about the purchase of machinery or what crops to plant are of this type. On the other hand, if the question is whether or not to borrow money, to buy insurance, or to buy a car, there is likely to be a family discussion before any final action is taken. These questions relate directly to family welfare rather than solely to the farm business. If money is borrowed, thrift is forced upon the members of the family. If insurance is purchased, other things will have to be foregone. A new car will be either liked or disliked by each family member, hence each one has an interest in the decision.

The tendency for husbands to be solely responsible for financial decisions is more marked in families on large than on small farm enterprises. There is a suggestion in the evidence that as standards of competitive business efficiency enter farming, the splitting of executive responsibility into home and farm divisions may become more pronounced. The prophets of chemurgy as well as the discoveries of conservative research point to extreme and imminent changes in the practices and skills of agricultural production. The rate and final extent to which country life may be deruralized, of course, cannot even be conjectured, but even now milk-dresses and bathtub-tomatoes are more than pure fantasy.

There are also some new problems of financial administration in the home. In the days of family self-sufficiency, production and consumption were one dual process, begun and finished on one farm unit. Cash was unimportant. Now, however, the medium of exchange has a new significance to each family member. How do families meet the problem of an equitable or satisfactory distribution of cash among the separate members? There is little evidence of any one answer sufficiently extant to be called a folkway, but it is likely that low income or disagreement over distribution of income is a new source of tension in the farm family.

Some phases of the parent-child relationship have been mentioned above. In the X family, we found each child attending school, the oldest boy in college. This illustrates an attitude frequently voiced by mothers who want to educate their children into white-collar strata. We found each child with a definite place in the work pattern, a place of responsibility increasing with age, and allocated to house or field and barn on the basis of sex. We found parents and children playing as well as working together. We found children included occasionally in family councils, we found the boys with property and money of their own, we found ritualistic observance of the children's birthdays. (The manner of giving money to children in the X family is not general in the area studied. Irregular amounts of spending money are more customary in other families.) We found both parents sharing some responsibilities for guidance and control, yet allowing children relative freedom in such matters as choice of vocation. However, we still find that unfailing obedience is expected.

"Mr. X, how do you get the children to do what you ask them to do?"
"Why, we just tell them." Mrs. X added, "We never believed in bribing them or paying them to do things." Mr. X continued, "We always calc'late that if they are told to do anything they are s'posed to do it."
"What methods of punishment do you use?"
"Oh, the whip and the strap. Often we deny them something they want. But we always make it clear to them just why we are doing it."

The changed status of wife and mother is at once cause and effect of changing family relationships. It has been only recently that a writer of syndicated newspaper features could, with impunity, advise a farm wife to "calmly announce to your husband that unless you can have a maid on your farm next year you will refuse to do any canning, gardening or chicken raising. Plan to cut down your work at least 40 percent."[11]

But women have been enfranchised. The law now recognizes their property rights. Educational levels have been raised. They participate freely in the organizational life of communities. Their roles with respect to household and farm work have changed. Some of these things have tended to give them a social status both within and out of the home that is more than ever like that of their husbands. The relationship of the mother's position to patterns within the family has been recognized in the foregoing discussion. It is related also to the role of the family in the community, a role that has changed since the days of pioneering. Mrs. X does not operate the family car but 43 percent of the farm wives in neighboring homes are licensed drivers.

The organized participation of the family centers largely in the church. All members of the family attend church and Sunday School regularly. They have not joined the Grange. Mr. X belongs to the Farm Bureau and Dairymen's League. Mrs. X is a faithful attendant at meetings of the Missionary Society. They have not been to a moving picture since they were married. Mr. X goes to the village or a nearby city about twice a week and Mrs. X not more than once a month. Entertainments take them out not more than once a month. Mrs. X visits with neighbors on the telephone from one to three times a day. Mr. X confesses, however, that he probably does just as much visiting if not more than his wife. He meets neighbors on the road and stops to chat with them or he exchanges work with his neighbors and gossips while he works. Once a year they have friends from the city who come to spend a week or a few days with them. Mr. X has been on the church board; he has been a church steward and has been on the church building committee. Mrs. X teaches Sunday School and is vice-

[11] Garry C. Meyers, *The Parent Problem.* (Syndicated newspaper column, Dec. 11, 1936.)

president of the Missionary Society. Mr. X is now collector and school trustee of the school district.

The proportion of husbands who did not participate in any organization (15 percent) is greater than for wives (12 percent). Similarly husbands attended an average of only 2.6 organizations while the wives attended an average of 3.2 organizations. This is a reversal of pioneer customs. Commercial recreation is infrequent. The husbands and wives attend moving pictures only about once in two months. Town contacts were twice as frequent for husbands as for wives because of business trips.

The local leadership of their communities comes from these families. Three-fourths of the men and over half of the women had a record of some past or present office in an organization. This changed community role of the farm wife, however, has not yet removed her from the family group enough to threaten the integration of the home, for wives with leadership records were found in those homes in which there are many shared activities. Furthermore, this activity in organizations often helps mothers to cope with current change. "Parent-teacher associations, child study clubs, and similar organizations render an important service in establishing norms of child behavior and strengthening the morale of the associated members in their efforts to maintain them."[12]

Although the strength of the great family as a rural social control has weakened materially we find the kinship group still important. One-fourth of the households studied included some relative whose only home was with the family group.

In many respects, then, the portrait of today's farm family in Central New York could well hang on the same wall with needlework samplers and framed hair flowers. In other respects, it would not be out of place in a modern mural. It is a modification of old patterns, a partial acceptance of new patterns. It is smaller than the pioneer family, yet it is still among our chief sources of population increase. The rural social organization of the area is no longer familistic, but it is at least "semi-familistic." [13] The roles of parent and child are less fixed in the mores. There is a definite heritage of paternal dominance, but the outlines of the heritage become progressively more dim. Obedience and subjection of children stand forth still as parental goals but with less and less filial recognition. Specialization and education have affected the division of labor, but shared work and shared leisure are still formative of the family pattern. Propinquity continues to foster solidarity, resisting the centrifugal effects of urbaniza-

[12] Dwight Sanderson, "Trends in Family Life Today," *Journal of Home Economics*, 24, April 1932, 317.

[13] Dwight Sanderson, *op. cit., Journal of Home Economics*, 29, April 1937, 223.

tion. There has been definite democratization in the changes of role and status. That is evidenced particularly in the joint executive function of mother and father. The rate at which this change occurs accelerates with the advance of business efficiency and industrialization in agriculture. Both rate and direction of future change in the farm family pattern are, therefore, quite as likely to depend upon larger economic and social influences affecting agriculture as upon the dictation of tradition.

THE ITALIAN FAMILY IN THE UNITED STATES

Paul J. Campisi

Adapted from "Ethnic Family Patterns: The Italian Family in the United States," *The American Journal of Sociology*, 53 (1948), 443-49, by permission of The University of Chicago Press.

The changes in the Italian family in America can be visualized in terms of a continuum which ranges from an unacculturated Old World type to a highly acculturated and urbanized American type of family. This transformation can be understood by an analysis of three types of families which have characterized Italian family living in America: the Old World peasant Italian family which existed at the time of the mass migration from Italy (1890-1910) and which can be placed at the unacculturated end of the continuum; the first-generation Italian family in America, which at the beginning of contact with American culture was much like the first but which changed and continues to change increasingly so that it occupies a position somewhere between the two extremes; and, finally, the second-generation Italian family which represents a cross-fertilization of the first-generation Italian family and the American contemporary urban family, with the trend being in the direction of the American type. Consequently, the position this family assumes is near the American-urban end of the continuum.

Since there are significant differences between the northern Italian and southern Italian families and since there are even greater differences between peasant, middle-class, and upper-class families, it seems expedient to single out one type of family for discussion and analysis, namely, the southern Italian peasant family. During the period of mass migration from Italy the bulk of the immigrants were from southern Italy (including Sicily).[1] These immigrants came mostly from small-village backgrounds as

[1] During the decade of 1900-1910, of the 2,045,877 Italians who came to America, the majority were from southern Italy.

peasant farmers, peasant workers, or simple artisans, and as such they brought with them a southern Italian folk-peasant culture. It is this type of background which the majority of Italian families in America have today.[2]

This paper cannot possibly present an adequate analysis of all the important changes observed in the Italian family. Therefore, a simple tabular form (see Table 1) is used to display the most important details.

The Southern Italian Peasant Family in America

At the time of the great population movement from Italy to America, beginning at the end of the nineteenth century, the southern Italian peasant family was a folk societal family. One of the chief characteristics of the folk society is that its culture is highly integrated, the separate parts forming a strongly geared and functionally meaningful whole.[3] This intimate interconnection between the various parts of a folk culture indicates that it would be artificial and fruitless to attempt to isolate, even for the sake of study and analysis, any one part, such as the family, and to proceed to discuss that as a discrete and distinct entity. All the characteristics of the Old World Italian peasant family are intimately tied in with such institutions and practices as religion, the planting and gathering of food, the celebrations of feasts and holidays, the education of the children, the treatment of the sick, the protection of the person, and with all other aspects of small-village folk culture. In the final analysis Old World peasant-family life meant small-village life, and the two were inseparable aspects of a coercive folk-peasant culture. This fact sharply distinguishes the Old World peasant family from the first- and second-generation families in America.

The First-Generation Southern Italian Peasant Family in America

By the first-generation Italian family is simply meant that organization of parents and offspring wherein both parents are of foreign birth and wherein an attempt is made to perpetuate an Italian way of life in the transplanted household. This is a family in transition, still struggling against great odds to keep alive those customs and traditions which were

[2] The observations in this paper are based on the literature in the field, on my own specific research in America on the acculturation of Italians, and, finally, on personal impressions and conclusions as a participant observer. A visit to southern Italy and Sicily three years ago gave me an opportunity to come in contact with the Old World peasant-type family. While this type of family has changed considerably from the time of the mass migration to America, enough structural and functional family lags exist to make the reconstruction of it in this paper reasonably valid.

[3] See Robert Redfield, "The Folk Society," *American Journal of Sociology*, LII, 1947, 293-308.

sacred in the Old World culture. As a result of many internal and external pressures which have cut it off from its Old World foundations, the first-generation family is marked by considerable confusion, conflict, and disorganization. The uncertain and precarious position of the first-generation Italian family today is further aggravated by the loss of that strong family and community culture which had been such an indispensable part of the Old World peasant family. It is this loss in the first-generation family which pushes it away from the unacculturated end of the continuum to a position somewhere in the middle.[4]

The Second-Generation Southern Italian Family in America

This refers to that organization of parents and offspring wherein both the parents are native American-born but have foreign-born parents who attempted to transmit to them an Italian way of life in the original first-generation family in America.

Among the significant characteristics of this type of family is the orientation which the American-born parents make to the American culture. This adjustment tends to take three forms. One is that of complete abandonment of the Old World way of life. The individual changes his Italian name, moves away from the Italian neighborhood and in some cases from the community, and has little to do with his foreign-born parents and relatives.[5] The ideal is to become acculturated in as short a time as possible. This type of second-generation Italian generally passes for an American family and is rare. A second form of second-generation Italian family is a marginal one. In this type there is a seriously felt need to become Americanized and hence to shape the structure and functions of the family in accordance with the contemporary urban American type of family. The parental way of life is not wholly repudiated, although there is some degree of rejection. This family is likely to move out of the Italian neighborhood and to communicate less and less with first-generation Italians, but the bond with the first-generation family is not broken completely. Intimate communication is maintained with the parental household, and the relationships with the parents as well as with immigrant relatives are affectionate and understanding. A third form which the second-generation family takes is of orientation inward toward an Italian way of life. This type of family generally prefers to remain in the Italian neighborhood,

---

[4] For an excellent analysis of the importance of a strong family and community culture see Margaret Park Redfield, "The American Family: Consensus and Freedom," pp. 23-37 above.

[5] See Carlo Sforza, *The Real Italians,* New York, Columbia University Press, 1942, for an interesting account of Italian-Americans who change their names.

close to the parental home. Its interaction with the non-Italian world is at a minimum, and its interests are tied up with those of the Italian community. Of the three, the second type is the most representative second-generation Italian family in America. This is the family depicted in Table 1.

Table 1 reveals the movement of the first- and second-generation Italian families away from the Old World peasant pattern and toward the contemporary American family type. In this persistent and continuous process of acculturation there are three stages: (1) the initial-contact stage, (2) the conflict stage, and (3) the accommodation stage.

The Initial-Contact Stage

In the first decade of Italian living in America the structure of the Old World family is still fairly well intact, but pressures from within and outside the family are beginning to crack, albeit imperceptibly, the Old World peasant pattern. Producing this incipient distortion are the following: the very act of physical separation from the parental family and village culture; the necessity to work and operate with a somewhat strange and foreign body of household tools, equipment, gadgets, furniture, cooking utensils, and other physical objects, in addition to making an adjustment to a different physical environment, including climate, urban ecological conditions, and tenement living arrangements; the birth of children and the increasing contact with American medical practices regarding child care; the necessity to work for wages at unfamiliar tasks, a new experience for the peasant farmer; the attendance of Italian children in American parochial and public schools; the informal interaction of the children with the settlement house, the church associations, the neighborhood clubs, the neighborhood gang, and other organizations; the continuing residence in America and increasing period of isolation from the Old World; the acceptance of work by the housewife outside the home for wages; the increasing recognition by both parents and children that the Italian way of life in the American community means low status, social and economic discrimination, and prejudice; and the increasing pressure by American legal, educational, political, and economic institutions for the Americanization of the foreigner.

Nonetheless, the first-generation Italian family in this phase is a highly integrated one, as in the Old World. The demands of the American community are not seriously felt in the insulated Italian colony, and the children are too young seriously to articulate their newly acquired needs and wishes. The Italian family is stabilized by the strong drive to return to Italy.

TABLE 1.   Differences between the Southern Italian Peasant Family in Italy and the First- and Second-Generation Italian Family in America

Southern Italian Peasant Family in Italy	First-Generation Southern Italian Family in America	Second-Generation Southern Italian Family in America
A. *General characteristics:*		
1. Patriarchal	Fictitiously patriarchal	Tends to be democratic
2. Folk-peasant	Quasi-urban	Urban and modern
3. Well integrated	Disorganized and in conflict	Variable, depending on the particular family situation
4. Stationary	Mobile	High degree of mobility
5. Active community life	Inactive in the American community but somewhat active in the Italian neighborhood	Inactive in the Italian neighborhood, but increasingly active in American community
6. Emphasis on the sacred	Emphasis on the sacred is weakened	Emphasis on the secular
7. Home and land owned by family	In the small city the home may be owned, but in a large city the home is usually a flat or an apartment	Ownership of home is an ideal, but many are satisfied with flat
8. Strong family and community culture	Family culture in conflict	Weakened family culture reflecting vague American situation
9. Sharing of common goals	No sharing of common goals	No sharing of common goals
10. Children live for the parents	Children live for themselves	Parents live for the children
11. Children are an economic asset	Children are an economic asset for few working years only and may be an economic liability	Children are an economic liability
12. Many family celebrations of special feasts, holidays, etc.	Few family celebrations of feasts and holidays	Christmas only family affair, with Thanksgiving being variable
13. Culture is transmitted only by the family	Italian culture is transmitted only by family, but American culture is transmitted by American institutions other than the family	American culture is transmitted by the family and by other American institutions
14. Strong in-group solidarity	Weakened in-group solidarity	Little in-group solidarity
15. Many functions: economic, recreational, religious, social affectional, and protective	Functions include semi-recreational, social, and affectional	Functions reduced to affectional, in the main

TABLE 1. (Continued)

Southern Italian Peasant Family in Italy	First-Generation Southern Italian Family in America	Second-Generation Southern Italian Family in America
B. *Size:*		
1. Large-family system	Believe in a large-family system but cannot achieve it because of migration	Small-family system
2. Many children (10 is not unusual)	Fair number of children (10 is unusual)	Few children (10 is rare)
3. Extended kinship to godparents	Extended kinship, but god-parent relationship is weakened	No extended kinship to godparents
C. *Roles and statuses:*		
1. Father has highest status	Father loses high status, or it is fictitiously maintained	Father shares high status with mother and children; slight patriarchal survival
2. Primogeniture: eldest son has high status	Rule of primogeniture is variable; success more important than position	No primogeniture; all children tend to have equal status
3. Mother center of domestic life only and must not work for wages	Mother center of domestic life but may work for wages and belong to some clubs	Mother acknowledges domestic duties but reserves time for much social life and may work for wages
4. Father can punish children severely	Father has learned that American law forbids this	Father has learned it is poor psychology to do so
5. Family regards itself as having high status and role in the community	Family does not have high status and role in the American community but may have it in the Italian colony	Family struggles for high status and role in the American community and tends to reject high status and role in the Italian community
6. Women are educated for marriage only	Women receive some formal education as well as family education for marriage	Emphasis is on general education with reference to personality development rather than to future marriage
7. The individual is subordinate to the family	Rights of the individual increasingly recognized	The family is subordinate to the individual
8. Daughter-in-law is subservient to the husband's family	Daughter-in-law is in conflict with husband's family	Daughter-in-law is more or less independent of husband's family
9. Son is expected to work hard and contribute to family income	Son is expected to work hard and contribute to family income, but this is a seldom-realized goal	Son expected to do well in school and need not contribute to family income

TABLE 1.   (Continued)

Southern Italian Peasant Family in Italy	First-Generation Southern Italian Family in America	Second-Generation Southern Italian Family in America
D. *Interpersonal relations:*		
1. Husband and wife must not show affection in the family or in public	Husband and wife are not demonstrative in public or in the family but tolerate it in their married children	Husband and wife may be demonstrative in the family and in public
2. Boys are superior to girls	Boys are regarded as superior to girls	Boys tend to be regarded as superior to girls, but girls have high status also
3. Father is consciously feared, respected, and imitated	Father is not consciously feared or imitated but is respected	Father is not consciously feared. He may be imitated and may be admired
4. Great love for mother	Great love for mother but much ambivalence from cultural tensions	Love for mother is shared with father
5. Baby indulgently treated by all	Baby indulgently treated by all	Baby indulgently treated by all with increasing concern regarding sanitation, discipline, and sibling rivalry
E. *Marriage:*		
1. Marriage in early teens	Marriage in late teens or early twenties	Marriage in early or middle twenties
2. Selection of mate by parents	Selection of mate by individual with parental consent	Selection of mate by individual regardless of parental consent
3. Must marry someone from the same village	This is an ideal, but marriage with someone from same region (*i.e.,* province) is tolerated; very reluctant permission granted to marry outside nationality; no permission for marriage outside religion	Increasing number of marriages outside nationality and outside religion
4. Dowry rights	No dowry	No dowry
5. Marriage always involves a religious ceremony	Marriage almost always involves both a religious and a secular ceremony	Marriage usually involves both, but there is an increasing number of marriages without benefit of religious ceremony
F. *Birth and child care:*		
1. Many magical and superstitious beliefs in connection with pregnancy	Many survivals of old beliefs and superstitions	Few magical and superstitious notions in connection with pregnancy

TABLE 1. (Continued)

Southern Italian Peasant Family in Italy	First-Generation Southern Italian Family in America	Second-Generation Southern Italian Family in America
2. Delivery takes place in a special confinement room in the home; midwife assists	Delivery takes place generally in a hospital; may take place in home; family doctor displaces midwife	Delivery takes place almost always in a hospital; specialist, obstetrician, or general practitioner assists
3. Child illnesses are treated by folk remedies; local physician only in emergencies or crises	Child illnesses are treated partially by folk remedies but mostly by the family doctor	Child illnesses are treated by a pediatrician; much use of latest developments in medicine (vaccines, etc.)
4. Child is breast-fed either by the mother or by a wet nurse; weaning takes place at about end of 2d or 3d year by camouflaging the breasts	Child is breast-fed if possible; if not, it is bottle-fed; same practice with variations regarding weaning	Child is bottle-fed as soon as possible; breast-feeding is rare; no weaning problems
5. No birth control	Some birth control	Birth control is the rule
G. *Sex attitudes:*		
1. Child is allowed to go naked about the house up to the age of 5 or 6; after this there is rigid enforcement of the rule of modesty	Variable, depending on the individual family's situation	This is variable, depending on the individual family; development of modesty is much earlier than in Old World peasant family
2. Sex matters are not discussed in family	Sex matters are not discussed in family	Sex matters increasingly discussed in family but not as freely as in "old" American family
3. Adultery is severely punished by the man's taking matters into his own hands	Adultery results in divorce or separation	Adultery may result in divorce or separation
4. Chastity rule rigidly enforced by chaperonage; lack of it grounds for immediate separation on wedding night	Attempts to chaperon fail, but chastity is an expectation; lack of it is grounds for separation, but there are few cases of this kind in America	No chaperonage; chastity is expected, but lack of it may be reluctantly tolerated
5. No premarital kissing and petting are allowed	No premarital kissing and petting are allowed openly	Premarital kissing and petting are allowed openly
6. Boys and girls attend separate schools	Schools are coeducational	Schools are coeducational

TABLE 1.   (Concluded)

Southern Italian Peasant Family in Italy	First-Generation Southern Italian Family in America	Second-Generation Southern Italian Family in America
H. *Divorce and separation:*		
1. No divorce allowed	No divorce allowed, but some do divorce	Religion forbids it, but it is practiced
2. Desertion is rare	Desertion is rare	Desertion is rare
I. *Psychological aspects:*		
1. Fosters security in the individual	Fosters conflict in the individual	Fosters security with some conflict lags
2. The family provides a specific way of life; hence, there is little personal disorganization	Family is in conflict, hence cannot provide a specific way of life; yields marginal American-Italian way of life	Family reflects confused American situation, does not give individual a specific way of life, but marginality is weakened
3. Recreation is within family	Recreation is both within and outside the family	Recreation is in the main outside the family; this is variable, depending on individual family situation

The Conflict Stage

In this period the first-generation family experiences its most profound changes and is finally wrenched from its Old World foundation. It is now chiefly characterized by the conflict between two ways of life, the one American and the other Italian, and by the incompatibility of parents and children. This phase begins roughly during the second decade of living in America—specifically, when the children unhesitatingly express their acquired American expectations and attempt to transmit them in the family situation and when the parents in turn attempt to reinforce the pattern of the Old World peasant family. Conflicting definitions of various family situations threaten to destroy whatever stability the family had maintained through the first period. This is the period of great frustration and of misunderstanding between parents and children. In this undeclared state of war between two ways of life it is the parents who have the most to lose, for their complete acceptance of the American way of living means the destruction of the Old World ideal.

The first-generation Italian family is also constantly made to feel the force of external pressures coming from outside the Italian colony. It is inevitable that the family structure should crumble under the incessant hammering. Not able to draw upon a complete culture and social system to support its position, the family pattern, already weakened, now begins to change radically: the father loses his importance, the daughters acquire unheard-of independence; in short, the children press down upon the first-generation family an American way of life.

Accommodation Stage

This period begins with the realization by parents and children that the continuation of hostility, misunderstanding, and contraventive behavior can result only in complete deterioration of the family. The ambivalent attitude of the children toward the parents, of great affection, on the one hand, and hostility, on the other, now tends to be replaced by a more tolerant disposition. This stage begins when the offspring reach adulthood and marry and establish households of their own, for by this time the control by the parents is greatly lessened.

Among the many factors which operate to bring about a new stability in the family are the realization on the part of the parents that life in America is to be permanent; the adult age of the offspring; the almost complete dependence of the parents on the offspring, including use of the children as informants, interpreters, guides, and translators of the American world; recognition on the part of the parents that social and economic success can come to the offspring only as they become more and more like "old" Americans; the conscious and unconscious acculturation of the parents themselves with a consequent minimizing of many potential conflicts; the long period of isolation from the Old World which makes the small-village culture and peasant family seem less real; the decision by the parents to sacrifice certain aspects of the Old World family for the sake of retaining the affection of the children; the acknowledgment by the children that the first-generation family is a truncated one and that complete repudiation of the parents would leave them completely isolated; the success of the first-generation family in instilling in the offspring respect and affection for the parents; and the gradual understanding by the children that successful interaction with the American world is possible by accepting marginal roles and that complete denial of the Old World family is unnecessary.

The accommodation between parents and offspring permits the second-generation Italians to orient themselves increasingly toward an American way of life. The second-generation household, therefore, tends to pattern itself after the contemporary urban American family. Considerable intermarriage, the advanced age of the parents, the loosening of ties with the Italian neighborhood, and the development of intimate relationships with non-Italians make the transition of the second-generation family comparatively easy.

IS A NEW FAMILY FORM EMERGING IN THE URBAN FRINGE?

E. Gartly Jaco and Ivan Belknap | Adapted from the *American Sociological Review*, 18:5 (1953), 551-57.

One modern school of family sociologists, represented chiefly by Zimmerman, considers the present urban American family an alarming instance of disintegration in the familial process. This disintegration is believed to have reached such extremes that the family can no longer adequately discharge vital functions such as reproduction and socialization. Writers of this school imply that American, and Western civilization generally, faces the dilemma of social collapse through failure of the family functions, or a return to some form of the large rural or semi-rural family system.[1]

Another school considers the modern urban family to be making a reasonably satisfactory adjustment to population density, secondary relations, and diversity of urban institutions.[2] Most demographers apparently agree that basic population trends have been in harmony with the assumptions of this latter school.[3] Urban sociologists also have generally accepted the present urban "companionship," or small family as necessarily

[1] See C. C. Zimmerman, *Family and Civilization,* New York: Harper, 1947, who concludes that "unless some unforeseen renaissance occurs, the family system will continue headlong its present trend toward nihilism" (p. 808). Also E. Schmiedeler, *An Introductory Study of the Family,* New York: Appleton-Century-Crofts, 1947 (revised). The dilemma consists in the fact that in neither case can the present structure of social organization be maintained.

[2] E. W. Burgess and H. J. Locke, *The Family,* New York: American Book Company, 1945; and J. K. Folsom, *The Family and Democratic Society,* New York: John Wiley, 1943. For a discussion of both schools, see R. F. Winch, *The Modern Family,* New York: Henry Holt, 1952, pp. 472-474, and W. Waller and R. Hill, *The Family,* New York: Dryden Press, 1952, pp. 17-20.

[3] The recent jump in fertility and marriages is generally regarded as a temporary phenomenon, holding that as knowledge and practice of contraception and other effects of urban life continue to reach the high-fertility segments of U.S. population, the downward trend will be resumed. See F. W. Notestein, "The Facts of Life," *The Atlantic Monthly,* 177 (June, 1946), pp. 75-83, and his "The Population of the World in the Year 2000," *Journal of American Statistical Association,* 45 (September, 1950), pp. 335-349. C. V. Kiser's review of P. K. Whelpton's *Cohort Fertility* supports this contention in general, though less emphatically. See "Fertility Trends and Differentials in the United States," *Journal of American Statistical Association,* 47 (March, 1952), pp. 25-48.

typical of urban communities.[4] Burgess and Locke consider this type of family as one "which seeks to combine the values of both the old rural and the modern urban situations." [5]

It is the purpose of this paper to point out certain trends that may be leading toward the emergence of a variant type of urban family which may be able to maintain sufficient fertility and integration to satisfy the Zimmerman requisites and yet function adequately in the urban community. This variant type of urban family seems to be locating in the urban "fringe," [6] as a product of changing ecological and demographic forces in metropolitan regions, and as a new functional adjustment of the family to the urban way of life. Because this family apparently represents primarily an adjustment to or a product of the peripheral metropolitan ecological area, it might be tentatively termed the "fringe family." [7]

This new family form may be only temporarily connected, however, with the urban fringe. The new family type should not be construed as being permanently or intrinsically bound up with the "fringe." The "fringe family" label is offered only as a provisional, heuristic term which implies that this family form is initially a product of contemporary urban fringe development. This form may eventually spread to other ecological areas. Further research is needed, therefore, before a precise term pertaining to the social structure of this new family type can be given. . . .

[4] See L. Wirth, "Urbanism as a Way of Life," *American Journal of Sociology,* 44 (July, 1938), pp. 1-24; N. Anderson and E. C. Lindeman, *Urban Sociology,* New York: Knopf, 1928; B. A. McClenahan, *The Changing Urban Neighborhood,* Los Angeles: University of Southern California Press, 1929; E. R. Mowrer, *Family Disorganization,* Chicago: University of Chicago Press, 1927; W. F. Ogburn, *Social Characteristics of Cities,* Chicago: 1937; S. A. Queen and L. P. Thomas, *The City,* New York: McGraw-Hill, 1939; C. F. Ware, *Greenwich Village,* Boston: Houghton-Mifflin, 1935; H. W. Zorbaugh, *The Gold Coast and The Slum,* Chicago: University of Chicago Press, 1929. This typical family is described as (1) small in size; (2) equalitarian in member relations; (3) individualistic in terms of family formation and functioning; (4) tending to be located in multiple-family dwellings; and (5) lacking many of the functions of the rural family.

[5] Burgess and Locke, *op. cit.,* p. 143.

[6] The fringe herein considered includes suburbs, satellite cities, and any other territory located immediately outside central cities whose labor force is engaged in non-farm activities.

[7] This is not to take issue with either school. The type of family Zimmerman holds to be basic we regard as functional in rural areas. The companionship type of family is associated with *central cities.* However, with the current growth of fringe areas around central cities, we feel that a new urban family form is developing within this fringe. If this is the case, it would be a serious oversimplification to continue to regard the "companionship" family as typical of the entire urban community. Some urban sociologists are suggesting that the urban family forms need restudy. See, for example, Svend Riemer, *The Modern City,* New York: Prentice-Hall, 1952, pp. 255-259.

Trends Toward the Fringe Family

The following trends in American society do not prove conclusively the existence and operation of a fringe family. They do, however, indicate a very strong probability of the development of this form. Demographic, ecological, labor force, and stratification data justify the inference that such concerted forces *must* be affecting and sustained by the family system existing in such an area.

Demographic. Birth order is obviously an index of increasing or decreasing family size. The boom in first order of births during the war years has evidently not been sustained recently.[8] However, an analysis of higher birth orders between 1942 and 1949 (comparative percentages computed by dividing the number of each birth order by total live births) reveals some noteworthy changes (Regional Summaries appear in Table 1.) [9] Every state reporting has an increase in third order of births in 1949 over 1942. For fourth order of birth, 43 out of the 47 states reporting had equal or greater percentages for 1949 over 1942; the four states having less in 1949 were Arkansas, New Mexico, Tennessee, and West Virginia, the latter having the greatest disparity of only one per cent. Apparently the fourth order of birth was the peak increase in 1949 over 1942, since only 17 states had equal or higher percentages of fifth orders in 1949. For the sixth order of birth, only nine states showed an increase in 1949. In sum, between 1942 and 1949, all states had higher third order of births in 1949, and 43 out of 47 for fourth order. Seventeen states showed a consistent gain in third through fifth orders in 1949 over 1942: Louisiana, Maryland, New Hampshire, Oregon, South Carolina, Texas, California, Connecticut, Idaho, Illinois, Michigan, Minnesota, New Jersey, New York, Ohio, Washington, and Wisconsin. The first six states were consistently higher in percentages from the third through the fifth birth orders. For the U.S. generally, Whelpton's cohort study substantiates our results in his 1925 cohort's experience.[10]

Kiser has pointed out the increase in fertility ratios since 1940 which are

proportionately heaviest among groups previously characterized by lowest fertility. Thus the percentage increases have been larger in the Northeast than in

[8] Kiser, *op. cit.,* p. 25, among others. Demographers have tended to concentrate fertility analysis to either first or second or "very high" birth orders, ignoring the middle range which is significant to our thesis. See, for example, F. W. Notestein, "The Population of the World in the Year 2000," pp. 337 ff.

[9] Massachusetts is omitted from the New England region because of an absence of birth order data.

[10] Kiser, *op. cit.,* Fig. 1, p. 30.

TABLE 1. Percentages of 3rd, 4th, 5th, and 6th Birth Orders in U.S. by Regions, 1942 and 1949*

Region	Birth Order							
	Third		Fourth		Fifth		Sixth	
	1942	1949	1942	1949	1942	1949	1942	1949
New England	12.3	16.1	6.2	7.4	3.4	3.5	2.0	1.9
Middle Atlantic	12.0	15.1	5.9	6.6	3.2	3.1	1.9	1.6
East North Central	12.9	16.5	6.7	7.8	3.6	3.8	2.2	2.1
West North Central	13.7	16.8	7.6	8.4	4.5	4.3	2.8	2.4
South Atlantic	13.2	15.4	8.3	8.6	5.6	5.3	4.0	3.7
East South Central	13.3	14.8	9.0	8.9	6.3	5.9	4.7	4.2
West South Central	12.5	16.1	7.2	8.9	4.6	5.4	3.1	3.5
Mountain	14.4	17.2	8.3	9.1	5.2	4.9	3.3	2.9
Pacific	11.9	17.1	5.5	7.3	2.8	3.1	1.6	1.6

* Compiled from *Vital Statistics of the United States,* U. S. Government Printing Office, Washington, D. C., Part II, 1942 and 1949.

the South, larger among whites than nonwhites, larger among urban than rural-farm populations, and probably larger in the "upper" than in the "lower" socioeconomic classes.[11]

Further data of a demographic order are suggested by Firey's finding of an excess of children under 10 years of age in the fringe of Flint, Michigan, in 1945.[12] Scaff's study of a California suburb showed that the "commuting population adds young families and comparatively larger families to this community." Furthermore, "without question, the presence of the commuter group in the community introduces younger adults and children and helps to balance an age distribution that is otherwise heavily weighted by elderly people." [13]

While the census area of rural non-farm is not strictly identical to that of the urban fringe area, it does include the latter and offers a crude index of trends in the fringe. In 1949, for the first time, the number of children under 5 years of age per 1,000 women was higher in the rural non-farm area than in both urban and rural farm areas, showing a steadier and higher rise than for urban areas, while in 1949, the rural farm amount dropped.[14] Further, the per cent distribution for 4 persons per household

[11] *Ibid.,* p. 38.

[12] W. I. Firey, *Social Aspects to Land Use Planning in the Country-City Fringe,* Michigan State College Agricultural Experiment Station Special Bulletin No. 339, East Lansing, June, 1946, pp. 17-18.

[13] A. Scaff, "The Effect of Commuting on Participation in Community Organizations," *American Sociological Review,* 17 (April, 1952), p. 217.

[14] Bureau of the Census, *Statistical Abstract of the United States,* Washington, D.C., 1951, Table 24, p. 20.

in 1950 was highest for rural non-farm areas as compared to both urban and rural farm areas.[15]

With a spurt in higher birth orders and excess numbers of young children associated with a jump in fringe in-migration, a trend toward increased family size in the fringe can be deduced.

Ecological. Perhaps the most significant single index of the increase in the fringe population is the comparative rates of growth between 1940 and 1950: central cities grew 13 per cent; the hinterland increased 5.7 per cent; but the outlying parts of central cities showed a jump of 34.7 per cent.[16]

Home ownership is directly related to large families, both in urban and rural areas.[17] Census figures show that home ownership has increased 53.9 per cent between 1940 and 1950,[18] a change which may stimulate or be stimulated by larger families. The rate of home ownership in rural non-farm areas from 1930 through 1950 has been closer to that of the rural farm than to that of the urban area.[19]

An increase in the processes of concentration of population and decentralization of services is mentioned by Hauser[20] and Blumenfield.[21] Therefore, the growth of the urban fringe need not represent a deconcentration of urban population from central cities, but may indicate rather an expansion of urban population into broader territory. Many fringe areas exist apart from central city political boundaries only for a brief time until they become incorporated into the central city. However, the data accumulating on the ecology of the urban community do indicate a continuing increase in the distribution of the United States population into the fringe areas.

Labor Force. Whetten and Mitchell's study of a Connecticut suburb showed that "white-collar" workers predominated among occupational groups.[22] However, movement to the fringe is not confined to the middle and upper economic levels. The spread of the concept of the guaranteed

15 *Ibid.*, Table 33, p. 25.

16 From P. K. Hatt and A. J. Reiss, Jr. (editors), *Reader in Urban Sociology,* Glencoe, Illinois, Free Press, 1951, p. 68.

17 M. Parten and R. J. Reeves, "Size and Composition of American Families," *American Sociological Review*, 2 (October, 1937), p. 664.

18 Bureau of the Census, *op. cit.*, Table 866, p. 723.

19 *Ibid.*, same table; also Table 871, p. 725.

20 P. M. Hauser, "The Changing Population Pattern of the Modern City," in Hatt and Reiss, *op. cit.*, pp. 165-182.

21 H. Blumenfield, "On the Growth of Metropolitan Areas," *Social Forces*, 28 (October, 1949), pp. 59-64.

22 N. L. Whetten and D. Mitchell, "Migration from a Connecticut Suburban Town, 1930-1937," *American Sociological Review*, 4 (April, 1939), pp. 173-179.

annual wage in American industry, taken with the increase in fringe population, indicates a drift to the fringe and to single-family dwellings by the so-called "blue-collar" workers.[23] The bearing of the guaranteed wage on the possibility of home ownership in this latter group is clear enough. Moreover, by "living out" and "working in," as Liepmann puts it, the blue-collar worker may paradoxically increase his family stability as he increases his job mobility. By living in the fringe, that is, the worker ceases to be tied to a particular factory or combination residential-occupational area in the city. He is free to change his jobs without the disruption of family stability caused by residential relocation.[24]

With the increasing employment of women, single as well as married, many women who eventually marry and become mothers return to their early forms of work after their children reach a more independent age (Table 2). Increased employment of married women is conducive to fringe living. Being employed in an area distant from residence minimizes

TABLE 2. Percentage of Women in U. S. Labor Force, by Marital and Familial Status, 1949*

Status	Per Cent in Labor Force
All women in labor force	22.5
Women without children	28.7
With children under 6	10.0
With children some 6-11 only	24.7
With children some 12-17 only	31.3

* Adapted from A. J. Jaffe and C. D. Stewart, *Manpower Resources and Utilization*, New York: Wiley, 1951, Table 4, p. 133.

the conflict between the mother's familial and non-familial roles. It seems highly probable that the increase of employment of women in the higher age groups is partly an index of employment of fringe family mothers.[25]

Liepmann has pointed out that fringe families encourage and even require "secondary earners," particularly in lower economic groups.[26] There

[23] For a discussion of this possibility, see A. J. Jaffe, "Population Trends and City Growth," in Hatt and Reiss, *op. cit.*, pp. 188-189.

[24] K. K. Liepmann, *The Journey to Work,* New York: Oxford University Press, 1944, pp. 10-12. Logically, this should apply to the white-collar worker as well.

[25] A. J. Jaffe and C. D. Stewart, *Manpower Resources and Utilization,* New York: Wiley, 1951, p. 133, show that the rate of married women in the labor force increases as the age of their children increases.

[26] In this process the family assures itself some safety through economic diversification, while at the same time it develops in what is probably an equalitarian direction. (Liepmann, *op. cit.*, pp. 19-25). Liepmann states further, ". . . The family as a whole benefits from the varied employment of its members. It is economically safer for the

are equally cogent reasons for the employment of higher status fringe mothers in view of the increasing costs of educating children, intensified by the inflation which has persisted since World War II. Lower age at marriage makes such employment of mothers consistent with a higher birth rate, while their superior education permits them to seek employment affording an economic surplus after they have paid for maids, kindergartens, and other maternal surrogates for the youngest children.

Social Stratification.   Scaff holds that "education and membership in a profession become a badge of acceptance" in suburbs.[27] Coupled with its matricentric orientation, social stratification in the fringe is probably more distinct and overt than is apparent in the central city, where the "elite" is composed of professionals and their wives or widows, while industrial workers occupy lower social strata. If so, then social cleavages in the fringe may be disparate, and fixed along occupational lines.

Furthermore, if industrial workers make up the lower social strata and participate less in the suburban community, as Scaff's study indicates, then commuting by such a population may be viewed as "escape from status." That is by "working in and living out," a worker may absent himself from his inferior social position during the working day. This would be especially true if the wife and older children are also employed. Indeed, such flexibility may make more tolerable the occupancy of a lower social position in the fringe.

The Structure of the Fringe Family

If one follows the clues suggested by current urban population and economic trends in the light of what is already known about the modern suburban family a number of inferences on the structure of the fringe family becomes possible. Some of the more important of these inferences involve the probable fringe family roles and the integration of this family at various stages of the institutional life cycle.

When the family selects the fringe for the sake of rearing children, this selection can be regarded as involving emphasis on the reproductive-socializing roles of father and mother. There is evidence that some movements to suburbs are carried out to improve the educational and recreational life of children.[28] This improvement is a reciprocal affair, since in its

---

family not to have all of its eggs in one basket, i.e., not to depend on one industry which may decline while others prosper. Domestic life, moreover, is enriched by a variety of occupational interests among the family," p. 24.

[27] Scaff, *op. cit.*, p. 220.

[28] Arthur Jones found in a study of a Philadelphia suburb that 80 per cent of its families had moved there to give their children better educational and recreational

very nature it involves not only a greater control over the children's environment but a strengthening of the significance of the parental roles and the associated roles of the siblings. This strengthening of parental roles has already been examined for the mother by Mowrer, and by Burgess and Locke in their studies, of the matricentric suburban family.[29] With the increasing employment of mothers, the shorter work day and week, and the spread of relatively higher incomes among many employed classifications, the father will come to play a more prominent role in the family and the community than was possible either in the companionship or suburban matricentric family. In general it seems likely that all the family roles in the fringe family may be enhanced by proximity of members, and by mutual functional significance.

The Burgess viewpoint of family sociology, with certain qualifications, holds that the compansionship family is more in line with other urban social institutions. In view of the powerful stress required by this form on the intrinsic husband-wife relationship rather than on the father-mother bond, the companionship family may represent less of an adaptation than a negation of the family's important functions. There is an implication in this position that as the family "gives in" and loses its historical functions, it becomes better adjusted to the urban environment.[30] Hence the fringe type of family offers at least a compromise between familistic and companionship forms while maintaining at least an apparent emerging adaptation to the urban environment.

Implicit in the strengthening of the parental bonds in the fringe family is an increased control over the courtship process, and perhaps solidarity in the old age family roles.[31] In the central city, anonymity and diversity of interaction minimizes parental control over children in the realm of courtship and dating. In the fringe, the courtship process can be confined in some degree to peer groups selected by parents in tacit agreement with

---

opportunities. (*Cheltenham Township*, Philadelphia: University of Pennsylvania Press, 1940, pp. 51-52.)

Strictly speaking, the parental roles of mother and father represent the center of the family functions, rather than the husband-wife roles. The central-city companionship family stresses the latter; the fringe family the former.

[29] E. R. Mowrer, *The Family*, Chicago: University of Chicago Press, 1932; Burgess and Locke, *op. cit.*, pp. 131-134.

[30] See Zimmerman, *op. cit.*, Ch. 2.

[31] It seems very likely that the extension of one's family of procreation into some supervision and control of his children's courtship process may give greater integration to his family of old age (gerontation). This would be an important development in view of the present trend toward an aging population, and the present isolation of grandparents in the terminal segment of the urban family life cycle.

other parents of like status. This represents a compromise between parental mate choice for the children, and the theoretical free choice implied in the dating pattern of the urban youth culture. This actuarial kind of control over the courtship process by fringe parents obviously gives greater continuity to the family process as experienced by both parents and children. This continuity, together with the heightened significance of the member relationships, may be one explanation of the greater number of children in the fringe.

Summary and Discussion

Burgess and Locke list six long-time family-related trends which have been disrupted by the recent war and speculate about their continuation after the war. These are: (1) The declining birth rate; (2) The consequent smaller size of the family; (3) The increase in proportion of the married to those of marriageable age; (4) The decrease in the age at marriage; (5) The increase in the proportion of all women, and of married women gainfully employed; (6) The decline in the historic functions of the family—economic, educational, recreational, religious, and protective.[32] The apparent assumption is that, should these trends be only temporarily disrupted by the recent war, and should they continue as in the past, the companionship family will become institutionalized in the United States.

Trends pointed out in the preceding discussion, however, indicate that particularly for the rapidly increasing U.S. fringe area, a somewhat different picture is appearing: (1) Sustained fertility through higher orders of birth; (2) A consequent increase in the size of the fringe family; (3) Marriage rates for males higher in the rural non-farm areas than in both urban and rural farm areas; (4) Decrease in the age at marriage continuing; (5) Employment of both single and married women increasing, particularly for the higher age groups and with mothers of children from 12 to 17 years of age; (6) The historic functions of the family seemingly better retained in the fringe—the economic, with employment of mothers as secondary workers; the educational, in the selection of "better" schools for children; the recreational, in the encouragement of participation of children in selected peer groups and social sets; the religious, in belonging to and supporting the "right" churches, and the protective, in addition to the preceding, in providing the best care and rearing practices of medical and mental science.

There is some indication, therefore, that the interruption of the long-term trends listed above by Burgess and Locke may have become sustained in the fringe family. We can say definitely, at least, that this inter-

[32] Burgess and Locke, *op. cit.,* p. 750.

ruption has been associated with an enormous increase in the fringe population between 1940 and 1950 and that the concept of the fringe family may serve sociology as one useful research hypothesis for the analysis, in a relatively unexplored area, of the demographic, ecological, community structure and working force trends we have mentioned.

If a family form of the type suggested here is beginning to appear in the urban fringe, several current postulates in the social-psychology of personality development will also have to be reconsidered.[33] To urban sociological study, the presence and operation of this family will mean new possibilities and problems if the present trends continue toward fringe expansion of population, decentralization and subsequent relocation of industry and services in the fringe.

For the determination of the structure and function of the fringe family and the community processes associated with it, nine hypotheses are here suggested for further research: [34]

(1) More Protestant than Catholic or Jewish families appear to live in the fringe. If so, Protestant fertility may be rising, Catholic fertility declining, and Jewish fertility destined to continue at a low rate.

(2) More whites than non-whites live in the fringe. If this is true, white fertility may rise, Negro fertility decline even further.

(3) Social stratification may be more fixed and disparate in the fringe than in the central city. Analysis of stratification in the fringe should show new class criteria, and should suggest answers to such questions as whether social classes are becoming more or less numerous and rigid in the United States.

(4) The inhabitants of fringe areas are experimenting with new forms of age and sex social organization. New perspectives on the urban cultural life cycle for childhood, youth, maturity and old age for both sexes should appear in systematic studies of the fringe.

(5) In the fringe the kinship system is assuming a more prominent function as the basis of status. Is the strengthening of kinship in this sense making the family as important as occupation in determining status in the urban community? If occupation is still maintaining major importance as a status-basis in the fringe, is the kinship system becoming more important in maintaining occupational lines?

(6) A strengthening of sibling as well as parental family roles occurs in the fringe family compared to the central city family. This changed significance of the sibling roles should extend the range of kinship association among age peers and modify current urban voluntary association practices.

[33] Particularly those theories centering on the consequences to the developing child of membership in a family unit which is relatively small in numbers, feeble in social extent and power, and self-centered. See for a sociological treatment of these theories, Bossard, J. H. S., *The Sociology of Child Development*, New York: Harper, 1948, Chapter III.

[34] Each of these hypotheses may have corollaries which can be deduced in the process of setting up the research. Each of them must also be comparatively primitive in view of the current status of data on the fringe.

(7) When contrasted to the central city, the fringe presents the following differential demography: higher fertility, larger families, more marriage, more children, greater number in the labor force, more home ownership, greater fluidity, less mobility, lower mortality, more aged persons, lower age at marriage.

(8) There is a higher rate of family participation in social institutions in the fringe than in the central city. Re-alignment of the family with urban institutions has tremendous significance for the study of contemporary social organization.

(9) More parental control of marriage occurs in the fringe than in the central city. The influence of parents on courtship in the fringe may be such as to bring about new types of mate selection and courtship patterns differing from those associated with the urban companionship family.

The foregoing propositions are not meant to be exhaustive. It seems certain, however, in view of the evidence now at hand, that the verification of a few of these hypotheses should begin a new chapter in urban family sociology. Moreover, this verification will have decided implications for demography, ecology, and social psychology.

7. PARENTHOOD AND SOCIALIZATION

In preceding chapters we were concerned with the family as a universal societal structure with its cultural and subcultural variations. We now begin an analysis of the American family in terms of those functions that are most characteristically performed within the family: reproduction and socialization. To provide a background for this presentation, the present chapter begins with some statistical data on the reproductive performance of the American population and certain of its segments.

One of the major beliefs in the psychology of personality is that the way in which a child's personality develops is determined to some degree by the way in which his parents, especially his mother, interact with him. (There is considerable variation in the *amount* of influence which maternal behavior is believed to exert on the child's personality.) The more educated strata of American society have long rejected as an old wives' tale the belief that during pregnancy mothers can "mark" the children they are carrying. Yet evidence consonant with such a formulation comes from Ferreira's study in which he follows up some work of Sontag and draws the tentative conclusion that the effect of the mother's attitudes on her child's behavior "has its zero hour before birth." Whereas much of the clinical literature emphasizes that the mother all too frequently engenders a neurosis in the child, Spitz concludes that the child who receives no "mothering" is much less privileged than the child who receives normal maternal care.

This chapter concludes with the classic formulation of G. H. Mead on the development of the self by means of play, the game, the significant gesture, and the generalized other.

PATTERNS OF FERTILITY IN THE UNITED STATES

The Editors

I. Trends in Fertility

During the past few centuries the occidental family has undergone many remarkable changes in structure and function. The most conspicuous of these has been the reduction in the size of the nuclear family. It has been estimated that at the close of the eighteenth century the average American wife bore more than eight children during her lifetime. In the depression decade of the 1930s the average was less than two.[1] Since the years of the depression and especially since World War II the birth rate has risen sharply. Thus the overall picture shows a gradual decline of great magnitude running for more than a century to the bottom of the depression followed by a considerable upswing, especially since 1945.

Figure 1 depicts the downward sweep in American fertility since 1800 and the upswing since the depression. The figure shows that correlated with the marked decrease in fertility rates has been a corresponding reduction in the proportion of rural people in the total population. This correlation will be discussed subsequently. It should be noted that the reduction in fertility rates underlies the decline in the average size of the American nuclear family. The declining fertility rates and the consequent diminution of family size constitute the demographic correlates of Ogburn's treatment of changing family functions.[2] Of course the fact that there is not a perfect correlation between fertility rates and the proportion of the population which is rural is seen in the fact that, since the depression, urbanization of the population has continued whereas the fertility rates have risen.

Fertility: 1940-1960. Although there is little doubt that between 1800 and the 1930s the number of children ever born to American wives dropped by 75 percent, there is considerable doubt as to whether the remarkable rise since World War II will prove to be quite temporary or of long duration. Expert opinion supported by attitudinal data inclines to the view that the rise is temporary. The actual fertility data, however, do

[1] Inter-Agency Committee for the National Conference on Family Life, *The American Family: A Factual Background,* Washington, U. S. Government Printing Office, 1948, p. 24.

[2] See "The Changing Functions of the Family."

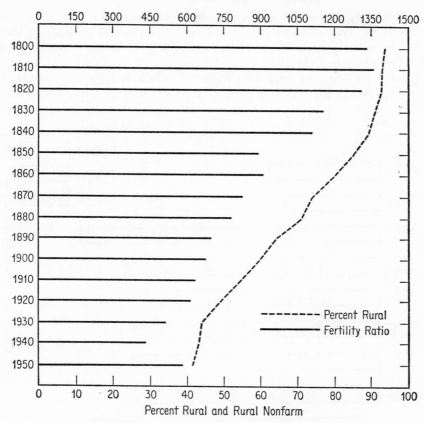

Figure 1.   Fertility ratio and percent of rural population by censal years, United States, 1800-1950.*

* Fertility data from P. K. Whelpton, *et al., Forecasts of the Population of the United States, 1945-1975.* Washington, D.C., U.S. Government Printing Office, 1947, p. 16. Data on percent rural are from Bureau of the Census, *1950 Census of Population, Preliminary Reports: General Characteristics of the Population of the United States: April 1, 1950.* Washington, D.C., February 25, 1951. Since final tabulations are not available, the percent rural for 1950 is an estimate. In 1950 the Bureau of the Census adopted a new definition of "rural," but, in order to make these data comparable, the pre-1950 definition was used throughout.

not seem to presage any decline or even immediate leveling of the numbers of children being born.

Because of the unemployment and reduced incomes during the depression many couples delayed marriage and married couples postponed having children. To the extent that the post-1940 rise was a reflection of deferred fertility, it would not be expected to contribute to a permanent rise in the rates. The "temporary" point of view is supported by a report that wives born in the period from 1936-1940 anticipate having fewer children than those born in 1931-1935.[3]

Actual fertility data appear in Figure 2, which is organized by "birth cohorts," that is, the surviving persons who were born in a specified period

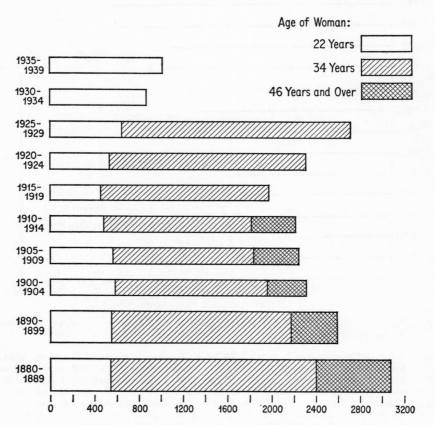

Figure 2. Number of children ever born per 1000 women, by age of woman, for white women born in 1880 to 1939, by year of birth of woman, for the non-institutional population of the United States: August 1959.*

*Source: Bureau of the Census, *Current Population Reports, Population Characteristics*, series P-20, no. 108, July 12, 1961, table 7, p. 23.

[3] Arthur A. Campbell, Pascal K. Whelpton, and Richard F. Tomasson, "The Reliability of Birth Expectations of U. S. Wives," to be published in the *Proceedings of the International Union for the Scientific Study of Population.* Pp. 7-8 in mimeographed manuscript.

of time. It is seen that women of completed fertility (those 46 years of age and older) show declining fertility from the cohort born in 1880-1889 through that born in 1910-1914. (Of course the subsequent cohorts had not yet reached the 46-and-over age-level.) Beginning with the cohort of 1910-1914, however, we can see the reversal, as each succeeding cohort at age 34 and at age 22 shows an increasing number of children. Indeed the women 34 years of age in the cohort born in 1925-1929 had more children than those of completed fertility in the four cohorts from 1890-1899 through 1910-1914. A lifetime average of about 2.2 children per woman would be required for the replacement of a stationary population under the mortality conditions of 1949-1951 and not affected by immigration or emigration. The women 30 to 39 years of age in 1959 had already born an average of 2.5 children per woman.[4] Such data indicate a continuing increase in the population, and no evidence of a decline or leveling off is visible in Figure 2.

The authors of the report on which Figure 2 is based, Wilson H. Grabill and Robert Parke, Jr., list some factors entering into the rise in the birth rate:

1. The historically less fertile segments of the population (urban, regions outside the South, wives in the labor force) have increased their fertility to approach the rates of the more fertile segments (rural, the South, and wives not in the labor force).
2. The proportion of women married has increased.
3. The median age of women at first marriage has declined.
4. Married women have born their first children at earlier ages than heretofore.

It should be pointed out that demographers were taken by surprise by the "baby boom" of the 1940s and, as indicated above, are uncertain about future trends. This is due largely to a lack of theory concerning fertility. In the past demographers have based their predictions upon extrapolations of historical fertility rates. For this reason they expected "more of the same" and did not foresee the sharp upswing in the 1940s and 1950s. An adequate science of demography would base its predictions upon theoretical considerations, which would presumably take account of such factors as economic conditions, changes in values, and such considerations as Grabill and Parke noted above. It is encouraging to note that some of the latest demographic analyses of fertility rates have attempted the theoretical approach.[5]

---

[4] Bureau of the Census, *Current Population Reports—Population Characteristics,* series P-20, no. 108, July 12, 1961, pp. 1-2.

[5] See especially Ronald Freedman, Pascal K. Whelpton, and Arthur A. Campbell, *Family Planning, Sterility and Population Growth,* New York, McGraw-Hill, 1959, pp. 7-8. See the editors' discussion of scientific theory in chapter 1 above.

II. Differentials in Fertility

To this point our focus has been on nation-wide rates. Rates for different segments of the population differ systematically along dimensions familiar in the literature of sociology—urbanism, social class, and race.

Urbanism, Suburbanism, and Fertility. As we have noted, the long-term trend in Figure 1 shows fertility to be declining as the rural proportion of the population has diminished. Table 1 shows that the urban-rural differential has been continuing although it also gives evidence that the gap may be narrowing. From 1950 to 1959 rural women of childbearing age bore more children per 1000 women than did urban women, but the proportionate increase in fertility was about three times as great in urban areas as in rural.

Within urban areas the birth rate is higher in suburbs than in central cities. A study in 1955 reports that the average number of children ever born to white married women 18-39 years old was 1.7 per woman in the twelve largest cities as compared with 2.1 in the suburbs of those cities.[6]

TABLE 1. Urban-Rural Differences in Fertility. Number of Children Ever Born per 1000 Women 15 to 44 Years Old (standardized for age), by Rural Farm, Rural Nonfarm and Urban Residence, 1950, 1957, 1959*

| Residence | Number Children Ever Born per 1000 Women 15-44 Years Old | | | Percent Increase 1950-1959 |
	1950	1957	1959	
Total	1395	1677	1777	27.4
Urban	1195	1504	1629	36.3
Rural farm	2064	2275	2298	11.3
Rural nonfarm	1681	1881	1911	13.7

* Source: Bureau of the Census, Current Population Reports-Population Characteristics, series P-20, no. 108, July 12, 1961, table 3, p. 19.

Social Class Differences. Whether a group is rural or urban it is subdivided by some form of class structure, and social class is another factor which we might expect to be correlated with differences in fertility. Social class is an abstract concept which cannot be measured directly. There are available, however, several commonly employed indices of social class which we can correlate with fertility. Here we shall be concerned with the three most commonly used indices: income, education and occupation. Table 2 reveals differences in fertility by differential income, while Table 3 reveals similar differences by educational levels. The patterns in these figures show that fertility is inversely related to social class. (The higher the level of education or income, the lower the fertility.) The most striking

[6] Clyde V. Kiser, "Fertility Rates by Residence and Migration," *Proceedings of the International Population Conference,* Vienna, 1959, p. 275.

feature about Table 4 is the absence of the decided relationship present in Tables 2 and 3. Since the occupational category also is a presumed index of social class, the absence of a clear pattern in this Table forces us to reassess the hypothesis that position in the class structure is inversely related to fertility.[7]

TABLE 2.   Number of Children Ever Born per 1000 Women 15-44 Years Old, Married and Husband Present (standardized for age) by Total Family Income, for the U.S., March 1957*

Family Income in Previous Calendar Year (1956)	Distribution of Women (%)	Children Ever Born per 1000 Women Married and Husband Present
Total reporting	100.0	2262
Under $2000	8.2	
$2000-$2999	8.5	2672
$3000-$3999	13.4	2483
$4000-$4999	17.3	2340
$5000-$6999	28.4	2159
$7000 and over	24.2	1842

* Source: Bureau of the Census, Current Population Reports-Population Characteristics, series P-20, no. 84, August 8, 1958, table 6, p. 12.

TABLE 3.   Number of Children Ever Born per 1000 Women 15-44 Years Old (standardized for age), by Educational Level, for the U.S., 1957*

Years of School Completed	Children Ever Born per 1000 Women		Percent Increase 1950-1957
	1950	1957	
Total	1395	1677	20.2
Elementary: Less than			
8 years	1972	2346	19.0
8 years	1642	1914	16.6
High school: 9-11 years	1501	1910	27.2
12 years	1141	1502	31.6
College: 13-15 years	1019	1360	33.5
16 years or more	807	1046	29.6
School years not reported	972		

* Source: Bureau of the Census, Current Population Reports-Population Characteristics, series P-20, no. 84, August 8, 1958, table 4, p. 10.

[7] If we disregard momentarily the findings based upon income and education and assume that occupation is *the* index of social class, we may explain the lack of pattern evident in Table 4 by one of the two following hypotheses: (1) Fertility is not related to social class. Kiser develops a hypothesis akin to this while discussing the lack of relation revealed in his earlier analysis of occupation and fertility. He suggests that class differentials are diminishing and that perhaps the apparent relations of fertility with education and income levels result more from lack of control over the residential

TABLE 4. Number of Children Ever Born per 100 Women 15-44 Years Old, Married with Husband Present (standardized for age), for the U.S., 1952 and 1957, by Occupational Group of Husband*

Categories of Occupational Group	Occupational Group of Husband	Children Ever Born per 1000 Women Married and Husband Present		Percent Increase 1952-1957
		1952	1957	
Total		1985	2313	16
Upper white collar workers	Professional, technical and kindred workers	1653	1939	17
	Managers, officials and proprietors except farm	1759	2085	18
Lower white collar workers	Clerical and kindred workers	1574	1920	22
	Sales workers	1535	2029	32
Upper blue collar workers	Craftsmen, foremen and kindred workers	1932	2285	18
Lower blue collar workers	Operators and kindred workers	2076	2454	18
	Service workers, including private household	1805	2229	24
	Laborers, except farm and mine	2380	2699	13
Farm workers	Farmers and farm managers	2704	3023	12
	Farm laborers and foremen	3153		

* Source: Bureau of the Census, Current Population Reports-Population Characteristics, series P-20, no. 84, August 8, 1958, table 5, p. 11.

If we conclude that social class and fertility are inversely related, the lack of a clear pattern in Table 4 means that we are still faced with the problem of describing more clearly the specific nature of this relationship.

There are at least two hypotheses which purport to do this. According to one, the relationship is linear (Figure 3), that is, for every increment in social class there is a constant decrement in fertility. According to the other hypothesis, the relationship is curvilinear (Figure 4). This means that starting at the bottom and running over most of the range of social class, as social class goes up, fertility goes down. At some class positions, however (possibly in the upper class or in the upper fringe of the middle class), the relationship shifts, and from here on, the higher the social class, the higher the fertility. Proponents of the latter hypothesis believe that in the upper social strata fertility is positively correlated with social class.

At this point we have three sets of relationships and two alternate hypotheses by which they may be interpreted. If we accept the linear hypothesis (Figure 3), two sets of data (education and income) appear to fit, but the third (occupation) does not. If we accept the curvilinear hypothesis (Figure 4) the occupational data begin to fit, but this leaves the linear relations unsatisfactorily explained. It may be that the first two relationships actually are curvilinear and that the curve does not appear because of inadequate categorization. It should be noted that all groups with annual money incomes of more than $7000 are lumped together in our data. This holds true as well for the upper educational groups, since college graduates are not distinguished from those who have gone on to professional schools. The few empirical studies which have made more refined break-downs in the upper income and educational levels have yielded evidence to support the curvilinear hypothesis.[8]

Racial Differences. The final social category which we shall consider in relation to human fertility is that of race, as reflected in differences between net reproduction rates of white and nonwhite Americans.[9] Table 5

factor than from a real relation between social class and fertility. See Clyde V. Kiser, "Fertility Trends and Differentials in the United States," *Journal of the American Statistical Association,* 47 (1952), 25-48, especially pp. 41-45. (2) Social class is related to fertility, and occupation is related to social class; the lack of pattern in the table results from inadequate occupational categorization and failure to control other factors. For a defense of this position, see Paul K. Hatt, "Occupation and Social Stratification," *American Journal of Sociology,* 55 (1950), 533-43.

[8] For a presentation of data which seem to support this hypothesis, see Robert F. Winch, *The Modern Family,* New York, Henry Holt and Co., 1952, pp. 118-21. See also, the forthcoming revised edition, Holt, Rinehart and Winston, ch. 7.

[9] Starting with a cohort of females at age zero, the net reproduction rate reflects the number of live females the first generation will bear. The net reproduction rate takes account of age-specific birth and death rates. A net reproduction rate of 1.00 signifies

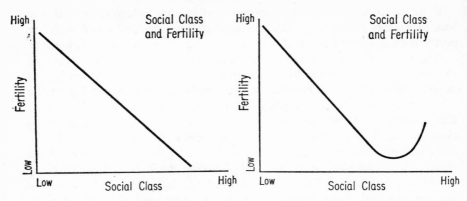

Figure 3. Hypothesis I: Linear rela-   Figure 4. Hypothesis II. Curvilinear re-
tion between social class and fertility.   lation between social class and fer-
    tility.

shows these differences from 1905 (estimated) to 1952. It is seen that the nonwhites have increased considerably more since the depression than have the whites although there was a short period after World War II when the increase in the net reproduction rate among whites exceeded that among nonwhites. In 1959 there were 1733 children ever born per 1000 white women 15-44 years old, as opposed to 1982 children ever born per 1000 nonwhite women 15-44 years old.[10]

TABLE 5.   Net Reproduction Rates for Whites and Nonwhites, 1905-1910, 1930-1935, 1935-1940, 1940, 1945, 1947, 1950, 1952*

Category	1905-10	1930-35	1935-40	Net Reproduction Rate 1940	1945	1947	1950	1952
Total	1.34	.98	.98	1.02	1.14	1.51	1.44	1.56
White	1.34	.97	.96	1.00	1.11	1.49	1.39	1.51
Nonwhite	1.33	1.07	1.14	1.20	1.38	1.59	1.78	1.89

* *Source:* Wilson H. Grabill, Clyde V. Kiser and Pascal K. Whelpton, *The Fertility of American Women,* New York, Wiley, 1958, tables 14 and 16, pp. 39 and 46.

III. Conclusions

From the early nineteenth century to the middle 1930s American reproductive rates declined remarkably. In the 1940s and 1950s a sharp rise occurred. In general, the greatest recent increases have been in the categories of the population previously having very low fertility.

---

that under the assumption of stable population the number of births equals just the number required for replacement.

[10] Bureau of the Census, *Current Population Reports,* series P-20, no. 108, *op. cit.,* table 1, p. 17.

Fertility rates vary with residence (urban, suburban, rural nonfarm, and rural farm), with race, and probably with social class—certainly with education and income.

THE PREGNANT WOMAN'S EMOTIONAL ATTITUDE AND ITS REFLECTION ON THE NEWBORN

Antonio J. Ferreira

Reprinted from *The American Journal of Orthopsychiatry*, 30:3 (1960), 553-561.

Throughout the centuries, in folklore and magic, there has been a prevalent belief that some specific events, happenings, or circumstances could ominously affect the pregnant woman, alter the course of her otherwise normal physiologic processes, and ultimately print indelible marks on the offspring. However, until recently, the subject of prenatal influences aroused but little scientific interest.

In 1896 Dabney [1] described 90 cases of fetal abnormalities which he believed were associated with and caused by "maternal impressions" incurred in the course of pregnancy. But it was only in the last twenty years, mostly through the efforts of Sontag and his co-workers, that our understanding of a prenatal environment began to take shape. Sontag pointed out that "deeply disturbed maternal emotion produces a marked increase in activity of the fetus" [6], and unearthed some evidence of the fact that "the psychophysiological state of the (pregnant) mother exerts an influence upon the behavior pattern of the normal fetus" [7]. A similar conviction was also expressed by Ernest Jones [2], who emphasized that the mother's attitude toward the unborn child influences the course of pregnancy and labor. In recent years, Turner [8], from a survey of 100 mothers and babies, was also led to the impression that "prenatal emotional stress might . . . alter the whole pattern of postnatal behavior."

The strength of these and other studies has not sufficed, however, to establish the existence and importance of a prenatal environment beyond the point of plausibility. And, perhaps also because of the difficulties inherent in a systematic study of the subject, the consideration of factors in the prenatal environment has remained speculative, if not unspoken.

This paper reports on an attempt to demonstrate the existence of a prenatal environment. Through a research study conducted at Letterman Army Hospital, San Francisco, and with the integrated participation of the Psychiatric, Obstetric and Pediatric Services, we set out to answer the question as to whether certain emotional attitudes of the pregnant mother were prenatally conveyed to the fetus and thus reflected upon the behavior of the newborn. We hoped to demonstrate experimentally the existence of a prenatal environment and, more specifically, to investigate the hypothesis that "upset" or deviant behavior in the newborn would bear a relationship with mother's prenatal "negative" attitude toward pregnancy and the baby-to-come.

The Design

The pregnant women were studied by means of a self-administered questionnaire. Utilizing the facilities of a Prenatal Clinic, every woman in her last four weeks of pregnancy (i.e., after the 36th week of pregnancy) was directed to fill out an attitude-inventory type of questionnaire. The questionnaire was based on the Parental Attitude Research Instrument (PARI) as developed by Schaefer and Bell [4]. It was composed of eight scales, seven from the PARI (constructed to measure the woman's Martyrdom, Dependency, Marital Conflict, Inconsiderateness-of-Husband, Irritability, Rejection of Homemaking Role, and Fear of Harming the Baby), and one other scale intended to measure Rejection of Pregnancy, in the same format as the other scales. Each of these scales consisted of eight items in the form of statements to which the individual was directed to respond with agreement or disagreement on a four-point scale. Weights of 3, 2, 1, 0 were assigned to the response of strong agreement, mild agreement, mild disagreement, and strong disagreement, respectively. A scale score was the sum of the item weights. The scores obtained from the scales of Fear of Harming the Baby (FHB) and Rejection of Pregnancy (RP) were to be considered as the criterion scores. They resulted from answers to the following statements:

Fear of Harming the Baby

3. A mother's greatest fear is that in a forgetful moment she may let something bad happen to the baby.
11. Mothers are fearful that they may hurt their babies in handling them.
19. All young mothers are afraid of their awkwardness in handling and holding the baby.
27. Mothers never stop blaming themselves if their babies are injured in accidents.
35. You must always keep tight hold of the baby during his bath, for in a careless moment he might slip.
43. Mothers often worry that people playing with the baby might be too rough.
51. There is no excusing the mother if a baby gets hurt.
59. A good mother must always be careful so that in a sleepy or busy moment she won't neglect the baby or hurt him.

Rejection of Pregnancy

8. A wise and intelligent woman avoids becoming pregnant.
16. Pregnancy makes a woman ugly.
24. Often a woman regrets having become pregnant.
32. A woman often doubts the wisdom of being pregnant.
40. There is nothing worse for a woman than being pregnant.
48. A pregnant woman is an unhappy woman.
56. Oftentimes a pregnant woman wishes her baby never to be born.
64. For natural reasons many times a woman resents having a baby.

The scores on the other six scales were to be regarded as not the focal part of the study. These six scales had been introduced mostly for the purpose of padding the questionnaire. However, they were chosen from among many other scales of the PARI with a view to their appropriateness for this group of women and with the faint hope of eliciting new leads for future investigation. As it developed, the introduction of these six other scales proved very useful as they ultimately offered us a rough measure of uncontrolled variables (degree of acquiescence, response "set," level of education, etc.) that, as it is well known, affect and strongly influence scores on this type of questionnaire.

In addition to answering the 64 items that made up the questionnaire, the individuals were instructed to answer a group of direct questions pertaining to background and other personal data: age, education, order in the family, parents' education, number of marriages, number and age of children, age of husband, whether they "believed" in planned pregnancies, whether they had planned "this" pregnancy, attitude toward breast and bottle feeding.

The expectant mothers under study were wives of people in the military (all ranks) in their last four weeks of pregnancy who utilized the service of the Prenatal Clinic during the period of time covered by this project (June 11-September 18, 1958). They filled out the self-administered questionnaire in an isolated room; they were instructed "not to sign it" and upon its completion, to place it in the envelope provided and to seal the envelope. A code number on the envelope permitted a later matching of mother and baby.

The babies were rated on the basis of the observation of their behavior during the first five days of life. The newborn's behavior was rated on five parameters: amount of crying, amount of sleep, degree of irritability, bowel movements, and feeding. Through daily interviews with the nurses who had been especially instructed to act as observers, an observational impression of each baby's behavior "for the past 24 hours" was obtained and rated in terms of each of the above-mentioned parameters. In each parameter, and for each day, the baby was described as normal (i.e. "like any average or usual newborn baby in the nurse's experience with babies"), somewhat deviant or markedly deviant (i.e., crying, sleeping, fussing, etc., "unusually more or less than the average newborn baby in the nurse's experience"). The frame of reference for such observations was carefully, repetitiously, and almost tediously emphasized by the investigator. The procedure was as follows: to the head nurse, at the end of her morning shift (about 3 P.M.), the following kind of question would be put for each baby and for each parameter. For instance, "Baby Smith . . . as a result of your observation during the morning shift . . . and the combined observations of the nurses in the two preceding shifts . . . and in terms of all of your experience with newborn babies . . . would you say that Baby Smith cried more or cried less than babies of this age usually and normally cry?"

From the ratings so obtained, a deviant day was operationally defined as any of those five days in the nursery in which the newborn baby displayed deviant behavior (scored either "somewhat" or "markedly" deviant) on any of the five parameters under consideration.

A deviant baby was then operationally defined as a baby who had two or more deviant days out of the first five days of life in the nursery. For the purpose of the study, the total baby population was thus divided into two groups operationally defined as nondeviant babies (with none or one deviant day) and deviant babies (with two or more deviant days).

For each baby, a record was also made of the sex, race, type of feeding (breast or bottle), length of labor, anesthetic used in delivery, type of delivery, birthweight, and the weight loss (or gain) at the end of the fifth day in the nursery. Twins, prematures, overdue, or physically ill babies (as determined by the recorded and totally independent opinion of the pediatric staff) were excluded from the study.

The hypothesis to be tested could now be formulated in its final form: that, as a group, mothers of deviant babies came from a population measurably different, in terms of the criterion variables (Fear of Harming the Baby and Rejection of Pregnancy), from mothers of nondeviant babies; and that this measurable difference was to be in the *direction* of higher criterion scores for mothers of deviant babies.

The Findings

A total of 268 mothers and 235 babies were studied. As expected, however, there were many babies "without" mothers, and mothers "without" babies. This is easily understood: Many mothers delivered outside (under the Medicare Program), others moved away before delivery, others delivered without having used and been tested at the Prenatal Clinic, a few did not fill out the questionnaire because of a language barrier, etc.

When babies and mothers were matched through code numbers, there were only 163 baby-mother pairs available for statistical treatment. Of the 163 babies involved, 28 were deviant babies (two or more deviant days out of five in the nursery) and 135 were nondeviant (none or one deviant day).

We shall now compare the two groups of mothers (mothers of deviant babies vs. mothers of nondeviant babies) in terms of the criterion variables under consideration: Fear of Harming the Baby and Rejection of Pregnancy.

Fear of Harming the Baby (FHB). Upon the collection of the first 50 cases, the hypothesis was tested by means of a point biserial correlation. The results appeared extremely encouraging, and displayed a statistically significant difference between the two groups of mothers. However, the observation of the FHB scores versus the total scores (sum total of the scores in the other seven scales), as plotted on a scattergram, brought a curious fact into evidence: that the individual score on the FHB scale (as in the other six scales from the PARI) seemed to be a linear function of the total score of the other scales! The higher the total score, the higher the FHB score seemed to be. From this observation we concluded that the FHB scores were being influenced by a number of factors or uncontrolled variables which similarly influenced the total scores. These uncontrolled variables were probably of the same type as those that ordinarily affect and influence scores on this sort of questionnaire: educational level, response "set," the degree of acquiescence, etc. It became apparent, therefore, that the raw FHB scores did not possess the desired comparative meaningfulness, and that they would have to be "adjusted" in terms of the uncontrolled variables involved. The uncontrolled variables were reflected in the total scores. Accordingly, and since the necessary statistical conditions seemed to be met [3], the FHB scores were treated by means of analysis of covariance using the total scores as a measurement of the uncontrolled variables. Thus treated, the data revealed that on the FHB scale the group of mothers of deviant babies had an adjusted mean score of $\overline{X}_D = 15.73$ versus an adjusted mean score of $\overline{X}_N = 14.07$ for mothers of nondeviant babies. These findings were statistically significant ($p < .025$, one-tailed).

It is important to mention at this point that we were also interested in investigating the construct validity of the FHB scale.* For this purpose, through consultation with obstetricians and other physicians, we established the soundness of a new hypothesis: that, as a group, primiparas

---

* Readers who are not conversant with the concept of construct validity may wish to consult Lee J. Cronbach and Paul E. Meehl, "Construct Validity in Psychological Tests," *Psychological Bulletin, 52,* 1955, 281-302—*Eds.*

would have a greater fear of harming the baby than multiparas. We proceeded then to test this hypothesis in terms of the criterion variable FHB for the two groups of primiparas and multiparas. Again the total score was used as a measure of the uncontrolled variables and analysis of covariance applied. The adjusted mean for the group of primiparas was $\overline{X}_P = 15.28$ and for the multiparas was $\overline{X}_M = 13.59$. This difference was statistically significant ($p < .01$, one-tailed) and thus indicated construct validity for the FHB scale.

We had anticipated that the FHB scores were to some extent a function of the educational level of the mothers. We found this to be so. When the mothers were divided into three subgroups according to the level of education attained, the obtained results were statistically significant ($p < .01$, one-tailed, analysis of variance) in terms of predicted decreased FHB scores with increased schooling. It is interesting to note that in each of these three subgroups, we found again that mothers of deviant babies scored appreciably higher on the FHB scale than mothers of nondeviant babies.

Rejection of Pregnancy (RP). In terms of the scores obtained on this scale, there was no statistically significant difference between the two groups of mothers. Worth noting is that seemingly the RP scale behaved quite differently from any of the other seven scales inasmuch as there was no appreciable correlation between scores on the RP scale and the total score.

However, the examination of the RP scores brought a very curious finding to our attention. It seemed that mothers of deviant babies had responded to the RP scale in such a way as to yield scores that belonged to either of the extreme portions of the range of RP scores. In other words, mothers of deviant babies seemingly scored either very high or very low on the range of scores of the RP scale. This had not been at all predicted and it should be eyed with great caution since it was brought to light by inspection of the data *after* its collection. Nevertheless, it appeared as an interesting point worthy of further exploration. Accordingly, the range of RP scores was broken down in four portions, each with approximately the same number of cases, and the two extreme portions compared with the two in-between. A comparison of the two groups so formed proved to be statistically significant ($p < .02$, chi square), allowing us to conclude *post facto* that mothers of deviant babies tended to score on the extreme range of the RP scale.

Baby's Deviancy. As previously mentioned, a deviant baby was operationally defined as a baby with two or more deviant days; and a deviant day was defined as any day when, on any of the five parameters, the baby was rated as deviant on the strength of the nursery nurses' observation. The nurses involved in the study were well experienced with the behavior of newborns; and throughout the study they maintained a very cooperative attitude with excellent understanding of the role they were playing as observers. Despite their excellence, it would have been desirable to know the reliability of their observations since on such observations rested the scores and the grouping of the babies. Such a reliability index was not available. However, a rough approach to an index of reliability could be obtained. The approach was based on the following reasoning: if a score of deviancy by the nurses had a better than chance probability of corresponding to a true over-all difference in the behavior of the newborn, then we would expect that, for any given baby scored deviant on a certain day, there would be a better-than-chance probability that he would again score deviant on any other day. Coarse as it was, we felt that we could look upon this approach as a rough index of the reliability of the nurses' observation of the newborns. Accordingly, for babies who were, for instance, deviant on the first day, we compared the expected (chance) and the occurring frequency with which deviancy was also scored on the second, third, fourth, and/or fifth day. The comparison revealed that the group of babies who were deviant on the first day (or any other day) were being scored as deviant on subsequent days with a frequency higher than chance ($p < .01$). This finding speaks well for the reliability of the baby's scores though it can of course also be interpreted as a "halo effect."

It is of interest to notice that there were just about as many babies who scored deviant on a given day as on any other day. The score of deviancy showed no visible preference for any day out of the five, and the results of a deviancy-by-days breakdown were very much within chance expectation.

An attempt was made to relate deviancy in the baby with other factors in the mother or in the baby, but no relationships were found. Deviancy in the baby did not relate with any of the scores obtained on the other six scales in the mother's questionnaire and, further, there was no relationship between deviancy in the baby and any one of the following factors: 1) race, 2) mother's age, 3) mother's education, 4) primiparity or multiparity, 5) length of labor, 6) type of delivery, 7) anesthesia during delivery, 8) breast or bottle feeding, 9) birthweight, 10) baby's weight loss (or gain) at the fifth day, 11) whether the pregnancy had been "planned" or not.

The group of deviant babies contained, however, more boys than girls, a discrepancy that when compared with the composition of the nondeviant group appeared, though unpredicted, statistically significant ($p < .05$).

Discussion

We had seen that an operationally defined deviancy in the newborn's behavior was statistically associated with nothing but a "negative" maternal attitude in evidence *prior* to delivery. The mothers of deviant babies scored significantly higher on an attitudinal scale of Fear of Harming the Baby; and on a Rejection-of-Pregnancy scale, the mothers of deviant babies again responded differently insofar as they tended to score either too high or too low through the whole range of scores.

The totality of the results obtained in this study speaks well for the existence of a prenatal environment. In terms of the variable Fear of Harming the Baby, mothers of deviant babies come from a population different from mothers of nondeviant babies. We feel that this observation is not only statistically significant, but psychologically meaningful as well. Having established construct validity for the criterion scale (FHB), we may conclude then that the "average" mother of a deviant baby had, during pregnancy, a conscious fear of harming her baby greater than the "average" mother of a nondeviant baby. As implied in our choice of the FHB scale as a criterion variable, we can assume further that this consciously "greater" fear of harming the baby corresponds to a "greater" *unconscious* hostility toward the baby-to-come.

On an *ad hoc* scale of Rejection of Pregnancy, the mothers of deviant babies again behaved in a significantly different fashion. We interpret this difference to mean that mothers of deviant babies have a conscious attitude of either extreme rejection or extreme nonrejection of their state of pregnancy. To the incautious reader, we must emphasize again the *post facto* nature of this finding, though its very occurrence cannot but add weight to the hypothesis of a prenatal environment. It bears repeating that no relationship was found between deviancy in the newborn and any such factor as race, age of mother, parity, length of labor, type of anesthesia, type of delivery, type of feeding, and whether or not pregnancy had been planned.

We have established that deviant behavior in the newborn was associated solely with "negative" attitudes in the mother, as expressed by a "higher" score on a scale of Fear of Harming the Baby and by an "either extreme" score on a scale of Rejection of Pregnancy. We may now ask: Is the association between newborn's deviant behavior and mother's negative attitude the result of a prenatal influence? Or, instead, is baby's deviancy the result of a direct and immediate contact with mother *after birth* during those five days in the nursery? This is a very crucial question. To answer it we return now to a further analysis and interpretation of the available data:

1. During the baby's five days in the nursery, mother and baby were together for a maximum of 10 hours. For the first 24 hours, nursery regulations allowed no contact between the mother and her baby. In the four subsequent days, the mother was to be with her baby five times a day for a feeding period not to exceed 30 minutes. Therefore, for those five days, mother and child were together for only 10 hours, the sum total of those feeding periods. During the rest of all those five days, i.e., 110 hours approximately, the newborns stayed in the nursery and totally away from mother.

2. However, we cannot so easily discount the possible importance of those ten hours during which mother and child were in contact for the first five days of the baby's life. For those ten hours occurred at a most crucial time—the feeding time. Fortunately, by means of a new assumption, the available data lends itself to further investigation. We assumed that if the observed deviancy were the result of mother's postnatal influence upon the newborn, and that if we could further assume a cumulative effect to such influence, there would be more babies deviant on the fourth and fifth days than on the first and second. The data revealed this not to be the case. In fact, there were about just as many babies deviant on the first and second days as there were on the fourth and the fifth. Therefore, we came to the conclusion that deviancy was related *not to postnatal factors,* but to prenatal ones.

3. Among the 163 mothers in the study, only one was unwed. Prior to delivery, this unwed mother had made arrangements to have her baby adopted. No postnatal contact whatsoever was allowed between this mother and her baby. It is interesting to note, therefore, that though there was no postnatal contact, her baby belonged to the deviant group (with three deviant days), while she herself had scored obviously "high" on the scale of Fear of Harming the Baby.

4. Another finding merits reporting, and further invites us to conclude for the existence of a prenatal environment. The finding in question has to do with the behavior of babies born of women who refused to fill out the questionnaire. From 268 mothers, only 4 opposed participation in the study and outrightly refused to fill out the questionnaire. Of these 4 women, 2 gave birth to babies who rated deviant in one day, one had a baby rated deviant in two days, and the other a baby deviant in three days out of five. On the basis of the data, the probability that this might have occurred by chance alone is approximately 2 in 10,000.

A number of interesting side findings emerged from this study. We found, for instance, that primiparas favored breast feeding, whereas multiparas preferred bottle feeding ($p < .02$, chi square). Also worth mentioning, at least as a somewhat humorous comment to the frailty of human intentions, was the finding that although 67 per cent of the mothers "believed" in planned pregnancies, only 32 per cent had actually planned "this" pregnancy! As the present study indicates, there is no relationship (fortunately!) between unplanned pregnancies and newborn's deviant behavior. As a matter of fact, there were more deviant babies from mothers who had planned their pregnancy than from those who did not. But the difference was not statistically significant.

If we are now to put together the weight of the results of this study, we come readily to conclude that they overwhelmingly tend to confirm the hypothesis of the existence of a prenatal environment. Of course we feel that the present study is in no way a final word on the subject. The coarseness of the measurements speaks for itself. And yet—it is conceivable that the influence of prenatal maternal attitudes upon the newborn may turn out to be such a grossly obvious phenomenon that even an approach as crude and blunt as ours was, stands a chance of bringing it into the light.

We look upon the results of this study as experimental confirmation of the existence and importance of a prenatal environment and for the present we find it reasonably safe to state that the influence of the emotional environment (mother's attitude) upon behavior has its zero hour before birth.

REFERENCES

1. Dabney, W. C., "Maternal Impressions," in *Diseases of Children*, pp. 191-216 (Edinburgh: Young J. Pentland, 1896). In M. E. Rogers, A. M. Lilienfeld and B. Pasamanick, *Prenatal and Paranatal Factors in the Development of Childhood Behavior Disorders*. Acta Psychiat. Neurol., Scand., Suppl. 102, 1955.

2. Jones, E., *Psychology and Childbirth*. Lancet, 1: 695-696, 1942.

3. Lindquist, E. F., *Design and Analysis of Experiments in Psychology and Education*. (Cambridge, Mass.: Riverside Press, 1953).

4. Schaefer, E. S., and R. Q. Bell, *Parental Attitude Research Instrument (PARI): Normative Data.* Unpublished manuscript. Library, National Institutes of Health, Bethesda, Md., 1955.

5. Siegel, S., *Nonparametric Statistics: For the Behavioral Sciences*, pp. 145-151 (New York: McGraw-Hill, 1956).

6. Sontag, L. W., "The Significance of Fetal Environmental Differences," *Am. J. Obst. & Gynec.*, 42: 995-1003, 1941.

7. ———, "Difference in Modifiability of Fetal Behavior and Physiology," *Psychosom. Med.*, 6: 151-154, 1944.

8. Turner, E. K., "The Syndrome in the Infant Resulting from Maternal Emotional Tension During Pregnancy," *Med. J. Australia*, 1: 221-222, 1956.

MOTHERLESS INFANTS

René A. Spitz

Adapted from "The Role of Ecological Factors in Emotional Development in Infancy," *Child Development,* 20 (1949), 145-55, Society for Research in Child Development, Inc., Purdue University.

In the following an extremely condensed report on our findings on psychosocial factors in infant development will be presented. To call attention to the function of such factors in infancy appears to us an urgent need, for it is not generally appreciated that at this age influences of a psychosocial nature are more startling in their consequences for development than at any other period of childhood in later life.

The reasons for this are manifold. At no later period is the development so rapid, so turbulent and so conspicuous. It involves, more obviously than at any later period, the somatic as well as the psychological aspects. Any variation in the development will be manifested in both these sectors, with the result, as will be shown further on, that such variations caused by psychosocial factors can literally become matters of life and death.

In a certain sense development at this age is facilitated by certain peculiarities which infancy does not share with later stages. In infancy development takes place from a quasi-animal level to the human level. This involves the problems of adaptation and communication which arise toward the end of the first year of life. Any anomaly in the solution of these problems will be particularly conspicuous.

Another peculiarity of infancy is that in it environment, in the widest sense of the word, is extraordinarily restricted. The radius of the infant's physical environment is extremely narrow. The social environment in the life of the normal infant is restricted practically to one single person: the infant's mother. This concentration of the infant's habitat and of its social contacts permits the investigator exceedingly close insight into the psychosocial factors operative in the infant's life.

A further facilitation in the study of our problem is offered by another peculiarity of infant life. The single infant can be studied in its family and also in large groups, in institutions. Nevertheless, our proposition that the infant's environment is restricted to one person, the mother or her substitute, remains substantially unchanged (unless the institution in question

should entrust the care of the infants to a number of different persons).
For during the largest part of the first year of life no interaction takes
place between one infant and the others, such as would be the case between the inmates of institutions at later ages. This makes it possible to
study infants of different races and of different hereditary endowment
under identical environmental conditions, keeping a large number of variables constant. . . .

A few brief statements will clarify our position on the question of emotional development as observed in the course of our work.

Emotions are not present ready-made from birth. Like any other sector
of the human personality they have to develop.

We usually conceive of emotions as paired: friendliness and anger,
love and hate, pleasure and displeasure, gay and sad, are the terms in
which we think of emotions. But at birth the first emotion visible is a state
of diffuse excitation in the nature of displeasure and no pleasurable emotion is observable as its counterpart, only a state of quiescence. A variety
of emotions develops from this beginning in the further course of the first
year of life. We have been able to correlate the progressive development
of specific emotions to definite age levels.

Two distinct emotional responses are differentiated in the course of the
first two months of life. They appear to correspond to pleasure and displeasure, and they seem to appear in reaction to physical stimulation.

A response to psychological stimulation seems to present itself for the
first time in the third month, when the infant smiles in response to a human partner's face [6].

Somewhat later displeasure also is manifested in response not only to
physical, but also to psychological stimulation. It can be observed in the
reaction of the infant to being left alone when its human partner goes
away.

After the sixth month negative emotions take the lead. Anxiety is differentiated from the displeasure reaction. We assume that a minimum of
ego development is the prerequisite for the development of anxiety and
find ourselves in agreement on this point with E. Hilgard [3].

In the following two months possessive emotions toward toys are manifested. Jealousy appears in the ninth and tenth months; between the
tenth and twelfth months disappointment, anger, love, sympathy, friendliness, enjoyment and a positive sense of property become observable.
The age levels mentioned should not be considered as definite limits.
They designate approximate ages at which these emotions appear and
may vary widely both according to individuals and circumstances.

The significant part of this emotional development is that during the
whole of the first year emotional discrimination is manifested approxi-

mately two months earlier than any other form of perception. The three months' smiling response, which is the infant's smiling recognition of the human partner's face, appears at an age at which no other object is recognized. Even food, the most familiar object in the baby's life, is recognized only more than two months later. The displeasure which the infant manifests at four months when left by its partner, appears two months earlier than the displeasure shown by the child when its toy is taken away. The eight months' anxiety shown by the child when confronted with strangers is a sign that it has achieved the capacity to discriminate between friend and stranger. This appears two months earlier than the child's capacity to differentiate toys and other objects from each other. Thus, emotional development acts as the trailbreaker for all other perceptive development during infancy. . . .

A brief summary of the first investigation made by us may serve as an illustration for the other ones of which only the results will be given.

The investigation in question [4, 5] was carried out in two institutions which we had the opportunity to observe simultaneously. Both institutions had certain similarities: the infants received adequate food; hygiene and asepsis were strictly enforced; the housing of the children was excellent; and medical care more than adequate. In both institutions the infants were admitted shortly after birth.

The institutions differed in one single factor. This factor was the amount of emotional interchange offered. In institution No. 1, which we have called "Nursery," the children were raised by their own mothers. In institution No. 2, which we have called "Foundlinghome," the children were raised from the third month by overworked nursing personnel: one nurse had to care for from eight to twelve children. Thus, the available emotional interchange between child and mother formed the one independent variable in the comparison of the two groups.

The response to this variable showed itself in many different ways. Perhaps the most comprehensive index of this response is offered by the monthly averages of the developmental quotients of these children.

The developmental quotient [1, 2] represents the total of the development of six sectors of the personality: mastery of perception, of bodily functions, of social relations, of memory and imitation, of manipulative ability and of intelligence. The monthly averages of the developmental quotients of the children in the two institutions over a period of twelve months are shown in Figure 1.

The contrast in the development of the children in the two institutions is striking. But this twelve months' chart does not tell the whole story. The children in "Foundlinghome" continued their downward slide and by the end of the second year reached a developmental quotient of 45. We have

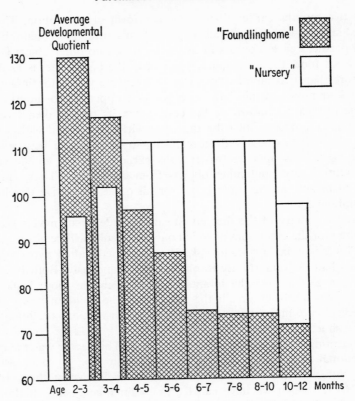

Figure 1.   Comparison of development in "nursery" and "found-linghome."

here an impressive example of how the absence of one psychosocial factor, that of emotional interchange with the mother, results in a complete reversal of a developmental trend. This becomes still clearer in Figure 2.

It should be realized that the factor which was present in the first case, but eliminated in the second, is the pivot of all development in the first year. It is the mother-child relation. By choosing this factor as our independent variable we were able to observe its vital importance. While the children in "Nursery" developed into normal healthy toddlers, a two-year observation of "Foundlinghome" showed that the emotionally starved children never learned to speak, to walk, to feed themselves. With one or two exceptions in a total of 91 children, those who survived were human wrecks who behaved either in the manner of agitated or of apathetic idiots.

The most impressive evidence probably is a comparison of the mortality rates of the two institutions. "Nursery" in this respect has an outstanding record, far better than the average of the country. In a five years' observation period during which we observed a total of 239 children, each for one year or more, "Nursery" did not lose a single child through death. In "Foundlinghome" on the other hand, 37 per cent of the children died during a two years' observation period (Figure 2).

The high mortality is but the most extreme consequence of the general decline, both physical and psychological, which is shown by children completely starved of emotional interchange.

We have called this condition marasmus, from the picture it shows; or

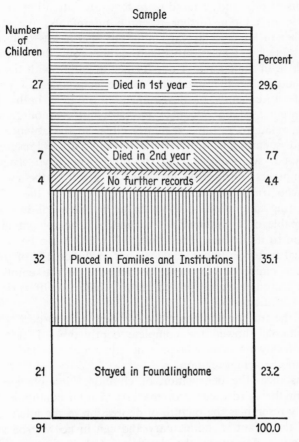

Figure 2. Mortality rate of children in "foundlinghome" during a two-years' observation period.

hospitalism according to its etiology. The ecological background of marasmus is the orphanage and the foundling home which were current in our
country in the last century. This ecological background leads to the picture
of a developmental arrest which progressively becomes a developmental
regression. Its earliest symptoms are developmental retardations in the
different sectors of personality. Changes occur in the emotional development, later the emotional manifestations become progressively impoverished, finally they give way to apathy. In those cases which survived
we have found, alternating with the apathetic children, a hyper-excitable
personality type. This personality type is on the lines described by Wallon
as "l'enfant turbulent" [8]. We have called this, for want of a better term,
the erethitic or agitated type.

The results of this study caused us to focus our attention on the mother-
child relation in all our further research on infants. We strove to examine whether in less spectacular conditions also it was truly such an all-
important influence. Closer investigation bore out this impression. We
could establish, in the course of our further research, with the help of statistical methods, that the regularity in the emergence of emotional response, and subsequently of developmental progress both physical and
mental, is predicated on adequate mother-child relations. Inappropriate
mother-child relations resulted regularly either in the absence of developmental progress, emotional or otherwise, or in paradoxical responses.

This is not a surprising finding for those who have observed infants
with their mothers; during the first year of life it is the mother, or her
substitute, who transmits literally every experience to the infant. Consequently, barring starvation, disease or actual physical injury, no other
factor is capable of so influencing the child's development in every field
as its relation to its mother. Therefore, this relationship becomes the central ecological factor in infant development in the course of the first year.
On the other hand, development, particularly in the emotional sector,
provides an extremely sensitive and reliable indicator of variations in the
mother-child relationship.

A few of the further findings made by us in this respect will follow. If
our first example showed the complete deprivation of emotional interchange, the following ones will present other and less striking modifications of the mother-child relation.

If, for instance, the deprivation of emotional interchange starts at a
later date, in the third quarter of the first year, a condition can develop
which greatly resembles the picture of depression in the adult. The psychic
structure of the infant is rudimentary and can in no way be compared to
that of the adult. Therefore, the similarity of the symptomatology should
not induce us to assume an identity of the pathological process. To stress

this difference we have called the condition anaclitic [7] depression.

As the name implies, the presenting symptom is a very great increase in the manifestations of the emotions of displeasure. This goes to the point where anxiety reactions in the nature of panic can be observed. Children in this condition will scream by the hour; this may be accompanied by autonomic manifestations such as tears, heavy salivation, severe perspiration, convulsive trembling, dilation of pupils, etc.

At the same time development becomes arrested (see Table 1). The arrest is selective: the least involved is the social sector which remains relatively advanced.

TABLE 1.

NURSERY	INFLUENCE OF SEPARATION FROM MOTHER ON DEVELOPMENTAL QUOTIENT
	Severe Depression
Duration of separation in months	Changes in points of DQ
Under 3	—12.5
3 to 4	—14
4 to 5	—14
Over 5	—25

One peculiarity of this condition is that the re-establishment of favorable emotional interchange will rapidly re-establish the developmental level. However, this is only true for separations which do not last longer than three months. If the deprivation lasts longer than five months no improvement is shown. On the contrary, the developmental quotient continues its decline, though at a slower rate, and it would seem that a progressive process has been initiated (see Table 2).

TABLE 2.

NURSERY	INFLUENCE OF SEPARATION FROM MOTHER ON DEVELOPMENTAL QUOTIENT
	Severe Depression
Duration of separation in months	Reversibility of decline in points of DQ
Under 3	+25
3 to 4	+13
4 to 5	+12
Over 5	— 4

Conclusion

. . . We believe . . . that the central psychosocial factor in the infant's life is its emotional interchange with its mother.

The particular ecological significance of this finding lies in the fact that this emotional interchange is largely governed by culturally determined mores and institutions on one hand, by social and economic conditions on the other. To give one example: marasmus was a frequent condition up to 1920 in our country, as foundling homes were still in general use. Today such conditions are difficult to find unless we look for them in countries where foundling homes are still the rule. In the United States placement in foster-homes has taken the place of foundling institutions.

Less extreme conditions, however, like anaclitic depression and the others described above can be readily found at present here too. Our social institutions do not encourage the mother to spend much time with her child—at least in the population at large. Industrial civilization tends to deprive the child in early infancy of its mother. I have recently learned that in another industrial country a financial premium has been introduced for mothers who return earlier to factory work after having delivered their child. The earlier they abandon their baby the more substantial the tax reduction they receive. We can be convinced that the consequences of these socio-economic measures will make themselves felt in a distortion of these children's psychological development, although a dozen or more years must pass before the change becomes evident.

Still less consideration is given in our social institutions and in our present-day mores to the question of how to prepare the future mother's personality for motherhood. From the influence on child development exerted by mood-swings in the mother, by the mother's infantile personality or by her neurosis, it is obvious that attempts to remedy such conditions should begin early. . . . I have stressed that preventive psychiatry should begin as early as possible, at birth at least, but preferably before delivery.

I have attempted to show in this paper that such preventive psychiatry will have to begin by applying measures which are largely of an ecological nature. These measures will have to include a re-arrangement of legislation, making it possible for mothers to stay with their children; of education to prepare our female population for motherhood; and they will have to comprise the introduction of social psychiatry to remedy in expectant mothers psychiatric conditions apt to damage their children.

REFERENCES

1. Buehler, Charlotte, and Hetzer, H., Kleinkinder Tests (Leipzig: Johann Ambrosius Barth, 1932).
2. Hetzer, H., and Wolf, K., "Babytests," Z. Psychol., 1928, 107, 62-104.
3. Hilgard, E., "Human motives and the concept of the self," Amer. Psychologist, 1949, 4, 375-382.
4. Spitz, R. A., "Hospitalism: an inquiry into the genesis of psychiatric conditions in early childhood," The Psychoanalytic Study of the Child (New York: Inter-

national Univ. Press, 1945), 1, 53-74.

5. Spitz, R. A., "Hospitalism, a follow-up report," *The Psychoanalytic Study of the Child* (New York: International Univ. Press, 1946), 2, 113-117.

6. Spitz, R. A., and Wolf, K. M., "The smiling response: a contribution to the ontogenesis of social relations," *Genet. Psychol. Monogr.*, 1946, 34, 59-125.

7. Spitz, R. A., and Wolf, K. M., "Anaclitic depression: an inquiry into the genesis of psychiatric conditions in early childhood," *The Psychoanalytic Study of the Child* (New York: International Univ. Press, 1946), 2, 313-342.

8. Wallon, H., *L'enfant turbulent* (Paris: Librairie Felix Alcan, 1925), p. 642.

THE DEVELOPMENT OF THE SELF

George Herbert Mead

Adapted and reprinted from *Mind, Self and Society* by George Herbert Mead, 146-156, by permission of The University of Chicago Press. Copyright 1934 by The University of Chicago. All rights reserved. Published 1934.

. . . Our symbols are all universal.[1] You cannot say anything that is absolutely particular; anything you say that has any meaning at all is universal. You are saying something that calls out a specific response in anybody else provided that the symbol exists for him in his experience as it does for you. There is the language of speech and the language of hands, and there may be the language of the expression of the countenance. One can register grief or joy and call out certain responses. There are primitive people who can carry on elaborate conversations just by expressions of the countenance. Even in these cases the person who communicates is affected by that expression just as he expects somebody else to be affected. Thinking always implies a symbol which will call out the same response in another that it calls out in the thinker. Such a symbol is a universal of discourse; it is universal in its character. We always assume

[1] Thinking proceeds in terms of or by means of universals. A universal may be interpreted behavioristically as simply the social act as a whole, involving the organization and interrelation of the attitudes of all the individuals implicated in the act, as controlling their overt responses. This organization of the different individual attitudes and interactions in a given social act, with reference to their interrelations as realized by the individuals themselves, is what we mean by a universal; and it determines what the actual overt responses of the individuals involved in the given social act will be, whether that act be concerned with a concrete project of some sort (such as the relation of physical and social means to ends desired) or with some purely abstract discussion, say the theory of relativity or the Platonic ideas.

that the symbol we use is one which will call out in the other person the same response, provided it is a part of his mechanism of conduct. A person who is saying something is saying to himself what he says to others; otherwise he does not know what he is talking about. . . .

. . . What is essential to communication is that the symbol should arouse in one's self what it arouses in the other individual. It must have that sort of universality to any person who finds himself in the same situation. There is a possibility of language whenever a stimulus can affect the individual as it affects the other. With a blind person such as Helen Keller, it is a contact experience that could be given to another as it is given to herself. It is out of that sort of language that the mind of Helen Keller was built up. As she has recognized, it was not until she could get into communication with other persons through symbols which could arouse in herself the responses they arouse in other people that she could get what we term a mental content, or a self.

Another set of background factors in the genesis of the self is represented in the activities of play and the game. . . .

We find in children . . . invisible, imaginary companions which a good many children produce in their own experience. They organize in this way the responses which they call out in other persons and call out also in themselves. Of course, this playing with an imaginary companion is only a peculiarly interesting phase of ordinary play. Play in this sense, especially the stage which precedes the organized games, is a play at something. A child plays at being a mother, at being a teacher, at being a policeman; that is, it is taking different rôles, as we say. We have something that suggests this in what we call the play of animals: a cat will play with her kittens, and dogs play with each other. Two dogs playing with each other will attack and defend, in a process which if carried through would amount to an actual fight. There is a combination of responses which checks the depth of the bite. But we do not have in such a situation the dogs taking a definite rôle in the sense that a child deliberately takes the rôle of another. This tendency on the part of the children is what we are working with in the kindergarten where the rôles which the children assume are made the basis for training. When a child does assume a rôle he has in himself the stimuli which call out that particular response or group of responses. He may, of course, run away when he is chased, as the dog does, or he may turn around and strike back just as the dog does in his play. But that is not the same as playing at something. Children get together to "play Indian." This means that the child has a certain set of stimuli which call out in itself the responses that they would call out in others, and which answer to an Indian. In the play period the child utilizes his own responses to these stimuli which he makes use of in building a self. The response which he has a tendency to make to these stimuli organizes them. He plays that he is, for instance, offering himself something, and he buys it; he gives a letter to himself and takes it away; he addresses himself as a parent, as a teacher; he arrests himself as a policeman. He has a set of stimuli which call out in himself the sort of responses they call out in others. He takes this group of responses and organizes them into a certain whole. Such is the simplest form of being another to one's self. It involves a temporal situation. The child says something in one character and responds in another character, and then his responding in another character is a stimulus to himself in the first character, and so the conversation goes on. A certain organized structure arises in him and in his other which replies to it, and these carry on the conversation of gestures between themselves.

If we contrast play with the situation in an organized game, we note the essential difference that the child who plays in a game must be ready to take the attitude of everyone else involved in that game, and that these different rôles must have a definite relationship to each other. Taking a very simple game such as hide-and-seek, everyone with the exception of the one who is hiding is a person who is hunting. A child does not require more than the person who is hunted and the one who is hunting. If a child is playing in the first sense he just goes on playing, but there is no basic organization gained. In that early stage he passes from one rôle to another just as a whim takes him. But in a game where a number of individuals are involved, then the child taking one rôle must be ready to take the rôle of everyone else. If he gets in a ball nine he must have the responses of each position involved in his own position. He must know what everyone else is going to do in order to carry out his own play. He has to take all of these rôles. They do not all have to be present in consciousness at the same time, but at some moments he has to have three or four individuals present in his own attitude, such as the one who is going to throw the ball, the one who is going to catch it, and so on. These responses must be, in some degree, present in his own make-up. In the game, then, there is a set of responses of such others so organized that the attitude of one calls out the appropriate attitudes of the other.

This organization is put in the form of the rules of the game. Children take a great interest in rules. They make rules on the spot in order to help themselves out of difficulties. Part of the enjoyment of the game is to get these rules. Now, the rules are the set of responses which a particular attitude calls out. You can demand a certain response in others if you take a certain attitude. These responses are all in yourself as well. There you get an organized set of such responses as that to which I have referred, which is something more elaborate than the rôles found in play.

Here there is just a set of responses that follow on each other indefinitely. At such a stage we speak of a child as not yet having a fully developed self. The child responds in a fairly intelligent fashion to the immediate stimuli that come to him, but they are not organized. He does not organize his life as we would like to have him do, namely, as a whole. There is just a set of responses of the type of play. The child reacts to a certain stimulus, and the reaction is in himself that is called out in others, but he is not a whole self. In his game he has to have an organization of these rôles; otherwise he cannot play the game. The game represents the passage in the life of the child from taking the rôle of others in play to the organized part that is essential to self-consciousness in the full sense of the term. . . .

The fundamental difference between the game and play is that in the latter the child must have the attitude of all the others involved in that game. The attitudes of the other players which the participant assumes organize into a sort of unit, and it is that organization which controls the response of the individual. The illustration used was of a person playing baseball. Each one of his own acts is determined by his assumption of the action of the others who are playing the game. What he does is controlled by his being everyone else on that team, at least in so far as those attitudes affect his own particular response. We get then an "other" which is an organization of the attitudes of those involved in the same process.

The organized community or social group which gives to the individual his unity of self may be called "the generalized other." The attitude of the generalized other is the attitude of the whole community.[2] Thus, for example, in the case of such a social group as a ball team, the team is the generalized other in so far as it enters—as an organized process or social activity—into the experience of any one of the individual members of it.

If the given human individual is to develop a self in the fullest sense, it

---

[2] It is possible for inanimate objects, no less than for other human organisms, to form parts of the generalized and organized—the completely socialized—other for any given human individual, in so far as he responds to such objects socially or in a social fashion (by means of the mechanism of thought, the internalized conversation of gestures). Any thing—any object or set of objects, whether animate or inanimate, human or animal, or merely physical—toward which he acts, or to which he responds, socially, is an element in what for him is the generalized other; by taking the attitudes of which toward himself he becomes conscious of himself as an object or individual, and thus develops a self or personality. Thus, for example, the cult, in its primitive form, is merely the social embodiment of the relation between the given social group or community and its physical environment—an organized social means, adopted by the individual members of that group or community, of entering into social relations with that environment, or (in a sense) of carrying on conversations with it; and in this way that environment becomes part of the total generalized other for each of the individual members of the given social group or community.

is not sufficient for him merely to take the attitudes of other human individuals toward himself and toward one another within the human social process, and to bring that social process as a whole into his individual experience merely in these terms: he must also, in the same way that he takes the attitudes of other individuals toward himself and toward one another, take their attitudes toward the various phases or aspects of the common social activity or set of social undertakings in which, as members of an organized society or social group, they are all engaged; and he must then, by generalizing these individual attitudes of that organized society or social group itself, as a whole, act toward different social projects which at any given time it is carrying out, or toward the various larger phases of the general social process which constitutes its life and of which these projects are specific manifestations. This getting of the broad activities of any given social whole or organized society as such within the experiential field of any one of the individuals involved or included in that whole is, in other words, the essential basis and prerequisite of the fullest development of that individual's self: only in so far as he takes the attitudes of the organized social group to which he belongs toward the organized, co-operative social activity or set of such activities in which that group as such is engaged, does he develop a complete self or possess the sort of complete self he has developed. And on the other hand, the complex co-operative processes and activities and institutional functionings of organized human society are also possible only in so far as every individual involved in them or belonging to that society can take the general attitudes of all other such individuals with reference to these processes and activities and institutional functionings, and to the organized social whole of experiential relations and interactions thereby constituted—and can direct his own behavior accordingly.

It is in the form of the generalized other that the social process influences the behavior of the individuals involved in it and carrying it on, i.e., that the community exercises control over the conduct of its individual members; for it is in this form that the social process or community enters as a determining factor into the individual's thinking. In abstract thought the individual takes the attitude of the generalized other[3] toward

[3] We have said that the internal conversation of the individual with himself in terms of words or significant gestures—the conversation which constitutes the process or activity of thinking—is carried on by the individual from the standpoint of the "generalized other." And the more abstract that conversation is, the more abstract thinking happens to be, the further removed is the generalized other from any connection with particular individuals. It is especially in abstract thinking, that is to say, that the conversation involved is carried on by the individual with the generalized other, rather than with any particular individuals. Thus it is, for example, that abstract concepts are concepts stated in terms of the attitudes of the entire social group or community; they

himself, without reference to its expression in any particular other individuals; and in concrete thought he takes that attitude in so far as it is expressed in the attitudes toward his behavior of those other individuals with whom he is involved in the given social situation or act. But only by taking the attitude of the generalized other toward himself, in one or another of these ways, can he think at all; for only thus can thinking—or the internalized conversation of gestures which constitutes thinking—occur. And only through the taking by individuals of the attitude of attitudes of the generalized other toward themselves is the existence or a universe of discourse, as that system of common or social meanings which thinking presupposes at its context, rendered possible. . . .

---

are stated on the basis of the individual's consciousness of the attitudes of the generalized other toward them, as a result of his taking these attitudes of the generalized other and then responding to them. And thus it is also that abstract propositions are stated in a form which anyone—any other intelligent individual—will accept.

8. PARENTHOOD AND SOCIALIZATION
(Continued)

Looking at the state of our knowledge about socialization, Maccoby asks which variables are most useful to orient research on this topic. She favors using such variables from experiments on animal learning as amount of reinforcement, proportion of responses reinforced, and the ratio of rewards to punishments. But in translation to the human level the problem becomes complicated by the question: "What do we mean by reward?" Experimenters can induce drive in laboratory animals by depriving them of food or water, but since most parents do not let their children get hungry, thirsty, wet or overtired, purely physiological rewards are not so relevant with children as with animals. The more complicated nature of the problem at the human level is seen in the fact that approval, affection, and attention become rewards, and the sex of the socializer may affect the outcome of a socializing episode.

In the sense that a culture denotes what is right, mothers in most societies "know" how to rear their young. Of course their "knowledge" is not necessarily valid in the scientific sense of having experimental verification that the procedures used will lead to the child's developing approved patterns of response. Rather their "knowledge" is secure in the sense that it is defined in the culture as the "right" and "proper" way to bring up children. Mothers pass the lore along to their daughters, and grandmothers are repositories of such wisdom. One of the concomitants of the industrial revolution has been the scientific revolution with its implication that science rather than the wisdom of the elders is the arbiter of the proper ways of meeting the gamut of human problems. As science succeeds folk wisdom among the *au courant*, the "child expert" displaces grandmother in knowing how to handle Junior. And mother learns how to raise children not from her mother, but from classes, books, and articles. The situation is further complicated by the fact that "expert" advice is subject to change practically without notice. One generation of mothers was indoctrinated in the idea of John B. Watson[1] that the proper maternal attitude is one of objective aloofness. But less than a generation later Margaret Ribble[2] and her followers were exhorting mothers to be warm and affectionate and to fondle their infants as much as possible. As the generation reared under Ribbleism entered parenthood, the literature began to emphasize that "parents too have rights" and that discipline (instead of fondling) gives the child security.

Like women's clothing, then, the subject of child-rearing in contemporary society has become subject to the swings of fashion. No doubt an important reason for the pendulum-swings in the advice of the "experts" is that their "knowledge" has never been grounded in solid scientific method.[3] As grandmothers became "old fogies," they clearly lost their authority. This loss created a market for expertise where there was not enough knowledge for anyone to be an expert. The selection from Brim's book assesses evidence concerning the effects of efforts to "educate" parents on how to raise children.

THE CHOICE OF VARIABLES IN THE STUDY OF SOCIALIZATION

Eleanor E. Maccoby

Adapted from Eleanor E. Maccoby, "The Choice of Variables in the Study of Socialization," *Sociometry*, 1961, 24, 357-371.

Perhaps the greatest change that has occurred in the field of child development in the past 15 years has been the increasing emphasis on socialization. The change may be traced by comparing the more traditional text books with recent ones. The scholarly child psychology text by Munn [15], for example, does not bring up the topic of parent-child interaction until the 16th chapter, and here devotes only eight pages to a topic called "environmental influences and personality," a heading under which he presents all that the book has to say on "mothering," on Freudian theory of developmental stages, on ordinal position—in fact, on socialization in general. Contrast this with a book such as Watson's [22], in which more than half the book is devoted to a discussion of socialization theory and a

[1] *The Psychological Care of Infant and Child,* New York, Norton, 1928, pp. 81-82.

[2] *The Rights of Infants,* New York, Columbia University Press, 1943.

[3] As pointed out in chapter 1, the more studies fall short of the fourfold design, the less scientifically conclusive they are. Most studies in this field have been of the clinical or one-cell design. For a criticism of Ribble's work see S. R. Pinneau, "A Critique on the Articles by Margaret Ribble," *Child Development,* 21, 1950, 203-228.

detailed consideration of the impressive amounts of research that have recently been done on the subject.

The same increasing emphasis on socialization may be seen in the child-development journals. And, of course, the widespread research interest in this topic has led to the development of several research instruments for the measurement of parental attitudes and behavior. There are the Fels scales [2], developed during the 40's, for the rating of parent behavior; the parent interview schedule developed by Sears and his associates at Harvard and Stanford [20], the parent attitude scales developed by Shoben at U.S.C. [21], and the widely-used Parent Attitude Research Instrument scales developed at the National Institute of Health by Schaefer and Bell [19], to mention only a few. Each investigator, when he sat down to make a first draft of his rating scale or interview schedule or attitude scale items, had to ask himself the question: what shall I measure? What are the important variables in parental behavior that ought to make a difference in the development of the child? The process of selecting and defining variables is, of course, the very heart of theory-making. There are as many possible variables as there are ideas about what causes what in human development. I cannot attempt here to give any sort of roster of variables; the task would be too great and might not prove very useful. I simply want to point out some of the major classes of variables that have been used and give a little of the history of the reasons why we have chosen to measure these things and not others and perhaps point to a few ways in which we could clarify the meaning of the dimension we are using.

Let us start with the traditional child psychologist, with his interests in motor development, emotional development, intelligence, concept formation, and personality development, all grounded in traditional principles of learning and maturation. He may look upon the current work in socialization with a jaundiced eye and inquire what the excitement is all about. He may feel that he has actually been studying socialization for years without calling it by this name. He might put his question this way: If it is true that socialization is the process of transmitting culture from one generation to another, and that the child acquires the modes of behavior prescribed by his culture through the process of learning, then how is the study of socialization any different from the study of learning itself? One might reply that in socialization studies, we study not only the child as learner but the parent as teacher. But a skeptic might still wonder how much difference this actually makes. For example, laboratory studies of learning have demonstrated that behavior which is followed by reward will be strengthened, and its probability of recurrence will be increased. Now, if a student of socialization does a study of dependency, and discovers that parents who reward their children for dependency have more dependent children, has he really found out anything that we didn't know already?

In my opinion, it is valuable to carry out at the human level studies which attempt to employ the standard variables that have grown out of laboratory studies on learning where most of the work has been done on sub-human species. But, in the process of applying such variables to socialization studies, the variables almost perforce undergo certain modifications and elaborations, with the result that translating traditional behavior theory variables into the socialization setting sometimes results in the addition of something new, and the possibility of getting new kinds of principles.

Let me give an example. Suppose we wanted to study the effects of a particular schedule of reward. What do we mean by reward? The traditional approach to reward has been to produce a physiological drive, such as hunger or thirst, through deprivation; and then to reinforce the desired behavior by presenting a drive-relevant reinforcing stimulus. But even in fairly young children, a rapid development of complex motivation occurs, and this changes the nature of the reinforcements to which children will be responsive. B. F. Skinner encountered this fact when he was developing his teaching machines. The early models were devised so as to emit little pieces of chocolate candy whenever a child made the correct response. But it was soon evident that a child progressed through a series of arithmetic or spelling problems just as readily without the candy; in fact, the giving of candy sometimes disrupted the learning process. Skinner, therefore, abandoned the candy rewards, and the current models of his machine rely upon no other reward than the child's interest in doing his work correctly—buttressed, no doubt, by a certain amount of pressure from the teacher and parents. This incident illustrates a major question about the definition of variables: what happens to the variable "amount of reward" when it is translated into situations of teacher-child, or parent-child, interaction? In modern societies, children's physiological drives are regularly and quite fully satisfied and are seldom used as a basis for training. That is, most parents do not let the child get hungry, thirsty, wet, or overtired, and then make the satisfaction of these needs conditional on good behavior. Rather, the rewards used are money, a trip to the zoo, being allowed to stay up for a special TV program, etc. A gift of candy for some children becomes symbolic of affection instead of vice versa. Very commonly, behavior is reinforced simply through the giving of approval, affection, or attention. So the concept "reward," when it refers to the rewards which parents use in socializing their children is not directly comparable to the concept as it was orginally developed in studies of animal learning. Of course, it is not really a new idea to point out that different kinds of organisms are capable of being rewarded by different kinds of things. It is clear enough that there are as many kinds of rewards as there are distinguishable motives, and that both motives and rewards vary between species and within species. But the new idea that has been added in socialization studies is that there may be distinguishable *classes* of rewards which may have different effects. The primary distinction made in studies so far has been between material reward and praise. Material reward covers all instances of giving the child some object or privilege that he wants, conditional upon good behavior. Praise depends to some degree upon the previous establishment of a relationship between the socializing agent and the child, such that the approval of this particular adult is something the child wants. That is, the effectiveness of praise ought to depend upon the identity of the person doing the praising and upon this person's being someone the child loves, fears, or must depend upon for the satisfaction of needs.

The same kind of differentiation of a variable has occurred with respect to punishment. Students of the socialization process have been working under the assumption that not all kinds of aversive events following a child's act will have the same effect. The distinction most commonly made is that between physical punishment and so-called love-oriented discipline, or withdrawal of love. There are other categories of punishment, too, such as withdrawal of privileges and ridicule, which are less interesting than the first two because there are fewer hypotheses about their probable effects. Let us concentrate for a moment on the distinction between physical punishment and withdrawal of love. Physical punishment is easy enough to define, although in rating its frequency and severity, the researcher is always troubled about the problem of how to weigh slaps and shakings in relation to formal spankings. More tricky by far is the matter of defining withdrawal of love. Sears and his associates [20] have defined it as any act or statement on the part of the parent that threatens the affectional bond between the parent and child. This would include the mother's turning her back on the child, refusing to speak to him or smile at him or be in the same room with him, saying she doesn't like him when he does disapproved things, etc. The system of classification of techniques of discipline presented by Beverly Allinsmith in her chapter in Miller and Swanson's book, *Inner Conflict and Defense,* [1] similarly emphasizes the distinction between "psychological" and "corporal" punishment, but defines psychological discipline somewhat differently. This classification for Allinsmith includes manipulating the child by shaming the child, appealing to his pride or guilt, and expressing disappointment over his misdeeds. But there is another dimension considered in the rating: namely, the amount of emotional control the mother displays in administering her discipline. Thus, if a mother shouts angrily at the child, "I hate you for doing that," Allinsmith would *not* classify this as psychological discipline, while Sears et al. would. But the mother who says calmly and perhaps coldly, "Now, dear, you know I don't like little boys who do that," would be classified as using psychological discipline in both systems. The difference in these two classification systems stems in part from two different views of the nature of the process which gives psychological discipline its effect. Sears et al. view it as a technique which arouses the child's anxiety over whether he is loved and approved of, and thereby elicits efforts on the child's part to regain his parents' approval by conforming, apologizing, or making amends. Allinsmith, on the other hand, emphasizes two things: (1) the *modeling* function of discipline, pointing out that a mother who loses her temper at the same time she is trying to teach the child to control his, will have a child who will do as the mother *does* rather than as she *says;* and (2) the target the child chooses for the aggressive impulses aroused in him as a consequence of punishment. The reasoning here is that the openly angry mother becomes a more legitimate target for the child's counter-aggression. The distinction between the two definitions of the dimension is further brought out when we consider the kinds of findings reported in the studies using them: Sears et al. found that withdrawal of love was associated with high development of conscience, physical punishment with low; Allinsmith found that psychological discipline, as she defined it, was associated with *indirect* fantasy expressions of aggression in the children they studied, corporal punishment with *direct* expression of aggression. All this illustrates the fact that fairly subtle differences in the definition of a dimension can affect the nature of child behavior that can be predicted from it. But more importantly, both these studies illustrate the fact that when we attempted to take over the variable "punishment" from the learning laboratories, we found it necessary to subdivide and differentiate the variable and gained predictive power by doing so.

I have been attempting to cite ways in which I think that socialization studies have improved upon some of the standard variables employed in laboratory studies. There are instances, alas, in which we have not taken note of the differences which exist between the laboratory and the standard socialization settings, and thus have failed to identify and make use of some potentially promising variables. For example, in laboratory studies, we can take it for granted that the experimenter is there during training sessions, administering either reinforcements or aversive stimuli in some orderly relationship to the subject's responses. In the parent-child relationship, the parent is by no means always functioning as a trainer, and parents differ greatly in the degree to which they do so. Some parents keep track quite continuously of what the child is doing, and engage in a constant flow of interaction, both verbal and non-verbal, with the child. Other parents, for a substantial portion of the time they are with their children, are bored, busy, withdrawn, intoxicated, watching television, or subject to some other state or activity which precludes their responding to the child unless he becomes very insistent. In such a household the children are, of course, in a very different learning situation than children growing up with more wholly attentive parents. I think the sheer amount of interaction may in some cases be a more important variable for predicting characteristics of the child than the nature of the interaction that does occur. Let me give you an example. In a study Dr. Lucy Rau and I are now doing at Stanford, we have selected groups of children who show discrepancies in their intellectual abilities. That is, we have one group of children who are good at verbal tasks but poor at number, another group who are good at spatial tasks but poor at verbal, etc. One of our students, Mrs. Bing, has interviewed the mothers of the children, and has also conducted some observation sessions in which the mother presents achievement tasks to the child while the observer records the kind and amount of the mother's involvement with the child's work. Mrs. Bing has found that it is the *amount*, rather than the *kind*, of mother-child interaction that best predicts what the child's pattern of intellectual skills will be. That is, the mothers of the highly verbal children use more praise, but also more criticism, than do the mothers of equally bright children whose area of special skill is non-verbal. Their total level of interaction with the child is greater, and this interaction includes the administration of what we would regard as aversive stimuli as well as reinforcements. The variable "amount of interaction" emerged in our factor analysis of the scales in the *Patterns of Child Rearing* study [20]—we titled this variable "responsible child-rearing orientation" for lack of a better name, but we never made much use of the variable because it did not fit in with the theoretical formulation of our study. But I suspect that for any future work in which we are trying to predict such things as the child's cognitive maturity level or his achievement motivation, we may find that this variable is a better predictor than the less global variables (such as amount of praise) that we have been relying on up till now.

So far, I have been discussing the process of translating variables from laboratory studies of learning to the socialization setting, and have pointed out that we have been successful in employing such variables as reward and punishment, but that in the process of using these variables, we have found useful ways of subdividing them. Let us consider the theoretical meaning of the elaborations of these variables that have occurred.

When we make the distinction between material reward and praise, and the distinction between love-oriented punishment and punishment that depends for its effect upon producing direct physical pain, we are really taking note of the fact that the effect of discipline, and in fact the very nature of the discipline that is possible to use with a child, depends upon the history of the relationship that has been developed between the child and the person who is training him. And here is a new class of variables that socialization studies have added to the list of variables derived from classical studies of learning. In laboratory studies of learning, it has not been found necessary (at least until very recently) to ask whether the experimental subject loved or hated the machine that was emitting pellets of food and drops of water, or whether the characteristics of the machine or person presenting the rewards made any difference in the effectiveness of the reinforcement. Socialization studies, on the other hand, have found the identity of the socializing agent, and certain of his personality characteristics, to be important.

The emphasis on the importance of the relationship between trainer and learner came, of course, out of psychodynamic theories of personality development.

Learning theory and psychoanalytic theory differ, I think, with respect to what they believe the basic nature of the socialization process is. This is an oversimplification, but I believe it would be reasonably accurate to say that a learning theorist would regard socialization as a learning process in which certain actions of the child's are selected out by virtue of reinforcement, others tried and dropped because they are in some way punished or non-reinforced. The parents have a primary role in administering the rewards and punishments for the child's actions, although they do not necessarily do this deliberately and consciously as a teaching effort. And, of course, there are other sources of reward and punishment than the parents' reactions which will help to determine what behavior the child retains.

The psychoanalytic approach, on the other hand, would emphasize not the detailed learning of specific actions on the basis of their outcome, but the providing of conditions which will motivate the child to take on spontaneously the socialized behavior the parent wants him to have. The terms introjection, internalization, learning through role-playing, and identification have been used in this connection; they all refer to the child's tendency to copy, to take on as his own, the behavior, attitudes, and values of the significant people in his life, even when the socializing agents have not said "that's a good boy" or given him a piece of candy for performing these acts or holding these values. I will not go into the controversy concerning which so much has been written as to whether the child is more likely to identify with the person who is powerful and feared or with the person who is loved; nor will I discuss the several thoughtful efforts by personality theorists to reconcile the two points of view. The only important point for our consideration here is that the psychoanalytic view of socialization has led to an exploration of such variables as the warmth or hostility of the socializing agent toward the child.

There can be no doubt that measures of the warmth of the parent-child relationship have turned out to be enormously useful in socialization studies, in a number of ways. In some studies, warmth has been found to have a direct relationship to some dependent variable. For example, McCord and McCord [13] have found that warmth in fathers was associated with low crime rate in sons. In other studies, warmth has turned out to be a useful crosscutting variable which interacts with other variables in such a way that other variables only begin to show their effects when the sample is first sub-divided into groups differing in parental warmth. For example, in the *Patterns of Child Rearing* study, Sears et al. [20] found that withdrawal of love is associated with rapid development of conscience, but only if this technique is employed by a warm mother; also that punishment for toilet accidents disrupts the toilet training process, but that the greatest disruption occurs if punishment is administered by a cold mother.

Warmth also occupies a central role in socialization studies in its relationship to other measures of child-training variables. There have been, to my knowledge, three factor analyses carried out on sets of socialization variables. One of these was on the Fels parent behavior rating scales [17], one on the PARI [26], and one on the dimensions employed by Sears et al. in the Patterns study [20]. In the latter two, warmth emerged as a fairly clear factor. In the first, there were two factors, one called "concern for the child" and the other called "parent-child harmony," which taken together are probably close to what is meant by warmth in the other two studies. It is clear, then, that both in terms of its predictive value for the child's behavior and its central place among the other interrelated child-training variables, warmth is a variable to be taken seriously. Why is it so important? I have already pointed out why the psychodynamic theorists believe it to be so—because of its role in producing identification. But the laboratory learning theorists can acknowledge its importance for another very simple reason: before a parent can socialize a child, he must have established a relationship with the child such that the child will stay in the vicinity of the parent and orient himself toward the parent. A warm parent keeps the child responsive to his directions by providing an atmosphere in which the child has continuous expectations that good things will happen to him if he stays near his parent and responds to his parent's wishes. Fear of punishment can also make the child attentive to the parent, of course, but it establishes as well the conflicting motivation to escape out of reach of the punisher.

I'm sure I needn't belabor any further the notion that warmth is an important variable. But to say this is not enough. We still are faced with considerable difficulty in definition. It has been the experience of a number of people working with child training data that they find themselves able to make reliable distinctions between mothers they call warm and mothers they call cold, and they find it possible to train others to make similar distinctions, but find it difficult indeed to define exactly what cues they are using to make the rating.

I suspect one source of difficulty is that the behavior we look for as indicating warmth varies with the age of the child the mother is dealing with. When the child is an infant, we are likely to label a mother as warm if she gives a good deal of the contact comfort that Harlow* has described. As the child grows older, the part played by the giving of contact comfort in the total constellation of warmth undoubtedly declines. When a child is ten, a mother seldom expresses her warm feelings for him by holding him on her lap. Rather, they are more likely to be expressed by the mother showing interest in the child and what he is doing, by helping unconditionally when help is needed, by being cordial and relaxed. Now warmth as expressed this way is not the same thing as giving contact comfort, and it is not to be expected that the same individuals would necessarily be good at both. Those who have read Brody's fascinating, detailed descriptions of mothers' behavior toward their infants [4] will perhaps have noted that the mothers who gave effective contact comfort, in the sense of holding the child comfortably and close, stroking it occasionally, imparting some rocking motion, handling it skillfully and gently in the process of caring for the child—the women who could do all these things well were not necessarily the same women who expressed delight and pride in their children, who noticed their little accomplishments, or who looked upon their infants as individuals. We should therefore not be surprised if there are low correlations between a mother's warmth toward her infant and her warmth toward the same child when it is older. If a primary ingredient of warmth is being able to gratify the child's needs unconditionally, and if the child's needs change from the infantile needs for being fed and being given contact comfort to the more mature needs for various kinds of ego support, then it is necessary for a mother to change considerably as her child changes, in order to be warm towards him at all ages. Some mothers make this change more easily than others.

* See H. F. Harlow, "The Nature of Love," *Amer. Psychologist*, 1958, 13, pp. 673-685—*Eds.*

It is true that Schaefer and Bayley [18], in their longitudinal study of a group of mothers, did find a substantial degree of continuity in the degree of warmth displayed by a given mother toward a given child as the child grew older. There were undoubtedly individual differences in the ways warmth was manifested, and in the appropriateness of a mother's particular style of warmth-giving to the needs of her child at each developmental stage.

From the standpoint of making use of the variable in research, it appears that we should recognize that measuring the mother's current warmth at the time the child is, say, in nursery school or in the primary grades, may not be an especially good index of how warm she was to the child as an infant. Furthermore, her warmth in infancy might predict quite different characteristics of the child than her warmth in middle childhood. If there is any relation at all between nurturance to an infant and its later personality traits, infant nurturance ought to relate only to those aspects of personality that presumably have their foundation in infancy—such as Erikson's dimension of trust [6], or various aspects of orality. Achievement motivation, on the other hand, if it is related to the mother's warmth at all, ought to be related to measures of this variable taken when the child is older. A finding of Bronfenbrenner's [5] seems to support this point about the importance of warmth-giving being appropriate to the developmental level of the child. He was studying high-school-aged children and employed several variables relating to the kind and amount of affectionate interchange between these adolescents and their parents. He measured the parents' affection-giving (in the sense of direct demonstrativeness), use of affective rewards, nurturance, and affiliative companionship. Among these variables, it was only the last one, affiliative companionship, that correlated with the child's current level of responsibility taking. We can speculate that this particular aspect of warmth is the one that fits in much better with an adolescent's needs than either giving him kisses or peanut butter sandwiches. All this means that warmth has to be defined in terms of parental responsiveness to the changing needs of the child.

I have referred to socialization variables that came originally from laboratory studies of learning, and that have been adapted for use in studying the socialization process. I have also referred to variables that originated in psychodynamic thinking. There is a set of variables that is difficult to classify in terms of these two theoretical systems; I am referring to the dimension "permissiveness vs. restrictiveness," which emerged in our factor analysis of the *Patterns* variables, and to the related dimension of "control vs. laissez-faire" which has come out of the factor analysis of the PARI scales. The theoretical status of these variables is confusing because they relate to both psychoanalytic and learning theory, but the predictions from the two theories as to the probable effects of "permissiveness" or "control" are sometimes quite different. To cite a familiar example, there is the issue of what ought to be the effects of permissive treatment of the infant's sucking responses. The question is complex, but a simplified version of the opposing positions would be this: the learning theorist would argue that if an infant is permitted extensive sucking, his sucking habit will be strengthened, and he will be more likely to suck his thumb, pencils, etc., at a later age. The psychodynamic theorist would argue that permitting extensive infantile sucking satisfies oral needs and reduces the likelihood of excessive oral behavior at a later age. The same kind of difference of opinion can be found concerning whether permissive treatment of a child's aggressive or dependent responses should increase or decrease those responses. Now, of course, the fact that different theories produce different predictions concerning the effects of a variable is no reason for abandoning the variable. On the contrary, it is cause for rejoicing, and we should by all means continue to use the variable so that we can get data which will bear upon the validity of the theories. The trouble is that when we arrive at the point of trying to get agreement on the interpretation of findings, it sometimes turns out that the two schools of thought did not mean the same thing by "permissiveness." If a study shows that the more permissive parents are toward their children's aggression the more aggressive the children become, the psychodynamic theorist may say, "Well, by permissiveness I didn't mean *license;* the child must have limits set for him but he must also be allowed to express his feelings." If, on the other hand, a study shows that children heavily punished for aggression are more aggressive on the playground, or prefer aggressive TV programs, the learning theorist may say, "Well, of course, if the parents' methods of stopping aggression are such as to provide additional instigation to aggression, then their non-permissiveness won't eliminate the behavior." We begin to see that there are some hidden meanings in such a term as "permissiveness" and that we are dealing with several dimensions. Continuing with the example of aggression, we can see that permissiveness for aggression could mean the following things:

1. The mother holds the attitude that aggression is an acceptable, even desirable, form of behavior.
2. The mother does not like aggressive behavior and expects to limit it in her children, but feels that it is natural and inevitable at certain ages and so does not react strongly when her young child displays anger. A related definition of permissiveness would be placing the demands for self control placed upon the child to correspond with his developmental level.
3. The mother is not especially interested in the child or is otherwise occupied,

and does not act to stop or prevent his aggression because she does not notice what he is doing unless his actions are aimed directly at her.
4. The mother does not act early in a sequence of her child's aggressive behavior, but waits till the behavior has become fairly intense.

And at the other end of the scale, the effect of *non*-permissiveness ought to depend upon how the non-permitting is done—whether by punishment, by reinforcing alternative behavior, by environmental control that removes the instigations to undesired behavior, or some other means. The basic point I wish to emphasize is that I believe "permissiveness" is not a unitary variable, and that we need to work more directly with its components.

So far I have discussed several classes of variables: the ones translated as directly as possible from laboratory studies of learning (e.g., amount and kind of reward and punishment), and variables such as warmth and permissiveness of the socializing agent, which have their origins more in psychodynamic theories. There is another class of variables which has been emerging as more and more important, namely the "social structure" variables. These variables have their origin largely in sociological thinking. I do not have time to give them more than the most cursory attention, but I do not believe they can be omitted if we are to do any sort of justice to the scope of significant variables employed in current socialization studies. One has only to list a few findings which have come out of the investigation of social structure factors to see how essential it has become to take them into account. Here is a brief sampling of such findings:

1. In adolescents, parents are most strict with children who are of the same sex as the dominant parent. [16].
2. A mother's use of strongly dominant child-rearing techniques (called "unqualified power assertion" in this study) is related to her husband's F score (authoritarian personality score), but not to her own. [10].
3. A mother's behavior toward her children is more closely related to her husband's education than her own, and her behavior is more closely related to her husband's education than is *his* behavior with his own education. Thus it appears that it is the family's social status, as indicated by the husband's education, that influences the mother's socialization practices. [5].
4. Sons are more intra-punitive if their mothers are primarily responsible for discipline than they are if their fathers are the primary disciplinarians. [9].
   Aspects of social organization such as whether residence is patrilocal, matrilocal, or neolocal, and whether marriage is polygamous or monogamous, determine such aspects of culture as the length of the post-partum sex taboo, the duration of exclusive mother-child sleeping arrangements, and the amount of authority the father has over the child; these factors in turn determine such socialization practices as the age of weaning, the severity of the socialization pressures which are directed toward breaking up the child's dependency upon the mother, and the existence and nature of puberty rites at adolescence.

These socialization practices then in their turn determine certain aspects of personality, including certain culturally established defense systems. [23, 24, 25].

6. When offered a choice between a small piece of candy now vs. a large one later, children from father-present homes can postpone gratification more easily than children from father-absent homes. [14].

These findings all represent efforts to put socialization practices into a cultural context. In each case, socialization practices are regarded as a link in a several-step chain, and consideration is given to the factors which determine the socialization practices themselves, as well as to the effects these practices in their turn have upon the child. It is clear that the way parents treat their children will be a function of their relationship to each other (especially of the distribution of authority between them), of the place the family has in the status system of the society in which the family resides, of the society's kinship system, etc. Of course, not every student of socialization need concern himself with all the steps in the complex sequence; he may, and often does, select a set of socialization practices and relate them to the child's behavior without going back to the conditions which led to these practices. But he needs to be aware of the degree to which socialization practices are embedded in a cultural context, and even needs to be alert to the possibility that the "same" socialization practice may have different effects when it is part of different cultural settings. So far, few studies have been planned or analyzed with this possibility in mind, but it might be worth some empirical examination.

It is time to make explicit an assumption that has been implicit so far about the constancy of personality from one situation to another and from one time to another. When we select aspects of parental behavior to study, and try to relate these to measured characteristics of the child, we usually measure what we believe to be reasonably pervasive, reasonably enduring "traits" of the parent and child. Orville Brim [3] in a recent paper, has leveled a direct attack at the notion of trait constancy. He has asserted that there is no such thing as a "warm" person, nor an "aggressive" person, nor a "dependent" person, but that behavior is specific to roles. This would mean that the same individual may be aggressive with his subordinates and dependent toward his boss; that a child may be emotionally expressive with his same-sexed age mates, but not with his teachers or his parents. The question of exactly how general personality traits are, is, of course, a matter that personality theorists have struggled with for many years. But our view of this matter will have some bearing upon our selection and definition of socialization variables. For if a child's behavior is going to be entirely specific to roles, then there is no point in trying to predict any generalized traits in the child; rather, we should be looking for those aspects of the socialization situation that will determine what behavior will be adopted by the child in each different role relationship in which he will find himself. If we wanted to find what socialization practices were associated with the child's becoming dominant or submissive, for example, we would have to study how his dominant behavior had been reacted to when he was playing with same-sexed siblings, and study this separately from the socialization of the same behavior when he was playing with opposite-sexed siblings; only thus could we predict, according to Brim, how dominant he would be with other boys in the classroom, and we would have to make a separate prediction of his dominance with girls in the classroom. We have already been following Brim's advice, in essence, when we do studies in which we test how the child's behavior varies with the role characteristics of the person with whom he is interacting. A good example is Gewirtz' and Baer's study on the interaction between the sex of the experimenter and the effects of interrupted nurturance [7]. But to follow Brim's point further, we would have to investigate the ways in which the child's behavior toward specific categories of "others" was conditioned by differential socialization in these role relationships.

I do not believe that either socialization or the child's reaction tendencies are as role-specific as Brim claims; but obviously role differentiation does occur, and he is quite right in calling our attention to the fact that for some variables at least, we should be studying socialization separately within roles. Actually, role is only one aspect of situational variability; we have known ever since the days of Hartshorne and May [8] that trait behavior like "honesty" is situation-specific; they found, for example, that the child who will cheat on the playground is not necessarily the same child who will cheat in the classroom, and that cheating is a function of the specific task presented to the child. This means that in studying the effects of socialization, we either have to abandon efforts to predict characteristics like "honesty" and attempt to study only those characteristics of the child that are at least somewhat constant across situations, or we have to choose socialization variables that are themselves much more situation-specific, and make much more detailed predictions. An example of the utility of making socialization variables more specific to the situations they are intended to predict is provided in a study by Levy [12], in which it was found that a child's adjustment to a hospital experience was *not* a function of the parents having trained the child generally to meet many different kinds of stress situations; rather, the child's response to hospitalization was predicted only from the amount of training the parent gave in advance for the meeting of this *particular* stress situation.

The same sort of situation prevails with respect to trait constancy over time. In their recent article on dependency, Kagan and Moss [11] were able to present repeated measurements of dependency in the same group of individuals—measurements which began at the age of three and continued into the late twenties. The most notable feature of their findings was the absence of continuity in this trait. The children who were dependent at age three and four were not the same individuals who emerged as dependent in adulthood. There was simply no continuity at all for boys, while there was some, but not a great deal, for girls. Let us consider Kagan's finding from the standpoint of efforts to study the socialization practices that are related to dependency. The first and obvious point is that we cannot expect to find any characteristic of the parent's behavior that will correlate with dependency in the young child and also correlate with dependency when the child is an adolescent or adult. This is not to say that the only correlations we can hope for are those between socialization practices and child characteristics measured at the same point in time. It is of course most likely that we will be able to find aspects of a parent's current behavior that correlate with characteristics his child is displaying at the same time. But it is also possible that we could find aspects of the parent's current behavior whose effects will not show up until later. That is, perhaps there were things the parents of Kagan's sample of children were doing when these children were three and four that had some bearing upon how dependent the children became at the age of ten or eleven. But it is clear enough that whatever these delayed-action variables are, they could hardly be the same variables as the ones which determined how dependent the children were at age three, since it was not the same children who were displaying large amounts of dependency behavior at the two ages.

I have pointed to the way in which different theoretical systems, and different social science disciplines, have converged to define and elaborate some of the variables which have been used in studies of socialization. In some cases this convergence has produced useful new knowledge; in others it has produced confusion over the meaning of variables. More importantly, it has produced a startling range of findings which have not yet been integrated into a theory of socialization. This is a major task that remains to be done.

REFERENCES

1. Allinsmith, Beverly, "The directness with which anger is expressed," in D. R. Miller and G. E. Swanson, *Inner Conflict and Defense* (New York: Holt, 1960), chapter 14.

2. Baldwin, A. L., Kalhorn, Joan, and Breese, Fay H., "The Appraisal of Parent Behavior," *Psychol. Monogr.*, 1949, 63.

3. Brim, O. G., Jr., "Personality development as role learning," in I. Iscoe and H. W. Stevenson, eds., *Personality Development in Children* (Austin: Univer. Texas Press, 1960).

4. Brody, Sylvia, *Patterns of Mothering* (New York: International Universities Press, 1957).

5. Bronfenbrenner, U., "Some Familial Antecedents of Responsibility and Leadership in Adolescents," unpublished manuscript, Cornell University, 1959.

6. Erikson, E. H., *Childhood and Society* (New York: Norton, 1950).

7. Gewirtz, J. L. and Baer, D. M., "Does Brief Social 'Deprivation' Enhance the Effectiveness of a Social Reinforcer ('Approval')?" *Amer. Psychologist*, 1956, 11, 428-429.

8. Hartshorne, H. and May, M. A., *Studies in Deceit* (New York: Macmillan, 1928).

9. Henry, A. F., "Family Role Structure and Self-blame," *Social Forces*, 1956, 35, 34-38.

10. Hoffman, M. L., "Power Assertion by Parents and Its Impact on the Child," *Child Develpm.*, 31, 1960, 129-144.

11. Kagan, J. and Moss, H. A., "The Stability of Passive and Dependent Behavior from Childhood through Adulthood," *Child Develpm.*, 1960, 31, 577-591.

12. Levy, E., "Children's Behavior under Stress and Its Relation to Training by Parents to Respond to Stress Situation," *Child Develpm.*, 1959, 30, 307-324.

13. McCord, W. and McCord, Joan, *The Origins of Crime* (New York: Columbia Univ. Press, 1959).

14. Mischel, W., "Preference for Delayed Reinforcement: An Experimental Study of a Cultural Observation," *J. Abnorm. Soc. Psychol.*, 1958, 56, 57-61.

15. Munn, N. L., *The Evolution and Growth of Human Behavior* (Boston: Houghton Mifflin, 1955).

16. Papanek, Miriam L., "Family Structure and Child-Training Practices," unpublished doctoral dissertation, Radcliffe College, 1954.

17. Roff, M., "A Factorial Study of the Fels Parent Behavior Scales," *Child Develpm.*, 1949, 20, 29-45.

18. Schaefer, E. S. and Bayley, Nancy, "Consistency of Maternal Behavior from Infancy to Preadolescence," *J. Abnorm. Soc. Psychol.*, 1960, 61, 1-6.

19. Schaefer, E. S. and Bell, R. Q., "Development of a parental attitude research instrument," *Child Develpm.*, 1958, 29, 339-361.

20. Sears, R. R., Maccoby, Eleanor E., and Levin, H., *Patterns of Child Rearing* (Evanston: Row Peterson, 1957).

21. Shoben, E. J., "The Assessment of Parental Attitudes in Relation to Child Adjustment," *Genet. Psychol. Monogr.*, 1949, 39.

22. Watson, R. I., *Psychology of the Child* (New York: Wiley, 1959).

23. Whiting, J. W. M., "Sin, sorcery and superego," in M. R. Jones, ed., *Nebraska Symposium on Motivation* (Lincoln: Univ. Nebraska Press, 1959), pp. 174-195.

24. Whiting, J. W. M., Chasdi, Eleanor H., Antonovsky, Helen F., and Ayres, Barbara C., "The Learning of Values," in C. Kluckhohn and E. Z. Vogt, eds., *The Peoples of Rimrock: A Comparative Study of Values Systems*. In press.

25. Whiting, J. W. M., Kluckhohn, R., and Anthony, A., "The Function of Male Initiation Rites at Puberty," in Elea-

nor E. Maccoby, T. M. Newcomb, and E. L. Hartley, eds., *Readings in Social Psychology* (New York: Holt, 1958), pp. 359-370.

26. Zuckerman, M., Barrett-Ribback, Bea-

trice, Monashkin, I., and Norton, J., "Normative Data and Factor Analysis on the Parental Attitude Research Instrument," *J. Consult. Psychol.*, 1958, 22, 165-171.

EVIDENCE CONCERNING THE EFFECTS OF EDUCATION FOR CHILD REARING

Orville G. Brim, Jr.

Adapted and reprinted from *Education for Child Rearing*, by Orville G. Brim, Jr., New York, Russell Sage Foundation, 1959, pp. 287-317.

What Does the Evidence Show?

In the section that follows we will review all the evidence on the effects of parent education programs, even though much of it is poor by the standards set forth above. Throughout we will endeavor to sound a note of caution, recognizing that poor studies may be dangerously misleading. Before we turn to the experimental studies, there are three other approaches to evaluation which require brief mention.

One of these [e.g., 33] is to ask persons responsible for the program for a subjective evaluation of the results which were produced. For example, leaders of child study groups when asked their opinion of a program may report that they think "the effects are good." This commands no attention as a serious research effort, since it involves the obvious problem of bias when one is judging one's own work; vast research literature on the distorting effects of motivation on judgment and perception indicates all too clearly that what people think happened, is happening, or will happen is colored by their own desires.

Second, there are a number of studies employing a case history approach. For example, Kinnis [25] and French and her colleagues [16] present cases in which brief educational counseling of a mother resulted in change in the mother's behavior which was beneficial to the child. Such studies are useful in the same general way that single case studies are ordinarily useful, namely, as a source of insight and suggestion for further theory and research. However, there is no way of estimating whether such results would occur in other similar cases, since the number is too small to justify any estimate of the contribution of random changes. In addition, no controls are present. The extension of such case findings to other cases, not to speak of parent education programs, is an unscientific and misleading procedure.

A third line of investigation which has become pertinent to the effects of parent education makes use of data suggesting historical shifts in child-care practices among American parents. The interpretation of the historical data is complicated and seems to permit no clear conclusion. The major line of argument is that parent education materials over the past two decades have urged the parent toward greater leniency in child care [36, 37]; therefore, any shifts in actual parent practice toward greater leniency during the past two decades can be interpreted as the effect of these parent education materials.

Some of the data on historical change come from comparisons of older with younger persons. Other data are from longitudinal studies of parents. Staples and Smith [35], for example, have shown that grandmothers are significantly more strict in child care than their daughters, now mothers themselves. However, in this and other longitudinal studies an equally probable explanation of the stricter attitudes and behavior of older parents is simply that they are older, and perhaps have less patience with children; in other words, this constitutes a maturational effect rather than an effect of the younger group being exposed to a different kind of parent education. Another interpretation of the same data is that there has been a general shift in the United States in all aspects of the society toward a more liberal attitude concerning other individuals, which would show in parent role behavior of younger persons and, indeed, if one wished to go farther, might even account for the changes in the advice transmitted to parents. In any event, the problems of interpretation of age difference data are serious ones.

Still other data on historical change come from the comparison of samples of parents studied at different times during the past two decades in respect to the degree of leniency of their child-rearing practices. Bronfenbrenner [8] has recently completed an analysis of all such studies dating from Anderson's [1], published in 1936, and including a number of recent and still unpublished studies. The author has done a notable service to our understanding of parent behavior. In this review the various subjects in the different samples of parents are classified as to whether they are white- or blue-collar workers in order to make the studies comparable. In reviewing the overall trends of the data he finds that reliance on breast feeding is decreasing while self-demand scheduling is becoming more common. With respect to class differences on the practices of weaning, of bowel and bladder training, and of both breast feeding and self-demand scheduling, he finds that while these were less common among the middle-class or white-collar wives before World War II, the direction has now been reversed and the middle-class mother is relatively more permissive than the lower-class. Bronfenbrenner then points out the relation between these trends and Wolfenstein's [40] analysis of changes in content. . . . The changes in the middle-class mothers parallel quite closely the changes noted in the Children's Bureau publication, *Infant Care*. Bronfenbrenner calls attention to another body of evidence . . . showing that middle-class mothers are more likely to read such publications on child care than are working-class mothers. He concludes, therefore, that his analysis suggests that these mothers not only read *Infant Care* (and other materials) but take them seriously and over time are influenced by them.

Bronfenbrenner's analysis of these historical data thus has the characteristics of an experimental evaluation in which the middle class is the experimental group and the working class is (comparatively speaking) the control group. Continuous appraisals of its self-reports on attitudes and overt behavior over time indicate that the experimental group changes significantly in the direction of attitudes and behavior advocated by parent education programs. While one can readily think of several other explanations of this pattern of results, they tend to be rather complicated and speculative and have neither the simplicity nor the common sense characteristics of Bronfenbrenner's interpretation.

If the interpretation is valid, then implications follow both for parent education and for its evaluation. First, one concludes that parent education has had some measurable effects on the American parent. Second, one surmises that the effects of parent education over any short period of time and for any given group of individuals may be so small as to escape notice in the ordinary experimental evaluation studies to be reviewed subsequently; but that the change over many years and in many people is such that after several decades significant effects are observed. In conclusion, it now appears that the strategy of evaluation research should include periodic appraisals of attitudes and behavior of parents and children, which are then related to data on the information presented and on exposure rates, so that the long term effects of parent education can be analyzed.

Introduction to the Research Studies

. . . only a few of the many studies undertaken in parent education are satisfactory from the standpoint of design and analysis. Unhappily, the majority of them have various characteristics which run the gamut of research deficiencies: no controls, failure to handle loss of subjects, procedures not specified clearly, use of inappropriate tests of significance, and so on. As already indicated, we will mention them all but will distinguish between them according to level of competence.

The studies are classified into three major groups. The first includes all studies evaluating a single method; the second, studies of multiple methods; while the third consists of studies comparing the effectiveness of different methods. Within each, the studies are further subdivided into those with either partially complete, or complete experimental designs. Finally, these subclasses are themselves subdivided according to the dependent variable (the "effect") being studied. The section closes with a brief overview of research in progress.

Table 1 summarizes this classification, and shows the result of each study. This table is helpful as a guide to the survey which follows.

TABLE 1.   A Classification of Studies Evaluating the Effects of Parent Education*

Method Evaluated	Effects Studied in Partially Complete Design			
	Parent Knowledge	Parent Attitudes	Parent Behavior	Child Behavior
Group procedures	22(+) 31(+)	9(+) 13(+) 22(+) 23(?)	24(+)	
Mass media				
Individual counseling			7(?)	7(?)
Combined methods		10(—) 38(—)	10(+) 21(+) 26(+) 38(—)	17(+)

Method Evaluated	Effects Studied in Complete Design			
	Parent Knowledge	Parent Attitudes	Parent Behavior	Child Behavior
Group procedures	29(+)	5(?) 6(?) 29(—) 32(+)		
Mass media	15(+)		18(—) 27(?+) 30(+)	
Individual counseling		19(?)		11(?+)
Combined methods	2(+)			

* The figures under the various headings refer to studies listed at the end of this article. The (+), (—), and (?) *after* the figures indicate, respectively, that the study reports significant improvement, no change, or results that cannot be interpreted clearly.

Note that in the table the few studies [2, 5, 6, 13, 31] comparing the effectiveness of different methods are reported in the rows for the actual methods concerned for the reasons set forth on pages 244-245.

Studies Evaluating the Effectiveness of a Single Method

1. Studies with Partially Complete Experimental Designs (No Control Group). The first set of studies with a partially complete experimental design evaluates the effects of group discussion procedures on increasing parents' knowledge. Studies by Hedrich [22] and Schaus [31] both report significant increases in factual knowledge.

Turning now to attitude as the criterion, Chandler [9] had as subjects 28 mothers exposed to an eight-week reading and group discussion course. He used a "traditional-developmental" measure [14] which was administered to the mothers before and after the study-group program. This test is based on responses to the questions: What are five things a good mother does? and What are five things a good child does? Chandler found that responses to the "good mother" questions were significantly more developmental after participation in the program. (The probability of the change occurring by chance was less than one in a hundred. Henceforth, such statistically relevant findings will be written as follows: $<.01$.) This difference was also true for responses to the "good child" question ($<.05$, that is, the probability of its chance occurrence was less than five in a hundred).

Another study [23, pp. 80-81], using only mothers who attended a series of six or eight meetings, investigated changes in the mother's ranking of six items, describing what she hoped to gain from the study group. The results show that the mother's preference for a general philosophy of child care and for general knowledge of children increased significantly, while the desire for reassurance about handling the child and for specific information on how to handle difficulties declined ($<.10$).

Hedrich [22] used as subjects four groups of parents with a total of 48 subjects. Each group met six times. The educational program was centered on teaching positive attitudes and practices for parents toward the development of self-reliance in their children. It was focused specifically on the four areas of eating, sleeping, toileting, and use of clothing. A before-and-after administration of Ojemann's self-reliance scale [28] showed a significant increase in favorable attitudes of the parents toward self-reliance practices. Attitudes also were significantly improved in child-rearing areas other than those dealt with in the groups, for example, play, thus suggesting that the change was generalized.

The largest study which utilized before-and-after measures with an experimental group, but had no control group, is the Davis and McGinnis research [13]; the subjects consisted of members of study groups under the auspices of the Institute of Child Welfare in Minnesota. The individual subjects totaled more than 1,000. The attitude instrument consisted of a fifty-item trait list pertaining to children, and the parents were asked to rate the degree of importance or seriousness of each of these traits on a four-point scale. Ratings were made separately for boys and girls, and for the ages five, nine, and fifteen. Comparisons of ratings before and after the study-group series show that there is an average reduction in the degree to which parents conceive the traits as serious. This result holds generally for ratings for all different sexes and ages. Separate analysis considers the shift of particular items into and out of the category of the "ten most serious" traits. Ratings after the study-group program resulted in the inclusion among the ten most serious items several which pertain to withdrawal types of behavior previously not perceived as serious.

Where parents' overt behavior, rather than attitudes, is concerned, only one study exists. This study by Jack [24] used 38 mothers as subjects and used home interviews before and after a study-group series to obtain information on the mothers' child-rearing practices. The items were scored on the basis of expert judgment as to the degree of favorableness of the practice. Comparison of before-and-after scores showed that the group as a whole improved in the direction of experts ($<.10$). Comparisons within the group between the two halves who were initially the low and high scorers show that the initially low scorers made a significant improvement in changing behavior ($.01$), while the latter did not ($.20$). However, this is likely the result of a regression of the extremes toward the average when tested a second time, a phenomenon well known and adequately explained by the theory of sampling.

Turning now to the next type of parent education method, mass media, there seem to be no evaluation studies with a partially complete experimental design which have been concerned with this method. We thus pass on to studies of individual counseling. One study [7] falls into this class and was designed to evaluate the effectiveness of a counseling procedure paralleling that used in numerous well-baby or child health conferences. The study had a before-and-after design with no control group, the focus being on internal comparisons between subjects. While in one sense it might be said the study had its own internal controls since the emphasis was on the relation of change to personality variables, the fact is that, strictly speaking, it is not a complete experimental design. The subjects were 50 mothers attending child health stations in New York City who reported feeding problems. Subjects were selected on the basis of age, birth order, and health of their children, and the fact that the subjects' current behavior in response to the feeding problem was nonpermissive.

The mothers were interviewed and counseled individually, with permissive handling of feeding practices described and recommended, and a pamphlet describing this practice was given to each. The mothers' own report of the handling of food refusals was obtained during the initial interview and then again in a second interview of the same type three to four months later. The results showed that 8 of the mothers adopted permissive practices in feeding, 16 tried the practice but rejected it, 26 made no attempt to try it.

Perhaps the unusual aspect of the study is that it attempted to relate the effects of educational counseling to several personality characteristics of the subjects. The results suggest that the differential effects of counseling of this type are in part predictable from knowledge of the mothers' characteristics. Some of the findings are the following: of the subjects who considered previous advice from physicians to be helpful, there were more who tried permissive feeding; more of the mothers who were primarily concerned about the child's diet, in contrast to his size, made attempts and subsequently adopted permissiveness; more of the subjects low in general dominance or authoritarianism in child-rearing practices adopted permissiveness than did subjects who were not. When the recommended practice was supported by other sources of parental education to which the subjects were exposed, subjects would more often both try the practice and adopt it.

2. *Studies with Full Experimental Designs.* Several studies fall into the category of research which evaluates the effects of group discussion methods. One by Owings [29], reported in Witmer [39], considered the effects of a course of instruction of parents in how to impart sexual knowledge to their children. This study made separate analyses of knowledge items and of attitudes toward sex instruction. Significant differences for the experimental group but not the control group are reported in gains in knowledge items.

Where attitudes are the focus of research, we find that a study by Balser and his colleagues [5, 6] used 12 parents as one of the experimental groups which participated in a series of group-centered, psychiatrically led, seminars concerning child development and parent-child relations. Two of the four control groups also consisted of parents, and we will be concerned here only with the comparison between the experimental parent group and the two parent control groups. Before-and-after measures for all three groups consisted of the Minnesota Multi-phasic Personality Inventory, of a sentence-completion test, and of a scale of parent attitudes [34]. The results showed no significant changes in any group in the sentence-completion test. On the attitude scale, the experimental group showed improved scores, the before-and-after difference for this group approaching statistical significance ($<.10$). However, one of the two control groups of parents also showed a significant ($<.01$) and even larger improvement in score on this measure. This finding should be kept in mind when we consider other studies which utilize no controls, because it demonstrates the possible error of attributing special effects to some educational program when actually none has occurred. In this instance, we need not necessarily conclude that the seminar had no effect on the parents, but rather that it had no more effect on them than did whatever miscellaneous experiences the control group was having at the same time. On the MMPI scales the experimental group tended to show an improvement on the family relation scale which was not paralleled by the control group; on the other MMPI scales, it showed no noticeable change.

Another study using both experimental and control groups was made by Shapiro [32]. He exposed 25 experimental subjects, carefully matched with control subjects, to a parent education group discussion program consisting of 12 sessions. The group leadership procedures were carefully controlled. The before-and-after measure consisted of five attitude scales based on the work of Shoben [34] and Harris, Gough, and Martin [20], and included measures of authoritarianism, parent-child integration, rigidity, fussiness, and good judgment. They thus parallel to some extent the attitude measures used in the Balser study. The attitude measures were administered both to experimental and control groups in such a way that the subjects did not relate them to the group education program. The results showed the experimental subjects to have improved to a significantly greater degree than the control subjects on the authoritarianism, good judgment, and possessiveness scales, but not on the other two. Moreover, Shapiro found that parents attending four or more meetings changed more than those attending three or fewer; that the change was the result of gains fairly evenly distributed throughout the experimental group; and that those experimental subjects who initially held more desirable attitudes on the scales changed more than those holding less desirable attitudes.

The Owings study [29] referred to above was concerned both with attitudes and with knowledge. In contrast to the Balser and Shapiro studies there were no significant differences on the attitude items in Owings' study.

We turn now to another group of studies, those evaluating the effects of mass media. There are four such studies. The first deals with effectiveness in increasing the knowledge of parents. The other three studies all are concerned with changes in self-reported behavior of parents.

In the first study [15] a pamphlet of the Minnesota Department of Health on *Getting Your Child Ready for School,* which included information on the importance of physical and dental examinations, immunization, safety training, and the like, was given to an experimental group of 21 parents and withheld from a control group of 14. Two days later 10 multiple-choice questions pertaining to the pamphlet were administered to both groups, and the experimental group was found to be superior to a significant degree to the control group in this knowledge.

The remaining three studies are all large-scale studies, and more importantly, they all investigate the effects of the same pamphlet series. The studies are in part comparable, and thus provide us a rare opportunity to consider the similarity of results of similar research. The pamphlet series evaluated was the *Pierre the Pelican* series of the Louisiana Society for Mental Health [30]. The series consists of 12 four-page pamphlets, customarily mailed to parents of first children at monthly intervals during the first year of the infant's life. These pamphlets discuss physical and emotional aspects of child care pertinent to the first year of life, and make concrete suggestions concerning parental care.

Of these large-scale studies, one made in New Orleans [30] utilized an experimental group (identified as "Group II" in the report and in Table 2 [1]) of 159 mothers whose children at the time of the study were sixteen months old and who had received the pamphlet series in the preceding year; a control group consisted of 227 mothers with children of the same age to whom the pamphlets had not been mailed. Another experimental group (identified as "Group I," omitted from Table 2) had been used earlier, but the children of these mothers were not comparable in age to those of the control group, and the results will not be considered here. Subjects were interviewed in their home on the basis of a sixty-item questionnaire. It is important to note that the interviewers knew prior to the interview which subjects were in the experimental and control groups.

On 54 comparisons the results showed 18 significant differences (critical ratios of 2.0 or more) favoring the experimental group, and on almost all remaining items there was a consistent but not significant difference favoring the experimental group. The items tend to be unrelated, so that their content is difficult to summarize.

The second of these large-scale studies [18] was conducted in North Carolina and used two experimental groups, whites and nonwhites, with 868 and 288 subjects, respectively, compared with two control groups, whites and nonwhites, of 765 and 278 subjects respectively. Experimental and control subjects were randomly selected from records of registration of first births in randomly selected counties. The experimental groups received through the mail the complete pamphlet series during the subsequent twelve months, while none was sent to the control groups. Carefully

[1] See p. 255.

trained interviewers visited mothers at their homes after the series was completed. The five key questions all concerned feeding practices (handling of food refusals, degree of concern over food schedules, changes in methods of giving milk, leaving the infant alone with a bottle on his pillow, and encouraging the child to feed himself). The questions were masked so that their evaluative characteristics were not apparent. The interviewers did not know whether a given subject was a member of the experimental or control group. The results show that differences on the five items between the experimental and control group members were not significant in any instance. Only in regard to the last-mentioned item did the difference approach statistical significance between the nonwhite experimental and control groups.

The third [27] in this group of large-scale studies was carried out in Michigan. It utilized experimental and control groups initially composed of 1,000 mothers each. Subjects were selected by sampling from experimental and control counties, based on registration of first births. The pamphlet series was then mailed to the experimental groups but withheld from the control groups. Following exposure to the pamphlets, a forty-three-item questionnaire based on the pamphlet materials was mailed to both groups. Returns (and thus the actual sample sizes) from the experimental group numbered 477, and from the control group 537. A comparison of experimental and control groups showed significant differences ($<.05$) on 10 of the 43 items. Two of these 10 differences favored the control group.

Three other measures of effects were used in that study. One treated all 43 items as if they constituted a test of information, and scored the items as to right answers. The average percentage of correct answers for the experimental group was greater than that for the control, and the difference approached statistical significance ($<.10$). The second considered only "concept items," those involving some understanding beyond simple factual information. These numbered 24 of the 43 items. Again using these as a "test," the average percentage correct for the experimental group was significantly greater ($<.03$) than for the control group. The third compared the effect of 11 background information variables, for example, education of the mother, upon subjects' responses. The results showed that 9 of these 11 background variables had a greater effect upon the answers of the control group than those of the experimental group. This suggests an interesting finding: that the pamphlet series reduced the individual variability in child-care knowledge and attitudes arising from differences in cultural and other background characteristics, by providing a new and common core of knowledge for all experimental subjects.

Another aspect of the Michigan study utilized public health nurse interviews with 30 experimental and 34 control mothers. A twenty-eight-item questionnaire was followed in the interview. Significant differences were found on 11 items, with 10 of these favoring the experimental group.

These three studies clearly show different results for this pamphlet series. In Table 2 we compare the results for those items which are identical or very similar for the studies. The comparisons are mainly between the Louisiana and Michigan investigations.

With respect to the *survey* results from the Michigan study, Table 2 shows that the studies agree in finding significant differences on one item (item 6) and possibly three others (items 4, 5, and 7). They agree in finding no significant differences on eight items (item 2 and items 17 through 23). The studies disagree in their findings (one study reporting significant differences, another none) on the remaining 11 items. Concerning the data from the nurses' interviews in the Michigan study, the studies agree in finding significant differences on three more items (items 10, 12, and 13), but now disagree in their findings where before they had agreed (items 21 and 22).

TABLE 2.   A Comparison of Results for Three Evaluation Studies of *Pierre the Pelican* Pamphlets

Subject Matter Identical or Similar in All Three Studies	Louisiana Study (30)* (Group II data only)	North Carolina Study (18)*	Michigan Study	
			Survey (27)*	Nurses' Interview (27)*
1. Parents do not leave baby with bottle and pillow.	S †	NS	NS	
2. Parents not concerned over scheduling.	NS	NS	NS	
3. Parents try later after child's food refusals.	S	NS		
4. Parents promote independence in feeding.	S	NS ‡		
5. Father changes diapers.	NS §		S	
6. Parents ask child's permission to use his things for new baby.	S		S	
7. Parents do not spank child reaching for breakables.	NS ¶		NS	S
8. Parents do not believe babies are afraid of the dark.	S		NS	
9. The baby goes to sleep in the dark.	S		NS	
10. Parents would take child to guidance center when problems can't be solved.	S		NS	S
11. Parents would give child allowance.	S		NS	
12. Parents believe baby should have separate room.	S		NS	S
13. Parents tell first child the second is coming.	S		NS	S

TABLE 2.   (Continued)

Subject Matter Identical or Similar in All Three Studies	Louisiana Study (30)* (Group II data only)	North Carolina Study (18)*	Michigan Study	
			Survey (27)*	Nurses' Interview (27)*
14. Parents have thought about what to say when child asks where he came from.	S		NS	
15. Parents believe if they tell child where he came from he will be less interested.	S		NS	
16. Make-believe friends indicate loneliness.	S		NS	
17. Use toilet chair for baby.	NS		NS	
18. Parents not bothered when children make house disorderly.	NS		NS	
19. Parents do not believe children under two years should borrow from each other.	NS		NS	
20. Child allowed to do things for self.	NS		NS	
21. Have family budget.	NS		NS	S
22. Parents know older child should have special attention after new baby.	NS		NS	S
23. Parents know children may be jealous of new sibling.	NS		NS	

* Reference number as listed at end of chapter.

† S and NS refer, respectively, to significant or nonsignificant differences between experimental and control groups in the study.

‡ Has borderline significance, nonwhites only.

§ All other items in the "father care" complex show significant differences.

¶ Borderline significance.

Rarely in social investigations are studies sufficiently similar to permit a comparison of their results. However, a comparison is very instructive because it makes apparent the limitation of any one study and the need for repetition of evaluation research. For example, if a generalization were made from the findings of either the Michigan or Louisiana study, the predictions would be wrong as to the outcome of the other study on about one-half of the items (approximately the results that would be obtained if a coin had been flipped). Since the disagreements in the results stem mainly from the Louisiana and Michigan comparison, one might consider the differences to arise from the use of an interview in the first study and a questionnaire in the second. However, this is doubtful for several reasons. The comparison of the results of the nurses' interviews in the Michigan study does not reduce the disagreement; the different methods were used in the Michigan and North Carolina studies, yet on the two items where they can be compared the results agree; moreover, the comparison of similar items in the Louisiana and North Carolina studies with interviewers used in both studies, shows disagreement. One more likely possibility is that the interviewers' knowledge of subjects' membership in the experimental or control groups in the Louisiana study may have resulted in some bias entering into the interview procedure. The most likely explanation of all may be that the parents in the separate studies had different kinds of interests, needs, and cultural traditions, producing differential acceptance and rejection of the material in the pamphlets.

Our next group of studies having complete experimental designs deals with the effects of counseling as a parent education procedure. While two studies fall into this category, in each case there is some difficulty in interpreting the nature of the results. In the first, Cooper [11] has evaluated the results of a counseling program attached to a well-baby clinic in Baltimore. Parents were interviewed and counseled on the general situation of the child, and general advice was given. Data on the child and his situation also were given by the counseling staff to pediatricians and general practitioners of the clinic and to nurses making home visits for their use in counseling.

The measure of effectiveness was the change in the behavior of the child of counseled parents. Use was made of the behavior records of 100 children whose parents had been seen at least three times in a year by the counseling service. The record at the last conference was used. These were compared with the behavior records of the *first* counseling session of 81 other children who were used as controls. The 81 control children were matched pair by pair with 81 of the experimental children on age, sex, and race. The behavior ratings were based on data from an interview with the mother and from observation of the child.

Comparisons of the records of the two groups show those of the experimental group to be generally better (more satisfactory behavior) than those of the controls. Also, 96 of the 100 experimental children improved in ratings from earlier to later sessions. While these results are encouraging, the study design does not in fact permit us to attribute the improvement to the counseling received. It may be that any or most undesirable behavior traits of a child run their course and improve with time; the control procedure used in Cooper's study compares records of behavior traits at possibly different stages of inception and thus does not permit us to rule out this possible explanation of changes.

The other study evaluating the results of a counseling program is by Hale [19], who reports that in a comparison of a group of parents who were counseled with three groups who were not, the former indicated subsequent to counseling that they had many more family life problems than the latter. This study design makes interpretation of findings difficult.

It may be, as someone suggested, that this simply represents what is commonly found in counseling situations, namely, that an emotionally healthy person is cautious about revealing himself and his problems until he has tested a situation and the people involved in it and is convinced that he can expose his problems without being ridiculed. On the other hand, the results may show that educational counseling increases the number of problems one perceives himself to have and reports.

Studies Evaluating the Effects of Multiple Methods in Parent Education

As we have pointed out above, several studies have evaluated multiple procedures in parent education and so cannot be classified under any of the major methods described. Only one such study has had a complete experimental design. This is a study by Andrew [2] which evaluated the overall effects of an educational workshop upon gains in knowledge. This workshop utilized methods of lectures, films, recordings, and various types of group discussions. Using a control group of college students not exposed to this overall program, the results show that the experimental group composed of a large number of parents made a significant increase in knowledge ($<.01$) in contrast to the student control group.

All of the other studies which evaluated multiple techniques have only a partially experimental design, in the sense that the use of control subjects is absent. Two of the studies have examined the effects of multiple methods upon the parents' behavior; two more have dealt with changes both in attitudes and behavior of parents; and a fifth has considered the effects upon children's behavior, one of the few to take this as a criterion variable. Considering the first, Hattendorf [21] concentrated on providing a group of mothers with factual information on sex, and on methods of educating children on sexual matters. One hundred and thirteen mothers were exposed to interviews and counseling, lectures, discussion groups, and printed material, as well as careful observation of programs of instruction of their children in their homes. The results suggest that on a before-and-after basis there was an increase in the use of casual incidental instruction of children in sexual matters as recommended by the program.

A report by Klatskin [26] of the effects upon child-care practices of those who had participated in the Yale Rooming-In Program might be viewed as a study of the effects of exposure to an educational program suggesting "flexible care," which uses counseling, reading, and, of course, some supervision during the rooming-in period in the actual care of the child. Two-hundred and twenty-nine mothers reported on their own child-care practices a year or more after the birth of the child. Klatskin used the Davis and Havighurst materials [12] describing child-care practices as the comparison data, on the assumption that the group had not been exposed to this type of educational program. Comparison on five child-care terms, including feeding when hungry, beginning bowel training before the child is six months old, and the like shows the 229 subjects to be substantially more flexible and/or more permissive in child care than were the Davis and Havighurst subjects. These differences are open to other interpretations, of course; for example, regional variations. Klatskin also reports in regard to differences between social classes in her sample that the middle- and lower-class subjects are approximately identical in their behavior. Thus, one might make the inference that exposure to educational programs reduces the effects of cultural variation on child care. This parallels the finding of the Michigan study [27] concerning the effects of the *Pierre the Pelican* pamphlet series reported above.

Of the two studies concerned both with attitudes and behavior of parents one [38] is described by Witmer [39, p. 74], in which the evaluation is of an unspecified course of instruction. The study reports that before-and-after interviews and tests of mothers' opinions and behavior regarding sex instruction show no significant differences with respect to their behavior toward their children.

The second and more recent study concerned with parental attitudes and behavior was made by Collins [10]. In this study the subjects were 17 mothers attending an annual two-week training program for parents of hearing-handicapped preschool children. The purpose of the program was to promote sound attitudes on child management. The study asked the subjects to report from their own experience examples of what they considered to be good or bad handling of their children during the preceding week at home; examples from this week were again selected during the last part of the workshop. The incidents reported as "good handling" were rated on eight of the Fels Parent Behavior Scales [4]. The results showed significant differences between the ratings of incidents reported before and after the training program, on the control-freedom scale, and the freegrowth-training scale. The Shoben scale [34] was also used as a before-and-after measure, but no significant changes were found.

The fifth study having only a partial experimental design in evaluating multiple methods is one made some years ago by Giblette and Macrae [17], utilizing as subjects mothers having feeding difficulties with their children. The mothers were exposed to a program of class instruction in nutrition, in child psychology, and to a period of observation of their children at nursery school under the direction of school personnel. The educational program also included individual counseling interviews. Subsequent to the program the children were rated as improving on 20 of 30 traits.

Studies Evaluating the Relative Effectiveness of Different Methods

This last group of studies consists of those which have sought to evaluate the relative effectiveness of two or more major methods, that is, group discussion, mass media, or individual counseling, or studies which have sought to evaluate the relative effectiveness of variations in procedure within each of these major methods.

Some of the studies in this group have employed a complete experimental design, in that control groups are added and the experimental groups are exposed to variations in method. All the studies which have this complete design have been referred to previously, for they have achieved two objectives in one research study. This follows from the fact that the answer to the question of whether or not parent education is effective can be obtained by comparing all experimental groups with the control groups receiving no education. If, in addition, information is obtained on the relative effectiveness of different educational procedures by virtue of having such experimental variations in the study, then one can ask what is the relative effectiveness of the different methods themselves? We now ask this additional question of some of the studies previously reported.

The first class of studies is concerned with comparisons of *major methods*. There is only one of this kind. It compares the effects of group discussion with the effects of lectures. We have pointed out in Chapter VII on methods that although there is a great deal of literature in other areas of social change dealing with the relative effectiveness of group discussion and other methods, the same studies had not been repeated in the area of parent education. The existing study is by Schaus [31]. Subjects consisted of members of three lecture groups and three study discussion groups. Each had eight sessions. Mimeographed summaries of the content material, which was the same for both the lecture and study groups, were given to all subjects and a comparison of the before-and-after gains in knowledge of subject matter showed the study group to have learned significantly more ($<.01$) than the lecture group.

The second class of studies deals with variations within major methods. The first two are expansions of studies reported previously which utilize a complete experimental design. In one of these studies Andrew [2] evaluated the effects of four different types of leadership. Eight groups of seven to ten each participated in discussion programs with the four following types of leadership procedure: group-oriented, authority, question-answer, and leaderless. Each of the four procedures was used with two groups, making eight groups in all. Members of the groups consisted of parents, teachers, and public health nurses. The groups met from two to three times. A thirty-item questionnaire, based on the content of the information discussed, was administered before and after the meetings. Comparisons of the groups on all items showed that all the procedures resulted in gain, but that there was a significantly greater increase *in knowledge* in the two groups which had no leader. The other groups did not differ significantly. Balser and his colleagues [5, 6] used a research design permitting a comparison between study groups using different leadership procedures. One was group-centered and led by a psychiatrist; a second was leadership-centered and led by a psychologist. While, as we have pointed out earlier, changes did occur as a result of the educational program, there were no apparent differences between groups attributable to the use of a psychologist or psychiatrist as leader, or to the leader-centered in contrast to group-centered process of discussion.

Remaining studies of variations within a method used no control groups. That of Davis and McGinnis [13] evaluated the effects of variations in leadership of discussion groups. The evaluation of the effects of different leaders employed a fifty-item list of children's behavior traits which were rated by parents as to seriousness. The parents' ratings were then compared with the ratings of experts, and the degree of agreement found. These ratings were obtained from parents both before and after the discussion groups. The amount of *improvement* in agreement with experts was compared for groups led by "specialists trained in child development and child psychology" and "groups . . . taught by local leaders." The results showed greater improvement for subjects taught by the specialists. The possible effects of other factors, such as education, were controlled. The authors' tentative conclusion is that experts can teach mental hygiene better than nonexperts.

Two reports by Andrew complete this class. Both come from her study [2] evaluating the effect of a workshop in parent education. As one experimental variation, an attempt was made to evaluate the relative effectiveness of four different techniques of presentation of information. This seems to pertain most closely to variations in the effectiveness of different types of mass media. All subjects were exposed to one general session of the following types: a panel discussion by lay persons, a lecture, a film, and two recordings. A thirty-item questionnaire based on the content covered in four general sessions was administered before and after the series of general sessions. The results show that there was significantly greater improvement in knowledge of the content covered in the sessions using the records and the lecture. However, the possible interaction between the type of content presented and the method of presentation makes it impossible to conclude whether the greater improvement results from the technique, or from the associated content being easier to learn.

Using the same general experimental situation, Andrew [3] investigated the effects of a variation which is hard to classify, but seems to pertain more to variation in content of the educational procedure than to method. If it is so construed, it is, of course, the only one like this. The same thirty-item information test was used as in the foregoing research, to test the learning in different groups. It was found that in groups which permitted the expression of individual "cathartic needs," and where subjects did in fact express them, there was a tendency for a greater learning of factual material to take place than in groups where catharsis was not permitted. . . .

In Table 1 on page 247 we have presented a simple classification of the existing evaluation studies. A brief summary of the nearly two dozen studies leaves little doubt that their results are inconclusive.

With respect to the method of group procedures, consider the studies using a complete experimental design. One of these [29] reports an increase in parent knowledge, and no change in parent attitudes. A more recent and more significant study [32], on the other hand, finds definite improvement in parent attitudes. The third study [5, 6], however, finds no improvement on one attitude measure, a significant improvement on subscales of another, and a change paralleled by change in the control group on still a third.

Five of the six studies evaluating group procedures which use no control groups, and hence have only a partially complete design, report significant change of some kind occurring in the parents involved in the program. One of these studies [24] reports an improvement in parent behavior; of four others concerned with attitudes [9, 13, 22, 23], three indicate a significant improvement; one of these [22] joins another [31] in reporting significant gains in parent knowledge.

At first glance, it would appear that the evidence points toward significant and desirable changes in parents resulting from participation in parent study or discussion groups. However, when the six not truly scientific evaluation studies are removed from the picture, the other three mentioned above present inconclusive results.

Turning to the research on the effects of mass media, all four of the studies in this field have employed control groups in a complete experimental design. One of these, a comparatively small study [15] concerned with improvement in factual information about children's health, reports a favorable change occurring in the parents involved. The other three major studies have all been concerned with changes in parent attitudes. One of these three [18] reports that no changes occurred; the other two [27, 30] report positive effects upon parent attitudes. However, in these two there are certain deficiencies in the selection of the control groups and, in addition, the positive changes reported differ from one study to the other; thus, the findings of each are open to various interpretations.

With respect to the method of individual counseling, there are only three studies in all. One of these [7] shows changes in both parent and child behavior, although the number of persons changing was few, and the study used no controls. Of the other two studies, which used a more complete experimental design, one [19] indicates improvement in children's behavior, but there was a methodological question involving changes due to maturation raised in connection with this study; the other study [11], really in the nature of a minor report, suggests that counseling increases the number of problems one perceives he has or is willing to report.

In the areas of both mass media and counseling, then, as was the case in the area of group procedures, the data delineate no clear conclusions. Nor is the situation better when we consider the studies evaluating the effectiveness of multiple procedures. Only one of these [2] uses an experimental design in the true sense and this study reports parents as increasing their factual information. The remaining studies in this category, of which there are five, all lack controls; and, in addition, present a quite mixed picture of changes. Two of these [10, 38] report no changes in parent attitudes. With respect to behavior, two [10, 21] report positive changes, one [38] no change, and for the other [26] the results are positive but subject to question: A final study [17] reports significant improvement in child behavior but this is a study having little value. Thus, while the one important study here indicates that parents improve in factual knowledge, this is hardly an impressive finding, and the other studies in this area are inconsistent in their results.

The issue of how effective is parent education in changing parents or children therefore remains unresolved at present. In the absence of conclusive evidence, the arguments favoring one side or the other of the issue doubtless will continue. On the one hand, there are those who argue that there is no reason to expect any important aspects of the adult personality to change as a result of parent education. Protagonists of this position would point out that we do not even know if the adult individual ever undergoes any important changes other than changes in factual knowledge even when exposed to educational experiences much more impressive than parent education, such as attending college or doing graduate work in child development over a two- or three-year period. Therefore, it is unreasonable to expect that participating one hour a week in a twelve-week seminar, or reading a pamphlet, or being counseled for half an hour during a monthly visit to a pediatrician could influence the parent-child relation.

On the other hand, an argument presented by many is that the educational programs for parents do produce changes but that for several reasons they will continue to escape detection, given the current measurement procedures. The changes which occur are alleged to be too subtle to be captured by other than clinical techniques, or too small to show up in the sparse samples of parents utilized in most evaluation studies; or are delayed in their occurrence so that they are not discernible except through a longitudinal study. The argument that changes occur but are too small to be detected has received support from Bronfenbrenner's review of the historical data on child care previously mentioned in detail. The supposition is that the changes in any given instance are too small to be measurable but that they are cumulative, so that exposure of parents to a variety of educational events, for example, not to just one pamphlet, or one type of counseling, or one study-group session, but to dozens of pamphlets and discussions over time, produces cumulative change in parent behavior.

Since the issue of effectiveness of parent education is still unresolved, one looks forward to future studies. The critic of programs who seeks to demonstrate that the efforts involved have little value, the educator himself who seeks to convince the public and other professionals that education for parents is justified, the independent social scientist interested in a test of theory, all will be involved in major evaluation research. This increased interest must result in a deeper understanding of the significance of this modern social movement.

REFERENCES

1. Anderson, John E., *The Young Child in the Home: A Survey of Three Thousand American Families.* Report of the Committee on the Infant and Preschool Child, White House Conference on Child Health and Protection (D. Appleton-Century Co., New York, 1936).

2. Andrew, Gwen, "A Study of the Effectiveness of a Workshop Method for Mental Health Education," *Mental Hygiene,* vol. 38, 1954, pp. 267-278.

3. Andrew, Gwen, "The Relationship Between Learning and Expression of Self-Orientation Needs at a Mental Health Education Workshop," *Mental Hygiene,* vol. 38, 1954, pp. 627-633.

4. Baldwin, Alfred L., Joan Kalhorn, and Fay H. Breese, "Patterns of Parent Behavior," *Psychological Monographs,* vol. 58, 1945, no. 3.

5. Balser, Benjamin H., Fred Brown, Minerva L. Brown, Edward D. Joseph, and Donald K. Phillips, "Preliminary Report of a Controlled Mental Health Workshop in a Public School System," *American Journal of Psychiatry,* vol. 112, 1955, pp. 199-205.

6. Balser, Benjamin H., Fred Brown, Minerva L. Brown, Leon Laski, and Donald K. Phillips, "Further Report on

Experimental Evaluation of Mental Hygiene Techniques in School and Community," *American Journal of Psychiatry*, vol. 113, 1957, pp. 733-739.

7. Brim, Orville G., Jr., "The Acceptance of New Behavior in Child Rearing," *Human Relations*, vol. 7, 1954, pp. 473-491.

8. Bronfenbrenner, Urie, "Socialization and Social Class Through Time and Space" in *Readings in Social Psychology*, edited by Theodore M. Newcomb, Eugene L. Hartley, and Eleanor E. Maccoby. 3d ed. (Henry Holt and Co., New York, 1958), pp. 400-425.

9. Chandler, Barbara A., "An Exploratory Study of the Professed Parent-Role Concepts and Standards of Child Behavior of Mothers in a Parent Education Project," *Dissertation Abstracts*, vol. 15, 1955, pp. 219-220.

10. Collins, Marjorie G., "A Study of Parent Attitudes on Child Management Before and After Training, Utilizing the Critical Incident Technique," *Dissertation Abstracts*, vol. 14, 1954, pp. 872-873.

11. Cooper, Marcia M., "Evaluation of the Mother's Advisory Service," *Monographs of the Society for Research in Child Development*, vol. 12, 1947, p. 1.

12. Davis, Allison, and Robert J. Havighurst, "Social Class and Color Differences in Child-Rearing," *American Sociological Review*, vol. 11, 1946, pp. 698-710.

13. Davis, Edith A., and Esther McGinnis, *Parent Education: A Survey of the Minnesota Program* (University of Minnesota Press, Minneapolis, 1939).

14. Duvall, Evelyn M., "Conceptions of Parenthood," *American Journal of Sociology*, vol. 52, 1946, pp. 193-203.

15. Ford, M., and E. E. Hartman, "Measuring Reader Comprehension of a Preschool Pamphlet," *Public Health Reports*, vol. 69, 1954, pp. 498-502.

16. French, Anne C., M. Levbarg, and H. Michal-Smith, "Parent Counseling as a Means of Improving the Perform-

ance of a Mentally Retarded Boy: A Case Study Presentation," *American Journal of Mental Deficiency*, vol. 58, 1953, pp. 13-20.

17. Giblette, C. T., and A. Macrae, "An Experiment in the Treatment of Feeding Problems Through Parental Education," *Mental Hygiene*, vol. 18, 1934, pp. 92-108.

18. Greenberg, B. G., M. E. Harris, C. F. MacKinnon, and S. S. Chipman, "A Method for Evaluating the Effectiveness of Health Education Literature," *American Journal of Public Health*, vol. 43, 1953, pp. 1147-1155.

19. Hale, Clara B., "Parent Need for Education and Help with Family Problems," *California Journal of Educational Research*, vol. 6, 1955, pp. 38-44.

20. Harris, Dale B., Harrison G. Gough, and William E. Martin, "Children's Ethnic Attitudes: II. Relationship to Parental Beliefs Concerning Child Training," *Child Development*, vol. 21, 1950, pp. 169-181.

21. Hattendorf, Kay W., "A Home Program for Mothers in Sex Education," *University of Iowa Studies in Child Welfare*, vol. 6, 1932, pp. 11-92.

22. Hedrick, Blanche E., "The Effectiveness of a Program of Learning Designed to Change Parental Attitudes Toward Self-Reliance," *University of Iowa Studies in Child Welfare*, vol. 10, 1934, pp. 249-268.

23. Institute of Child Study, Toronto, *Well Children: A Progress Report* (University of Toronto Press, Canada, 1956).

24. Jack, Lois M., "A Device for the Measurement of Parent Attitudes and Practices," *University of Iowa Studies in Child Welfare*, vol. 6, 1932, pp. 137-149.

25. Kinnis, Gladys C., "Emotional Adjustment of the Mother to the Child with a Cleft Palate," *Medical Social Work*, vol. 3, 1954, pp. 67-71.

26. Klatskin, Ethyln H., "Shifts in Child Care Practices in Three Social Classes

under an Infant Care Program of Flexible Methodology," *American Journal of Orthopsychiatry*, vol. 22, 1952, pp. 52-61.

27. Michigan State Department of Mental Health, *A Report of Some Aspects of the Effectiveness of the Pierre the Pelican Mental Health Pamphlets*. Lansing, Mich., 1952, mimeographed.

28. Ojemann, Ralph H., "The Measurement of Attitude Toward Self-Reliance," *University of Iowa Studies in Child Welfare*, vol. 10, 1934, pp. 101-111.

29. Owings, Chloe, *The Effectiveness of a Particular Program in Parental Sex Education* (University of Minnesota Press, Minneapolis, 1931).

30. Rowland, Loyd W., *A First Evaluation of the Pierre the Pelican Health Pamphlets*. Louisiana Mental Health Studies, No. 1, Louisiana Society for Mental Health, New Orleans, 1948.

31. Schaus, Hazel W., "An Experimental Investigation of Methods in Parent Education," *University of Iowa Studies in Child Welfare*, vol. 6, 1932, pp. 117-134.

32. Shapiro, Irving S., "Is Group Parent Education Worthwhile? A Research Report," *Marriage and Family Living*, vol. 18, 1956, pp. 154-161.

33. Shirley, May, *Can Parents Educate One Another? A Study of Lay Leadership in New York State*. National Council of Parent Education, New York, 1938.

34. Shoben, Edward J., Jr., "The Assessment of Parental Attitudes in Relation to Child Adjustment," *Genetic Psychological Monographs*, vol. 39, 1949, pp. 101-148.

35. Staples, Ruth, and June W. Smith, "Attitudes of Grandmothers and Mothers Toward Child-Rearing Practices," *Child Development*, vol. 25, 1954, pp. 91-97.

36. Stendler, Celia B., "Sixty Years of Child-Training Practices," *Journal of Pediatrics*, vol. 36, 1950, pp. 122-134.

37. Vincent, Clark E., "Trends in Infant Care Ideas," *Child Development*, vol. 22, 1951, pp. 199-209.

38. Witmer, Helen L., *The Attitudes of Mothers Toward Sex Education*. University of Minnesota Press, Minneapolis, 1929.

39. Witmer, Helen L., *The Field of Parent Education: A Survey from the Point of Research*. National Council of Parent Education, New York, 1934.

40. Wolfenstein, Martha, "Trends in Infant Care," *American Journal of Orthopsychiatry*, vol. 23, 1953, pp. 120-130.

9. SOCIAL STRUCTURE AND SOCIALIZATION

In this chapter we present research that shows that socialization is affected by two sets of structural variables. In the first paper Barry, Bacon and Child ask whether, in the differential rearing of the sexes, our society makes "an arbitrary imposition on an infinitely plastic biological base," or if the cultural imposition is to be found "uniformly in all societies as an adjustment to the real biological differences between the sexes." They find that societies differ in the degree of emphasis upon sex-typed behavior and that such differences are correlated with the nature of the economy. In particular, an economy in which physical strength is important tends to emphasize large sex differences in the process of socialization.

The second article deals not with the macrostructures of societies but with the microstructures of families, and more precisely with the variable of same-sex versus opposite-sex siblings in two-child families. Brim has reanalyzed data gathered previously by Koch and finds that in these data: (1) cross-sex siblings have more traits of the opposite sex than do same-sex siblings and (2) this effect is greater for the younger rather than the older sibling.

A CROSS-CULTURAL SURVEY OF SOME SEX DIFFERENCES IN SOCIALIZATION

Herbert Barry III, Margaret K. Bacon, and Irvin L. Child

Adapted from the *Journal of Abnormal and Social Psychology*, 55, 1957, 327-332. Most of the societies used in this article are listed in "Relation of Child Training to Subsistence Economy" by Barry, Child and Bacon, *American Anthropologist*, 1959, 61, 51-63.

In our society, certain differences may be observed between the typical personality characteristics of the two sexes. These sex differences in personality are generally believed to result in part from differences in the way boys and girls are reared. To the extent that personality differences between the sexes are thus of cultural rather than biological origin, they seem potentially susceptible to change. But how readily susceptible to change? In the differential rearing of the sexes does our society make an arbitrary imposition on an infinitely plastic biological base, or is this cultural imposition found uniformly in all societies as an adjustment to the real biological differences between the sexes? This paper reports one attempt to deal with this problem.

Data and Procedures

The data used were ethnographic reports, available in the anthropological literature, about socialization practices of various cultures. One hundred and ten cultures, mostly nonliterate, were studied. They were selected primarily in terms of the existence of adequate ethnographic reports of socialization practices and secondarily so as to obtain a wide and reasonably balanced geographical distribution. Various aspects of socialization of infants and children were rated on a 7-point scale by two judges (Mrs. Bacon and Mr. Barry). Where the ethnographic reports permitted, separate ratings were made for the socialization of boys and girls. Each rating was indicated as either confident or doubtful; with still greater uncertainty, or with complete lack of evidence, the particular rating was of course not made at all. We shall restrict the report of sex difference ratings to cases in which both judges made a confident rating. Also omitted is the one instance where the two judges reported a sex difference in opposite directions, as it demonstrates only unreliability of judgment. The number of cultures that meet these criteria is much smaller than the total of 110; for the several variables to be considered, the number varies from 31 to 84.

The aspects of socialization on which ratings were made included:

1. Several criteria of attention and indulgence toward infants.
2. Strength of socialization from age 4 or 5 years until shortly before puberty, with respect to five systems of behavior; strength of socialization was defined as the combination of positive pressure (rewards for the behavior) plus negative pressure (punishments for lack of the behavior). The variables were:

(a) Responsibility or dutifulness training. (The data were such that training in the performance of chores in the productive or domestic economy was necessarily the principal source of information here; however, training in the performance of other duties was also taken into account when information was available.)

(b) Nurturance training, i.e., training the child to be nurturant or helpful toward younger siblings and other dependent people.

(c) Obedience training.

(d) Self-reliance training.

(e) Achievement training, i.e., training the child to orient his behavior toward standards of excellence in performance, and to seek to achieve as excellent a performance as possible.

Where the term "no sex difference" is used here, it may mean any of three things: (a) the judge found separate evidence about the training of boys and girls on this particular variable, and judged it to be identical; (b) the judge found a difference between the training of boys and girls, but not great enough for the sexes to be rated a whole point apart on a 7-point scale; (c) the judge found evidence only about the training of "children" on this variable, the ethnographer not reporting separately about boys and girls.

Sex Differences in Socialization

On the various aspects of attention and indulgence toward infants, the judges almost always agreed in finding no sex difference. Out of 96 cultures for which the ratings included the infancy period, 88 (92%) were rated with no sex difference by either judge for any of those variables. This result is consistent with the point sometimes made by anthropologists that "baby" generally is a single status undifferentiated by sex, even though "boy" and "girl" are distinct statuses.

On the variables of childhood socialization, on the other hand, a rating of no sex difference by both judges was much less common. This finding of no sex difference varied in frequency from 10% of the cultures for the achievement variable up to 62% of the cultures for the obedience variable, as shown in the last column of Table 1. Where a sex difference is reported, by either one or both judges, the difference tends strongly to be in a par-

TABLE 1.   Ratings of Cultures for Sex Differences on Five Variables of Childhood Socialization Pressure

Variable	Number of Cultures	Both Judges Agree in Rating the Variable Higher in		One Judge Rates No Difference, One Rates the Variable Higher in		Percentage of Cultures with Evidence of Sex Difference in Direction of		
		Girls	Boys	Girls	Boys	Girls	Boys	Neither
Nurturance	33	17	0	10	0	82%	0%	18%
Obedience	69	6	0	18	2	35%	3%	62%
Responsibility	84	25	2	26	7	61%	11%	28%
Achievement	31	0	17	1	10	3%	87%	10%
Self-reliance	82	0	64	0	6	0%	85%	15%

ticular direction, as shown in the earlier columns of the same table. Pressure toward nurturance, obedience, and responsibility is most often stronger for girls, whereas pressure toward achievement and self-reliance is most often stronger for boys.

For nurturance and for self-reliance, all the sex differences are in the same direction. For achievement there is only one exception to the usual direction of difference, and for obedience only two; but for responsibility there are nine. What do these exceptions mean? We have reexamined all these cases. In most of them, only one judge had rated the sexes as differently treated (sometimes one judge, sometimes the other), and in the majority of these cases both judges were now inclined to agree that there was no convincing evidence of a real difference. There were exceptions, however, especially in cases where a more formal or systematic training of boys seemed to imply greater pressure on them toward responsibility. The most convincing cases were the Masai and Swazi, where both judges had originally agreed in rating responsibility pressures greater in boys than in girls. In comparing the five aspects of socialization we may conclude that responsibility shows by far the strongest evidence of real variation in the direction of sex difference, and obedience much the most frequently shows evidence of no sex difference at all. . . .

The observed differences in the socialization of boys and girls are consistent with certain universal tendencies in the differentiation of adult sex role. In the economic sphere, men are more frequently allotted tasks that involve leaving home and engaging in activities where a high level of skill yields important returns; hunting is a prime example. Emphasis on training in self-reliance and achievement for boys would function as preparation for such an economic role. Women, on the other hand, are more frequently allotted tasks at or near home that minister most immediately to the needs of others (such as cooking and water carrying); these activities have a nurturant character, and in their pursuit a responsible carrying out of established routines is likely to be more important than the development of an especially high order of skill. Thus training in nurturance, responsibility, and, less clearly, obedience, may contribute to preparation for this economic role. These consistencies with adult role go beyond the economic sphere, of course. Participation in warfare, as a male prerogative, calls for self-reliance and a high order of skill where survival or death is the immediate issue. The childbearing which is biologically assigned to women, and the child care which is socially assigned primarily to them, lead to nurturant behavior and often call for a more continuous responsibility than do the tasks carried out by men. Most of these distinctions in adult role are not inevitable, but the biological differences between the sexes strongly predispose the distinction of role, if made, to be in a uniform direction.[1]

The relevant biological sex differences are conspicuous in adulthood but generally not in childhood. If each generation were left entirely to its own devices, therefore, without even an older generation to copy, sex differences in role would presumably be almost absent in childhood and would have to be developed after puberty at the expense of considerable relearning on the part of one or both sexes. Hence, a pattern of child training which foreshadows adult differences can serve the useful function of minimizing what Benedict termed "discontinuities in cultural conditioning" [1].

The differences in socialization between the sexes in our society, then, are no arbitrary custom of our society, but a very widespread adaptation of culture to the biological substratum of human life.

Variations in Degree of Sex Differentiation

While demonstrating near-universal tendencies in direction of difference between the socialization of boys and girls, our data do not show perfect uniformity. A study of the variations in our data may allow us to see some of the conditions which are associated with, and perhaps give rise to, a greater or smaller degree of this difference. For this purpose, we classified cultures as having relatively large or small sex difference by two different methods, one more inclusive and the other more selective. In both methods the ratings were at first considered separately for each of the five variables. A sex difference rating was made only if both judges made a rating on this variable and at least one judge's rating was confident.

In the more inclusive method the ratings were dichotomized, separately for each variable, as close as possible to the median into those showing a large and those showing a small sex difference. Thus, for each society a large or a small sex difference was recorded for each of the five variables

---

[1] For data and interpretations supporting various arguments of this paragraph, see Mead [2], Murdock [3], and Scheinfeld [6].

on which a sex difference rating was available. A society was given an over-all classification of large or small sex difference if it had a sex difference rating on at least three variables and if a majority of these ratings agreed in being large, or agreed in being small. This method permitted classification of a large number of cultures, but the grounds for classification were capricious in many cases, as a difference of only one point in the rating of a single variable might change the over-all classification of sex difference for a culture from large to small.

In the more selective method, we again began by dichotomizing each variable as close as possible to the median; but a society was now classified as having a large or small sex difference on the variable only if it was at least one step away from the scores immediately adjacent to the median. Thus only the more decisive ratings of sex difference were used. A culture was classified as having an over-all large or small sex difference only if it was given a sex difference rating which met this criterion on at least two variables, and only if all such ratings agreed in being large, or agreed in being small.

TABLE 2. **Culture Variables Correlated with Large Sex Difference in Socialization, Separately for Two Types of Sample**

Variable	More Selective Sample		More Inclusive Sample	
	∅	N	∅	N
Large animals are hunted	.48*	(34)	.28*	(72)
Grain rather than root crops are grown	.82**	(20)	.62**	(43)
Large or milking animals rather than small animals are kept	.65*	(19)	.43*	(35)
Fishing unimportant or absent	.42*	(31)	.19	(69)
Nomadic rather than sedentary residence	.61**	(34)	.15	(71)
Polygyny rather than monogamy	.51*	(28)	.38**	(64)

* $p < .05$.
** $p < .01$.

Note.—The variables have been so phrased that all correlations are positive. The phi coefficient is shown, and in parentheses, the number of cases on which the comparison was based. Significance level was determined by χ^2, or Fisher's exact test where applicable, using in all cases a two-tailed test.

We then tested the relation of each of these dichotomies to 24 aspects of culture on which Murdock has categorized the customs of most of these societies[2] and which seemed of possible significance for sex differentiation. The aspects of culture covered include type of economy, residence pattern, marriage and incest rules, political integration, and social organization. For each aspect of culture, we grouped Murdock's categories to make a dichotomous contrast (sometimes omitting certain categories as irrele-

[2] These data were supplied to us directly by Professor Murdock.

vant to the contrast). In the case of some aspects of culture, two or more separate contrasts were made (e.g., under form of marriage we contrasted monogamy with polygyny, and also contrasted sororal with nonsororal polygyny). For each of 40 comparisons thus formed, we prepared a 2×2 frequency table to determine relation to each of our sex-difference dichotomies. A significant relation was found for six of these 40 aspects of culture with the more selective dichotomization of over-all sex difference. In four of these comparisons, the relation to the more inclusive dichotomization was also significant. These relationships are all given in Table 2, in the form of phi coefficients, along with the outcome of testing significance by the use of χ^2 or Fisher's exact test. In trying to interpret these findings, we have also considered the nonsignificant correlations with other variables, looking for consistency and inconsistency with the general implications of the significant findings. We have arrived at the following formulation of results:

1. Large sex difference in socialization is associated with an economy that places a high premium on the superior strength, and superior development of motor skills requiring strength, which characterize the male. Four of the correlations reported in Table 2 clearly point to this generalization: the correlations of large sex difference with the hunting of large animals, with grain rather than root crops, with the keeping of large rather than small domestic animals, and with nomadic rather than sedentary residence. The correlation with the unimportance of fishing may also be consistent with this generalization, but the argument is not clear.[3] Other correlations consistent with the generalization, though not statistically significant, are with large game hunting rather than gathering, with the hunting of large game rather than small game, and with the general importance of all hunting and gathering.

2. Large sex difference in socialization appears to be correlated with customs that make for a large family group with high cooperative interaction. The only statistically significant correlation relevant here is that with polygyny rather than monogamy. This generalization is, however, supported by several substantial

[3] Looking (with the more inclusive sample) into the possibility that this correlation might result from the correlation between fishing and sedentary residence, a complicated interaction between these variables was found. The correlation of sex differentiation with absence of fishing is found only in nomadic societies, where fishing is likely to involve cooperative activity of the two sexes, and its absence is likely to mean dependence upon the male for large game hunting or herding large animals (whereas in sedentary societies the alternatives to fishing do not so uniformly require special emphasis on male strength). The correlation of sex differentiation with nomadism is found only in nonfishing societies; here nomadism is likely to imply large game hunting or herding large animals, whereas in fishing societies nomadism evidently implies no such special dependence upon male strength. Maximum sex differentiation is found in nomadic nonfishing societies (15 with large difference and only 2 with small) and minimum sex differentiation in nomadic fishing societies (2 with large difference and 7 with small difference). These findings further strengthen the argument for a conspicuous influence of the economy upon sex differentiation.

correlations that fall only a little short of being statistically significant. One of these is a correlation with sororal rather than nonsororal polygyny; Murdock and Whiting [4] have presented indirect evidence that co-wives generally show smoother cooperative interaction if they are sisters. Correlations are also found with the presence of either an extended or a polygynous family rather than the nuclear family only; with the presence of an extended family; and with the extreme contrast between maximal extension and no extension of the family. The generalization is also to some extent supported by small correlations with wide extension of incest taboos, if we may presume that an incest taboo makes for effective unthreatening cooperation within the extended family. The only possible exception to this generalization, among substantial correlations, is a near-significant correlation with an extended or polygynous family's occupying a cluster of dwellings rather than a single dwelling.[4]

In seeking to understand this second generalization, we feel that the degree of social isolation of the nuclear family may perhaps be the crucial underlying variable. To the extent that the nuclear family must stand alone, the man must be prepared to take the woman's role when she is absent or incapacitated, and vice versa. Thus the sex differentiation cannot afford to be too great. But to the extent that the nuclear family is steadily interdependent with other nuclear families, the female role in the household economy can be temporarily taken over by another woman, or the male role by another man, so that sharp differentiation of sex role is no handicap. . . .

Both of these generalizations contribute to understanding the social background of the relatively small difference in socialization of boys and girls which we believe characterizes our society at the present time. Our mechanized economy is perhaps less dependent than any previous economy upon the superior average strength of the male. The nuclear family in our society is often so isolated that husband and wife must each be prepared at times to take over or help in the household tasks normally assigned to the other. It is also significant that the conditions favoring low sex differentiation appear to be more characteristic of the upper segments of our society, in socioeconomic and educational status, than of lower segments. This observation may be relevant to the tendency toward smaller

[4] We think the reverse of this correlation would be more consistent with our generalization here. But perhaps it may reasonably be argued that the various nuclear families composing an extended or polygynous family are less likely to develop antagonisms which hinder cooperation if they are able to maintain some physical separation. On the other hand, this variable may be more relevant to the first generalization than to the second. Occupation of a cluster of dwellings is highly correlated with presence of herding and with herding of large rather than small animals, and these economic variables in turn are correlated with large sex difference in socialization. Occupation of a cluster of dwellings is also correlated with polygyny rather than monogamy and shows no correlation with sororal vs. nonsororal polygyny.

sex differences in personality in higher status groups [cf. Terman and Miles, 8].

The increase in our society of conditions favoring small sex difference has led some people to advocate a virtual elimination of sex differences in socialization. This course seems likely to be dysfunctional even in our society. Parsons, Bales, *et al.* [5] argue that a differentiation of role similar to the universal pattern of sex difference is an important and perhaps inevitable development in any social group, such as the nuclear family. If we add to their argument the point that biological differences between the sexes make most appropriate the usual division of those roles between the sexes, we have compelling reasons to expect that the decrease in differentiation of adult sex role will not continue to the vanishing point. In our training of children, there may now be less differentiation in sex role than characterizes adult life—so little, indeed, as to provide inadequate preparation for adulthood. This state of affairs is likely to be especially true of formal education, which is more subject to conscious influence by an ideology than is informal socialization at home. With child training being more oriented toward the male than the female role in adulthood, many of the adjustment problems of women in our society today may be partly traced to conflicts growing out of inadequate childhood preparation for their adult role. This argument is nicely supported in extreme form by Spiro's analysis of sex roles in an Israeli kibbutz [7]. The ideology of the founders of the kibbutz included the objective of greatly reducing differences in sex role. But the economy of the kibbutz is a largely nonmechanized one in which the superior average strength of men is badly needed in many jobs. The result is that, despite the ideology and many attempts to implement it, women continue to be assigned primarily to traditional "women's work," and the incompatibility between upbringing or ideology and adult role is an important source of conflict for women. . . .

Summary

A survey of certain aspects of socialization in 110 cultures shows that differentiation of the sexes is unimportant in infancy, but that in childhood there is, as in our society, a widespread pattern of greater pressure toward nurturance, obedience, and responsibility in girls, and toward self-reliance and achievement striving in boys. There are a few reversals of sex difference, and many instances of no detectable sex difference; these facts tend to confirm the cultural rather than directly biological nature of the differences. Cultures vary in the degree to which these differentiations are made; correlational analysis suggests some of the social conditions influencing these variations, and helps in understanding why our society has relatively small sex differentiation.

REFERENCES

1. Benedict, Ruth, "Continuities and discontinuities in cultural conditioning," *Psychiatry*, 1938, *1*, 161-167.
2. Mead, Margaret, *Male and female* (New York: Morrow, 1949).
3. Murdock, G. P., "Comparative data on the division of labor by sex," *Social Forces*, 1937, *15*, 551-553.
4. Murdock, G. P., and Whiting, J. W. M., "Cultural determination of parental attitudes: The relationship between the social structure, particularly family structure and parental behavior," in M. J. E. Senn, ed., *Problems of infancy*

and childhood: *Transactions of the Fourth Conference, March 6-7, 1950.* New York: Josiah Macy, Jr. Foundation, 1951. Pp. 13-34.
5. Parsons, T., Bales, R. F., *et al.*, *Family, socialization and interaction process* (Glencoe, Ill.: Free Press, 1955).
6. Scheinfeld, A., *Women and men* (New York: Harcourt, Brace, 1944).
7. Spiro, M. E., *Kibbutz: Venture in Utopia* (Cambridge: Harvard Univ. Press, 1956).
8. Terman, L. M., and Miles, Catherine C., *Sex and personality* (New York: McGraw-Hill, 1936).

FAMILY STRUCTURE AND SEX ROLE LEARNING BY CHILDREN: A FURTHER ANALYSIS OF HELEN KOCH'S DATA

Orville G. Brim, Jr.

Adapted from *Sociometry*, 21 (1958), 1-15.

The structure of a social group, delineated by variables such as size, age, sex, power, and prestige differences, is held to be a primary influence upon the patterns of interaction within the group, determining in major part the degree to which any two group members interact. It is held, second, that social roles are learned through interaction with others, such interaction providing one with the opportunity to practice his own role as well as to take the role of the other. On this basis one may hypothesize that group structure, by influencing the degree of interaction between group members, would be related to the types of roles learned in the group: one would learn most completely those roles which he himself plays, as well as the roles of the others with whom he most frequently interacts. This argument is applied in this paper specifically to the relation between family structure, described in terms of age, sex and ordinality of children, and the sex role learning by the children.

The process of role learning through interaction, which has been described in detail by Mead [15], Cottrell [2], and others, can be sketched as follows. One learns the behavior appropriate to his position in a group through interaction with others who hold normative beliefs about what his role should be and who are able to reward and punish him for correct and incorrect actions. As part of the same learning process, one acquires expectations of how others in the group will behave. The latter knowledge is indispensable to the actor, in that he must be able to predict what others expect of him, and how they will react to him, in order to guide his own role performance successfully. Accurate or erroneous understanding and prediction are respectively rewarding and punishing to the actor, and learning proceeds systematically through the elimination of incorrect responses and the strengthening of correct ones.

It has been the distinctive contribution of sociology to demonstrate that learning the role of others occurs through the actor's taking the role of the other, i.e., trying to act as the other would act. While this role-taking of the other can be overt, as with children who actively and dramatically play the role of the parent, it is commonly covert in adults, as with the husband who anticipates what his wife will say when he returns home late, or the employee who tries to foresee his employer's reaction when he asks for a raise.

It follows that, whether taking the role of others is overt or covert, certain responses (belonging to the role of the other) are in fact made, run through, completed, and rewarded if successful, i.e., accurate, and that this process adds to the repertoire of possible actions of a person those actions taken by others in their own roles. Such actions, as part of one's repertoire or pool of learned responses, are available for performance by an actor, not now simply in taking the role of the other, but as resources which he can use as part of his *own* role performances.

The critical fact is that the actor not only can, but *does,* make use of responses learned in role-taking in his own role performances. There are two senses in which this happens. The first, which does not concern us in this paper, involves the direct transfer of the role of the other to a new and parallel status of one's own, where there is a straightforward adoption of the other's role. Such transfer may be appropriate and rewarded, as where the oldest child performs the role of the parent to his sibs, or simply interesting and tolerated, as where the new assistant professor plays the department chairman to the graduate students.

The second sense, which is our major concern here, involves a more complex process of convergence between one's own role and that of the other which he takes, where there is a spill-over of elements belonging to another's role into one's own performance when it is not necessarily appropriate. Our basic hypothesis, set forth by Cottrell [2] and others, is that interaction between two persons leads to assimilation of roles, to the incorporation of elements of the role of the other into the actor's role.

Thus, one says, husbands and wives grow more alike through time, and long-time collaborators in research begin to think alike.

While not pretending to a full analysis of the process underlying assimilation, several causes can be described. First, the actor may note that the other is successful to a high degree in some of his behavior and consciously transfer to his own role such behavioral elements for trial. To the extent that they prove successful for him, in his performance, and are not eliminated through punishment from others for being inappropriate, he will adopt them. Second, faced with novel situations where his "own" behavior fails, the elements of others' roles are already learned and available for trial and hence would tend to be tried prior to the development of totally new responses; again, if successful, they tend to be assimilated to the role. Third, the actions learned by taking the role of others are ordinarily performed implicitly and under limited conditions, e.g., in interaction with the other. However, the cues which guide and elicit one's own role performance may be difficult to differentiate from cues eliciting taking the role of the other. It would appear that for the young child this is especially difficult, and data indeed show that the child has difficulty discriminating between reality and fantasy, between what his role is or even what his self is, and what belongs in the category of the "other." In this way, behavior learned through role-taking and appropriate to the other is confused with and undifferentiated from behavior learned as part of one's own role. The latter becomes tinged or diluted with characteristics belonging to someone else's role.

Among the hypotheses which are derivative of the general hypothesis of assimilation through interaction, two are pertinent here. First, the process of discrimination between what belongs to oneself and what belongs to the other is aided by the guidance of other persons. Thus, the parent helps the son differentiate between what belongs to him and what belongs to his sister; the fledgling nurse is assisted in a proper demeanor and in separating her duties from those of the physician. Rewards and punishments administered by others govern the discrimination process. Where the process of assimilation comes primarily from inability to discriminate between roles, it follows that where greater attention is paid to helping the learner discriminate, the process of assimilation is to a greater degree arrested.

Second, given two other persons with whom one interacts and who differ in power over the actor, i.e., differ in the degree to which they control rewards and punishments for the actor, one would predict that the actor would adopt more of the characteristics of the powerful, as contrasted to the less powerful, other person. This follows from the fact that it is more important to the actor to predict the behavior of the powerful figure, that he is motivated more strongly to take his role, that the rewards and punishments are more impressive and the learning consequently better. Interaction between two figures of unequal power should give a parallel result, namely, there would be a greater assimilation of the role of the other into the actor's role for the less powerful figure, for the same reasons as above. Thus the employee gravitates toward the boss more than the reverse, and the child becomes more like the parent than the other way round. However, this is not to imply that the more powerful figure need not take the role of the other, nor that he does not assimilate (to a lesser degree) elements from the other's role. The weaker figure always has some control over rewards and punishments, requiring therefore that his reaction be considered. The displeased employee can wound his boss through expressions of dislike, and the angry child can hurt his parents in a variety of ways, from refusing to eat to threatening to leave home.

Turning now to a consideration of sex-role learning specifically, pertinent reviews [1, 17] of the data show that sex-role prescriptions and actual performance begin early. The accepted position is that children in a family learn their appropriate sex roles primarily from their parents. There is remarkably little data, other than clinical materials, on this topic, perhaps because of its obviousness. What systematic data there is, is not inconsistent with the role learning propositions set forth above. Sears, Pintler, and Sears [14] have shown that in families where the father is absent the male child is slower to develop male sex-role traits than in families where the father is present, a finding predictable from the fact that there is no father whose role the child needs to take. Both Sears [13] and Payne and Mussen [12] have shown that father role-playing, identification with the father, and masculinity of attitudes are positively related to the father's being warm, affectionate, and rewarding. This strikes one as the same type of finding as the first, but at the other end of the interaction range; insofar as warm, affectionate, and rewarding fathers interact more with their sons, or are perceived as such because they interact more, it follows that the sons have more experience in taking their role.

In regard to the effects of sibling characteristics upon sex-role learning, there is again almost no information. Fauls and Smith [3] report that only children choose sex-appropriate activities more often than do children with older same-sex siblings, a finding which seems to fit none of our role-learning propositions. While one might hold that the only child has more interaction, because of sibling absence, with his same-sex parent, hence learns his sex role better, one might equally say, especially for the young boys, that it is the cross-sex parent with whom the child interacts and hence the only child should not learn his sex role well. In any case, the finding serves to stress the limitations of the data we are to report, namely, that they pertain to variations within two-child families, and that generalization to families of varying sizes is unwarranted. We return to this point later.

Even with respect to theory concerning the effects of siblings on sex-role learning, we have not noted any systematic predictions in the literature. It seems to us implicit in Parsons' recent analysis [11] of sex-role learning in the nuclear family that when the child begins his differentiation between the father and mother sex roles he would be helped in making the differentiation if he had a cross-sex sibling; this is not formally stated, however, and we may be guilty of misinterpretation.

It is against this background of comparative absence of research and theory on the effects of siblings on sex-role learning that our own report must be viewed. The very valuable data on personality traits of children presented in recent publications by Helen Koch [4, 5, 6, 7, 8, 9, 10] provide the opportunity to apply several of the general hypotheses set forth above to the substantive area of sibling effects on sex-role learning. The specific application of these hypotheses can be summarized as follows:

First, one would predict that cross-sex, as compared with same-sex, siblings would possess more traits appropriate to the cross-sex role. When taking the role of the other in interaction, cross-sex siblings must take the role of the opposite sex, and the assimilation of roles as delineated above should take place.

Second, one would predict that this effect would be more noticeable for the younger, as compared with the older, sibling in that the latter is more powerful and is more able to differentiate his own from his sibling's role.

Third, on the assumption that siblings close in age interact more than those not close in age, one would predict that this effect would be more noticeable for the siblings who are closest together in age. This is in essence an extension of the first hypothesis to deal with variations in interaction within the cross-sex sibling groups.

Procedures

Our description of procedures must of necessity be broken into two parts. The first consists of a brief description of the procedures in Helen Koch's original study; complete details are available in the publications cited previously. The second consists of our mode of further analysis of the reported data.

In her series of papers Helen Koch has reported results from a major research project concerned with the relation between structural characteristics of the family, namely, sex of child, sex of sibling, ordinal position of child, and age difference between siblings, and the child's ratings on more than fifty personality traits. In her study, all subjects were obtained from the Chicago public schools and one large private school. The characteristics of the children used as subjects can be summarized as follows. All children were from unbroken, native-born, white, urban, two-child families. The children were five- and six-year-olds, free of any gross physical or mental defect. In most cases only one sibling in a family was a subject in the study.

The subjects numbered 384. "The experimental design included three sibspacing levels, two ordinal positions, subjects of two sexes and siblings of two sexes. There were 48 children in each of the following categories —male with a male sib older, male with a male sib younger, male with a female sib older, male with a female sib younger, female with a male sib older, female with a male sib younger, female with a female sib older, and female with a female sib younger. Each of these groups of 48 children was composed of three subgroups of 16 children, representing the following three sibling-age-difference levels: siblings differed in age by under two years, by two to four years, and four to six years, respectively. Hence our basic subgroups of 16 numbered 24" [7, p. 289]. The groups were matched, approximately, on an individual subject basis with respect to age of child and father's occupational status.

Teachers' ratings were made for each child on 58 traits. The teachers, all of whom were women, were trained in a conference or two to make the ratings. No teacher rated a child with whom contact had been less than three months, and in most cases the contact ranged from six to nine months. The 58 traits included 24 of the Fels Child Behavior Scales, and 34 items from the California Behavior Inventory for Nursery School Children. All ratings were made on line scales, converted later to 9-point scales. Ratings on each trait were subsequently normalized, prior to analysis of the data.

The relation between personality trait ratings and the structure of the family from which the children came was assessed by analysis of variance for each of the 58 traits. Helen Koch presents in her publications the findings from the variance analyses. It is this data on which we made our further study.

The procedures for the further analysis involved several steps. First, the writer, with the assistance of three professional persons as additional judges,[1] judged each of the 58 traits in terms of its pertinence to either a masculine or feminine role. Our conception of the characteristics of the two sex roles was based on recent empirical studies describing sex-role differences in small problem-solving groups [16] and in the nuclear fam-

[1] Dr. John Mann, Mr. David Glass, and Mr. David Lavin.

ily [18], and on the major theoretical treatment of such differences by Talcott Parsons [11]. In these studies the now-familiar distinction between the instrumental or task role and the expressive or social-emotional role in a social group is shown to be related to sex-role differentiation, particularly in the family, with the male customarily taking the instrumental role and the female the expressive role. Hence in the judging process our decision as to whether a trait was masculine or feminine was essentially dependent on whether we believed the trait to belong to the instrumental or expressive role respectively.

Substantial descriptive data are available on sex-role differences in children for some of the traits which we judged. These findings, summarized by Terman and Tyler [17], were consulted after the judging was completed and strongly corroborate our assignment of traits: e.g., male children are judged higher on traits we believed instrumental, such as dominance and aggression, and lower on traits we judged to pertain to the expressive role, such as affection and absence of negativism.

In judging the traits it was recognized that many of them would be part of the role requirements for both roles. However, it was clear that there exists for each of the roles what is essentially a rank order of characteristics in terms of their importance for the role. Hence the basis for our judgments was whether the trait appeared to be higher in the rank order of requirements for the instrumental or the expressive role. Traits which seemed pertinent to neither, e.g., stammering, or for which no judgment of greater importance could be made, e.g., curiosity, were not ascribed to either role and were omitted from subsequent steps in the analysis. It was possible to assign 31 of the 58 traits to either the instrumental or expressive role. Twenty of the 31 traits pertain to the expressive role, the children evidently having been rated on a predominantly female cluster of traits.

Some of the traits were stated in a negative way which made them, while pertinent to the role, incongruent with the role conception. Thus, "uncooperativeness with group" seemed clearly to be relevant to the expressive role but as an incongruent trait. In like manner, both affectionateness and jealousy seemed most important as aspects of the expressive role, the former being congruent with the role conception, the latter incongruent. It therefore was necessary to make a second judgment regarding each trait, namely, whether it was a congruent or incongruent aspect of the role to which it pertained.

Table 1 lists the 31 traits, the role to which they seemed most pertinent, and the indication of whether the trait was a congruent or incongruent characteristic of the role.

With the judging of the traits completed, the next step was a careful

TABLE 1. Traits Assignable to Male (Instrumental) or Female (Expressive) Roles

Trait name	Pertains primarily to instrumental (I) or expressive (E) role	Trait is congruent (+) or incongruent (−) characteristic of role
1. Tenacity	I	+
2. Aggressiveness	I	+
3. Curiosity	I	+
4. Ambition	I	+
5. Planfulness	I	+
6. Dawdling and procrastinating	I	−
7. Responsibleness	I	+
8. Originality	I	+
9. Competitiveness	I	+
10. Wavering in decision	I	−
11. Self-confidence	I	+
12. Anger	E	−
13. Quarrelsomeness	E	−
14. Revengefulness	E	−
15. Teasing	E	−
16. Extrapunitiveness	E	−
17. Insistence on rights	E	−
18. Exhibitionism	E	−
19. Uncooperativeness with group	E	−
20. Affectionateness	E	+
21. Obedience	E	+
22. Upset by defeat	E	−
23. Responds to sympathy and approval from adults	E	+
24. Jealousy	E	−
25. Speedy recovery from emotional disturbance	E	+
26. Cheerfulness	E	+
27. Kindness	E	+
28. Friendliness to adults	E	+
29. Friendliness to children	E	+
30. Negativism	E	−
31. Tattling	E	−

reading of Helen Koch's findings. A tabulation was made of all differences on the 31 traits between the 16 basic subgroups reported by her as significant (close to or at the .05 level, based on the separate analyses of variance). Such differences involved single structural characteristics, e.g., first-born versus second-born; single interactions of characteristics, e.g., girls with brothers versus girls with sisters; and multiple interactions, e.g., first-born boys with sisters versus first-born boys with brothers. These significant differences in traits were then entered in some preliminary forms of Tables 2 and 3. The procedure for entering differences was somewhat complicated and is described as follows:

TABLE 2.   Instrumental and Expressive Traits for Five- and Six-year-old Girls

Subjects	Sib Age Difference	Male (or Instrumental) Traits		Female (or Expressive) Traits	
		High Masculinity Ratings	Low Masculinity Ratings	High Femininity Ratings	Low Femininity Ratings
Older girl with younger sister	0-2 years	2,5,7	4,6,9,10	13,14,15,16,17,18, 19,20,21,24,30	22,23,25,26, 27
	2-4 years	7	2,4,9,10,11	13,14,15,16,17,18, 19,20,21,24,30	22,23,26,27, 28
	4-6 years	7	2,4,6,9,10, 11	13,14,15,16,17,18, 19,20,21,24,30	22,23,25,26, 27,28
Older girl with younger brother	0-2 years	1,2,3,4,5, 9,10	6	13,14,15,16,19,20, 21,25,26,27,30	22,24
	2-4 years	1,2,3,4,9, 10	6	13,14,15,16,19,20, 25,26,27,30,31	22,24
	4-6 years	1,2,4,6,7, 9,10	6	13,14,15,19,20,21, 25,26,27,28,30	22,24,31
Younger girl with older sister	0-2 years	2,5,6,7,8	3,4,9,10,11	12,13,14,15,16,18, 19,20,21,22,23,30	17,25,26,27
	2-4 years		3,4,5,8,9, 10,11	12,13,14,15,16,18, 19,20,21,22,23,30	17,25,26,27, 28
	4-6 years	6,7	2,3,4,9,10, 11	12,13,14,15,16,18, 19,20,21,22,23,30	17,25,26,27 28
Younger girl with older brother	0-2 years	1,2,3,4,7, 8,9,10,11		12,13,14,15,16,18, 19,20,21,22,23,25, 26,27,28,30	
	2-4 years	1,4,6,7,10, 11		12,13,14,15,16,18, 19,20,21,22,23,25, 26,27,28,30	
	4-6 years	2,4,5,10,11		12,13,14,15,16,18, 19,20,21,22,23,25, 26,27,28,30	

Note: Trait numbers refer to listing in Table 1. Traits entered in high masculinity rating column are male-congruent traits with high ratings, male-incongruent traits with low ratings. The reverse is true for low masculinity rating column. Female trait entries are made in the same manner.

First, with respect to a trait judged pertinent to the male or instrumental role, and considered a *congruent* aspect of that role: when any subgroup or groups were rated significantly higher than others on that trait, the number of the trait was entered in the high masculinity column for such a group; the subgroup or groups they were higher than, i.e., the low groups, had the number of the trait entered in the low masculinity column. Second, with respect to a male trait considered an *incongruent* aspect of the role: when any subgroup was rated higher than another on such a trait, the trait number was entered in the low masculinity column for such a group; for the group it was higher than, i.e., the low group, the trait number was entered in the high masculinity column. The procedure for the female or expressive traits was identical, except the female columns were used.

This procedure means that for any subgroup, entries in the high masculinity column consist of congruent male traits on which the group is high, and incongruent male traits on which it is low; entries in the low masculinity column consist of incongruent male traits on which the group is high, and congruent male traits on which it is low. Female column entries are read the same way. An example may be helpful at this point. Consider in Table 3 the subgroup "Younger Boy with Older Brother" at the four- to six-year age difference. In the high masculinity column the entry of trait number 2 means that the group was rated significantly *high* on aggressiveness; the entry of trait number 10 means that the group was rated significantly *low* on wavering in decision. In the low masculinity column, trait number 6 indicates a *high* rating on dawdling and procrastinating, while trait number 7 indicates a *low* rating on responsibleness.

The preliminary forms of Tables 2 and 3 were complicated and two further steps toward simplification were taken before reaching the present form. The initial tables were marred by the occurrence of duplicate trait-number entries in the cells, arising primarily from the multiple reporting of the original data and the multiple differences emerging between the various subgroups. Hence, where duplicate trait-entries occurred, only one entry was kept. The result is to make each entry read that that subgroup is significantly higher (or lower) than some other group *or groups* on that particular trait. Second, the tables were complicated by the fact that for all subgroups there were at least some trait numbers which appeared in *both* the high and low subdivisions of either the male or female column. This indicated, of course, that a subgroup was higher (or lower) than some other group on that trait, but also lower (or higher) than still another group; i.e., on the ranking of mean ratings on the trait, the subgroup would have differed significantly from both the top and bottom ranks. To clarify the tables, and also substantially to increase the reliability of the subgroup differences reported here, all traits on which a subgroup had both high and low entries were dropped for that subgroup. In summary, the result of this step, combined with the one above, is to make *each entry in the final tables read that that subgroup is significantly higher (or lower) than one or more groups on that trait, and is significantly lower (or higher) than none.*

TABLE 3. Instrumental and Expressive Traits for Five- and Six-year-old Boys

Subjects	Sib Age Difference	Male (or Instrumental) Traits		Female (or Expressive) Traits	
		High Masculinity Ratings	Low Masculinity Ratings	High Femininity Ratings	Low Femininity Ratings
Older boy with younger brother	0-2 years	9,10	1,2,7,11		12,13,14,15,16,19, 22,23,25,26,27,30
	2-4 years	4,9,10	1,2,5,7,11		12,13,14,15,16,19, 20,21,22,23,25,26, 27,28,30
	4-6 years	2,4,9,10	1,7,11		12,13,14,15,16,19, 20,22,23,25,26,27, 30,31
Older boy with younger sister	0-2 years	11	2,4,7,9,10	25,26,27,31	12,13,14,15,16,17, 18,19,20,21,22,23, 24,30
	2-4 years	2,3,5,11	4,7,9,10	25,26,27,28	12,13,14,15,16,17, 18,19,20,21,22,23, 24,30
	4-6 years	3	2,4,7,9,10	25,26,27,28	12,13,14,15,16,17, 18,19,20,21,22,23, 24,30
Younger boy with older brother	0-2 years	4,9,10	1,2,3,5,6,7, 8	22,23,24	13,16,18,19,21, 25,26,27,28,30
	2-4 years	4,9,10	1,3,6,7	22,23,24	13,16,18,19,20,21, 25,26,27,28,30
	4-6 years	2,4,5,8,9, 10	6,7	22,23,24,29	13,16,18,19,20,21, 25,26,27,28,30
Younger boy with older sister	0-2 years		2,4,5,6,7,8, 9,10	17,22,23,24, 25,26,27	13,19,30
	2-4 years		2,4,6,9,10	17,22,23,24, 25,26,27,28	13,16,19,20,21,30
	4-6 years		2,4,6,7,9, 10	17,22,23,24, 25,26,27,28	13,16,19,20,21,30

Note: See note to Table 2.

Results and Discussion

The data presented in Tables 2 and 3 can be brought to bear upon our hypotheses by considering the distribution by subgroups of the traits indicating high or low masculinity or femininity. Our concern is with the frequency of trait entries of the four types, rather than with the descriptive content of any particular trait. Essentially we give each separate trait an equal weight, then summarize in terms of masculinity (many high rating, few low rating entries) and of femininity, associated with each sub-group.

With respect to our first hypothesis, that through interaction and taking the role of the other the cross-sex sibs would have more traits of the opposite sex than would same-sex sibs, an examination of the distribution in Table 2 shows that this is clearly the case. Controlling for ordinality, the older girl with a younger brother has more high masculinity traits and fewer low masculinity traits, than does her counterpart, the older girl with a younger sister. This distribution of traits is even more pronounced for the girls in the second ordinal position, the younger girl with older brother being substantially higher on masculinity than her counterpart with an older sister. One will note that the acquisition of male traits does not seem to reduce the number of feminine traits of the girls with brothers. The more accurate interpretation is that acquisition of such traits adds to their behavioral repertoire, probably with a resultant dilution of their femininity in behavior, but not a displacement.

Examination of Table 3 with respect to this first hypothesis indicates that it holds for boys also. While not pronounced for the boys in the eldest child position, the boy with the sister is feminine to a greater degree than the boy with the brother. For the boys who are second-born, the difference is clear: the boy with the elder sister is substantially more feminine than his counterpart with an older brother. For the boy with the older sister the acquisition of feminine traits would seem to have displaced, rather than simply diluted, his masculinity and he thus contrasts with the girls for whom this did not occur. We can offer no explanation for this, but it may provide a lead for further study in this area.

In connection with this result, the role of the parent requires attention. While all would agree that parents actively assist cross-sex sibs in separating their sex roles, the data show they are unsuccessful in completely arresting the process of assimilation. Perhaps in earlier times, when children's sex roles were stressed more strongly, and perhaps today for some parents making an extreme effort, the effects of interaction would be reduced. However, it certainly appears that the average parent today cannot completely avoid the effects of such sib interaction. Even were more attention given by parents to cross-sex as opposed to same-sex sibs in this matter, we believe that the tremendously greater cross-sex interaction of the former would leave its mark.

With respect to our second hypothesis, that because of differences in control of rewards and punishments and in ability to discriminate between self and other roles the effects of role-taking would be more pronounced for the younger child, an examination of Tables 2 and 3 again seems to support the hypothesis. While the younger, as contrasted with the older, girl with a brother manifests only a slightly greater degree of masculinity, this difference for boys is quite striking: the younger, as contrasted with the older, boy with a sister is substantially more feminine.

With respect to our third hypothesis, that on the assumption of interaction varying inversely with age-gap and greater interaction producing greater role-taking, the effects of role-taking would be largest for the sibs closest in age, the results in both tables are negligible. One might discern some such relationship for the boy with an older sister, and the girl with an older brother, but even here it is tenuous. Because the assumption that interaction varies with sib age differences may in fact be untenable, we cannot in this instance say we have made a direct test of the hypothesis that more frequent interaction produces more role assimilation. Since the first hypothesis, which in essence states the same point, was so strongly confirmed, our inclination is to reject our assumption that interaction varies with age difference, at least to a degree sufficient to produce differences in role-taking.

There are two further aspects of Tables 2 and 3 which are quite noticeable and which need comment. We refer first to the fact that girls with brothers appear to be masculine to a greater degree than do any of the males themselves. The simplest and most likely explanation, hence the one which we favor, is that this result occurs because of certain biases in the teachers' ratings. We submit that teachers implicitly rated boys and girls on different scales, i.e., girls were implicitly rated on a girls' scale, boys on a boys' scale. The girl with an extreme masculine trait—extreme, that is, for a girl—receives a very high rating; a boy with the same absolute degree of such a trait, or even more of it, would on the boys' scale not be extreme and his rating consequently would be reduced. In the subsequent analysis of variance, where the male and female ratings are treated as if on the same absolute scale, certain girls extremely high for girls would score significantly higher than even certain boys high on the trait. To some extent we see the same effect in reverse for the younger boys with an older sister; while not being more feminine than girls, they almost tie certain girls, e.g., older girls with younger sisters. The probable use of different implicit rating scales, the implausibility of any group of girls being more masculine than all boys and the important fact that when girls and boys are assuredly rated on the same absolute scale [e.g., 3, 17] boys regularly outscore girls on masculine traits, all tend to support this interpretation.

The second additional aspect of the tables which merits discussion is that all girls seem to be more feminine than the boys are masculine; indeed, the major characteristic of the boys is to be antifeminine, not masculine. In part this is explained by the assumed bias in the ratings mentioned above; boys are outscored on their own traits by some girls. In part also this is explained by the preponderance of feminine traits used in the ratings, so that boys could only express their masculinity, as it were, by being rated low on such traits. In part, and an intriguing part indeed, it may be explained by certain developmental processes commonly assumed in clinical theory and recently put in a role theory context by Parsons [11, pp. 95-101]. Parsons points out that both boy and girl first identify with the mother and tend to play an expressive role. In development the boy must break away and establish a new identification with the father, which is difficult and involves much new learning, in the role-taking sense. At the same time, the boy must "push far and hard to renounce dependency." Girls, continuing identification with the mother and the expressive role, face neither of these problems. It may be, then, that the girls' femininity and the boys' antifemininity and yet lack of masculinity which shows itself in Tables 2 and 3 arises in part because the children have been caught by the raters at an age where the boy is trying to shift his identification from mother to father.

To conclude, our analysis of Helen Koch's data indicates that cross-sex siblings tend to assimilate traits of the opposite sex, and that this effect is most pronounced in the younger of the two siblings. These findings support the role-learning theory presented here, and also stand as a substantive contribution to the area of sex-role learning. We wish now to stress two points mentioned earlier.

First, these findings must be subject to strict limitations to two-child families. Not only does the Fauls and Smith study demonstrate this limitation with regard to only-child families, but observation suggests that in larger families other variables come into play; e.g., in the four-child family with a three and one sex split, parents may actively help the solitary child in differentiating sex roles; or in the four-child family with a two and two split, siblings may pair off by sex and the cross-sex role-taking effect is minimized.

Second, with respect to the substantive value of these results, we would point out that even though parents must remain as the major source of sex-role learning, almost every child has a mother and father to learn from. Hence the *variations* in type and amount of sex-role learning occur on top of this base, so to speak, and in this variability the effect of a same or a cross-sex sib may play as large or larger a role than variations in parental behavior, mixed versus single-sexed schooling, sex of neighborhood playmates, and the like. Speculations on the durable and considerable effects of sex of sib on sex-role learning thus seem warranted and lead one to consider problems such as the effect of sex of sibling on one's later role in the marital relation, on career choices, and on other correlates of the adult sex role.

Summary

This paper reports some relations between ordinal position, sex of sibling, and sex-role learning by children in two-child families. The findings are based on a further analysis of Helen Koch's data relating personality traits of children to their sex, sex of sibling, ordinal position, and age difference from sibling. In this analysis the personality traits were classified as pertaining either to the instrumental (masculine) role or the expressive (feminine) role. The distribution of such traits in children as a correlate of family structure was then assessed.

General propositions describing role learning in terms of interaction with others, including taking the role of the other, leads to hypotheses that cross-sex siblings will have more traits of the opposite sex than will same-sex siblings, and that this effect will be greater for the younger, as contrasted with the older, sibling. Both hypotheses are confirmed by the data presented.

REFERENCES

1. Brim, O. G., Jr., "The Parent-Child Relation as a Social System: I. Parent and Child Roles," *Child Development,* 1957, 28, 344-364.

2. Cottrell, L. S., Jr., "The Analysis of Situational Fields in Social Psychology," *American Sociological Review,* 1942, 7, 370-382.

3. Fauls, L. B., and W. D. Smith, "Sex Role Learning of Five-Year-Olds," *Journal of Genetic Psychology,* 1956, 89, 105-117.

4. Koch, H. L., "The Relation of 'Primary Mental Abilities' in Five- and Six-Year-Olds to Sex of Child and Characteristics of His Sibling," *Child Development,* 1954, 25, 210-223.

5. Koch, H. L., "Some Personality Correlates of Sex, Sibling Position, and Sex of Sibling Among Five- and Six-Year-Old Children," *Genetic Psychology Monographs,* 1955, 52, 3-50.

6. Koch, H. L., "The Relation of Certain Family Constellation Characteristics and the Attitudes of Children Toward Adults," *Child Development,* 1955, 26, 13-40.

7. Koch, H. L., "Attitudes of Children Toward Their Peers as Related to Certain Characteristics of Their Sibling," *Psychological Monographs,* 1956, 70, No. 19 (whole No. 426).

8. Koch, H. L., "Children's Work Attitudes and Sibling Characteristics," *Child Development,* 1956, 27, 289-310.

9. Koch, H. L., "Sibling Influence on Children's Speech," *Journal of Speech and Hearing Disorders,* 1956, 21, 322-328.

10. Koch, H. L., "Sissiness and Tomboyishness in Relation to Sibling Characteristics," *Journal of Genetic Psychology,* 1956, 88, 231-244.

11. Parsons, T., "Family Structure and the Socialization of the Child," in T. Parsons and R. F. Bales, *Family, Socialization and Interaction Process* (Glencoe, Illinois: Free Press, 1955).

12. Payne, D. E., and P. H. Mussen, "Parent-Child Relations and Father Identification Among Adolescent Boys," *Journal of Abnormal and Social Psychology,* 1956, 52, 359-362.

13. Sears, P. S., "Child-Rearing Factors Related to Playing of Sex-Typed Roles,"

American Psychologist, 1953, 8, 431 (abstract).

14. Sears, R. R., M. H. Pintler, and P. S. Sears, "Effect of Father Separation on Preschool Children's Doll Play Aggression," *Child Development,* 1946, 17, 219-243.

15. Strauss, A., *The Social Psychology of George Herbert Mead* (Chicago: Phoenix Books, University of Chicago Press, 1956).

16. Strodtbeck, F. L., and R. D. Mann, "Sex Role Differentiation in Jury De-liberations," *Sociometry,* 1956, 19, 3-11.

17. Terman, L. M., and L. E. Tyler, "Psychological Sex Differences," in L. Carmichael (ed.), *Manual of Child Psychology* (2d ed.) (New York: Wiley, 1954).

18. Zelditch, M., Jr., "Role Differentiation in the Nuclear Family: A Comparative Study," in T. Parsons and R. F. Bales, *Family, Socialization and Interaction Process* (Glencoe, Illinois: Free Press, 1955).

10.   SOCIAL CLASS AND SOCIALIZATION

This chapter and the next consider the impact of social class on socialization. After showing that middle-class children stay in school longer and perform better than do lower-class children, Toby attributes such differences to the sub-cultures of these classes and to the family as the conduit through which the values are transmitted from parents to children.

Kohn finds a broadly common set of values in middle- and working-class parents with respect to the kind of persons their children ought to be. He also finds differences between the two classes, which he says are related to their living conditions. Kohn interprets values of high priority in either situation as those which the parents view as being both important and problematic of attainment. Middle-class parents are more likely than are working-class parents to value "internal" standards (for example, honesty, self-control) and less likely to value obedience.

ORIENTATION TO EDUCATION AS A FACTOR IN THE SCHOOL MALADJUSTMENT OF LOWER-CLASS CHILDREN

Jackson Toby

Adapted by permission of the author and publisher from *Social Forces*, 35 (1957), 259-266. Copyright, University of North Carolina Press, 1957.

Even taking an extremely crude index of school achievement, that of grade placement, *for every age level* the average grade of middle-class urban children is higher than that of lower-class children. (See Tables 1, 2, and

3.) These differences can be observed at 7 and 8 years of age as well as at 17. Apparently whatever produces the difference starts operating to differentiate lower-class from middle-class children from the early grades. Another way of looking at class selectivity of the educational process is to observe the proportion of lower-class boys in high school a generation ago (Tables 4 and 5) or in college today.[1]

Why are middle-class children more successful in their studies? Why do lower-class children drop out at younger ages and complete fewer grades? One hypothesis is that school teachers are middle-class in their values, if not in their origins, and penalize those students who do *not* exhibit the middle-class traits of cleanliness, punctuality, and neatness or who *do* exhibit the lower-class traits of uninhibited sexuality and aggression.[2] Some social scientists believe that lower-class children, even though they may have the intellectual potentialities for high levels of academic achievement, lose interest in school or never become interested because they resent the personal rejection of their teachers. Such rejection is, they say, motivated by the teachers' mistaken notion that lower-class children are deliberately defying them. Davis and Havighurst show that children are the prisoners of their experience and that lower-class children behave the way they do, not because of any initial desire to defy school authorities, but rather because of their lower-class childhood training.[3]

According to this hypothesis, teacher rejection makes the lower-class boy resentful and rebellious. His attitude is, "If you don't like me, I won't cooperate." Unfortunately for him, however, school achievement is related to later occupational advancement. Failure to cooperate with the teacher cuts off the lower-class boy from a business or professional career. Professor August Hollingshead describes what happens to lower-class boys from a small town in Illinois who withdraw from school to escape the psychic punishment meted out by the teachers and upper-class children.

The withdrawees' job skills are limited to what they have learned from contact with parents, relatives, friends, and through observations and personal experience, largely within the community; no withdrawee has any technical training for any type of job; furthermore, few have plans to acquire it in the future. . . . The boys have some acquaintance with working on farms, washing cars, loading and unloading grain, repairing cars, driving trucks, doing janitor work, clerking in stores, and odd jobs, but their lack of training, job skills, and experience combined with their youth and family backgrounds severely limit their job opportuni-

[1] Helen B. Goetsch, *Parental Income and College Opportunities* (New York: Teachers College, Columbia University, Contributions to Education, No. 795, 1940).

[2] W. L. Warner, R. J. Havighurst, and M. B. Loeb, *Who Shall Be Educated?* (New York: Harper & Brothers, 1944).

[3] Allison Davis and Robert J. Havighurst, *Father of the Man* (Boston: Houghton Mifflin, 1947).

ties. These factors, along with need, force them to take whatever jobs they can find. . . . Menial tasks, long hours, low pay, and little consideration from the employer produces discontent and frustration, which motivate the young worker to seek another job, only to realize after a few days or weeks that the new job is like the old one. This desire for a more congenial job, better pay, shorter hours, and a better employer gives rise to a drift from job to job.[4]

The association between education, job levels, and prestige in the social structure is so high that the person with more education moves into the high-ranking job and the person with little education into the low-ranking job. Furthermore, and this is the crucial fact from the viewpoint of the person's relation to the social structure, each tends to remain in the job channel in which he starts as a young worker. This is especially true if he has less than a high school education; then he starts as an unskilled menial and has few opportunities in later years to change to skilled labor, business, or the professions. Therefore, his chances to be promoted up through the several levels of the job channel in which he functions are severely limited. As the years pass, his position in the economic system becomes fixed, and another generation has become stable in the class structure.[5]

In other words, Professor Warner and his colleagues point out that the American public school teacher is suspicious of lower-class children and unwilling to give them a chance. If they withdraw from school to escape the pressures, they must surrender their chance to realize the American dream: social mobility.

Another hypothesis attributes the inferior performance of lower-class children at school *directly* to the economic disabilities of their families.

TABLE 1. Median Years of School Completed by Native White Boys by Monthly Rental Value of Home and by Age in Cities of 250,000 Inhabitants or More, 1940

Age	Monthly Rental Value of Home						
	Under $10	$10-$14	$15-$19	$20-$29	$30-$49	$50-$74	$75 and over
7 years	1.3	1.5	1.6	1.7	1.7	1.7	1.7
8 years	2.1	2.4	2.4	2.5	2.6	2.6	2.7
9 years	2.8	3.2	3.3	3.4	3.5	3.7	3.7
10 years	3.6	4.0	4.2	4.4	4.5	4.6	4.7
11 years	4.4	4.9	5.1	5.3	5.5	5.6	5.6
12 years	5.4	5.7	6.0	6.2	6.5	6.6	6.7
13 years	6.0	6.7	7.1	7.2	7.5	7.7	7.8
14 years	7.2	7.8	7.9	8.2	8.5	8.7	8.8
15 years	8.3	8.5	8.8	9.2	9.4	9.6	9.8
16 years	8.6	9.3	9.6	9.8	10.3	10.5	10.6
17 years	9.4	9.9	10.2	10.7	10.7	11.3	11.5

Source: Bureau of the Census, *Sixteenth Census of the United States* (1940), *Monograph on Population Education: Educational Attainment of Children by Rental Value of Home* (Washington: Government Printing Office, 1945), p. 3.

[4] August B. Hollingshead, *Elmtown's Youth* (New York: John Wiley & Sons, 1949), p. 369.

[5] *Ibid.*, p. 388.

John is a poor student because he lacks the nourishing food for sustained effort or because he is compelled to work after school instead of doing his homework; or he is a truant because he is ashamed to appear at school in ragged clothes or torn shoes. Like the rejecting teacher hypothesis, the economic disability hypothesis treats the child as essentially passive. According to both, he is victimized by a situation over which he has no control, in the one case by teachers who reject him, in the other by an economic system which does not allow him the opportunities to realize his ambitions.

TABLE 2. **Distribution of Retarded and Nonretarded Pupils According to Occupational Status of Father (Sims' Scale) in the New York City Public Schools, 1931-32**

Father's Occupational Status	Total	Slow Progress	Normal Progress	Rapid Progress
Total	100.0	100.0	100.0	100.0
Professional	3.7	1.3	4.4	6.2
Clerical	19.8	11.2	19.4	31.9
Artisan	24.0	22.0	25.5	24.8
Skilled laborer	36.9	43.8	35.1	29.8
Unskilled laborer	15.6	21.7	15.6	7.3

TABLE 3. **Percentage Distribution of Pupils According to Father's Occupational Status and Pupils' Progress Status, 1931-32**

Father's Occupational Status	Pupils' Progress Status			
	Total	Slow	Normal	Rapid
Professional	100.0	13.2	39.7	47.1
Clerical	100.0	21.3	32.7	46.0
Artisan	100.0	34.6	35.6	29.8
Skilled laborer	100.0	45.0	31.9	23.1
Unskilled laborer	100.0	53.0	33.6	13.4

Source for Tables 2 and 3: Eugene A. Nifenecker, *Statistical Reference Data Relating to Problems of Overageness, Educational Retardation, Non-Promotion, 1900-1934* (New York: Board of Education, 1937), p. 233.

But it is not at all clear that the average lower-class child has academic aspirations which are thwarted by his teachers or his economic circumstances. Studies of withdrawees from high school show that the majority leave school with no regrets; some volunteer the information that they hate school and are delighted to get through with it.[6] These data suggest

[6] Howard C. Seymour, The Characteristics of Pupils Who Leave School Early—A Comparative Study of Graduates with Those Who are Eliminated Before High School Graduation, unpublished Ph.D. dissertation, Harvard University, 1940; Harold J. Dillon, *Early School Leavers* (New York: National Child Labor Committee, 1949).

that some lower-class children view the school as a burden, not an opportunity. Perhaps it is not only teacher prejudice and his parents' poverty that handicap the lower-class child at school. *He* brings certain attitudes and experiences to the school situation just as his teacher does.

TABLE 4.   **High School Attendance of the Children of Fathers Following Various Occupations, Seattle, St. Louis, Bridgeport, and Mount Vernon, 1919-1921**

Parental Occupation	Number in High School for Every 1,000 Men 45 Years of Age or Over
Proprietors	341
Professional service	360
Managerial service	400
Commercial service	245
Building trades	145
Machine trades	169
Printing trades	220
Miscellaneous trades	103
Transportation service	157
Public service	173
Personal service	50
Miners, lumber workers, and fishermen	58
Common labor	17

TABLE 5.   **Percentage of Students in Each of Two High School Years from Each of the Occupational Groups, 1919-1921**

Parental Occupation	Freshman Class	Senior Class
Proprietors	17.7	22.9
Professional service	7.7	12.5
Managerial service	15.4	19.1
Commercial service	8.6	11.1
Clerical service	5.9	5.9
Agricultural service	2.3	2.3
Artisan-proprietors	4.4	3.5
Building trades	8.8	5.3
Machine trades	8.3	4.6
Printing trades	1.0	0.8
Miscellaneous trades	4.8	2.3
Transportation service	6.2	3.6
Public service	1.7	1.1
Personal service	1.4	0.9
Miners, lumber workers, and fishermen	0.5	0.3
Common labor	1.8	0.6
Unknown	3.5	3.2

Source for Tables 4 and 5: George S. Counts, *The Selective Character of American Secondary Education* (Chicago: University of Chicago Press, 1922), pp. 33, 37.

Whereas the middle-class child learns a socially adaptive fear of receiving poor grades in school, of being aggressive toward the teacher, of fighting, of cursing, and of having early sex relations, the slum child learns to fear quite different social acts. His gang teaches him to fear being taken in by the teacher, of being a softie with her. To study homework seriously is literally a disgrace. Instead of boasting of good marks in school, one conceals them, if he ever receives any. The lower-class individual fears not to be thought a street-fighter; it is a suspicious and dangerous social trait. He fears not to curse. If he cannot claim early sex relations his virility is seriously questioned.[7]

Of course, not all lower-class children have a hostile orientation to the school. As a matter of fact, the dramatic contrast between the educational attainments of drafted enlisted men in the two World Wars show that the public schools are being used more and more; and some of this increase undoubtedly represents lower-class youths who eagerly take advantage of educational opportunities.[8] Still, many lower-class children do *not* utilize the educational path to social advancement.[9] Apparently, one reason for this is a chronic dissatisfaction with school which begins early in their academic careers. Why should middle-class children "take to" school so much better?

To begin with, it should not be taken for granted that any child, whatever his socio-economic origin, will find school a pleasant experience from the very first grade. On the contrary, there is reason to believe that starting school is an unpleasant shock. The average child cannot help but perceive school as an invasion of his freedom, an obligation imposed on him by adults. Forced to come at set times, to restrain his conversation so that the teacher may instruct the class as a group, he may not see any relation-

[7] Allison Davis, *Social Class Influences on Learning* (Cambridge, Massachusetts: Harvard University Press, 1949), p. 30.

[8] 41 percent of the selectees of World War II were high school graduates or better, as contrasted with only 9 percent in World War I. Samuel A. Stouffer and others, *The American Soldier* (Princeton: Princeton University Press, 1949), I, 59. Compulsory school attendance laws may have something to do with this difference, but the average age of high school graduation is beyond the age of compulsory attendance in most states.

[9] The assumption here is that the goal of success is sufficiently widespread in the American ethos and the penalties for criminal deviance sufficiently great that the failure to utilize a legitimate channel of social mobility can usually be explained as due (1) to a failure on the part of the individual to *perceive* that channel as feasible for him and to define it as an opportunity, (2) to objective disabilities which cannot be overcome by effort, or (3) to his perception of other and better opportunities. We assume, therefore, that the lower-class subculture (uncongenial to social mobility) has its roots in a sour-grapes reaction. This does *not* mean that every lower-class boy yearns for higher socio-economic status at some time or other in his life. Some of them have been socialized into the sour grapes tradition before having the experience on which they might personally conclude that the grapes are sour.

ship between what she asks him to learn and what he might be interested in doing. And in terms of maximizing his pleasure at the time, he is quite right. Except for kindergarten and ultra-progressive schools, the curriculum is a discipline imposed on the pupil rather than an extension and development of his own interests. This is not to condemn the school system, But it does point up the problematic nature of school adjustment.

Middle-class parents make it quite clear that school is nothing to be trifled with. They have probably graduated at least from high school, and their child is aware that they *expect* him to do the same or better. If he has difficulty with his studies, they are eager (and competent) to help him. And not only do his *parents* expect him to apply himself to his studies, so do his *friends* and *their* parents. He is caught in a neighborhood pattern of academic achievement in much the same way some lower-class boys are caught in a neighborhood pattern of truancy and delinquency. This concern with education is insurance against the child's fall in social status. Middle-class parents convey to their children subtly or explicitly that they must make good in school if they want to go on being middle-class. This may be phrased in terms of preparation for a "suitable" occupation (an alternative to a stigmatized occupation such as manual labor), in terms of a correlation between a "comfortable" standard of living and educational level, or in terms of the honorific value of education for its own sake.

Middle-class parents constantly reinforce the authority and prestige of the teacher, encouraging the child to respect her and compete for her approval. The teacher makes a good parent-surrogate for him because his parents accept her in this role.[10] They urge him to value the gold stars she gives out and the privilege of being her monitor. But although the middle-class child's initial motivation to cooperate with the teacher may spring from his parents, motivation functionally autonomous of parental pressure usually develops to supplement it.[11] Part of this new motivation may be the intrinsic interest of the subject matter, or at least some of it, once he has gotten well along in his course. *Learning* to read may be a disagreeable chore; but the time soon comes when interesting stories are made accessible by the development of reading skill. An even more important

[10] Professor Green maintains that the middle-class boy is more closely supervised by his mother than the lower-class boy and that this "personality absorption" creates a dependence on adult authority much greater than that of the less well supervised lower-class boy. If this theory were accepted, we would thus find additional reason for the relative tractability and cooperativeness of the middle-class boy in school. Arnold W. Green, "The Middle Class Male Child and Neurosis," *American Sociological Review*, XI (1946), 31-41.

[11] See Gordon W. Allport, *Personality* (New York: Henry Holt and Company, 1937), pp. 191-206, for a discussion of functional autonomy.

source of motivation favorable to school is the recognition he gets in the form of high marks. He learns that scholastic competition is somewhat analogous to the social and economic competition in which his parents participate. The object of scholastic competition is to win the approving attention of the teacher, to skip grades, and to remain always in the "bright" classes. (In grade school the "bright" and the "dull" classes take approximately the same work, but pupils and teachers have no difficulty in separating the high prestige groups. In high school, "commercial," "trade," and "general" courses have different curricula from the high prestige "college" course. Again, there is consensus among the students as well as the teachers that the non-college courses are for those who are not "college material.")[12]

Of course it is not competition alone that gives the middle-class child an emotional investment in continued scholastic effort; it is the *position* he achieves in that competition. Apparently his preschool training prepares *him* much better for scholastic competition than his lower-class class-mate.[13] His parents mingle with lawyers, accountants, businessmen, and others who in their day-to-day activities manipulate symbols. In the course of conversation these people use a sizeable vocabulary including many ab-stractions of high order. He unconsciously absorbs these concepts in an effort to understand his parents and their friends. He is stimulated in this endeavor by the rewards he receives from his parents when he shows verbal precociousness. These rewards are not necessarily material or con-scious. The attention he receives as a result of a remark insightful beyond his years, the pride his mother shows in repeating a bright response of his to her friends, these are rewards enough. This home background is valu-able preparation for successful competition in school. For, after all, school subjects are designed to prepare for exactly the occupational level to which his parents are already oriented. Hence he soon *achieves* in school a higher than average status. (See Tables 1, 2, and 3.) To maintain this status intact (or improve it) becomes the incentive for further effort, which involves him deeper and deeper in the reward and punishment system of the school. Thus, *his success cumulates and generates the condi-tions for further success.*

[12] George S. Counts, *The Selective Character of American Secondary Education* (Chicago: University of Chicago Press, 1922), shows the middle-class orientation of the "college" course; see also R. E. Eckert and T. O. Marshall, *When Youth Leaves School* (New York: The Regents' Inquiry, McGraw-Hill, 1938), p. 67.

[13] Millie C. Almy, *Children's Experiences prior to First Grade and Success in Be-ginning Reading* (New York: Teachers College, Columbia University, Contributions to Education, No. 954, 1949); Dorris M. Lee, *The Importance of Reading for Achiev-ing in Grades Four, Five, and Six* (New York: Teachers College, Columbia Univer-sity, Contributions to Education, No. 556, 1933).

A similar conclusion was reached after a study of the success and failure of children in certain nonacademic activities. Dr. Anderson concluded that success and practice mutually reinforce one another, producing remarkable differentiations in performance.

. . . a child is furnished from early life with the opportunity to hammer nails. In the course of the next ten or fifteen years, the child has 100,000 opportunities to hammer nails, whereas a second child in the same period of time has only ten or fifteen opportunities to hammer nails. At the age of twenty, we may be tremendously impressed with the ease and accuracy with which the first child hammers nails and likewise with the awkwardness and incapacity of the second child. We speak of the first child as an expert and the second child as a boob with respect to the nail hitting situation, and we may naïvely ascribe the ability of the first child to an inherited ability because its appearance is so inexplicable in comparison with the lack of ability of the second child.[14]

The most significant fact which comes out of these observations is the fact that if we take a particular child and record his relationship to the group, we find that in ninety-five percent of the situations with which he is presented in the play situation, he is the dominating or leading individual, whereas another child under the same conditions is found to be in the leading position only five percent of the time.

. . . the social reactions of these particular children . . . may be the product of hereditary factors, environmental factors, more rapid rate of development, or a large number of factors combined. The important fact for our discussion is that within a constant period one child is getting approximately twenty times as much specific practice in meeting social situations in a certain way as is a second child. Life is something like a game of billiards in which the better player gets more opportunity for practice and the poorer player less.[15]

For the average middle-class child, the effective forces in his life situation form a united front to urge upon him a favorable orientation to school. Of course, this may not be sufficient to produce good school adjustment. He may not have the native intelligence to perform up to the norm. Or he may have idiosyncratic experiences that alienate him from scholastic competition. But, apparently, for the *average* middle-class child, this favorable orientation, combined with the intellectual abilities cultivated in his social milieu, results in satisfactory performance in the school situation.

The other side of the coin is the failure of some lower-class children to develop the kind of orientation which will enable them to overcome the initial frustration of school discipline.[16] To begin with, the parents of the

[14] John E. Anderson, "The Genesis of Social Reactions in the Young Child," *The Unconscious: A Symposium,* ed. by E. S. Dummer (New York: Alfred A. Knopf, 1928), pp. 83-84.

[15] *Ibid.,* pp. 81-82.

[16] At this point we are abstracting from such situational considerations as teacher

lower-class child may not support the school as do middle-class parents. His parents probably do not have much education themselves, and, if not, they cannot very well make meaningful to him subjects that they do not themselves understand. Neither are they able to help him surmount academic stumbling blocks. Even more important, they lack the incentive to encourage him in and praise him for school accomplishment at that critical early period when he finds school new and strange and distasteful. Almost the same reasoning can be applied to the inculcation of a cooperative attitude toward school in the child as has been applied to an acceptant attitude toward toilet training. If the parents convey to the child their eagerness to have him adjust to irksome school discipline, he will probably accept it to please them and retain their love just as he learned to urinate and defecate at appropriate times and places. But toilet training and school adjustment training differ in an important particular. Parents *must* toilet train the child because permitting him to soil himself at will is a constant and immediate nuisance.

The consequences of a child's disinterest in school may also be unpleasant, both for him and for his parents, but it is not immediate. In the short run, allowing him to neglect school may be the least troublesome course for his parents to take. If they are neutral or antagonistic toward school, a result (1) of the esoteric nature of the curriculum from the point of view of skills cultivated and appreciated in the lower-class milieu and (2) of their failure to see the relevance of education to occupational advancement into a higher socio-economic class, they do not *have* to give the kind of support to the school given by middle-class parents. There is no reason to assume that the value of education is self-evident. For those lower-class people who have lost hope in social mobility, the school is a symbol of a competition in which they do not believe they can succeed. If they themselves have given up, will they necessarily encourage their children to try to be better?

Moreover, coming as he does from a social stratum where verbal skills are not highly developed, the lower-class child finds school more difficult than does his middle-class contemporary. His father, a carpenter or a factory worker, manipulates concrete objects rather than symbols in his occupational role. In so far as he learns from his father, he is more likely to learn how to "fix things" than the importance of a large vocabulary.[17] This learning does not help him with his school work, for school tends to give a competitive advantage to those with verbal facility.

---

rejection, the economic resources of the family and native capacity. We are considering only the orientations of the boy himself.

[17] Of course this is a matter of degree. The lower-class boy acquires verbal skills but not on so high a level as the middle-class boy.

This disadvantage with respect to verbal skills may account for the poorer showing of lower-class children on standard intelligence tests.[18]

. . . the cultural bias of the standard tests of intelligence consists in their having fixed upon only those types of mental behavior in which the higher and middle socio-economic groups are superior. In those particular areas of behavior, the tests might conceivably be adequate measures of mental differences among individual children within the more privileged socio-economic groups. But they do not measure the comparative over-all mental behavior of the higher and lower socio-economic groups, because they do not use problems which are equally familiar and motivating to all such groups.[19]

In other words, middle-class children have an advantage because they are more familiar with the sort of problems that occur on the tests. This does not necessarily mean that the intelligence tests are invalid. It depends upon what the investigator thinks he is measuring. If he believes he is getting at "innate" ability, abstracted from cultural milieu and idiosyncratic learning, he is naïve. An intelligence test is a valid measure of the native intellectual ability of an individual only under special circumstances, one of these being that the respondent's experience is similar to that of the group on which the test was standardized. Thus, a Navaho boy who scores 80 on the Stanford-Binet (Revised Form) may be unusually intelligent. Until a test is designed to tap the experiences of Navahos, there exists no reference point about which to assess superiority and inferiority.[20]

However, it is not only the *content* of the intelligence test that gives middle-class urban children a better chance at high scores. It is the *structure* of the test situation. Even if we could find items equally familiar or unfamiliar to everyone taking the test, differential interest in solving abstract problems would work against the lower-class student.

. . . finding completely unfamiliar problems is not a possible choice, because such problems (namely, those involving some relationship between esoteric geometrical figures) do not arouse as great interest or as strong a desire to achieve a solution among low socio-economic groups as among high groups. The reason is clear: such an unrealistic problem can arouse the child's desire to achieve a solution only if the child has been trained to evaluate highly any and all success in tests. No matter how unreal and purposeless the problem may seem, the average child in a high socio-economic group will work hard to solve it, if his parents, his teacher, or other school officers expect him to try hard. The average slum child, however, will usually react negatively to any school test, and especially to a test whose problems have no relation to his experience.[21]

[18] Walter S. Neff, "Socio-economic Status and Intelligence: a Critical Survey," *Psychological Bulletin,* XXXV (1938), 727-757.

[19] Allison Davis, *op. cit.,* p. 48.

[20] Dorothy Leighton and Clyde Kluckhohn, *Children of the People* (Cambridge, Massachusetts: Harvard University Press, 1947), pp. 148-155.

[21] *Ibid.,* pp. 68-69.

However justified the criticisms of the intelligence test as an instrument measuring native intellectual ability, it is highly predictive of academic accomplishment. A student with a high I.Q. score does better in his studies, on the average, than one with a low I.Q. score.[22] Hence the discrepancy between the scores of lower-class students and of middle-class students is an index of the former's disadvantage in the school situation.

One possible response of the lower-class child to his disadvantages in the school situation is to increase his efforts. But his initial orientation drives him in the opposite direction. He is more likely to respond to competitive failure by going on strike psychologically, neglecting his homework, paying no attention in class, annoying the teacher. Uninterested in the curriculum, he learns as little as he can. Instead of a situation where the student and the teacher work toward a common goal, the development of the student's understanding of certain ranges of problems, he and his teacher are oriented antagonistically to one another. The teacher tries to stuff into his head as much of the curriculum as possible; he tries to absorb as little as is consistent with his own safety, in terms of sanctions mobilized by the school and his parents.

But school subjects are cumulative. Within a few years he is retarded in basic skills, such as reading, absolutely necessary for successful performance in the higher grades. Whether he is promoted along with his agemates, "left back," or shunted into "slow" programs makes relatively little difference at this point. For whatever is done, he finds himself at the bottom of the school status hierarchy. He is considered "dumb" by the more successful students and by the teachers. This makes school still more uninteresting, if not unpleasant, and he neglects his work further. Eventually he realizes he can never catch up.

Without realizing what he was doing, he had cut himself off from the channels of social mobility. In those crucial early grades where the basis for school adjustment was being laid, he had not yet known that he wanted more out of life than his parents. Or, if he knew, he did not realize that school achievement and high occupational status are related. And he was not lucky enough to have parents who realized it for him and urged him on until he was old enough to identify with the school through choice. There is a certain irreversibility about school maladjustment. The student can hardly decide at 18 that he wants to become a lawyer if he is five years retarded in school. It is no longer possible for him to "catch up" and use school as a means to realize his ambitions. Sometimes lower-class men will rue their failure to take advantage of the opportunities presented by the

[22] Eugene A. Nifenecker, *Statistical Reference Data Relating to Problems of Overageness, Educational Retardation, Non-Promotion,* 1900-1934 (New York: Board of Education, 1937), p. 111.

school. James T. Farrell captures the flavor of this regret in the following passage from one of his novels:

Walking on, seeing the lights of Randolph Street before him, he wondered if they were college football players [referring to the young men walking in front of him]. That was what Studs Lonigan might have been. Even if he did admit it, he had been a damn good quarterback. If he only hadn't been such a chump, bumming from school to hang around with skunky Weary Reilley and Paulie Haggerty until he was so far behind at high school that it was no use going. It wouldn't have been so hard to have studied and done enough homework to get by, and then he could have set the high school gridiron afire, gone to Notre Dame and made himself a Notre Dame immortal, maybe, alongside of George Gipp, the Four Horsemen, Christie Flannagan and Carideo. How many times in a guy's life couldn't he kick his can around the block for having played chump.[23]

If on the other hand, the social milieu of the lower-class boy supported the school and encouraged him to bend every effort to keep up with his work, he would finish high school whether he enjoyed it or not—the way middle-class boys do. At graduation he might decide that he would like to become a plumber. That is, he might not crave middle-class status enough to suffer the discipline of continued education. But if he were not content with a lower-class status, if he wanted above all things to "be somebody," the educational route to high status would still be open. He would still have a *choice;* he would not be forced to accept a menial occupational role whether he liked it or not. As it is, the crucial decision is made before he is old enough to have a voice in it; it is made by his parents, his neighbors, and his friends.

To sum up, the middle-class child has the following advantages in school compared with the lower-class child: (1) his parents are probably better educated and are therefore more capable of helping him with his school work if this should be necessary; (2) his parents are more eager to make his school work seem meaningful to him by indicating, implicitly or explicitly, the occupational applications of long division or history; (3) the verbal skills which he acquires as part of child training on the middle-class status level prepare him for the type of training that goes on in school and give him an initial (and cumulating) advantage over the lower-class child in the classroom learning situation; and (4) the coordinated pressure of parents, friends, and neighbors reinforce his motivation for scholastic success and increase the probability of good school adjustment.

[23] James T. Farrell, *Judgment Day* (New York: Vanguard Press, 1935), p. 24.

SOCIAL CLASS AND PARENTAL VALUES

Melvin L. Kohn

Adapted from *The American Journal of Sociology*, 64:4 (1959), 337-351, by permission of The University of Chicago Press.

We undertake this inquiry into the relationship between social class and parental values in the hope that a fuller understanding of the ways in which parents of different social classes differ in their values may help us to understand why they differ in their practices.[1] This hope, of course, rests on two assumptions: that it is reasonable to conceive of social classes as subcultures of the larger society, each with a relatively distinct value-orientation, and that values really affect behavior.

Sample and Method of Data Collection

Washington, D.C.—the locus of this study—has a large proportion of people employed by government, relatively little heavy industry, few recent immigrants, a white working class drawn heavily from rural areas, and a large proportion of Negroes, particularly at lower economic levels. Generalizations based on this or any other sample of one city during one limited period of time are, of course, tentative.

Our intent in selecting the families to be studied was to secure approximately two hundred representative white working-class families and another two hundred representative white middle-class families, each family having a child within a narrowly delimited age range. We decided on fifth-grade children because we wanted to direct the interviews to rela-

---

[1] There now exists a rather substantial, if somewhat inconsistent, body of literature on the relationship of social class to the ways that parents raise their children. For a fine analytic summary see Urie Bronfenbrenner, "Socialization and Social Class through Time and Space," in Eleanor E. Maccoby *et al.*, *Readings in Social Psychology* (New York: Holt; 1958). Bronfenbrenner gives references to the major studies of class and child-rearing practices that have been done.

For the most relevant studies on class and *values* see Evelyn M. Duvall, "Conceptions of Parenthood," *American Journal of Sociology*, LII (November, 1946), 193-203; David F. Aberle and Kaspar D. Naegele, "Middle Class Fathers' Occupational Role and Attitudes toward Children," *American Journal of Orthopsychiatry*, XXII (April, 1952), 366-78; Herbert H. Hyman, "The Value Systems of Different Classes," in Reinhard Bendix and Seymour M. Lipset (eds.), *Class, Status, and Power* (Glencoe, Ill.: Free Press, 1953), pp. 426-42.

tionships involving a child old enough to have a developed capacity for verbal communication.

The sampling procedure involved two steps: the first, selection of census tracts. Tracts with 20 per cent or more Negro population were excluded, as were those in the highest quartile with respect to median income. From among the remaining tracts we then selected a small number representative of each of the three distinct types of residential area in which the population to be studied live: four tracts with a predominantly working-class population, four predominantly middle-class, and three having large proportions of each. The final selection of tracts was based on their occupational distribution and their median income, education, rent (of rented homes), and value (of owner-occupied homes). The second step in the sampling procedure involved selection of families. From records made available by the public and parochial school systems we compiled lists of all families with fifth-grade children who lived in the selected tracts. Two hundred families were then randomly selected from among those in which the father had a "white-collar" occupation and another two hundred from among those in which the father had a manual occupation.

In all four hundred families the mothers were to be interviewed. In every fourth family we scheduled interviews with the father and the fifth-grade child as well.[2] (When a broken family fell into this sub-sample, a substitute was chosen from our over-all sample, and the broken family was retained in the over-all sample of four hundred families.)

When interviews with both parents were scheduled, two members of the staff visited the home together—a male to interview the father, a female to interview the mother. The interviews were conducted independently, in separate rooms, but with essentially identical schedules. The first person to complete his interview with the parent interviewed the child.

Indexes of Social Class and Values

Social Class. Each family's social-class position has been determined by the Hollingshead Index of Social Position, assigning the father's occupational status a relative weight of 7 and his educational status a weight of 4. We are considering Hollingshead's Classes I, II, and III to be "middle class," and Classes IV and V to be "working class." The middle-class sample is composed of two relatively distinct groups: Classes I and II are

[2] We secured the co-operation of 86 per cent of the families where the mother alone was to be interviewed and 82 per cent of the families where mother, father, and child were to be interviewed. Rates of non-response do not vary by social class, type of neighborhood, or type of school. This, of course, does not rule out other possible selective biases introduced by the non-respondents.

almost entirely professionals, proprietors, and managers with at least some college training. Class III is made up of small shopkeepers, clerks, and salespersons but includes a small number of foremen and skilled workers of unusually high educational status. The working-class sample is composed entirely of manual workers but preponderantly those of higher skill levels. These families are of the "stable working class" rather than "lower class" in the sense that the men have steady jobs, and their education, income, and skill levels are above those of the lowest socioeconomic strata.

Values. We shall use Kluckhohn's definition: "A value is a conception, explicit or implicit, distinctive of an individual or characteristic of a group, of the desirable which influences the selection from available modes, means, and ends of action." [3]

Our inquiry was limited to the values that parents would most like to see embodied in their children's behavior. We asked the parents to choose, from among several alternative characteristics that might be seen as de-

TABLE 1. Proportion of Mothers Who Select Each Characteristic as One of Three "Most Desirable" in a Ten- or Eleven-Year-Old Child

Characteristics	For Boys Middle Class	For Boys Working Class	For Girls Middle Class	For Girls Working Class	Combined Middle Class	Combined Working Class
1. That he is honest	0.44	0.57	0.44	0.48	0.44	0.53
2. That he is happy	.44*	.27	.48	.45	.46*	.36
3. That he is considerate of others	.40	.30	.38*	.24	.39*	.27
4. That he obeys his parents well	.18*	.37	.23	.30	.20*	.33
5. That he is dependable	.27	.27	.20	.14	.24	.21
6. That he has good manners	.16	.17	.23	.32	.19	.24
7. That he has self-control	.24	.14	.20	.13	.22*	.13
8. That he is popular with other children	.13	.15	.17	.20	.15	.18
9. That he is a good student	.17	.23	.13	.11	.15	.17
10. That he is neat and clean	.07	.13	.15*	.28	.11*	.20
11. That he is curious about things	.20*	.06	.15	.07	.18*	.06
12. That he is ambitious	.09	.18	.06	.08	.07	.13
13. That he is able to defend himself	.13	.05	.06	.08	.10	.06
14. That he is affectionate	.03	.05	.07	.04	.05	.04
15. That he is liked by adults	.03	.05	.07	.04	.05	.04
16. That he is able to play by himself	.01	.02	.00	.03	.01	.02
17. That he acts in a serious way	0.00	0.01	0.00	0.00	0.00	0.01
N	90	85	84	80	174	165

* Social-class differences statistically significant, 0.05 level or better, using chi-squared test.

[3] Clyde Kluckhohn, "Values and Value Orientations," in Talcott Parsons and Edward A. Shils (eds.), *Toward a General Theory of Action* (Cambridge, Mass.: Harvard University Press, 1951), p. 395.

sirable, those few which they considered *most* important for a child of the appropriate age. Specifically, we offered each parent a card listing 17 characteristics that had been suggested by other parents, in the pretest interviews, as being highly desirable. (These appear down the left margin of Table 1. The order in which they were listed was varied from interview to interview.) Then we asked: "Which three of the things listed on this card would you say are the *most* important in a boy (or girl) of (fifth-grade child's) age?" The selection of a particular characteristic was taken as our index of value.

Later in this report we shall subject this index to intensive scrutiny.

Class and Values

Middle- and working-class mothers share a broadly common set of values—but not an identical set of values by any means (see Table 1). There is considerable agreement among mothers of both social classes that happiness and such standards of conduct as honesty, consideration, obedience, dependability, manners, and self-control are highly desirable for both boys and girls of this age.

Popularity, being a good student (especially for boys), neatness and cleanliness (especially for girls), and curiosity are next most likely to be regarded as desirable. Relatively few mothers choose ambition, ability to defend one's self, affectionate responsiveness, being liked by adults, ability to play by one's self, or seriousness as highly desirable for either boys or girls of this age. All of these, of course, might be more highly valued for children of other ages.

Although agreement obtains on this broad level, working-class mothers differ significantly[4] from middle-class mothers in the relative emphasis they place on particular characteristics. Significantly fewer working-class mothers regard happiness as highly desirable for *boys*. Although characteristics that define standards of conduct are valued by many mothers of both social classes, there are revealing differences of emphasis here too. Working-class mothers are more likely to value obedience; they would have their children be responsive to parental authority. Middle-class mothers are more likely to value both consideration and self-control; they would have their children develop inner control and sympathetic concern for other people. Furthermore, middle-class mothers are more likely to regard curiosity as a prime virtue. By contrast, working-class mothers put the emphasis on neatness and cleanliness, valuing the imaginative and exploring child relatively less than the presentable child.[5]

[4] The criterion of statistical significance used throughout this paper is the 5 per cent level of probability, based, except where noted, on the chi-square test.

[5] Compare these results with Bronfenbrenner's conclusion, based on an analysis of

TABLE 2. Proportion of Fathers Who Select Each Characteristic as One of Three "Most Desirable" in a Ten- or Eleven-Year-Old Child

Characteristics	For Boys Middle Class	Working Class	For Girls Middle Class	Working Class	Combined Middle Class	Working Class
1. That he is honest	0.60	0.60	0.43	0.55	0.52	0.58
2. That he is happy	.48	.24	.24	.18	.37	.22
3. That he is considerate of others	.32	.16	.38	.09	.35*	.14
4. That he obeys his parents well	.12*	.40	.14	.36	.13*	.39
5. That he is dependable	.36*	.12	.29*	.00	.33*	.08
6. That he has good manners	.24	.28	.24	.18	.24	.25
7. That he has self-control	.20	.08	.19	.00	.20*	.06
8. That he is popular with other children	.08	.16	.24	.45	.15	.25
9. That he is a good student	.04	.12	.10	.36	.07	.19
10. That he is neat and clean	.16	.20	.14	.09	.15	.17
11. That he is curious about things	.16	.12	.10	.00	.13	.08
12. That he is ambitious	.20	.12	.14	.00	.17	.08
13. That he is able to defend himself	.04	.16	.00*	.18	.02*	.17
14. That he is affectionate	.00	.04	.05	.18	.02	.08
15. That he is liked by adults	.00	.08	.00	.09	.00	.08
16. That he is able to play by himself	.00	.08	.05	.00	.02	.06
17. That he acts in a serious way	0.00	0.04	0.00	0.00	0.00	0.03
N	25	25	21	11	46	36

* Social-class differences statistically significant, 0.05 level or better, using chi-square test.

Middle-class mothers' conceptions of what is desirable for boys are much the same as their conceptions of what is desirable for girls. But working-class mothers make a clear distinction between the sexes: they are more likely to regard dependability, being a good student, and ambition as desirable for boys and to regard happiness, good manners, neatness, and cleanliness as desirable for girls.

What of the *fathers'* values? Judging from our subsample of 82 fathers, their values are similar to those of the mothers (see Table 2). Essentially the same rank-order of choices holds for fathers as for mothers, with one major exception: fathers are not so likely to value happiness for their daughters. Among fathers as well as mothers, consideration and self-control are more likely to be regarded as desirable by the middle class;

reports of studies of social class and child-rearing methods over the last twenty-five years: "In this modern working class world there may be greater freedom of emotional expression, but there is no laxity or vagueness with respect to goals of child training. Consistently over the past twenty-five years, the parent in this group has emphasized what are usually regarded as the traditional middle class virtues of cleanliness, conformity, and (parental) control, and although his methods are not so effective as those of his middle class neighbors, they are perhaps more desperate" (*op. cit.*).

middle-class fathers are also more likely to value another standard of conduct—dependability. Working-class fathers, like their wives, are more likely to value obedience; they are also more likely to regard it as desirable that their children be able to defend themselves.[6]

We take this to indicate that middle-class parents (fathers as well as mothers) are more likely to ascribe predominant importance to the child's acting on the basis of internal standards of conduct, working-class parents to the child's compliance with parental authority.

There are important differences between middle- and working-class parents, too, in the way in which their choice of any one characteristic is related to their choice of each of the others.[7]

[6] A comparison of the values of the fathers in this subsample with those of the mothers in this same subsample yields essentially the same conclusions.

We do not find that fathers of either social class are significantly more likely to choose any characteristic for boys than they are to choose it for girls, or the reverse. But this may well be an artifact of the small number of fathers in our sample; Aberle and Naegele (*op. cit.*) have found that middle-class fathers are more likely to value such characteristics as responsibility, initiative, good school performance, ability to stand up for one's self, and athletic ability for boys and being "nice," "sweet," pretty, affectionate, and well-liked for girls.

[7] A logical procedure for examining these patterns of choice is to compare the proportions of parents who choose any given characteristic, B, among those who do and who do not choose another characteristic, A. But since a parent who selects characteristic A has exhausted one of his three choices, the a priori probability of his selecting any other characteristic is only two-thirds as great as the probability that a parent who has not chosen A will do so. (A straightforward application of probability considerations to the problem of selecting three things from seventeen when one is interested only in the joint occurrence of two, say, A and B, shows that we can expect B to occur 2/16 of the time among those selections containing A and 3/16 of the time among those not containing A.) This, however, can be taken into account by computing the ratio of the two proportions: p_1, the proportion of parents who choose B among those who choose A, and p_2, the proportion who choose B among those who do *not* choose A. If the ratio of these proportions (p_1/p_2) is significantly larger than two-thirds, the two are positively related; if significantly smaller, they are negatively related.

The test of statistical significance is based on the confidence interval on a ratio, originally given by Fieller, with the modification that we deal here with the ratio of two independent proportions whose variances under the null hypothesis (chance) are known and whose distribution we assume to be normal. The 95 per cent confidence interval on the true ratio, R, of the two proportions, p_1 and p_2, that hold for any given A and B, is given by:

$$R = \frac{r \pm (1/8p_2) \ \sqrt{(28 \ / \ n_1) + (39r^2 \ / \ n_2) - [\ (28 \times 39) \ / \ 64n_1n_2p_2^2]}}{[\ 1 - (39 \ / \ 64n_2p_2^2)\ ]}$$

where p_1 and p_2 are the observed sample proportions, $r = p_1/p_2$, $n_1 =$ the number of persons selecting A, and $n_2 =$ the number of persons who do not select A.

The logic of the testing procedure is as follows: If the interval contains the null

We have already seen that parents of both social classes are very likely to accord *honesty* first-rank importance. But the choice of honesty is quite differently related to the choice of other characteristics in the two classes (see Table 3). Middle-class mothers[8] who choose honesty are more likely than are other middle-class mothers to regard consideration, manners, and (for boys) dependability as highly desirable; and those mothers who regard any of these as desirable are more likely to value honesty highly. Consideration, in turn, is positively related to self-control, and manners to neatness. Honesty, then, is the core of a set of standards of conduct, a set consisting primarily of honesty, consideration, manners, and dependability, together with self-control and neatness. As such, it is to be seen as one among several, albeit the central, standards of conduct that middle-class mothers want their children to adopt.

This is not the case for working-class mothers. Those who regard honesty as predominantly important are not especially likely to think of consideration, manners, or dependability as comparable in importance; nor are those who value any of these especially likely to value honesty. Instead the mothers who are most likely to attribute importance to honesty are those who are concerned that the child be happy, popular, and able to defend himself. It is not that the child should conduct himself in a considerate, mannerly, or dependable fashion but that he should *be* happy, *be* esteemed by his peers, and, if the necessity arise, *be* able to protect himself. It suggests that honesty is treated less as a standard of conduct and more as a quality of the person; the emphasis is on being a person of inherent honesty rather than on acting in an honest way.

Note especially the relationship of popularity to honesty. For middle-class mothers these are *negatively* related. To value honesty is to forego valuing popularity; to value popularity is to forego valuing honesty. One must choose between honesty "at the risk of offending" and popularity at

---

hypothesis value of $R = \frac{2}{3}$ implied by chance selection, then we assume no association between B and A. If the interval excludes $\frac{2}{3}$ such that the lower limit is larger than $\frac{2}{3}$, we conclude that the true R is greater than we expect on the basis of randomness and hence that B is positively associated with A. On the other hand, if the upper limit of the interval is smaller than $\frac{2}{3}$, then we conclude that the true R is smaller than $\frac{2}{3}$ and hence B and A are negatively related.

This procedure was suggested by Samuel W. Greenhouse. For the derivation of the test see E. C. Fieller, "A Fundamental Formula in the Statistics of Biological Assay, and Some Applications," *Quarterly Journal of Pharmacy and Pharmacology,* XVII (1944), 117-23; see also Pandurang V. Sukhatme, *Sampling Theory of Surveys with Applications* (Ames, Iowa: Iowa State College Press, 1954), pp. 158-60.

[8] This analysis and those to follow will be limited to the mothers, since the sample of fathers is small. For simplicity, we shall present data separately for boys and for girls only where the relationship under discussion appears to differ for the two sexes considered separately.

the sacrifice of absolute honesty. The exact opposite obtains for working-class mothers: those who accord high valuation to either are *more* likely to value the other. The very mothers who deem it most important that their children enjoy popularity are those who attribute great importance to honesty. Honesty does not interfere with popularity; on the contrary, it enhances the probability that one will enjoy the respect of one's peers.

However, working-class mothers who value obedience, manners, or consideration are distinctly unlikely to value popularity, and vice versa. They do see each of these standards of conduct as inconsistent with popularity.[9] This further substantiates the view that working-class mothers are more likely to view honesty as a quality of the person, a desideratum of moral worth, rather than as one among several highly valued standards of conduct.

Happiness, in distinction to honesty, implies neither constraints upon action nor a moral quality; rather, it indicates a desired goal, achievable in several different ways. One way of specifying what is implied when happiness is regarded as a major value is to ascertain the other values most likely to be related to the choice of happiness.

The two choices positively related to the choice of happiness by middle-class mothers are curiosity and (for boys) ambition. Those middle-class mothers who deem it exceedingly important that their children aspire for knowledge or success are even more likely than are middle-class mothers in general to value their children's happiness highly.

Working-class mothers who value these, however, are no more likely to value happiness. Instead, curiosity is related to consideration, to the child's concern for others' well-being, and ambition to dependability, to his being the type of person who can be counted on. The values that are positively related to happiness by working-class mothers are honesty, consideration (for boys), and popularity (for girls). Not aspirations for knowledge or for success, but being an honest—a worthy—person; not the desire to outdistance others, but, for boys, concern for others' well-being and for girls, enjoyment of the respect and confidence of peers: these are the conceptions of the desirable that accompany working-class mothers' wishes that their children be happy.

[9] It may be that these three characteristics have more in common than that they are all standards of conduct. The fact that working-class mothers who value consideration for their *daughters* are especially likely to value manners, and the converse, suggests the possibility that consideration may be seen as a near-equivalent to manners by at least a sizable portion of working-class mothers. If so, all three values negatively related to popularity can be viewed as reflecting close conformance to directives from parents—as contrasted to directives from within. (Note, in this connection, that working-class mothers who would have their daughters be mannerly are distinctly unlikely to deem it important that they be dependable.)

TABLE 3. **All Cases* Where Mothers' Choice of One Characteristic as "Desir-able" Is Significantly Related to Their Choice of Any Other Character-istic as "Desirable"**

| | | | Proportion Who Choose B among Those Who: | | |
			Choose A (p_1)	Do Not Choose A (p_2)	p_1/p_2
	Middle-Class Mothers				
	Characteristic A	B			
Positive relationships:					
1. Honesty	Consideration		0.42	0.37	1.14
2. Honesty	Manners		.22	.16	1.38
3. Honesty	Dependability (boys)		.33	.22	1.50
4. Consideration	Honesty		.47	.42	1.12
5. Manners	Honesty		.52	.43	1.21
6. Dependability	Honesty (boys)		.54	.41	1.32
7. Consideration	Self-control		.24	.22	1.09
8. Self-control	Consideration		.41	.39	1.05
9. Manners	Neatness		.24	.08	3.00
10. Neatness	Manners		.42	.16	2.63
11. Curiosity	Happiness		.58	.43	1.35
12. Happiness	Curiosity		.23	.14	1.64
13. Happiness	Ambition (boys)		.13	.06	2.17
Negative relationships:					
1. Honesty	Popularity		.04	.24	0.17
2. Popularity	Honesty		.12	.50	0.24
3. Curiosity	Obedience		.03	.24	0.13
4. Obedience	Consideration		0.17	0.45	0.38
	Working-Class Mothers				
Positive relationships:					
1. Happiness	Honesty		0.51	0.55	0.93
2. Popularity	Honesty		.62	.51	1.22
3. Honesty	Popularity		.20	.14	1.43
4. Honesty	Defend self		.07	.05	1.40
5. Consideration	Manners (girls)		.42	.30	1.40
6. Manners	Consideration (girls)		.31	.20	1.55
7. Consideration	Curiosity		.11	.04	2.75
8. Ambition	Dependability		.29	.19	1.53
9. Happiness	Consideration (boys)		.35	.27	1.30
10. Consideration	Happiness (boys)		.32	.25	1.28
11. Happiness	Popularity (girls)		.25	.16	1.56
Negative relationships:					
1. Obedience	Popularity		.05	.24	0.21
2. Manners	Popularity		.00	.23	0.00
3. Consideration	Popularity		.02	.23	0.09
4. Popularity	Obedience		.10	.38	0.26
5. Popularity	Manners		.00	.29	0.00
6. Popularity	Consideration		.03	.32	0.09
7. Manners	Dependability (girls)		0.00	0.20	0.00

* Where it is not specified whether relationship holds for boys or for girls, it holds for both sexes. In all the relationships shown, p_1 and p_2 are each based on a minimum of 20 cases.

Still the perhaps equally important fact is that no choice, by mothers of either social class, is negatively related to the choice of happiness.

The final bit of information that these data provide concerns the conception of *obedience* entertained in the two classes. Middle-class mothers who value curiosity are unlikely to value obedience; those who value obedience are unlikely to value consideration. For middle-class mothers, but not for working-class mothers, obedience would appear to have a rather narrow connotation; it seems to approximate blind obedience.

TABLE 4. Mothers' Socioeconomic Status and Their Choice of Characteristics as "Most Desirable" in a Ten- or Eleven-Year-Old Child

Characteristic	I	Proportion Who Select Each Characteristic Socioeconomic Stratum (on Hollingshead Index)			
		II	III	IV	V
Obedience	0.14	0.19	0.25	0.35	0.27
Neatness, cleanliness	.06	.07	.16	.18	.27
Consideration	.41	.37	.39	.25	.32
Curiosity	.37	.12	.09	.07	.03
Self-control	.24	.30	.18	.13	.14
Happiness	.61	.40	.40	.38	.30
Boys		.48	.40	.27	
Girls		.54	.40	.45	
Honesty	0.37	0.49	0.46	0.50	0.65
N	51	43	80	128	37

Class, Subculture, and Values

In discussing the relationship of social class to values we have talked as if American society were composed of two relatively homogeneous groups, manual and white-collar workers, together with their families. Yet it is likely that there is considerable variation in values, associated with other bases of social differentiation, *within* each class. If so, it should be possible to divide the classes into subgroups in such a way as to specify more precisely the relationship of social class to values.

Consider, first, the use we have made of the concept "social class." Are the differences we have found between the values of middle- and working-class mothers a product of this dichotomy alone, or do values parallel status gradations more generally? It is possible to arrive at an approximate answer by dividing the mothers into the five socioeconomic strata delineated by the Hollingshead Index (see Table 4). An examination of the choices made by mothers in each stratum indicates that variation in values parallels socioeconomic status rather closely:

a) The higher a mother's status, the higher the probability that she will choose consideration, curiosity, self-control, and (for boys)[10] happiness as highly desirable; curiosity is particularly likely to be chosen by mothers in the highest stratum.

b) The lower her status, the higher the probability that she will select obedience, neatness, and cleanliness; it appears, too, that mothers in the lowest stratum are more likely than are those in the highest to value *honesty*.

Mothers' values also are directly related to their own occupational positions and educational attainments, independently of their families' class status. (The family's class status has been indexed on the basis of the husband's occupation and education.) It happens that a considerable proportion of the mothers we have classified as working class hold white-collar jobs.[11] Those who do are, by and large, closer to middle-class mothers in their values than are other working-class mothers (see Table 5). But those who hold manual jobs are even further from middle-class mothers in their values than are working-class mothers who do not have jobs outside the home.

So, too, for mothers' educational attainments: a middle-class mother of *relatively* low educational attainment (one who has gone no further than graduation from high school) is less likely to value curiosity and more likely to value (for girls) neatness and cleanliness (see Table 6). A working-class mother of *relatively* high educational attainment (one who has at least graduated from high school) is more likely to value self-control for boys and both consideration and curiosity for girls. The largest differences obtain between those middle-class mothers of highest educational attainments and those working-class mothers of lowest educational attainments.

Even when we restrict ourselves to considerations of social status and its various ramifications, we find that values vary appreciably within each of the two broad classes. And, as sociologists would expect, variation in values proceeds along other major lines of social demarcation as well. Religious background is particularly useful as a criterion for distinguishing subcultures within the social classes. It does *not* exert so powerful an effect that Protestant mothers differ significantly from Catholic mothers of the same social class in their values.[12] But the combination of class and reli-

---

[10] The choice of happiness is, as we have seen, related to social class for boys only. Consequently, in each comparison we shall make in this section the choice of happiness for *girls* will prove to be an exception to the general order.

[11] No middle-class mothers have manual jobs, so the comparable situation does not exist. Those middle-class women who do work (at white-collar jobs) are less likely to value neatness and cleanliness and more likely to value obedience and curiosity.

[12] The index here is based on the question "May I ask what is your religious background?"

Even when the comparison is restricted to Catholic mothers who send their children

gious background does enable us to isolate groups that are more homogeneous in their values than are the social classes *in toto*. We find that there is an ordering, consistent for all class-related values, proceeding from middle-class Protestant mothers, to middle-class Catholic, to working-class Protestant, to working-class Catholic (see Table 7). Middle-class Protestants and working-class Catholics constitute the two extremes whose values are most dissimilar.

TABLE 5.   Working-Class Mothers' Own Occupations and Their Choice of Characteristics as "Most Desirable" in a Ten- or Eleven-Year-Old Child

Characteristic	Proportion Who Select Each Characteristic		
	White-Collar Job	No Job	Manual Job
Obedience	.26	.35	.53
Neatness, cleanliness	.16	.18	.42
Consideration	.39	.21	.05
Curiosity	.10	.04	.00
Self-control	.13	.14	.11
Happiness	.33	.40	.26
Boys	.32	.21	
Girls	.36	.59	
N	69	77	19

Another relevant line of social demarcation is the distinction between urban and rural background.[13] As we did for religious background, we can arrange the mothers into four groups delineated on the basis of class and rural-urban background in an order that is reasonably consistent for all class-related values. The order is: middle-class urban, middle-class rural, working-class urban, working-class rural (see Table 8). The extremes are middle-class mothers raised in the city and working-class mothers raised on farms.

Several other variables fail to differentiate mothers of the same social class into groups having appreciably different values. These include the

---

to Catholic school versus Protestant mothers of the same social class, there are no significant differences in values.

Jewish mothers (almost all of them in this sample are middle class) are very similar to middle-class Protestant mothers in their values, with two notable exceptions. More Jewish than Protestant mothers select popularity and ability to defend one's self —two values that are not related to social class.

[13] We asked: "Have you ever lived on a farm?" and then classified all mothers who had lived on a farm for some time other than simply summer vacations, prior to age fifteen, as having had a rural background.

Ordinarily, one further line of cultural demarcation would be considered at this point—nationality background. The present sample, however, is composed predominantly of parents who are at least second-generation, United States-born, so this is not possible.

mother's age, the size of the family, the ordinal position of the child in the family, the length of time the family has lived in the neighborhood, whether or not the mother has been socially mobile (from the status of her childhood family), and her class identification. Nor are these results a function of the large proportion of families of government workers included in the sample: wives of government employees do not differ from other mothers of the same social class in their values.

TABLE 6. Mothers' Education and Their Choice of Characteristics as "Most Desirable" in a Ten- or Eleven-Year-Old Child

	Middle-Class Mothers			
	Proportion Who Select Each Characteristic			
	Male Child		Female Child	
Characteristic	At Least Some College	High-School Graduate or Less	At Least Some College	High-School Graduate or Less
Obedience	0.11	0.22	0.13	0.29
Neatness-cleanliness	.03	.09	.03*	.23
Consideration	.47	.35	.41	.37
Curiosity	.31*	.13	.31*	.06
Self-control	.33	.19	.19	.21
Happiness	0.50	0.41	0.59	0.40
N	36	54	32	52

	Working-Class Mothers			
	Proportion Who Select Each Characteristic			
	Male Child		Female Child	
Characteristic	At Least High-School Graduate	Less than High-School Graduate	At Least High-School Graduate	Less than High-School Graduate
Obedience	0.29	0.43	0.28	0.32
Neatness-cleanliness	.12	.14	.21	.35
Consideration	.32	.27	.33*	.14
Curiosity	.07	.05	.12*	.00
Self-control	.22*	.07	.16	.08
Happiness	0.27	0.27	0.47	0.43
N	41	44	43	37

* Difference between mothers of differing educational status statistically significant, 0.05 level or better, using chi-squared test.

In sum, we find that it is possible to specify the relationship between social class and values more precisely by dividing the social classes into subgroups on the basis of other lines of social demarcation—but that social class seems to provide the single most relevant line of demarcation. . . .

This study does not provide disinterested observations of the parents' behavior. Our closest approximation derives from interviews with the parents themselves—interviews in which we questioned them in considerable detail about their relevant actions. Perhaps the most crucial of these data

TABLE 7.   Mothers' Religious Background and Their Choice of Characteristics as "Most Desirable" in a Ten- or Eleven-Year-Old Child

| Characteristic | Proportion Who Select Each Characteristic | | | |
	Middle-Class Protestant	Middle-Class Catholic	Working-Class Protestant	Working-Class Catholic
Obedience	0.17	0.25	0.33	0.36
Neatness, cleanliness	.08	.15	.17	.27
Consideration	.36	.38	.26	.29
Curiosity	.24	.12	.07	.05
Self-control	.28	.15	.15	.09
Happiness	.47	.42	.38	.30
Boys	.48	.32	.35	.13
Girls	0.45	0.52	0.42	0.54
N	88	52	107	56

TABLE 8.   Rural versus Urban Background of Mothers and Their Choice of Characteristics as "Most Desirable" in a Ten- or Eleven-Year-Old Child

| Characteristic | Proportion Who Select Each Characteristic | | | |
	Middle-Class Urban	Middle-Class Rural	Working-Class Urban	Working-Class Rural
Obedience	0.19	0.24	0.29	0.42
Neatness, cleanliness	.11	.12	.17	.25
Consideration	.42	.27	.31	.18
Curiosity	.19	.12	.07	.04
Self-control	.20	.33	.15	.11
Happiness	.47	.42	.41	.25
Boys	.44	.47	.28	.25
Girls	0.50	0.37	0.57	0.26
N	141	33	110	55

are those bearing on their actions in situations where their children behave in *disvalued* ways. We have, for example, questioned parents in some detail about what they do when their children lose their tempers. We began by asking whether or not the child in question "ever really loses his temper." From those parents who said that the child does lose his temper, we then proceeded to find out precisely what behavior they consider to be "loss of temper"; what they "generally do when he acts this way"; whether they "ever find it necessary to do anything else"; if so, what else they do, and "under what circumstances." Our concern here is with what the parent reports he does as a matter of last resort.[14]

Mothers who regard *self-control* as an important value are more likely

[14] This comparison and those to follow are limited to parents who say that the child does in fact behave in the disvalued way, at least on occasion. (Approximately equal proportions of middle- and working-class mothers report that their children do behave in each of these ways.)

TABLE 9.   Choice of "Self-Control" as "Most Desirable" Characteristic and Most Extreme Actions That Mothers Report They Take when Their Children Lose Their Tempers

| | Middle Class | | Proportion Working Class | | Both | |
	Choose Self-control	Don't Choose Self-control	Choose Self-control	Don't Choose Self-control	Choose Self-control	Don't Choose Self-control
Punish physically	0.26	0.20	0.44	0.26	0.32	0.23
Isolate	.20	.11	.11	.12	.17	.11
Restrict activities, other punishments	.06	.05	.17	.14	.10	.10
Threaten punishment	.06	.03	.00	.02	.04	.02
Scold, admonish, etc.	.31	.40	.17	.31	.26	.36
Ignore	0.11	0.21	0.11	0.15	0.11	0.18
	1.00	1.00	1.00	1.00	1.00	1.00
N	35	113	18	113	53	226

to report that they punish the child—be it physically, by isolation, or by restriction of activities; they are unlikely merely to scold or to ignore his loss of temper altogether (see Table 9).

To punish a child who has lost his temper may not be a particularly effective way of inducing self-control. One might even have predicted that mothers who value self-control would be less likely to punish breaches of control, more likely to explain, even ignore. They do not, however, and we must put the issue more simply: mothers who assert the value are more likely to report that they apply negative sanctions in situations where the child violates that value. This response would certainly seem to conform to their value-assertion.

A parallel series of questions deals with the mother's reactions when her child "refuses to do what she tells him to do." Mothers who assert that they regard *obedience* as important are more likely to report that they punish in one way or another when their children refuse.[15] There is also evidence that mothers who value *consideration* are more likely to respond to their children's "fighting with other children," an action that need not necessarily be seen as inconsistent with consideration, by punishing them, or at least by separating them from the others.[16]

[15] The figures are 47 versus 29 per cent for middle-class mothers; 36 versus 18 per cent for working-class mothers.

[16] The figures are 42 versus 29 per cent for middle-class mothers; 61 versus 37 per cent for working-class mothers.

There is also some indication that *working-class* mothers who value *honesty* have been more prone to insist that their children make restitution when they have "swiped" something, but the number of mothers who say that their children have ever swiped something is too small for this evidence to be conclusive. (The figures for working-

In all three instances, then, the reports on parental reactions to behavior that seem to violate the value in question indicate that mothers who profess high regard for the value are more likely to apply negative sanctions.

Interpretation

Our first conclusion is that parents, whatever their social class, deem it very important indeed that their children be honest, happy, considerate, obedient, and dependable.

The second conclusion is that, whatever the reasons may be, parents' values are related to their social position, particularly their class position.

There still remains, however, the task of interpreting the relationship between parents' social position and their values. In particular: What underlies the differences between the values of middle- and of working-class parents?

One relevant consideration is that some parents may "take for granted" values that others hold dear. For example, middle-class parents may take "neatness and cleanliness" for granted, while working-class parents regard it as highly desirable. But what does it mean to say that middle-class parents take neatness and cleanliness for granted? In essence, the argument is that middle-class parents value neatness and cleanliness as greatly as do working-class parents but not so greatly as they value such things as happiness and self-control. If this be the case it can only mean that in the circumstances of middle-class life neatness and cleanliness are easily enough attained to be of less immediate concern than are these other values.

A second consideration lies in the probability that these value-concepts have differing meanings for parents of different cultural backgrounds. For example, one might argue that honesty is a central standard of conduct for middle-class parents because they see honesty as meaning truthfulness; and that it is more a quality of the person for working-class parents because they see it as meaning trustworthiness. Perhaps so; but to suggest that a difference in meaning underlies a difference in values raises the further problem of explaining this difference in meaning.

It would be reasonable for working-class parents to be more likely to see honesty as trustworthiness. The working-class situation is one of less

---

class mothers are 63 versus 35 per cent; for middle-class mothers, 38 versus 33 per cent.)

The interviews with the children provide further evidence that parents have acted consistently with their values—for example, children whose mothers assert high valuation of dependability are more likely to tell us that the reason their parents want them to do their chores is to train them in responsibility (not to relieve the parents of work).

material security and less assured protection from the dishonesty of others. For these reasons, trustworthiness is more at issue for working-class than for middle-class parents.

Both considerations lead us to view differences in the values of middle- and working-class parents in terms of their differing circumstances of life and, by implication, their conceptions of the effects that these circumstances may have on their children's future lives. We believe that parents are most likely to accord high priority to those values that seem both *problematic,* in the sense that they are difficult of achievement, and *important,* in the sense that failure to achieve them would affect the child's future adversely. From this perspective it is reasonable that working-class parents cannot afford to take neatness and cleanliness as much for granted as can middle-class parents. It is reasonable, too, that working-class parents are more likely to see honesty as implying trustworthiness and that this connotation of honesty is seen as problematic.

These characteristics—honesty and neatness—are important to the child's future precisely because they assure him a respectable social position. Just as "poor but honest" has traditionally been an important line of social demarcation, their high valuation of these qualities may express working-class parents' concern that their children occupy a position unequivocally above that of persons who are not neat or who are not scrupulously honest. These are the qualities of respectable, worthwhile people.

So, too, is obedience. The obedient child follows his parents' dictates rather than his own standards. He acts, in his subordinate role as a child, in conformity with the prescriptions of established authority.

Even in the way they differentiate what is desirable for boys from what is desirable for girls, working-class mothers show a keen appreciation of the qualities making for respectable social position.

The characteristics that middle-class parents are more likely to value for their children are internal standards for governing one's relationships with other people and, in the final analysis, with one's self. It is not that middle-class parents are less concerned than are working-class parents about social position. The qualities of person that assure respectability may be taken for granted, but in a world where social relationships are determinative of position, these standards of conduct are both more problematic and more important.

The middle-class emphasis on internal standards is evident in their choice of the cluster of characteristics centering around honesty; in their being less likely than are working-class parents to value obedience and more likely to value self-control and consideration; and in their seeing obedience as inconsistent with both consideration and curiosity. The child is to act appropriately, not because his parents tell him to, but because he wants to. Not conformity to authority, but inner control; not because you're told to but because you take the other person into consideration—these are the middle-class ideals.

These values place responsibility directly upon the individual. He cannot rely upon authority, nor can he simply conform to what is presented to him as proper. He should be impelled to come to his own understanding of the situation.[17] He is to govern himself in such a way as to be able to act consistently with his principles. The basic importance of relationship to self is explicit in the concept of self-control. It is implicit, too, in consideration—a standard that demands of the individual that he respond sympathetically to others' needs even if they be in conflict with his own; and in the high valuation of honesty as central to other standards of conduct: "to thine own self be true."

Perhaps, considering this, it should not be surprising that so many middle-class mothers attribute first-rank importance to happiness, even for boys. We cannot assume that their children's happiness is any less important to working-class mothers than it is to middle-class mothers; in fact, working-class mothers are equally likely to value happiness for *girls*. For their sons, however, happiness is second choice to honesty and obedience. Apparently, middle-class mothers can afford instead to be concerned about their son's happiness. And perhaps they are right in being concerned. We have noted that those middle-class mothers who deem it most important that their sons outdistance others are especially likely to be concerned about their sons' happiness; and even those mothers who do not are asking their children to accept considerable responsibility.

[17] Curiosity provides a particularly interesting example of how closely parents' values are related to their circumstances of life and expectations: the proportion of mothers who value curiosity rises very slowly from status level to status level until we reach the wives of professionals and the more highly educated businessmen; then it jumps suddenly (see Table 4). The value is given priority in precisely that portion of the middle class where it is most appropriate and where its importance for the child's future is most apparent.

11. A CRITIQUE OF STUDIES RELATING SOCIAL CLASS TO METHODS OF CHILD-REARING AND THEIR CONSEQUENCES IN THE PERSONALITIES OF CHILDREN

In certain intellectual circles during the period between the two World Wars the middle classes were viewed with distaste and were represented as psychologically rigid, Puritanically prejudiced against fun, intellectually dull, and esthetically insensitive. Such writers as Sinclair Lewis and such painters as Grant Wood were fluent in their denunciation of middle-class Philistinism. For many this outlook carried the corollary that people in the lower social strata were generally more easy-going, more generous, had greater *joie de vivre*, and were generally more attractive. Whether this *Zeitgeist* gave coloration to research results or whether there has been a profound change in the two subcultures has been a moot question. In any case not only have subsequent studies failed to confirm the early findings that middle-class mothers are more strict and less giving than those in the lower class, but actually the more recent papers report that the situation is just the opposite.

Sewell reviews the literature on this topic, finds that the level of research and theoretical sophistication has been "appallingly low," and offers suggestions as to fruitful directions for future research.

SOCIAL CLASS AND CHILDHOOD PERSONALITY

William H. Sewell

Adapted from William H. Sewell, "Social Class and Childhood Personality," *Sociometry*, 1961, 24, 340-356.

Introduction

During the past twenty-five years there has been a great deal of interest in the relationship between social class and personality—particularly in the bearing of social class on the personality of the child and the relationship between social class and adult mental illness. This paper will concentrate on influences of social class on childhood personality and will not be concerned with the literature on youth and adults. Numerous books, monographs, research articles and essays have been published on this topic. Often they have contradictory emphases and conclusions depending on the convictions, theoretical orientations, and research styles of the authors.

The theoretical basis for expecting a substantial relationship between social class and personality rests on three major assumptions upon which there seems to be widespread agreement among social scientists. The first is that in all societies some system of social stratification exists whereby the members of the society are differentiated into subgroups or classes which bear to one another a relationship of social inequality. It is further generally acknowledged that persons in the society can be more or less located in the stratification system in terms of the characteristic social roles they play. Consequently, it is possible to infer, crudely at least, the social class position of most individuals in terms of readily ascertainable criteria. The particular criteria will be dependent on the culture of the society in question.[1] There are rather wide differences among writers as to the origins of stratification, the functions of stratification, the criteria of

[1] A number of books dealing with social stratification have appeared in recent years which summarize contemporary theory and research. These include: R. Bendix and S. M. Lipset, (eds.) *Class, Status and Power: A Reader in Social Stratification* (Glencoe, Ill.: The Free Press, 1953), J. A. Kahl, *The American Class Structure* (New York: Rinehart and Co., 1953), K. B. Mayer, *Class and Society* (New York: Doubleday, 1951), B. Barber, *Social Stratification* (New York: Harcourt, Brace and Co., 1957), S. M. Lipset and R. Bendix, *Social Mobility in Industrial Society* (Berkeley: University of California Press, 1959), and L. Reissman, *Class in American Society* (Glencoe, Ill.: The Free Press, 1959).

social classes, the meaning of the term class,[2] the number of classes, the rigidity of any particular stratification system, and almost any other aspect of theory, substance, or measurement which could possibly be raised, but almost everyone seems agreed that some system of stratification based on social inequality is an inevitable product of organized group life. The empirical basis of this proposition is strong in that no society has yet been studied in which a stratification system, fulfilling at least the minimum requirements stated above, has not been found.

The second assumption is that the position of the child's family in the stratification system determines in considerable measure not only the social learning influences to which he will be subjected during the early period of his life, and in later life for that matter, but greatly affects also the access that he will have to certain opportunities that are socially defined as desirable. Certainly, there seems to be ample evidence that this is true even in societies in which the stratification structure is not particularly rigid or the differences between the social classes extreme. While many social scientists would deny that American society has fixed classes each with its own distinctive subculture, none would claim that the learning environment of the child whose family is highly placed in the stratification structure does not differ materially from that of the child whose social class position is low. Also it is readily apparent that the styles of life, the material comforts, the value systems and the instruction, both intentional and unintentional, which the child receives about the roles available to him in society differ depending on the social class position of his family. And finally even his treatment in the neighborhood, community and larger society will depend for some time, at least, on his social status origins.

The third assumption on which there is general agreement is that the early experiences of the individual will be of considerable importance in determining his later social behavior. To be sure, there is rather massive disagreement about the particular psychodynamics of the relationship between early experience and later behavior, the specific or patterned experiences which produce other patterns or traits of later personality, or even the critical periods in terms of days, months and years in which the individual is most susceptible to influence. However, these details and dif-

---

[2] The writer is not convinced that social class is the best term for describing the socioeconomic levels treated in most of the literature covered in this paper. Actually the term social class implies much more than has been established concerning the existence of classes with distinctive boundaries and subcultures. What is meant operationally by social class in most studies is simply a convenient category of socioeconomic status. While the writer would prefer to use the more accurate term socioeconomic status or simply social status he bows to the trend in the literature and will use the term social class in this paper except in referring to those studies where the authors have themselves used socioeconomic status or social status.

ferences of theory and commitment have not led to any widespread rejection of the basic notion of the primary importance of early experiences
in shaping later personality. The experimental evidence on animal behavior and the somewhat more inferential knowledge about human learning furnish the empirical foundation for this assumption.

On the basis of these assumptions the reasonable expectation would be
that some distinct personality traits, configurations, or types might be
found which would differentiate the children of the several social classes,
or at least that the incidence of certain personality characteristics would
be different for the children of the various social classes. The results of
research efforts to elucidate these relationships have been disappointing
for a number of methodological and theoretical reasons. It would be
impossible and is unnecessary to review each of the numerous writings
which have direct bearing on the problem, but it does seem worthwhile
to examine some of the most important of them to see if it is possible to
reach any valid conclusions on the extent and nature of the relationship between social class and childhood personality, to point out some of
the weaknesses of the research in the field, and to make some suggestions
for future research. This is the purpose of the present paper.

An Examination of Selected Studies

A convenient point of departure might be to look at examples of studies
which illustrate various approaches to the problem. As a minimum these
would seem to include (1) work based primarily on typological and informal observational procedures, (2) those in which detailed observations on class-related child-training procedures have been made and personality characteristics inferred, observed or systematically assessed, and
finally (3) studies in which some measure of social class position has
been related directly to some independent assessments of personality.

Perhaps the best known example of the first type of study mentioned is
Arnold Green's "The Middle-Class Male Child and Neurosis" which was
originally published in 1946 and has been republished in numerous col-
lections of readings.[3] Green, stimulated by the neo-Freudian writers
Horney and Fromm, and on the basis of his recollections of his childhood
and young adulthood in a Massachusetts industrial community of about
3,000 persons, delineated a set of social psychological conditions that he
had observed in middle-class families which he believed predisposed
middle-class male children to neurosis. He observed that the middle-class
parent is caught up in a life-long struggle for improvement of personal
position in the class structure. The father's work takes him away from the

[3] Arnold Green, "The Middle-Class Male Child and Neurosis," *American Sociological Review,* 11 (1946), 31-41.

home and involves the manipulation of others around him to further his personal career. He is ambivalent toward his son because the child takes time, money and energy that could be used for the father's social advancement and also interferes with his role as a partner and companion to his wife. The mother, too, is ambivalent toward her child. He interferes with her career aspirations and her individual pleasures. Also, he causes worries and demands great care and attention. Despite the socially structured ambivalence of both parents toward their son, they train him to love them for the care and sacrifices they have made for him and force him to feel lost without their love. Thus, the middle-class boy suffers "personality absorption" to such an extent that he cannot turn to others for genuine emotional satisfaction. Moreover, he is faced with the constant threat of withdrawal of parental love. Little wonder, then, that he feels small, insignificant, unworthy, inferior, helpless and anxious! He can never escape his parents' norms at home, in school or in his play groups —always he must try to live up to their high expectations of him, or he will lose their love. Thus, he lives "alone and afraid in a world he never made." The lower-class (Polish-American) child suffers no such fate. Although parental authority is often harsh and brutal, it is also casual and external to the "core of the self." The children avoid their parents, in fact have contempt for them and band together in common defense against their cruelty. Consequently the parents do not have the opportunity or the techniques to absorb the personalities of their children. Thus, the lower-class boys do not suffer from the guilt, anxiety and extreme sense of insecurity from which the middle-class boy suffers as a result of his extreme dependency on his parents.

This is possibly an all too brief portrayal of Green's argument, but it summarizes his main points. Although the paper purports to be based on careful observation, no indication is given about the number of observations made of the socialization practices of either lower- or middle-class families, nor is there any indication of the frequency of neurosis or neurotic behavior among either lower- or middle-class boys—much less any direct evidence on the incidence of neurotic behavior among those middle-class boys (or lower-class boys) who have, as against those who have not, been socialized in the "middle-class way." Consequently the article might well be dismissed as a provocative and speculative essay except for the fact that it has served as one of the principal supports for the currently widely-held stereotype of the neurotic middle-class child and has fostered the idea that the lower-class child in our culture is relatively less subject to neurotic tendencies and symptoms. It also illustrates something of the current state of the field in that a paper which is based essentially on speculation and retrospection should be widely accepted as portraying an accurate account of the influence of social class on childhood personality.

The second type of study is perhaps most conveniently illustrated by the research done by members of the Committee on Human Development at the University of Chicago and originally reported in 1948 in two articles, one by Allison Davis and Robert J. Havighurst,[4] and another by Martha C. Ericson.[5] These studies were the first to report systematic empirical findings indicating that child-rearing practices of middle-class parents differ significantly from those of lower-class families. The findings of the Davis-Havighurst study were based on interviews with 98 middle-class (48 white and 50 Negro) and 102 lower-class (52 white and 50 Negro) mothers and dealt with a wide variety of child-training questions and the mothers' expectations concerning their children. Perhaps the most important finding of the study was the restrictiveness of the middle-class mothers in the critical early training of the child. They were shown to be less likely to breast-feed, more likely to follow a strict nursing schedule, to restrict the child's sucking period, to wean earlier and more sharply, to begin bowel and bladder training earlier and to complete toilet training sooner than were lower-class mothers. In addition, they generally followed stricter regimes in other areas of behavior and expected their children to take responsibility for themselves earlier. From these results the inference was drawn that middle-class children encounter more frustration of their impulses and that this is likely to have serious consequences for their personalities. Their findings regarding the differences in nursing and toilet training between the middle and lower classes were widely heralded and served to strengthen the conviction, especially of psychoanalytically oriented workers in the field—particularly those at the forefront of the culture and personality movement—that the socialization of the middle-class child in America was producing neurotic middle-class children and adults.[6] Davis and Havighurst themselves did not make this assertion. Their own conclusions were rather equivocal concerning the supposed consequences of these differences in training for the middle-

---

[4] Allison Davis and Robert J. Havighurst, "Social Class and Color Differences in Child Rearing," *American Sociological Review,* 11 (1946), 698-710. See also their *Father of the Man* (Boston: Houghton-Mifflin Co., 1947).

[5] Martha C. Ericson, "Child Rearing and Social Status," *American Journal of Sociology,* 52 (1946), 190-192.

[6] For a summary and critique of personality and culture literature see A. Inkeles and D. J. Levinson, "National Character: The Study of Modal Personality and Sociocultural Systems," in G. Lindzey, *Handbook of Social Psychology* (Cambridge, Mass.: Addison-Wesley Publishing Co., 1954), Ch. 26, and A. R. Lindesmith and A. L. Strauss, "A Critique of Culture-Personality Writings," *American Sociological Review,* 15 (1950), 587-600.

class child. Actually, the inference they drew regarding personality effects was that the training influences to which middle-class children are subjected are likely to produce an orderly, conscientious, responsible, tame but frustrated child. The only direct evidence they presented about the personalities of the children studied was that thumbsucking, which may be seen as an evidence of oral deprivation, and masturbation, which may indicate general frustration, are both much more frequently reported for middle-class than for lower-class children.

The findings of the Chicago group and the inferences made from their findings as to the personality consequences of class-related child-training practices were widely accepted and held sway without competition for some time. However, they were finally challenged by the results of two carefully designed empirical studies with quite different research objectives. The first of these was the attempt by the present writer to determine the consequences of a variety of infant-training practices on independently assessed childhood personality characteristics and the second was the careful study of patterns of child-rearing made by a group of behavioral scientists at Harvard under the leadership of Robert R. Sears.

The study of infant training and personality, published in 1952, was based on interviews conducted in 1947 with the mothers of 165 rural Wisconsin children concerning the practices they followed in rearing their children and subsequently relating the data thus obtained to the personality characteristics of the same children as these were determined from scores on both paper-and-pencil and projective tests of personality and ratings of the children's behavior by their mothers and teachers.[7] The specific infant-training practices studied were those most stressed in the psychoanalytic literature including: feeding, weaning, nursing schedule, bowel training, bladder training and punishment for toilet accidents. These experiences were not found to be significantly related to childhood personality characteristics as assessed in the study. Moreover, two carefully constructed factor-weighted indexes measuring permissiveness in toi-

---

[7] Wm. H. Sewell, "Infant Training and the Personality of the Child," *American Journal of Sociology*, 58 (1952), 150-159. Another paper dealing with the effects of feeding techniques on oral symptoms was published by the writer and Paul H. Mussen, "The Effects of Feeding, Weaning and Scheduling Procedures on Childhood Adjustment and the Formation of Oral Symptoms," *Child Development*, 23 (1952), 185-191. Other papers reporting on theoretical, methodological and substantive aspects of the study include: Wm. H. Sewell, "Field Techniques in Social Psychological Study in a Rural Community," *American Sociological Review*, 14 (1949), 718-726, Wm. H. Sewell, P. H. Mussen, and C. W. Harris, "Relationships among Child Training Practices," *American Sociological Review*, 20 (1955), 137-148, Wm. H. Sewell, "Some Observations on Theory Testing," *Rural Sociology*, 21 (1956), 1-12.

let training and feeding produced even less positive results.[8] In all only 18 out of 460 relationships tested in the study were significant at the .05 level and, of these, seven were opposite from the predicted direction.[9] These results, along with evidence from studies not so directly focused on the problem, tended to undermine the confidence of many who had made the inferential leap from class-determined early training practices to class-linked childhood personality characteristics and types.

Equally upsetting evidence came in 1954 with the publication of a preliminary report from the Harvard study by Eleanor E. Maccoby and P. K. Gibbs,[10] and later when the more complete report of the study was published by Robert Sears, Eleanor Maccoby and Harry Levin.[11] Their results, based upon careful interviews with 379 New England middle-class and lower-class mothers (labeled "upper-middle" and "upper-lower" by Maccoby and Gibbs), clearly indicated no differences in infant-feeding practices between the two social classes, more severity in toilet training in the lower-class families, less permissiveness in sex training in the lower-class families, more restriction of aggression toward parents and peers (and more punitiveness where such aggression took place) in lower-class families, greater imposition of restrictions and demands on the child in the lower-class family, more physical punishment, deprivation of privileges and ridicule by lower-class parents, but no differences between the two groups on isolation and withdrawal of love. Needless to say, these results were in important respects directly contradictory to the findings of the Chicago group and provided little factual basis for continued acceptance of the stereotyped version of the middle-class mother as a rigid, restrictive, demanding and punitive figure whose behavior can but result in frustrated, anxious, conforming and overly dependent children.[12] Neither was there any evidence whatever to support Green's contention about personality absorption of the child in the middle-class family or its supposed consequent—the neurotic middle-class child.

As might well be expected, the findings of the Harvard group provoked considerable debate and Havighurst and Davis did a comparison of

[8] For the indexes, see Sewell, Mussen and Harris, *op. cit.,* p. 144.

[9] A replication of the study in Ceylon resulted in the same conclusions. M. A. Straus, "Anal and Oral Frustration in Relation to Sinhalese Personality," *Sociometry,* 20 (1957), 21-31.

[10] E. E. Maccoby and P. K. Gibbs, "Methods of Child Rearing in Two Social Classes," in W. E. Martin and C. B. Standler (eds.), *Readings in Child Development* (New York: Harcourt, Brace & Co., 1954).

[11] R. R. Sears, E. E. Maccoby, and H. Levin, *Patterns of Child Rearing* (Evanston, Ill.: Row, Peterson and Co., 1957).

[12] *Ibid.,* Ch. 12.

the data of the two studies after adjusting to make the age groups more comparable but they still found substantial and large differences between the results of the Chicago and Harvard studies.[13] A number of other studies have appeared in recent years that generally confirm the findings of the Harvard group.[14] Finally, Urie Bronfenbrenner, on the basis of an examination of a whole battery of studies, both published and unpublished, found a basis for explaining some of the differences, particularly in infant feeding and toilet training, in terms of a trend toward greater permissiveness in these areas on the part of lower-class mothers up to World War II but with a reversal since then, middle-class mothers subsequently becoming more permissive in infant training.[15] The data gathered over the 25-year period on the training of the young child seem to him to show that middle-class mothers have been consistently more permissive towards the child's expressed needs and wishes, less likely to use physical punishment and more acceptant and equalitarian than have lower-class mothers. Finally, he sees indications that the gap between the social classes may be narrowing. While one might disagree with some of his interpretations and question some of the data on which his trends are based, it is clear from his review that in the present situation the evidence clearly supports the findings of the Harvard group and furnishes little basis for the belief that the training practices of middle-class parents are more likely than those of lower-class parents to produce neurotic personalities in their children.

One other important study which carries the analysis of class-related child-rearing a step forward has recently been reported by Daniel R. Miller and Guy E. Swanson.[16] In their study of child-rearing in Detroit, Michigan, they add to the stratification position represented by social class a second

[13] R. J. Havighurst and A. Davis, "A Comparison of the Chicago and Harvard Studies of Social Class Differences in Child Rearing," *American Sociological Review,* 20 (1955), 438-442.

[14] These include: R. A. Littman, R. A. Moore, J. Pierce-Jones, "Social Class Differences in Child Rearing: A Third Community for Comparison with Chicago and Newton, Massachusetts," *American Sociological Review,* 22 (1957), 694-704, M. S. White, "Social Class, Child Rearing Practices and Child Behavior," *American Sociological Review,* 22 (1957), 704-712, and M. L. Kohn, "Social Class and the Exercise of Parental Authority," *American Sociological Review,* 24 (1954), 352-366, and M. L. Kohn, "Social Class and Parental Values," *American Journal of Sociology,* 64 (1959), 337-351.

[15] Urie Bronfenbrenner, "Socialization and Social Class through Time and Space," in E. E. Maccoby, T. M. Newcomb, and E. L. Hartley, *Readings in Social Psychology,* (New York: Henry Holt and Co., 1958), pp. 400-425.

[16] D. R. Miller and G. E. Swanson, *The Changing American Parent* (New York: John Wiley and Co., 1958). See also their *Inner Conflict and Defense* (New York: Holt, Rinehart and Winston, 1960) which appeared after this paper was written.

variable dealing with integrative position in the social structure which they have called "entrepreneurial-bureaucratic integration." Families with entrepreneurial orientations are those in which the husband works in organizations that are relatively small in size, with a simple division of labor, have relatively small capitalization, and provide for mobility and income through individual risk-taking and competition. Families with bureaucratic orientations are those ·in which the father works in a large and complex organization employing many specialists, paying fixed wages or salaries for particular jobs, and, in place of reward for individual risk-taking, provides security in continuity of employment and income for those who conform with organizational demands. Miller and Swanson feel there is reason to believe that this aspect of status interpenetrates the family and influences child-rearing practices. Consequently, in their analysis they classify their families not only by social class but also by entrepreneurial-bureaucratic position. The addition of this new dimension of status produced results which were not nearly as clear-cut and definite as they had expected. In keeping with their predictions, entrepreneurial middle-class mothers were not less permissive than entrepreneurial lower-class mothers and there were no differences between bureaucratic lower-class and middle-class families in this regard. Their predictions that entrepreneurial middle-class mothers would be more likely to train their children in an active and manipulative view of the world was not supported. Moreover, entrepreneurial and bureaucratic lower-class mothers did not differ to any appreciable extent in the way they trained their children. If only the class differences are considered their results are quite similar to those of the Harvard group. It seems quite probable that the relative failure of the new dimension to add much to the predictive power of social class was to some extent due to the inadequacy of their scheme for determining entrepreneurial-bureaucratic orientation.[17] Consequently in future studies better categorization and assessment of this dimension may produce greater associations.[18] In any event the idea of introducing other dimen-

[17] Apparently the use of this variable was something of an afterthought and consequently its operational definition for purposes of the research had to be based on data available from the interview rather than what might have been more pertinent information. (*Ibid.,* pp. 67-70.)

[18] Possibly it would be a better test of the hypothesis to simply compare the personality characteristics, for a large number of cases, of children brought up in families more clearly representing the entrepreneurial and bureaucratic ends of the continuum, i.e. children of owner-operators of independent retail establishments vs. children of government clerks. It might even be more rewarding to drop the bureaucratic-entrepreneurial orientation entirely and to examine the influence of specific occupations on socialization norms and practices on the assumption that occupations differ in the extent to which they interpenetrate family life and influence of the behavior of its members.

sions of status than social class position seems to be a good one and should be tested in other studies.

A third type of research bearing directly on the relationship between social class and personality involves the correlation between measures of socioeconomic status (henceforth referred to as SES) and children's scores on personality tests and is perhaps well illustrated by a study by the present writer and A. O. Haller.[19] A comprehensive review of the studies in which SES had been measured objectively and correlated with independent assessments of the personality of the child indicated that middle-class children consistently made a better showing than lower-class children. For the most part the correlations were low or the differences were small and often there was no indication that the association was statistically significant, that sampling was adequate, that the tests of status and personality were dependable, or that variables known to be related to status or personality or both were controlled.[20]

Consequently it was decided to make a rigorous test of the hypothesized relation between SES and personality using a design in which both variables were measured objectively and independently for a large sample (1,462) of grade-school children in a culturally homogeneous community with a fairly wide range of SES. Correlation analysis techniques were used to determine the relationship between SES, as measured by father's occupation and a rating of the prestige of the family in the community, and personality adjustment as indicated by a factor-weighted score on the California Test of Personality. The zero order correlation coefficients between the two status measures and the personality scores were determined. Then the multiple correlation coefficient of the two status measures and personality scores was computed, and, finally, the relationship was determined with sibling position, intelligence, and age controlled. The results indicated a low but significant association between status and measured personality (.16 for father's occupation and child's personality score, .23 for prestige position and child's personality score, .25 for the multiple correlation of the two status measures and child's personality

---

[19] Wm. H. Sewell and A. O. Haller, "Social Status and the Personality Adjustment of the Child," *Sociometry*, 19 (1956), 114-125. A number of earlier studies of this type are given in the bibliography attached to the present paper. Illustrative recent studies include L. G. Burchinal, "Social Status, Measured Intelligence, Achievement and Personality Adjustment of Rural Iowa Girls," *Sociometry*, 22 (1959), 75-80, L. G. Burchinal, B. Gardner, and G. R. Hawkes, "Children's Personality Adjustment and the Socioeconomic Status of Their Families," *Journal of Genetic Psychology*, 92 (1958), 149-159, and H. Angelino, J. Dollins, and E. V. Mech, "Trends in the 'Fears and Worries' of School Children as Related to Socioeconomic Status and Age," *Journal of Genetic Psychology*, 89 (1956), 263-277.

[20] Sewell and Haller, *op. cit.*, pp. 114-115.

score). The combined effect of the two status measures was not significantly reduced when the controls were introduced. The direction of the correlations indicated that the lower the SES of the child's family the less favorable his personality test score.

Certainly these results indicate that only a relatively small amount of the variance in measured personality found in this group of children can be accounted for by their SES. However, the test of the hypothesis was stringent and the correlations might well be higher in communities with more distinct stratification systems, and if more refined measures of status and personality were used. In any event the correlations, particularly since they are not markedly different from those reported by others who have followed similar methods, should not be dismissed. They at least help to explain some of the variance in measured personality—an area in which little measured variance has been explained by other measured variables. However, the results do not provide much encouragement for the view that social class is a major determinant of childhood personality and offer still another instance of evidence against the claim that middle-class children suffer greater personality maladjustment than lower-class children.

In an attempt to explore further the relationship between SES and personality, the writers next did a factor analysis of the 30 personality test items which had been found to be most highly correlated with SES.[21] The results of this analysis indicated that four factors explained approximately 90 per cent of the common variance among the items. These factors were tentatively identified as (1) *concern over status,* (2) *concern over achievement,* (3) *rejection of family,* and (4) *nervous symptoms.* Each factor was negatively correlated with SES, their respective correlations being —.31, —.18, —.12 and —.26, indicating that the lower the status of the child the greater the tendency to score high (unfavorably) on each of the factors. The intercorrelation between the factors range from +.25 to +.59. Thus, there seems to be a tendency for children who are concerned about their social status to worry about their achievements, to reject their families and to display nervous symptoms. The evidence from this study points to the fact that these characteristics are more common among lower- than higher-status children. Again the correlations between SES and the personality characteristics indicated by the factors, although statistically significant, are low and offer only limited support for the notion that the position of the child in the stratification system has bearing on his personality pattern. They are, however, sug-

[21] Wm. H. Sewell and A. O. Haller, "Factors in the Relationship between Social Status and the Personality Adjustment of the Child," *American Sociological Review,* 24 (1959), 511-520.

gestive of a line of attack on the problem which may be somewhat more rewarding than some of the approaches employed thus far.

Conclusions Regarding Social Class and Childhood Personality

On the basis of this brief review of studies of the bearing of social class on the personality of the child, the following conclusions seem justified:

First, there is a growing body of evidence from empirical studies of several types indicating a relatively low correlation between the position of the child in the stratification system (social class) and some aspects of personality, including measured personality adjustment. The relationship has not been shown to be nearly as close as might have been expected, but there is mounting evidence that at least some of the variance in childhood personality can be explained by the social status position of the child. Possibly when better measures are used the relationship will prove to be higher. The present crude techniques of measuring both variables doubtless result in underestimation of the correlation.

Second, the direction of the relationships found offer absolutely no support for the notion that middle-class children more commonly exhibit neurotic personality traits than do children of lower-class origins. Indeed all of the empirical evidence points to the opposite conclusion.

Third, the studies of child-rearing in relation to social class, made since the publication of the Chicago studies, have found fewer class-related differences in infant training than might have been expected and those differences that have been found tend to indicate greater permissiveness in feeding and toilet training on the part of middle-class mothers rather than lower-class mothers. The findings in relation to early childhood training indicate less impulse control, less punitiveness, less reliance on strict regime, less restrictiveness in sex behavior and less restriction on aggression—in other words, generally greater permissiveness on the part of middle-class mothers.

Fourth, empirical studies of the consequences of child training have given a great deal of attention to such aspects of infant discipline as manner of nursing, weaning, scheduling, bowel and bladder training, but have found very little or no relationship between these experiences and childhood personality traits and adjustment patterns. Much less attention has been given to the consequences of other aspects of child training but some low correlations have been found between such factors as patterns of punishment, permissiveness for aggression and mother's affectional warmth for the child and such aspects of personality as feeding problems, dependency and aggression. Although these correlations explain only a small portion of the variance in childhood personality, they cannot be entirely dismissed and, to the extent that the child-training practices are class-linked, they must be credited with having some bearing on the relationship between social class and personality. Certainly, however, the empirical evidence does not permit any lavish claims regarding the influence of the child-training variables studied on the personality of the child.

Fifth, a final inescapable conclusion from reading these and other writings on social class and childhood personality is that, with a few notable exceptions, the level of research and theoretical sophistication in this area has been appallingly low. Some of the most influential work has had little or no acceptable empirical basis. The evidence upon which widely accepted claims have been founded is sometimes from samples that are so small or so clearly biased that no reliable conclusions could possibly be reached. In fact there is not a single study that can claim to be representative of the whole society or any region of the county and only a small handful are clearly representative of any definable social system. The statistical techniques in some of the studies are clearly inappropriate for the data. The theoretical guide-lines for most of the studies are seldom specified and often are not even discernible. The chain of inference from theory to data-to conclusions-to wider generalizations is sometimes unclear and instances can be cited in which links in the chain are entirely missing. Great lack of conceptual clarity, particularly concerning the two principal variables, social class and personality, is generally apparent. Thus, statistical categories of socioeconomic status measured by crude techniques are treated as social classes in the broader meaning of that term, and inferences are drawn about subcultures, learning environments, value systems and other social class characteristics without the necessary empirical evidence of their existence. Likewise the term personality is used in a variety of ways but with little attention to definition and specification. Often inferences are made about deeper levels of personality from more or less surface variables. Because of these weaknesses in theory and method, more definitive conclusions about the relationship between social class and childhood personality must await better designed studies.

Suggested Direction for Future Research

In the light of the present situation in the field and because of the basic importance of the problem, both from the theoretical and practical points of view, a few suggestions regarding future research may be helpful.

First, although the available evidence concerning the relationship between child training and personality does not provide much of a basis for explaining personality variation and despite the fact that the present evidence concerning the relationship between social class and child training does not seem to indicate a very close correspondence, further studies of social class and child-rearing are desirable and necessary. With our present knowledge of sampling procedures, data gathering methods, and analysis techniques, a carefully designed large-scale study using a sample of sufficient size to permit racial and ethnic breakdowns and other needed controls and concentrating on various aspects of child-rearing ranging over the whole period of childhood is the indicated next step. In such a study additional attention should be given to assessment of the behavioral correlates of child-training practices and to appropriate delineation of larger personality configurations. Additional studies of small local communities with relatively narrow stratification systems or studies in larger communities with samples that are inadequate to represent the full range of the stratification system are not likely to add much to the knowledge already available from existing studies and could well be dispensed with.

Second, despite the fact that many studies have been made of the relationship between the SES and the measured personality of school children, there is still need for a definitive study in a large community with a heterogeneous population. Since it would be an overwhelming task to map the actual class structure of such a community, several objective SES indicators might be employed singly and in combination, using modern multivariate statistical analysis techniques to determine their relationship to measured personality as indicated by personality test results. With large samples a number of variables could be controlled and some definitive conclusion might be reached regarding the relationship between SES and childhood personality. Further analysis, following the general model used by Sewell and Haller in their most recent study of test items having high correlations with SES, could be done to determine the factors which account for variance in responses to personality tests.

Third, there is need for an intensive study of the relationship between social class and personality in some community or society in which a functionally existing class system with well established subcultures is present, if indeed such can be found. It is clear from the literature that classes in this sense have been assumed to be operative but in no case have they been properly delineated, validated and sampled in any study of social class and personality. Such a task would be a major undertaking and probably could be carried out only in a modest-sized community, but, if an appropriate community were found, this would provide a more critical test of the theoretical relationship between social class and personality than any yet attempted. Unless and until such a study is done, no one is really justified in implying that social class is more than a convenient statistical category in discussing its relation to personality.

Fourth, it now seems clear that scientific concern with the relation between social class and personality has perhaps been too much focused on global aspects of personality and possibly too much on early socialization. Therefore, it is suggested that the more promising direction for future research will come from a shift in emphasis, toward greater concern with those particular aspects of personality which are most likely to be directly influenced by the position of the child's family in the social stratification system such as attitudes, values and aspirations, rather than with deeper personality characteristics. It also is suggested that instead of focusing so much attention on the very young child, more research should be done on older children and adolescents—on the assumption that whatever differences one's social class position makes to the above-mentioned aspects of personality are likely to be the product of later and more gradual socialization experiences rather than the more proximate effect of specific aspects of early experiences.[22] Evidence already available from a number of studies seems to clearly indicate that adolescents' belief systems and values differ rather clearly in relation to their social status positions: adolescents of lower-class backgrounds appear to have lower need-achievement, lower achievement values,[23] are much less likely to place high value on a college education, less frequently aspire to high level achievement in educational pursuits and occupational activities[24] and are less willing to defer their gratifications than are middle- and upper-class adolescents.[25] Recent studies by the writer and others have shown that there is a marked negative correlation between the socioeconomic status and the educational and occupational aspirations of high school seniors and that relationship remains even when

[22] For elaboration of this point of view, see Fred L. Strodtbeck, "Family Interaction Values and Achievement," in D. C. McClelland and others, *Talent and Society* (New York: D. Van Nostrand Co., 1958), Ch. II.

[23] B. C. Rosen, "The Achievement Syndrome: A Psychocultural Dimension of Social Stratification," *American Sociological Review*, 21 (1956), 203-211, and "Race, Ethnicity and the Achievement Syndrome," *American Sociological Review*, 24 (1959), 47-60.

[24] There is much evidence on this point but see particularly H. H. Hyman, "The Value Systems of Different Classes: A Social Psychological Contribution to the Analysis of Social Classes," in Bendix and Lipset, *op. cit.*, pp. 426-442, and Wm. H. Sewell, A. O. Haller, and M. A. Straus, "Social Status and Educational and Occupational Aspirations," *American Sociological Review*, 22 (1957), 65-73. Many additional studies are referred to in the latter article.

[25] A. Davis and J. Dollard, *Children of Bondage* (Washington, D.C.: American Council on Education, 1940), and L. Schneider and S. Lysgaard, "The Deferred Gratification Pattern: A Preliminary Study," *American Sociological Review*, 18 (1953), 142-149.

sex and intelligence are controlled.[26] As yet unpublished, results from a more recent study which the writer is currently conducting on a statewide sample of Wisconsin high school seniors indicate that other variables such as parental pressure, rural-urban background, community and peer group influences may also be introduced without negating the influencing of SES on these aspirations. It may well be that it is precisely in the area of attitudes, values, and aspirations that social class influences are most pronounced. If so, it would be profitable to study children of different social class backgrounds to determine when and how these characteristics of personality develop and how responsive they may be to other influences.

Fifth, it is suggested that it might be more revealing and more promising in terms of the knowledge that may be gained about social influences on personality to focus some attention on intraclass differences instead of being concerned exclusively with interclass variations. The data of most empirical studies of social class in relation to personality and related variables indicate the existence of considerable intraclass variation. Thus in the writer's studies of social status and personality there were many children in each social status level who made favorable scores on the personality measures and some at each level who made unfavorable scores. Even in the study of differences in educational and occupational aspirations, where sizable differences were found between SES groups, there were important differences within each status group.[27] Obviously, it is not intelligence or sex which accounts for the within class differences found in this study because these variables have been controlled, but it may well be that family attitudes and values, peer group influences, and community forces will be found to explain a sizable portion of the variance. Another interesting series of questions suggested by these results is: What are the personality effects of having values and aspirations that are deviant from those of one's social class? Does it mean that the lower-class child has to reject the values of his family and neighborhood in order to be socially mobile? If so, what are the dimensions and what is the nature of the stress experienced by the upwardly mobile lower-class child and what personality consequences flow from such striving? If he is successful in his mobility aspirations, will the lower-class child find it possible to internalize the values of his new status position or will he be constantly plagued by the conflict between his old values and the new? These are just a few of the kinds of questions that could be studied in relation to intraclass differences in personality.

Sixth, it is suggested that, as basic a variable as social class is for social behavior, there are other important aspects and dimensions of social

[26] Sewell, Haller and Straus, *op. cit.,* pp. 70-72.
[27] *Ibid.*

structure that cut across the social stratification system which should not be neglected in the study of the personality development of the child.[28] Among the more important of these are the mobility and the occupational orientations of the family. In addition, there are other traditional social structure variables such as age, sex, family size, sibling position, race, ethnic background and religion which probably play a significant role but have been taken into account insufficiently in studies of social structure and childhood personality. Moreover, much theoretical and analytical work is needed on the possible influence of various combinations of social structure variables and their joint as well as independent influence on the personalities of children.

[28] Some provocative suggestions along these lines are given in R. T. Morris and R. J. Murphy, "The Situs Dimension in Occupational Structure," *American Sociological Review,* 24 (1959), 231-239.

12. ADOLESCENTS AND THEIR PARENTS

Believing that adolescence is not necessarily filled with "storm and stress," Hollingshead presents a sociological interpretation of the age category which the term "adolescence" denotes. While holding that the conflict between adolescents and their parents may be overemphasized, Kingsley Davis analyzes the grounds for whatever antagonisms may exist. Allison Davis describes important differences between the middle and lower classes in the expression of aggressive and sexual impulses among adolescents.

To study orientation and motivation toward achievement among adolescent boys Strodtbeck began by looking for values relevant to occupational mobility in Italian and Jewish cultures. He then carried out what the statistician calls a $3 \times 2 \times 2$ factorial design (three levels of socioeconomic status, two categories of ethnicity, and two levels of achievement). With this design it is possible to hold socioeconomic status constant while examining variation due to ethnicity, and vice versa. It develops that much of the variation associated with the Italian-Jewish difference in ethnicity (among New Haven subjects) washes out when socioeconomic status is held constant, or in his phrasing, "ethnic differences in family interaction are not of great relevance in explaining [differential achievement]." It also turns out that whereas responses of adolescent sons and their fathers show significant variation with socioeconomic status but not with ethnicity, responses of mothers vary significantly with ethnicity but not with socioeconomic status. Commenting on this finding, Strodtbeck observes that it is the fathers rather than the mothers whose workaday efforts contribute directly to familial mobility.

ADOLESCENCE: A SOCIOLOGICAL DEFINITION

August B. Hollingshead

Adapted with permission from August B. Hollingshead, *Elmtown's Youth,* New York, 1949, John Wiley & Sons, Inc., pp. 5-7.

In the past half-century physiologists, psychologists, educators, clergymen, social workers, and moralists have turned their attention to the physical and psychological phenomena connected with adolescence, and . . . of the millions of words written on the subject most have had a worried tone. This interest in the adolescent, with its emphasis on the "problems" of adolescence, can be traced to the monumental work of G. Stanley Hall.[1] Hall blended evolutionary theory, the facts of physical growth (as they were then known), instinct psychology, and a liberal sprinkling of ethnographic facts taken out of their cultural context with a set of strong moral judgments. He assumed that the individual in the course of his life recapitulates the evolutionary development of the human species.[2]

Hall conceived of adolescence as the period in the life cycle of the individual from age 14 to 24, when the inexorably unfolding nature of the organism produces a "rebirth of the soul" which brings the child inevitably into a conflict with society. This was believed to be a period of "storm and stress," of "revolution," in the individual. In the "new birth" the "social instincts undergo sudden unfoldment." [3] Hall also asserted that "the adolescent stage of life" is marked by a struggle between the needs of the organism and the desires of society, which is "biologically antagonistic to genesis."[4] "All this is hard on youth . . ."[5] This psychology of adolescence included in its scope physiology, anthropology, sociology, sex, crime, religion, and education. Needless to say, this is a broad area which only a system maker who ignored facts as they exist in society could cover in a single sweep.

Hall's prestige as a psychologist, educator, and university president was so great and his influence over students so dominant that his theories were accepted widely by psychologists and educators. Gradually, how-

[1] G. Stanley Hall, *Adolescence, Its Psychology and Its Relations to Physiology, Anthropology, Sociology, Sex, Crime, Religion, and Education,* New York, D. Appleton and Company, 1904, two volumes.

[2] *Ibid.,* vol. I, p. viii.

[3] *Ibid.,* p. xv.

[4] *Ibid.,* p. xvi.

[5] *Ibid.,* p. xviii.

ever, the weight of empirical information indicated that these views were largely doctrinal. But, even now, the idea that adolescence is a period of "storm and stress," of conflict between individual and society, is held by many people, in spite of the fact that this has never been demonstrated to be true. On the contrary, common-sense observation will cast grave doubts upon its validity. Nevertheless, a recent summary of the field of adolescent psychology insisted upon the "casual" connection between the physical manifestations of adolescence and social behavior.[6]

Eventually, the conclusion was reached that, from the viewpoint of the sociologist, adolescence is distinctly different from psychologists', physiologists', and educators' concepts of it. *Sociologically, adolescence is the period in the life of a person when the society in which he functions ceases to regard him* (male or female) *as a child and does not accord to him full adult status, roles, and functions.* In terms of behavior, it is defined by the roles the person is expected to play, is allowed to play, is forced to play, or prohibited from playing by virtue of his status in society. It is not marked by a specific point in time such as puberty, since its form, content, duration, and period in the life cycle are differently determined by various cultures and societies. Sociologically, the important thing about the adolescent years is the way people regard the maturing individual. The menarche, development of the breasts, and other secondary manifestations of physical adolescence in the female, and the less obvious physical changes in the male connected with sex maturation, such as rapid growth, voice changes, the appearance of labial, axial, and pubic hair, derive their significance for the sociologist from the way they are regarded by the society in which the adolescent lives.

THE SOCIOLOGY OF PARENT-YOUTH CONFLICT

Kingsley Davis

Adapted and reprinted from the *American Sociological Review,* 4 (1940), 523-35.

It is in sociological terms that this paper attempts to frame and solve the sole question with which it deals, namely: Why does contemporary western civilization manifest an extraordinary amount of parent-adolescent

---

[6] Wayne Dennis, "The Adolescent," in *Manual of Child Psychology,* edited by Leonard Carmichael, John Wiley and Sons, Inc., New York, 1946, pp. 633-666. We would agree with Dennis if he or any other psychologist demonstrated any "causal" connection between the physical phenomenon of puberty and the social behavior of young people during the adolescent period, irrespective of cultural milieu.

conflict?[1] In other cultures, the outstanding fact is generally not the rebelliousness of youth, but its docility. There is practically no custom, no matter how tedious or painful, to which youth in primitive tribes or archaic civilizations will not willingly submit.[2] What, then, are the peculiar features of our society which give us one of the extremest examples of endemic filial friction in human history?

Our answer to this question makes use of constants and variables, the constants being the universal factors in the parent-youth relation, the variables being the factors which differ from one society to another. Though one's attention, in explaining the parent-youth relations of a given milieu, is focused on the variables, one cannot comprehend the action of the variables without also understanding the constants, for the latter constitute the structural and functional basis of the family as a part of society.

The Rate of Social Change

The first important variable is the rate of social change. Extremely rapid change in modern civilization, in contrast to most societies, tends to increase parent-youth conflict, for within a fast-changing social order the time-interval between generations, ordinarily but a mere moment in the life of a social system, becomes historically significant, thereby creating a hiatus between one generation and the next. Inevitably, under such a condition, youth is reared in a milieu different from that of the parents; hence the parents become old-fashioned, youth rebellious, and clashes occur which, in the closely confined circle of the immediate family, generate sharp emotion. . . .

[1] In the absence of statistical evidence, exaggeration of the conflict is easily possible, and two able students have warned against it. E. B. Reuter, "The Sociology of Adolescence," and Jessie R. Runner, "Social Distance in Adolescent Relationships," both in *Amer. J. Sociol.,* November 1937, 43: 415-16, 437. Yet sufficient nonquantitative evidence lies at hand in the form of personal experience, the outpour of literature on adolescent problems, and the historical and anthropological accounts of contrasting societies to justify the conclusion that in comparison with other cultures ours exhibits an exceptional amount of such conflict. If this paper seems to stress conflict, it is simply because we are concerned with this problem rather than with parent-youth harmony.

[2] Cf. Nathan Miller, *The Child in Primitive Society,* New York, 1928; Miriam Van Waters, "The Adolescent Girl Among Primitive Peoples," *J. Relig. Psychol.,* 1913, 6: 375-421 (1913) and 7: 75-120 (1914); Margaret Mead, *Coming of Age in Samoa,* New York, 1928, and "Adolescence in Primitive and Modern Society," 169-88, in *The New Generation* (ed. by V. F. Calverton and S. Schmalhausen, New York, 1930; A. M. Bacon, *Japanese Girls and Women,* New York and Boston, 1891 and 1902.

The Birth-Cycle, Declerating Socialization, and Parent-Child Differences

Note, however, that rapid social change would have no power to produce conflict were it not for two universal factors: first, the family's duration; and second, the decelerating rate of socialization in the development of personality. "A family" is not a static entity but a process in time, a process ordinarily so brief compared with historical time that it is unimportant, but which, when history is "full" (*i.e.,* marked by rapid social change), strongly influences the mutual adjustment of the generations. This "span" is basically the birth-cycle—the length of time between the birth of one person and his procreation of another. It is biological and inescapable. It would, however, have no effect in producing parent-youth conflict, even with social change, if it were not for the additional fact, intimately related and equally universal, that the sequential development of personality involves a constantly decelerating rate of socialization. This deceleration is due both to organic factors (age—which ties it to the birth-cycle) and to social factors (the cumulative character of social experience). Its effect is to make the birth-cycle interval, which is the period of youth, the time of major socialization, subsequent periods of socialization being subsidiary. . . .

Physiological Differences

Though the disparity in chronological age remains constant through life, the precise physiological differences between parent and offspring vary radically from one period to another. The organic contrasts between parent and *infant,* for example, are far different from those between parent and adolescent. Yet whatever the period, the organic differences produce contrasts (as between young and old) in those desires which, at least in part, are organically determined. Thus, at the time of adolescence the contrast is between an organism which is just reaching its full powers and one which is just losing them. The physiological need of the latter is for security and conservation, because as the superabundance of energy diminishes, the organism seems to hoard what remains.

Such differences, often alleged (under the heading of "disturbing physiological changes accompanying adolescence") as the primary cause of parent-adolescent strife, are undoubtedly a factor in such conflict, but, like other universal differences to be discussed, they form a constant factor present in every community, and therefore cannot in themselves explain the peculiar heightening of parent-youth conflict in our culture. . . .

Psychosocial Differences: Adult Realism versus Youthful Idealism

Though both youth and age claim to see the truth, the old are more conservatively realistic than the young, because on the one hand they take Utopian ideals less seriously and on the other hand take what may be called operating ideals, if not more seriously, at least more for granted. Thus, middle-aged people notoriously forget the poetic ideals of a new social order which they cherished when young. In their place, they put simply the working ideals current in the society. There is, in short, a persistent tendency for the ideology of a person as he grows older to gravitate more and more toward the status quo ideology, unless other facts (such as a social crisis or hypnotic suggestion) intervene. With advancing age, he becomes less and less bothered by inconsistencies in ideals. He tends to judge ideals according to whether they are widespread and hence effective in thinking about practical life, not according to whether they are logically consistent. Furthermore, he gradually ceases to bother about the *untruth* of his ideals, in the sense of their failure to correspond to reality. He assumes through long habit that, though they do not correspond perfectly, the discrepancy is not significant. The reality of an ideal is defined for him in terms of how many people accept it rather than how completely it is mirrored in actual behavior.[3] Thus, we call him, as he approaches middle age, a realist.

The young, however, are idealists, partly because they take working ideals literally and partly because they acquire ideals not fully operative in the social organization. Those in authority over children are obligated as a requirement of their status to inculcate ideals as a part of the official culture given the new generation. The children are receptive because they have little social experience—experience being systematically kept from them (by such means as censorship, for example, a large part of which is to "protect" children). Consequently, young people possess little ballast for their acquired ideals, which therefore soar to the sky, whereas the middle-aged, by contrast, have plenty of ballast. . . .

Sociological Differences: Parental Authority

Because of his strategic position with reference to the new-born child (at least in the familial type of reproductive institution), the parent is given considerable authority. Charged by his social group with the responsibility of controlling and training the child in conformity with the mores

---

[3] When discussing a youthful ideal, however, the older person is quick to take a dialectical advantage by pointing out not only that this ideal affronts the aspirations of the multitude, but that it also fails to correspond to human behavior either now or (by the lessons of history) probably in the future.

and thereby insuring the maintenance of the cultural structure, the parent, to fulfill his duties, must have the privileges as well as the obligations of authority, and the surrounding community ordinarily guarantees both.

The first thing to note about parental authority, in addition to its function in socialization, is that it is a case of authority within a primary group. Simmel has pointed out that authority is bearable for the subordinate because it touches only one aspect of life. Impersonal and objective, it permits all other aspects to be free from its particularistic dominance. This escape, however, is lacking in parental authority, for since the family includes most aspects of life, its authority is not limited, specific, or impersonal. What, then, can make this authority bearable? Three factors associated with the familial primary group help to give the answer: (1) the child is socialized within the family, and therefore knowing nothing else and being utterly dependent, the authority of the parent is internalized, accepted; (2) the family, like other primary groups, implies identification, in such sense that one person understands and responds emphatically to the sentiments of the other, so that the harshness of authority is ameliorated;[4] (3) in the intimate interaction of the primary group control can never be purely one-sided; there are too many ways in which the subordinated can exert the pressure of his will. When, therefore, the family system is a going concern, parental authority, however inclusive, is not felt as despotic.

A second thing to note about parental authority is that while its duration is variable (lasting in some societies a few years and in others a lifetime), it inevitably involves a change, a progressive readjustment, in the respective positions of parent and child—in some cases an almost complete reversal of roles, in others at least a cumulative allowance for the fact of maturity in the subordinated offspring. Age is a unique basis for social stratification. Unlike birth, sex, wealth, or occupation, it implies that the stratification is temporary, that the person, if he lives a full life, will eventually traverse all of the strata having it as a basis. Therefore, there is a peculiar ambivalence attached to this kind of differentiation, as well as a constant directional movement. On the one hand, the young person, in the stage of maximum socialization, is, so to speak, *moving into* the social organization. His social personality is expanding, *i.e.,* acquiring an increased amount of the cultural heritage, filling more powerful and numerous positions. His future is before him, in what the older person is leaving behind. The latter, on the other hand, has a future before him only in the sense that the offspring represents it. Therefore, there is a dis-

---

[4] House slaves, for example, are generally treated much better than field slaves. Authority over the former is of a personal type, while that over the latter (often in the form of a foreman-gang organization) is of a more impersonal or economic type.

parity of interest, the young person placing his thoughts upon a future which, once the first stages of dependence are passed, does not include the parent, the old person placing his hopes vicariously upon the young. This situation, representing a *tendency* in every society, is avoided in many places by a system of respect for the aged and an imaginary projection of life beyond the grave. In the absence of such a religio-ancestral system, the role of the aged is a tragic one.[5]

Let us now take up, point by point, the manner in which western civilization has affected this *gemeinschaftliche* and processual form of authority.

1. Conflicting Norms. To begin with, rapid change has, as we saw, given old and young a different social content, so that they possess conflicting norms. There is a loss of mutual identification, and the parent will not "catch up" with the child's point of view, because he is supposed to dominate rather than follow. More than this, social complexity has confused the standards *within* the generations. Faced with conflicting goals, parents become inconsistent and confused in their own minds in rearing their children. The children, for example, acquire an argument against discipline by being able to point to some family wherein discipline is less severe, while the parent can retaliate by pointing to still other families wherein it is firmer. The acceptance of parental attitudes is less complete than formerly.

2. Competing Authorities. We took it for granted, when discussing rapid social change, that youth acquires new ideas, but we did not ask how. The truth is that, in a specialized and complex culture, they learn from competing authorities. Today, for example, education is largely in the hands of professional specialists, some of whom, as college professors, resemble the sophists of ancient Athens by virtue of their work of accumulating and purveying knowledge, and who consequently have ideas in advance of the populace at large (*i.e.*, the parents). By giving the younger generation these advanced ideas, they (and many other extrafamilial agencies, including youth's contemporaries) widen the intellectual gap between parent and child.[6]

3. Little Explicit Institutionalization of Steps in Parental Authority. Our society provides little explicit institutionalization of the progressive

---

[5] Sometimes compensated for by an interest in the grandchildren, which permits them partially to recover the role of the vigorous parent.

[6] The essential point is not that there are other authorities—in every society there are extrafamilial influences in socialization—but that, because of specialization and individualistic enterprise, they are *competing* authorities. Because they make a living by their work and are specialists in socialization, some authorities have a competitive advantage over parents who are amateurs or at best merely general practitioners.

readjustments of authority as between parent and child. We are intermediate between the extreme of virtually permanent parental authority
and the extreme of very early emancipation, because we encourage release in late adolescence. Unfortunately, this is a time of enhanced sexual
desire, so that the problem of sex and the problem of emancipation occur
simultaneously and complicate each other. Yet even this would doubtless
be satisfactory if it were not for the fact that among us the exact time
when authority is relinquished, the exact amount, and the proper ceremonial behavior are not clearly defined. Not only do different groups and
families have conflicting patterns, and new situations arise to which old
definitions will not apply, but the different spheres of life (legal, economic,
religious, intellectual) do not synchronize, maturity in one sphere and immaturity in another often coexisting. The readjustment of authority between individuals is always a ticklish process, and when it is a matter of
such close authority as that between parent and child it is apt to be still
more ticklish. The failure of our culture to institutionalize this readjustment by a series of well-defined, well-publicized steps is undoubtedly a
cause of much parent-youth dissension. The adolescent's sociological exit
from his family, via education, work, marriage, and change of residence,
is fraught with potential conflicts of interest which only a definite system
of institutional controls can neutralize. The parents have a vital stake in
what the offspring will do. Because his acquisition of independence will
free the parents of many obligations, they are willing to relinquish their
authority; yet, precisely because their own status is socially identified with
that of their offspring, they wish to insure satisfactory conduct on the latter's part and are tempted to prolong their authority by making the decisions themselves. In the absence of institutional prescriptions, the conflict of interest may lead to a struggle for power, the parents fighting to
keep control in matters of importance to themselves, the son or daughter
clinging to personally indispensable family services while seeking to evade
the concomitant control.

4. *Concentration within the Small Family.* Our family system is peculiar in that it manifests a paradoxical combination of concentration and
dispersion. On the one hand, the unusual smallness of the family unit
makes for a strange intensity of family feeling, while on the other, the fact
that most pursuits take place outside the home makes for a dispersion of
activities. Though apparently contradictory, the two phenomena are really
interrelated and traceable ultimately to the same factors in our social
structure. Since the first refers to that type of affection and antagonism
found between relatives, and the second to activities, it can be seen that
the second (dispersion) isolates and increases the intensity of the affectional element by shearing away common activities and the extended kin.

Whereas ordinarily the sentiments of kinship are organically related to a number of common activities and spread over a wide circle of relatives, in our mobile society they are associated with only a few common activities and concentrated within only the immediate family. This makes them at once more instable (because ungrounded) and more intense. . . . Consequently, a great deal of family sentiment is directed toward a few individuals, who are so important to the emotional life that complexes easily develop. This emotional intensity and situational instability increase both the probability and severity of conflict.

In a familistic society, where there are several adult male and female relatives within the effective kinship group to whom the child turns for affection and aid, and many members of the younger generation in whom the parents have a paternal interest, there appears to be less intensity of emotion for any particular kinsman and consequently less chance for severe conflict.[7] Also, if conflict between any two relatives does arise, it may be handled by shifting mutual rights and obligations to another relative.[8]

5. *Open Competition for Socioeconomic Position.* Our emphasis upon individual initiative and vertical mobility, in contrast to rural-stable regimes, means that one's future occupation and destiny are determined more at adolescence than at birth, the adolescent himself (as well as the parents) having some part in the decision. Before him spread a panorama of possible occupations and avenues of advancement, all of them fraught with the uncertainties of competitive vicissitude. The youth is ignorant of most of the facts. So is the parent, but less so. Both attempt to collaborate on the future, but because of previously mentioned sources of friction, the collaboration is frequently stormy. They evaluate future possibilities differently, and since the decision is uncertain, yet important, a clash of wills results. The necessity of choice at adolescence extends beyond the occupational field to practically every phase of life, the parents having an interest in each decision. A culture in which more of the choices of life were settled beforehand by ascription, where the possibilities were fewer and the responsibilities of choice less urgent, would have much less parent-youth conflict.[9]

6. *Sex Tension.* If until now we have ignored sex taboos, the omission has represented a deliberate attempt to place them in their proper

[7] Margaret Mead, *Social Organization of Manua,* 84, Honolulu, Bernice P. Bishop Museum Bulletin 76, 1930. Large heterogeneous households early accustom the child to expect emotional rewards from many different persons. D. M. Spencer, "The Composition of the Family as a Factor in the Behavior of Children in Fijian Society," *Sociometry* (1939), 2: 47-55.

[8] The principle of substitution is widespread in familism, as shown by the wide distribution of adoption, levirate, sororate, and classificatory kinship nomenclature.

[9] M. Mead, *Coming of Age in Samoa,* 200 ff.

context with other factors, rather than in the unduly prominent place usually given them.[10] Undoubtedly, because of a constellation of cultural conditions, sex looms as an important bone of parent-youth contention. Our morality, for instance, demands both premarital chastity and postponement of marriage, thus creating a long period of desperate eagerness when young persons practically at the peak of their sexual capacity are forbidden to enjoy it. Naturally, tensions arise—tensions which adolescents try to relieve, and adults hope they will relieve, in some socially acceptable form. Such tensions not only make the adolescent intractable and capricious, but create a genuine conflict of interest between the two generations. The parent, with respect to the child's behavior, represents morality, while the offspring reflects morality *plus* his organic cravings. The stage is thereby set for conflict, evasion, and deceit. For the mass of parents, toleration is never possible. For the mass of adolescents, sublimation is never sufficient. Given our system of morality, conflict seems wellnigh inevitable.

Yet it is not sex itself but the way it is handled that causes conflict. If sex patterns were carefully, definitely, and uniformly geared with nonsexual patterns in the social structure, there would be no parent-youth conflict over sex. As it is, rapid change has opposed the sex standards of different groups and generations, leaving impulse only chaotically controlled.

The extraordinary preoccupation of modern parents with the sex life of their adolescent offspring is easily understandable. First, our morality is sex-centered. The strength of the impulse which it seeks to control, the consequent stringency of its rules, and the importance of reproductive institutions for society, make sex so morally important that being moral and being sexually discreet are synonymous. Small wonder, then, that parents, charged with responsibility for their children and fearful of their of their own status in the eyes of the moral community, are preoccupied with what their offspring will do in this matter. Moreover, sex is intrinsically involved in the family structure and is therefore of unusual significance to family members *qua* family members. Offspring and parent are not simply two persons who happen to live together; they are two persons who happen to live together because of past sex relations between the parents. Also, between parent and child there stand strong incest taboos, and doubtless the unvoiced possibility of violating these unconsciously intensifies the interest of each in the other's sexual conduct. In addition, since sexual behavior is connected with the offspring's formation of a new family of his own, it is naturally of concern to the parent. Finally, these factors taken in combination with the delicacy of the authoritarian relation,

[10] Cf., e.g., L. K. Frank, "The Management of Tensions," *Amer. J. Sociol.*, March 1928, 33: 706-22; M. Mead, *op. cit.*, 216-217, 222-23.

the emotional intensity within the small family, and the confusion of sex standards, make it easy to explain the parental interest in adolescent sexuality. Yet because sex is a tabooed topic between parent and child,[11] parental control must be indirect and devious, which creates additional possibilities of conflict.

Summary and Conclusion

Our parent-youth conflict thus results from the interaction of certain universals of the parent-child relation and certain variables the values of which are peculiar to modern culture. The universals are (1) the basic age or birth-cycle differential between parent and child, (2) the decelerating rate of socialization with advancing age, and (3) the resulting intrinsic differences between old and young on the physiological, psychosocial, and sociological planes.

Though these universal factors *tend* to produce conflict between parent and child, whether or not they do so depends upon the variables. We have seen that the distinctive general features of our society are responsible for our excessive parent-adolescent friction. Indeed, they are the same features which are affecting *all* family relations. The delineation of these variables has not been systematic, because the scientific classification of whole societies has not yet been accomplished; and it has been difficult, in view of the interrelated character of societal traits, to seize upon certain features and ignore others. Yet certainly the following four complex variables are important: (1) the rate of social change; (2) the extent of complexity in the social structure; (3) the degree of integration in the culture; and (4) the velocity of movement (*e.g.,* vertical mobility) within the structure and its relation to the cultural values.

Our rapid social change, for example, has crowded historical meaning into the family time-span, has thereby given the offspring a different social content from that which the parent acquired, and consequently has added to the already existent intrinsic differences between parent and youth, a set of extrinsic ones which double the chance of alienation. Moreover, our great societal complexity, our evident cultural conflict, and our emphasis upon open competition for socioeconomic status have all added to this initial effect. We have seen, for instance, that they have disorganized the important relation of parental authority by confusing the goals of child control, setting up competing authorities, creating a small

---

[11] "Even among the essentially 'unrepressed' Trobrianders the parent is never the confidant in matters of sex." Bronislaw Malinowski, *Sex and Reproduction in Savage Society,* London, 1927, p. 36n. Cf. the interesting article, "Intrusive Parents," *The Commentator,* September 1938, which opposes frank sex discussion between parents and children.

family system, making necessary certain significant choices at the time of adolescence, and leading to an absence of definite institutional mechanisms to symbolize and enforce the progressively changing stages of parental power.

If ours were a simple rural-stable society, mainly familistic, the emancipation from parental authority being gradual and marked by definite institutionalized steps, with no great postponement of marriage, sex taboo, or open competition for status, parents and youth would not be in conflict. Hence, the presence of parent-youth conflict in our civilization is one more specific manifestation of the incompatibility between an urban-industrial-mobile social system and the familial type of reproductive institutions.[12]

CLASS DIFFERENCES IN SEXUAL AND AGGRESSIVE BEHAVIOR AMONG ADOLESCENTS

Allison Davis

Adapted from Allison Davis, "Child Rearing in the Class Structure of American Society," *The Family in a Democratic Society: Anniversary Papers of the Community Service Society of New York,* New York, Columbia University Press, 1949, pp. 56-59. Originally appeared in *Social-Class Influences upon Learning,* Inglis Lecture, Cambridge, Harvard University Press, 1948, pp. 32-37. Copyrighted by the President and Fellows of Harvard College, by whom permission has also been granted.

Before comparing middle-class and lower-class adolescents, a warning must be interjected. We recall that the long, indulgent nursing period of lower-class infants does not prevent their developing marked fear of starvation in later childhood and adulthood. This fact means that new situations, if strongly organized physically or socially, make new behavior. This is a cardinal principle of the new integrated science of social psychology. Basic learning can and does appear at any age level, provided that society or the physical environment changes the organization of its basic rewards and punishments for the individual.

[12] For further evidence of this incompatibility, see the writer's "Reproductive Institutions and the Pressure for Population," (*Brit.*) *Sociol. Rev.;* July 1937, 29: 289-306.

Secondly, we should not be so naïve as to think that lower-class life is a happy hunting ground, given over to complete impulse expression. Slum people must accept all the basic sexual controls on incest, on homosexuality, on having more than one mate at a time, and on marital irresponsibility. In fact, there is evidence to indicate that slum people are more observant of the taboos upon incest and homosexuality than are those in the upper class. Furthermore, the same pattern which holds in their food intake—deprivation, relieved by peaks of great indulgence—is typical of lower-class sexual life. Lack of housing, lack of a bed for oneself, frequent separations of mates or lovers, the hard daily work of mothers with six to fourteen children, the itinerant life of the men, all make sexual life less regular, secure, and routine than in the middle class. In the slum, one certainly does not have a sexual partner for as many days each month as do middle-class married people, but one gets and gives more satisfaction, over longer periods, when one does have a sexual partner. With this reservation in mind, we may proceed to examine adolescent behavior in the two classes.

The aggressive behavior of adolescents is a crucial case in point. In the middle class, aggression is clothed in the conventional forms of "initiative," or "ambition," or even of "progressiveness," but in the lower class it more often appears unabashed as physical attack, or as threats of, and encouragement for, physical attack. In general, middle-class aggression is taught to adolescents in the form of social and economic skills which will enable them to compete effectively at that level. The lower classes not uncommonly teach their children and adolescents to strike out with fist or knife and to be certain to hit first. Both girls and boys at adolescence may curse their father to his face or even attack him with fists, sticks, or axes in free-for-all family encounters. Husbands and wives sometimes stage pitched battles in the home; wives have their husbands arrested; and husbands, when locked out, try to break in or burn down their own homes. Such fights with fists or weapons, and the whipping of wives, occur sooner or later in most lower-class families. They may not appear today, nor tomorrow, but they will appear if the observer remains long enough.

The important consideration with regard to physical aggression in lower-class adolescents is, therefore, that it is learned as an approved and socially rewarded form of behavior in their culture. An interviewer recently observed two nursery school boys from lower-class families; they were boasting about the length of their fathers' clasp knives! The parents themselves have taught their children to fight, not only children of either sex, but also adults who "make trouble" for them. If the child or adolescent cannot whip a grown opponent, his father or mother will join the fight. In such lower-class groups, an adolescent boy who does not try to be a good fighter will not receive the approval of his father, nor will he be acceptable to any play group or gang. The result of these cultural sanctions is that he learns to fight and to admire fighters. The conception that aggression and hostility are neurotic or maladaptive symptoms of a chronically frustrated adolescent is an ethnocentric view of middle-class individuals. In lower-class families, physical aggression is as much a normal, socially approved, and socially inculcated type of behavior as it is in frontier communities.

There are many forms of aggression, of course, which are disapproved by lower-class as well as by middle-class adolescents. These include, among others, attack by magic or poison, rape, and cutting a woman in the face. Yet all these forms of aggression are fairly common in some lower-class areas. Stealing is another form of aggression which lower-class parents verbally forbid, but which some of them in fact allow—so long as their child does not steal from his family or its close friends. The model of the adolescent's play group and of his own kin, however, is the crucial determinant of his behavior. Even where the efforts of the parent to instill middle-class mores in the child are more than half-hearted, the power of the street culture in which the child and adolescent are trained overwhelms the parental verbal instruction. The rewards of gang prestige, freedom of movement, and property gain all seem to be on the side of the street culture.

Like physical aggression, sexual relationships and motivation are more direct and uninhibited in lower-class adolescents. The most striking departure from the usual middle-class motivation is that, in much lower-class life, sexual drives and behavior in children are not regarded as inherently taboo and dangerous.

There are many parents in low-status culture, of course, who taboo these behaviors for their girls. Mothers try to prevent daughters from having children before they are married, but the example of the girl's own family is often to the contrary. At an early age the child learns of common-law marriages and extra-marital relationships of men and women in his own family. He sees his father disappear to live with other women, or he sees other men visit his mother or married sisters. Although none of his siblings may be illegitimate, the chances are very high that sooner or later his father and mother will accuse each other of having illegitimate children; or that at least one of his brothers or sisters will have a child outside marriage. His play group, girls and boys, discuss sexual relations frankly at the age of eleven or twelve, and he gains status with them by beginning intercourse early.

With sex, as with aggression, therefore, the social instigations and reinforcements of adolescents who live in these different cultures are opposites. The middle-class adolescent is punished for physical aggression and for physical sexual relations; the lower-class adolescent is frequently rewarded, both socially and organically, for these same behaviors. The degree of anxiety, guilt, or frustration attached to these behaviors, therefore, is entirely different in the two cases. One might go so far as to say that in the case of middle-class adolescents such anxiety and guilt, with regard to physical aggression and sexual intercourse, are proof of their normal socialization in their culture. In lower-class adolescents in certain environments, they are evidence of revolt against their own class culture, and possibly of incipient personality difficulties.

The point which these considerations seems to make clear, and which seems to be borne out by many detailed life histories of adolescents of each class, is as follows: The social reality of individuals differs in the most fundamental respects according to their status and culture. The individuals of different class cultures are reacting to different situations. If they are realistic in their responses to these situations, their drives and goals will be different. This basic principle of comparative psychology implies that in order to decide whether an individual in American society is normal or neurotic, one must know his social class and likewise his ethnic culture. He may be quite poorly oriented with regard to middle-class culture, simply because he has not been trained in it and, therefore, does not respond to its situations. If his behavior is normal for lower-class culture—which clinicians, teachers, and guidance workers do not usually know—he may appear to them to be maladjusted, unmotivated, unsocialized, or even neurotic. In dealing with such cases, the reference points of social reality of the teacher or psychologist must be set up with regard to the basic demands of lower-class culture upon its members.

FAMILY INTERACTION, VALUES, AND ACHIEVEMENT

Fred L. Strodtbeck

Adapted from David C. McClelland, Alfred L. Baldwin, Urie Bronfenbrenner and Fred L. Strodtbeck, *Talent and Society,* pp. 135-194. Copyright 1948, D. Van Nostrand Company, Inc., Princeton, New Jersey.

. . . we [chose] those values which appeared most likely to have accounted for the differences in occupational achievement after these two groups [Jews and Southern Italians] came to the United States. This task entailed . . . a comparison of Italian-Jewish values. . . . Finally the problem narrowed to a comparison at five points, as follows:

(1) *Man's Sense of Personal Responsibility in Relation to the External World.* . . . For the present-day achiever in the United States, rational mastery of the situation has taken the place of the "hard work" of the Calvinists, and the threat of almost continuous review of his record has been equated with anxiety over eventual salvation. There is no necessary personal deprivation which must be endured; indeed, one's accomplishment can be facilitated by "breaks." But the breaks are now of the individual's own making; it is a matter of being available with what is needed at the right place and at the right time. Just as the breaks are not doled out by a beneficent power, neither are failures. Whatever failure an individual has suffered could always have been foreseen and circumvented if the individual had been sufficiently alert. For the modern achiever there is no legitimate excuse for failure. His sense of personal responsibility for controlling his destiny is enormous.

Old-culture Jewish beliefs appear to be congruent in many, if not all, respects with such a belief in a rational mastery of the world. For the Jew, there was always the expectation that everything could be understood, if perhaps not controlled. Emphasis on learning as a means of control was strong. Neither religious nor secular learning, once attained (unlike the Protestant's salvation and the achiever's status), was in continual jeopardy. For men who were learned in trades but not religious scholars, the expectations of charity to others of the community who were less fortunate was a continuing goad to keep working; but if misfortune befell a former benefactor, the community understood. The sense of personal responsibility existed along with a responsibility of the community for the individual which eased somewhat the precariousness associated with "all or none" expectations of the individual.

For the Italian, there was no real logic in striving; the best-laid plans of man might twist awry. Misfortune originated "out there," out beyond the individual. *Destino* decreed whether a particular event would or would not come to pass. A sort of passive alertness was thus inculcated. Although no one knew when he might be slated for a lucky break, at the same time there was no motivation for any rational undertaking of heroic proportions; such an undertaking might be *destined* to fail.

(2) *Familism versus Loyalty to a Larger Collectivity.* . . . The old Jewish pattern sanctioned separation from the family for purposes of business and education, and there was a distinct consciousness that a man's first responsibility was toward his children. That is, obligations were primarily from those who have more to those who have less—from which, practically speaking, it followed that children need not always stay to nur-

ture parents who might be better off than they were. Although the Jews did not go so far as the present American achiever in weakening the ties to parents, the pattern contrasts sharply with that of the Southern Italians who put loyalty upward to the extended family first.

(3) *Perfectability of Man.* An aspect of Calvinism perhaps best captured for popular consumption in *Poor Richard's Almanac* by Benjamin Franklin is the insistence that at every moment of every day a man must work to improve himself. The old Jewish culture also, with its emphasis on religious scholarship and study, represented a similar belief in the responsibility for self-improvement. For the achiever in the United States, this perfectability has, in one sense, been relaxed; but insofar as it remains, it has become even more stringent. Now, we are told, the improvement should be acquired in a relaxed manner, with no apparent effort; self-improvement is something to be "enjoyed" not "endured" as earlier. But in any case, an interest in education should be (and has been) high because it is so obviously one of the ways in which man perfects himself.

For the Southern Italian there has always been considerable doubt as to whether man could perfect himself or, indeed, whether he need try. According to his interpretation of Catholicism, he must conscientiously fulfill his duties, but his "good works" do not form a rationalized system of life. Good works may be used to atone for particular sins, or, as Weber points out, stored up as a sort of insurance toward the end of one's life; but there is no need to live in every detail the ideal life, for there is always the sacrament of absolution. Furthermore, the Southern Italian sees man as living in an uneasy peace with his passions, which from time to time must be expected to break through. Man is really not perfectable—he is all too human. So he would do well not to drive himself or his mind too relentlessly in trying to reach that impossible goal, perfection.

(4) *Consciousness of the Larger Community.* The Calvinist's dictum that "each man is his brother's keeper" has given way in the United States to a less moralistic rationale based upon a recognition of the interdependencies in modern society. Just as the whole Jewish community could vicariously participate in the charities of its wealthiest members, there is a sense in which the strengthening of various aspects of American society is recognized as contributing to the common good.

The Jew from the older culture, enabled by his success to assume a responsibility for the community, had little choice in the matter. The social pressures were great, and they were ordinarily responded to with pride and rewarded by prominence in the community forum. The identification went beyond the extended family. The giver was not to be rewarded in kind; his reward came from community recognition. Such community identification—as contrasted with family identification—has not been highly developed among Southern Italians. Reduced sensitivity to community goals is believed to inhibit the near altruistic orientations which in adolescence and early maturity lead individuals to make prolonged personal sacrifices to enter such professions as medicine or the law.

(5) *Power Relations.* . . . The old-culture Jew, on the other hand, did not see power in the context of some external system of pre-established impersonal relationships. He tended, like the Calvinist, to translate power questions into other terms—to the equity of a particular bargain, for example; but unlike the Calvinist, he saw these relationships always as specific, both as to persons and content, and not part of a larger system. His primary concern was to make his relationships good with others with whom he was in close contact over a particular issue. The specificity of his relations with others, including his separation of business and family matters, is also like the functional specificity of modern bureaucratic society, but again unlike it in overlooking the *system* of such functional relationships.

The old-culture Italian tended to see power entirely in immediate interpersonal terms. Power was the direct expression of who can *control* the behavior of another rather than who knows more for a job in an impersonal system. "Who's boss?" was his constant inquiry. Every relationship he turned into a "for me-against me" or "over me-under me" polarity.

The New Haven Sample

In the process of developing the sampling frame in New Haven, further data were obtained which bear upon Italian-Jewish cultural differences. A questionnaire was administered to 1151 boys between the ages of 14 and 17 (and a somewhat larger number of girls) in the New Haven public and parochial schools. Data obtained on this questionnaire were utilized primarily to identify a set of third-generation Italian and Jewish boys, who were in turn stratified by their school performance and socio-economic status. The questionnaire touched generally upon values and more particularly upon materials relating to occupational choice, parental expectations, parental control, educational aspirations, and balance of power within the family.[1]

Boys from Catholic families who reported one or more paternal and one or more maternal grandparent born in Italy were considered Italian.

---

[1] A more detailed form of this questionnaire has been deposited as Document number 5501 with the ADI Auxiliary Publications Project, Photoduplication Service, Library of Congress, Washington 25, D.C. A copy may be secured by citing the Document number and by remitting $2.50 for photoprints, or $1.75 for 35 mm. microfilm. Advance payment is required. Make checks or money orders payable to: Chief, Photoduplication Service, Library of Congress.

Boys who reported the religion of both their parents as Jewish were considered Jewish. Socio-economic status was determined from information provided by the son relating to his parents' education and his father's occupation. . . . In terms of these two criteria the following frequencies were obtained:

Socio-economic Status	Italian	Jewish	Other
High (classes 1 and 2; owners of large businesses; major and minor professionals)	8	24	52
Medium (classes 3 and 4; owners of small businesses; white-collar workers; supervisors)	80	66	213
Low (classes 5, 6, and 7; skilled workers; laborers)	182	17	455
Unclassified	15	2	59
	285	109	779

To demonstrate even more clearly the differential status distribution of the two groups, one may construct an index number using the distribution of "Others" as a base. For example, 52 out of the total 720 in column 3 (excluding the unclassified "Others") are of high socio-economic status. On a pro rata basis, 19.5 Italians of high status would be expected. Significantly fewer than this—only 8, or 41 per cent of the expected—turn up. For the Jews of high status, 310 per cent of the expected are observed. The full set of indices is as follows:

Percentage of Expectation

Socio-economic Status	Italian	Jewish
High	41	310
Middle	100	209
Low	107	25

We used the boy's achievement in school as a criterion of his own performance, just as the status of the family might be used as a criterion of the father's performance. Toward this end, each boy's performance on intelligence and achievement tests was inspected, and his grade performance in terms of the norms of the particular school predicted. When the boy's school grades exceeded the expected performance, he was considered an over-achiever; when his grades fell short, he was classified as an under-achiever. The different standards and testing systems of the various schools made it necessary to adjust slightly the degree to which the boy had to depart from expectation before he was considered an over- or under-achiever.

Being an over-achiever proved to be positively related to higher socio-economic status. This may be illustrated with the 674 "other" students for whom full information was available.

Socio-economic Status	Percentage of Over-achievers	
High	47%	(47)
Medium	35%	(201)
Low	27%	(426)

It thus becomes apparent that since socio-economic status is not an analytic element of central interest, provision must be made for controlling or removing its effect if other variables are to be understood. The standard procedure for making this correction is a factorial design. Forty-eight boys, according to our estimate, could be studied intensively, and they were selected from the larger frame of cases to be allocated as follows.[2]

Socio-economic Status	Italian Boys School Achievement		Jewish Boys School Achievement	
	"Over"	"Under"	"Over"	"Under"
High	4	4	4	4
Medium	4	4	4	4
Low	4	4	4	4
		Total 48		

. . . To initiate our relations with the families, each of the 48 boys was first contacted in the school during his study period and told that he had been selected by a random process to assist with the development of a new kind of test. The "test" consisted of a set of six 8 × 10 pictures, similar in appearance to the TAT cards, designed to elicit n Achievement scores.[3] These pictures were presented to the boy one at a time. with instructions to make up a good story around the picture "about real people and real problems." The administration procedure adopted was comparable to what McClelland and his co-workers have described as "neutral" [1, pp. 100 ff.], and it was not assumed that the boy's achievement motivation was any more mobilized than it would ordinarily be in a school situation. The girl psychologist who administered the pictures was young and attractive; the atmosphere was casual and businesslike.

After the session, the boy's cooperation was sought in arranging a visit to his home at a time when it would be possible to talk with him and his

[2] In making the final selection, it was necessary in scattered instances to use families with parents who were born elsewhere, but who had come to this country as very young children. The socio-economic status classification is in all cases based upon the interviewer's notes obtained in the interviews with the parents. One Italian family was obtained from a residential community adjacent to New Haven.

[3] Briefly described, the pictures are as follows: (1) boy in classroom; (2) operation in background, boy in foreground; (3) man and boy in foreground, horses in background; (4) young man in foreground, crossroads in background; (5) two male figures in workshop; and (6) boy with broom in foreground and several teen-agers in background.

parents. On that same day, a letter was sent from the principal explaining the investigation and stating a hope that the parents would cooperate. The experimenter then phoned the parents and completed arrangements to visit the home. The objective of the investigation was explained to the parents as an effort to illuminate ways in which parents and sons go about making occupational decisions.

The parents were almost unanimously cooperative (as soon as they were assured that we did not have anything to sell). Our only refusals came from two families—one where there was illness and one in which the father would not participate.

The Experimental Procedure

In addition to the questionnaire administered to the boy at school, questionnaires were given to the father, mother, and son in the home. Some questions were asked of the son both in school and at home so that instances of shift in response might be checked against other family information.

The team visiting each home consisted of an experimenter and an assistant, who carried portable sound equipment. As soon as the answer sheets had been completed, the assistant compiled a set of items for discussion. These he selected, if the distribution of original responses made it possible, with an eye to making three coalitions of the following type:

(a) Mother and son agree, father disagrees;
(b) Father and mother agree, son disagrees;
(c) Father and son agree, mother disagrees.

While this collation of responses was being carried out by the assistant, the experimenter gave the family other forms to fill out and subsequently moved them into position around the recorder. He then presented the first item to the family with the following instructions:

We've looked over your responses to the first set of items and, in many cases, all three of you answered the items in the very same way. In some cases, two of you agreed, but the third person picked a different alternative. What we would like to do is ask the three of you as a group to consider again some of these items on which the agreement was not complete. We would like you to talk over the item until you understand clearly why each person marked the item as he did. We want you to try to agree on one choice which would best represent the opinion of the family, if this is possible.

The experimenter then read the item in question saying, roughly:

Mr. ―――― said ――――, and Mrs. ―――― said, and (calling the son by his first name) said ――――. Talk this over and see if it's possible to agree on one of the choices. When you are finished, call me.

. . . The details of the revealed-difference routine were evolved by a series of trial-and-error modifications which may be briefly described. If one contrasts conversations between husbands and wives obtained by concealed recording devices with those obtained by a recorder in full view, one finds no striking differences. Evidently (a) the importance of resolving a difference of opinion with a person with whom one had a solid relationship, and (b) the concurrent requirement of having each member act so that his behavior is consistent with the expectations developed in previous interaction, combine to give a measure which is not greatly influenced by the recording paraphernalia. At the heart of the process is the necessity for "revealing a difference," as has been most clearly demonstrated in a Cornell study by Arthur J. Vidich [7]. Vidich attempted to have married couples discover and discuss whatever differences they might have about disposing of a legacy. In this he encountered great resistance, with a tendency for couples to be most interested in explaining their respective thinking to the experimenter instead of to one another. Vidich's experience suggests that the group cohesiveness which, when a difference is revealed, creates the motivation for interaction operates to conceal and resist differences when they arise under conditions which the group can control. . . .

To illustrate the experimental procedure concretely, we will quote from the discussion in one Italian home, along with scattered background information. Michael's father, a machinist who stopped attending school just before graduating from high school, conceived of himself as a strict disciplinarian.

I probably should be ashamed to say it, but up to a few months ago I used to beat him, I really let him have it. I still believe that sparing the rod spoils the child. I still do let him have it every so often. I wore out a strap on that boy. You can't overlook badness. It's got to be nipped in the bud.

In his discussion with the interviewers, Michael's father gave this picture of his own discussions with Michael:

Sometimes I feel he keeps quiet when I want him to put up an argument, especially when I look at things the wrong way; maybe I misunderstand the whole situation. I may be wrong, maybe I came home crabby, the kid may have an argument on his hands with me. He may be right; I may be wrong. Well, my tone of voice, my manner makes him keep quiet. Maybe he had all the right in the world in his argument, and he keeps quiet about the whole situation, and then he gets heck from me for not putting up an argument.

. . . The protocol of their discussion of the first revealed difference question is as follows:

Michael's Family

Experimenter: Two fathers were discussing their boys, one of whom was a brilliant student and the other a promising athlete. Some people believe that one father was more fortunate than the other father. Do you think that the father with the athletic son or the father with the studious son was more fortunate? Michael said that the father with the athletic son was the more fortunate and (the father and mother) said that the father with the brilliant son was more fortunate. We would like you to discuss this.

Father: Why do you say the ah, ah, father of an athlete? (8)[4]

Michael: Because if the son is an athlete he must be getting good marks in order to play sports. (5) He must be getting good marks (6) and—

Father: Not necessarily. (10) Not necessarily. (10)

Mother: While he's out playing, he doesn't get his studies. (5)

Father: No! (10) No! (10) That's not it either. (10) Let's look at it this way. (6) Forget about the school part. (6) Don't attach the athletic life to the school life. (6) Don't make it— Don't make it that the boy in order to be an athlete has to have good marks. (6) We know that. (5) But take it as a kid's life; (6) as a guy's life. (6) Would you think that a guy who was a good athlete would get more out of life; (8) get ahead in life more than a kid who was smart in his studies and made every grade just like that? (5)

Michael: Well, the way you're asking the question, you're putting it a little different than the way it reads on the paper, I think. (10)

Father: No! (10) No! (10) I'm not. (10) It means the same thing. (10) It's just that I probably made it a little longer. (5)

Michael: Well, what is the last sentence on the paper exactly? (7)

Father: Look. Do you . . . (sternly) . . . ? (12) I'll read the whole thing. (6)

Michael: (Attempts to protest that rereading is not necessary.) (11)

Father: Two fathers were discussing their boys; (6) one of whom was a brilliant student and the other an athlete of great promise. (6) (continues to read question given above.)

Michael: (inaudible remark) Athletic son . . . (11)

Father: Well. (6) I think if ah, ah, my son were studious and he pursued any vocation at all, (6) Michael (6), I wouldn't worry as much as I would even if I knew he were a brilliant football player. (5) What good is that? ah (8)

Michael: Well, it's like I said before. (10) If he's good in sports, he must be good in marks. (5)

Father: Yes, Michael. (3) What good is being a football player, ah, towards helping you to become something? (8) An engineer or draftsman or something? (6) Football and baseball, there's a limit to it. (5) You've got to live with it and make something out of it. (5)

Michael: I don't know. (10) What do you think, Maw? (8)

Mother: I'd still say the studious type. (5)

Father: Try to make your son understand, Mother, that even if he were a great football or basketball player, after he's 35 or 40 he can't play any more. (4)

Mother: Play any more. (3) That's right, Michael. (5)

Father: What are you going to do then? (8) Live on your laurels? (12)

---

[4] The scores in parentheses are Bales' Interaction Process categories [2].

Michael: No! (3) No! (3) You'd have to quit by then (3) but I mean, I mean you'd have to have good marks before. (10)

Mother: Yes, but—(10)

Father: In other words you agree. (5) You agree you have to be studious first? (8)

All protocols were scored directly from the recordings and were not transcribed. The subsequent processing of the data may be illustrated with Michael's family's protocol. In Table 1 the number of acts by each family member is shown for each decision in each of the three coalition patterns. Previous research [4] leads one to expect that persons who talk most should have most power in the sense of winning the most decisions; and that an isolate role, necessitating an explanation of one's position to two others, should also increase participation.

In Michael's family, the differentiation in participation is marked, with the father accounting for more than half of the total acts originated. Even in instances where others are the isolates, he continues to dominate. To anticipate the statistical analysis, the acts originated are converted to percentage values, then transformed to angular readings in this way:

	Father	Mother	Son
Original acts	432	141	240
Percentage	53%	17%	30%
Arc sine	47	24	33

TABLE 1. Acts by Person by Decision for Michael's Family

Type of Decision	Originator			Total
	Father	Mother	Son	
	47	16	28	91
Fa vs. Mo-So	65	19	37	121
	76	23	41	140
	188	58	106	352
	39	16	17	72
Mo vs. Fa-So	31	16	10	57
	52	39	43	134
	122	71	70	263
	52	6	17	75
So vs. Fa-Mo	23	4	21	48
	47	2	26	75
	122	12	64	198
Total	432	141	240	813

Throughout the statistical analysis and in subsequent tables, arc sine values are used to stabilize the variance.

To form a power score based upon decisions won, it is convenient to assign arbitrary scores, so that winning, or holding one's position when in the minority, is weighted more heavily when one is an isolate than when one is a member of the larger coalition. The conventions are as follows:

Nature of Decision	Coalition Members		Minority Member
Coalition Wins	1	1	0
Minority Wins	0	0	2
No Decision	.5	.5	1

In Michael's family, the resultant measure of power is markedly differentiated:

	Father	Mother	Son
Original Score	9.5	5.0	3.5
Percentage	53%	28%	19%
Arc Sine	47	32	26

and it may be noted that Michael's father, who participated most heavily, also demonstrated the highest power. Michael had the second highest participation, but ranked third in power. . . .

The V-scale and Other Attitudinal Differences

Fifteen items were included in the original screening questionnaire. These items, adapted from research of the Harvard Seminar in Social Mobility,[5] dealt very generally with the types of value differences which have been previously described as characterizing older Italian-Jewish differences. Not all points in the value analysis were covered in the questionnaire. The analysis was completed late in the study, and the questionnaire had been the original device for selecting subjects for the study by the revealed difference technique.

In the first stage of the analysis, we were looking for items which would discriminate at the .05 level between over-achieving and under-achieving

V-score	Percentage above Average	Number
0	0	2
1	0	6
2	17	46
3	20	82
4	23	146
5	26	207
6	30	226
7	42	220
8	51	76

[5] The assistance of Florence Kluckhohn, Talcott Parsons, and Samuel A. Stouffer, joint directors of this seminar, is gratefully acknowledged.

students (both Italians and Jews being excluded from this comparison). The original set of 15 items was reduced to 8 (see Table 2). Although in this process items of uneven coverage resulted, it was nonetheless apparent that these scores could be combined (1 for achievement-related responses, 0 for the alternate responses) to provide a moderately efficient discrimination of students receiving above average grades.

TABLE 2.   V-scale Items, Factor Loadings and Italian-Jewish Response Levels

Factor Loading			Percentage Who Disagree	
Factor I "Mastery"	Factor II "Independence of Family"	Items	Jews	Italians
.64	.00	(1) Planning only makes a person unhappy since your plans hardly ever work out anyhow.	90	62
.49	.28	(2) When a man is born, the success he's going to have is already in the cards, so he might as well accept it and not fight against it.	98	85
.58	.15	(3) Nowadays, with world conditions the way they are, the wise person lives for today and lets tomorrow take care of itself.	(80)*	(79)
.04	.60	(4) Even when teen-agers get married, their main loyalty still belongs to their fathers and mothers.	64	46
.21	.60	(5) When the time comes for a boy to take a job, he should stay near his parents, even if it means giving up a good job opportunity.	91	82
.29	.68	(6) Nothing in life is worth the sacrifice of moving away from your parents.	82	59
—.02	.28	(7) The best kind of job to have is one where you are part of an organization all working together even if you don't get individual credit.	54	28
—.05	.00	(8) It's silly for a teen-ager to put money into a car when the money could be used to get started in business or for an education.**	(65)	(63)

* The difference is not significant at the .05 level for pairs of values in parentheses; for the remaining values the differences are significant at the .05 level or greater.
** Per cent "Agree" reported for this item.

Since neither the Italians nor the Jews had been involved in the original computations, Italian-Jewish differences provide an independent check on the distribution of one type of "achievement potential" in the two populations. From inferences made on the basis of status mobility, it was predicted that Jews would have higher achievement-related responses than Italians. Table 2 shows that this prediction was significantly confirmed for six of the eight items, with no differences observed in the other two cases. . . .

As to the validity of the scale so developed, three bits of evidence are relevant.

(1) The first is based upon the way the fathers responded to the V-items on the questionnaires administered in the home. One assumes that second-generation fathers of higher status have by their own work personally accounted for some appreciable part of their mobility. In terms of such an assumption, one might predict that fathers of higher status would have higher V-scores than those of lower status. These data, presented in Table 3, may be analyzed so that each effect associated with the factorial design (including status) is isolated. The form of the analysis is as follows:

Source of Variation	Degrees of Freedom
Corrected sum of squares	47
between groups	11
1. Linear SES	1
2. Quadratic SES	1
3. Italian (I) v Jews (J)	1
4. (O) v (U) Achievers	1
5. I v J $\times$ O v U	1
6. I v J $\times$ Linear SES	1
7. I v J $\times$ Quadratic SES	1
8. O v U $\times$ Linear SES	1
9. O v U $\times$ Quadratic SES	1
10. I v J $\times$ O v U $\times$ Linear SES	1
11. I v J $\times$ O v U $\times$ Quadratic SES	1
Residual	36

It will be our practice throughout the analysis to examine the variance associated with each degree of freedom. In this instance three significant effects are observed:

Primary Sources of Variation for Fathers' V-scores

Line 1.	Higher SES groups have higher values	$F =$	10.85
		$p \leqq$	0.01
Line 4.	Fathers of over-achievers are higher than fathers of under-achievers	$F =$	4.74
		$p \leqq$	0.05
Line 6.	There is a greater linear SES trend for Jews than for Italians	$F =$	4.16
		$p \leqq$	0.05

Of primary interest is the relation between the father's class position and the V-score. This effect is significant and in keeping with the hypothesis that persons who have achieved higher status have higher V-scores. We must not, of course, lean too heavily upon this finding, because we have not demonstrated that the higher V-scores preceded the attainment of higher status; the opposite might well be the case. But if there had been no relationship, or a reversed relationship, then there would have been less ground for believing that a high V-score in high school would necessarily be associated with status mobility. The observed finding leaves open the possibility that the higher V-scores of the higher-status fathers may have been continuously operative and contributed to the status attained.

From line 6 one learns that there is a greater difference between the V-score of high-status and low-status Jewish fathers than there is between high- and low-status Italian fathers. This finding, which in itself appears to be of little consequence, serves merely to draw our attention to the fact that, save for this exception, there were *no* Italian-Jewish differences. That is, the two items—stratification by class and educational achievement of son—remove the Italian-Jewish differences found originally in the school population.

TABLE 3.   Fathers' V-score by SES, Ethnicity, Over- and Under-Achievement

Socio-economic Status SES	Italians		Jews	
	Over-Achievers	Under-Achievers	Over-Achievers	Under-Achievers
High	6	6	8	7
	6	6	8	6
	8	5	8	7
	6	7	8	7
Medium	5	6	6	8
	7	2	7	6
	8	6	8	8
	7	5	7	6
Low	5	5	5	6
	6	8	6	3
	4	5	6	4
	7	6	6	6

(2). From line 4 one obtains a second, slightly different, validation of the significance of the V-score: fathers of over-achievers have higher V-scores than fathers of under-achievers.

Would the same effects be present for mothers, or is the pattern of their relationship different? To conserve space, the table of actual values for mothers' V-score is omitted, and the results of an analysis of variance examined directly.

Primary Sources of Variation for Mothers' V-scores

Line 3.	Jewish mothers are higher than Italian mothers	$F = 4.46$
		$p \leq 0.05$

In this case, mothers of higher socio-economic status are not differentiated from those of lower socio-economic status. As an after-the-fact speculation, one might say that the status of a family is primarily established by the husband's occupation; therefore there is less reason to believe that higher-status wives personally contributed by extra-familial efforts to the mobility. Hence the lack of SES effects would not controvert the finding in the case of the fathers. Equally interesting is the fact that the mothers of over-achieving boys do not show disproportionately higher V-scores. Again the mother's contribution to a highly achieving son might involve something other than parallel attitudes about the universe, family ties, work relations, and the like. The ethnic difference in the case of the mothers is not removed by the stratification; the expected cultural relationship persists: Jewish women have higher V-scores than Italian women. These data are provocative. Yet the one instance of V-score variation which goes toward validation—that is, the ethnic difference—is counterbalanced by the absence of higher V-scores for mothers of over-achievers.

(3) There remains, of course, the matter of particular interest—the sons' V-scores:

Primary Sources of Variation for Sons' V-scores

Line 4.	Over-achievers have higher V-scores than under-achievers	$F = 5.17$
		$p \leq 0.05$

For sons, as for fathers, there are, after stratification, no ethnic differences, but over-achieving boys are significantly higher than under-achieving boys. In so far as both Italian and Jewish boys were excluded from the sample at the time the eight items were selected, this finding constitutes, on an independent sub-population, a third instance of validation of the V-scale as a measure of values which are associated with actual achievement. When both parents are in agreement on the positive alternative of the V-score item—or other attitudinal points, for that matter—then the son may be prevented from playing the parents against each other. It is notable that instances of joint V-score agreement in the positive direction are significantly more frequent among Jewish parents than Italian parents.

Primary Sources of Variation for Joint Parental "Achievement Positive" Responses to V-items

Line 1. Parents from higher SES groups agree more	$F =$	9.78
	$p \leqq$	0.01
Line 3. Jewish parents agree more than Italians	$F =$	13.43
	$p \leqq$	0.01

Choice of occupation for the son is another point at which the value structure of the family members is obviously apparent. Data on this point were obtained from the questionnaire. All the boys in our high-school sample, as well as the parents of the 48 boys in the intensive sample, were asked whether they would be pleased or disappointed if the sons chose the following occupations (listed by status rank):

1. Doctor, advertising executive
2. Druggist, jewelry store owner
3. Bank teller, bookkeeper
4. Carpenter, auto mechanic
5. Mail carrier, bus driver
6. Night watchman, furniture mover

The results have been reported in full elsewhere [6]. What is relevant here is that in the total sample, the slope of self-reported pleasure in the occupations by Jewish boys was significantly steeper ($p < .01$) than for Italian boys, meaning that the Jewish boys rejected the occupations of lower status more decidedly. The same result was obtained for the parents; Jewish parents rejected lower-status occupations for their sons more decidedly ($p < .05$) than Italian parents. Finally, there was more agreement among parents and sons ($p < .05$) in the Jewish than in the Italian families.

The difference in emphasis upon education also stands out. For example, the percentage of the respondents in the large sample who "want to" and "expect to" go to college is sharply differentiated between Italians and Jews; but, interestingly enough, Italians are not differentiated from "others."

SES	Italians	Jews	Other
High	(75%)*	83%	77%
Middle	45%	83%	51%
Low	38%	(71%)*	31%

* Values in parentheses are based on low frequencies. See p. 359 *supra*.

Some of the same factors differentiate over- and under-achievers. In cases where boys differed from their parents, the over-achieving boys preferred the higher status occupations significantly more frequently than under-achieving boys ($p < .01$). Also there was more initial consensus among the three family members over all the "revealed differences" in the families of over-achievers. In short, these data support strongly the conclusion based on V-scale results: Jews have values more likely to promote high achievement than Italians do, and there is greater agreement among family members. The additional findings agree with the V-scale also in that they show higher occupational aspiration and greater family consensus among over-achievers than among under-achievers.

n Achievement Scores

A point of articulation between the V-score and prior research arises in connection with the n Achievement scores. The scores for each boy in the sample, based on the presence or absence of achievement imagery in the stories written about the pictures shown him, have been analyzed in the manner illustrated with the V-scores.

Primary Sources of Variation for Sons' n Achievement Scores		
Line 4. Over-achievers have higher n Achievement than under-achievers	$F = 4.79$	$p \leq .05$

In view of previously reported differences between Italians and Jews as to "age of mastery," the absence of an Italian-Jewish difference is surprising, notwithstanding the stratification. The small difference present, an average of 3.2 stories with n Achievement imagery for Italians to 3.7 for Jews, is in the expected direction but *not* significant. The difference between over-achievers (3.9) and underachievers (3.0) is significant at the 0.05 level and constitutes an additional confirmation of the relationship of n Achievement to high-school grades [3].

Family Interaction and Power

Examination of the family patterns of interaction in terms of Bales interaction process categories showed no significant relationships to socioeconomic status, to ethnicity, to over- and under-achievement, to V-scores, or to n Achievement. Only two significant effects emerged for the supportiveness index. The first was a greater supportiveness toward their sons by Italian than by Jewish fathers. The interpretation seems to be that there was a tendency for the Italian father to look upon his son as a less mature person; hence it suggests a denial of near-adult status. Second, mothers were more supportive to fathers as the status of the fathers improved.

The point to be emphasized is the very great similarity of Italian and Jewish interaction patterns. If there has been differential achievement—and according to our data this is indeed the case—then one must conclude that ethnic differences in family interaction are not of great relevance in explaining it. . . .

[With respect to power] the significant trends within the family are as follows:

Primary Sources of Variation in Family Power Scores

Fathers:	Line 1.	Fathers from higher SES groups have higher power scores	$F =$	11.82
			$p \leqq$	0.01
	Line 8.	There is a greater linear SES trend for fathers of over-achievers than for fathers of under-achievers	$F =$	5.09
			$p \leqq$	0.05
Mothers:	Line 3.	Jewish mothers have higher power scores than Italian mothers	$F =$	4.19
			$p \leqq$	0.05
Sons:	Line 1.	Sons from higher SES groups have lower power scores	$F =$	6.47
			$F \leqq$	0.05

The higher the status of the families, the less the power of the sons and the greater the power of the fathers (just as the sons had reported). The mothers' power scores do not seem to be influenced by status, but Jewish mothers have more power than Italian mothers. One significant interaction is found; namely, the trend over status is steeper for fathers of over-achievers than for fathers of under-achievers. There were no ethnic differences in fathers' and sons' power scores and no differences in the sons' power scores related to school achievement.

To test the assertions that there were more departures from equality among Italians than Jews, a coefficient was formed by squaring the mean deviations of the power scores within each family. Analysis of this measure by the standard techniques reveals:

Primary Sources of Variation for Coefficient of Dispersion of Family Power
(Transformed to Rankits)

Line 3.	Power dispersion is greater in Italian than Jewish families	$F = 3.15$
		$p \leqq 0.05$

In short, our data show less equality among the family members in Italian than in Jewish families. This fact agrees with the ratings of power in Jewish-Italian families in the Greater Boston area reported earlier, as well as with the distribution of parental power as reported by Jewish and Italian boys in our own larger sample.

Although participation scores, like the other Bales categories, are not related to any of the variables in this study, they were significantly associated with power scores, as in the author's previous study of husband-wife interaction [5]. The residual correlations (after effects of classificatory variables are removed) are for the father .57 ($p < .001$), for the mother .48 ($p < .01$), for the son .56 ($p < .001$). In short, he who talks most wins most.

Summary

. . . There is . . . evidence . . . that the following three values contained in the V-scale are important for achievement in the United States.

1. *A belief that the world is orderly and amenable to rational mastery; that, therefore, a person can and should make plans which will control his destiny* (*three items in the V-scale*). The contrary notion, that man is subjugated to a destiny beyond his control, probably impeded Southern Italians in their early adjustment to the United States, just as in this study it impeded boys in school of less successful fathers in their choice of occupations. Unfortunately, we cannot say with any assurance in which direction the curse worked—whether the poor performance of the Italians and of the less successful fathers or sons was the result of their belief in fate, or whether the belief in fate was the result of their poor performances. But since we do know, in the case of the Italians, that the belief was part of their earlier culture and therefore antedated their performance, we may feel justified in concluding that the belief came first so far as the adjustment of Southern Italians to the United States is concerned.

2. *A willingness to leave home to make one's way in life.* Again, the South Italian stress on "familism," for which we found evidence in the V-scale, may well have interfered with upward mobility and contributed to the lower occupational achievement of Italians as compared with Jews. Family balance of power also affects willingness to leave home, as we shall see in a moment—a fact which demonstrates that one's position in life can produce a value disposition as well as the reverse. But whether the willingness to break up the family comes from an "old culture," from the power balance in the family, or from the father's or son's relative lack of success in job and school, it is certainly a value of importance in the achievement complex.

3. *A preference for individual rather than collective credit for work done.* Because our evidence is based upon only one item of the V-scale, it must be interpreted with caution. We have earlier argued that for achievement to arise from a heightened desire for individual credit, a certain basic competence and discipline within a larger relationship system (i.e., a profession or modern bureaucracy) is required. Familistic organization with emphasis upon collateral rewards has not historically fitted the requirements for achievement of intermediate status in the United States—particularly not as well as more individualistic orientations. Our finding that Jews are more inclined toward individual credit than Italians has positive implications for achievement, but this ethnic difference is less important to our argument than the more general emphasis that individual credit must be sought within a framework of norms which, like the Calvinist's, are pointed toward the betterment both of society and the particular actor.

Beyond the V-scale results, which are impressive because they reflect differential achievement of cultures (Jews versus Italians), of fathers (high versus low SES) and of sons (over- versus under-achievement in school), there are two facts from the larger questionnaire study which relate to a fourth expected value difference between Italians and Jews—

namely, the value placed on the *perfectability of man*. The Jews definitely had higher educational and occupational expectations for their sons. Practically speaking, this would mean they believed that man could improve himself by education and that no one should readily submit to fate and accept a lower station in life, the way the Italians were prepared to do.

The fifth and final expected Italian-Jewish value difference had to do with power relationships. From ethnographic reports and other studies, we had been led to believe that Italians would be more concerned than Jews with establishing dominance in face-to-face relationships. Such indeed turned out to be the case. Both in the boys' reports of who was dominant at home and in the actual decision-winning in the homes we studied intensively, the Italians showed greater variations from equality of power than the Jews. While this finding is probably of less importance than those presented above, it nonetheless sharpens our curiosity about the effects of power balance on the son's achievement. Is it perhaps true that when relatively equalitarian relations exist in the home, the son can move to new loyalties for larger systems of relationship, such as those provided by college or a job, without an outright rupture of family controls? Is such an adjustment to new institutions outside the home harder the more the home has tended to be dominated by one parent or the other? Furthermore, what would be the cost to the son of such a rupture—both in performance and in motivation to continue on his own? One wonders, of course, whether the conflict would not be less, the frustration less, when the break came—and consequently the emotional and intellectual adjustment more efficient—if the son had come from a home where controls were already diffuse and equalitarian as they are in many situations in life? The present research involved only a single visit with the families; in subsequent research it is to be hoped that more contact can be arranged as the child is growing up. Thus one could follow the effects of a balance of power on the child's adjustment inside the family and subsequently to life outside it.

So we come back to one of the most persistent and important themes of this study: what have power and the adjustment to power to do with achievement? Let us review the steps of the argument briefly. We held that, to achieve on the American scene, one must adjust to a more or less impersonal, bureaucratic system where power lies not with the individual but with the system, and is used to reward and punish according to the way individuals live up to impersonal specialized standards of performance. In addition, we argued that the family is also a "power system" and that the son's adjustment to it should generalize to his life outside. Of course, the reverse should also be true: performance outside should generalize (at least for the fathers) to performance inside the family. Our data confirm this expectation. Fathers who have adjusted successfully to the American scene (and therefore have high SES in our terms) are significantly more powerful in their homes too. Interestingly enough, the same is not true of sons: those who have done well outside (the over-achievers) do not necessarily have more power at home. This is because the family consists of two adults whose largely complementary roles tend to create a strong coalition working on and for the child; the latter could wield influence only if both the other members were very weak or disunited. The father's occupational success is something else again. He is a key member of the parental coalition; if he fails in his function of "bringing home the bacon"—of adapting successfully outside—his power is reduced at home, too.

But to return to our main concern here—the generalization of the boy's experience with power at home to his possible future achievement—our data on this point are especially striking. They point most clearly to a link between family "democracy" (that is, a relatively powerful mother) and the V-scale. Now, since we have just shown that the V-scale contains values (belief in control of one's destiny, willingness to leave home) related to three types of achievement, we can feel justified in assuming that power balance in the family is of importance in giving a child ideas which will bear on his later success or failure. And oddly enough, it is the power balance that is correlated with the ideas and not whether those same ideas are held by the parents or not. A clear case of the children believing what the parents do and not what they say! For example, a father may have a high V-score and believe that one can control his destiny, as perhaps he himself has done in achieving a high-status occupation. But is his son likely to accept this belief if his father pushes him around all the time? Apparently not, to judge by our data. The son is more likely, at least in this stage of his life, to resign himself to the notion that there are forces beyond his control—in this instance, father.

This analysis immediately suggests the popular notion that there is alternation of generations in the production of great men in a family, or Franz Alexander's analysis of "chronic" achievers as persons who experience guilt for having usurped their father's role.[6] At least it provides a

---

[6] Franz Alexander, in the *Age of Unreason* (1), explains ultra-aggressive, ruthless, and belligerently self-centered personality types produced by impoverished immigrant families in terms of the failure of the parental coalition: "A common solution is that the son usurps the father's place in the mother's affection as well as in economic importance and acquires an inordinate ambition. He wants to justify all his mother's hopes and sacrifices and thus appease his guilty conscience about his father. He can do this only by becoming successful at whatever cost. Success becomes the supreme value and failure the greatest sin because it fails to justify the sacrifice of the father. In consequence of this all other defects such as insincerity in human rela-

fairly solid ground for such theories to build on: father's power is *inversely* related to V-scale values. It also adds one further item to the growing body of evidence that power relations in the family are an important determinant of personality development. Finally, it provides an interesting example of how the theoretical analysis of family structure, so ably made by Parsons and Bales [2], can be tested by empirical studies of the sort reported here.

REFERENCES

1. McClelland, D. C., Atkinson, J. W., Clark, R. A., and Lowell, E. L., *The Achievement Motive* (New York: Appleton-Century-Crofts, 1953).
2. Parsons, T., and Bales, R. F., *Family, Socialization and Interaction Process* (Glencoe, Illinois: Free Press, 1955).
3. Ricciuti, H. N., and Sadacca, R., "The Prediction of Academic Grades with a Projective Test of Achievement Motivation: II. Cross-Validation at the High School Level," Princeton, N. J.: Educational Testing Service, 1955.
4. Shannon, J., "Early Detachment and Independence in a Study of Creativity,"
unpublished manuscript, Univ. of Chicago, 1957.
5. Strodtbeck, F. L., "Husband-Wife Interaction over Revealed Differences," *Amer. Soc. Rev.*, 1951, 16, 468-473.
6. Strodtbeck, F. L., McDonald, M. R., and Rosen, B. C., "Evaluation of Occupations: A Reflection of Jewish-Italian Mobility Differences," *Amer. Soc. Rev.*, 1957, 22, 546-553.
7. Vidich, A. J., "Methodological Problems in the Observations of Husband-Wife Interaction," unpublished manuscript, Cornell Univ., 1957.

tionships, unfairness in competition, disloyalty, disregard of others, appear comparatively slight, and the result is a ruthless careerist, obsessed by the one idea of self-promotion, a caricature of the self-made man, and a threat to Western civilization, the principle of which he has reduced to absurdity."

13. YOUNG ADULTS AND THEIR PARENTS

Let us think of young adults as being around twenty-five years old and their parents as around fifty; in the next chapter we shall be looking at the relations between these generations some twenty-five years later, that is, when they are about fifty and seventy-five respectively. The first of the two articles in this chapter reports a study to determine some of the conditions under which young married couples do or do not maintain active contact with their parents. The second paper concerns authority relations between these two generations and the consequence for relations between grandparents and grandchildren.

In seeking correlates of the degree to which young married couples maintain active contact with their parents, Sussman finds that relatively high contact (or "family continuity," in his phrasing) is associated with developmental (rather than traditional) child-rearing methods, with similarity between both members of the young couple and the parents (and of course parents-in-law) in such background characteristics as ethnoreligious identity and social class, and with traditional courtship and wedding practices.

Dorrian Sweetser tests two hypotheses on ethnographic materials and comes to the conclusion that, if grandparents exercise authority over parents, the relations of the grandparents to their grandchildren tend toward formality, whereas if the grandparents do not exercise such authority, they have indulgent, close and warm relationships with their grandchildren.

FAMILY CONTINUITY: SELECTIVE FACTORS WHICH AFFECT RELATIONSHIPS BETWEEN FAMILIES AT GENERATIONAL LEVELS

Marvin B. Sussman

Adapted from *Marriage and Family Living*, 16 (1954), 112-120.

. . . in this study we have attempted to establish the importance of selective factors which affect relationships between families at two generational levels, i.e., family continuity. The factors investigated are: (1) the socio-cultural background of marriage mates; (2) type of courtship and marriage ceremony; (3) family and child rearing philosophy and practice; (4) the development of a help pattern between parents and their married children; and (5) the residential location of the parental and child's family after marriage.[1]

Sample and Methods

To determine the importance of these factors, parents of ninety-seven families of New Haven, Connecticut and suburbs who were middle-class, white, Protestant, whose children had married and left home were intensively interviewed. From these interviews 195 parent-child relationships were selected which comprise a final sample. A case study approach was employed and the statistical device of chi-square (χ^2) was used to test the significance of relationships of specific items of behavior to the factors which affect the continuity of family relationships after a child marries and leaves home. Then also, χ^2 was used to test the relationship of the five factors to a family continuity scale. This was thought necessary to check the importance and the degree to which these factors affected

---

[1] For a full explanation of the factors and their assigned ratings and use in a rating scales see the author's Ph.D. dissertation, *Family Continuity: A Study of Factors Which Affect Relationships Between Families at Generational Levels* (New Haven: Yale University Library, 1951). Plus (+) values were assigned to each of the factors of every case when a child married a mate of similar cultural background; where parents have knowledge of impending marriage; where children have been reared developmentally (an explanation of this concept can be found in Evelyn M. Duvall, "Conceptions of Parenthood," *American Journal of Sociology*, LII: 193-203, November, 1946); where a help pattern develops between parents and married child's family; and where both families were located in the same or nearby communities. A minus (—) value was assigned wherever opposite conditions existed.

family continuity. This was accomplished first by constructing a five point scale which indicated levels of intergenerational family relationships and each case was rated according to it.

After families had been rated according to this continuity scale, they were assigned a qualitative rating for each of the factors, e.g., in the first factor, if similarity of background was present a plus value was assigned and a minus value if absent (See footnote 2). By this technique it was possible to correlate the factors with the levels of family continuity, thus determining their importance to continued relationships between families of two generations.

Family Continuity Scale

In constructing the family continuity scale the completed case histories were read and rated by the investigator and one other person. The basis for scaling was the raters' judgment and interpretation of the existing configuration of family relationships in terms of the following criteria:

Rating one (1) indicates *no* family continuity marked by complete mutual rejection by the parents and child's family of each other. (Since no families were rated in level one it was no longer considered.)

Rating two (2) indicates *poor* family continuity marked by partial or complete mutual acceptance by parents and their son or daughter, but mutual rejection of parents and their son- or daughter-in-law.

Rating three (3) indicates *fair* family continuity in which parents and married child's family partially accept each other.

Rating four (4) indicates *good* family continuity in which parents and their son or daughter accept one another, and parents and their child's spouse partially accept each other.

Rating five (5) indicates *high* family continuity, marked by complete mutual acceptance by the parents and child's family of each other.

TABLE I. Levels of Family Continuity as Rated by Author and One Judge

Level Assigned by Judge	Level Assigned by Author			
	2	3	4	5
2	19	3	1	
3	2	13	7	4
4		8	15	2
5		1	4	116

$r = 0.87$.

The results indicated a high and significant coefficient of correlation, $r = .87$ between the two sets of scores. (See Table 1.) Because of this the author decided to use his own ratings of families.

Discussion

Socio-cultural Background Factor. Our findings indicated that when children married individuals of a background similar to their own—in terms of ethnic origin, church membership, class position, and educational attainment—their relationships with parents-in-law tended to be harmonious. When difficulties occurred between the families they were generally solved on a personal basis. Most differences between them fell within a range of tolerated behavior. In contrast were the relationships of parents and their child's family when the spouse was of dissimilar background. Here differences were projected in terms of out-group stereotypes and family relationships were strained.[2]

Frequently because of differences in cultural background parents and their children-in-law are uncertain as to how to anticipate one another's responses and tend to magnify small and petty differences into a formidable barrier against continued friendly relationships. This is clearly illustrated in the Rogers family. The Rogers have two children, both girls. The eldest chose a marriage partner, acceptable to the parents, from a family similar in background to her own. Sally, the youngest, an impetuous girl, according to her mother, eloped and married "out of her background" a young man from the Near East.

Mrs. Rogers explained how their son-in-law differed from their American one and was unpredictable in his response to them and their activities. Obviously these cultural differences become emotionally defined and relationships between family members become tense and uneasy. She illustrated: "We enjoy having Mary's husband around (Mary is the eldest and married according to family and clique expectations). He used to practically live here before he married Mary and I knew his mother well. . . . He is just like our son; he likes bread pudding, and likes to play cards with us; he is crazy about tools and gadgets. He and Bill (husband) are quite a pair; they both go in for gadgets and wood working. He is a

---

[2] One of several social scientists who established the importance of common frames of reference as consequence of group interaction is Muzafer Sherif. In his autokinetic experiments and later researches in intergroup relations he has aptly illustrated the effects of shared experience upon the formation of group norms and the development of negative attitudes and aggressive behavior towards out-groups. See *Psychology of Social Norms* (New York: Harper and Brothers, 1936) and *Groups in Harmony and in Tension* (New York: Harper and Brothers, 1953). When a group shares common frames of reference and expectations they also possess a number of alternative behavior responses which are understood and accepted by members. These come within a range of toleration of acceptable differences, thus making for ease in resolving difficulties and in adjusting the relationships of the group's members. See Ralph Linton, *The Study of Man* (New York: D. Appleton Century Co., 1936), especially pages 273-79.

wholesome American boy. . . . With Herim it is altogether different. We don't know what to expect of him when he is here. He is very emotional and may burst out at any particular time—that makes for unpleasant relations. We can't act naturally when he is here. . . ."

On one occasion Herim disturbed a family ritual, group singing, and the Rogers bitterly resented it.[3] She related the incident, "When Mary and Clyde would come over for an evening, Bill would suggest that we all gather around the piano and sing some songs. He would get out his banjo and Mary would play the piano and all join in and sing as we used to do before the girls married. When Herim would come here he would take over the piano and play some 17th century French sonatas that no one would know enough about to be able to say whether it was well played or not or know what to do about it. He would monopolize the piano and never join in our group singing. . . ."

According to Mrs. Rogers, Herim raised the grandchild in not "quite an American way," "not exactly like ours." She illustrated, "For instance, we cannot understand, nor do we like the idea of Herim calling the boy 'Fritzie.' You know what that word means to us . . . that is the name of an enemy soldier and you know how we have been feeling about them for the last twenty-five years. I am not sure but according to his native country, Fritzie is a very nice pet name. We just don't like it."

Mr. Rogers was also bitter about his son-in-law's "peculiar philosophy." His wife said: "Bill gets riled up at Herim. He is always telling Bill that a wife should be on call for her husband's needs and desires. Nothing burns him up more than to hear this kind of talk. It is so much against what Bill believes in." (Case 58)

These differences among the Rogers appear to be small and petty. However, they add up to an impervious obstacle against continued harmonious relationships.[4] If Herim had been of the same cultural background as

[3] Family rituals are folkways or practices which often effectively control the family members' behavior and also assist in the development of family cohesion. By participating in family rituals members develop closer understandings with one another, and learn to share similar values and ideals. Often, as a consequence of this, parents and children generate in-group feelings, and further resent any efforts by outsiders to modify or interfere with these rituals. For an excellent analysis of the functions of rituals in family life see James H. S. Bossard and Eleanor Boll, *Ritual in Family Living* (Philadelphia: University of Pennsylvania Press, 1950).

[4] Harriet R. Mowrer suggests, "Cultural differences of this sort give rise to conflict largely because they symbolize obstacles to, or lack of, identification of the individuals concerned. . . . These conflicts are largely a matter of early prejudices and aversions. In this emotional realm conflict takes on a symbolic character. The individual responds not only to the immediate situation but to the whole complex of past experiences and associations which are revived." *Personality Adjustment and Domestic Discord* (New York: American Book Company, 1935).

the Rogers and had called his child by a "good American" nickname such as Skipper or Sandy, much of the present difficulty would have been avoided. Herim may in time learn these nuances of socially approved behavior and the Rogers may also gain understanding to expect the unexpected, but until then, friendship between the generational families will continue to be unstable.

Of the causal factors, socio-cultural background appears to be the most important in affecting family continuity, and proved to be as important to it as it is to marital happiness as demonstrated by the Terman, Wallin, Burgess and Cottrell researches. The chi-square test was used to relate similarity and dissimilarity to the family continuity scale. The results indicated that 118 of 121 families who had been given a rating of level five (high continuity) were those where marriage mates were of similar background. In the lowest level, two (poor continuity), only 2 out of 23 families involved mates of similar background. Only 18 of the 174 families which were rated within the range of high family continuity were those in which mates were of dissimilar background.

TABLE II. Family Continuity by Background

Level of Intergenerational Continuity	Socio-Cultural Background of Marriage Mates	
	Similar	Dissimilar
High 5	118	3
Fair 4	15	10
Average 3	21	5
Poor 2	2	21

Chi square = 83.65; 3 df; P less that .01.

Type of Courtship and Marriage Ceremony Factor. When traditional courtship and marriage practices were followed, intergenerational family continuity was increased.[5] In providing the church wedding and reception for their child, parents conformed to clique expectations, thus ensuring their position within it.

Conversely, when a child eloped, his family was denied the benefits of these social occasions. Frequently the parents became subjects of gossip which created embarrassment and unhappiness for them in their personal relationships, and in some instances reduced their prestige among friends.[6]

[5] For a further discussion of parental influences in the choice of marriage mates see Marvin B. Sussman, "Parental Participation in Mate Selection and Its Effect Upon Family Continuity," *Social Forces,* 32:76-81, October, 1953.

[6] A. B. Hollingshead indicates that parents attempt to limit their children's contacts to those within the same class and usually bring pressure upon them to drop friends

Parents indicated that when children had courtships over which they had some control they had the opportunity to become better acquainted with their future children-in-law. Then also, they often developed a pattern of cooperative activities with them, which usually carried on after marriage. Furthermore, they were able in this pre-nuptial period to instruct their children in their future marital roles.[7] By participation in the nuptial events their own emotional upset was lessened. . . .[8]

In order to check these conclusions derived from the case data, observation and non-observation of traditional courtship patterns were correlated with the various levels of the intergenerational family continuity scale. The results indicate that 115 of the 121 cases, where observation of such sequence occurred, were rated with the highest level of family continuity, and only seven of such cases were graded at the very lowest.

TABLE III.   Family Continuity by Courtship

Level of Intergenerational Continuity	Traditional Courtship Pattern	
	Observation	Non-Observation
High 5	115	6
Fair 4	17	8
Average 3	14	12
Poor 2	7	16

Chi square = 51.906; 3df; P less than .01.

Family Child-rearing Factor. Parents who raised their children developmentally, i.e., growing with them as they passed through successive age periods, and who encouraged them to become independent and self-reliant, but to continue affectionate ties with family members, had an easier time in adjusting to their child's marriage status than those who reared their progeny traditionally. Elders in using the latter method tended to keep a strict control over their children, and to retard their emancipation by creating a dependent relationship between them and parents. Upon the children's marriage they had difficulty in severing the emotional ties with them, and as a result were prone to interfere unduly in their married life. . . .

---

who differ in class position. He intimates parents may "lose face" within their class or clique if they don't act firmly in this matter. *Elmtown's Youth* (New York: John Wiley and Son, 1949), pp. 429-33.

[7] M. H. Kuhn suggests that daughters more than ever need preparation for marriage, especially training to be a homemaker and housekeeper, and parents may in part meet this need. H. Becker and R. Hill, eds., *Marriage and the Family* (Boston: D. C. Heath and Company, 1942).

[8] Emotional stress and readjustment during this period of leave-taking of children are concisely described by Ruth S. Cavan, *The American Family,* Chapter 20.

A statistical check was made of the influence of the rearing pattern upon the speed with which new behavior is learned after a child's marriage. To accomplish this the two rearing patterns were correlated with the pattern of parental suggestion after marriage. A breakdown of the cases in this sample indicated that in 110 of the 136 cases categorized as developmental, and in 18 of 59 traditional, parents did not give suggestions to the newly-weds unless they were asked. This indicates that where the developmental process of child-rearing is observed, parents are more likely to interfere less in the child's new family, thus enhancing family continuity.

TABLE IV.　Rearing Pattern by Parental Suggestion

Child-Rearing Pattern	Parents Unsolicited Suggestions to Children After Marriage	
	Yes	No
Developmental	26	110
Traditional	41	18

Chi Square = 47.68; P less than .01.

When developmental and traditional rearing patterns were correlated with the levels of the intergenerational family continuity scale the results indicate that 100 of the 121 cases where developmental practices were used were rated with the highest level of family continuity and only 8 [out of 23] such cases were graded at the very lowest.

TABLE V.　Continuity by Rearing Pattern

Level of Intergenerational Continuity	Child-Rearing Pattern	
	Developmental	Traditional
High 5	100	21
Fair 4	15	10
Average 3	12	14
Poor 2	8	15

Chi Square = 28.89; 3 df; P less than .01.

Help Factor. Intergenerational family continuity was also encouraged when parents and children maintained economic and other assistance between each other. Parents do best in maintaining harmonious relationships with their child's family, however, when they exercise self-restraint in their gift giving, and do not override the independence and self-respect of the younger couple. As might be expected, the flow of financial aid was from the parents to children. Parents wanted to help children, as they expressed it, "to get started" at the same class level as their own, and to use their wealth to accomplish this. Most parents did not permanently subsidize their children by providing them with a steady income. Rather, they assisted in such large purchases as an automobile, house, household furnishings and so forth. In turn for this aid, elders expected the affection of their children and grandchildren.[9]

Mr. Light illustrated the pattern of indirect giving used successfully by other parents in saying: "When we visit Babs and her family we take things for the grandchildren and foods which we know they can't afford. We try to do something to give them pleasure, something which we know they can't afford or won't go out to get by themselves. Here is what I mean: I once wrote a letter to Johnny asking about the cost of baby sitters. He replied with a two-page letter giving complete details perhaps thinking that I wanted to go into the business! I realized that they were having difficulty in getting out by themselves so I sent them $35, with each dollar bill in a separate envelope. I addressed it to the granddaughter with a note, 'You can use this when your mother wants to get a baby sitter.' I did not want them to turn down invitations. I wouldn't tell them this but had to give the gift in a certain way so that they would take it.

"Our life has changed in that we are trying to do things that will mean more in the long run to them. They are very proud; they don't want to be helped. It has to be done with a certain amount of subtlety." (Case 98)

Residential Factor. The location of children's residence is of itself a variant influence upon family continuity. Distance appears to be associated with other factors, such as approval of the marriage by parents and presence of grandchildren, rather than to have a single constant effect upon family relationships. When parents are satisfied with their child's choice of spouse, they desire his family to be located in the same or nearby community, generally within a radius of fifty miles, so that they can develop joint activities. On the other hand when children have married persons of whom their parents disapprove, family harmony is best maintained by residence not too close together. Probably, the location of the breadwinner's employment is a more important influence upon location of households than is the status of intergenerational family relationships. The importance of this factor needs further study.

The Coopers maintain close contact with their married daughter who lives fifty miles away. Ever since a grandchild arrived about a year ago their visits to her have increased. She said: "I'm surprised how Mr. Cooper, after a hard day's work, will say to me, 'Let's hop into the car and run up to see Bonnie.' I am really amazed because I never used to be able to get him to move out of his chair once he sat down for the evening." (Case 23)

[9] For a fuller explanation of this pattern see Marvin B. Sussman, "The Help Pattern in the Middle Class Family," *American Sociological Review.* 18: 22-28, February, 1953.

The Moodys do not approve of their son-in-law and are satisfied to have their daughter some distance from them. Mrs. Moody related: "It seems strange for a mother to say these things but after her marriage I was so heartbroken that I didn't want her near me. I never told her this. Her husband who is an engineer had an offer to go to Texas. We urged them to go. I think every mother wants her daughter nearby, but in this situation where we didn't accept her husband it was best for us that we didn't see them too often. They are now living in Texas and she will come back about once a year. It has proved very satisfactory though we do miss her and the grandchild." (Case 73)

These generalizations are given added support when the factor of residential nearness is correlated with the levels of the continuity scale. The results indicate an almost equal distribution of location of households on all levels of the scale. Of 122 families rated on level five (high continuity), in 72 both parents and children lived in the same community, and in 50 they were located at a distance from one another. On level two (low continuity), 14 families were residents of the same community and 9 were separated at some distance. The χ^2 of this distribution suggests that there is no functional relationship between the degree of continuity between families of two generations and the location of their respective households.

TABLE VI. Continuity by Residential Location

Level of Intergenerational Continuity	Location of Households	
	In Same or Neighboring Community	Distantly Separated
High 5	71	50
Fair 4	16	9
Average 3	11	15
Poor 2	14	9

Chi Square = 6.12; 3 df; P more than .01.

Analysis

In analyzing these factors we have assumed that they are of equal importance in affecting intergenerational relationships between families. Actually, in any given parent-child relationship, two or more of these factors form a configuration, that is, they are associated with each other in a patterned relationship and do not act singly. What these configurations are needs examination.

The first three factors which are considered, socio-cultural background, type of courtship, and child-rearing practices, may be said to be causal ones in that they determine the pattern of intergenerational family relationships. The other two, help pattern and residential location emerge at or soon after the marriage, and are more incidental than determinative of such relationships.

These first three causal factors are shown to be related when examined in terms of levels of family continuity in Table II.

This figure has been constructed by assigning plus (+) values to factors in each case when the child married a mate of similar background, observed the traditional courtship and was reared developmentally; and a minus (−) value was assigned for the converse of this. For any particular relationship involving families of two generations there may be as many as three plus (+) characteristics present or there may be none at all. In the figure it will be noted that in level five which represents the highest continuity, 91.8 per cent of the characteristics have been appraised as positive. The positive percentages decrease with each level and for the lowest continuity, level two, only 24.3 per cent of the characteristics are positive.

Since the differentiation by continuity level in the percentage of positive characteristics is so markedly graduated, we conclude that family continuity and these factors are closely related.

Conclusions

In this study our interest has centered upon parents and their roles in the families of their married children. We have assumed that if intergenerational family continuity is to be maintained they must assist constructively, if at all, in the establishment of their children's marriages. The children need their services and resources at least at the start of their mar-

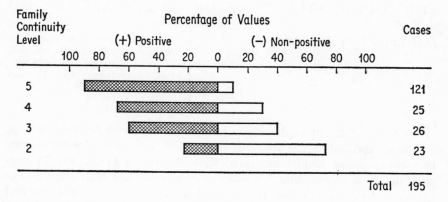

Figure 1. Percentage of positive factors (sociocultural similarities, traditional courtship, child-rearing pattern) by family continuity level.

riages. Most often, parents give freely of their help, expecting nothing in return but affection and appreciation. When the two sets of families establish mutually reinforcing relationships, stability is secured to both. . . .

In this study three factors emerge as determinative of intergenerational family behavior and these appear to operate in association with one another in any given parent-child relationship. Thus, family continuity between generational levels is greater when a child marries a mate of similar cultural background, meets parental expectations by having a traditional courtship and marriage and has been reared developmentally.

Endogamous choice of a marital partner and observing the norms regarding the manner of marriage are the two most important causal factors affecting family continuity.

A child reared developmentally is more likely to seek, and secure financial and other assistance from his parents than if he had been reared traditionally. Parents, in turn, while expecting some affectional response from children and the right to exercise the grandparent role, are less likely to interfere in the lives of the newly-weds if they have practiced developmental methods of child care.

Family continuity is increased when a pattern of mutual aid exists, in moderation, between the two sets of families.

Family continuity can best be continued if parents and their child's family, who have established harmonious relationships, can be located within a radius of fifty miles or so of each other. Conversely, when relationships are strained, intergenerational ties can best be preserved if parents and their married children's households are distantly separated.

THE SOCIAL STRUCTURE OF GRANDPARENTHOOD

Dorrian Apple Sweetser

Adapted from *American Anthropologist*, 58 (1956), 656-663. (Paper originally published under the authorship of Dorrian Apple.)

Two rather different explanations have been advanced for the warmth, closeness, and indulgence which is proverbially expected of grandparents in many societies besides our own. Radcliffe-Brown (1952:96-97) suggests that "friendly equality" between grandparents and grandchildren exists as a relieving reaction to the tension caused between parents and children by parental authority and by the obligations each has toward the other. This explanation, for whose adequacy he reports no test, relates friendly equality to a constant on the parent-child relation, namely, authority with accompanying tension. By this hypothesis, grandparents would be universally disposed to friendly equality with grandchildren. Radcliffe-Brown does not advance any ideas about other factors in family relations which might either hinder or be required for the manifestation of this predisposition. His is a psychological explanation, in that it focuses on the inner effects of experiences. Nadel (1951:234-236), on the other hand, suggests that informality between grandfather and grandson is associated with lack of family authority by the grandfather; as evidence he cites data from ten Nuba tribes on the presence or absence of a joking relationship between grandfather and grandson. This is a structural analysis, since it relates the rights and duties of one role to those of another.

This study will report a structural analysis which confirms and expands Nadel's hypothesis that friendly equality between grandparents and grandchildren appears only with certain patterns of authority in the family. Such a structural analysis does not rule out psychological analysis. It rather states conditions which permit psychological predispositions to manifest themselves in behavior with such clarity and regularity that they become generally expected.

Data used in this study are classifications of expected modal kinship behavior described in ethnological reports. In some societies, grandchildren have the same relationship with both paternal and maternal grandparents, though there is intersociety difference in whether or not friendly equality is present. Other societies show an intrasociety difference, in that relations of grandchildren with one set of grandparents are more friendly and equal than with the other set of grandparents. Since there are two kinds of variation in the grandparent-grandchild relationship, we need two hypotheses.

Both hypotheses are based on two general assumptions or postulates. One we will call the postulate of structural congruity in kinship behavior: that given a certain quality in the relationship of x and y, the possible qualities in the relationship of x and z and of y and z are thereby limited. This postulate underlies the ideas of Radcliffe-Brown and Nadel which were cited, and also underlies a number of other analyses of kinship behavior; see for example Homans and Schneider (1955:15-16) and Eggan (1955). The other postulate we will call the compatibility of authority and formality: that the more closely associated a person is to the status possessing the most jural authority in a hierarchy, the less likely he is to have an informal and equal relationship with subordinates in the hierarchy. This postulate has also been used in other analyses of kinship behavior; see for example the discussion of the mother's brother by Radcliffe-Brown (1952:15-31) and the analysis of unilateral cross-cousin marriage by Homans and Schneider (1955).

For societies in which relationships of grandchildren are the same with both sets of grandparents, we will test the following hypothesis, which we will hereafter refer to as Hypothesis I: when the grandparental generation continues to exercise considerable authority over the parental generation after the grandchildren are born, the relation of the grandchildren to the grandparents will not be one of friendly equality. The opposite is of course implied; when there is no such authority, there will be friendly equality. For societies in which grandchildren have a higher degree of friendly equality with one set of grandparents than with the other, we will test the following hypothesis, which we will hereafter refer to as Hypothesis II: if the grandchildren's relation with one set of grandparents has less friendly equality than with the other, the former grandparents will be those related to the grandchildren through the parent who possesses (or whose lineage member possesses) more household authority in the nuclear family. The phrase, "lineage member possesses," is put in to refer to the MoBro in avuncupotestal societies.

Information for testing the hypotheses was obtained from ethnological reports on 75 societies, 54 of which are among those used by Murdock (1949) and/or are in the Yale Human Relations Area Files. The sample is not representative of all societies, for there is no way to define the universe so that a representative sample could be drawn. Some areas, Africa for one, are over-represented, while others such as South America are under-represented. Lacking precise knowledge of the representativeness of the sample, we can generalize from it only on plausibility, not probability, to say that it is likely that the patterns found in this sample occur often enough to be important, though other patterns might be found in other samples which are not disclosed in this one. One support for the plausibility of generalization from the sample comes from a test of whether any major cultural area contributed significantly more or less than its share of successes, that is, cases which fit the hypotheses. A test of each area, using the formula for the standardized deviate,

$$z = \frac{|\, x - np \,| - .5}{\sqrt{npq}}$$

with x = number of successes in the area, n = number of cases in the area, and p = proportion of successes among all cases, showed that no area contributed significantly more or less than its share of successes.

Reliability of the categorizations of kinship behavior was tested by two independent raters on a separate random sample for each dichotomous variable. A rater would be given the definitions of both dichotomies of a variable and asked to sort excerpted information on kinship behavior into two stacks containing the same number of societies. All agreements between each rater's judgments and the writer's had a probability of occurring by chance which was less than .01. The formula for the "matching problem" (Mosteller and Bush 1954:I:307-311) was used to calculate the probabilities of agreement. All coefficients of association were over .90. As for validity of the data, we can only say that it has face validity, since there exist no independent measures for validating it.

Hypothesis I requires a distinction between societies in which grandparents continue to exercise a considerable amount of authority over the parental generation after the grandchildren are born and societies in which grandparents do not retain such authority. The raters for the reliability test were instructed that the former condition is defined thus: "relatively high degree of authority by the grandparental generation (fathers, fathers-in-law, or mother's brothers) over the parental generation (grown sons, sons-in-law, or nephews) is indicated by statements to the effect that the family head of an extended family possesses authority over the persons and property of married children, or that the head is the guardian or trustee of the extended family and receives respect and obedience, or that respect and obedience is expected from grown sons and daughters. It can also be inferred from the details of the kinds of decisions that a lineage or extended family head must make for other adults in the generation below." The latter condition was defined thus: "a lack of authority by the grandparental generation (fathers, fathers-in-law, or mother's brothers) over the parental generation (grown sons, sons-in-law, or nephews) is indicated by statements to the effect that children at adulthood take on the right to authority which their parental generation possessed, that children may live and work where and with whom they please, that the father (or mother's brother) controls the household where the nuclear family is the basic familial unit, that lineage authority is low, that parents may not interfere in their grown children's affairs, that seniority is little emphasized in determining status, and/or that parents do not control or command their children."

Hypothesis I also requires a distinction between societies in which friendly equality is present between grandparents and grandchildren and societies in which it is absent. The former was defined as: "there is friendly equality, permitted disrespect on the part of the grandchild, closeness and informality, or the grandchild is less respectful and reserved with the grandparent in contrast to the demeanor required toward parents." The latter was defined as: "the grandchild is submissive, reservedly respectful, not on easy and friendly terms with the grandparents owing to the discipline they exert, and/or must show a high degree of formal esteem to grandparents in comparison with the formal esteem shown to other relatives."

Table 1 presents a test of Hypothesis I. It will be seen that we can confidently reject the null hypothesis of no relationship between the dichotomous variables.

TABLE 1. Authority of Grandparents and Formality of Grandchildren

		Grandparental Authority over Parents		
		present		absent
Friendly Equality between Grandparents and Grandchildren	absent	Ashanti Tanala Moro Nama Koalib Hottentots Ibo Cow Creek Marki Verre Seminole Bachama Atsugewi Yaqui Jicarilla Apache		Shilluk Nyima Ramkokamekra Lakhers
	present	Kurama Korea Tallensi		Tarahumara Papago Saramaccaner Dakota Baiga Kiowa-Apache Bhuiya Cheyenne Burmese Fox Woleaians Kaska Fiji, Lau Islands Sanpoil Lesu Ojibwa Abelem Oto Mountain Arapesh Lozi Ontong Java Korongo Tikopia Dilling Chiricahua Apache Otoro Malabu Katab Kare-Kare Jukun Nyakyusa Azande

Chi-square corrected for continuity $= 20.17$; $p < .001$.

Kinship positions referred to in the ethnological reports are, for grandchildren, SoSo, SoDa, DaSo, and DaDa in 44 societies, SoSo, and SoDa for 6 societies, and DaSo and DaDa for one society; grandparents referred to are FaFa, FaMo, MoFa, and MoMo, for 43 societies, FaFa and MoFa in one society, FaFa for 5 societies, FaFa and FaMo in one society, and MoFa and MoMo for one society. It will be noted that for seven societies we have information on relations of grandchildren with only one grandparent or set of grandparents; these cases could not be used in testing Hypothesis II but can be used in testing Hypothesis I.

For Hypothesis II we need data on which parent, or parent's lineage member, has more household authority, and on which set of grandparents it is with whom the grandchildren have the less friendly and equal relationship. Possession of more household authority was defined as: " 'household authority' means the right to make decisions about work and about the upbringing of children. 'Father's side' means father or father's patrilineal lineage member such as father's father or father's older brother. 'Mother's side' means mother, mother's matrilineal lineage member such as mother's brother or mother's mother's brother, or mother's father (where residence is matrilocal and authority patripotestal instead of avuncupotestal)." A less friendly and equal relationship was defined as: "(1) behavior toward one grandparent can be more familiar, while the other is more strict and elicits reservedly respectful behavior; (2) both are relatively close and familiar, but one is said to be closer or more indulgent than the other; (3) both are relatively close and familiar, but joking is permitted with one and not with the other, or less restrained joking is permitted with one than with the other."

In 19 societies the discrimination in formality fell between the maternal and paternal grandparents. In 5 societies only one member of a set of grandparents was distinguished by more formality. In 2 of the 5 societies, the discrimination fell between the FaFa, FaMo, and MoFa on the one hand and MoMo on the other. In one society relations with FaFa were more formal than with FaMo, MoFa, and MoMo; in one society MoMoBro was distinguished from MoMo, MoFa, FaFa, and FaMo; and in one society MoMoBro and MoMo received more formal behavior than MoFa, FaFa, and FaMo. In 5 societies information was available on grandsons only; for the others the behavior of both grandsons and granddaughters was described. Where information was available about him, MoMoBro was defined as a member of the set of maternal grandparents because in avuncupotestal societies the MoMoBro is the functional equivalent of the FaFa in patripotestal societies, in that he is the man who has, or had, jural authority over the man who has jural authority over the grandchild.*

Table 2 presents the test of Hypothesis II. We can confidently reject the null hypothesis of no relationship between the dichotomous variables.[1]

---

* FaFa refers to father's father, FaMo to father's mother, etc. The term "avuncupotestal," etymologically, refers to the power of the uncle. In this context, it can be seen that the reference is to mother's brother. "Patripotestal," of course, refers to power of the father—*Eds.*

[1] One problem in cross-cultural studies such as this is that inclusion of societies which are too alike may falsely inflate the numbers in the cells of the tables. Precaution against such inflation was taken while the sample was being collected by requiring that a society have had independent existence for about six generations and self-definition as a separate people. For a more stringent test, however, we can make

TABLE 2.   Parental Authority and Formality of Grandchildren

		Less Friendly Equality with:		
		paternal grandparents		maternal grandparents
Household Authority on:	father's side	Lemba Venda Thonga Dahomey Nankanese Bena Swazi Lenge Comanche Zulus Tswana Hehe Hera Tonga	Kurtatchi Murngin Eastern Cherokee	Western Apache
	mother's side			Haida Hopi Navaho Truk Marshallese Manus

Fisher's exact test, p = .00005.

While the data for these hypotheses were being collected, data were also tabulated on the occurrence of sexual joking between grandparents and grandchildren. In many societies, some relatives who may marry or who may have sexual relations are expected to joke about sex with each other. In 34 societies in our sample, sexual joking is present between some relatives. In the 18 societies in which there is sexual joking between grandparents and grandchildren, there are 7 instances in which people may marry those who stand in a kind of grandparental relationship to them, either because some persons who are potential spouses are called by the kinship term for grandparents, or because a man can inherit his

---

further eliminations of societies with some influence on each other. For Hypothesis I, we will eliminate Koalib and leave Moro, eliminate Bachama and leave Marki Verre, eliminate Abelem and leave Arapesh, eliminate Dakota and leave Cheyenne, eliminate Dilling and Otoro and leave Korongo, and eliminate Malabu and Kare-Kare and leave the Katab. For the remaining cases, Chi-square corrected for continuity is 14.107 and p <.001. For Hypothesis II, we will eliminate Nankanese and leave Dahomey, eliminate Swazi, Lemba, Tswana, Lenge, and Venda and leave Thonga. Fisher's exact test on the remaining cases shows that the null hypothesis can be rejected at the .004 level. The findings are evidently not the result of inflating the numbers with over-similar societies.

paternal grandfather's widows. In the 16 societies in which sexual joking between grandparents and grandchildren is not present, there are only two instances of possible marriage with a kind of grandparent. This association between sexual joking and marriage is not strong enough to be statistically significant. The point of interest is that the trend in the data is the same as that found by Brant (1948: 160-162), who tested with cross-cultural data the association between sexual joking and marriage in the relationships of a male and WiSi, BroWi, MoBroDa, and FaSiDa. Brant found that the rare cases are those in which a sexual relation is possible but sexual joking is absent, but his data show, as do ours, that the possibility of a sexual relationship is not a sufficient cause of the sexual joking.

One further elaboration is possible for the two main hypotheses of this study. They can be united into a single and more abstract hypothesis: formality between grandparents and grandchildren is related to association of grandparents with family authority. The phrase, "association of grandparents with family authority," covers both those grandparents who exercise family authority (see Hypothesis I) and those grandparents who are related to the parent associated with authority in the nuclear family (see Hypothesis II).

This more abstract formulation of the findings of Hypotheses I and II suggests the question of whether the amount of authority of parents over children has any effect. Some evidence on this question can be had from the ratings of Bacon and Barry on several aspects of the obedience required of children in various societies.[2] They have rated 28 of the societies in this study's sample on a 14-point scale. A mean obedience score for each society was obtained from the separate ratings for boys and for girls on reward for obedience, punishment for disobedience, and frequency of demands for obedience. These scores were then dichotomized at their median. Table 3 presents a test of whether the relationship between the association of grandparents with family authority and formality between grandparents and grandchildren holds within the categories of high and low parental authority as measured by obedience scores. Since the relationship does hold within both categories of parental authority, we can conclude that the amount of parental authority over children has no effect on our findings.

In summary, this study has confirmed and expanded Nadel's idea that friendly equality between grandparents and grandchildren appears only when family social structure permits it. It may be true, as Radcliffe-Brown suggests, that tensions between parent and child tend to draw

[2] I am indebted to Dr. Margaret Bacon, Mr. Herbert Barry, and Professor Irvin L. Child of Yale University for their kind permission to use these unpublished ratings.

TABLE 3. Relationship between Authority of Grandparents and Formality with Grandchildren, Within Categories of Parental Authority

		Association of Grandparents with Family Authority: Obedience Score above the Median	
		present	absent
Formality between grandparents	present	7	
and grandchildren	absent	1	6
	Fisher's exact test, p = .002		
		Obedience Score below the Median	
		present	absent
Formality between grandparents	present	6	1
and grandchildren	absent	1	6
	Fisher's exact test, p = .02		

grandparent and grandchild together, but such closeness is no universal in kinship behavior. In essence, the hypotheses of this study show that formality between grandparents and grandchildren is related to association of grandparents with family authority, while the indulgent, close, and warm relationship is fostered by dissociation of grandparents from family authority.

REFERENCES

Brant, C. S., "On joking relationships," American Anthropologist (1948) 50:160-162.

Eggan, F. (ed.), Social anthropology of North American tribes, enlarged edition (Chicago: The University of Chicago Press, 1955).

Homans, G. C., and D. M. Schneider, Marriage, authority, and final causes (Glencoe, Illinois: The Free Press, 1955).

Mosteller, F., and R. R. Bush, Selected quantitative techniques. The handbook of social psychology, ed. by G. Lindzey (Cambridge, Mass.: Addison-Wesley, 1954).

Murdock, G. P., Social structure (New York: Macmillan, 1949).

Nadel, S. F., The foundations of social anthropology (Glencoe, Illinois: The Free Press, 1951).

Radcliffe-Brown, A. R., Structure and function in primitive society (London: Cohen and West, 1952).

14. AGED PARENTS AND THEIR OFFSPRING

In this chapter the two generations are some twenty-five years older, that is, the older generation is now around seventy-five. According to American culture, parents and their adult offspring can hardly expect to live together amicably. One reason for this, according to Cavan, is that the adult roles of father and son and of mother and daughter tend to be competitive rather than complementary. She goes on to remark that social change and cultural divergence are other potential sources of conflict.[1] That this may be a class-linked view, however, is suggested by Willmott and Young, who note the solidarity of English working-class women with their families of orientation, especially with their mothers, and who speak of extended kin as the "woman's trade union."

It is the increasing dependency of older men and women as they age and become more feeble that creates pressure for close contact and perhaps a common household with their children. Schorr presents data on opinion polls in this country and on actual performance with respect to the amount and kinds of support children give their aged parents, and reports that "in living together families prefer to organize themselves around a mother-daughter tie." He also cites an English study which reveals among the aged a disposition to compromise the conflict between the neolocal custom and the desire for contact whereby the older generation wishes to live separately from but near their children and to see them frequently.

[1] It will be recalled that Kingsley Davis made the same general point concerning relationships between adolescents and their parents.

SOME STATISTICS CONCERNING OUR AGED POPULA-TION

Adapted from 87th Congress,
Special Committee on Aging,
"New Population Facts on Older
Americans, 1960," Washington,
D.C., United States Government
Printing Office, 1961, pp. 1-4, 7-
12 and 20-23.

A. NATIONAL TRENDS

In the 10 years since 1950, the aged population of the United States (persons aged 65 and over) increased by nearly 35 percent, to 16,559,580. This rate of growth was exceeded only in the case of children aged 5 to 14. In contrast, the total population increased by only 18.5 percent. In 1950, the aged constituted 8.1 percent of the total U.S. population but by 1960 they made up 9.2 percent of the total. The growth in this segment of our population is dramatized by the fact that in 1920 the aged as a proportion of the total U.S. population was only 4.7 percent. Thus, while only 1 out of every 20 Americans was 65 or older in 1920, the ratio has increased to 1 in every 11 in 1960.

The gradual but definite increase in average life expectancy taking place over the past several decades is reflected in the detailed composition of the aged population, as indicated in Table I. The 65-plus population is not one homogeneous aggregate of persons of the same age. Within this 65-and-over population, furthermore, the very oldest group, those aged 85 and over, has increased in number since 1950 by more than 60 percent, and now number (as of 1960) more than 929,000.

TABLE I. Population 65 and Older, 1950 and 1960

	1950	1960	Percent increase, 1950 to 1960
Total aged population	12,294,698	16,559,580	34.7
65 to 69 years	5,013,490	6,257,910	24.8
70 to 74 years	3,419,208	4,738,932	38.6
75 to 79 years	} 3,284,061	{ 3,053,559 }	41.1
80 to 84 years		{ 1,579,927 }	
85 years and over	577,939	929,252	60.8

Table II below indicates the dramatic increase in the number of aged persons since 1920. The 236-percent increase in the number of persons 65 years and older and the 279-percent increase of those 75 years and older in the 40 years between 1920 and 1960 is extraordinary. The 920-percent increase in those 85 years and older is phenomenal.

TABLE II. **Number and Increase of Persons 65, 75, and 85 Years of Age, 1920-60**

	Number		Increase, 1960 over 1920	
	1920	1960	Number	Percent
65 and older	4,929,000	16,559,580	11,630,580	236
75 and older	1,469,704	5,562,733	4,093,029	279
85 and older	91,085	929,252	838,167	920

In 1920, 30 percent of the aged were 75 years and older; in 1960, 34 percent were 75 years and older. Those 85 years and older in 1920 constituted 1.8 percent of the aged population; in 1960, 5.6 percent were 85 years or older.

Life expectancy for persons born in 1920 was 54.1 years. Those born in 1959 (the latest year for which estimates are available) may expect to live 69.7 years, on the average . . .

Future Trends

Population projections made by the Department of Health, Education, and Welfare indicate that in the 40 years between 1960 and 2000 the aged population will more than double in number, bringing the total to over 30 million people. This would be an increase of more than 25 million aged persons since 1920. From the first census in 1790 when the total population was about 4 million, it was just about 70 years (in 1860) before the total U.S. population numbered over 30 million people. This projection to the year 2000 of 30 million persons aged 65 or more, it should be emphasized, does not take into account any of the effects of modern medical research in the chronic diseases characterizing an aging population, such as cancer and atherosclerosis.[1]

[1] A report from Wayne State University ("Inside Wayne," Apr. 26, 1961) notes a breakthrough in the use of artificial plastic arteries to prevent crippling caused by diseases of the blood vessels. There has been dramatic progress in the last 7 years in replacing arteries, which had become plugged as their lining thickened from atherosclerosis, the cause of most heart disease and strokes. Dr. Herbert S. Robb, of the Wayne College of Medicine, has said: "If we could cure atherosclerosis, the ordinary lifespan, except for cancer, might be 120 to 130 years."

Women Outlive Men

The fact that women are outliving men is reconfirmed by the 1960 census. There are nearly 1.6 million more women than men aged 65 and over. Stated another way, of all persons 65 and over, 55 percent are women. More strikingly, women make up 61 percent of the 85-year-and-older population.

TABLE III.   **Percent Increase in Population 65 and over by Sex, 1950-60**

Age	Percent Increase, 1950 to 1960		
	Total	Men	Women
65 and over	34.7	29.1	39.7
65 to 69 years	24.8	20.6	28.8
70 to 74 years	38.6	33.8	43.0
75 to 84 years	41.1	34.0	47.1
85 and over	60.8	52.6	66.5

The ratio of women to men is 121 women for every 100 men in the 65-year-and-older group. The ratio increases with each 5-year age group, from 114 women for every 100 men in the 65-to-69-year age group to 157 women per 100 men in the 85-year-and-older group. Women as a percent of the total aged population increases from 53.2 percent up to 61 percent for the 85-year-and-older group . . .

B. MARITAL STATUS OF THE AGED, 1950-60

The ramifications of the larger proportion of women in the aged population are far greater than statistics alone might suggest. For example, the marital status of the aged becomes significant in terms of both social and economic consequences.

Aged Men. In 1950, approximately two-thirds of all men aged 65 and over were married. Slightly less than one-fourth were widowed. Ten years later, more than seven-tenths were married, and slightly less than one-fifth (19 percent) were widowed.

The Census Bureau now provides more detailed information than previously available, and the data indicate that among men aged 65 to 74,

TABLE IV.   **Marital Status of Aged by Sex, 1960, by Percent**

	65 to 74 Years		75 Years and Older	
	Male	Female	Male	Female
Married	78.9	45.6	59.1	21.8
Widowed	12.7	44.4	31.6	68.3
Divorced	1.7	1.7	1.5	1.2
Single	6.7	8.4	7.8	8.6

nearly 8 out of every 10 were married, and less than 1 out of 8 were widowers. On the other hand, for men 75 and older, less than 6 out of 10 were married, and nearly a third were widowers . . .

Aged Women. In 1950, slightly more than one-third of aged women were married. More than one-half were widowed. By the end of 10 years, in 1960, the proportion married remained relatively unchanged, 37 percent, and there was a slight decrease in the proportion who were widows, to 53 percent. Among aged women in the 65-to-74 age group, about 46 percent were married, and 44 percent were widows. Among those 75 and older, only about one-fifth were married, and nearly 7 out of every 10 were widows . . .

Greater Number of Unmarried Aged Women. The United States is witnessing a decreasing number of all aged men per 100 aged women— starting with 101 men per 100 women in 1920 and declining to 83 per 100 in 1960. At the same time, the numbers of widowed, divorced, and single women over the age of 65 have increased sharply in the past 10 years, from 4.1 million in 1950 to nearly 5.7 million in 1960. This is an increase of nearly 40 percent.

On the other hand, the number of widowed, divorced, and single aged men has grown from 1.96 million to only 2.06 million in the same period of time. This is hardly any increase at all, in sharp contrast to the 40-percent increase in the number of aged women who are widowed, divorced, or single. Furthermore, the difference in absolute numbers is significant.

The crucial point is that nearly one-half of all the aged in this country, about 8 million persons, are widowed, single, or divorced. This fact has important implications for problems of public policy, family responsibility, and income maintenance, etc., especially in the case of aged women, most of whom have not been wage earners in their productive years, and therefore, as a case in point, with less favorable protection under our system of social security and private pension benefits.

TABLE V. Marital Status of All Aged, by Sex, 1960, by Percent

Marital status	Percent
Total	100.0
Married	53.1
Male	32.7
Female	20.4
Widowed, divorced, or single	46.9
Male	12.4
Female	34.5

Source: Census Bureau, "Current Population Reports, Population Characteristics," March 1960, Series P-20, No. 105.

As table V indicates, widowed, divorced, or single women constitute the largest group of aged in the United States: nearly 35 percent of all the aged are in this category. Only one-fifth of all the aged are married women. One-third of the aged are married men, and the remainder, about one-eighth, are widowed, divorced, or single men.

Household Characteristics of the Aged

From the preceding sections on the proportion of men and women in the aged population and their marital status, it is obvious that household and family characteristics of the older population should significantly reflect the basic statistics.

In March 1960, there were 4.8 million husband-wife families with the head of the family aged 65 or older. The average family size for this group was 2.47 persons, as compared to 3.77 for all husband-wife families . . .

Nearly three-fourths of the aged are living in "primary" families— that is, families that include among their members the head of a household. Slightly more than one-fifth are unrelated individuals, numbering about 3.6 million individuals. Nearly 40 percent of these aged unrelated persons are 75 years old and over. Seventy percent are women. "Unrelated individuals," by Bureau of the Census definition, are persons (other than inmates of institutions) who are not living with any relatives.

The remaining proportion, of 4 to 5 percent, are mostly in institutions such as homes for the aged, mental hospitals, and other group quarters, as well as secondary families (families not including among the members the head of a household, such as lodgers) and subfamilies (married couples living in a household and related to, but not including the head of the household).

TABLE VI.   Family Status of Aged Population, Males and Females, March 1960

		Both Sexes		Male		Female	
		65 to 74	75+	65 to 74	75+	65 to 74	75+
Total	thousands	10,307	5,334	4,778	2,280	5,529	3,054
In families	do	7,943	3,681	4,074	1,738	3,869	1,943
Unrelated individuals	do	2,188	1,444	613	458	1,575	986
Inmates of institutions	do	176	209	91	84	85	125
Percent		100.0	100.0	100.0	100.0	100.0	100.0
In families		77.1	69.1	85.3	76.2	70.0	63.6
Unrelated individuals		21.2	27.1	12.8	20.1	28.5	32.3
Inmates of institutions		1.7	3.9	1.9	3.7	1.5	4.1

C. THE AGED POPULATION: AN INCREASING PERCENTAGE OF THE TOTAL ADULT POPULATION

The aged as a percent of the population 21 years and older is now 15.4 percent, as compared to 12.6 percent 10 years ago. These figures indicate the greater actual or potential weight the aged have in the total voting population. The figures do not exactly represent the voting age population since they include aliens and others not eligible to vote, and exclude younger persons permitted to vote in Alaska, Georgia, Hawaii, and Kentucky.

The aged have special problems and special demands flowing from their social, economic, and physical conditions. The chances of effective action on these special problems may well be affected by the extent and nature of their representation in the total electorate.

TABLE VII. Aged as a Portion of the Population 21 Years and Older [In thousands]

Year	Number of persons 65 years and older	Number of persons 21 years and older	Aged as a percent of population 21 years and older
1920	4,933	60,887	8.1
1930	6,634	72,944	9.1
1940	9,019	83,997	10.7
1950	12,295	97,403	12.6
1960	16,560	107,377	15.4

The aged population since 1920 has increased 236 percent while the total adult population has increased by only 76 percent in the same period of time.

More than two-fifths of the increase between 1950 to 1960 in the total adult population is attributable to the increase in the aged population

TABLE VIII. Percent of Total Adult Population Growth Attributable to Aged Population and Percent Increase in Aged and Adult Population, 1920-60 (Percentages rounded off.)

	Decennial increase in aged population as percent of total adult population decennial increase	Percent increase in aged population	Percent increase in total adult population
1920 to 1930	14	35	20
1930 to 1940	22	36	15
1940 to 1950	24	36	16
1950 to 1960	43	35	10
1920 to 1960	25	236	76

alone. This decade witnessed a growth of slightly less than 10 million adult persons, with nearly 4.3 million of them in the 65-and-older age group.

D. THE AGED IN THE STATES

The national data provide a composite picture of diverse trends in the individual regions, States, and localities. Every State experienced in the last 10 years an increase in the population 65 years and over, but this increase ranged from low values of 10.6, 13.6, 13.9, and 17.2 percent in Vermont, Alaska, Maine, and New Hampshire, respectively, to high values of 55.1, 65.4, 103.9, and 132.9 in New Mexico, Nevada, Arizona, and Florida, respectively. The proportion of increase in the former four states was below the national average of 34.7 percent. . . .

When changes are presented in percentages, the absolute numbers involved is not evident. California has had an increase of more than 480,-000 aged persons since 1950, and now has a total of more than 1,376,-000. New York witnessed an increase of nearly 430,000, and now has the largest population of aged persons, 1,687,590, as of 1960. . . .

E. THE AGED IN THE RURAL POPULATION

Of the total population 65 years and older, more than 5.3 million live in rural areas, either on farms or in small towns with less than 2,500 population. In other words, one-third of the aged live in rural areas, 8 percent on farms, and nearly 25 percent in nonfarm areas. Ten years ago, the total number of the aged in the rural population was only 4.4 million, with 36 percent of the aged living in rural areas. Therefore, while the number of aged in the rural population has increased by nearly 1 million, the percent of the aged in the rural population has declined slightly.[2] . . .

The important point to be recalled is that over 5 million people aged 65 and over live in rural areas—33 percent of the aged population. This is not an insignificant statistic since an overall shortage of medical facilities and personnel and other community programs is most frequently present in the rural areas of the Nation.

The ratio of aged women per 100 aged men in rural areas is less than

[2] . . . in 1960 a new definition of farm population was adopted as a result of an increasing number of families living in the open country who are not obtaining their livelihood from agriculture. The new definition applies a more restrictive criterion in classifying places as farms. The aged make up 33 percent of the total rural population in terms of both the 1950 and 1960 definitions. Under the 1950 definitions, however, 11 percent of the aged live on farms and 22 percent in nonfarm areas, as of 1960. . . .

TABLE IX.   1960 Rural Population by Age and Aged in Rural Population, by Percent, 1960 and 1950

Age and Year	Farm	Nonfarm	Total Rural
1960 definition:			
Total, 1960	100.0	100.0	100.0
Under 14 years	31.9	34.0	33.5
14 to 64 years	59.6	58.5	58.8
65 years and over	8.5	7.5	7.7
1950 definition:			
65 years and over, 1950	7.5	8.6	8.2
65 years and over, 1960	9.1	7.1	7.6

for the total population. Nationally, there are 121 aged women for every 100 aged men; in rural areas, there are 108 aged women for every 100 aged men. In the farm areas, there are only 84 aged women for every 100 aged men. This is the only major segment of the aged population in which men outnumber women. . . .

F. THE AGED IN THE LABOR FORCE

The U.S. Department of Labor counts 3.3 million persons 65 years and over as part of the total labor force. This means that one-fifth of the aged are in the labor force, compared with one-third in 1920 and slightly more than one-fourth in 1950. The percentage decline of the aged in the labor force is a long-term development as seen below.

TABLE X.   Percentage of Aged 65 and over in the Labor Force, Male and Female, 1890-1960

Year	Total	Male	Female	Year	Total	Male	Female
1890	39.9	70.0	8.5	1940	22.4	43.3	6.7
1900	37.4	64.9	9.3	1950	26.3	45.0	9.5
1920	32.8	57.1	8.2	1960	20.0	33.6	11.2
1930	31.4	55.5	8.2				

These trends reveal that the labor force participation rate among aged men has been reduced by more than half and that the rate among aged women increased somewhat since 1890.

The 1940-50 increase reflects the general increase in labor demand during the war period.

The rate of unemployment among the aged in the labor force has increased from 4.8 to 5.4 percent since 1951.

While the number of aged has increased 34.7 percent and their labor force participation rate has decreased 6.3 percent in 10 years, the absolute number of aged in the labor force has increased 9.4 percent in

10 years. Employment in agricultural industries has actually declined by 11 percent, however.

Of the persons 65 years and over who are employed, more than a third work part time (less than 35 hours per week). Slightly more than one-fourth work more than 40 hours per week. . . .

TABLE XI.   Full- and Part-Time Employment of Persons 65 Years and Older, 1960

Hours worked per week:	Percent
Total employed	100.0
1 to 34 hours	34.6
35 to 40 hours	38.1
41 hours or more	27.3

G. SUMMARY OF POPULATION FACTS ON THE AGED

1. In 1900, 1 person out of 25 was 65 years and older; today 1 out of every 11 is 65 years and older.

2. In 10 years, 1950-60, the aged population increased nearly 35 percent. The general population increased only 19 percent.

3. In contrast to the 65-plus population, the population aged 45-64 has only grown at about the same rate as the total population, 17.4 percent, between 1950-60.

4. In 1920, 1.8 percent of the aged were 85 years and older. In 1960, 5.6 percent of the aged were 85 years and older. Today the number of persons 85 years and older is 920 percent greater than it was in 1920.

5. In 1920, 8.1 percent of the voting age population was 65 years and older. In 1960, 15.4 per cent of the voting age population was 65 years and older.

6. Nationally, there are 121 aged women for every 100 aged men. On farms, however, there are only 84 aged women for every 100 aged men.

7. Nearly one-half of the aged are widowed, single, or divorced. Seven out of every ten aged men are married, while only one out of every three aged women is married.

8. Between 1930 and 1940, an average of 758,000 persons were reaching the aged status annually. Between 1950 and 1959, an average of 1,198,000 persons a year reached their 65th birthday.

9. The net increase each year in the past decade has been more than 426,000 on the average, that is, each year the population 65 and older increased by this amount. The average annual net increase for the 75 and older population alone has been more than 170,000 during the past decade.

10. The aged population is expected to more than double in the 40 years between 1960 and 2000.

11. One-fourth of the aged live in rural nonfarm areas. Only 8 percent of the aged live on farms.

12. One-fifth of the aged are in the labor force. In 1920, one-third of the aged were in the labor force.

FAMILY TENSIONS BETWEEN THE OLD AND THE MIDDLE-AGED

Ruth Shonle Cavan

Adapted from *Marriage and Family Living,* 18 (1956), 323-327.

Over the past thirty years special needs of old people have gradually been recognized and met. Financial needs, better health care, suitable recreation, adequate housing, and retirement jobs have all been at least partially provided.

Primary Group Needs. Perhaps the most fundamental need of old people, however, is for the intimate and affectionate contacts in a primary group. This need is either not recognized or is passed over lightly in discussions of old age.

At every age level except old age, the need for intimate contacts is recognized—in fact, stressed. The significance of marriage is said to lie in its close companionship; the child's need for loving care by the parents has never been more strongly emphasized. But suddenly, in old age, men and women long accustomed to primary group life are assumed no longer to need it.

Actually, old people are in great need of belonging to some intimate group. Many old people have lost the mate who supplied affection. Approximately one third of all men and two thirds of all women aged sixty-five and over are without wife or husband, the great majority being widowers or widows.[1] The decades add years and deaths, until among the small number who remain at age eighty-five and over, about 60 per cent of the men and 85 per cent of the women are widowed.[2] These stranded old people are the olds who speak of loneliness, of no one caring, of neglect by children, and of the uselessness of living.

[1] *Statistical Abstract of the United States, 1954* (Washington: Government Printing Office, 1954), p. 263.

[2] Ruth Shonle Cavan, *The American Family* (Thomas Y. Crowell Co., 1953), p. 590.

In addition to these solitary individuals are old couples. The need for love is not so apparent as long as an old couple is able to remain physically self-sufficient. But the old couple in ill health or with failing strength feels much the same need that a child feels for the sympathetic and protective care of someone who is attached to him by affection and loyalty.

The family is the group in our society designated to supply love. The living arrangements of old people give some indication of the accessibility of the family as a source of affection. Among men aged seventy-five and over, out of every hundred, forty-one live with the wife, thirty-four live with other relatives, chiefly children, in either their own or the relatives' home, and twenty-five live alone or in institutions, hotels, lodging houses, and the like.[3] Of every hundred women aged seventy-five and over, seventeen live with the husband and fifty-one with other relatives, primarily children; thirty-two live alone, in institutions, hotels, and the like.

Mutual Rejection. When old people live with children or other relatives, the relationships are not always happy, nor are they always cordial between adult children and old parents who live separately from each other. Respectable middle-aged people have been heard to make such remarks as these about their parents: "You ought to take every person over sixty-five out and shoot him," or, rhetorically, "Why don't the old fools die," or more generally, "What we need is more homes for the old." Social workers whose clientele includes old people sometimes share this rejecting attitude. In one published article, perhaps somewhat extreme, a social worker speaks of people "sacrificing" their lives for their parents and becoming "bitter, frustrated, warped" individuals.[4] The same article concedes that the child owes his parents a definite responsibility but not so "overwhelming that it stifles his own life" or "jeopardizes his own family relationships"—meaning relationships to spouse and children, but excluding parents from the concept of the family.

At the same time that middle-aged children reject their parents, they also feel a sense of duty toward them and of guilt if they are not cared for or if they are placed in an institution instead of being cared for in the family.

Old people also vary in their feelings. They want independence and resent interference from their children. At the same they complain of neglect and yearn for the love and affection of their children.

Social Change and Conflict. Some writers have assumed that conflict

[3] Based on *United States Census of Population, Special Report PE No. 2D* (Washington: Government Printing Office, 1953), pp. 21-22.

[4] Margaret B. Ryder, "Case Work with the Aged Parent and His Adult Children," *The Family,* 26: 243-50, November, 1945.

between old and middle-aged is inevitable. It seems, however, that much of the conflict is due to contemporary social changes and types of family organization resulting from these changes. If it is true that conflict grows out of temporary social changes, then, as these situations become stabilized or pass into social history, it seems probable that the present strained relations between old parents and middle-aged children may also change. This paper will examine some of the probable social causes for the conflict between old and middle-aged; personality conflict per se are not included in the discussion.

The basis for many seeming personality conflicts is, in reality, some form of cultural conflict, of which there are three types, all based on cultural transition. These three types of transition are from rural to urban, from foreign culture to American, and from lower to higher class status. Each of these transitions is typically made by one small family unit, usually the younger generation moving to the new position, while the parent generation remains in the original cultural position.

1. Rural to urban transition. Young people who move from farm to city readily adapt themselves to urban values and modes of behavior. Present-day middle-aged people who were born and reared on farms usually are thoroughly urbanized. Their parents who remained on their farms tend to be rural in outlook on life, chiefly because their attitudes were firmly set before present means of mass communication and easy transportation had developed. The clash of rural-urban values between the two generations was clearly demonstrated by Dinkel in his study of fifty Minnesota families, reported in 1943.[5] The conflict was increased by the authoritarianism of the old parents who felt that their position entitled them to insist upon compliance with their values, whereas their children felt entitled to follow their own more urbanized values.

2. Foreign to American culture. The great influx of foreign-born people between 1890 and 1914 is now part of the social history of the United States. The young foreigners who came then are old people now. For example, in one small industiral city which drew many immigrants, when 16 per cent of the total population was foreign-born, 45 per cent of persons aged sixty-five and over were foreign-born.[6]

The old age group among the foreign-born is often still loyal to many of the precepts and customs of the foreign culture, whereas their middle-aged children have pulled away from their parents and tend to identify themselves with their more Americanized children. Tensions between old

[5] Robert M. Dinkel, "Parent-Child Conflict in Minnesota Families," *American Sociological Review,* 8: 412-19, 1943.

[6] Ruth Shonle Cavan, "Old Age in a City of 100,000," *Illinois Academy of Science Transactions,* 40: 156-70, 1947. Figures are for 1940 and not available for later dates.

parents and their children tend to originate in cultural conflicts. An interesting experiment in family-life education for a group of aged Jews in St. Louis revealed that the subject of most interest was the old people's feeling of rejection by their children.[7] The leaders of the group concluded that in addition to the usual adjustments of the old, this group had also to adjust to the cultural alienation from their Americanized children with its attendant feeling of isolation and rejection.

3. Upward social mobility. In the United States upward mobility is taken for granted. Sons enter occupations requiring more education and commanding higher salaries than the occupations of their fathers; for their sons they plan still more education and a further push upward. In the upward climb, it is not sufficient that the climber should affiliate himself with the class level above him, assimilating their culture; he must also break his identification with the class level left behind, most often represented by his parents. In an exploratory study, LeMasters tested the hypothesis that social class differences are a factor in creating tension between old people and their upwardly mobile children. His cases tended to show that rural-urban differences and differences in educational and social class level were all related to difficulties in joint living of parents and their married children. When these differences were not present, "parents seemed to live quite harmoniously with their married children, barring such personal factors as personality problems" (not covered by the study).[8]

Upward social mobility often is closely related to one of the other two types of cultural transition—rural to urban or foreign to American culture. The chasm between old parents and middle-aged children is wider and deeper when two, rather than one, types of cultural transition are involved.

The Effect of Small Family Units. Adjustment to the cultural conflicts just discussed and to the geographic mobility accompanying cultural change has included the breaking of the family into generational units, with each unit consisting of the parent generation and their dependent children. As the children mature, they in turn establish independent households and detach themselves from their parents. When we say "family," we customarily refer to husband, wife, and dependent children; rarely do we include grandparents in the concept of the family. This limited concept of the family has many implications.

1. Each small family unit becomes autonomous. It not only lives in its own domicile but jealously guards its privacy and its right to determine

---

[7] Sidney Hurwitz and Jacob C. Guthartz, "Family Life Education with the Aged," *Social Casework,* 33: 382-87, 1952.

[8] E. E. LeMasters, unpublished study; quotation from personal letter.

its own family pattern of living. The autonomy of the generations is established when the younger couple are about twenty-five years of age and the parental couple about fifty. The question of joint living—or even close interdependence—does not arise until some twenty-five years later when the couples are respectively fifty and seventy-five years old. In the meantime a quarter of a century of separate living has caused the two couples to become firmly set in their disparate ways of family life.

2. Separate family units call for complementary roles within each unit, which become competitive roles between the units. Each daughter becomes a homemaker and each son an independent head of his family. Thus daughter and mother have identical and hence competitive roles; son and father likewise have identical and competitive roles. These roles are necessary as long as the two families remain separate. If in the old age of the parents, the two households merge or even increase their inter-family contacts, the roles compete. Mother and daughter compete, and father and son. Sometimes instead of peaceful competition, there is conflict.

3. Small family units conceal latent conflict over authority. The older authoritarian type of family life, fitted to a culturally stable rural society, has given way to a more equalitarian type of family life. The present middle-aged generation is the critical generation in the transition from authoritarian to the equalitarian family where old and young alike have a voice in decision-making. The present old generation is the last of the truly authoritarian generations of parents. The present middle-aged generation waged the battle for freedom in the 1920's and outwardly at least won the battle. The interaction between old and middle-aged, observed in many cases and indirectly revealed in many studies, raises the question whether the parents really changed their authoritarian attitudes; and whether the children really achieved equality and independence. Perhaps the old attitudes remain but have been kept in abeyance by the geographical separation of households. When the old age of the parents brings them into closer contact with their middle-aged children, the old struggle often seems to revive. The middle-aged alternately submit and rebel, sometimes bursting into tears or temper tantrums in true adolescent fashion. The old, for their part, try to dominate and bend their sons and daughters to their will. At the same time, the adult children feel protective toward their parents, and the parents want affection and sympathy. Thus the chances are great that both old and middle-aged will act with inconsistent behavior because of the inconsistencies within themselves and the struggle between the generations.

Decline of Conflict Situations. The preceding discussion tends to show that the alienation of children from parents is transitional in nature, the concomitant of temporary social changes. Some at least of the changes are already disappearing. The passage from rural to urban will undoubtedly continue as long as the rural birth rate exceeds needs for rural labor, and as long as urban industries expand. However, the cultural rift in this movement is now being nullified by the urbanization of rural life. Ethnic conflicts may remain acute for foreign-born groups, but because of reduction in immigration they will affect a much smaller proportion of the population. Upward social mobility will be made less painful with the decline in the other two types of culture conflict, and with the continued spread of education into the lower class levels. There will continue, however, to be some cultural differences between the generations, due to the speed of cultural change now typical of the United States, which shows no signs of abatement. Only a static society could eliminate all cultural differences between generations.

The conflict between authoritarian oldsters and their middle-aged children may also be expected to decrease. Part of this authoritarianism was fostered by rural and foreign backgrounds. As ruralism and foreign culture disappear, authoritarianism may be expected to decline in favor of a more equalitarian type of family living. Also, the middle-aged people who struggled to break the bonds are less authoritarian with their children, and this hard struggle between the generations may perhaps never be repeated with future generations.

As long as the family breaks off into generational units, the competition of roles will continue if the units try to reunite. Separate dwelling units over a quarter of a century will also continue to lead to the formation of distinctive patterns of family living. However, with reduction in cultural change, the differences in family patterns should be less sharp.

Is Separation the Best Solution? The chief solution offered at present for generational conflicts is separate living quarters, with the institutional nursing home supplanting the independent dwelling as old age decline advances. Old people are urged not to identify themselves closely with their children but to seek friends of their own age. These suggestions are directed at reduction of conflict on a superficial basis by keeping the antagonists apart. They overlook the need of the old for affection and primary group contacts; and they do not explore the benefits that might accrue to old and young alike from closer contacts maintained over the years. . . .

Conclusion. This paper has tried to show that there are specific, identifiable causes for present tensions between old parents and their middle-aged children; that these causes are to some degree temporary in nature; and that they are decreasing in force. In the meantime, attitudes of rejection are becoming fixed and are supported by such rationalizations as the supposition that in-laws cannot live peaceably together, that the middle generation owes sole attention to the oncoming generation, and that old people are better off in institutions than in close association with their children. Separation into small family units and mutual rejection are becoming crystallized into a permanent pattern. Such fixation will hinder family adjustment to less conflicting cultural patterns of the future and will reduce the number of mutual benefits that closer family contacts can bring.

FAMILY, CLASS, AND GENERATION IN LONDON

Peter Willmott and
Michael Young

Adapted from *Family and Class in a London Suburb*, London, Routledge & Kegan Paul Ltd., 1960, pp. 123-127.

Woodford is a middle-class suburb of London. Bethnal Green is a working-class "village in the middle of London." There follows an excerpt from the conclusions of a study comparing the family in these two settings—Eds.

In one important way, Woodford has turned out to be much as we expected. Kinship ties there are much looser than in Bethnal Green. When a couple marry they set up a genuinely independent household; relatives' homes are more often connected by occasional missions, not by the continuous back and forth which make two homes into one in Bethnal Green. Kinship matters less—friendship more. That is no surprise. But we also asked other questions, whose answers were not as we expected. This applies particularly to four important points—old age, the feminine core to the kinship system, the friendliness of the suburb and the difference between the working classes of the two districts.

1. Old Age. Money apart, Bethnal Green could almost have been designed expressly for aged parents, so well is it suited to their needs, providing them with company, care, and, at least if they are mothers, a place of eminence in the family. How could Woodford reach such a standard? Was it not inevitable that more independence for the young would mean less security for the old? It sounds plausible, but it is not true, not wholly so at any rate . . . the generations do come together, but not so much on marriage as on bereavement. Bethnal Green daughters give up adolescent freedoms and return to their mother's hearth when they marry; Woodford parents come to their children's door when one of them is left widowed or when both are too old or too infirm to take care of themselves any longer. From a purely physical point of view, Woodford seems to look after its old as well as Bethnal Green.

This is not to say there is nothing to worry about. Physical care is only one of the needs of old age. Others are not so well satisfied in Woodford—the need for respect, the need to give as well as receive, the independence which is necessary to a sense of identity. To some extent this is inevitable. Woodford belongs more fully to a modern technological world than does Bethnal Green, and is therefore less in awe of tradition. Woodford wives do not pay great heed to their mothers' opinions about baby-care, as they do in Bethnal Green; they are more likely to consult the clinic, their doctor or each other. The wisdom of age is not valued in the same way. The generations have been apart for many years and, this being so, they are almost bound to have grown apart too. It may be a shock to both sides when they reunite, and find each other so prickly.

But if the friction of which we saw evidence is partly unavoidable, it is not wholly so. A good deal of the trouble arises because the generations so often have to live under the same roof. Here Woodford is at a disadvantage compared with Bethnal Green. The latter has more variety of accommodation, one- and two-room flats and floors, tiny houses as well as large, and Mums can use their influence with rent-collectors to get suitable quarters for each segment of their family. The old can live in their own little place across the road from their daughter. But things change as one goes towards the periphery of the city—at the centre there are more small dwellings in large buildings, flats and old houses, and in the suburbs more large dwellings in small houses. In Woodford the housing is as nearly uniform as the most hide-bound municipal estate—it has almost nothing but three- or four-bedroom houses, for families with children, ill-designed for the elderly. The generations move in together because they have no alternative—there is no suitable accommodation nearby.

In the long run more small houses will presumably be built, mixed up with the larger, so that whole families can more often live *near* to each other. There is scope for a private counterpart to the Council dwellings which are becoming increasingly popular with the old. Meanwhile, most people will have to make do as best they can in houses not designed with the needs of three generations in mind. It seemed to us that wherever a semblance of a separate dwelling for the old had been carved out inside the house, a cooking-stove and a wash-basin installed, or best of all a separate bathroom and kitchen, the aged were more content because they enjoyed more independence and suffered less wound to their pride. The younger people benefited through being on their own too, for at least some of the time. What can be done depends upon the size and layout of the house, and how many people have to fit into it. But some sort of physical conversion should often be possible and worthwhile. It is now obligatory, under the House Purchase and Housing Act of 1959, for local authorities to make grants for conversions and improvements. This could be a most useful measure for old people. But perhaps Councils could, in co-operation with voluntary bodies, do something more—not just to publicize the existence of this new Act (which is probably unknown to almost all the citizens of Woodford, as of everywhere else), but to provide a source of architectural advice. The husbands of Woodford, as we noticed in Chapter II, are many of them formidable handymen already. But this does not mean they would be averse to suggestions made on the spot by a knowledgeable person about ways of reorganizing rooms, plumbing and heating to suit the needs of old people without impairing the eventual value of the house. This new architectural service would naturally draw together what is already known—and it is considerable—about the special housing needs of old people. Where a local authority was already building bungalows for them it would be particularly easy to add on the additional function of advising private owners and landlords about conversions and adaptations.

There is another service which would be especially helpful to the young. One of our informants was an aged woman who had recently spent six weeks in the nearby Langthorne Hospital so that her children could have an urgently needed break. She was exceptional. Far more help of this sort is needed. Langthorne Hospital has been a pioneer in providing for temporary admissions of aged patients in order to relieve their relatives.[1] Sheldon[2] suggested short-stay hostels to help meet the same objective; other measures include the fuller use of day hospitals and home helps. Relief of this sort not only helps hard-pressed relatives; by assisting old people to stay well, it also eases the pressure of demand for beds in chronic wards.

The social services generally can do much to relieve the financial pressure on old people. They can usually manage at the moment, without being an intolerable burden on the younger generation, because they have some support from their pension. This is not to say it is adequate. . . . we saw how much the income of manual workers falls upon retirement, and how much scope there is for an improved pension which would do a little more to maintain their standard of living in old age. There is another vital job for the social services, State and voluntary, and that is to care for the large minority of people who have no families to aid them. . . . we would say that the task of supporting them adequately has only been started. The kind of old people who were isolated in Woodford needed more frequent visits from somebody—a few of them saw practi-

[1] DeLargy, J., "Six Weeks In: Six Weeks Out."
[2] Sheldon, J. H., *The Social Medicine of Old Age*, pp. 197-8.

cally nobody. They also needed more help with housing, small old people's homes with a family atmosphere about them, "foster children" and, generally, the sense that society had not deserted them.

2. Mothers and Daughters. In the light of our research in Bethnal Green, and that in other working-class districts of Britain, we ventured to expect "the stressing of the mother-daughter tie to be a widespread, perhaps universal, phenomenon in the urban areas of all industrial countries." But we went on to qualify this statement by saying "at any rate in the families of manual workers." [3] Manual workers seemed to have more need for the extended family as the "woman's trade union." Working-class mothers suffered more from insecurity. They received uncertain and ungenerous housekeeping allowances even when their husbands were not out of a job. Death rates were high, and if husbands were not killed in war or in peace, they were always liable to desert. Woman's only protection, we argued, was herself. The wife clung to the family into which she was born, and particularly to her mother.

Since middle-class wives were not beset by such insecurity we did not expect them to be so dependent on their own families of origin. We imagined that the closeness of husband and wife . . . might exclude any special bond between a wife and her mother. We were not completely wrong—the mother-daughter tie is not so tight as in Bethnal Green. But it is there all the same; the feminine relationship is still stressed; it is very often still the axis of such extended family relationships as there are . . . occupational mobility did not seem to set up barriers between mothers and daughters as it did between fathers and sons. Insecurity or not, married women have a common interest in children and housekeeping, in their common occupation, which is stronger than the interest shared by fathers and sons engaged in quite different occupations.[4] We would now remove the qualification from the above statement.

[3] *Family and Kinship in East London,* p. 163.
[4] An as-yet-unpublished study Michael Young and Hildred Geertz in a San Francisco suburb suggested that the same tie between mothers and daughters could be noticed there.

CURRENT PRACTICE OF FILIAL RESPONSIBILITY

Alvin L. Schorr

Adapted from *Filial Responsibility in the Modern American Family,* United States Government Printing Office, 1960, pp. 5-18.

Financial Assistance in Separate Living

Although the aged (or those over 65 years old) do not include all the parents of adult children, they are the group whose income has been most carefully studied. On the whole, one may assume that the financial assistance they receive from their children is at peak level, for the need of the over-65's is greatest and the earnings of their children has approached its highest level. Almost three out of five of those over 65 had less than $1,000 cash income in 1958 [101]. When evaluating such a figure, one must take into account certain differences from younger people—income that is not in cash, for example, more home ownership, higher medical costs. The results of the Steiner and Dorfman analysis for 1951 [75] are illuminating. It found that the following percentages of the aged were living below "subsistence budget" levels—27 percent of the couples, 33 percent of the nonmarried [1] men, and 50 percent of the nonmarried women. In any philosophy, these are large numbers of people who are in need of additional income. Roughly three out of four of the aged have living children [38, 69A].

The attitudes that people express about financial support by adult children have been surveyed from time to time. The results that are produced are at least superficially confusing. For example, in answer to the general question of what a "good" son or daughter will do for his parents, more than 75 percent of both parents and children thought he should provide financial assistance [73]. A National Opinion Research Center study, on the other hand, provides a brief story about a widower whose children, in modest circumstances, have families of their own [69]. Only 30 percent of the over-65-year-olds thought the widower should ask his children to help him. The results of some surveys suggest that old people assert a right to support that, however, they will not voluntarily exercise. They make contradictory statements; they say children should support but when they are in need, do not ask for help [77, 73, 4]. When one inquires whether

[1] In the material that follows, figures and percentages about nonmarried men and women refer to current status. That is, the divorced and widowed are included in the nonmarried.

children should help a great deal or "some," only 5 percent will take the former [4]. Cultural influence is evident in one set of findings [4]: 42 percent of Negroes thought that children should help their parents and 37 percent of Caucasians, but only 24 percent with Spanish background. The low figure for people with Spanish background is interesting for, in fact, theirs is much more deeply an extended family culture [66]; however, they do not share "Anglo" guilt at seeking public assistance.

It is difficult to summarize the results of opinion surveys in the area of filial responsibility. The better surveys are so carefully sampled and analyzed that they produce a wholly misleading sense of precision, for the values they seek to measure are in conflict and colored by deep feeling. In general, surveys yield respectable percentages in favor of filial responsibility as long as the question is put quite simply and in ethical terms. But responses may be manipulated almost at will by framing questions that introduce the responsibility of the adult child to his wife and children, his responsibility for his own advancement, choice between the adult child and other sources of assistance, choice between money and other things children can give, and the sense that one speaks of one's self rather than in general. Many surveys introduce one or more of these considerations, with the result that a majority oppose and a fairly large minority (a third or more) affirm the responsibility of children to support. There is little doubt that a question heavily stacked with these conflicting values would produce a smaller percentage (perhaps 5-15 percent) of aged people who would say their children should support them. Then one recalls that in many areas Americans do not feel constrained to practice what we profess, or vice versa, and it becomes important to turn to what is our actual practice.

The percentage of aged people who receive cash contributions from one or more children is probably between 5 and 10 percent (that is, about 1-1.5 million parents); the proportion may be lower but it is not higher. There are studies that arrive at lower percentages [63, 75]. Other studies arrive at figures up to 20 or 25 percent [4], but higher figures invariably include help from other relatives or friends, or the estimated value of free rent and other help in kind. Both are apt to add materially to percentages. Though there is some impression that children who do not ordinarily contribute will pay for emergency or special needs, in a given year 10 percent or fewer of the aged report doctor bills paid by a relative [32, 63]. Presumably, if data were available for younger (*i.e.*, less than 65 years old) parents of adult children, all these percentages would be lower.

In short, the pattern of cash contribution by adult children is not significant in terms of numbers. For the 5 to 10 percent who receive some contribution, the amount does become meaningful. For example, a national study [89A] found in 1957 that the median yearly contribution for those aged people who received contributions from a child or other relative was $300 for couples ($150 per person) and $240 for nonmarried individuals. As the median total money income was, respectively, $2,250 and $1,070, these contributions take on importance. (Contributions that are not made in cash are treated in the next section.)

In general, contributions by children are more often made to nonmarried women,[2] less often to nonmarried men, and least often to couples. Undoubtedly this difference reflects the circumstances of each group. Married couples have higher incomes, their health is likely to be better, and various studies have indicated that they are in a more favorable state of mind. The nonmarried women, on the other hand, are likely to be the group with the lowest earnings and the poorest health. Their average age will be higher and their chances of remarriage not so favorable as for the men. It has been observed, further, that personal relationships are easier when mothers get help from married daughters than from married sons [14, 82].

The percentage who contribute to parents increases regularly with the income of the giver. Conversely, it is the parents in the lowest income group who more often receive contributions [102, 42, 77, 91]. That those who have most should give most often to those who have least could, perhaps, have been predicted. But it is inconsistent with the observation in a number of sociological studies [35, 14] that working-class mores require help to parents and other relatives in a way that middle-class mores do not. Cash contribution is not the complete story, however, as financial help may be given in other forms, notably living together.

It is interesting that investigations of filial contributions treat them as if the flow were one way, from the adult child to the parent. Perhaps this is in the nature of studies that focus on the income of the aged and, after all, are not concerned with the income of younger groups. Or perhaps we assume, because we are acutely conscious of the problem of the aged, that what flows the other way will be insignificant. Where the question has been asked, however, the balance of financial aid proves to be "greater in the direction of helping children" [79, 77]. Moreover, this held true in each income category. While one must suppose that there is a decrease of assistance from parent to child with decreasing income, individuals and two-person families in the lowest income categories still show small percentages making contributions for someone else's support [102].

One has only to reflect on how students attend college to appreciate the extent to which parental financial assistance operates beyond children's minority. Seventy-four percent of unmarried full-time college students

2 See footnote, p. 417.

received assistance from their parents in 1952-53, in a median amount of $764 [36]. Two out of five college students are 20 years of age or older [85]. Though parental help may be concentrated among the younger students, obviously a substantial number of older students are also being helped. Other areas that might reflect similar investment are cash contributions when young adults establish a home and are rearing children. It is entirely representative of these developments that a scale developed to measure performance in the parent role [30] rates highest the parent who "has no need to give [adult children] everything. If he supports them financially, does so unobtrusively and in a matter-of-fact way. . . ." Here is a development that may represent more change in voluntary patterns than the contributions up the age ladder that we have been discussing—that is, the extension of parental support into their children's majority.

Living Together

A special survey [75], which provides the most usable information for our purpose, indicates that 33 percent of the aged in the country in 1952 were living with children. Fifty-eight percent were living alone or as couples only; the remaining 10 percent had other shared living arrangements. Higher percentages of aged living with others are provided by measures that combine other relatives as well as children in one figure. Such figures are roughly a third higher than those that count only aged people living with children. Lower percentages are provided if one counts as living together only the aged who are not heads of households [70]. Who is named head of the household may reflect courtesy, however, more than the actual circumstances [22, 75]. The long-term trend appears to be towards a declining percentage of aged people living with children [51, 26]. By 1957, about 28 percent were living with children [69A].

It is the currently nonmarried and especially the women who are most likely to share living arrangements. The third who lived with children in 1957 include the following: 23 percent of all aged couples, 27 percent of the nonmarried men, and 37 percent of the nonmarried women [69A]. Other factors that increase the likelihood of living with children are advancing age, poor health, and low income. Obviously these factors are interrelated and, as has been indicated, they are related to sex and marital status as well. Nevertheless, one should not conclude that only poor, sick, old widows live with children, for though the numbers of men, of couples, and of well-to-do who live with children are smaller, they are noticeable. Women prove to be pivotal in live-together arrangements. Aged parents are more likely to live with their daughter and her husband and children, if any, than with a son and his family. Young couples starting out are more likely to live with the wife's family than with the husband's [26, 45]. Aged women with low incomes will tend, if they have children, to live with them. Men's choices are apparently dependent on other factors; they are as likely to live alone, whether or not they have children [11].

It has been noted that cash contributions are more frequently given by those with higher incomes, and more frequently received by those with lower incomes. This, as far as it goes, is inconsistent with sociological and anthropological studies that conclude that working-class, but not middle-class, mores require assistance to relatives beyond husband and children. Living together is somewhat more in accordance with this expectation, as the following table indicates.

Distribution by Income Class of All Families Compared with Families That Include an Aged Member, 1951

| | | Families That Include Aged Member | |
Money Income Class	All Families in United States (Percent)	Families with Aged Couples (Percent)	Families with Aged Nonmarried Individuals (Percent)
All income	100.0	100.0	100.0
Less than $2,000	20.5	27.0	31.8
$2,000 to $2,999	15.4	14.5	12.3
$3,000 to $4,999	35.3	27.1	29.3
$5,000 or more	28.7	31.3	26.5

Sources: (86), table 1; (22), table 8.

Thus, among the families with less than $2,000 income, there is substantially more living together (this includes other relatives as well as children) than their representation in the population. Conversely, families with more than $2,000 income tend less to have aged relatives living with them.

How is one to interpret this? The more that people have, the more likely they are to give in cash. But having more will not lead them (except perhaps in the highest income brackets) to share their living quarters more readily. The overrepresentation in the lowest income brackets must be a result, at least in part, of the need of the parent rather than the income of the family. For example, a retired laborer is five times as likely to live with relatives as a retired corporation executive [13]. Undoubtedly, family income among the laborers' families is lower, but the retired laborer's own income is lower than the retired executive's. The laborer's *need* for a living arrangement that will in part support him presents the question of help to his family. The executive's family is less likely to have a question to consider. Thus, the fact that more low-income families live together is, to some extent, simply a reflection of the fact that more of these parents need help. Why does not starker need of their parents also produce more cash contributions from the lowest income families? Presumably because their own plight is so difficult that they must select the most efficient way of sharing, which is living together. Possibly too, the reciprocal services that the parent may offer—babysitting, housekeeping—are more meaningful to the low income families. To summarize perhaps oversimply, the *giving of cash* reflects a feeling of having some excess that may be shared (a feeling, apparently, that few Americans at any income level have). *Living together* reflects much more simply a response to the need of the parent.

Aside from sheer need, there may be a degree of readier receptivity to living together among lower income families, but such a statement must be used guardedly. Recent English exploration of the relationship between kinship intimacy and class concludes that class is not determining. Rather, certain factors are determining that may be (or may not be) associated with class: economic ties within families (helping one another get a job in the same factory, for example), whether their neighborhood tends to throw them together, social and geographic mobility, and opportunity to make other contacts [5]. Moreover, it seems possible that it was not so much a working-class as an immigrant characteristic that older studies reported. It has been observed that loyalty to an extended family is an immigrant tradition, particularly among those from Eastern and Southern European countries [28, 42, 12]. That working-class and immigrant groups were in many places identical would make such confusion unavoidable. There is an ironic note here: as immigration declines and Americans become more thoroughly Americanized, alarm arises that a tradition of extended-family solidarity (thought by some to be American) is vanishing. That such a development is in fact involved tends to be indicated by findings that the lower their socioeconomic position, the more likely aged people living alone are to become isolated with advancing age, *except where immigrant cultural traditions intervene* [42].

Obviously, it is more difficult to assess the value of living together than of cash contributions. An estimate based on 1951 data found 10 percent of the aged couples and 38 percent of the nonmarried living without payment in quarters that they did not own. The guess was made that the value of free quarters to the old person was $360 a year [22], a substantial addition to his income. In addition to free rent, 3 percent of the aged couples and 8 percent of the aged nonmarried said that they contributed less than their full share to household expenses, or had bills paid for them by relatives or friends. In any case, value cannot be measured in financial terms alone. The relationship of advanced age and poor health to living together strongly suggests that protection or nursing care is often involved. Less tangible factors, such as attention and company, are certainly involved too.

There are also the possibilities that living together is for the children's benefit, for mutual benefit, or for no benefit at all, but simply the way family living is working out. These are possibilities that, as with cash contributions, are often obscured by the assumption that living together is for the parent's benefit. Upward of a third of the aged who live with children or other relatives own their own home [92, 91, 63]. Though this means that the parents are giving something, one can only conjecture how many are therefore contributing more than their share. When they are specifically asked for whose benefit they are living together, the majority of parents say they are helping the children [77]. The Rhode Island Governor's Commission [63] concluded that "in many instances the arrangement is undoubtedly one of mutual economic necessity." Aside from occasional wry observations that the positive attributes of parents in their families "must be taken as a possibility also," [73, 77, 37] the literature of aging has little more to contribute. It is curious that one must turn from the field of aging, where one is so often urged to stress the positive, to other technical fields in the United States and to English research for evidence of the utility of living together to adult children.

To begin with, the living together of parents and adult children is influenced by parental concern about children just as is parental contribution of cash. In 1951, one in five married couples postponed setting up a separate home during the first year of marriage, one in eight for 3 years. Though the housing shortage may have been involved at that time, 3 years later 12 percent of couples under 25 did not have their own household [26]. In 1959, one out of four or five couples in their middle years had a young adult child living with them [84]. That parents are turned to in later years in the face of emergencies—death, divorce, desertion, and migration—has also been noted [105] but not counted. In 1950, there were 4.8 million people 25 years and over living with parents who were the head of the household, accounting for about half the living-together households [70, 63]. We have noted, of course, that who is head is an uncertain indication of actual circumstances.

To turn to nonmonetary services to adult children, Dr. Ernest Burgess has described contributions that are well known in folklore, basing his comments, however, on studies of engagement and marriage [8]. "Where both parents and children elect to live together," he writes, "the arrangement may work out more or less satisfactorily. Where the wife is working, the mother-in-law often takes on the major charge of the household responsibilities. She may be very happy to function as a babysitter. . . . Although there may be some disagreements, these tend to be minor, and both generations report the relationship as satisfying." The recent survey of working mothers documents the role of grandparents in caring for children [43]. The major role of aged parents in household help has also been verified [72, 71, 104]. Finally, disaster and stress studies, which have been carried on since the Second World War, find uniformly that families which include grandparents are more flexible and resilient in the face of the father's induction into the Armed Forces, and of flood, fire, or other disaster than are parent-child families [55, 112, 33].

It is unfortunate that description and documentation appear to segment the relations of adult children and their parents, as if living together were separate from the giving of cash and as if the benefit of the parent were separate from the benefit of the adult child. Actually, as Dr. Burgess' statement suggests and as studies in England make clear, at least for England, the dominant note is one of "reciprocal services being freely performed" [82]. In view of this and of the fact that so many adult children and parents do live together, it is important to examine the general opposition to this arrangement. From half to 90 or 95 percent of the population will say that it is better for parents and children to live separately [73, 42]. The smaller percentages are produced when the query is addressed to a group such as aged nonmarried women who, it may be assumed, do not see the question as entirely abstract [4]. Even so the percentage opposed to living together is high. Though there is some dissent in the general population, specialists in family relations and in aging are nearly unanimous. "It is impossible to say," reads a typical statement, "that with us it is 'natural' for any other group than husband and wife and their dependent children to maintain a common household. . . . It is, of course, common for other relatives to share a household with the conjugal family, but this scarcely ever occurs without some important elements of strain. For independence is preferred. . . ." [52]

It must be noted, without implying that this is the reason social scientists hold it, that such a point of view serves several functions in our society. It has been amply demonstrated [109, 50, 67] that the conjugal family, permitting emphasis on individual merit and mobility, is the most suitable for an industrial society. It also seems clear, or at least it is widely believed [65], that young adults in our society need support in making a clean break from their parents. This emancipation function is evident as one follows the later phases of child-parent relationship—children break away in late adolescence or early adulthood, only to reestablish ties with parents after marriage and particularly after their own children arrive [5]. Finally, the American ethic that the individual (and his child) comes first has been supported in the face of the poignant appeal that growing numbers of retired people were bound to exercise. The type of sweeping endorsement of separate living that has been cited has served these functions, and it has the advantage of simplicity. It is time to recognize, however, that the evidence for such a statement is entirely anecdotal and a priori.

At the moment, clinical and theoretical evidence is virtually all we have and it is not to be disregarded; but we do not have field studies such as the English have conducted (which imply rather different conclusions[3]) and we do not have sufficiently deep and adequately controlled studies of living together. Possibly we have been too much involved with schedules and questionnaires, which by their nature simplify, and too little involved in direct observations of living patterns. Pending better evidence we must retain perspective. The anecdotal evidence is made up of repeated reports of the problems that occur when parents live with children [16, 12]. "Severe family tensions," these say in one way or another, "most painful conflicts to old and young, are reported by psychiatrists and caseworkers when families are trying to combine arrangements satisfying for two or three generations in the same home" [40]. More recently, people reporting strain have been asked to define the chief problem in living together, and they frequently speak of privacy, space, and conflict over raising children [42, 46, 73]. The consistency with which such reports appear makes it quite clear that the sharing of households by parents and adult children *may* cause problems. They do not establish that it necessarily causes problems.

Theoretical evidence moves beneath the complaints that are voiced, to indicate the tensions that modern psychology recognizes in the living together of adult children and parents. Otto Pollak, in a brief review of this evidence [59], touches on the added financial burden a parent represents, the possibility that a child may be caught with an unresolved need to be dependent on his parent, or that the parent may resent being dependent on a child whom he has never really wished to see adult. He points out the possibilities for working out forgotten resentments between parent and child, the consequent guilt, and the possibility that either parent or child will find himself with less status in this new situation than he has had. "The economic and psychological mining of such situations leads easily to explosions for reasons implied in our culture," Dr. Pollak concludes. There can be no question that there are potential strains when parents and adult children live together. But potential strains are inherent in any living situation—in work, in rearing children, in marrying. If technical and popular

[3] For example [5]: "It seems to me that many clinical workers, doctors, and family research workers take it for granted that [the conjugal family] is the natural and normal form for familial behavior to take. Advice based on this assumption must be rather bewildering."

literature confined themselves to the strains intrinsic to each of these activities, would we conclude that we should give them up?

An analogy may make this point clearer. One may suppose that a visitor from Israel in the year 2000, third or fourth generation of a line of
children raised communally in a kibbutz, would inquire whether Americans find it a strain to raise children *in the family home.* Do they not interfere with their parents' privacy; are they not insensitive and demanding; they must be an enormous physical and financial drain? Does not
rearing children provoke unresolved conflict about dependency, and so
forth? What can one answer to these questions but "yes," adding if we
will that we find compensating satisfactions or that we do not like the alternative of raising our children communally. Moreover, a healthy family
organizes itself so that these strains do not mount up as much as, to the
uninitiate, it might seem. The useful questions about raising children, as
about living together, are not categorical (that is, is there strain? or are
there failures?) but discriminatory: when is strain greater and lesser? how
are strains handled by those who handle them? what are the situations in
which they cannot normally be handled?

We have looked into these questions very little indeed. Dr. Burgess
has noted that in-law conflict is almost always with the mother-in-law [7],
and we have seen that in living together families prefer to organize themselves around a mother-daughter tie. There has been some speculation
about this, but we hardly know, for example, whether we may say that
such arrangements do often work out well. Havighurst has suggested that
constructive plans for living together may be made if, first, parents have
truly freed their children [31]. It is in England that the mechanisms of
intergenerational living have begun to be identified. In a major step forward [82], it is concluded that when the mother is related to the adult
son, living together is workable if she can yield authority to her daughter-
in-law. With unmarried children, a mother may without special difficulty
retain more authority. People seem willing to acknowledge a special bond
of mother and daughter, though too close a tie of father and daughter or
son and mother seems to create a problem. This work also describes the
manner in which husbands and wives segregate certain financial and domestic roles so that one or the other is shielded from relationships that
may be painful. This material points, of course, to the importance of further research, but more than that it indicates that living together may be
workable and, indeed, satisfying *in certain circumstances.* The significant
point appears to be not that parents and children should or should not
live together, but that they should know whether they can and select
whether they wish to.[4] It is now a decade since Dr. Burgess wrote [8]

[4] Recent writing for lay audiences presents such a point of view somewhat more

that "particularly significant is the hypothesis that mutually satisfying relations are much more likely to be maintained where living with or away from parents is a matter of choice and not of necessity both for the elderly and for the younger couple." Since then, in this country, the significance of choice has received inadequate attention.

Care and Affection

Many older people say, often when their financial means are modest or less than modest, that the important thing they wish from their children is affection. If they are asked to rank material and affectional support to indicate importance, material help turns up a poor second [77, 69]. They want to live alone but at the same time they want to be close to their children [82]. Adult children make complementary statements [42]. The Havighurst scale for role of adult child of aging parents [30] which, in its own way, states a consensus, rates highest the child who "keeps in close personal touch . . . by visits, letters, or actually living together." Ranked high is the child who "has no responsibility for caring financially for parents but feels a real responsibility for maintaining satisfactory relations with them—visiting them, keeping in touch. . . ." The adult child is rated medium who "feels that their lives are fairly separate from his."

There has been widespread alarm, though in recent technical material there is effort to dispel it, that older people are isolated from their children and other realtives. The stereotype is abroad that the aged person is a lonely old woman, querulous or embittered, needy, and either demanding or overly proud, with no one to care for her. It is suggested occasionally that this stereotype springs from the guilt of the young who write about the aging. Possibly it is a result of strenuous efforts to create interest in a growing problem in which there was, for a time, no interest. Unquestionably, also, there has been a tendency to assume that physical isolation means social isolation. For example, analysis by Talcott Parsons [52] demonstrated that the effect of conjugal family living, of an economy that maintains a man at full employment and then abruptly retires him, and of the connection between work and where one lives—the effect of these forces is physically to separate the aged from the families of their children. It was more or less assumed that physical separation meant that the children no longer cared about or kept in touch with their parents.

A body of material has accumulated that makes it clear that in Eng-

than technical material, perhaps because the writer must face the fact that a third of his audience does live with children. For example (18): "For one reason or another it may be best to live with your married children. This need not be the fate above all to be avoided that folklore says it is. It may work out very pleasantly for all of you. Of course, it may be unbearable."

land this assumption is quite inaccurate. The United States has substantial social differences from England, but our recent investigations yield findings that are consistent with the English. In general, their studies find that old people prefer to live alone, but they live near to children or other relatives, whom they see quite frequently. Paying special attention to feelings of loneliness, the English found that four out of five made no complaint and only 8 percent were very lonely. Moreover, it was not those who had lived alone all their lives (the "isolated") who were lonely. It was those who had lost a spouse (the "desolated") who felt lonely, and with time their grief might diminish. There were single men who "lived completely lonely lives without any trace of loneliness." The parents got a great deal of help from their children, and they reciprocated to a marked degree [71, 111, 5, 82, 72].

Similarly, studies at Duke University and the National Opinion Research Center find that "while some older people are isolated from their families, most older people are a part of family configuration" [69]. Upwards of two-thirds of aged parents see their children at least weekly [77, 63]. Even when there are not visits, there may be daily telephone contact [45]. Study of middle-class families concludes that the generations "have a desire to help one another . . . parents had established a pattern of giving moderate help and service . . . parents and children take care of one another during illness regardless of distance. . . ." [79, 78]

In an area of give and take where intangibles are the chief currency, it is particularly difficult to say that parents or children have, in general, the net advantage. Deeper attachments are involved than may be understood in remaining overly focused on "responsibility" in filial relations. For example, although isolation of parents from adult children is frequently attributed to children's ambition to move into a higher social class, there is evidence that parents will say they are closer to children who have been more successful [77]. This can be puzzling unless one realizes that parents too are ambitious for their children. Thus one sees that parents take pleasure from the success of adult children, and find in it a means of continued expansion of their own lives, even if its visible result is to reduce physical contact between parents and children [77, 41].

Similarly folklore recognizes, though middle-class mores may reject, a tacit tie between grandparents and grandchildren [14, 106], which is the reverse face of the common complaint that grandparents interfere with the raising of children. Grandparents do often seem to mitigate parental anger at children, in the course of which naturally they may become the object of anger themselves. Grandparents may indulge their grandchildren in a way parents cannot—in a way, in fact, that parents cannot approve.

But this is a relationship from which grandparents get satisfaction, and it may serve salutary purposes for the grandchildren. Moreover, the relief that the parents, the middle generation, may find in the mediating influence of grandparents has been less adequately explored than the resentment that they express. The outward evidences of this tie of alternate generations—babysitting, child care, and nursing—have been noted.

To give recognition to the positive tone of filial relations for so many people is not to imply that there are not aged people in need of protection, care, and social contact. We know that, when a retired person is incapable of handling his insurance benefit, it is often difficult to find someone who will help him [89]. That there is conflict between generations, particularly when circumstances thrust living together upon them, has been noted above. Further, there are old people who do not have and who miss social contact [42, 77, 11]. To many the advance of years unavoidably brings illness, declining income, death of contemporaries, senility, or declining responsibility and status. For the three out of four who have grown children, they are often a satisfaction and a support. When they are not, this may be particularly disappointing.

Current Practice, in Closing

Adults with children of their own or only with personal wants of their own face now for the first time, in substantial numbers, the free choice of sacrificing these to the needs of their parents. We deal with this in the abstract by expressing allegiance to all moral values at once—that is, to the rights of children, the independence of adults, and responsibility to parents.[5] Our practice in relation to cash, however, is tersely described by an old German proverb: "One father takes better care of 10 children than 10 children take care of one father." When parents' need is stark, and often then for protection and care as well as for income, adult children do in substantial numbers live together with their parents. As for the intangibles —attention, affection, and so forth—there is no value conflict, or at any rate less, for the giving of these things to parents does not diminish what remains for a spouse and children. On the contrary, a warm relationship with grandparents may mean more to be shared by everyone. The evidence is that these ties are usually to some extent satisfying, though unquestionably there are also lonely and isolated parents.

Two key points are implicit in the material that has been reviewed.

[5] For example, [17] p. 32, "It is not the function of children of small families to support the older members of families in which they were reared. That the parent and child families are related by blood and bound together by one common member is more or less immaterial." P. 327, "One of the deep-seated traditions in this country is that adult children will care for their aged parents when necessary."

First is the reciprocal nature of filial relations. In the net, parents may give more to adult children in cash, though less in living together. Living together is frequently a situation in which *both* parties benefit, though a set of scales would tip one way or the other. And whatever the material balance, the receiver often renders service or repays with fondness, an important coin in itself. Conflict between such drives as advancement of one's own family and ties to the older generation is not necessarily intergenerational conflict; parents are as much wedded to their children's advancement as are the children. Though to the onlooker, on occasion, adult children may be living disproportionately well compared with their parents, it may be that the parents would not have it otherwise and take pleasure in their children's success.,

The other key point is the spontaneous nature of filial relations. What children and their parents give to each other has little connection with law or compulsion. The money contribution, which is the only gift that can actually be compelled, is a relatively unimportant pattern. Helping each other with chores, visiting, and showing concern, which cannot be compelled, is the dominant pattern. Moreover, it is not only children but other relatives and friends who engage in these activities. We have noted that other relatives and friends add materially when they are included in the statistics of cash contributions to the aged, and these groups add about 25 percent when they are included in statistics of living together. They illuminate a point about filial responsibility that is perhaps elementary—the help or exchange that takes place between people is a function of the way they feel about each other. In the presence of fondness or gratitude, the exchange will take place even if individuals are not thought to be responsible; in their absence, it is unlikely to take place.

That this is the pattern of filial relations should be useful to us. Many have confused the family support that was assured by the pre-Civil War economy with the voluntary support that they wish would be forthcoming today. They have equated the absence of filial contributions with the absence of filial ties. In consequence, they point to general deterioration of filial relations. This exaggerates the problem and, in any case, it is inaccurate; we should take heart. On the other hand, pointing to deterioration may be a way of evading problems that need to be faced. (Since this is an evasive course that often invokes the Fifth Commandment, one might label it as "taking the Fifth. . . .") Plugging up the gaps in our income-maintenance programs, voluntary protective agencies for the aged [48], and the homemaker services that have seemed to move so slowly are among the needs that may be faced. The need for broader counseling and research activities is implicit in our discussion. We have dwelt here particularly on the need to deal realistically with living together, which has so far been treated as if it were a very poor arrangement indeed. We are old enough as a country and in our social sciences to take a more sophisticated position. It appears that children should, at some point, live independently. Later on, if it suits them and if it suits their parents, perhaps they can satisfactorily share a household again. The task before us is to learn when this may be sound and when it may be disastrous, and the techniques that work in living together, so we may equip our families to make sound choices.

REFERENCES

(Page references are to direct quotations)

1. Abbott, Edith, "Abolish the Pauper Laws," *Social Service Review*, March 1934.
2. Abbott, Edith, *Public Assistance*, Vol. 1, 1940, University of Chicago Press, Chicago, Ill.
3. Beard, Belle Boone, "Are the Aged Ex-Family?" *Social Forces*, March 1949.
4. Bond, Floyd A., et al., *Our Needy Aged*, 1954, Henry Holt & Co., New York, pp. 256-259.
5. Bott, Elizabeth, *Family and Social Network*, 1957, Tavistock Publications, Limited, London W. 1., p. 218.
6. Breckenridge, Elizabeth, *How Public Welfare Serves Aging People*, July 1955, American Public Welfare Association, Chicago, Ill., p. 11.
7. Burgess, Ernest W., and Wallin, Paul, with Shultz, Gladys Denny, *Courtship, Engagement, and Marriage*, 1954, Lippincott, Philadelphia, Pa.
8. Burgess, Ernest W., "Family Living in the Later Decades," in *The Annals of the American Academy of Political Science*, January 1952, "Social Contribution by the Aging," pp. 111-112.
9. Burns, Eveline M., *Social Security and Public Policy*, McGraw-Hill, New York, 1956.
10. Calhoun, Arthur W., *A Social History of the American Family*, 1917, 3 vols., Vol. II, pp. 51-54; Vols. I, II, and III, passim; Vol. I, II, pp. 67-68, A. H. Clark, Cleveland, Ohio.
11. Cavan, Ruth Shonle, "Family Life and Family Substitutes in Old Age," *American Sociological Review*, August 1943.
12. Cavan, Ruth Shonle, et al., *Personal Adjustment in Old Age*, 1949, Science Research Associates, Chicago, Ill.
13. Corson, John J., and McConnell, John W., *Economic Needs of Older People*, 1956, Twentieth Century Fund, New York.
14. Davis, Allison; Gardner, Burleigh B.; and Gardner, Mary R., *Deep South*, 1941, University of Chicago Press, Chicago, Ill.
15. De Tocqueville, Alexis, *Democracy in America*, 1899, Introduction to Vol. I, Appleton & Co., New York.
16. Dinkel, Robert M., "Parent-Child Conflict in Minnesota Families," *American Sociological Review*, August 1943.
17. Drake, Joseph T., *The Aged in America's Society*, 1958, Ronald Press Co., New York.
18. Duvall, Evelyn Millis, *In-Laws—Pro and Con*, 1954, American Book-Stratford Press, Inc., New York, p. 363.
19. Epler, Elizabeth, "Old-Age Assistance: Determining Extent of Children's Ability to Support," *Social Security Bulletin*, May 1954.
20. Epler, Elizabeth, "Old-Age Assistance: Plan Provisions on Children's Responsibility for Parents," *Social Security Bulletin*, April 1954.
21. Epstein, Abraham, *The Challenge of the Aged*, 1928, The Vanguard Press, New York.

22. Epstein, Lenore A., "Economic Resources of Persons Aged 65 and Over," *Social Security Bulletin,* June 1955.

23. Epstein, Lenore A., "Money Income of Aged Persons: A Ten-Year Review, 1948 to 1958," *Social Security Bulletin,* June 1959.

23A. Epstein, Lenore A., "Money Income Sources of Aged Persons, December 1959," *Social Security Bulletin,* July 1960.

24. Florida Legislative Council and Reference Bureau, *Florida Welfare Services: A Research Report of the Select Committee on Welfare,* 1959, Tallahassee.

25. Florida Department of Public Welfare, Division of Research and Statistics, "Effect of the Migration of Oldsters to Florida on the OAA Program," *Florida Public Welfare News,* February 1955, p. 4.

26. Glick, Paul C., *American Families,* 1957, John Wiley & Sons, New York.

27. Greenfield, Margaret, *Administration of Old Age Security in California,* May 1, 1950, University of California, Berkeley.

28. Handlin, Oscar, *The Uprooted,* 1951, Little, Brown and Co., Boston.

29. Hart, Ethel J., "The Responsibility of Relatives Under the State OAA Laws," *Social Service Review,* March 1941.

30. Havighurst, Robert J., "Research Memorandum . . . ," in Anderson, John E., ed., *Psychological Aspects of Aging,* 1956, George Banta Co., Inc., Menasha, Wisconsin, p. 294 and p. 296.

31. Havighurst, Robert J., "Social and Psychological Needs of the Aging," in *The Annals of the American Academy of Political Science,* January 1952, "Social Contribution by the Aging."

32. Health Information Foundation, "Use of Health Services by the Aged," *Progress In Health Services,* April 1959, New York.

33. Hill, Reuben, *Families Under Stress,* 1949, Harper, New York.

34. Hitrovo, Michael V., "Responsibility of Relatives in the Old-Age Assistance Program in Pennsylvania," *Social Service Review,* March 1944, p. 69 and p. 75.

35. Hollingshead, August B., "Class Differences in Family Stability," reprinted from *The Annals of the American Academy of Political Science,* November 1950, in Herman D. Stein and Richard A. Cloward, *Social Perspectives on Behavior,* 1958, The Free Press, Glencoe, Ill.

36. Hollis, Ernest V., *Costs of Attending College,* U.S. Department of Health, Education, and Welfare, Office of Education Bulletin No. 9, 1957.

37. Kaplan, Saul, "Old-Age Assistance: Children's Contributions to Aged Parents," *Social Security Bulletin,* June 1957.

38. Kaplan, Saul, Technical Supplement to (37), mimeographed, p. 1, Bureau of Public Assistance, July 1957.

39. Kent, James, *Commentaries on American Law,* 1836, Vol. 2, New York.

40. Kraus, Hertha, "Housing Our Older Citizens," in *The Annals of the American Academy of Political Science,* January 1952, "Social Contribution by the Aging," p. 127.

41. Kuhlen, Raymond G., "Changing Personality Adjustment During the Adult Years," in Anderson, John E., ed., *Psychological Aspects of Aging,* 1956, George Banta Co., Inc., Menasha, Wis.

42. Kutner, Bernard, et al., *Five Hundred Over Sixty,* 1956, Russell Sage Foundation, New York.

43. Lajewski, Henry C., "Working Mothers and Their Arrangements for Care of Their Children," *Social Security Bulletin,* August 1959.

44. Lazarus, Esther, "Economic Functioning of Older People: Implications for Social Work," in *Toward Better Understanding of the Aged,* 1958,

Council on Social Work Education, New York, pp. 98-99.

45. Leichter, Hope J., *Kinship and Casework*, Groves Conference, Chapel Hill, N.C., Apr. 6, 1959, mimeographed.

46. Lenzer, Anthony, with Pond, Adele S., and Scott, John, *Michigan's Older People, Six Hundred Thousand Over Sixty-Five*, 1958, State of Michigan Legislative Advisory Council.

47. Leyendecker, Hilary M., *Problems and Policy in Public Assistance*, 1955, Harper, New York.

48. McCann, Charles W., "Guardianship and the Needy Aged," *Public Welfare*, July 1958.

49. Minnesota Commission on Aging, *Minnesota's Aging Citizens*, January 1953.

50. Moore, Wilbert E., "The Aged in Industrial Societies," in Milton Derber, ed., *The Aged and Society*, December 1950.

51. Ogburn, William F., with Tibbitts, Clark, "The Family and Its Functions," in *Recent Social Trends in the United States*, Report of the President's Research Committee on Social Trends, 1934.

52. Parsons, Talcott, "Age and Sex in the Social Structure of the U.S.," reprinted from the *American Sociological Review*, October 1942, in Herman D. Stein and Richard A. Cloward, *Social Perspectives on Behavior*, 1958, The Free Press, Glencoe, Ill., p. 200.

53. Pennsylvania Department of Public Welfare, *Composition of Shelter Groups, Latter Half of January 1958*, Special Analysis, mimeographed, June 9, 1958.

54. Pennsylvania Department of Public Welfare, *Effect of Length-of-Residence Requirement on Eligibility for Public Assistance in Pennsylvania*, mimeographed, Mar. 5, 1959.

55. Perry, Helen Swick, and Perry, Steward E., *The Schoolhouse Disasters*, National Academy of Sciences, 1959, Washington.

56. Pleak, Janet, "Reports of Payments to Out-of-State Recipients," *Public Welfare*, May 1951.

57. Polemis, Zane M., "Public Assistance Payments to Out-of-State Recipients," *Public Welfare*, October 1955.

58. Pollak, Otto, *Social Adjustment in Old Age*, 1948, Bulletin No. 59, Social Science Research Council, New York.

59. Pollak, Otto, *The Social Aspects of Retirement*, 1956, Richard D. Irwin, Inc., Homewood, Ill., p. 13.

60. The President's Commission on Veterans' Pensions, *Survivor Benefits for Service-Connected Deaths and Veterans' Insurance*, Staff Report No. VII, May 22, 1956, p. 30 and p. 49.

61. The President's Commission on Veterans' Pensions, *The Historical Development of Veterans' Benefits in the U.S.*, 1956.

62. Remmers, Mable J., "Meditations of a Visitor En Route," *Kansas Welfare Digest*, May 1945, p. 2.

63. Rhode Island Governor's Commission to Study Problems of the Aged, *Old Age in Rhode Island*, July 1953, Providence, p. 7.

64. Riesenfeld, Stefan, with Maxwell, Richard C., *Modern Social Legislation*, 1950, Foundation Press, Brooklyn, N.Y., pp. 692-700. Regarding legal base of filial responsibility, see also, *Corpus Juris Secundum*, vol. 67, "Parent and Child," section 24, and vol. 70, "Paupers," section 60c.

65. Rose, Arnold M., "Acceptance of Adult Roles and Separation from Family," *Marriage and Family Living*, May 1959.

66. Saunders, Lyle, "English-Speaking and Spanish-Speaking People of the Southwest," reprinted from *Cultural Differences and Medical Care: The Case of the Spanish Speaking People of the Southwest*, in Herman D. Stein and Richard A. Cloward, *Social Perspectives on Behavior*, 1958, The Free Press, Glencoe, Ill.

67. Schorr, Alvin L., "Families on Wheels," *Harper's,* January 1958.

68. Schorr, Alvin L., "Problems in the ADC Program," *Social Work,* April 1960.

69. Shanas, Ethel, "Some Sociological Research Findings About Older People Pertinent to Social Work," in *Toward Better Understanding of the Aged,* Council on Social Work Education, 1958, New York, p. 52.

69A. Shanas, Ethel, *The Living Arrangements of Older People in the United States,* a paper prepared for delivery at the Fifth Congress, International Association of Gerontology, August 1960.

70. Sheldon, Henry D., *The Older Population of the U.S.,* 1958, John Wiley & Sons, New York.

71. Sheldon, J. H., "Old-Age Problems in the Family," *Milbank Memorial Fund Quarterly,* April 1959, p. 123.

72. Sheldon, J. H., "Social Aspects of the Aging Process," in Milton Derber, ed., *The Aged and Society,* December 1950.

73. Smith, William M., Jr.; Britton, Joseph H.; and Britton, Jean O., *Relationships Within 3-Generation Families,* Pennsylvania State University, College of Home Economics Research Publication 155, April 1958, p. 7.

74. State of Maine, unpublished studies, 1956 and 1958.

75. Steiner, Peter O., and Dorfman, Robert, *The Economic Status of the Aged,* University of California Press, 1957, pp. 68-85, 109-111.

76. Stevens, David H., and Springer, Vance G., "Maine Revives Responsibility of Relatives," *Public Welfare,* July 1948.

77. Streib, Gordon F., "Family Patterns in Retirement," *Journal of Social Issues,* Vol. XIV, 1958, No. 2, issue on "Adjustment in Retirement," edited by Gordon F. Streib and Wayne E. Thompson with Ernest W. Burgess, Edward A. Suchman, and Ethel Shanas.

78. Sussman, Marvin B., "Family Continuity: Selective Factors Which Affect Relationships Between Families at Generational Levels," *Marriage and Family Living,* May 1954.

79. Sussman, Marvin B., "The Help Pattern in the Middle Class Family," *American Sociological Review,* February 1953, p. 23 and p. 25.

80. Tawney, Richard H., "Economic Virtues and Prescriptions for Poverty," from *Religion and the Rise of Capitalism,* in Herman D. Stein and Richard A. Cloward, *Social Perspectives on Behavior,* pp. 266-287, 1958, The Free Press, Glencoe, Ill.

81. Tierney, Brian, *Medieval Poor Law,* 1959, University of California Press, Berkeley.

82. Townsend, Peter, *The Family Life of Old People,* 1957, The Free Press, Glencoe, Ill., p. 165.

83. U.S. Department of Commerce, Bureau of the Census, *Current Population Reports,* Series P-20, No. 73, Mar. 12, 1957.

84. ———, ———, ———, Series P-20, No. 100, Apr. 13, 1960.

85. ———, ———, ———, Series P-20, No. 101, May 22, 1960.

86. ———, ———, ———, Series P-60, No. 12, June 1953.

87. U.S. Department of Health, Education, and Welfare, *Trends,* 1959 edition, p. 55.

88. ———, Advisory Council on Public Assistance, *Public Assistance,* January 1960, Appendixes A, G, and H.

89. ———, Bureau of Old-Age and Survivors Insurance (Social Security Administration), *Adults Who Are Incapable of Handling Their Own Benefit Funds or Who Are Marginally Capable of Doing So,* mimeographed, May 1958.

89A. ———, ———, "Income of Old-Age and Survivors Insurance Beneficiaries: Highlights from Preliminary

Data, 1957 Survey," *Social Security Bulletin,* August 1958.

90. ———, ———, *Interstate Mobility of Aged Beneficiaries Under OASI in 1957, . . . in 1958,* mimeographed.

91. ———, ———, *National Survey of OASI Beneficiaries, 1957,* unpublished tables.

92. ———, ———, *National Survey of OASI Beneficiaries, 1951,* unpublished tables.

93. ———, ———, *Program Simplification Memorandum,* Nov. 24, 1958, "Dependency Requirements for Auxiliary Beneficiaries," also *Report of the Work Simplification Group,* August 1959.

94. ———, Bureau of Public Assistance (Social Security Administration), *Recipients of Old-Age Assistance in Early 1953,* Part I—State Data, tables 8 and 18.

95. ———, ———, *Methods Used by States in Determining Need and Amount of Payment for Assistance in the Aid to Dependent Children Program,* Gladys O. White, October 1959, p. 24.

96. ———, Office of Vocational Rehabilitation, *Facts in Brief,* Rehabilitation Service Series No. 450, May 1958, reissued May 1959.

97. ———, ———, *Measuring the Disabled Individual's Economic Need,* Rehabilitation Service Series No. 248, p. 15.

98. ———, Region II, *Administrative Cost Study,* State of New York, Department of Social Welfare, July-October, 1959.

99. ———, ———, *Administrative Study of New Jersey Department of Institutions and Agencies,* 1958.

100. ———, ———, *Study of Special Support and Services,* Office of Public Assistance, Pennsylvania Department of Public Welfare, 1959.

101. ———, Social Security Administration, Division of Program Research, *Family Status of the Aged in 1959 and Money Income in 1958,* Research and Statistics Note No. 5, mimeographed Jan. 25, 1960.

102. U.S. Department of Labor, Bureau of Labor Statistics, *Study of Consumer Expenditures, Incomes and Savings, Urban U.S.—1950,* Vol. XI, 1957.

103. ———, *Population and Labor Force Projections for the U.S., 1960-1975,* Bulletin No. 1242 (work-life expectancy is cited from a paper in July 1957 by Seymour L. Wolfbein).

104. Varchaver, Catherine, *Older People in the Detroit Areas and the Retirement Age,* 1956, Wm. B. Eerdmans Publishing Co.

105. Von Hentig, Hans, "The Sociological Function of the Grandmother," *Social Forces,* May 1946.

106. Wentworth, Edna C., *How Persons Receiving Social Security Benefits Get Along,* Address June 23, 1959, to American Library Association, Washington, D.C.

107. West, James, *Plainsville, U.S.A.,* 1945, Columbia University Press, New York, pp. 58-59.

108. Wickenden, Elizabeth, *The Needs of Older People,* 1953, American Public Welfare Association, Chicago.

109. Wilensky, Harold L., and Lebeaux, Charles N., *Industrial Society and Social Welfare,* 1958, Russell Sage Foundation, New York.

110. Wisconsin Legislative Council, *Problems of the Aged,* 1953, Vol. I, Part II.

111. Young, Michael, "The Extended Family Welfare Association," *Social Work* (London, England), January 1956, p. 150.

112. Young, Michael, "The Role of the Extended Family in a Disaster," *Human Relations,* 1954, No. 3.

15. CRITERIA FOR SELECTING A MATE

As we saw in chapters 5 and 6, industrialization and the decline of traditions are quite as visible in the Orient as in the West. The first of Sugimoto's vignettes illustrates the point that marriages are sometimes contracted not only between a man and woman who are not in love with each other but who are not aware of the other's identity or existence. The second of Sugimoto's vignettes illustrates the conflict of the transition.

In considering marriage with and without love Freeman views ethnocentrism and the avoidance of incest as two pressures which govern mate-selection, and he sets up a classification on the basis of degree of arrangement and of preferential mating.

Noting that marriage is a link between kin lines and that it affects the ownership of property and the exercise of influence, Goode asserts that mate-selection and love are "too important to be left to the children." He goes on to distinguish several methods of "love-control," such as child marriage and segregation of the sexes. The upper strata of a society have much more at stake in the maintenance of the social structure and thus are more strongly motivated to control mate-selection. Also, their young have much more to lose than is the case in the lower strata. It therefore follows, according to Goode, that the upper strata will maintain stricter control over love and courtship to keep mate-selection within the field of eligibles and that, as a result, there will be less free choice of mates in the upper strata than in the lower strata.

OLD LOVE AND NEW

Etsu Inagaki Sugimoto

Adapted from *A Daughter of the Samurai,* by E. I. Sugimoto. Copyright 1925 by Doubleday and Company, Inc., pp. 55-58, 88-89. Reprinted by permission of the publisher.

The Honourable All Ordains

This family council was the largest that had been held since Father's death. Two gray-haired uncles were there with the aunts, besides two other aunts, and a young uncle who had come all the way from Tokyo on purpose for this meeting. They had been in the room for a long time, and I was busy writing at my desk when I heard a soft "Allow me to speak!" behind me, and there was Toshi at the door, looking rather excited.

"Little mistress," she said with an unusually deep bow, "your honourable mother asks you to go to the room where the guests are."

I entered the big room. Brother was sitting by the *tokonoma,* and next to him were two gray-haired uncles and the young uncle from Tokyo. Opposite sat Honourable Grandmother, the four aunts, and Mother. Tea had been served and all had cups before them in their hands. As I pushed back the door they looked up and gazed at me as if they had never seen me before. I was a little startled, but of course I made a low, ceremonious bow. Mother motioned to me, and I slipped over beside her on the mat.

"Etsu-ko," Mother said very gently, "the gods have been kind to you, and your destiny as a bride has been decided. Your honourable brother and your venerable kindred have given much thought to your future. It is proper that you should express your gratitude to the Honourable All."

I made a long, low bow, touching my forehead to the floor. Then I went out and returned to my desk and my writing. I had no thought of asking, "who is it?" I did not think of my engagement as a personal matter at all. It was a family affair. Like every Japanese girl, I had known from babyhood that sometime, as a matter of course, I should marry, but that was a faraway necessity to be considered when the time came. I did not look forward to it. I did not dread it. I did not think of it at all. The fact that I was not quite thirteen had nothing to do with it. That was the attitude of all girls. . . .

The Honourable All Defied

I must have been very young when my brother went away, for though I could distinctly recall the day he left, all memory of what went before or came immediately after was dim. I remember a sunny morning when our house was decorated with wondrous beauty and the servants all wore ceremonial dress with the Inagaki crest. It was the day of my brother's marriage. . . .

Ishi and I wandered from room to room, she explaining that the bride for the young master would soon be there. . . .

. . . The "seven-and-a-half-times" messenger in his stiff-sleeved garment . . . had returned from his seventh trip to see if the bridal procession was coming, and though the day was bright with sunshine, was lighting his big lantern for his last trip to meet it halfway—thus showing our eagerness to welcome the coming bride. . . .

Then suddenly something was wrong. Ishi caught my shoulder and pulled me back, and Brother came hurriedly out of Father's room. He passed us with long, swinging strides, never looking at me at all, and stepping into his shoes on the garden step, he walked rapidly toward the side entrance. I have never seen him after that day.

The maiden my brother was to have married did not return to her former home. Having left it to become a bride, she was legally no longer a member of her father's family. This unusual problem Mother solved by inviting her to remain in our home as a daughter; which she did until finally Mother arranged a good marriage for her.

In a childish way I wondered about all the strangeness, but years had passed before I connected it with the sudden going away at this time of a graceful little maid named Tama, who used to arrange flowers and perform light duties. . . . Tama was not a servant. In those days it was the custom for daughters of wealthy tradesmen to be sent to live for a short time in a house of rank, that the maiden might learn the strict etiquette of *samurai* home life. The position was far from menial. . . .

The morning after my brother went away I was going, as usual, to pay my morning greetings to my father when I met Tama coming from his door, looking pale and startled. She bowed good morning to me and then passed quietly on. That afternoon I missed her and Ishi told me that she had gone home.

Whatever may have been between my brother and Tama I never knew; but I cannot but feel that, guilt or innocence, there was somewhere a trace of courage. . . . In that day there could be only a hopeless ending to such an affair, for no marriage was legal without the consent of parents, and my father, with heart wounded and pride shamed, had declared that he had no son.

It was not until several years later that I heard again of my brother. One afternoon Father was showing me some twisting tricks with a string. . . . A maid came to the door to say that Major Sato, a Tokyo gentleman whom my father knew very well, had called. . . . I shall never forget that scene. Major Sato, speaking with great earnestness, told how my brother had gone to Tokyo and entered the Army College. With only his own efforts he had completed the course with honour and was now a lieutenant. There Major Sato paused.

My father sat very still with his head held high and absolutely no expression on his stern face. For a full minute the room was so silent that I could hear myself breathe. Then my father, still without moving, asked quietly, "Is your message delivered, Major Sato?"

"It is finished," was the reply.

"Your interest is appreciated, Major Sato. This is my answer: I have daughters, but no son."

MARRIAGE WITHOUT LOVE: MATE-SELECTION IN NON-WESTERN SOCIETIES

Linton C. Freeman

Adapted from Robert F. Winch, *Mate-Selection*, New York, Harper & Bros., 1958, 20-39.

Virtually all societies consider married life the most desirable type of existence for adults. The details may vary from one society to another, but all societies encourage marriage; all advocate a relatively stable union between two or more persons. And in most cases the marital union involves such activities as living together, working together, having children, and rearing them. This is marriage; it is the rule in every known human group, past or present, "primitive" or modern.

If marriage is the rule in every society it is also the practice. Most people marry. In America Bill Brown marries Alice Jones because he loves her. But a couple of centuries ago in feudal Japan it was different. There, Ito Satake took Haru Taira as his wife, although he had never met her, because she was selected by his family, and he by hers. Among the Yaruro Indians of Venezuela, José Miguel was wed to Anita because she was his mother's brother's daughter. But in Southwest Africa, Gogab

Garub, a Hottentot man, could not marry Hoaras Garis even though he wanted to because she was his seventh cousin, a very distant relative indeed! People marry everywhere, but the person whom they select to marry and their reasons for selecting that person show wide variation. It is some of this variation which we shall explore in the present chapter.

Nowhere is mate-selection a random activity. It is always guided by two principles which underlie the process of selection. These are (1) *preferential mating,* which serves to define and delimit a field of eligible mates—sometimes narrow, sometimes wide—into which a person is encouraged to marry, and (2) *marriage arrangement,* which refers to the degree to which persons other than those marrying participate in the process of selection. These principles will be discussed in turn below.[1]

Preferential Mating

All societies have rules prohibiting certain classes of persons from marrying one another. One part of these rules results from the prohibition of incest, or the tendency of all peoples to prohibit sexual activities— and hence marriage—among people who are close kinsmen. In mate-selection incest prohibition is expressed as the universal tendency to prohibit marriage between mother and son, father and daughter, or brother and sister. And many societies extend these prohibitions. Sometimes they include all known blood relatives and even persons with whom kinship ties are very distant. Some Australians, for example, prohibit marriage between persons of similarly named clans living a hundred miles away, and most probably entirely unrelated.[2] From this point of view, then, the ideal mate for any person would be a stranger, an outsider, an individual not related to him in any real or imagined way.

In every society there is also a principle of mate-selection which runs counter to the incest prohibition, namely, the force of ethnocentrism, or group-conceit, which is common to all human groups. More or less universally the outsider is suspect; people tend to distrust and to dislike persons of different races, creeds, or cultural backgrounds. On the other hand, members of one's immediate family and community are accepted; they share the same traditional cultural values. In mate-selection ethnocentrism is expressed through prohibiting marriage with outsiders. Almost all societies, for example, oppose marriages between persons of markedly different racial or cultural backgrounds. Very often marriage with in-

---

[1] This discussion is based in part upon G. P. Murdock's "Social Laws of Sexual Choice," as outlined in his *Social Structure,* New York, Macmillan, 1949. The present scheme represents a modification and simplification of Murdock's analysis.

[2] Robert H. Lowie, *An Introduction to Cultural Anthropology,* New York, Farrar & Rinehart, 2nd ed., 1940, p. 233.

dividuals outside the village or tribe is prohibited, and those from other social classes are considered unacceptable. From this point of view, the ideal mate for a man would be his mother or sister or some other member of his immediate family, since it is with these persons that he shares the most social relationships, that he feels the most cohesion. But as we have seen in the preceding paragraph, selection of a close relative as a mate, which would be a case of extreme ethnocentrism, is prohibited by the incest taboo.

From this discussion we can observe two conflicting tendencies which govern preferential mate-selection. They may be viewed as forces operating in opposite directions to locate a field of eligible mates. The location of the field of eligibles in any society will be determined by the relative strength of these two principles in that particular situation. For some groups ethnocentrism may be particularly forceful, and the result will be a high proportion of marriages among persons who are closely related to each other. Such was the case among the rulers of ancient Hawaii, Egypt, and the Incas, where brother-sister marriage was common. In our own country the intermarriage of cousins has been remarked in two very different settings: among the highlanders of the southeastern states and among the "proper Bostonians." [3] Other groups, however, have stressed incest prohibitions. Traditionally, among the Chinese, for example, one could not marry a person with the same family name as oneself, no matter how distant the relationship, and in a nation of China's size such a relationship might be distant indeed! In certain African tribes persons with a common ancestor cannot marry even though they may be quite remotely related. Just so, certain societies have prohibited marriage between persons of the same village or even of the same tribe.

In most cases neither of these forces is expressed in the extremes described above. Rather, the forces of ethnocentrism and of incest prohibition are both applied to isolate a field of eligibles within which mate-selection is expected to occur. The field of eligible or approved mates may be large or it may be small, its limits may be vague or they may be well defined, but in any case all societies have rules which tend to establish such a field. Even in modern American society we are expected to marry a person from our own race, religion, class, and age group, but one who is not too closely related to us by blood ties. In other societies the rules of preferential mating may take different forms, but some rules will always be present.

[3] J. S. Brown, "The Conjugal Family and the Extended Family," *American Sociological Review* (1952), *17*:297-306; Cleveland Amory, *The Proper Bostonians*, New York, Dutton, 1947, chap. 1.

Marriage Arrangement

The second principle guiding mate-selection refers to the process of choosing a person within the field of eligibles. The field is established according to the rules for preferential mating, but very often a choice must be made among several eligibles within the field; this is where the principle of marriage arrangement comes into play. This principle is a reflection of the fact that almost every society has rules according to which persons other than those marrying participate in the process of mate-selection to a greater or lesser degree. As we shall see, some societies specify extreme intervention; they encourage the practice of fully arranged marriages. Others, such as our own, allow very little outside intervention and permit the person marrying to make his or her own choice. But in any case, the principle of marriage arrangement constitutes a real dimension of marriage choice once the field of eligibles has been established through the principle of preferential mating.

The Family and Mate-Selection[4]

Variations in the bases upon which mates are selected can perhaps best be understood through examining the setting in which marriage occurs in various societies. Marriage takes place in every society, and in each case it results in the formation of a marriage group or nuclear family. Such families consist of mates and their offspring.

Some societies, such as our own, consider this nuclear family as *the* family. To us, the family includes a married couple and their unmarried children. They usually live in a separate residence, often far from other relatives. Although a relationship with others, say grandparents, uncles, aunts, and cousins, is recognized, these relatives are almost outsiders; they are not included in the intimate family interaction. The family, then, is small and independent of other ties, but even this small unit is somewhat loosely organized and is seldom the center of the activities of its members. This is especially true of the father, who spends the larger part of his waking hours away from home and family at his place of work. Then too, after infancy the children are busy outside of the home most of the time, either in school or in recreational activities. The mother is more likely to be involved with home and family. Mothers with young children are by necessity more home-bound and family-oriented than others in our

[4] This discussion of the contemporary American family is, for the most part, based upon the following sources: John Sirjamaki, *The American Family in the Twentieth Century,* Cambridge, Harvard University Press, 1955, and Robert F. Winch, *The Modern Family,* New York, Holt, 1952.

society, but today, with the working wives and the clubwomen, even more mothers show progressive emancipation from family ties.

Interaction and intimacy are certainly not absent in the small family group, but contacts outside that core tend to be made with age-group friends rather than with relatives. Such contacts may even preclude activities within the small family—the poker party or bowling night removing the father, the bridge club taking the mother, and the soda bar or other neighborhood hangout demanding the presence of the youngster.

Such is the typical family life of the contemporary American middle class. It is a small family and a loosely integrated one. It is a family which reflects the individualistic stress of our way of life. Individuals are taken as such; a person is known—is understood—by his job, his income, his education, his personal achievements, and, above all, his individual personality. Questions concerning his race, his religion, and his neighborhood, may be important in classifying him, but his background, his grandfather's occupation, and so on, are relatively unimportant. He may move to a position of greater or lesser prestige in the society; in fact, he is encouraged to succeed." He is an individual, and his successes and failures are his own. They reflect little upon his antecedents or his descendants—they tend to be his and his alone. It is true that his family, his wife and children, will be affected by a man's success or failure in the economic world. But a working wife may mediate this position considerably, and then the children are encouraged to go out on their own to succeed or fail for themselves.

This same individualism is reflected in American patterns of mate-selection. Bill Brown, it was said above, marries Alice Jones because he loves her. This is the ideal American middle-class pattern. They meet, "love strikes," and they marry. The basis for their feeling is undetermined—unquestioned; love is enough. However, it is likely that she will be selected from a field of eligibles which is determined according to the principle of preferential mating. The incest prohibitions will force him to choose a mate among those outside of his immediate family group. Ethnocentrism, on the other hand, dictates that he marry within his own race, his own social class, his own general age group, and usually from his own neighborhood or at least his own town. This provides him with a field of eligibles, people who are pretty much like him, sharing the same values, attitudes, and interests. Geographical migration may have led him away from his family spatially, and his own success or failures may have led him away in terms of attitudes and interests. But he will probably have more in common with the women in his field of eligibles than with his more distant relatives. Thus, though the range of acceptable mates is wide, it serves to provide him with a potential wife with whom he has much in common, and still someone who is outside of his own family group.

Ideally, within this field of eligible mates, American society gives the person marrying great leeway in selecting a spouse. Parental opposition may provide a minor setback, but since the parents have no direct personal interest in the union other than their children's happiness, their recalcitrance can usually be overcome or ignored. After all, the children have already been largely independent of their parents for some years. They have run with their own peers, made their own decisions, and somewhat governed their own lives. So selecting their own mates is merely another instance of their independence. Parents do not feel that they should interfere in the selection process, and probably they frequently doubt that they could intervene effectively if they wished to. Sometimes parents are not even consulted, for love is a personal affair between the two who experience it. Marriage, too, is their affair. When young people marry they usually leave their parents both spatially and emotionally; they go off by themselves and start their own family in their own home. Marriage usually destroys the last vestiges of dependence upon their parents. They are free—their new family is an independent unit in an individualistic world.

Most societies, however, are organized along very different lines. Imagine a society without specialists in various occupations, without schools or teachers, without churches or clergymen, without jails or police, without courts or judges. This is a society in which there are no governmental bureaucrats, no physicians or nurses, no farmers or merchants, one in which the only division of labor is made along age and sex lines. In such a society everyone would be pretty much like everyone else. Within age and sex groups all would do the same things and have the same rights and the same duties. But how would such a society be organized? Without teachers, who would educate the children? Without clergymen, who would guide the religious life of the community? Without police, who would apprehend criminals, and without judges, who would determine their guilt or innocence? Who would run the society, cure the ill, grow the food, sell the basic necessities? The same answer may be given for all of these questions: the family.

Instead of the individual, the family is the key unit of social organization. An individual's position comes not from his occupation or income, or even from his accomplishments, but rather from his family. An individual has identity only in his family setting. Aginsky provides some notion of the importance of the family in his report of a statement of an old Pomo Indian:

A man must be with his family to amount to anything with us. If he had nobody else to help him, the first trouble he got into he would be killed by his enemies because there would be no relatives to help him fight the poison of the other group. No woman would marry him because her family would not let her marry a man with no family. He would be poorer than a newborn child; he would be poorer than a worm, and the family would not consider him worth anything. He would not bring renown or glory with him. He would not bring support of other relatives either. The family is important. If a man has a large family and a profession and upbringing by a family that is known to produce good children, then he is somebody and every family is willing to have him marry a woman of their group. It is the family that is important. In the white ways of doing things the family is not so important. The police and soldiers take care of protecting you, the courts give you justice, the post office carries messages for you, the school teaches you. Everything is taken care of, even your children, if you die; but with us the family must do all of that.[5]

Thus, people living in such societies view the family very differently from the way we do. The family to them is a larger kinship group consisting of the nuclear marriage group and various other relatives including grandparents, uncles, aunts, cousins, grandchildren, and so on. In these societies the larger kinship group or extended family often shares a common residence and is a self-supporting coöperative unit. It is the center of activities for the individuals comprising it, and the various marriage groups within it are submerged into the larger system.

The Yaruros of Venezuela[6]

Earlier we said that among the Yaruros of Venezuela José Miguel married Anita because she was his mother's brother's daughter. Let us explore the background for such a marriage in order to determine its basis. The Yaruros are a tribe of river nomads who inhabit a vast plain southeast of the Andes in inland Venezuela. Throughout most of the year the climate is hot and dry, the temperature climbing well above 100 degrees each day. This suffocating heat is made bearable only by the trade winds, which blow incessantly with a gale force over the level sands.

During the dry season the land is a veritable desert—miles of flat sand with very little animal or plant life. But during the rainy season the land is overrun with flora and fauna, food is plentiful, and life is relatively easy. Easy, that is, except for the insects. Insects are more than plentiful; they are omnipresent, and only the constant winds save the people from these pests.

[5] B. W. Aginsky, "An Indian's Soliloquy," *American Journal of Sociology* (1940), 46:43-44.

[6] This description is based upon Vincenzo Petrullo, "The Yaruros of the Canpanaparo River, Venezuela," *Smithsonian Institution, Bureau of American Ethnology, Bulletin 123,* Anthropological Papers, No. 11, 1938, pp. 161-290.

The Yaruros are a short, dark, small-boned people who have a definite Mongoloid appearance. They consider themselves to be one people but divide their lands according to recognized territorial limits. Each band has a large area along a river over which its members roam in quest of food.

The life of the Yaruros is simple to the extreme. Their economy is based upon fishing, hunting, and gathering, and although they travel far and wide in search of food, they never venture far from the river. They migrate in canoes and their only possessions can be carried in this fashion. Aside from a few baskets and pots, and some hunting equipment, they own nothing. They move on whenever food becomes scarce and settle again, but only temporarily, in any likely-looking spot.

Men do all the hunting and fishing, using bows and arrows and fish-hooks. Women gather plants and herbs and roots, which they carry in baskets slung from their heads. The diet is variable, but young crocodiles are the basic staple. Crocodile eggs, turtles, turtle eggs, and various plants and roots are also common. And honey and tobacco are continually sought. Except for dogs, the Yaruros have no domesticated animals.

The clothing worn by men consists only of a loincloth, while the women wear more elaborate girdles fashioned from foliage. Due to their sparse clothing and their constant exposure to the bright sun these people are tanned a rich brown most of the time.

The Yaruros are a peaceful people, given to religious musings. They travel in small hunting groups, and shamans or religious leaders are common. A typical camp might consist of an old man and his wife, their unmarried sons and daughters, their married daughters and husbands, their grandchildren, and sometimes their unmarried brothers and sisters. This family travels and hunts and camps as a group.

Campsites may be found anywhere along the river. The only shelters are temporary windbreaks fashioned out of branches, but these must not be too elaborate or effective or they will allow insects to gather. People find shelter from the sun and wind by sitting behind them, but they keep warm at night by burying themselves in the sand.

Yaruro society is divided into two moieties or halves on the basis of kinship. Each person belongs to one or the other moiety. Descent is traced in the female line; one inherits his moiety from his mother. And one must marry into the opposite moiety.

For the Yaruro, this means that a man must marry one of his cross-cousins, that is, one of his mother's brother's daughters or one of his father's sister's daughters. Each hunting group is made up of a single family, and other hunting groups nearby are usually closely related to it. Contacts with other more distant groups are very infrequent. It would seem, therefore, that a Yaruro man might marry only among his close relatives. However, it is considered incest to marry his sisters, his mother, and his mother's sisters, whom he calls mother also. Likewise, his mother's sister's daughters are called sister, and marriage with them is prohibited. And his father's brother's daughters belong to his own moiety. So this seems to leave only his mother's brother's daughters and his father's sister's daughters, that is, his cross-cousins. But since cross-cousin marriage probably took place in the parental generation also, his mother's brother's daughters *are* his father's sister's daughters. And it is among these cross-cousins that a Yaruro man is expected to marry.

When a young man among the Yaruros reaches marriageable age he talks with his father. The father takes him to the shaman—the religious leader—who instructs him in the obligations of marriage. Then the shaman goes to one of the boy's uncles, who selects one of his daughters, and the boy simply moves into his uncle's household. Thereafter he is obligated to work and hunt with his uncle. In effect, he takes the place of the uncle's sons who go off to live in the camps of their own wives.

Yaruro marriage practices, therefore, typify a procedure which serves to delimit an extremely narrow field of eligibles. Taken together, the incest taboos and the attitudes of ethnocentrism restrict the field of eligibles for the typical Yaruro man to one of his cross-cousins. Such an arrangement solves the problems raised by poor communication and sparse population. It affords access to a potential mate in the immediate vicinity but requires that the mate be obtained from another camp. This promotes interaction between camps and tends to maintain interfamilial solidarity.

While the dictates of preferential mating tend to limit a person's freedom to select a mate in Yaruro society, marriage arrangement carries this restriction farther still. Preferential mating requires that a person marry his cross-cousin; marriage arrangement allows his uncle to select which particular cross-cousin it shall be. Such a pattern differs from the American one in the two important respects which we are considering here. First, one must select a mate from a small field of eligibles, i.e., from among one's cross-cousins. Persons other than cross-cousins are not defined as suitable marriage partners. Second, the choice of spouse from the field of eligibles is made not by the person to be married but by another—the uncle of the boy, who is also father of the girl.

Feudal Japan[7]

Some societies are permissive in both preferential mating and marriage arrangement, like the Americans; and some, like the Yaruros, are re-

[7] This description is based principally upon the following sources: Lafcadio Hearn, *Japan, an Attempt at Interpretation,* New York, Macmillan, 1904; Kenneth S. La-

strictive in both. Others, however, permit selection from a wide field of eligibles but restrict the persons marrying from participating in the choice. To illustrate this, we must remember that in feudal Japan Ito Satake married Haru Taira because she was selected by his family and he by hers. Let us examine the setting in which this marriage took place and try to uncover some of the factors involved in this type of marriage choice.

Arranged marriage was found in its purest form among the aristocracy of Japan during the feudal era. Let us say, therefore, that our couple were aristocrats and that they lived during the early 1700's, the height of Japanese feudalism.

Japan is a group of about three thousand islands lying off the east coast of continental Asia. The four main islands are Hokkaido, Honshu, Shikoku, and Kyushu. These form a long chain running for 1500 miles from northwest to southeast. The terrain consists chiefly of mountainous areas covered with vast forests, and only 15 percent of it is arable.[8] The climate shows great variation from region to region, and no single description will hold for all of Japan; northern Japan is subject to long and bitter winters while the south is typically subtropical.

The people of Japan are also varied. They are predominantly Mongoloid, but much Malayan and some Caucasian Ainu ancestry is evident. They are short-legged people with straight black hair and dark eyes. Skin color ranges from an almost Nordic pink and white to the light brown of the Pacific Islanders. Since they are a heterogeneous group, few generalizations could be made which would accurately describe the physical characteristics of the people of Japan.

During the feudal era Japan was divided into small duchies, each ruled by a local lord backed by armed knights. A hereditary military leader governed the entire nation and relegated the Emperor to comparative

---

tourette, *The Development of Japan,* New York, Macmillan, 1920; George Bailey Sansom, *Japan: A Short Cultural History,* New York, Appleton-Century, 2nd ed., 1943; Etsu Sugimoto, *Daughter of the Samurai,* New York, Doubleday, Page, 1925.

The following sources were also consulted: Alice Mabel Bacon, *Japanese Girls and Women,* Boston, Houghton Mifflin, 1902; Ruth Benedict, *The Chrysanthemum and the Sword,* Boston, Houghton Mifflin, 1946; Frank Brinkely, *A History of the Japanese People from the Earliest Times to the End of the Meiji Era,* New York, Encyclopaedia Britannica, Inc., 1915; John Embree, *The Japanese Nation,* New York, Farrar & Rinehart, 1945; Douglas G. Haring, "Japan and the Japanese," in Ralph Linton (ed.), *Most of the World,* New York, Columbia University Press, 1949; Arthur M. Knapp, *Feudal and Modern Japan,* Boston, J. Knight Co., 1897, 2 vols.; Inazo Ota Nitobe, *The Japanese Nation,* New York, Putnam, 1912; George Bailey Sansom, *The Western World and Japan,* New York, Knopf, 1950.

[8] H. A. Meyerhoff, "Natural Resources in Most of the World," in Ralph Linton, *op. cit.,* p. 63.

insignificance. Social classes were defined rigidly, and one's class was reflected in his dwelling, his style of dress, his food, etc. Both commoners and knights owed allegiance to the local lord.

However, more important even than his responsibility to his lord was a person's responsibility to his family. His thoughts, his feelings, his behavior were so completely bound up in family life that understanding him is unthinkable without first examining his family.

Family life in feudal Japan was organized around the patriarch. A typical household might include the patriarch, his wife, all his sons and their wives and children, his unmarried daughters, servants, and younger brothers of the patriarch and their wives and children. Such a family was a relatively large-scale coöperative unit under the direction of the patriarch. He was the director of the household. He handled all household business except for the purchase of supplies, and even there he might intervene and make the final decision. He was the priest in family worship and the manager of family properties. In short, the patriarch was the almost absolute ruler of all members of the Japanese family.

The power of the patriarch, however, was not unlimited. Important decisions were usually made by the family council, which included most of the mature males and the old women in the family. This group met to decide such things as adoption of a child or marriage of a family member, but day-to-day decisions were made by the family patriarch.

Children were reared into a pattern of male dominance and rigid conformity. Infants were fondled and petted, but after the first few years permissive treatment gradually gave way to a demand for conformity. This gradual increase in restriction was manifest in a range of techniques. The emphasis, however, was never upon forcing the child to express himself within the vague framework of abstractions like "good" or "bad." Rather, self-expression was denied; children learned to conform to a rigid pattern of expectations which demanded almost ritual performance of even the simplest tasks. In the Japanese desire to create and maintain "face," etiquette played a major role. Children were taught polite customs by their mothers, who actually manipulated the child's body into the desired positions. This way a child learned to bow, and to sit on the floor attentively, for hours if necessary, while studying. And young girls were taught to sleep flat on their back, legs straight together and arms straight and at their sides.

Children were forced to succumb to these rigid demands through the continual threat of humiliation or withdrawl of family affection. Regular teaching and emphasis upon "family honor" tended to center the child's attention upon the family. The threat of denial of the satisfactions of the family was, therefore, a real and effective mechanism in insuring conforming behavior.

The relative social positions of men and women were learned quite early, as was the position of each family in the social hierarchy. Most important, the child learned the intricate system of duties and obligations which characterized Japanese culture. He learned to think and act, not as an individual, but as a family member. His position in the family and the position of his family in the larger society determined his fate. He grew to accept that fate—to behave always in terms of his proper station in life.

Ideally, by the time they reached adulthood the Japanese had learned to view each other, not as individuals at all, but almost completely as stereotypes. If two people were members of the same family they treated each other in terms of their relationship. They met neither as personalities nor as persons, but only as representatives of particular relationships. All fathers treated, and were treated by, their sons in much the same way. Their interaction was based upon their kinship, not upon personal feelings.

Relationships with persons outside the family were governed by the importance of the families involved. Even here, personalities were relatively unimportant; people met as representatives of their families, not as individuals. It was more important to know that a person was a member of the Nakamura family than to know that he was Kiyoshi Nakamura. As a person he did not count; as a representative of a certain family he became important. In short, in traditional Japanese society individuality was almost completely submerged in the family system.

In studying traditional Japan we see a society in which the family was central to all activities. Everything an individual did—each of his interpersonal relationships—was organized around his position in the family. Marriage, then, like any other interpersonal relationship was an affair of the entire family. People married, not to start a new family, but rather to perpetuate the already existing one. In a sense, a marriage was not a union between two persons but a technique for adding another person in order to keep the family going. Selecting the right person was important to everyone in the family since he or she might either contribute to the general welfare or be an active source of disorganization. Moreover, marriage established a long-term bond between two families. When considering marriage they looked for an eligible person from a family whose position might enhance their prestige and security.

Thus marriage among the traditional Japanese aristocracy was too important for all members of both families for them to allow the choice to be made by two young, inexperienced people. It was too big even for the patriarch to handle alone. So marriages were arranged by the family council.

When the family of a young man considered their son at an appropriate age for marriage they engaged the services of a family friend to act as go-between. This was necessary in order to save "face." For if they themselves started negotiations with the family of some prospective bride and were rejected, their loss of face or public esteem would be irreparable. It would be a personal affront to them—one from which recovery would be impossible. So some trusted family friend was elected, who might suggest several young ladies of suitable social standing and after discussion proceed to contact the family of one of these. After preliminary negotiations the girl's family usually appointed their own go-between, and thereafter followed a long period of careful investigation and negotiation. Considerations of social class, and honor and health among each family turned out to be healthy, honorable, and socially acceptable, agreement was reached and the betrothed was informed of the decision. The prospective bride and groom had little or no voice in the proceedings and often did not meet until the wedding ceremony.

The wedding ceremony was held at the groom's home. At the start of the ceremony the bride was often dressed in white (to signify her sorrow at losing her family), but during the proceedings she changed to a colorful kimono. Sake was drunk by the bride and groom, and the wedding ceremony was complete.

Marriage in traditional Japan thus presented a picture of almost complete subjugation of individual initiative in the familial group. The field of eligibles was relatively wide, but the marrying individuals were not even consulted in preparing for their marriage, the family council making all decisions. This pattern is understandable in light of the importance of family ties for the Japanese individual. It affords an illustration of the degree to which persons other than the ones being married may dominate the process of mate-selection.

Some societies have gone further still. Among villagers in India, the medieval English, and the inhabitants of the island of Buka in Melanesia child bethrothal was quite common. The parents of young children would get together and contract a marriage for their immature offspring. Such a contract was binding even though the union could not be consummated for several years. This practice was common also among the Kazaks of central Asia. In some cases two Kazak fathers might agree to unite their as yet unborn children. Such a practice represents an extreme of parental intervention in the mate-selection of their offspring.

The Hottentots[9]

A fourth form of mate-selection is typified by the pattern among the Hottentots of Southwest Africa. Here, you will recall, Gogab Garub, a Hottentot man, could not marry Hoaras Garis because she was his seventh cousin. Although the Hottentots allow the persons marrying a more or less free choice of a mate, the mate must be chosen from within a relatively small field of eligibles. Thus, for these people the rules of preferential mating are restrictive, but marital arrangement plays only a minor part in the selection process.

The Hottentots are a group of nomadic herders who live in a great grassy plateau in the southwestern corner of Africa. The climate is cool and dry; it provides little in the way of vegetation except for grass, but animal life is abundant. Animals of all sizes and kinds—from the elephant to the tiniest insect—literally cover the plain, but a few roots, berries, bulbs, and melons are the only edible vegetation.

Today the Hottentots number about twenty thousand. Once they were far more numerous, but warfare and disease have steadily diminished their number ever since European contact. They are extremely small people—men average about five feet three inches; women, four feet eleven. Their skin is light brown in color, and it is often very wrinkled. Their bodies are relatively hairless, and the hair on their heads is usually short, black, and quite kinky.

Hottentot economy is based upon hunting and herding. The people hunt game, large and small, and gather roots and berries. They raise large herds of cattle and sheep, but these are seldom slaughtered. Instead, they milk the cows and ewes, and use oxen as beasts of burden. Thus, meat and milk constitute the basic part of the Hottentot diet.

Women do most of the work, and the men are characterized as lazy. The women milk the animals, gather the edible plants, carry water and firewood, cook, maintain the houses, and make clothing and pottery. The herds are tended by young boys or captive servants, while the men spend their days hunting. Sometimes men prepare skins, or work with wood or metal, but for the most part they confine their activities to the hunt.

The Hottentots are divided into twelve tribes. Although each tribe is associated with a general territorial location, the members of a tribe do not live together in a single community. Instead, each tribe is broken

[9] This description is based upon the following sources: S. S. Dornan, *Pygmies and Bushmen of the Kalahari,* London, Seeley, Service, 1925; George P. Murdock, *Our Primitive Contemporaries,* New York, Macmillan, 1934; I. Schapera, *The Khoisan Peoples of South Africa,* London, Routledge, 1930; I. Schapera and B. Farrington (eds.), *The Early Cape Hottentots,* Cape Town, The Van Rebeeck Society, 1933.

down into a number of clans—groups of persons united by a common ancestor. In general, the members of a clan form a single community, or sometimes, if a clan is large, it may split up into several extended families, each of which goes off to live by itself.

The Hottentots are a seminomadic people. A typical community consists of a number of light portable dome-shaped huts. They are made of sticks and covered with rush mats and are constructed in such a manner that when the water supply runs low, or hunting or grazing becomes difficult, they may be torn down, loaded on oxen, and transported to a new location where water, grass, or game is more plentiful. A single hut is occupied by a nuclear family—a man, his wife, and their unmarried children.

In any encampment the huts are arranged according to age seniority. The Hottentots place great value upon age, and the eldest male member of any clan is the chief. But political and judicial authority are vested in the chief only with reservation. A clan is governed by a council consisting of the older men of the clan. This council directs the activities of clan members, settles quarrels within the clan, and punishes minor offenders. On the tribal level these activities are performed by the tribal council composed of the chiefs of all the clans in the tribes.

Marriage for the Hottentots is usually monagamous. However, some of the wealthier men take more than one wife. Each wife lives in a separate hut along with her children. Later or secondary wives give precedence to the first wife, whose children enjoy preferred rights of inheritance. Women own their huts and their own herds of animals. They also control the distribution of household provisions. A husband must respect his wife's rights in these matters, and if he violates them he is obliged to pay a penalty of sheep or cattle.

Neither boys nor girls are considered marriageable until they have passed through a series of rites at puberty. Each boy must also demonstrate his proficiency as a hunter by killing some big-game animal. Once these rites have been completed, however, the young people are allowed considerable sexual freedom. There is also marked freedom in the choice of mate, the principal limitation on which is the prohibition against marrying a person of their own clan or a person bearing their clan name even though from another tribe. On the other hand, they are discouraged from marrying a person too alien; they are—like the Yaruros—required to marry a cross-cousin, that is, a daughter of either the mother's brother or the father's sister. Thus among the Hottentots the range of potential mates is severely restricted from the start.

From this description we can see that the field of eligibles for the Hottentots is established on the basis of kinship. Incest prohibitions are strong—they are extended to include every member of a person's clan—everyone in his local encampment. The Hottentots camp in clan groups like the Yaruros, and most interpersonal contacts are with kinsmen. And like the Yaruros, isolation and ethnocentrism force them to seek a spouse from a neighboring encampment. In both cases the person sought is a cross-cousin. But here the resemblance ends. For while Yaruro custom dictates that the choice among cross-cousins be made by the uncle, the Hottentots allow the persons marrying to make their own choice.

Thus within the field of eligibles—his cross-cousins—a young man is free to choose his own mate. When he finds a desirable prospect, a young Hottentot man speaks to his parents. They, in turn, send emissaries to the parents of the girl to ask her hand. Tradition dictates that they refuse. But the young man is undaunted by this seeming opposition. He attempts to enlist the support of the girl. He watches her house at night until he determines the location where she sleeps. Then, when everyone has retired, he goes in and lies down next to her. Usually she gets up and moves to another part of the hut but he remains in her place until morning. The next night he returns to her hut. If he finds her in the same spot he knows that his suit is favored. She may leave again, but sooner or later she will stay and the marriage is consummated.

The marriage feast is conducted the same day. He provides a cow for slaughter and goes off to hunt game for a feast. He presents gifts of cattle to her family but receives an equal number in return. The couple resides in the camp of the bride for about a year or until their first child is born. Thereafter they move to the camp of the groom's parents and establish their permanent residence in their own hut.

For the Hottentots, then, the process of actually selecting a mate is one of great freedom and individual initiative. Both bride and groom have a voice, and there is usually little or no outside interference. They are in this respect like the modern Americans, but for them, unlike the Americans, the range of potential mates from which the choice is made is relatively small.

We have defined two principles which govern the process of mate-selection: (1) preferential mating and (2) marriage arrangement. And we have indicated that each of these principles may vary from society to society. Some societies may delimit a narrow field of eligibles and some a wide one. Some may allow extreme intervention—by parents, for example—in the choice of a mate while others allow no intervention whatsoever. In order to illustrate their operation we have described four extreme types.

The first type is familiar to all of us who live in American society. Here we are allowed to choose a mate within a relatively wide field of eligibles, and outside intervention in our choice by parents or others is minimal.

The opposite of the American form in both respects was typified by the Yaruros of Venezuela. Here the field of eligibles is extremely narrow, and choice within that field is made by persons other than the ones being married.

A third extreme form is exemplified by the Japanese of feudal times. In this case the field of eligibles was wide, but the person marrying had little voice in the proceedings.

The fourth and final extreme may be illustrated by the Hottentots of Southwest Africa. This society prohibits marriage except within a relatively narrow field of eligibles, but within that field the persons marrying make their own choice.

We may summarize these points as follows:

Degree of Arrangement of the Marriage	Preferential Mating	
	Highly Specified Preferences Leading to Narrow Field of Eligibles	Little Specification of Preferred Mate Leading to Wide Field of Eligibles
High: Parents or others select one's spouse	Yaruros	Feudal Japan
Low: Principal selects own spouse	Hottentots	Middle-class U.S.A.

THE THEORETICAL IMPORTANCE OF LOVE

William J. Goode

Adapted from the *American Sociological Review*, 24 (1959), 38-47.

Because love often determines the intensity of an attraction[1] toward or away from an intimate relationship with another person, it can become one element in a decision or action.[2] Nevertheless, serious sociological attention has only infrequently been given to love. Moreover, analyses of

[1] On the psychological level, the motivational power of both love and sex is intensified by this curious fact: (which I have not seen remarked on elsewhere) Love is the most projective of emotions, as sex is the most projective of drives; only with great difficulty can the attracted person believe that the object of his love or passion does not and will not reciprocate the feeling at all. Thus, the person may carry his action quite far, before accepting a rejection as genuine.

[2] I have treated decision analysis extensively in an unpublished paper by that title.

love generally have been confined to mate choice in the Western World, while the structural importance of love has been for the most part ignored. The present paper views love in a broad perspective, focusing on the structural patterns by which societies keep in check the potentially disruptive effect of love relationships on mate choice and stratification systems.

Types of Literature on Love

For obvious reasons, the printed material on love is immense. For our present purposes, it may be classified as follows:

1. Poetic, humanistic, literary, erotic, pornographic: By far the largest body of all literature on love views it as a sweeping experience. The poet arouses our sympathy and empathy. The essayist enjoys, and asks the reader to enjoy, the interplay of people in love. The storyteller—Bocaccio, Chaucer, Dante—pulls back the curtain of human souls and lets the reader watch the intimate lives of others caught in an emotion we all know. Others—Vatsyayana, Ovid, William IX Count of Poitiers and Duke of Aquitaine, Marie de France, Andreas Capellanus—have written how-to-do-it books, that is, how to conduct oneself in love relations, to persuade others to succumb to one's love wishes, or to excite and satisfy one's sex partner.[3]

2. Marital counseling: Many modern sociologists have commented on the importance of romantic love in America and its lesser importance in other societies, and have disparaged it as a poor basis for marriage, or as immaturity. Perhaps the best known of these arguments are those of Ernest R. Mowrer, Ernest W. Burgess, Mabel A. Elliott, Andrew G. Truxal, Francis E. Merrill, and Ernest R. Groves.[4] The antithesis of romantic love, in such analyses, is "conjugal" love; the love between a settled, domestic couple.

A few sociologists, remaining within this same evaluative context, have

---

[3] Vatsyayana, *The Kama Sutra,* Delhi: Rajkamal, 1948; Ovid, "The Loves," and "Remedies of Love," in *The Art of Love,* Cambridge, Mass.: Harvard University Press, 1939; Andreas Capellanus, *The Art of Courtly Love,* translated by John J. Parry, New York: Columbia University Press, 1941; Paul Tuffrau, editor, *Marie de France: Les Lais de Marie de France,* Paris L'edition d'art, 1925; see also Julian Harris, *Marie de France,* New York: Institute of French Studies, 1930, esp. Chapter 3. All authors but the first *also* had the goal of writing literature.

[4] Ernest R. Mowrer, *Family Disorganization,* Chicago: The University of Chicago Press, 1927, pp. 158-165; Ernest W. Burgess and Harvey J. Locke, *The Family,* New York: American Book, 1953, pp. 436-437; Mabel A. Elliott and Francis E. Merrill, *Social Disorganization,* New York: Harper, 1950, pp. 366-384; Andrew G. Truxal and Francis E. Merrill, *The Family in American Culture,* New York: Prentice-Hall, 1947, pp. 120-124, 507-509; Ernest R. Groves and Gladys Hoagland Groves, *The Contemporary American Family,* New York: Lippincott, 1947, pp. 321-324.

instead claimed that love also has salutary effects in our society. Thus, for example, William L. Kolb[5] has tried to demonstrate that the marital counselors who attack romantic love are really attacking some fundamental values of our larger society, such as individualism, freedom, and personality growth. Beigel [6] has argued that if the female is sexually repressed, only the psychotherapist or love can help her overcome her inhibitions. He claims further that one influence of love in our society is that it extenuates illicit sexual relations; he goes on to assert: "Seen in proper perspective, [love] has not only done no harm as a prerequisite to marriage, but it has mitigated the impact that a too-fast-moving and unorganized conversion to new socio-economic constellations has had upon our whole culture and it has saved monogamous marriage from complete disorganization."

In addition, there is widespread comment among marriage analysts, that in a rootless society, with few common bases for companionship, romantic love holds a couple together long enough to allow them to begin marriage. That is, it functions to attract people powerfully together, and to hold them through the difficult first months of the marriage, when their different backgrounds would otherwise make an adjustment troublesome.

3. Although the writers cited above concede the structural importance of love implicitly, since they are arguing that it is either harmful or helpful to various values and goals of our society, a third group has given explicit if unsystematic attention to its structural importance. Here, most of the available propositions point to the functions of love, but a few deal with the conditions under which love relationships occur. They include:

(1) An implicit or assumed descriptive proposition is that love as a common prelude to and basis of marriage is rare, perhaps to be found as a pattern only in the United States.
(2) Most explanations of the conditions which create love are psychological, stemming from Freud's notion that love is "aim-inhibited sex." [7] This idea is expressed, for example, by Waller who says that love is an idealized passion which develops from the frustration of sex.[8] This proposition, although rather crudely stated and incorrect as a general explanation, is widely accepted.
(3) Of course, a predisposition to love is created by the socialization experience. Thus some textbooks on the family devote extended discussion to the ways in which our society socializes for love. The child, for example, is told that

[5] William L. Kolb, "Sociologically Established Norms and Democratic Values," *Social Forces,* 26 (May, 1948), pp. 451-456.
[6] Hugo G. Beigel, "Romantic Love," *American Sociological Review,* 16 (June, 1951), pp. 326-334.
[7] Sigmund Freud, *Group Psychology and the Analysis of the Ego,* London: Hogarth, 1922, p. 72.
[8] Willard Waller, *The Family,* New York: Dryden, 1938, pp. 189-192.

he or she will grow up to fall in love with some one, and early attempts are made to pair the child with children of the opposite sex. There is much joshing of children about falling in love; myths and stories about love and courtship are heard by children; and so on.

(4) A further proposition (the source of which I have not been able to locate) is that, in a society in which a very close attachment between parent and child prevails, a love complex is necessary in order to motivate the child to free him from his attachment to his parents.

(5) Love is also described as one final or crystallizing element in the decision to marry, which is otherwise structured by factors such as class, ethnic origin, religion, education, and residence.

(6) Parsons has suggested three factors which "underlie the prominence of the romantic context in our culture": (a) the youth culture frees the individual from family attachments, thus permitting him to fall in love; (b) love is a substitute for the interlocking of kinship roles found in other societies, and thus motivates the individual to conform to proper marital role behavior; and (c) the structural isolation of the family so frees the married partners' affective inclinations that they are able to love one another.[9]

(7) Robert F. Winch has developed a theory of "complementary needs" which essentially states that the underlying dynamic in the process of falling in love is an interaction between (a) the perceived psychological attributes of one individual and (b) the complementary psychological attributes of the person falling in love, such that the needs of the latter are felt to be met by the perceived attributes of the former and *vice versa*. These needs are derived from Murray's list of personality characteristics. Winch thus does not attempt to solve the problem of why our society has a love complex, but how it is that specific individuals fall in love with each other rather than with someone else.[10]

(8) Winch and others have also analyzed the effect of love upon various institutions or social patterns: Love themes are prominently displayed in the media of entertainment and communication, in consumption patterns, and so on.[11]

4. Finally, there is the cross-cultural work of anthropologists, who in the main have ignored love as a factor of importance in kinship patterns. The implicit understanding seems to be that love as a pattern is found only in the United States, although of course individual cases of love are sometimes recorded. The term "love" is practically never found in indexes of anthropological monographs on specific societies or in general anthropology textbooks. It is perhaps not an exaggeration to say that Lowie's comment of a generation ago would still be accepted by a substantial number of anthropologists:

[9] Talcott Parsons, *Essays in Sociological Theory,* Glencoe, Ill.: Free Press, 1949, pp. 187-189.

[10] Robert F. Winch, *Mate-Selection,* New York: Harper, 1958.

[11] See, e.g., Robert F. Winch, *The Modern Family,* New York: Holt, 1952, Chapter 14.

But of love among savages? . . . Passion, of course, is taken for granted; affection, which many travelers vouch for, might be conceded; but Love? Well, the romantic sentiment occurs in simpler conditions, as with us—in fiction. . . . So Love exists for the savage as it does for ourselves—in adolescence, in fiction, among the poetically minded.[12]

A still more skeptical opinion is Linton's scathing sneer:

All societies recognize that there are occasional violent, emotional attachments between persons of opposite sex, but our present American culture is practically the only one which has attempted to capitalize these, and make them the basis for marriage. . . . The hero of the modern American movie is always a romantic lover, just as the hero of the old Arab epic is always an epileptic. A cynic may suspect that in any ordinary population the percentage of individuals with a capacity for romantic love of the Hollywood type was about as large as that of persons able to throw genuine epileptic fits.[13]

In Murdock's book on kinship and marriage, there is almost no mention, if any, of love.[14] Should we therefore conclude that, cross-culturally, love is not important, and thus cannot be of great importance structurally? If there is only one significant case, perhaps it is safe to view love as generally unimportant in social structure and to concentrate rather on the nature and functions of romantic love within the Western societies in which love is obviously prevalent. As brought out below, however, many anthropologists have in fact described love *patterns*. And one of them, Max Gluckman,[15] has recently subsumed a wide range of observations under the broad principle that love relationships between husband and wife estrange the couple from their kin, who therefore try in various ways to undermine that love. This principle is applicable to many more societies (for example, China and India) than Gluckman himself discusses.

The Problem and Its Conceptual Clarification

The preceding propositions (except those denying that love is distributed widely) can be grouped under two main questions: What are the consequences of romantic love in the United States? How is the emotion of love aroused or created in our society? The present paper deals with the first question. For theoretical purposes both questions must be reformulated, however, since they implicitly refer only to our peculiar system of romantic love. Thus: (1) In what ways do various love patterns fit into the social structure, especially into the systems of mate choice and strati-

[12] Robert H. Lowie, "Sex and Marriage," in John F. McDermott, editor, *The Sex Problem in Modern Society*, New York: Modern Library, 1931, p. 146.

[13] Ralph Linton, *The Study of Man*, New York: Appleton-Century, 1936, p. 175.

[14] George Peter Murdock, *Social Structure*, New York: Macmillan, 1949.

[15] Max Gluckman, *Custom and Conflict in Africa*, Oxford: Basil Blackwell, 1955, Chapter 3.

fication? (2) What are the structural conditions under which a range of love patterns occurs in various societies? These are overlapping questions, but their starting point and assumptions are different. The first assumes that love relationships are a universal psychosocial possibility, and that different social systems make different adjustments to their potential disruptiveness. The second does not take love for granted, and supposes rather that such relationships will be rare unless certain structural factors are present. Since in both cases the analysis need not depend upon the correctness of the assumption, the problem may be chosen arbitrarily. Let us begin with the first.[16]

We face at once the problem of defining "love." Here, love is defined as a strong emotional attachment, a cathexis, between adolescents or adults of opposite sexes, with at least the components of sex desire and tenderness. Verbal definitions of this emotional relationship are notoriously open to attack; this one is no more likely to satisfy critics than others. Agreement is made difficult by value judgments: one critic would exclude anything but "true" love, another casts out "infatuation," another objects to "puppy love," while others would separate sex desire from love because sex presumably is degrading. Nevertheless, most of us have had the experience of love, just as we have been greedy, or melancholy, or moved by hate (defining "true" hate seems not to be a problem). The experience can be referred to without great ambiguity, and a refined measure of various degrees of intensity or purity of love is unnecessary for the aims of the present analysis.

Since love may be related in diverse ways to the social structure, it is necessary to forego the dichotomy of "romantic love—no romantic love" in favor of a continuum or range between polar types. At one pole, a strong love attraction is socially viewed as a laughable or tragic aberration; at the other, it is mildly shameful to marry without being in love with one's intended spouse. This is a gradation from negative sanction to positive approval, ranging at the same time from low or almost nonexistent institutionalization of love to high institutionalization.

The urban middle classes of contemporary Western society, especially in the United States, are found toward the latter pole. Japan and China, in spite of the important movement toward European patterns, fall toward the pole of low institutionalization. Village and urban India is farther toward the center, for there the ideal relationship has been one which at least generated love after marriage, and sometimes after betrothal, in contrast with the mere respect owed between Japanese and Chinese spouses.[17] Greece after Alexander, Rome of the Empire, and perhaps the

---

[16] I hope to deal with the second problem in another paper.

[17] Tribal India, of course, is too heterogeneous to place in any one position on such

later period of the Roman Republic as well, are near the center, but somewhat toward the pole of institutionalization, for love matches appear to have increased in frequency—a trend denounced by moralists.[18]

This conceptual continuum helps to clarify our problem and to interpret the propositions reviewed above. Thus it may be noted, first, that individual love relationships may occur even in societies in which love is viewed as irrelevant to mate choice and excluded from the decision to marry. As Linton conceded, some violent love attachments may be found in any society. In our own, the Song of Solomon, Jacob's love of Rachel, and Michal's love for David are classic tales. The Mahabharata, the great Indian epic, includes love themes. Romantic love appears early in Japanese literature, and the use of Mt. Fuji as a locale for the suicide of star crossed lovers is not a myth invented by editors of tabloids. There is the familiar tragic Chinese story to be found on the traditional "willowplate," with its lovers transformed into doves. And so it goes—individual love relationships seem to occur everywhere. But this fact does not change the position of a society on the continuum.

Second, reading both Linton's and Lowie's comments in this new conceptual context reduces their theoretical importance, for they are both merely saying that people do not *live by* the romantic complex, here or anywhere else. Some few couples in love will brave social pressures, physical dangers, or the gods themselves, but nowhere is this usual. Violent, self-sufficient love is not common anywhere. In this respect, of course, the U.S. is not set apart from other systems.

Third, we can separate a *love pattern* from the romantic love *complex*. Under the former, love is a permissible, expected prelude to marriage, and a usual element of courtship—thus, at about the center of the continuum, but toward the pole of institutionalization. The romantic love complex (one pole of the continuum) includes, in addition, an ideological prescription that falling in love is a highly desirable basis of courtship and marriage; love is strongly institutionalized.[19] In contemporary United States,

---

a continuum. The question would have to be answered for each tribe. Obviously it is of less importance here whether China and Japan, in recent decades, have moved "two points over" toward the opposite pole of high approval of love relationships as a basis for marriage than that both systems as classically described viewed love as generally a tragedy; and love was supposed to be irrelevant to marriage, i.e., non-institutionalized. The continuum permits us to place a system at some position, once we have the descriptive data.

[18] See Ludwig Friedländer, *Roman Life and Manners under the Early Empire* (Seventh Edition), translated by A. Magnus, New York: Dutton, 1908, Vol. 1, Chapter 5, "The Position of Women."

[19] For a discussion of the relation between behavior patterns and the process of institutionalization, see my *After Divorce*, Glencoe, Ill.: Free Press, 1956, Chapter 15.

many individuals would even claim that entering marriage without being in love requires some such rationalization as asserting that one is too old for such romances or that one must "think of practical matters like money." To be sure, both anthropologists and sociologists often exaggerate the American commitment to romance;[20] nevertheless, a behavioral and value complex of this type is found here.

But this complex is rare. Perhaps only the following cultures possess the romantic love value complex: modern urban United States, Northwestern Europe, Polynesia, and the European nobility of the eleventh and twelfth centuries.[21] Certainly, it is to be found in no other major civilization. On the other hand, the *love pattern,* which views love as a basis for the final decision to marry, may be relatively common.

Why Love Must Be Controlled

Since strong love attachments apparently can occur in any society and since (as we shall show) love is frequently a basis for and prelude to marriage, it must be controlled or channeled in some way. More specifically, the stratification and lineage patterns would be weakened greatly if love's potentially disruptive effects were not kept in check. The importance of this situation may be seen most clearly by considering one of the major functions of the family, status placement, which in every society links the structures of stratification, kinship lines, and mate choice. (To show how the very similar comments which have been made about sex are not quite correct would take us too far afield; in any event, to the extent that they are correct, the succeeding analysis applies equally to the control of sex.)

Both the child's placement in the social structure and choice of mates are socially important because both placement and choice link two kin-

---

[20] See Ernest W. Burgess and Paul W. Wallin, *Engagement and Marriage,* New York: Lippincott, 1953, Chapter 7 for the extent to which even the engaged are not blind to the defects of their beloveds. No one has ascertained the degree to which various age and sex groups in our society actually believe in some form of the ideology.

Similarly, Margaret Mead in *Coming of Age in Samoa,* New York: Modern Library, 1953, rates Manu'an love as shallow, and though these Samoans give much attention to love-making, she asserts that they laughed with incredulous contempt at Romeo and Juliet (pp. 155-156). Though the individual sufferer showed jealousy and anger, the Manu'ans believed that a new love would quickly cure a betrayed lover (pp. 105-108). It is possible that Mead failed to understand the shallowness of love in our own society: Romantic love is, "in our civilization, inextricably bound up with ideas of monogamy, exclusiveness, jealousy, and undeviating fidelity" (p. 105). But these are *ideas* and ideology; *behavior* is rather different.

[21] I am preparing an analysis of this case. The relation of "courtly love" to social structure is complicated.

ship lines together. Courtship or mate choice, therefore, cannot be ignored by either family or society. To permit random mating would mean radical change in the existing social structure. If the family as a unit of society is important, then mate choice is too.

Kinfolk or immediate family can disregard the question of who marries whom, only if a marriage is not seen as a link between kin lines, only if no property, power, lineage honor, totemic relationships, and the like are believed to flow from the kin lines through the spouses to their offspring. Universally, however, these are believed to follow kin lines. Mate choice thus has consequences for the social structure. But love may affect mate choice. Both mate choice and love, therefore, are too important to be left to children.

The Control of Love

Since considerable energy and resources may be required to push young-sters who are in love into proper role behavior, love must be controlled *before* it appears. Love relationships must either be kept to a small num-ber or they must be so directed that they do not run counter to the ap-proved kinship linkages. There are only a few institutional patterns by which this control is achieved.

1. Certainly the simplest, and perhaps the most widely used, structural pattern for coping with this problem is child marriage. If the child is be-trothed, married, or both before he has had any opportunity to interact intimately as an adolescent with other children, then he has no resources with which to oppose the marriage. He cannot earn a living, he is physi-cally weak, and is socially dominated by his elders. Moreover, strong love attachments occur only rarely before puberty. An example of this pattern was to be found in India, where the young bride went to live with her husband in a marriage which was not physically consummated until much later, within his father's household.[22]

2. Often, child marriage is linked with a second structural pattern, in which the kinship rules define rather closely a class of eligible future spouses. The marriage is determined by birth within narrow limits. Here, the major decision, which is made by elders, *is when* the marriage is to oc-cur. Thus, among the Murngin, *galle,* the father's sister's child, is scheduled to marry *due,* the mother's brother's child.[23] In the case of the "four-class"

---

[22] Frieda M. Das, *Purdah,* New York: Vanguard, 1932; Kingsley Davis, *The Popu-lation of India and Pakistan,* Princeton: Princeton University Press, 1951, p. 112. There was a widespread custom of taking one's bride from a village other than one's own.

[23] W. Lloyd Warner, *Black Civilization,* New York: Harper, 1937, pp. 82-84. They may also become "sweethearts" at puberty; see pp. 86-89.

double-descent system, each individual is a member of *both* a matri-moiety and a patri-moiety and must marry someone who belongs to neither; the four-classes are (1) ego's own class, (2) those whose matri-moiety is the same as ego's but whose patri-moiety is different, (3) those who are in ego's patri-moiety but not in his matri-moiety, and (4) those who are in neither of ego's moieties, that is, who are in the cell diagonally from his own.[24] Problems arise at times under these systems if the appropriate kinship cell—for example, parallel cousin or cross-cousin—is empty.[25] But nowhere, apparently, is the definition so rigid as to exclude some choice and, therefore, some dickering, wrangling, and haggling between the elders of the two families.

3. A society can prevent widespread development of adolescent love relationships by socially isolating young people from potential mates, whether eligible or ineligible as spouses. Under such a pattern, elders can arrange the marriages of either children or adolescents with little likelihood that their plans will be disrupted by love attachments. Obviously, this arrangement cannot operate effectively in most primitive societies, where youngsters see one another rather frequently.[26]

Not only is this pattern more common in civilizations than in primitive societies, but is found more frequently in the upper social strata. *Social segregation* is difficult unless it is supported by physical segregation—the harem of Islam, the zenana of India[27]—or by a large household system with individuals whose duty it is to supervise nubile girls. Social segregation is thus expensive. Perhaps the best known example of simple social segregation was found in China, where youthful marriages took place between young people who had not previously met because they lived in

[24] See Murdock, *op. cit.*, pp. 53 ff. *et passim* for discussions of double-descent.

[25] One adjustment in Australia was for the individuals to leave the tribe for a while, usually eloping, and then to return "reborn" under a different and now appropriate kinship designation. In any event, these marital prescriptions did not prevent love entirely. As Malinowski shows in his early summary of the Australian family systems, although every one of the tribes used the technique of infant betrothal (and close prescription of mate), no tribe was free of elopements, between either the unmarried or the married, and the "motive of sexual love" was always to be found in marriages by elopement. B. Malinowski, *The Family Among the Australian Aborigines,* London: University of London Press, 1913, p. 83.

[26] This pattern was apparently achieved in Manus, where on first menstruation the girl was removed from her playmates and kept at "home"—on stilts over a lagoon—under the close supervision of elders. The Manus were prudish, and love occurred rarely or never. Margaret Mead, *Growing Up in New Guinea,* in *From the South Seas,* New York: Morrow, 1939, pp. 163-166, 208.

[27] See Das, *op. cit.*

different villages; they could not marry fellow-villagers since ideally almost all inhabitants belonged to the same *tsu*.[28]

It should be emphasized that the primary function of physical or social isolation in these cases is to minimize informal or intimate social interaction. Limited social contacts of a highly ritualized or formal type in the presence of elders, as in Japan, have a similar, if less extreme, result.[29]

4. A fourth type of pattern seems to exist, although it is not clear cut; and special cases shade off toward types three and five. Here, there is close supervision by duennas or close relatives, but not actual social segregation. A high value is placed on female chastity (which perhaps is the case in every major civilization until its "decadence") viewed either as the product of self-restraint, as among the 17th Century Puritans, or as a marketable commodity. Thus love as play is not developed; marriage is supposed to be considered by the young as a duty and a possible family alliance. This pattern falls between types three and five because love is permitted before marriage, but only between eligibles. Ideally, it occurs only between a betrothed couple, and, except as marital love, there is no encouragement for it to appear at all. Family elders largely make the specific choice of mate, whether or not intermediaries carry out the arrangements. In the preliminary stages youngsters engage in courtship under supervision, with the understanding that this will permit the development of affection prior to marriage.

I do not believe that the empirical data show where this pattern is prevalent, outside of Western Civilization. The West is a special case, because of its peculiar relationship to Christianity, in which from its earliest days in Rome there has been a complex tension between asceticism and

[28] For the activities of the *tsu*, see Hsien Chin Hu, *The Common Descent Group in China and Its Functions*. New York: Viking Fund Studies in Anthropology, 10 (1948). For the marriage process, see Marion J. Levy, *The Family Revolution in Modern China*, Cambridge: Harvard University Press, 1949, pp. 87-107. See also Olga Lang, *Chinese Family and Society*, New Haven: Yale University Press, 1946, for comparisons between the old and new systems. In one-half of 62 villages in Ting Hsien Experimental District in Hopei, the largest clan included 50 per cent of the families; in 25 per cent of the villages, the two largest clans held over 90 per cent of the families; I am indebted to Robert M. Marsh who has been carrying out a study of Ching mobility partly under my direction for this reference: F. C. H. Lee, *Ting Hsien. She-hui K'ai-K'uang t'iao-ch'a*, Peiping: Chung-hua p'ing-min Chiao-yu ts'u-chin hui, 1932, p. 54. See also Sidney Gamble, *Ting Hsien: A North China Rural Community*, New York: International Secretariat of the Institute of Pacific Relations, 1954.

[29] For Japan, see Shidzué Ishimoto, *Facing Two Ways*, New York: Farrar and Rinehart, 1935, Chapters 6, 8; John F. Embree, *Suye Mura*, Chicago: University of Chicago Press, 1950, Chapters 3, 6.

love. This type of limited love marked French, English, and Italian upper class family life from the 11th to the 14th Centuries, as well as 17th Century Puritanism in England and New England.[30]

5. The fifth type of pattern permits or actually encourages love relationships, and love is a commonly expected element in mate choice. Choice in this system is *formally* free. In their 'teens youngsters begin their love play, with or without consummating sexual intercourse, within a group of peers. They may at times choose love partners whom they and others do not consider suitable spouses. Gradually, however, their range of choice is narrowed and eventually their affections center on one individual. This person is likely to be more eligible as a mate according to general social norms, and as judged by peers and parents, than the average individual with whom the youngster formerely indulged in love play.

For reasons that are not yet clear, this pattern is nearly always associated with a strong development of an adolescent peer group system, although the latter may occur without the love pattern. One source of social control, then, is the individual's own 'teen age companions, who persistently rate the present and probable future accomplishments of each individual.[31]

Another source of control lies with the parents of both boy and girl. In our society, parents threaten, cajole, wheedle, bribe, and persuade their children to "go with the right people," during both the early love

[30] I do not mean, of course, to restrict this pattern to these times and places, but I am more certain of these. For the Puritans, see Edmund S. Morgan, *The Puritan Family,* Boston: Public Library, 1944. For the somewhat different practices in New York, see Charles E. Ironside, *The Family in Colonial New York,* New York: Columbia University Press, 1942. See also: A. Abram, *English Life and Manners in the Later Middle Ages,* New York: Dutton, 1913, Chapters 4, 10; Emily J. Putnam, *The Lady,* New York: Sturgis and Walton, 1910, Chapter 4; James Gairdner, editor, *The Paston Letters, 1422-1509,* 4 vols., London: Arber, 1872-1875; Eileen Power, "The Position of Women," in C. G. Crump and E. F. Jacobs, editors, *The Legacy of the Middle Ages,* Oxford: Clarendon, 1926, pp. 414-416.

[31] For those who believe that the young in the United States are totally deluded by love, or believe that love outranks every other consideration, see: Ernest W. Burgess and Paul W. Wallin, *Engagement and Marriage,* New York: Lippincott, 1953, pp. 217-238. Note Karl Robert V. Wikman, *Die Einleitung Der Ehe. Acta Academiae Aboensis (Humaniora),* 11 (1937), pp. 127 ff. Not only are reputations known because of close association among peers, but songs and poetry are sometimes composed about the girl or boy. Cf., for the Tikopia, Raymond Firth, *We, the Tikopia,* New York: American Book, 1936, pp. 468 ff.; for the Siuai, Douglas L. Oliver, *Solomon Island Society,* Cambridge: Harvard University Press, 1955, pp. 146 ff. The Manu'ans made love in groups of three or four couples; cf. Mead, *Coming of Age in Samoa, op. cit.,* p. 92.

play and later courtship phases.³² Primarily, they seek to control love relationships by influencing the informal social contacts of their children:
moving to appropriate neighborhoods and schools, giving parties and helping to make out invitation lists, by making their children aware that certain individuals have ineligibility traits (race, religion, manners, tastes,
clothing, and so on). Since youngsters fall in love with those with whom
they associate, control over informal relationships also controls substantially the focus of affection. The results of such control are well known and
are documented in the more than one hundred studies of homogamy in
this country: most marriages take place between couples in the same
class, religious, racial, and educational levels.

As Robert Wikman has shown in a generally unfamiliar (in the United
States) but superb investigation, this pattern was found among 18th
Century Swedish farmer adolescents, was widely distributed in other Ger-
manic areas, and extends in time from the 19th Century back to almost
certainly the late Middle Ages.³³ In these cases, sexual intercourse was
taken for granted, social contact was closely supervised by the peer group,
and final consent to marriage was withheld or granted by the parents
who owned the land.

Such cases are not confined to Western society. Polynesia exhibits a
similar pattern, with some variation from society to society, the best
known examples of which are perhaps Mead's Manu'ans and Firth's Tiko-
pia.³⁴ Probably the most familiar Melanesian cases are the Trobriands
and Dobu,³⁵ where the systems resemble those of the Kiwai Papuans of
the Trans-Fly and the Siuai Papuans of the Solomon Islands.³⁶ Linton
found this pattern among the Tanala.³⁷ Although Radcliffe-Brown holds
that the pattern is not common in Africa, it is clearly found among the

³² Marvin B. Sussman, "Parental Participation in Mate Selection and Its Effect
upon Family Continuity," *Social Forces*, 32 (October, 1953), pp. 76-81.

³³ Wikman, *op. cit.*

³⁴ Mead, *Coming of Age in Samoa, op. cit.*, pp. 97-108; and Firth, *op. cit.*, pp.
520 ff.

³⁵ Thus Malinowski notes in his "Introduction" to Reo F. Fortune's *The Sorcerers
of Dobu,* London: Routledge, 1932, p. xxiii, that the Dobu have similar patterns, the
same type of courtship by trial and error, with a gradually tightening union.

³⁶ Gunnar Landtman, *Kiwai Papuans of the Trans-Fly,* London: Macmillan, 1927,
pp. 243 ff.; Oliver, *op. cit.*, pp. 153 ff.

³⁷ The pattern apparently existed among the Marquesans as well, but since Linton
never published a complete description of this Polynesian society, I omit it here. His
fullest analysis, cluttered with secondary interpretations, is in Abram Kardiner, *Psychological Frontiers of Society,* New York: Columbia University Press, 1945. For
the Tanala, see Ralph Linton, *The Tanala,* Chicago: Field Museum, 1933, pp. 300-
303.

Nuer, the Kgatla (Tswana-speaking), and the Bavenda (here, without sanctional sexual intercourse).[38]

A more complete classification, making use of the distinctions suggested in this paper, would show, I believe, that a large minority of known societies exhibit this pattern. I would suggest, moreover, that such a study would reveal that the degree to which love is a usual, expected prelude to marriage is correlated with (1) the degree of free choice of mate permitted in the society and (2) the degree to which husband-wife solidarity is the strategic solidarity of the kinship structure.[39]

Love Control and Class

These sociostructural explanations of how love is controlled lead to a subsidiary but important hypothesis: From one society to another, and from one *class* to another within the same society, the sociostructural importance of maintaining kinship lines according to rule will be rated differently by the families within them. Consequently, the degree to which control over mate choice, and therefore over the prevalence of a love pattern among adolescents, will also vary. Since, within any stratified society, this concern with the maintenance of intact and acceptable kin lines will be greater in the upper strata, it follows that noble or upper strata will maintain stricter control over love and courtship behavior than lower strata. The two correlations suggested in the preceding paragraph also apply: husband-wife solidarity is less strategic relative to clan solidarity in the upper than in the lower strata, and there is less free choice of mate.

Thus it is that, although in Polynesia generally most youngsters indulged in considerable love play, princesses were supervised strictly.[40] Similarly, in China lower class youngsters often met their spouses before

38 Thus, Radcliffe-Brown: "The African does not think of marriage as a union based on romantic love, although beauty as well as character and health are sought in the choice of a wife," in his "Introduction" to A. R. Radcliffe-Brown and W. C. Daryll Forde, editors, *African Systems of Kinship and Marriage*, London: Oxford University Press, 1950, p. 46. For the Nuer, see E. E. Evans-Pritchard, *Kinship and Marriage Among the Nuer*, Oxford: Clarendon, 1951, pp. 49-58. For the Kgatla, see I. Schapera, *Married Life in an African Tribe*, New York: Sheridan, 1941, pp. 55 ff. For the Bavenda, although the report seems incomplete, see Hugh A. Stayt, *The Bavenda*, London: Oxford University Press, 1931, pp. 111 ff., 145 ff., 154.

39 The second correlation is developed from Marion J. Levy, *The Family Revolution in China*, Cambridge, Harvard University Press, 1949, p. 179. Levy's formulation ties "romantic love" to that solidarity, and is of little use because there is only one case, the Western culture complex. As he states it, it is almost so by definition.

40 E.g., Mead, *Coming of Age in Samoa, op. cit.*, pp. 79, 92, 97-109. Cf. also Firth, *op. cit.*, pp. 520 ff.

marriage.[41] In our own society, the "upper upper" class maintains much greater control than the lower strata over the informal social contacts of their nubile young. Even among the Dobu, where there are few controls and little stratification, differences in control exist at the extremes: a child betrothal may be arranged between outstanding gardening families, who try to prevent their youngsters from being entangled with wastrel families.[42] In answer to my query about this pattern among the Nuer, Evans-Pritchard writes:

You are probably right that a wealthy man has more control over his son's affairs than a poor man. A man with several wives has a more authoritarian position in his home. Also, a man with many cattle is in a position to permit or refuse a son to marry, whereas a lad whose father is poor may have to depend on the support of kinsmen. In general, I would say that a Nuer father is not interested in the personal side of things. His son is free to marry any girl he likes and the father does not consider the selection to be his affair until the point is reached when cattle have to be discussed.[43]

The upper strata have much more at stake in the maintenance of the social structure and thus are more strongly motivated to control the courtship and marriage decisions of their young. Correspondingly, their young have much more to lose than lower strata youth, so that upper strata elders *can* wield more power.

Conclusion

In this analysis I have attempted to show the integration of love with various types of social structures. As against considerable contemporary

[41] Although one must be cautious about China, this inference seems to be allowable from such comments as the following: "But the old men of China did not succeed in eliminating love from the life of the young women. . . . Poor and middle-class families could not afford to keep men and women in separate quarters, and Chinese also met their cousins. . . . Girls . . . sometimes even served customers in their parents' shops." Olga Lang, *op. cit.,* p. 33. According to Fried, farm girls would work in the fields, and farm girls of ten years and older were sent to the market to sell produce. They were also sent to towns and cities as servants. The peasant or pauper woman was not confined to the home and its immediate environs. Morton H. Fried, *Fabric of Chinese Society,* New York: Praeger, 1953, pp. 59-60. Also, Levy (*op. cit.,* p. 111): "Among peasant girls and among servant girls in gentry households some premarital experience was not uncommon, though certainly frowned upon. The methods of preventing such contact were isolation and chaperonage, both of which, in the 'traditional' picture, were more likely to break down in the two cases named than elsewhere."

[42] Fortune, *op. cit.,* p. 30.

[43] Personal letter, dated January 9, 1958. However, the Nuer father can still refuse if he believes the demands of the girl's people are unreasonable. In turn, the girl can cajole her parents to demand less.

opinion among both sociologists and anthropologists, I suggest that love is a universal psychological potential, which is controlled by a range of five structural patterns, all of which are attempts to see to it that youngsters do not make entirely free choices of their future spouses. Only if kin lines are unimportant, and this condition is found in no society as a whole, will entirely free choice be permitted. Some structural arrangements seek to prevent entirely the outbreak of love, while others harness it. Since the kin lines of the upper strata are of greater social importance to them than those of lower strata are to the lower strata members, the former exercise a more effective control over this choice. Even where there is almost a formally free choice of mate—and I have suggested that this pattern is wide-spread, to be found among a substantial segment of the earth's societies—this choice is guided by peer group and parents toward a mate who will be acceptable to the kin and friend groupings. The theoretical importance of love is thus to be seen in the sociostructural patterns which are developed to keep it from disrupting existing social arrangements.

16. PROPINQUITY AND HOMOGAMY IN AMERICAN MATE-SELECTION

Who marries whom? Students of marriage have formulated two responses to the question: (1) like seeks like and/or (2) opposites attract. The two formulations have been labelled homogamy and heterogamy respectively.

The emphasis in the present chapter is on homogamy. Residential propinquity is an ecological form of homogamy. Koller's study shows that spouses tend before marriage to live near each other and that the lower the prestige of the husband's occupation, the greater the degree of the couple's residential propinquity. This suggests, Koller adds, that "the parents of boys and girls of marriageable age have unconsciously helped select their son's or daughter's mates by choosing to live in a given urban area."

Hollingshead presents data to show homogamy with respect to age, religion, and social class. Locke, Sabagh, and Thomes use data from the United States and from Canada to support the hypothesis that the rate of interfaith marriage of Catholics increases as the proportion of that group in the population decreases. (Incidentally the Locke hypothesis is supported by Hollingshead's data on Catholics and Protestants, but his data on people of Jewish faith show more homogamy than Locke's hypothesis would predict.)

According to Golden, social controls serving to discourage Negro-white intermarriage are visible not only in the family but also in the clergy, the military, and among government functionaries.

RESIDENTIAL AND OCCUPATIONAL PROPINQUITY

Marvin R. Koller

Adapted from the *American Sociological Review*, 13 (1948), 613-616.

Pioneer work by James Bossard in Philadelphia, 1931, suggested that residential propinquity of mates at the time of marriage is a factor in mate selection. Dr. Bossard examined five thousand consecutive marriage licenses in which one or both of the applicants were residents of Philadelphia. The study found that one out of four couples lived within two city blocks of each other; one-third of the couples lived within five blocks or less of each other. In an apt statement, Dr. Bossard concludes: "Cupid may have wings, but apparently they are not adapted for long flights." [1] Follow-up studies by Maurice R. Davie and Ruby Reeves in New Haven, Connecticut, 1931; Dr. W. A. Anderson in Genessee County, New York, 1934, and Carmella Frell in Warren, Ohio, 1947, confirmed Dr. Bossard's original hypothesis.

It was the purpose of the present study to apply more rigorous techniques to determine the validity of residential propinquity as a factor in mate selection in Columbus, Ohio.

The first refinement of Bossard's study was to shift the years of study away from the "Depression" of the early thirties. The possibility that the economic conditions of 1931 were responsible for Dr. Bossard's findings had to be eliminated. By selecting 1938 as a year well removed from the trough of the economic cycle and yet not a year closely connected with World War II and 1946 as the post-war year, this possibility was eliminated. The war years themselves were not used, since residence was often waived as a licensing requirement. If a high degree of residential propinquity showed itself in these years, then its long run effect could be presumed.

The second change from Bossard's study was to discard information relating to Negroes as far as possible. Dr. Bossard had included them in his study, admitting that one-ninth of the city's population was Negro. The elimination of Negroes from our sample was done because Negroes do not have the freedom to move about a city compared to whites and hence are forced to live in segregated areas.

---

[1] J. H. S. Bossard, *Marriage and Family,* Chapter Four. Philadelphia, University of Pennsylvania Press, 1940, "Residential Propinquity as a Factor in Marriage," pp. 79-92.

A third refinement was to employ the standard city block equal to one-eighth of a mile as the unit of measurement rather than to count individual city blocks "as if" they were equal. City blocks vary in length within cities and between cities. The use of the standard city block enables future studies to secure comparative data upon which we shall finally base our conclusions concerning the importance of residential propinquity as a factor in mate selection.

Dr. Bossard's study dealt with the first five months of the year and did not deal with those people who married during the last seven months of the year. To make sure that all couples in all twelve months would have a chance to be selected for study, a sampling technique of selecting the first fifty couples each month, starting with the first of the month and an additional fifty couples starting with the fifteenth of the month, or one hundred couples per month was employed.

A fifth refinement was to study solely the couples who were both residents of the city of Columbus, Ohio, and adjacent incorporated suburbs. A pilot study preliminary to defining the universe was run to determine how many of the couples marrying in Franklin County, the county in which Columbus is located, were both residents of the city. Over 70 per cent of those applying for a marriage license at the Franklin County Courthouse were both city residents. It was therefore decided to study only those couples who were both city residents rather than mix an urban and a rural study of distance between the homes of couples about to be married.

Lastly, Dr. Bossard's work was carried a bit further by investigating two factors which might possibly help explain whatever residential propinquity was found. These were the age and occupation of the male. The age of the man was used because there is the greater probability of accuracy with men due to the alleged tendency of women to under- or over-state their age. Occupation for men was regarded as more important since the occupation of the woman tends either to end with marriage or to be less permanent.

Summarizing the hypotheses implemented by the study:

1. Residential propinquity is a factor in mate selection for white mates who were both residents of the city of Columbus, Ohio, 1938 and 1946.
2. Residential propinquity is in part a function of (a) age and (b) occupation of males in Columbus, Ohio, 1938 and 1946.

It was possible to succeed in the original plan to take twelve hundred cases in 1946 as there were adequate numbers. In 1938, however, the numbers were considerably smaller and therefore the sampling in 1938 yielded only one thousand, one hundred and thirty-two cases. The total cases handled by the study amounted to two thousand, three hundred and thirty-two cases. . . .

The residences of both the male and female on each license were pinpointed on a large map of Columbus from the Office of the City Engineer. The closest possible distance in standard city blocks was desired. In the face of obstacles such as rivers, railroad tracks, golf courses, large estates, undeveloped areas, and large state institutions it was necessary to measure around them. For example, a couple whose residences faced each other across the river could not be measured in city blocks directly across the river but rather around the river to the nearest bridge and thence to the residence of the individual concerned.

The assembled data indicated very clearly that the refinements of Dr. Bossard's study when applied to Columbus, Ohio, for 1938 and 1946 sustained Dr. Bossard's original findings. The men tended to select women in the city who lived near the men's homes. Criticism that Bossard's study might have been influenced by the "Depression," by the inclusion of Negroes, by the treatment of all city blocks as equal, or by the sampling technique is not supported by this study in Columbus. In 51 per cent of the 1,132 cases studied in Columbus, Ohio, 1938, the men selected a girl living within 12 standard city blocks. In 1946, 50 per cent of 1,200 men selected a girl living within 15 standard city blocks. Combining the cases of 1938 and 1946, a total of 1,205 or 51 per cent of 2,332 cases chose a mate living within 14 standard city blocks.

Analysis of the frequency distributions found for both years into quartiles yields the following table.

TABLE 1.   Residential Propinquity of Mates in Standard City Blocks in Columbus, Ohio, 1938, 1946, and Total, by Quartiles

Quartiles	1938	1946	Total
Q_1	2.85	3.34	3.14
Q_2	11.75	14.69	13.33
Q_3	31.00	31.53	31.72
Q_D*	14.08	14.05	14.29

* Q_D (semi-interquartile range) $= \dfrac{Q_3 - Q_1}{2}$ —*Eds.*

There appears to be little difference in the degree of residential propinquity in the samples for 1938 and 1946. . . .

The potential criticism, that residential propinquity as found by this study might be nothing more than the operation of chance due to the selected nature of the sample universe, needs to be answered. Would any group of women and men with the common characteristics of being white, both city residents, and marriage license applicants in a given year on the basis of chance also select each other as future mates? If a close degree of residential propinquity was found for people who had these characteristics and yet had not selected each other as mates, the operation of the residential propinquity factor by sheer chance would be established.

The case records used in the study were thoroughly shuffled and two hundred cases were taken at random for each year. By plotting the residence of the male of the first card and the residence of the female on the second card, one hundred men and one hundred women who did not select each other as mates and yet who possessed the three common characteristics mentioned above were measured relative to their degree of residential propinquity. The degree of residential propinquity of these "chance couples" varied markedly from the degree of residential propinquity found for those couples who had actually selected each other as future mates. One-fourth of the "chance couples" lived within 17 standard city blocks whereas one-fourth of the actual couples lived within 3 standard city blocks of each other's homes. The median for the chance sample was about 29 blocks whereas for the couples who selected each other the median was about 13 blocks. The upper quartile for the chance sample was about 43 blocks whereas the couples about to be married had an upper quartile of 32 blocks. These findings indicate that something other than chance explains the close degree of residential propinquity found between the couples studied.

Analysis of the frequency distributions of the various age groups and occupations of the men studied in 1938 and 1946 indicates possible factors operative in the degree of residential propinquity.

TABLE 2.   Rank Order of Medians of Age Groups by Standard City Blocks, 1938 and 1946

1938		1946	
Age Group	Median	Age Group	Median
24-27	16.21	24-27	18.00
32-35	15.00	20-23	15.30
28-31	14.83	28-31	13.88
20-23	10.64	32-35	13.25
Over 35	6.66	Over 35	7.83

Dividing the frequency distributions for age groups in quartiles reveals again the close degree of residential propinquity for each age group for each year . . . the results varying with each age group. The age group 24-27 consistently demonstrated the greatest distance in standard city blocks. The age group Over 35 consistently demonstrated the closest degree of residential propinquity. The group 28-31 consistently remained in the middle between the age group 24-27 and the age group Over 35. Other age groups were erratic in their behavior. . . .

TABLE 3.   Rank Order of Medians of Occupational Groups by Standard City
            Blocks, 1938 and 1946

1938 Occupational Group	Median	1946 Occupational Group	Median
Clerical and sales	17.38	Professional and managerial	18.50
Professional and managerial	15.94	Clerical and sales	15.25
Skilled	10.30	Skilled	13.90
Semi-skilled	9.33	Service	13.33
Service	8.63	Semi-skilled	11.17
Unskilled	2.83	Unskilled	4.79

A similar study of the occupations of the males yielded more significant data. In general, the higher a man ranged on the occupational scale
the greater the distance in standard city blocks in which he selected
his future mate. The lower the occupational position, the greater the degree of residential propinquity found. Professional and Managerial men
had a median frequency of about 16 to 18 standard city blocks between themselves and their future wives, whereas the Unskilled men married girls living within three to five standard city blocks.

The findings of this study sustain the original hypothesis, namely, that
residential propinquity is a factor in mate selection for white mates who
were both city residents of Columbus, Ohio, 1938 and 1946 and that
residential propinquity is explainable, in part, as a function of (a) age
and (b) occupation of males in Columbus, Ohio, 1938 and 1946. Generalizations applied to age and occupational groups are not wholly correct unless one specifies which age group or which occupational group in
a given year he means.

What interpretation should be given these findings? What social inference can we find here? If our findings are correct, then some of Bossard's
original ideas that there are "social types in urban communities" who
tend to marry may be correct. It is further suggested that because residential propinquity is operative in the city, the parents of boys and girls
of marriageable age have unconsciously helped select their son's or daughter's mates by choosing to live in a given urban area. There appears to be
a stronger than fifty-fifty chance that a young boy or girl in the city
will marry someone living very close to his residence. Here, we might have
a predictive device of great value.

We must be very cautious, however, before we generalize too freely
about residential propinquity. Thus far the studies have dealt with
marriage license applications. The findings must be supplemented by additional research, such as interviews, to determine if the residential propinquity reported in the documents is more apparent than real. More research using similar methods to this one should be undertaken to check

these findings in Columbus. More researches using different methods are also welcomed as they might reveal discrepancies that cannot appear in a statistical study.

With Bossard then, we repeat, "Yes, Cupid has wings but he doesn't fly very far. . . ."

CULTURAL FACTORS IN THE SELECTION OF MARRIAGE MATES

August B. Hollingshead

Adapted from the *American Sociological Review,* 15 (1950), 619-27.

The question of who marries whom is of perennial interest, but only during the last half-century has it become the subject of scientific research. Throughout American history there has always been a romantic theory of mate selection, supported by poets, dramatists, and the public at large. Social scientists, however—a group of jaundiced realists, by and large—have little faith in this pleasant myth as an explanation for the selection of marriage mates.[1] Their theories can be divided between (1) the homogamous and (2) the heterogamous.[2] The theory of homogamy postulates that "like attracts like"; the theory of heterogamy holds that "opposites attract each other."

Certain aspects of each theory have been investigated by psychologists and sociologists. The psychologists have confined their attention almost exclusively to individual physical [3] and psychological [4] characteristics. Sociologists have focused, in the main, upon factors external to the in-

[1] For a discussion of this theory and some facts to refute it see A. B. Hollingshead, "Class and Kinship in a Middle Western Community," *American Sociological Review,* 14 (August, 1949), 469-475.

[2] E. W. Burgess and Paul Wallin, "Homogamy in Social Characteristics," *American Journal of Sociology,* 49 (September, 1943), 109-124.

[3] J. A. Harris, "Assortive Mating in Man," *Popular Science Monthly,* 80 (1912), 476-492. This is the earliest review in the literature that tries to give a scientific explanation of the question of who marries whom. The studies reviewed primarily dealt with physical characteristics: deafness, health, longevity, age, stature, cephalic index, hair and eye color.

[4] Harold E. Jones, "Homogamy in Intellectual Abilities," *American Journal of Sociology,* 35 (1929), 369-382; E. L. Kelly, "Psychological Factors in Assortive Mating," *Psychological Bulletin,* 37 (1940), 493 and 576; Helen M. Richardson, "Studies of Mental Resemblance Between Husbands and Wives and Between Friends," *Psychological Bulletin,* 36 (1939), 104-120.

dividual. As a consequence, sociological research has stressed such things as ethnic origin,[5] residential propinquity,[6] race,[7] religion,[8] socioeconomic status,[9] and social characteristics in general.[10] While all of these researches have used empirical data, only a few of them have attempted to measure the significant cutural factors that impinge upon mate selection against the background of the theories of homogamy and heterogamy. We shall attempt to do this in this paper.

My attack upon this problem will be to state the theoretical limits within

[5] Bessie B. Wessel, "Comparative Rates of Intermarriage Among Different Nationalities in the United States," *Eugenical News,* 15 (1930), 105-107; Bessie B. Wessel, *An Ethnic Survey of Woonsocket, R.I.,* Chicago, University of Chicago Press, 1931; James H. S. Bossard, "Nationality and Nativity as Factors in Marriage," *American Sociological Review,* 4 (December, 1939), 792-798; Ruby Jo Reeves, *Marriages in New Haven since 1870 Statistically Analyzed and Culturally Interpreted,* doctoral dissertation Yale University (unpublished), 1938; Ruby Jo Reeves Kennedy, "Single or Triple Melting-Pot? Intermarriage Trends in New Haven, 1870-1940," *American Journal of Sociology,* 39 (January, 1944), 331-339; Milton L. Barron, *Intermarriage in a New England Industrial Community,* Syracuse, Syracuse University Press, 1946. Barron has a good bibliography of studies in this area, pp. 355-366.

[6] James H. S. Bossard, "Residential Propinquity as a Factor in Marriage Selection," *American Journal of Sociology,* 38 (1932), 219-224; Maurice R. Davie and Ruby Jo Reeves, "Propinquity in Residence Before Marriage," *American Journal of Sociology,* 44 (1939), 510-517; Ruby Jo Reeves Kennedy, "Pre-Marital Residential Propinquity and Ethnic Endogamy," *American Journal of Sociology,* 48 (March, 1943), 580-584; John S. Ellsworth, Jr., "The Relationship of Population Density to Residential Propinquity as a Factor in Marriage Selection," *American Sociological Review,* 13 (August, 1948), 444-448.

[7] Romanzo Adams, *Interracial Marriage in Hawaii,* New York, The Macmillan Co., 1937; Otto Klineberg, *Characteristics of the American Negro,* New York, Harper, 1944, especially Part V where Negro-white intermarriage and the restrictions on it imposed by law are discussed; U. G. Weatherly, "Race and Marriage," *American Journal of Sociology,* 15 (1910), 433-453; Robert K. Merton, "Intermarriage and the Social Structure," *Psychiatry,* 4 (August, 1941), 371-374; Constantine Panunzio, "Intermarriage in Los Angeles, 1924-1933," *American Journal of Sociology,* 47 (March, 1942), 399-401.

[8] Reuben R. Resnick, "Some Sociological Aspects of Intermarriage of Jew and Non-Jew," *Social Forces,* 12 (October, 1933), 94-102; J. S. Slotkin, "Jewish-Gentile Intermarriage in Chicago," *American Sociological Review,* 7 (February, 1942), 34-39; Ruby Jo Reeves Kennedy, "Single or Triple Melting-Pot?" *op. cit.*

[9] Richard Centers, "Marital Selection and Occupational Strata," *American Journal of Sociology,* 54 (May 1949), 530-535; Donald M. Marvin, "Occupational Propinquity as a Factor in Marriage Selection," *Publications of the American Statistical Association,* 16 (September, 1918), 131-156; Meyer F. Nimkoff, "Occupational Factors and Marriage," *American Journal of Sociology,* 49 (November 1943), 248-254.

[10] Walter C. McKain, Jr., and C. Arnold Anderson, "Assortive Mating in Prosperity and Depression," *Sociology and Social Research,* 21 (May-June, 1937), 411-418; E. W. Burgess and Paul Wallin, "Homogamy in Social Characteristics," *American Journal of Sociology,* 49 (September, 1943), 109-124.

which mate selection may take place, then turn to a body of data to determine how, and to what extent, specific factors influence the selection of marital partners.[11]

Viewed in the broadest theoretical perspective of democratic theory, the choice of marriage mates in our society might be conceived of as a process in which each unattached biologically mature adult has an equal opportunity to marry every other unattached biologically mature adult of the opposite sex. Viewed from the narrowest perspective of cultural determinism, biologically mature, single males or females have only limited opportunity to select a marital partner. The first proposition assumes complete freedom of individual choice to select a mate; the second assumes that mates are selected for individuals by controls imposed on them by their culture. If the first assumption is valid we should find no association between cultural factors and who marries whom; if the second is descriptive of the mate selection process we should expect to find a strong association between one or several cultural factors and who marries whom. The second proposition, however, allows for individual choice within limits of cultural determinism; for example a Jew is expected to marry a Jew by the rules of his religion; moreover, he is more or less coerced by his culture to marry a Jewess of the same or a similar social status, but he has a choice as to the exact individual.

In the remainder of this paper I shall test five factors—race, age, religion, ethnic origin, and class—within the limits of the theories of homogamy and heterogamy and the abstract model I have outlined. The data utilized to measure the influence of these factors on the selection of marriage mates were assembled in New Haven, Connecticut, by a research team during the last year through the cooperation of the Departments of Vital Statistics of the State of Connecticut and the City of New Haven. All marriage license data on marriages in New Haven during 1948 were copied. Then parents, relatives, in some cases neighbors, were asked in February, 1949, to supply the addresses of each newly married couple. Addresses were obtained for 1,980 couples out of a total of 2,063 couples married in the city in 1948. Nine hundred and three couples, 45.8 per cent, had moved from the city, and 1,077, 54.4 per cent, were living in it in February, 1949. A 50 per cent random sample, drawn by Census Tracts from the 1,077 couples resident in New Haven, was interviewed with a schedule. The interview, which lasted from about an hour and a quarter to three hours, took place in the home of the couple, usually with both the husband and wife present, and occurred most

[11] For purposes of this paper we shall rely upon tests of significance and measures of association to tell us what cultural factors are of greater or lesser importance in the determination of who marries whom.

generally in the evening or late afternoon.[12] In addition, twenty-eight census-like items such as age, occupation, birthplace, residence, and marital status, were available on all of the 1,980 couples.

The 523 interviewed couples were compared with the 1,457 non-interviewed couples, census item by census item, to determine if the interviewed group differed significantly from the non-interviewed group. No significance of difference was found at the 5 per cent level for any item, except where the husband and wife were both over 50 years of age.[13] Having satisfied ourselves that the interviewed group was representative of the total group, we proceeded with a measure of confidence to the analysis of our data.

Race

Our data show that the racial mores place the strongest, most explicit, and most precise limits on an individual as to whom he may or may not marry. Although interracial marriages are legal in Connecticut, they are extremely rare; none occurred in New Haven in 1948. Kennedy's analysis of New Haven marriages from 1870 through 1940 substantiates the rule that Negroes and whites marry very infrequently. Thus, we may conclude that a man's or woman's marital choice is effectively limited to his or her own race by the moral values ascribed to race in this culture. Race, thus, divides the community into two parts so far as marriage is concerned. Because there were no interracial marriages in 1948, and because of the small percentage of Negroes in New Haven, we will confine the rest of our discussion to whites.

Age

Age, like race, is a socio-biological factor that has a definite influence on marital choice. The effects of cultural usages and values on the selection of a marriage partner may be seen by a study of Table 1. While there is a very strong association between the age of the husband and the age of the wife at all age levels, it is strongest when both partners are under 20 years of age. Men above 20 years of age tend to select wives who are in the same 5 year age group as they are, or a younger one. After age 20 the percentage of men who marry women younger than themselves increases until age 50. After 50 the marital partners tend to

---

12 Eighty-seven per cent of the interviewing was done by senior undergraduates and graduate students, 5 per cent by an assistant, and 8 per cent by the writer. Six per cent of the interviews were checked for reliability from one month to four months after the original interview.

13 The principal reasons for this deviation were (1) twice as many older couples refused to be interviewed as those below fifty years of age, and (2) the age gap between interviewers and potential interviewees influenced the situation.

be nearer one another in age. Table 1 indicates further that controls relative to age rather effectively limit a man's choice to women of his age or younger, but that the woman cannot be too much younger or counter controls begin to operate. Evidence accumulated in the interviews shows it is widely believed that a young woman should not marry "an old man." The effects of this belief and practice are reflected in the lower left hand section of Table 1. There we see that only 4 men above 45 years of age, out of a total of 144, married women under 30 years of age.

TABLE 1.   Age of Husband and Wife by Five-Year Intervals for New Haven Marriages, 1948

Age of Husband	Age of Wife								Total
	15-19	20-24	25-29	30-34	35-39	40-44	45-49	50 & up	
15-19	42	10	3						55
20-24	153	504	51	10	1				719
25-29	52	271	184	22	7	2			538
30-34	5	52	87	69	13	5			231
35-39	1	12	27	29	21	2	3		105
40-44		1	9	18	17	8	2	1	56
45-49	1		3	6	16	16	7	1	49
50 & up			1	4	11	15	21	43	95
Total	254	850	365	168	86	47	33	45	1848

$\chi^2 = 2574.8905$ $P < .01$ $C = .76$ $\bar{C} = .80$.

C = The coefficient of contingency.
$\bar{C}$ = The corrected coefficient of contingency corrected for broad grouping by the formula given in Thomas C. McCormick, *Elementary Social Statistics*, McGraw-Hill, 1941, p. 207.

The age-sanctions that impinge on a woman with reference to the age of a potential husband narrow her marital opportunities to men her age, or to slightly older men. This usage is reflected in the upper right corner of Table 1, where marriages between older women and younger men are conspicuous by their absence. In short, differences in the customs relative to age and marital partners place greater restrictions on a woman's marital opportunities than a man's. Nevertheless, it is clear that the values ascribed to age restrict an individual's marital opportunities within narrow limits; and a woman's more than a man's.

Religion

The effects of religious rules on an individual's marital choices were very clear.[14] Next to race, religion is the most decisive factor in the segregation of males and females into categories that are approved or disapproved with respect to nuptiality, Ninety-one per cent of the mar-

[14] R. J. R. Kennedy, "Single or Triple Melting-Pot? Intermarriage Trends in New Haven, 1870-1940," *op. cit.*

riages in this study involved partners from the same religious group. In the case of Jews, this percentage was 97.1, among Catholics it was 93.8 per cent; it fell to 74.4 per cent for Protestants. The differences in percentage, we believe, are a reflection of the relative intensity of in-group sanctions on the individual in the three religious groups. A striking point that emerged from our data is that the effects of religion on marital choice has not changed between the parental and present generation.[15] Table 2 shows that the number of Catholics who married Catholics, and Jews who married Jews, was almost the same in both generations. The number of Protestants who married Protestants dropped in the present generation, but not significantly in terms of the numbers involved.[16] The influence of religious affiliation on the selection of a marriage mate is obviously strongest in the Jewish group and weakest in the Protestant. This is reflected in the number of mixed marriages. On this point, we would remark that there is no consistent bias between sex and mixed Catholic-Protestant marriages; either partner is likely to be a Catholic or a Protestant. On the other hand, in Jewish-Gentile marriages it has been a Jewish male who has married a Gentile female.

I shall point out, in passing, that the very high association we found between religion and marriage is not unique. Burgess and Wallin reported a coefficient of contingency of .75 for the 1,000 engaged couples they studied in Chicago;[17] our data revealed a coefficient of contingency of .77 in the present generation. This is not essentially different from theirs. Because religion is so effective a control in the selection of marriage mates I shall hold it constant and analyze other factors in terms of it.

Ethnic Origin

New Haven remained almost wholly Protestant religiously, and British ethnically, from its settlement in 1638 until the late 1830's. Between 1830 and 1880 Irish arrived by the hundred; Germans and Scandinavians by the score. The Irish and a minority of the Germans were Catholic and they soon established themselves in this burgeoning railroad and manufacturing center. An expanding economy, coupled with political and economic unrest in Southern and Eastern Europe, resulted in the in-

---

[15] Our discussion on this and subsequent points includes only white marriages where the religion of the couple and of their four parents was known. Moreover, the tabular materials include only white cases where the specific data called for by the table were complete. "Unknown" cases were eliminated in particular instances.

[16] The religious affiliation of marital partners in the present and parental generations was tested for significance; none was found; $\chi^2 = 6.7015$ with 8 degrees of freedom.

[17] E. W. Burgess and Paul Wallin, *op. cit.*, p. 115.

TABLE 2. Religious Affiliation in the Parental and Present Generations

	A. Wife's Father and Mother*		
		Wife's Mother	
Wife's Father	Catholic	Protestant	Jewish
Catholic	274	11	0
Protestant	9	75	0
Jewish	2	1	65
Total	285	87	65

$$\chi^2 = 522.4592 \quad P < .01 \quad C = .74$$

	B. Husband's Father and Mother		
		Husband's Mother	
Husband's Father	Catholic	Protestant	Jewish
Catholic	273	12	0
Protestant	14	70	0
Jewish	0	0	68
Total	287	82	68

$$\chi^2 = 494.4359 \quad P < .01 \quad C = .73$$

	C. Husband and Wife		
		Wife	
Husband	Catholic	Protestant	Jewish
Catholic	271	20	0
Protestant	17	61	0
Jewish	1	1	66
Total	289	82	66

$$\chi^2 = 636.0297 \quad P < .01 \quad C = .77$$

* The religious affiliation claimed by the interviewees is used here.

flux of thousands of Polish and Russian Jews, and tens of thousands of Italians between 1890 and 1914. After 1914, the stream of immigration became a trickle that has never again been allowed to run freely. Thus, today, New Haven is composed mainly of three large religious groups and seven European-derived ethnic stocks: British, Irish, German, Scandinavian, Italian, Polish, and Polish Jewish.[18]

We cannot discuss how ethnicity is related to the selection of a marriage mate apart from religion, because religion and ethnic origin are so closely related. Observation of . . . Tables 3 through 6 . . . will show that ethnicity within a religious group has been a very potent factor in influencing the mate selection process in both the parental and the present generations, but it was stronger a generation ago than it is now. Although

[18] We are excluding Negroes from our discussion.

ethnic lines are crossed within the Catholic and the Protestant faith more frequently in the present than in the parental generation, this is not true for the Jews. Furthermore, ethnic lines in both generations were crossed, for the most part, within religious groups. This means that the Catholics are becoming a mixture of Irish, Polish, and Italian as a result of intermarriage between these groups, but there is still a large block of unmixed Italian stock in New Haven and smaller blocks of Irish and Polish. The Protestants, on the other hand, select marriage partners mainly from the

TABLE 3.†   Per cent of Interethnic and Intra-ethnic Marriages in the Present and Parental Generations among Catholics*

	Couples	Parents
Intra-ethnic	56	84
Interethnic	44	16
	100	100
	(N = 271)	(N = 542)

TABLE 4.†   Per cent of Interethnic and Intra-ethnic Marriages in the Present and Parental Generations among Protestants*

	Couples	Parents
Intra-ethnic	34	68
Interethnic	66	32
	100	100
	(N = 61)	(N = 122)

TABLE 5.†   Per cent of Interethnic and Intra-ethnic Marriages in the Present and Parental Generations among Jewish*

	Couples	Parents
Intra-ethnic	100	100
Interethnic	0	0
	100	100
	(N = 66)	(N = 132)

TABLE 6.†   Per cent of Interethnic and Intra-ethnic Marriages in the Present and Parental Generations among Mixed Religions*

	Couples	Parents
Intra-ethnic	18	86
Interethnic	82	14
	100	100
	(N = 39)	(N = 78)

* The religious affiliation claimed by the interviewees is used here.
† Adopted from table 3 in original—*Eds.*

British segment of the city's population; a minority chose a partner from a Northwestern European group, and in some cases both partners will be of German or Scandinavian descent. Kennedy discovered this process in her study of New Haven marriage records from 1870 to 1940, and developed her theory of the triple melting-pot in terms of it.[19]

. . . Table 6 . . . indicates that, in most cases, marriages across religious lines involve the mixing of ethnic stocks. This is true whether Catholics and Protestants marry, or Jews and Gentiles, because the members of each religious group came from such different parts of Europe. From the viewpoint of assimilation, marriages across religious lines are crucial if the triple melting-pot is to become a single melting-pot. But as Kennedy's and our data show, we are going to have three pots boiling merrily side by side with little fusion between them for an indefinite period. Furthermore, if the rules relative to mixed marriages in the Roman Catholic and Jewish churches were followed strictly there would be no mixing of the contents of one pot with those of another. To be sure, ethnic intermixture would occur, but within each respective religious group.

Class

Our discussion of the relationship between social class and marriage will be based on cases where the husband, the wife, and both parental families were *de facto* residents of New Haven.[20] The analysis of 1,008 marriages where the husband, the wife, and their families were residents of New Haven revealed that the class of residential area in which a man's or a woman's family home is located has a very marked influence on his or her marital opportunities. In 587 of these 1,008 marriages, or 58.2 per cent (see Table 7), both partners came from the same class of residential area. When those that involved a partner from an adjacent class area were added to the first group the figure was raised to 82.8 per cent of all marriages.

Careful study of the data presented in Table 7 will reveal that the residential class in which a family has its home has a different effect on a woman's marital opportunities in comparison with a man's. While the modal, as well as the majority, of marriages at all levels united class

[19] For a discussion of this theory see Ruby Jo Reeves Kennedy, "Single or Triple Melting-Pot? Intermarriage Trends in New Haven, 1870-1940," *op. cit.*

[20] The index of class position used here was developed by Maurice R. Davie on the basis of the ecological analysis he had made of the city of New Haven. Davie has ranked the 22 natural ecological areas that are primarily residential into six classes. Class I is the best and class VI the worst type of residential area. For a discussion of the project on which these ratings are made, see Maurice R. Davie, "The Patterns of Urban Growth," *Studies in the Science of Society,* G. P. Murdock, *ed.,* New Haven, 1937, pp. 133-161.

TABLE 7.   **Residential Class of Husband and Wife for Residents of New Haven**

Class of Husband	Class of Wife						Total
	I	II	III	IV	V	VI	
I	13	7	1	0	3	1	25
II	8	56	8	12	13	8	105
III	1	4	15	5	7	7	39
IV	0	8	4	55	35	38	140
V	0	12	8	30	252	87	389
VI	0	5	9	40	60	196	310
Total	22	92	45	142	370	337	1008

$$\chi^2 = 1045.0605 \quad P < .01 \quad C = .71 \quad \overline{C} = .77$$

equals, when class lines were crossed the man selected a woman from a lower class far more frequently than was true for women. For instance, if you look at Table 7 you will see that 12 men from class I married women from lower ranking areas, and four of the twelve married girls from class V and class VI areas. On the other hand, 9 women from class I

TABLE 8.   **Residential Class of Husband and Wife by Religious Groups**

	A. Catholic	
Residential Class of Husband	Residential Class of Wife	
	I-III	IV-VI
I-III	16	7
IV-VI	12	161
Total	28	168

$$\chi^2 = 74.8413 \quad P < .01$$

	B. Protestant	
Residential Class of Husband	Residential Class of Wife	
	I-III	IV-VI
I-III	12	4
IV-VI	1	18
Total	13	22

$$\chi^2 = 18.0923 \quad P < .01$$

	C. Jewish	
Residential Class of Husband	Residential Class of Wife	
	I-III	IV-VI
I-III	24	2
IV-VI	3	15
Total	27	17

$$\chi^2 = 26.6687 \quad P < .01$$

areas married men from lower ranking areas, but 8 of the 9 came from a class II area and 1 from a class III area. No man from class IV, V, or VI areas married a woman from a class I area. If you follow down the successive class levels on Table 7 you will see that this tendency is repeated all the way to class VI. It is clearest, however, in classes IV and V. In class IV, only 12 women from classes II and III combined married men from class IV. On the other hand, class IV men married 35 class V and 38 class VI women, for a total of 73. Fifty class V men married women

TABLE 9. Years of School Completed by Husband and Wife by Religion

A. Catholic

Years of School Husband	Years of School Wife		
	9 & less	10-12	13 & more
9 & less	35	19	1
10-12	33	128	27
13 & more	5	15	19
Total	73	162	47

$$\chi^2 = 80.9784 \quad P < .01$$

B. Protestant

Years of School Husband	Years of School Wife		
	9 & less	10-12	13 & more
9 & less	11	3	0
10-12	10	26	7
13 & More	3	6	16
Total	24	35	23

$$\chi^2 = 38.9932 \quad P < .01$$

C. Jewish

Years of School Husband	Years of School Wife		
	9 & less	10-12	13 & more
9 & less	0	0	0*
10-12	0	22	11
13 & more	0	8	26
Total	0	30	37

$$\chi^2 = 12.6033 \quad P < .01$$

* The zero cells were not included in the χ^2.

from classes II, III, and IV, but 87 married class VI women. These figures reveal that the man has a wider range of choice than a woman, but he tends, when he goes outside of his own class, to marry a woman in a lower class. From whatever way we view Table 7, it is evident that the class position of a family is a factor that exerts a very important influence on the marriage choice of its children.

Now that we have seen the larger picture, we will look at it from the special perspective of a combination of religion and residential class. Because the number of cases where we knew both religion and class level was small in some residential areas, we have combined classes I through III, and classes IV through VI in Table 8. Table 8 indicates very clearly that the class factor operates independently of religion, and with about equal force in each religious group. What is especially significant is that the effects of class position on who marries whom are so strong in each religious group.

Education operates in the same way as residence to sort potential marriage mates into horizontal status groups within the confines of religion. Within each religious group men with a particular amount of education married women with a comparable amount of education in very significant numbers. This tendency was strongest in the Jewish and weakest in the Catholic group. The strong association between the educational level of the husband and the wife, so evident in Table 9, is not a new development. We compared the education of husbands and wives in the parental generation by religious groups and found that for both the husband's parents and the wife's parents the association held. Moreover, the coefficients of contingency for each set of parents by religion were almost the same, as the following tabulation shows:

Religion	Husband's Parents'	Wife's Parents'
Catholic	.57	.58
Protestant	.58	.59
Jewish	.59	.59

These coefficients indicate that education, along with religion, has influenced the mate selection process for at least two generations.

In summary, this paper has attempted to throw light on three questions: *first,* does a biologically mature unattached adult have an equal opportunity to marry an unattached mature adult of the opposite sex? *Second,* what restrictions are placed on his choice by society, and *third,* how effective are certain selected restrictions in limiting his choice? These questions become meaningful only when we relate them to the two propositions outlined in the introduction. There I set up a model with theoretical limits of absolute freedom of individual choice in the selection of a marital partner at one pole, and no choice at the other.

The data presented demonstrate that American culture, as it is reflected in the behavior of newly married couples in New Haven, places very definite restrictions on whom an individual may or may not marry. The racial mores were found to be the most explicit on this point. They divided the community into two pools of marriage mates and an individual fished for a mate only in his own racial pool. Religion divided the white race into three smaller pools. Persons in the Jewish pool in 97.1 per cent of the cases married within their own group; the percentage was 93.8 for Catholics and 74.4 for Protestants. Age further subdivided the potential pool of marriage mates into rather definite age grades, but the limits here were not so precise in the case of a man as of a woman. The ethnic origin of a person's family placed further restrictions on his marital choice. In addition, class position and education stratified the three religious pools into areas where an individual was most likely to find a mate. When all of these factors are combined they place narrow limits on an individual's choice of a marital partner. At the moment we cannot go beyond this point and assign a proportionate probable weight to each one.

In conclusion, I think the data we have presented strongly support the proposition that one's subculture, and one's race, age, and class positions in the society effectively determine the kind of a person one will marry, but not the exact individual. In a highly significant number of cases the person one marries is very similar culturally to one's self. Our data clearly support the theory of homogamy, rather than that of heterogamy, *but* a generalized theory of the precise influence of cultural and individual factors on the selection of marriage mates remains to be formulated. This is an objective for sociologists to work toward.

INTERFAITH MARRIAGES

Harvey J. Locke, Georges Sabagh, and Mary Margaret Thomes

Adapted from *Social Problems,* 4 (1957), 333-340.

K ennedy and Hollingshead, in their studies of New Haven, Connecticut, arrived at the conclusion that interfaith marriage rates are so small that religion is the chief barrier to assimilation in the United States. Kennedy found that the per cent of Catholics who married non-Catholics was consistently low, ranging from 14 to 18 per cent for three decades beginning with 1900. [6] [1] She concluded that "religious endogamy is persisting and the future cleavage will be along religious lines rather than along nationality lines as in the past." [6, p. 332] She held that the concept of the single melting pot of assimilation should be replaced by

---

[1] Catholics are equated with three nationality groups: Italians, Irish, and Poles; Protestants with British Americans, Germans, and Scandinavians; Jews were recorded separately. Kennedy secured her data from marriage-license records.

the idea of the triple melting pot. While she applied this idea to the United States as a whole, she emphasized that her data were exclusively for New Haven.

Hollingshead secured data on this question from interviews with 437 white couples in New Haven in 1949. He reported that of all married couples involving Jews, Catholics, and Protestants the per cent of interfaith marriages was about 3 for Jews, 6 for Catholics, and 27 for Protestants. [3] He concluded: "Kennedy's and our data show we are going to have three pots boiling merrily side by side with little fusion between them for an indefinite period." [3, p. 624]

John Thomas, using 1949 data, found that the triple-melting-pot idea did not apply even to the state of Connecticut, which had a relatively high rate of Catholic interfaith marriage. He reported that 40 per cent of all Catholic marriages in Connecticut were mixed. [9] The data given below tends to emphasize the inadequacy of the triple-melting-pot theory of assimilation.

The specific purpose of this paper is to test systematically the hypothesis that the rate of interfaith marriage of a given religious group increases as the proportion of that group in the population decreases. We shall also examine the factors involved when cases deviate from the hypothesis. The two variables in the hypothesis are interfaith marriages and the proportion of a religious group in the population. An interfaith marriage is defined as one in which either the bride or groom is a member of a given religious group while the spouse is not. The interfaith marriage rate used here is the per cent which interfaith marriages are of all marriages involving members of a given religious group. The hypothesis will be tested by data from the United States and Canada.

United States Data

Our analysis for the United States will be confined to interfaith marriages of Catholics from data in the *Official Catholic Directory.* [7] This information is fairly complete and reliable because each parish submits an annual report which includes total marriages and the number of interfaith marriages. From this we computed the interfaith marriage rate. Obviously the interfaith marriages included are only those which the Church sanctions and considers valid.[2] [9, p. 488] Figures in the *Direc-*

2 John Thomas has shown that the number of invalid interfaith marriages is high. (9, p. 488) He studied about 30,000 interfaith marriages and found that 40 per cent were not sanctioned by the Church. He indicated that the interfaith marriage rate would be considerably higher if both valid and invalid marriages were included. However, there are no adequate data on the number of invalid marriages.

tory on the number of Catholics, as in the case of most religious enumerations, are not entirely reliable since they are based on various Church censuses and estimates. They do, however, provide an approximation of the actual Catholic population. The general population figures of the various states are estimates of the United States Census.

The interfaith marriage rate and per cent of Catholics in the population of each state were computed for 1955 and also for 1945. The range of these variables may be illustrated by the 1955 figures. In that year Catholic interfaith marriage rates ranged from a low of 13 per cent in New Mexico to a high of 70 per cent in North Carolina. Only two states, New Mexico and Rhode Island, had rates of less than 20 per cent. On the other hand, 20 states had interfaith marriage rates between 40 and 70 per cent. The per cent of Catholics in the population ranged from 1 per cent in North Carolina to 61 per cent in Rhode Island.

High negative correlations were found between the interfaith marriage rates and the per cent of Catholics in the population for the 48 states.[3] The correlation for 1955 was -.86 and that for 1945 was -.76. These correlations indicate that the lower the per cent of a religious group in the population, the higher the interfaith marriage rate—at least for Catholics.

TABLE 1. **Per Cent Catholic of the Total Population and Per Cent Mixed Marriages, Regions of the United States** *

Region	Per Cent Catholic	Per Cent Interfaith Marriages
New England	47	22
Middle Atlantic	33	24
East North Central	25	26
Mountain	21	29
Pacific	20	34
West South Central	18	23
West North Central	18	30
South Atlantic	5	50
East South Central	4	47

* Computed from *The Official Catholic Directory,* New York, J. P. Kenedy and Sons, 1955.

This inverse relationship is further documented by regional data for the United States for 1955. In Table 1 the regions are ranked from highest to lowest per cent Catholic in the population. As the per cent Catholic (column 1) decreases the interfaith marriage rate (column 2) increases. The lowest interfaith marriage rate, 22 per cent, was in New England which had the highest proportion of Catholics, 47 per cent. The other

[3] Rank Order Correlations were used.

extreme, the East South Central region (Kentucky, Tennessee, Alabama, and Mississippi), had an interfaith marriage rate of 87 per cent but a Catholic population accounting for only 4 per cent of the total.

On the basis of the evidence presented above, one would not be surprised to find a low interfaith marriage rate in New Haven. Since two thirds of Hollingshead's sample of 437 white couples were Catholics, his low interfaith marriage rate was to be expected. It is apparent that findings for New Haven would not indicate a low rate of interfaith marriage for the United States.

Canadian Data

The hypothesis may be tested by data from official Canadian vital statistics and census publications. Brides and grooms state their religious affiliations at the time of marriage and from tabulations of this information we have computed interfaith marriage rates. The calculation of the per cent of a religious group in the population is possible because the Canadian census includes a question on religious affiliation.

TABLE 2.   Per Cent Catholic of the Total Population and Per Cent Interfaith Marriages, Provinces of Canada*

Province	Per Cent Catholic	Per Cent Interfaith Marriages
Quebec	88	2
New Brunswick	51	8
Prince Edward Island	46	9
Nova Scotia	34	17
Newfoundland	34	17
Ontario	25	22
Saskatchewan	24	26
Manitoba	20	32
Alberta	20	33
British Columbia	14	46
Canada	43	11

* Computed from Dominion Bureau of Statistics, *Vital Statistics, 1951*, Ottawa, 1954, pp. 400-409; and Dominion Bureau of Statistics, *The Canada Yearbook 1954*, Ottawa, 1954, p. 137. The rates were calculated for brides and bridegrooms taken together.

In Table 2, the 10 provinces of Canada are ranked according to the per cent Catholic in the population in 1951. It will be noted that there is a perfect negative relationship between the per cent Catholics and the interfaith marriage rates.

Anglican interfaith marriage rates and the per cent of Anglicans in the population are given in Table 3. While in general interfaith marriage rates increase with a decrease in the proportion of Anglicans, this relationship is not a perfect one.

Thus data on Catholics in the United States and on Catholics and Anglicans in Canada support the hypothesis that the rate of interfaith marriage of a given religious group increases as the proportion of that group in the population decreases.

There are, of course, many other factors that are related to the level of intermarriage in a given area. One way of finding out what these factors are is to consider cases that deviate greatly from what can be expected on the basis of our hypothesis.

For example, Table 4 gives three pairs of states with about the same proportion of Catholics in the population. This table shows that there is a wide difference between the states of each pair in the rate of interfaith marriage. In fact, the interfaith marriage rate of one state in each pair is about twice as large as that of the other state.

Social distance may be one variable affecting the low interfaith marriage rates of Texas and New Mexico as compared with the high rates in Nevada and Connecticut. In Texas and New Mexico, Catholics are predominantly Mexican-Americans, while Catholics in Connecticut and Nevada have a different cultural background. In Connecticut a very large proportion of Catholics are of Irish, Polish, or Italian origin. The social-distance studies of Bogardus show that there is a much greater distance between the general population and Mexican-Americans than between the general population and Irish, Poles, and Italians. [1] Richards, investigating attitudes of 1,672 white college students in Arkansas, Louisiana, Oklahoma, and Texas, found that the social distance between these students and Mexicans was as great as between them and Negroes. [8] Additional research might test the hypothesis that the greater the social distance between ethnic groups the lower the intermarriage rate.

The cohesiveness of religious groups may be another factor affecting the level of interfaith marriage. For example, a study of religion in Ellis County, Kansas, reports the presence of five homogeneous and cohesive Russian-German Catholic communities. [5] If a similar situation exists in other counties in Kansas, these relatively isolated Catholic groups would have little contact with Protestants and thus the interfaith marriage rate would be reduced. The hypothesis is that the greater the homogeneity and cohesiveness of a religious group, the lower the intermarriage rate.

The cohesiveness hypothesis may be applicable also to the case of Quebec. It will be seen from the data given in Table 3 that Quebec has the lowest per cent of Anglicans in the population and also a relatively

TABLE 3.   Per Cent Anglican of the Total Population and Per Cent Interfaith Marriages, Provinces of Canada *

Province	Per Cent Anglican	Per Cent Interfaith Marriages
Newfoundland	30	36
British Columbia	27	51
Ontario	20	49
Nova Scotia	18	47
Manitoba	16	54
Alberta	13	57
New Brunswick	12	57
Saskatchewan	11	59
Quebec	4	47
Canada	15	50

* Computed from Dominion Bureau of Statistics, *Vital Statistics, 1951*, Ottawa, 1954, pp. 400-409; and Dominion Bureau of Statistics, *The Canada Yearbook 1954*, Ottawa, 1954, p. 137. The rates were calculated for brides and bridegrooms taken together.

low rate of Anglican interfaith marriage. There is evidence indicating that the Catholic French-Canadians, who constitute the majority of the population, are a cohesive group. [4]

The level of economic status of a religious group may be related to the extent of interfaith marriage practiced by that group. The hypothesis would be that the higher the economic status, the higher the interfaith marriage rate. An example is the case of Arizona and Texas. There is evidence to indicate that Arizona Catholics have a much higher level of income than those in Texas, and this may account for the higher rate of Catholic interfaith marriage in Arizona.[4] Also Fichter's report of interfaith marriage in two parishes in New Orleans tends to support this idea. [2] One parish had a much higher socio-economic status than the other. Interfaith marriage was significantly greater in the parish having the higher status. This leads to the hypothesis that the higher the economic status of a religious group, the higher the intermarriage rate.

It appears that religion is not a major barrier to assimilation in the United States. At any rate, the proportion of Catholics who engage in interfaith marriage is relatively large in the United States as a whole. In 1955, of all Catholic marriages 27 per cent were valid interfaith marriages.

[4] Persons with Spanish surnames constitute about 70 per cent of the Catholic population in both Arizona and Texas. The median annual income of persons with Spanish surnames in Arizona is 44 per cent higher than of similar persons in Texas (1950 census data). We assume that persons with Spanish surnames are Catholics. On this basis we infer that Catholics in Arizona have a higher economic status than those in Texas.

If those not sanctioned by the Church were added to these, there would be an even higher per cent of interfaith marriage.

We have shown that the per cent of interfaith marriage of a religious group increases as the proportion of that group in the population decreases. In addition, we suggested that in cases where interfaith marriage rates differ from the expected, additional cultural factors such as social distance between groups, cohesiveness, and economic status may be related to such variations in interfaith marriage rates.

TABLE 4. **Per Cent Catholic of the Total Population and Per Cent Interfaith Marriages for Selected States***

States	Per Cent Catholic	Per Cent Interfaith Marriages
Kansas	13	27
Washington	11	46
Texas	20	20
Nevada	20	49
New Mexico	47	13
Connecticut	47	24

* Computed from *The Official Catholic Directory,* New York, J. P. Kenedy and Sons, 1955.

REFERENCES

1. Bogardus, Emory S., "Changes in Racial Distance," *International Journal of Opinion and Attitude Research,* 1 (1947), 58.
2. Fichter, Joseph H., *Dynamics of a City Church* (Chicago: University of Chicago Press, 1951), p. 109.
3. Hollingshead, August B., "Cultural Factors in the Selection of Marriage Mates," *American Sociological Review,* 15 (1950), 622.
4. Hughes, Everett C., *French Canada in Transition* (Chicago: University of Chicago Press, 1943), Chapters 2 and 15.
5. Johannes, Sister Mary Eloise, "A Study of the Russian-German Settlements in Ellis County, Kansas," *The Catholic University of America Studies in Sociology* (Washington: Catholic University of America Press, 1946).
6. Kennedy, Ruby Jo Reeves, "Single or Triple Melting-Pot? Intermarriage Trends in New Haven, 1870-1940," *American Journal of Sociology,* 49 (1944), 333.
7. *Official Catholic Directory,* New York, Kenedy, yearly publications.
8. Richards, Eugene S., "Attitudes of College Students in the Southwest Toward Ethnic Groups in the United States," *Sociology and Social Research,* 33 (1950), 27-28.
9. Thomas, John L., "The Factor of Religion in the Selection of Marriage Mates," *American Sociological Review,* 16 (1951), 489. Also, "Mixed Marriages in the United States," *Lumen Vitae,* 6 (1951), 173-186.

SOCIAL CONTROL OF NEGRO-WHITE INTER-MARRIAGE

Joseph Golden

Adapted by permission of the author and publisher from *Social Forces,* 36 (1958), 267-269. Copyright, University of North Carolina Press, 1958.

When one is reckoning up the various forces which contribute toward preventing the intermarriage of whites and Negroes, it is well to keep in mind some facts which are so obvious that they may be lost sight of. The two groups are separated, particularly in the South, by a system of segregation which effects most areas of living—employment, family life, residence, recreation, even transportation. Whether or not this system of segregated living is expressed in legislation, it is public sentiment which gives meaning to, and which enforces, the system. The structure is significant since it allows few opportunities for men and women of the two races to meet at all, fewer opportunities for them to engage in sexual contact, and even fewer for the kind of equalitarian contacts which lead to marriage.

The laws which prohibit Negro-white intermarriage have been described and analyzed in several publications.[1] It is pointed out that 29 [2] states possess laws which prohibit intermarriage. These states include all the southern states, five North Central states, all the Mountain states, except New Mexico, and Oregon as the only one of the Pacific states. The definition of "Negro" in these laws ranges from "one-fourth or more Negro blood" to "persons with any trace of Negro blood whatsoever." Punishment may be as high as $1,000 fine and/or ten years of imprisonment.

[1] Albert E. Jenks, "Legal Status of Negro-White Amalgamation in the United States," *American Journal of Sociology,* XXI (March 1916), 666-78; Charles S. Mangum, Jr., *The Legal Status of the Negro* (Chapel Hill, 1940), chap. X; Geoffrey May, *Marriage Laws and Decisions in the United States* (New York, 1929), passim; Gilbert T. Stephenson, *Race Distinctions in American Law* (New York, 1910), chap. VI; Chester G. Vernier, *American Family Laws* (Stanford, 1938), I, passim; Louis Wirth and Herbert Goldhamer, "The Hybrid and the Problem of Miscegenation," *Characteristics of the American Negro,* edited by Otto Klineberg (New York, 1944), chap. VIII; Justine W. Wise, "Intermarriage with Negroes—A Survey of State Statutes," *Yale Law Journal,* XXXVI (April 1927), 858-66.

[2] Twenty-eight at time of writing; see infra.

Both Wirth and Wise have shown that all states with more than five percent Negro population have such laws, as well as 12 states with less than three percent Negro population. West Virginia punishes only the white person.[3] Punishments are also provided in a number of states for ministers or officials who knowingly perform a ceremony, and for officials who knowingly issue a license, in violation of the law.

Since the publication of the most recent work on legislation, California has repealed its law. This is the first instance of repeal of a state law on miscegenation since the 1880's when four states[4] repealed their laws. The California law was called into question when Sylvester S. Davis, Jr., classified as Negro, and Andrea D. Perez, classified as white, filed a petition for a writ of mandamus against Los Angeles County authorities in an effort to compel the latter to issue a marriage license. The case was carried directly to the state's Supreme Court which held, on October 1, 1948, that the law was unconstitutional, since "marriage is something more than a civil contract, subject to regulation by the state. It is a fundamental right of free men." [5]

Even in states which do not prohibit intermarriage, it is reported that functionaries of the courts and of municipal and county governments put difficulties in the way of those mixed couples who seek to obtain a marriage license.[6] Pennsylvania and Rhode Island have no law prohibiting miscegenation, but require applicants for a marriage license to state their race ("color" in Pennsylvania). Some of the 50 couples interviewed in Philadelphia in 1949 and 1950 [7] stated that the clerk issuing the license had attempted to dissuade them from marrying. Such attempts were also reported on the part of clergymen of several denominations. Some of the Negro husbands reported a great deal of difficulty, while they were in the Army, in obtaining permission to marry from their superior officers and the army hierarchy.

In those states where intermarriage is illegal, the state itself attempts to prohibit miscegenation and to punish those who violate the proscrip-

[3] Charles S. Mangum, Jr., *op. cit.*, p. 241.

[4] Rhode Island, 1881; Maine, 1883; Michigan, 1883; Ohio, 1887. Gilbert T. Stephenson, *op. cit.*, p. 90.

[5] P. Murray, *States' Laws on Race and Color* (Cincinnati, 1951), p. 77; Editorial, *The Nation*, CLXXVI (October 16, 1948), p. 415.

[6] A clerk in the marriage bureau in Harrisburg, Pa., refused to issue a license to a mixed couple, saying that it was against the policy of his office. The bridegroom took the matter to a court, which ordered the marriage bureau to grant the license. They made out the marriage application, but the clerk tossed the pen they had used into the waste basket. P. Murray, *op. cit.*, p. 76.

[7] See Joseph Golden, "Characteristics of the Negro-White Intermarried in Philadelphia," *American Sociological Review*, XVIII (April 1953), 177-183.

tion. In those states, however, where no such law exists, there is no official agency or institution to uphold the mores of racial endogamy. There it is the family which, more than other institutions, assumes the responsibility for preventing miscegenation, especially in the form of intermarriage. Although many of the persons interviewed had few and not very strong family ties, a significant proportion related that their interracial marriage was preceded by family discussion. Their parents, siblings, and other relatives offered advice, pointed out the dangers of mixed marriage, and warned of the prejudice they would encounter. Some appealed to their loyalty—the marriage would injure their parents' social standing or economic status, or they would find it impossible to continue living in the community if the marriage became known. Few of the white spouses were living in the same community as their parents; it was, perhaps, this lack of strong family ties and the consequent anonymity which helped to make their marriage possible. Several white wives stated that they had never informed their parents of the marriage.

The wedding of the Thayers[8] (Negro husband—white wife) was preceded by discussions between the parties and their families and friends. Everyone advised against it. Her family seemed to be mainly concerned with the effect on them. Her sisters believed that if the community in St. Louis heard about this, they and their children would be completely ostracized. His children were concerned about what the marriage would do to his business, but told him to go ahead if that was what he wanted. They were married in the home of his daughter by the minister of the Negro church which she attended. Only his immediate family was present.

The parents of Mrs. Brown, a German-Jewish refugee, objected to the marriage, although they had no objection to her dates with Negroes. Her father was afraid that the marriage would hurt her career, while her mother feared that her father's jewelry-repair business would be injured when news of the marriage became known in the neighborhood. Mr. Brown feels that they were not "prejudiced," but were really concerned about her welfare.

Few of these persons met their in-laws before marriage. One couple tried it. "We engineered several occasions. They were impressed. Mom thought Jim was very intelligent, quite a speaker, etc., until they found him to be their future son-in-law. Then all hell broke loose. Pop was going to shoot us, shoot Jim, have us arrested and all kinds of things. Nothing ever happened. During the time that we were going together, I wanted to bring him home and say 'This is it.' Jim said, 'No, try to get them used to interracial company and then go ahead.' But it didn't happen. I doubt if it would have, for Pop brought home a picture of an interracial dance in England

---

[8] This name, as well as all other names of persons interviewed, is fictitious.

and during a lull in any conversation, when company was about, would whip it out and say, 'Look at those niggers, etc.' " [9]

There is no way of knowing how many couples are prevented from marrying by this sort of pressure. Miss Stone became friendly with a young Negro man whom she met at an artists' party. They courted over one year. When she told her parents that they were contemplating marriage, her father and her brother came to see him and tried to break up the affair. Her father died soon after and her mother became ill. Miss Stone felt that she was the cause of these events. After postponing the impending marriage for two years, they found they did not wish to marry.

As was pointed out at the beginning of the discussion, the primary force inhibiting interracial marriages is the social structure which discourages such contacts between the two races as may lead to marriage. Perhaps just as important are the attitudes which buttress the system of segregation, and the myths whose function it is to strengthen the prohibition against interracial marriage. Along with these attitudes goes the belief that interracial marriage is doomed to fail, that successful marriage is impossible for a Negro-white couple. Whether the belief is based on reality or not is, for the purpose of discussion, irrelevant. The belief itself is a factor in keeping the number of interracial marriages so small that they do not represent a threat to the existing sex mores. An examination was made of several texts currently used in functional college courses in the field of marriage and family life. Bowman states that racially mixed marriage "presents unusually difficult problems, which in some cases are hopelessly insoluble." [10] Christensen suggests that interracial marriages "are usually inadvisable on social and cultural grounds." [11] Duvall and Hill, discussing American research on interracial marriages, find that "Without exception the findings from this research argue against intermarriage." [12]

We have discussed, so far, a number of social controls which function to prevent or inhibit Negro-white intermarriage. Among them are: the segregated social structure of our culture; the system of attitudes, beliefs, and myths which grow out of the social system and which serve to strengthen it; the laws which express the sex and marriage customs of the culture; institutional functionaries, such as clergymen, army officers, and governmental employees who attempt to discourage interracial marriage even in those states which have not legislated on the subject; the family, especially the immediate family, which uses affectional ties to prevent

[9] Letter to author from Mrs. White.
[10] Henry A. Bowman, *Marriage for Moderns* (New York, 1950), p. 180.
[11] Harold T. Christensen, *Marriage Analysis* (New York, 1950), p. 264.
[12] Evelyn M. Duvall and Reuben Hill, *When You Marry* (Boston, 1945), p. 117.

intermarriage. In view of the institutional and moral forces arrayed against such intermarriages, why is it that some do take place? It is suggested that there are factors in our culture which operate to lessen the effectiveness of the social controls described. These factors will be discussed in a future study.

17.   DATING: LOVE AS A CRITERION IN MATE-SELECTION

Where marriages are agreed upon by the persons directly involved rather than by their families (as in traditional China), the society must make some provision for premarital association and for mate-selection. In its context the American practice of dating (eventually leading to marriage for most persons) makes good societal sense.

The diffidence of Elmtown adolescents on their first dates and the age pattern in dating are described by Hollingshead. Winch analyzes the dating pattern in a quest for its marriage-oriented and other functions.

Hunt summarizes Christian ambivalence about love and sex, and Beigel relates contemporary conceptions of love to courtly and romantic love. In the theory of complementary needs, Winch and Ktsanes and Ktsanes sought to conceptualize love so that it would better lend itself to research than it had previously. Here the latter two authors present a summary of the theory and an illustrative case. According to this theory, the homogamy in social characteristics, which was observed in chapter 16, creates for each individual a field of eligible spouse-candidates; within the field of eligibles it is believed that mate-selection proceeds on the basis of complementary needs. This means that rather than select a mate psychically similar to oneself, a person tends to select a mate whose pattern of needs is complementary to one's own.

Some years ago a sociologist set up for research purposes his own lonely-hearts-introduction-by-mail-service. Wallace presents some characteristics of his clients and discusses the functions of his (now defunct) service.

DATING IN ELMTOWN

August B. Hollingshead

Adapted by permission from August B. Hollingshead, *Elmtown's Youth,* New York, John Wiley & Sons, Inc., 1949, pp. 223-27.

Local folkways define picnics, dances, parties, and hayrides as date affairs at which a boy is expected to pair with a girl. The testimony of many students demonstrates that the vast majority have their first formal date on these occasions. Individuals recalled vividly whom the first date was with, where they went, who was there, and other details which marked this important step in the transition from childhood to adolescent life. The first date is often a cooperative enterprise which involves the members of two cliques of the opposite sex. Two illustrations will be given to illuminate the process; both were taken from autobiographies of seniors. The first was written by a class III girl.*

I began to date in the eighth grade with boys I had played with all my life. I ran with a group of girls, and there was also a group of fellows we liked. At all social functions where boys and girls mixed, these two groups came together. We started running around in the fifth grade at Central School, but we did not date yet. We were always together at all school parties, and by the eighth grade we were having our own private parties to which these two groups and no one else was invited. We held these parties at the homes of the girls fairly frequently, usually on a Friday or Saturday night, sometimes on Sunday, but not often. We still did not have any regular dates until the end of the eighth grade.

[Then] Marion Stowe's mother had a party for us. She invited all the kids in both our groups. The fellows got together and decided they would have dates. Tom Biggers asked me to go to the party with him. I was so thrilled and scared I told him to wait until I talked it over with Mother. Mother thought I was too young to start having dates. I argued with her for two days. Dad couldn't see anything wrong with me going to the party with Tom; so Mother let me say "Yes." Tom's dad came by for us in their car and took us to the party, but we walked home. My next date was with Joe Peters during the summer before I started to high school.

---

* Hollingshead divides Elmtown families into five social classes. Class I carries the most prestige, and class V, the least. Class III families, then, tend to cluster around the midpoint of Elmtown's prestige spectrum, while the generally "poor but honest" families of class IV are presented as distinctly lower in prestige than those of class III. For a description of the sub-cultural characteristics of Elmtown's five classes, see A. B. Hollingshead, *op. cit.,* chap. 5—*Eds.*

Eddie Parker, a class IV junior, believed that his interest in dating went back to the seventh grade, when he and his friends began "to feel shy in the presence of girls," whereas they had been indifferent, aloof, or hostile to them before. His clique talked "a lot" about dates, girls they would like to date, and "women" in general, but no one was bold enough to make a date. This went on until the spring "we were in the eighth grade when all of us [his clique] decided one Saturday we would make dates with the girls in our class who lived in the neighborhood. We went around to their houses and asked them if they would go to the show with us. We made dates with five of them, and that night we all went to the show together." In this case, it would appear that the boys, and probably the girls as well, derived support from one another. If we accept student reports, common characteristics of these first dates are shyness, fear of doing the wrong thing, of making statements the other person will resent, and overcautiousness in the physical approaches of one partner to another. Both persons have been filled with so much advice by parents, usually the mothers, about how to act and what to expect from the date that both play their roles clumsily. They are told precisely what to say and do, when to come home, what they should not do, and what the consequences will be if they violate their instructions. As one class III girl said:

I was so scared by what Mother told me Jim might do I did not like the experience at all. He did not even try to hold my arm. I knew I was supposed to "freeze up" if he did and I was so ready to "freeze up" we walked all the way home without saying much. I knew he was afraid of me so we just walked along. I was so disappointed in that first date I did not have another for a year. I had several crushes on boys, but I couldn't bring myself to say "Yes" when they asked me for a date. In the latter part of my sophomore year, I had a crush on Larry Jacobs, and when he asked me for a date I said "Yes." We went together a few times when Frank Stone asked me for a date. I went to the Junior Play with him. After the play we went to Burke's [a popular restaurant] with the rest of the kids and then home. Oh, we had fun! Since then, I have had a lot of dates, and now I really enjoy them.

The more adventurous youngsters begin to date when they are 12 years of age—at picnics and family group get-togethers—and the parents are usually present. A definite dating pattern becomes clear during the fourteenth year; 20 per cent of the girls and 15 per cent of the boys report that they had their first dates when they were 13. A much larger number begins to date in the fifteenth year, and by the end of it approximately 93 per cent of both sexes are dating with some regularity. Among the sixteen-year-olds, dating is the accepted procedure, and the boy or girl who does not date is left out of mixed social affairs. Our data make it clear that between the beginning of the fourteenth and the end of the sixteenth years the associational pattern of these adolescents changes from almost exclusive interaction with members of their own sex to a mixed associational pattern similar to that found in adult life. In this period, certain activities, such as girls' "hag parties" and hunting and baseball among the boys, are organized on a single sex basis; and others, such as dances and parties, are almost exclusively mixed.

Forty-three per cent of the boys and 58 per cent of the girls report that they experienced the thrill of their "first date" before they entered high school. Dating before entry into high school is not related significantly to age, town or country residence, or class. On the contrary, it is associated with clique membership. Some cliques have a much higher ratio of dates than others, but we did not search for an explanation of this fact either within the cliques which dated or those which did not. The discrepancy between boys and girls with dating experience prior to entry into high school continues throughout the freshman year. This differential disappears in the sophomore year, and by the time the junior year is reached more boys than girls report dates. At this level only 1 boy out of 13 and 1 girl out of 10 claim they have never had a date. All senior boys report they have had dates, but 3 girls are still looking forward to this event.

About 51 per cent of 553 dates the students reported during April, 1942, were with other students who belonged to the same school class; that is, freshman with freshman, and so on.[1] When the dating partners belong to different school classes, the pattern is significantly different between the boys and the girls. One-third of the boys' dates are with girls who belong to a class *below* them in school, whereas 31 per cent of the girls' dates are with boys *above* them in school. This gives the freshman girl a wider opportunity for dates than the freshman boy, for she can be dated by a freshman or a boy from the sophomore, junior, or senior classes. Freshmen and sophomore boys are reluctant to ask a girl who belongs to a class above them in school for a date. Many girls do not like to date younger boys unless they possess specific prestige factors, such as athletic prowess, "family background," or "good looks." Only 15

---

[1] All statistical data on the dating pattern unless otherwise indicated, such as the figures on dates and no dates before high school, are derived from the analysis of the dates the students reported they had during April, 1942. April was selected as our sample month because by that time we knew the students personally and had asked them so many questions that we assumed, and correctly, that they would give us information about their dating behavior. A second reason for choosing April was the belief that by this late in the year the dating pattern of the student group would be well established. We also believed that it was better to attempt a complete study of dating behavior for a single month than to trust student memories over a longer period.

per cent of the boys' dates are with girls from a higher school class than theirs. Almost two-thirds of these mixed dates (62 per cent) are between senior girls and junior boys; the remainder are between sophomore boys and junior and senior girls.

These figures bring out the effects of two customs on the dating relations of these young people. In the first place, the folkways of courtship encourage a boy to date a girl younger than himself. The complement of this is that the girl expects to date a boy older than herself. The operation of this rule results in boys dating girls either the same age as themselves or younger. With the school classes graded principally along age lines, this means that the boys date girls from their own class or a lower class in school. Thus, the freshman, sophomore, and junior girls have more opportunities for dates than the senior girls. In the second place, the senior girls' dating chances are limited still further by an administrative rule which restricts to high school students any high school party at which there is dancing. This rule severely restricts the senior girls' dating field, and to a less extent the juniors', particularly in class IV, because boys at this level drop out of school sooner than girls. Thus, a shortage of senior boys, combined with the school rule that only students may attend high school dances, forces the senior girls to ask junior boys for dates or let it be known that they would like to go with a junior boy or not date at school affairs. Another effect of this aspect of the dating system is the limited opportunity open to the freshman boy to date girls. Within the high school, the only girls he can date readily are freshmen, and here he competes with sophomore and junior boys who have more prestige in the eyes of the girls than he does. Then, too, the older boys are more sophisticated, more experienced in the arts of love, usually have more money, and give the girls more status in their own eyes than a "green kid" whom they have known through years of close contact in elementary school. The net effect of these factors on dating is a significantly lower ratio of dates among freshman boys in comparison with freshman girls, and of more junior and senior girls dating younger boys or boys outside the student group.

THE FUNCTIONS OF DATING IN MIDDLE-CLASS AMERICA

Robert F. Winch

Adapted from *The Modern Family,* New York, Holt, Rinehart and Winston, 1952, and from the forthcoming revision of it.

Morton Hunt speaks of dating as an American social invention of the 1920s and as

"the most significant new mechanism of mate selection in many centuries. In place of the church meeting, the application to father, and the chaperoned evenings in the family parlor, modern youth met at parties, made dates on the telephone and went off alone in cars to spend their evenings at movies, juke joints, and on back roads." [1]

Insofar as it is related to marriage, dating is the "window-shopping" period—it carries no commitment to buy the merchandise on display. Dating in American culture has a number of functions, some of which are only remotely related to marriage. In the first place, dating is a popular form of recreation, and thereby an end in itself. In the current setting, at least where the urban ethos prevails, a date carries no future obligation on the part of either party except, perhaps, for some reciprocation in entertainment.

A second function, especially in the school situation concerns the status-grading and status-achieving function. As Mead expresses it, "the boy . . . longs for a date [but not] . . . for a girl. He is longing to be in a situation, mainly public, where he will be seen by others to have a girl, and the right kind of girl, who dresses well and pays attention." [2] In his famous article on "The Rating and Dating Complex" [3] Waller made the point that in campus dating there was exploitation in two senses: (a) each party tried to make the other fall in love "harder" and earlier, and (b) each was interested in the other for status considerations. So far as fraternity men and women are concerned, the latter point has been corroborated on the campus of one midwestern university by Ray, who asserts that dating is "one of the ways of gaining, maintaining or losing

[1] Morton M. Hunt, *The Natural History of Love,* New York, Alfred Knopf, 1959, p. 356.

[2] Margaret Mead, *Male and Female,* New York, Morrow, 1949, pp. 286-287.

[3] Willard Waller, *American Sociological Review,* 2, 1937, 727-737.

prestige for the house." [4] He found this to be considerably more true of women's than of men's organizations and pointed out that while dating was one of the principal ways in which a sorority accumulated and maintained prestige, a fraternity had other avenues to prestige, such as athletics. Both in sororities and fraternities dating was found to be quite homogamous with respect to social status, but the men's dates tended to diverge more from their own statuses than did those of the women.

In such a setting as the coeducational campus considerable pressure to date is exerted upon those who would prefer not to. It is consistent with the directness and cruelty of adolescence that such pressure is often expressed as group ridicule. The social conditions of college life, then, stimulate one (a) to date, and (b) to date the type of person approved by one's social group. It follows that a date does not always signify a man's spontaneous and voluntary affectional interest in a girl. It should not be thought, however, that fraternities and peer groups generally represent the only source of such pressure. Families are frequently interested in the courtship progress of their young people, and are traditionally disposed to encourage the mating interest of their spinster daughters, irrespective of whether the age of spinsterhood be defined as beginning at eighteen or twenty, as in Colonial times, or a decade or so later, as in the college-trained groups of today.

A third function of the phenomenon of dating is that of socialization. It provides males and females with an opportunity to associate with each other, and thus to learn proper deportment and the social graces. It serves to eliminate some of the mystery which grows up about the opposite sex —a mystery which is fostered by the small-family system in which many children have no siblings of the opposite sex near their own age.

There is a fourth function, which is a corollary of the third. The opportunity to associate with persons of the opposite sex gives a person the chance to try out his own personality and to discover things about the personalities of others. In our discussion of the adolescent we noted that the social situation caused persons of both sexes to be somewhat uncertain as to how successfully they would work through their various tasks of self-validation, especially that of achieving the appropriate sex-type. The dating process is a testing ground—both in the sense of providing repeated opportunities for the adolescent to ascertain his stimulus value to persons of the opposite sex, and of providing learning situations so that he can improve his techniques of interaction. Dating allows him an opportunity to discover that potential love-objects are also insecure; thus the adolescent can universalize his insecurity and thereby reduce his own feel-

[4] J. D. Ray, *Dating Behavior as Related to Organizational Prestige,* Department of Sociology, Indiana University, 1942, p. 42 (unpublished master's thesis).

ings of inadequacy. In the process of learning about the personalities of the opposite sex, the male, for example, ceases to react to all females as "woman" and discovers that there are "women," i.e., that females too are individuals and have idiosyncrasies. Another way of speaking of this function is to interpret it as a means of defining the dater's identity:

"To a considerable extent adolescent love is an attempt to arrive at a definition of one's identity by projecting one's diffuse ego images on one another and by seeing them thus reflected and gradually clarified. This is why many a youth would rather converse, and settle matters of mutual identification, than embrace." [5]

The third and fourth functions of dating facilitate the fifth—mate-selection. Dating enables young men and women to test out a succession of relationships with persons of the opposite sex. One finds that one "gets along nicely" with some, not so well with others, that some relationships are thrilling—at least for a time—that others are satisfying, and still others are painful and laden with conflict. Through dating one can learn to interpret the behavior and thereby to diagnose the personalities of persons of the opposite sex.

By noting with what kind of person one's interaction is most gratifying one can learn something about the personality and values one would find desirable in a spouse. Armed with this experience one is in a vastly improved position to set about the task of selecting a mate. In the urban setting there is an emphasis on the need-meeting aspect of the marital relationship (in terms of affection, security, etc.) and because of the heterogeneity of the population there is considerable variation in values, life styles, etc. Accordingly, because of the complexity of the checks to be made, two or three dates barely allow the testing function to get under way.

A sixth function of dating is that of intensifying the anticipatory socialization of the dating individual into marital and other adult familial roles. Such a process is begun of course in the person's family of orientation for he learns a version of the content of the roles of husband and of wife, of father and of mother in his parental home. It is to be expected, however, that in a dynamic, non-traditional society like ours each generation will believe that there are ways to improve upon their parents' conceptions of familial roles, and in the dating relationship there is opportunity for discussion of other versions. The quest for more satisfactory definitions of familial roles can be illuminated by the explorations of the dating couple into each other's values, opinions, life style and life plans,

---

[5] Erik H. Erikson, *Childhood and Society,* New York, Norton, 1950, p. 228. Hunt quotes D. H. Lawrence as making roughly the same point. Cf. Morton H. Hunt, *op. cit.,* p. 355.

desire as to number of children, location and kind of community to live in, devoutness of religious orientation, occupational goals, and standard of living.

When viewing dating as a procedure for selecting a mate and for beginning the adjustment to marriage, it is important to recall the ways in which the dating relationship is *not* a rehearsal for marriage. Dating provides an opportunity to explore the personality and values of another human being in a situation of erotically tinged, fun-oriented recreation. Much of the content of marital and parental roles, on the other hand, involves the task-oriented activities of making and financing purchases, keeping a house clean, orderly, and stocked, and of tending children. For this reason the efficiency of dating as a procedure for mate-selection and particularly as a context for anticipatory socialization into marital roles and adjustment to marriage can be only partial.

THE HEART'S PARADOX

Morton M. Hunt

Adapted from *The Natural History of Love,* New York, Alfred A. Knopf, Inc., Copyright 1959 by Morton M. Hunt, pp. 126-127.

The influence of Christianity upon love and marriage is . . . an impossible tangle of opposites—purification and contamination, the rebuilding of the family and the total flight from the family, the glorification of one woman and the condemnation of womankind. But the paradoxical is not meaningless, for it is a reflection of a basic mechanism of human personality; in the human heart paradox has its equivalent in the form of the paired and warring drives of love and hate, selfishness and altruism, submissiveness and rebelliousness. It is, in a word, ambivalence—the ability to feel two ways about one thing: the most perplexing, best-hidden, and most pervasive aspect of human nature.

Christianity accentuated ambivalence in the area of love far beyond any previous social system. Formerly, men and women accepted sex as a human appetite, considered love a pastime and sometimes a torment, and viewed marriage as an inescapable social duty. They had some conflicting feelings about each, but not too many, and consequently felt at ease about taking their several satisfactions jointly or separately. Christianity forbade the separate gratification of the desires for sex, love, and marriage, but made it impossible to enjoy them together. Sex became so sinful and disgusting that even within marriage, as an act allowed by God, it seemed shameful and more ignoble than elimination. Men and women hungered for it and hated the hunger, enjoyed it and felt guilty for the enjoyment. Marriage itself was the only state in which the body's yearnings could legitimately be satisfied, and through which God's commandment to be fruitful was realizable, yet it was expressly said to be an inferior condition, and one suited for weaklings. The more elevated emotions of love, since they could hardly be linked with the shameful act of sex, were limited to the painful and dangerous artifice of continent marriage. An ever-larger fraction of the European population, unable to find any tolerable solution to these dilemmas, voluntarily deprived themselves of sex and marriage, and fled to a celibate life in monasteries and convents; there they could love God, the saints, and the Pope without complications.

Saint Augustine epitomized the Christian conflicts in a single terrible sentence: "Through a woman we were sent to destruction; through a woman salvation was sent to us." In comparison, the *"odi et amo"* of Catullus looks almost puerile, and the inner conflicts of Greek men concerning women seem like childish make-believe. Yet paradox has the final word, after all: out of ambivalence, out of sexual restraint, out of the virgin-mother image of woman, man would first fashion that new notion of romantic love that has seemed so enormously valuable to him ever since.

LOVE: COURTLY, ROMANTIC, AND MODERN

Hugo G. Beigel

Adapted from "Romantic Love," the *American Sociological Review*, 16 (1951), 326-34.

Three phases of formalized love are discernible in Western culture. The first encompasses the origin of courtly love in the twelfth century, the second its revival at the turn of the nineteenth century, and the third its present state and significance for marital selection. . . .

Courtly love was the conventionalization of a new ideal that arose in the feudal class and institutionalized certain aspects of the male-female relationship *outside marriage*. In conformity with the Christian concept of and contempt for sex, the presupposition for courtly love was chastity. Being the spiritualization and the sublimation of carnal desire, such love was deemed to be impossible between husband and wife. By application of the religious concept of abstract love to the "mistress," the married woman of the ruling class, who had lost her economic function, was endowed with higher and more general values: gentleness and refinement. Unselfish service to the noble lady became a duty of the knight, explicitly sworn to in the oath the young nobleman had to take at the dubbing ceremony.[1] Part of this service was ritualized; by means of such formalization the aggressiveness of unfulfilled cravings was channeled into codes and causes. In this manner sexual covetousness was deflected and the marital rights of husbands were—theoretically at least—safeguarded. This was obviously an important provision in an age in which social rules prevented free choice of a mate for marriage with the result that basic human needs were left unsatisfied.

Courtly love—in retrospect called romantic love—consequently was not a whimsical play. In spite of the surface appearance of its aesthetic formulation, it sprang from vital needs, from a deeply felt desire for the ennoblement of human relations, and from culture-bred frustrations. It made *māze* (moderation) a masculine virtue.

The fact that it is in the first place the sexual drive that was frustrated in this love relationship suggests an anology with adolescent love. We can assume that certain features in the development of an adolescent brought up in an earlier phase of our culture coincide with tendencies observable nowadays. Those produced by the physiological maturation of the organism, for instance, are universal, and medieval literature gives some evidence of the emotions involved in self-discovery and the experience of change at this age.[2]

While the sexual drive rises to its greatest intensity during adolescence, it is denied satisfaction. Abstinence and celibacy being among the highest religious ideals and sexual immorality being threatened with hellfire, conflicts are created that lead to feelings of guilt, depreciation of the ego, and a heightening of the ego ideal. The phantasy is quickened and the suppression of the intensified desires results in a high emotionality which seeks for vicarious outlets.[3] While sexual relations cannot be established before marriage, there is sufficient erotic stimulation from talk, from visual stimuli, and an occasional trespassing with females outside one's class to feed the hope for more. Unless hope is realized or relinquished, the adolescent strains his resources to impress any members of the opposite sex and one female in particular whose behavior allows anticipation of possible acceptance. The means are display of masculine skill and prow-

---

[1] "Monumenta Germaniae" (leges II, 363) in E. Sturtevant, *Vom guten Ton im Wandel der Jahrhunderte*, Berlin, Bong, 1917.

[2] Chretiens de Troyes, *Percival* (Conte del graal); Wolfram von Eschenbach, *Parzival*; Hartmann von Aue, *Der arme Heinrich*; Wirnt von Gravenberg, *Vigalois*.

[3] K. C. Garrison, *The Psychology of Adolescence*, New York, Prentice-Hall, 1950; A. H. Arlitt, *Adolescent Psychology*, New York, American Book Co., 1933.

ess which, under the influence of religious teachings, the group code, and the masculine ideal, are subordinated to socially acknowledged causes or such feats as can be interpreted as good causes. The striving to prove one's independence and manliness finds expression in the search for adventures. The female, being at the same time the weaker competitor, the object to be obtained, and the substitute for the mother, grows to be the ideal audience and the representative for the super-go; this has the effect that softer virtues often take precedence over coarser forms of behavior. While, in general, the adolescent does not aim at permanent possession of the female, any sign of approval by her is interpreted as accepted and props up the wavering self-esteem. For this service she is idealized; even the refusal of sexual gratification is taken as an indication of greater self-control and moral strength. Such greatness, on the other hand, reflects favorably on the quality of the one accepted, who tries to live up to moral perfection and thus to the beloved's assumed higher standard. Vows of self-improvement alternate with feelings of unworthiness and moments of expansive self-feeling.

The adolescent's showing-off attitude has its counterpart in the medieval knight's search for adventures and in the tournaments he fought for his mistress. Love tests are frequent. Certain feats like those of Ulrich von Lichtenstein, who sent his little finger to his mistress and drank the water in which she had washed, or of Peire Vidal, who had himself sewn into a bear's hide and hunted,[4] have their parallels in the adolescent's obsessional yearning to impress the chosen female by valiance, self-sacrifice, and self-punishment. As do adolescent relations, courtly love provided partial satisfactions of the sexual desire. The lover having become a *drutz*[5] had the right to accompany his lady to her bedchamber, to undress her to the skin, and to put her to bed. Sometimes he was even allowed to sleep with her if he promised to content himself with a kiss. The love symbols are similar; the adolescent feels the one-ness with the beloved by wearing a lock of her hair or a ribbon near his heart as the knight felt it when he tied her veil around his armor; and as the mistress wore her gallant's blood-stained shirt so may a girl today wear her boy's pin, blazer, or baseball hat.

Such and many more similarities provoke the conclusion that courtly love represents the aesthetization of adolescent feelings which, though recognized as precious, are rarely experienced in adulthood with the same

---

[4] H. Jantzen (ed.), *Dichtungen aus mittelhochdeutscher Fruehzeit; Goeschen,* 137, Leipzig, 1910.

[5] A. von Gleichen-Russwurm, *Kultur und Sittengeschichte aller Zeiten und Voelker,* Zurich, Gutenberg Verlag, 1920. The lover who has reached the fourth and highest state in the ritual of courtly love and is accepted.

ardor. Under the influence of the cherished tales of oriental love refinement, the pyre of adolescent emotions was artificially kept burning, producing that subtler form of male-female relations that exploited the elations and depressions of enforced chastity for the ennoblement of the mind and gave the newly consolidated ruling class moral distinction over the crude indulgence of the masses.

The cultural significance of this concept lies in the fact that the idealization of the female initiated her social elevation and that it introduced voluntary fidelity, restraint, and the magnanimous gentleness of the male consciously into the relation between the sexes, qualities that were not considered essential or even possible in a marriage based on the semi-patriarchal concept of the Middle Ages. As the idea spread, it influenced greatly the emotional development of the group as a whole. This penetration became evident when romantic love, the bourgeois adaptation of courtly love, was propagated by the Romanticists.

Presupposing the knowledge of the historic and socio-economic roots of the Romantic movement,[6, 7] we limit ourselves again to an outline of those trends that have direct bearing on our subject.

In formulating the idea of romantic love, the Romanticists merely propounded a concept that had become a socio-psychological necessity. Starting in the fourteenth century, the dissolution of the broader family had progressed to the point where its economic, religious, and political functions were gone. With increasing urbanization the impact of social isolation made itself felt upon the individual. As a result of industrialization and mercantilization the father's authority had decreased and the children remained longer under the more emotionally-oriented care of the mother, a fact that, together with the child's loss of economic function, effected a gradual change in personality, especially in the male personality. Reformation, revolutions, and wars had shaken the foundations of beliefs and traditions. Being the first to feel the pinch of the technological development on the treasured ideology of individualism, the Romanticists rebelled against the progressing de-humanization, the all-devouring materialism and rationalism, and sought escape from these dangers in the wonders of the emotions. In the basic feelings of humanity they hoped to find security and a substitute for the eliminated cultural values.

Under the increasing discomfort in a changing civilization, the aristocratic class had found a way to alleviate the defects of a family-prescribed monogamous marriage by dividing duty and satisfaction; the women reserved her loyalty for her husband and her love for her gallant. Continuing on the tracks laid by the concept of courtly love, the nobles of the seven-

6 R. M. Meyer, *Die Literatur des 19. und 20. Jahrhunderts*, Berlin, Bondi, 1921.

7 L. Walzel, *Romantik; Natur und Geisteswelt*, Leipzig, 1915, vols. 232, 233.

teenth and eighteenth centuries in Austria, Spain, France, the Netherlands, etc. still adhered to the tenet that love and marriage were irrecon-
cilable. Yet, love had dropped its cloak of sublimation. The medieval
concept had drawn a line between the spiritual and the animalic-sexual,
between love and marriage. The court society of the Baroque[8] and the
Rococo periods, by rewarding the gallant's deeds and duels with carnal
favors, actually integrated sex and love—though only outside marriage.
The adaptation noticeable in the ascending bourgeois class followed the
same line—integration of sex and love—with the important difference
that their economic struggle, their tradition of thrift, their religious ideas
(which, reformed to further their purposes, gave them moral support in
their ultimate contest with the group in power),[9] did not permit them to
accept illicit relationships as a solution of the problem. Yet, they had not
remained unaffected by the ideology of earthly love. The refined concept
had filtered down from the castles to the cities. Marriage, to be sure, was
still arranged on a family basis with an eye on business, and the status
of the wife was by no means enviable. But the verbiage of courtly love
had entered the relation of the sexes. However, it was addressed not to
the married woman, but, for the first time, to the marriageable maiden.
Of course, this was hardly possible before the betrothal since, as an anonymous writer, Ursula Margareta, wrote in her diary published posthu-
mously in 1805,[10] "the association with the opposite sex was not yet invented then (about 1760) . . . and we were shielded from them as
from chicken pox." But during the months between engagement and marriage the betrothed was expected to "court" the girl and to display his
emotional fervor in conversation, gifts, and poetry.

Preceded by the English novelist Samuel Richardson (1689-1761), who
is credited with having said first that love is needed for marriage, the men
of letters of those days pointed out both the immorality of the aristocratic
solution and the sterility of the bourgeois pattern. Visualizing love as an
antidote to the insecurity produced by social and technological changes,
they propagated its legitimization and thus its perpetuation in marriage.
The model for the bond between the sexes was the complex of feelings so
graciously depicted in medieval romances, and its realization was henceforth called romance or romantic love.

We thus encounter a third stage in the development of love relations.
The first admitted certain formalized features of adolescent feelings into

[8] M. Carrière, "Barocque," in Gleichen-Russwurm, *op. cit.,* vol. 11.

[9] R. H. Tawney, *Religion and the Rise of Capitalism,* Penguin Books, Harmonds-
worth, England, 1938.

[10] "Alte und neue Zeit; Taschenbuch zum geselligen Vergnuegen, 1805," reprinted
in Sturtevant, *op. cit.*

the adult relationship to bridge the dichotomy between sublimated sex desires and the prevailing sex-hostile ideology; the second justified with love adulterous sex relations to ease the burden of an unreformed monogamy; the third aimed at the integration of love and marriage. It was promulgated by the first spokesmen of the bourgeois culture, who pleaded for the right of the young people to make their own choice for marriage on the basis of their feelings. No longer was there to be a cleavage between the spirituality of love and the marital sex relation but the latter was to be sanctified by the former. This combination raised—though only ideologically at first—the woman of the middle class to the status which heretofore only the aristocratic lady had achieved in relation to the man.

Like courtly love, the concept of the Romanticists leaned noticeably on adolescent experiences. Though less ritualized than courtly love, romantic love acknowledged the value of certain pre-adult emotions. It established a hierarchy of characteristics that marked predestined affection. Foremost among them was emotional instead of rational evaluation, an attitude that contrasts clearly with the adult behavior normally aspired to, but is typical of adolescence, in which the rational powers do not operate at their optimum. Economic and status considerations were belittled. The female was idealized because of her "natural" kindness, her intuition, and her nearness to nature. The male conceived of himself as a restless, striving, and erring deviate, spoiled by civilization, who, inspired by the female's love, might find the way back to his better self. This tendency corresponds to the adolescent's moments of magnified feelings of inferiority in the face of the female's greater poise and virtue and the elation when he is accepted nevertheless. While in the romantic concept the adventures of the mind were valued over fighting and fencing, conflict, self-recognition, sensitivity and the preservation of one's "true" and original self were elevated to moral qualities. The analogy to the adolescent's defensive attitude toward practical adult goals is evident.

The romantic love relationship itself was pervaded by melancholy and *Weltschmerz* (world-woe), another trend that is generally encountered in adolescence when the young person, having severed his emotional ties with his protective elders and craving new attachments, finds himself abandoned and, in comparison with the still child-like ego ideal, inadequate. From the same experience, on the other hand, results the claim to uniqueness and originality. Owing to his maturing mental powers, his broadening experience and knowledge, the adolescent frequently senses suddenly some of the discrepancies between reality and the moral teachings of his group, especially those which are antagonistic to the fulfillment of his desires, In this whirl of contradictions, wishes, rebellious emotions, and thoughts he feels like a castaway or like a revolutionary, chosen for the fight against either the traditions or the temptations, like a hero or like a sinner, full of defiance or full of resolutions to prove himself better than anyone else. Simultaneously proud and afraid of his discoveries, he seeks reassurance, someone to confide to, a companion who confirms the value of his ideas and thus of his personality.

Unable to turn to his parents, who in a quickly changing world are no longer considered revered guides but old-fashioned antagonists, he can find assurance only with a friend who seems to be shaken by similar convulsions and consequently "understands." After a period of homosexual friendships, the social conventions, the ideal of masculinity, and the sex drive usually direct the choice toward heterosexual relations, the same relations whose secrecy, mood of conspiracy, exuberations and depressions were the raw material for romantic love which, minimizing the sexual aspect, introduced friendship between the sexes.

By the end of the nineteenth century love had won its battle along the whole line in the upper sections of the middle class. It has since been regarded as the most important prerequisite to marriage. The American concept that considers individual happiness the chief purpose of marriage is based entirely on this ideology. . . . Is this love identical with the formalized concept of romantic love?

Certainly, the all-pervading melancholy is relatively rare among young adults; the mood of lovers, though still vacillating between joy and depression, is, on the whole, less sentimentally sad and, owing to their greater independence and the diminishing outside interference, is based more often on anticipation of marital joys, cooperation, "having fun together," and pursuit of common interests. As contact between the sexes is freer, partial sexual outlets are frequently provided. And while such activities may still be followed by feelings of guilt, these seem to be greatly attenuated by a presumed necessity caused by a socially cultivated sexual competition. Sex competition, on the other hand, particularly potent among girls, tends to blur the line between the excitations of love and those of an aggressive ambition. As a result of the prevailing dating convention and its concomitant early initiation of the sexes on a social basis, the over-idealization of the female (the keynote in both courtly and romantic love) is curbed. The love conventions of the twelfth and the nineteenth centuries were grants made by the man to the female; love in our day and in this country, conversely, has become a demand of the female, who is in the privileged position to extend or withhold sexual favors. Her own desire probably being lessened by culturally necessitated repressions, she frequently uses such favors to reward or stimulate emotional expressions without regard to her own sex drive. Thus, it appears that the modern love concept is not identical with romantic love, but is a derivative, modified in concord with the conditions of our age and based more on ego demands than on ideal demands.

But whatever form it takes, love is rarely the only consideration upon which marriage is contracted. Rather, it is one selective factor operating within the controls imposed upon the mates by our culture. These controls involve age, race, religion, ethnic origin, and class,[11] and the thus defined field is furthermore narrowed by regional proximity. . . .

THE THEORY OF COMPLEMENTARY NEEDS IN MATE-SELECTION

Thomas Ktsanes and
Virginia Ktsanes

Original manuscript. The theory of complementary needs was first set forth by Robert F. Winch. For a more detailed exposition of the theory see his book *Mate-Selection,* New York, Harper, 1958.

Who Marries Whom?

The question of "who marries whom" is one which has aroused "common sense" as well as well as scientific interest. The common sense answer is paradoxical, for while everyone knows that "like marries like" and that "birds of a feather flock together," it is also equally clear that "opposites attract." As is frequently the case in folk wisdom, both assertions are probably true depending upon the characteristics considered. If by "like" one means similarity in regard to a variety of social characteristics such as ethnic origin, religion, occupation, residential location, and social status, then indeed the view that mates tend to be similar seems correct. If, on the other hand, "like" is used to denote similarity in a variety of psychological attitudes, traits, tendencies, or needs, then the situation is by no means clear. This being the case, it is in order to take a brief look at some studies which have attempted to answer the question of the degree to which homogamy or heterogamy prevails in marital choice. The tendency of persons to select mates who have certain characteristics similar to their own is called homogamy or assortative mating. Conversely, heterogamy refers to the selection of mates who are opposites or are merely different. We shall begin with a brief review of the research literature on homogamy Later we shall present the theory of complementary needs as a special type of heterogamy.

[11] A. B. Hollingshead, "Cultural Factors in the Selection of Marriage Mates."

Homogamy in Social Characteristics.   Interest in the problem of assortative mating is probably an analogical extension out of the field of biology where for lower animals there seems to be a trend toward similarity in size and vitality. On the human level also there is some slight evidence for homogamy in physical characteristics.[1] With human beings, however, physical similarity has not been the principal concern. Most work on assortative mating has concerned a variety of social characteristics. We shall now briefly examine some of this evidence.

In an early study by Marvin[2] it was noted that there was a greater than chance tendency for marriages to occur between persons with similar occupations. More recently Centers[3] has pointed out that there tend to be no wide differences in the occupational statuses of spouses. Burgess and Wallin[4] have shown that there is homogamy in educational level. Further, basing their conclusions on the ratings by the couple of the social status of their parents and on their report of the present income of their fathers, Burgess and Wallin state ". . . it is clear that there is a considerable excess over chance for young people to fall in love and become engaged to those in the same social and economic class." [5] Kennedy [6] has indicated that there is a strong trend toward homogamy in regard to religious affiliation and a tendency, though less marked, toward homogamy in ethnic origin.

Bossard,[7] in a study repeated by subsequent researchers, showed that people usually select their mates from those who live nearby. In Bossard's classic study more than half of the marriages in his sample were between persons living within twenty blocks of each other. However, the effect of this factor of mere spatial propinquity must not be over-emphasized for it overlaps with the factors discussed before. The various ecological areas of the city are characterized by heavy concentrations of certain socio-eco-

[1] In Mary Schooley, "Personality Resemblance Among Married Couples," *Journal of Abnormal and Social Psychology,* 31 (1936), 340-47, some low positive correlations were found to exist between mates on height, weight, visual acuity, and appearance.

[2] Donald Marvin, "Occupational Propinquity as a Factor in Marriage Selection," *Journal of the American Statistical Association,* 16 (1918-19), 131-50.

[3] Richard Centers, "Marital Selection and Occupational Strata," *American Journal of Sociology,* 54 (1949), 530-35.

[4] E. W. Burgess and Paul Wallin, "Homogamy in Social Characteristics," *American Journal of Sociology,* 49 (1943), 109-24.

[5] *Ibid.,* p. 114.

[6] R. J. R. Kennedy, "Single or Triple Melting-Pot? Intermarriage Trends in New Haven, 1870-1950," *American Journal of Sociology,* 63 (1952), 56-59.

[7] J. H. S. Bossard, "Residential Propinquity as a Factor in Marriage Selection," *American Journal of Sociology,* 38 (1932), 219-24.

nomic classes, ethnic and religious groups; and these groups as noted above tend to be endogamous.[8]

In summary, the studies reviewed indicate that persons who marry tend to be similar in regard to a variety of characteristics such as social class, ethnic background, educational level, religion, occupation, and area of residence. However, these findings actually bear little direct relationship to our problem. They are of some interest in that they give us a notion of the limits within which another principle of selection may operate. As we interpret them, these factors tend to define a field of eligibles from which a mate may be selected on psychological grounds.

Homogamy in Psychological Characteristics. Psychological character-istics which have been studied with respect to homogamy include a long and varied list. Characteristics investigated by means of "paper-and-pen-cil" personality inventories include neuroticism, dominance, self-suffi-ciency, etc. One early study[9] found moderately high correlations between mates on neurotic tendency and dominance. Burgess and Wallin[10] in their more recent study of 1000 engaged couples found homogamy in regard to a few traits. Their correlations, however, were of a rather low order and are therefore not too convincing. In regard to various "content" attitudes, *e.g.,* religious and political attitudes, there is some evidence for similarity.[11] These similarities, however, may have developed after marriage. The re-sults in this area are thus considerably short of being definitive. Stagner in reviewing the studies on homogamy in psychological characteristics has pointed out that correlations indicating similarity are higher with respect to intellectual, interest, and attitude scores, but that measures of tempera-ment do not show this tendency as clearly.[12] The measures of tempera-ment referred to by Stagner are those estimates of various traits such as dominance, self-sufficiency, etc., which are arrived at by means of paper-and-pencil tests. Confidence in paper-and-pencil tests is vitiated by the fact that subjects can "fake" their responses and thereby create what they regard as favorable impressions.[13] When we try to get behind the pic-ture of personality which the subject wants us to accept, and more particu-

[8] Endogamy refers to marriage within the group.

[9] E. L. Hoffeditz, "Personality Resemblances Among Married Couples," *Journal of Abnormal and Social Psychology,* 5 (1934), 214-27.

[10] E. W. Burgess and Paul Wallin, "Homogamy in Personality Characteristics," *Journal of Abnormal and Social Psychology,* 39 (1944), 475-81.

[11] T. M. Newcomb, and G. Svehla, "Intra-family Relationships in Attitude," *Sociometry,* 1 (1937), 180-205.

[12] Ross Stagner, *Psychology of Personality,* New York, McGraw-Hill, 1948, p. 387.

[13] Cf. Albert Ellis, "The Validity of Marriage Prediction Tests," *American Sociological Review,* 13 (1948), 710-718.

larly, when we want to understand a subject's motivational patterns of which he may be only partially aware, we find no systematic research on the question of homogamous *vs.* heterogamous mate-selection.[14] In the absence of experimental evidence various writers have been theorizing on this problem.

Toward a More Adequate Theory.   Ideas about types of harmonic intermeshing of needs have been suggested by various theorists and researchers. Many of these owe a debt to Freud, who made a distinction between "anaclitic" and "narcissistic" love.[15] By the anaclitic type Freud meant a love which was expressed in attitudes of self-derogation and reverential admiration toward the love-object. In this type of love one is dependent on the loved one toward whom he can express his need to revere and admire. Narcissistic love is essentially self-love but the narcissist has a great need to be admired by others as well as himself. Thus in his formulation of the narcissistic-anaclitic typology, Freud posited a type of complementary relationship, *i.e.,* the dependent person who has the need to revere and admire is attracted to the narcissistic person who has a great need to be admired and receive adulation.

Following the suggestion that persons with complementary psychic make-ups are attracted to each other, several psychoanalysts have proposed that matching occurs between those who are complementarily neurotic.[16] According to this hypothesis, for example, a dependent male with unresolved emotional ties to his mother would be attracted to an aggressive and dominant woman burdened with conflicts over her sex role. As a general theory of mate-selection, however, this literature is inadequate because the writers have explained attraction only in terms of the highly individualized neurotic patterns of their patients. What we are seeking is a theory which will be generally applicable, not merely to Freud's anaclitic and narcissistic types of persons, not merely to dependent people who marry nurturant people, not merely to neurotics, but to all kinds of personalities.

Gray [17] has used a broader approach to this problem. He hypothesized

14 A few individual cases have been reported at this "deep" level of analysis, but they have been neurotic patients and the authors' reports have lacked experimental control. Cf., *e.g.,* C. P. Oberndorf, "Psychoanalysis of Married Couples," *Psychoanalytic Review,* 25 (1938), 453-57.

15 Sigmund Freud, "On Narcissim: An Introduction," in *Collected Papers,* vol. 4, London, Hogarth, 1925, pp. 30-59.

16 Cf., *e.g.,* C. P. Oberndorf, *op. cit.;* Edmund Bergler, *Unhappy Marriage and Divorce,* New York, International Universities Press, 1946; and Bela Mittleman, "Complementary Neurotic Reactions in Intimate Relationships," *Psychoanalytic Quarterly,* 13 (1944), 479-91.

17 Cf., *e.g.,* H. Gray, "Psychological Types in Married People," *Journal of Social*

that mate-selection would be complementary with respect to the types of personality formulated by Jung (extrovert-introvert, etc.). His empirical findings, however, were not convincing.[18]

Other theorists have tried to identify various motivation-linked aspects of interaction. Bernard, for example, suggests various dimensions of love.[19] She notes the usual dimension of dominance and also dwells upon the desire for response or acceptance and on the differential ability of persons to "give" as she calls it. As we shall see later, these are similar to some of the "needs" in our conceptual scheme. Bernard did not systematically state that attraction occurred between persons who were complementary in regard to these dimensions. Others, however, have come very close to this notion. Ohmann[20] stated this idea by saying that we are attracted to those who complete us psychologically. We seek in a mate those qualities which we do not possess.

Taking leads from all of the foregoing, Winch attempted to pull them together. He began by defining love in terms of needs:

Love is the positive emotion experienced by one person (the person loving, or the lover) in an interpersonal relationship in which the second person (the person loved, or love-object) either (a) meets certain important needs of the first, or (b) manifests or appears (to the first) to manifest personal attributes (*e.g.*, beauty, skills, or status) highly prized by the first, or both.[21]

Then he hypothesized that mate-selection would take place according to what he called the theory of complementary needs:

In mate-selection each individual seeks within his or her field of eligibles for that person who gives the greatest promise of providing him or her with maximum need gratification.[22]

Perhaps this can be phrased more simply by hypothesizing that the personality needs of marriage partners tend to be complementary rather than similar. Two points require further clarification: (a) What are per-

---

Psychology, 29 (1949), 189-200; and "Jung's Psychological Types in Men and Women," *Stanford Medical Bulletin,* 6 (1948), 29-36.

[18] Winch applied tests of significance to some of Gray's data. These tests showed that the selection of mates in terms of Jung's types was not significantly greater than might have been expected by chance.

[19] Jessie Bernard, *American Family Behavior,* New York, Harper and Brothers, 1942, pp. 435-56.

[20] Oliver Ohmann, "The Psychology of Attraction," in Helen Jordan (*ed.*), *You and Marriage,* New York, Wiley, 1942, chap. 2.

[21] Robert F. Winch, *The Modern Family,* New York, Holt, 1952, p. 333.

[22] *Ibid.,* p. 406. In the phrase "field of eligibles" Winch takes account of the previously noted homogamy with respect to such social characteristics as race, religion, and social class.

sonality needs and which needs are germane to our problem? and (b) What exactly is meant by the term "complementary"?

Needs. One can think of the term "need" as meaning a goal-oriented drive. Goal in this sense refers not only to such things as material objects and status in the social structure but more particularly to such things as the quality and kind of response desired in interpersonal situations. Examples of the latter are the desire to give help or adulation to others, the desire to take care of others, the desire to control, etc. When these goals are attained, the need is gratified. However, gratification is a dynamic process, and a need once gratified does not cease to function. Patterns of behavior which are tension-reducing tend rather to be reinforced. In a marriage, for example, a woman who finds in her interaction with her spouse gratification for a need to control will continue to want to control him. One further characteristic of needs should be noted. Needs function at both the conscious and unconscious levels. A person may be conscious, partly conscious, or not at all conscious of the goals he desires.

Henry A. Murray has defined "need" in a more formal way:

> A need is a construct . . . which stands for a force . . . which organizes perception, apperception, intellection, conation, and action in such a way as to transform in a certain direction an existing, unsatisfying situation.[23]

Further, he has elaborated an extensive list of emotional needs. However, because Murray's list is so detailed, we found it necessary to depart from it in a number of ways. The following list of needs[24] is nevertheless based upon Murray's scheme.

Needs

n Abasement[25]	To accept or invite blame, criticism or punishment. To blame or harm the self.
n Achievement	To work diligently to create something and/or to emulate others.
n Approach	To draw near and enjoy interaction with another person or persons.
n Autonomy	To get rid of the constraint of other persons. To avoid or escape from domination. To be unattached and independent.
n Deference	To admire and praise a person.
n Dominance	To influence and control the behavior of others.
n Hostility	To fight, injure, or kill others.
n Nurturance	To give sympathy and aid to a weak, helpless, ill, or dejected person or animal.
n Recognition	To excite the admiration and approval of others.

[23] H. A. Murray, *et al., Explorations in Personality,* New York, Oxford University Press, pp. 123-24.

[24] R. F. Winch, *op. cit.,* pp 408-409.

[25] The notation "n" before the name of a variable is used as a shorthand form for the term "need," and where it is found on following pages, that is what it represents.

n Sex	To develop an erotic relationship and engage in sexual relations.
n Status Aspiration	To desire a socio-economic status considerably higher than one has. (A special case of achievement.)
n Status Striving	To work diligently to alter one's socio-economic status. (A special case of achievement.)
n Succorance	To be helped by a sympathetic person. To be nursed, loved, protected, indulged.

General Traits

Anxiety	Fear, conscious or unconscious, of harm or misfortune arising from the hostility of others and/or social reaction to one's own behavior.
Emotionality	The show of affect in behavior.
Vicariousness	The gratification of a need derived from the perception that another person is deriving gratification.

A study to test this theory has been undertaken with a group of middle-class subjects. Because striving for upward mobility (or higher socio-economic status) is so central to the middle-class value system, it was decided to include two variables pertaining to status.

Complementariness. To explain this theory let us imagine two person, *A* and *B,* interacting with each other. Let us assume that both are deriving gratification from this interaction. Then the interactional sequence will be in accordance with the theory of complementary needs if:

I. the same need is gratified in both *A* and *B* but at very *different* levels of *intensity;* or

II. *different needs* are gratified in *A* and *B.*

An example of I is the interaction between a person who wants others to do his bidding (high n Dominance) and one lacking the ability to handle his environment who is looking for someone to tell him what to do (low n Dominance). An example of II is found in the case of a person desirous of attention and recognition (n Recognition) who finds gratification in relationship with a person who tends to bestow admiration on the former (n Deference). These are referred to as Type I and Type II complementariness respectively and constitute two forms of heterogamy.

Illustration of the Theory

To illustrate the theory of complementary needs we have chosen a case from a sample of middle-class married couples and have attempted to show how these two partners complement each other need-wise. It will be noted that in this case the male shows some dependent trends. We do not feel that this case is atypical of our middle-class sample. Dependent needs in the personality of the middle-class male are probably more frequent than is popularly supposed.[26] It is to be emphasized that the man and wife discussed here are a normally functioning couple.

The Case of Anne and Frank Hamilton.[27]   Before we can understand how individual needs function for mutual gratification in a marital relationship, it is first necessary to present the personalities involved. We shall consider first the wife and then the husband before we attempt to understand their relationship to each other.

Anne Hamilton is best described in build as "hefty." Her outstanding features facially are her large mouth and rather prominent teeth. That her mouth is so noticeable the interviewer attributes to the fact that "it never seems to be still." She talks loud and fast. She punctuates her words by dramatic use of her hands and facial expressions. Even when she is listening, her face does not relax. She smiles broadly or raises her eyebrows or in some other way responds aggressively to what is said.

Anne's energy is also evident in her capacity to work. To finish college in three years, she carried extra courses each term and still sailed through her undergraduate work. She earned most of the money to pay her college expenses even though her family was able and willing to pay them. But she just liked to keep busy, so not only did she work and keep up her grade average, but she also held responsible positions in numerous extra-curricular affairs. She was so efficient in getting ads for the school yearbook that for the first time that publication had a financial surplus.

Going along with this terrific need to achieve, there is a high need to dominate others, which Anne describes as "a certain element of bossiness in me." She feels that her way of doing things is best and she wants people to do things "in the manner I so designate." [28]

She does not like to be "stepped on" nor does she admire people who

26 For further elaboration on this point, *cf.*, for example, Arnold Green, "The Middle Class Male Child and Neurosis," *American Sociological Review*, 11 (1946), 31-41; and Talcott Parsons, *The Social System*, Glencoe, Ill., The Free Press, 1951, esp. pp. 262-69.

27 The material upon which the case analysis was done consists of a case-history type interview, Thematic Apperception Test protocols, and a second type of interview designed to get at the more behavioral aspects of personality. The full case analysis was made by the research staff of this project which consists of Dr. Robert F. Winch, Mrs. Sandra K. Oreck, Dr. Oliver J. B. Kerner, and the authors of this article. The present report is a synopsis of their findings, which cannot be presented in their entirety because the analysis runs to about two hundred pages of manuscript. Much of the documentation for generalizations must be omitted. All names and identifying characteristics have been changed in order to preserve the anonymity of the couple without impairing the crucial facts of the case. It is our desire to present the case as simply as possible for the purpose of illustrating the theory.

28 Shortly we shall note that this domination of others occurred very early in her life in her relationship to her parents and other members of the household.

can be pushed around. Such people she cannot respect. "People that I cannot look up to, I have a tendency to shove out of my way or to trample on, just shove, push." Thus we see in Anne little need to feel sympathy for other persons (n Nurturance) but rather a hostile attitude towards them.

She tends to be critical of other people and apparently because of this she has encountered some difficulty in forming close friendships. She says that people usually like her if they can overcome their first impression which frequently is one of antagonism. She says on this point, "I'm very quick spoken and rarely stop to think that I may be hurting somebody's feelings or that they are not going to take it just the way I meant it." But she needs people and she wants them to like her.

The competitiveness and the need to manipulate people undoubtedly indicate compensatory behavior for feelings of insecurity at some level. There is some evidence to indicate that these feelings stem from her doubts about her being a feminine person. She tends to be jealous of pretty women. She is contemptuous towards them when their attractiveness and "poise" win them positions of prestige which they are not equipped to handle because of a lack of the "executive ability" that she possesses. All her life she states that she wanted to be like her mother who is pretty and sweet and "gives a lot, perhaps too much." She feels, however, that she has not succeeded in becoming this sort of woman. She regards herself as a person who is "quick, uneven-tempered and impatient, ambitious . . . ready to tell others how to do things." Evidence that she rejects this "masculine" component in her personality is her view that she would not want a daughter to be like herself, but "more like Mother."

The postulation of such a conflict helps to explain why Anne did not continue with her career plans. She took a master's degree in advertising the year following her undergraduate work. She then set out to make a career in this field, but there were no jobs immediately available. Employers did not want college graduates who had their own bright ideas about the business, and, according to her account, they were unwilling to employ her for menial jobs which she was willing to take because they felt she was too intelligent and soon would become disinterested.

At this point Anne's career drive began to fluctuate. She took a job in an office. While there and while formally engaged to another man, she met Frank. She and Frank were married six months after their meeting, and they moved to a city where she had obtained a good job and where he enrolled in college. At the end of a year she became pregnant and stopped working for awhile. By the third month of her pregnancy, however, she became bored with "sitting around home" and took a job as a waitress, much against the doctor's orders. She lost the child three months later. She stated that she wanted the child very badly and that she was broken up over her loss. This wish would be consistent with the feminine desire to be a "mother." In addition to the conscious desire to be feminine, it seems probable that she had an unconscious wish to abort and to deny willingness to play a feminine (maternal) role.

Perhaps if we look into Anne's background for a moment we can see more clearly the circumstances which led to the development of her pattern of aggressive behavior and the confusion over appropriate sex-role behavior.

Anne was the only child in a family of four adults. Her father was a self-made man, one who built up a trucking business to the point where it netted him an income of around $700 monthly even during the depression years. She describes him as being a short man, one who was hot-tempered and stubborn. He was 30 when Anne was born and her mother was only 18. The mother is described as being even-tempered, calm and dependent. The third adult was Anne's maternal grandmother who came to live with the family shortly after Anne was born. She managed the house and Anne's mother and apparently Anne's father as well. Anne says her grandmother often warned the father against his outbreaks of wrath in front of the child. The grandmother brought with her one of her sons who was about the age of Anne's father and who was similar to Anne's mother in temperament. He was very good to Anne and gave her everything she wanted. He married for the first time and left the household when he was 50 years old.

Anne was the center of attention for these four persons. What she could not get from one, she could get from another. This pattern of relationships was conducive to her manipulation of persons and the need for recognition from them which we have noted earlier.

Grounds for the competitiveness may also be found in this network of relationships. Anne's mother was very young and still dependent upon her mother who looked upon Anne as "her youngest child." Thus the relationship between mother and daughter resembled sibling rivalry, not only for the "mutual mother's" love but for the husband-father's love as well. Here were two bases for Anne to dislike her mother, but her mother was such a sweet young thing that she never gave Anne any rationalization for hating her. This left Anne with an unexpressed hostility which apparently has been partially sublimated into an achievement drive and partially displaced onto "feminine" women like her mother. Her mother was better looking than she, so Anne could not compete with her on these grounds but had to seek other means of achieving superiority.

To strive in an aggressive manner was satisfactory in another way too because the father, who wanted a son, approved of such behavior in his little tomboy. Further, grandmother was a model of aggressive behavior. Anne's gratifying relationship with her fostered an identification. The aggressive pattern was fairly well set by the time Anne reached adolescence as is evident in her report that, in junior high school, teachers commented on it. One teacher advised her to change her ways or she would never get a husband. Father also changed his mind about what he wanted and began to look upon her as "feminine" and wanted her to become dependent on him while she was in college. These undoubtedly are the sources of some of the ambivalence we note in her picture, especially concerning career and motherhood.

Although she had doubts about her "feminine appeal," Anne apparently had little trouble in finding dating relationships. Though she confesses she was not the most popular girl on campus and that her weekend calendar was not always filled, she dated from the time she first entered high school. She had only one serious relationship before meeting Frank. This was an engagement to a man described as "suave and smooth . . . and with nice manners." It apparently was a stormy affair, off and on several times. The engagement was broken finally over the issue of whether or not there should be a formal wedding. Anne wanted one, but her fiancé's family did not.

Frank is unlike Anne in many ways. Whereas she gets much gratification from work and positions of responsibility, he much prefers just loafing and being with people. He is now in college, at Anne's request, and very much looks forward to the time when he will be through. College is just a means to an end for him; the less work he has to do to get through, the happier he will be. He wants the degree, however, because it will facilitate his getting a good job. He looks to the job to bring him status and prestige and to provide a large income so that he can buy sports cars and a big house. Nevertheless, he does not like to work for such a position and is just as content if someone gets it for him.

Frank likes people and he gets along with them very well. It is important to him that they like him and give him attention. He loves to talk and to joke, and generally he is successful in winning friends. "I'm an easy person to get along with . . . I do a fair job of amusing people although I feel that people don't regard me as entirely full of nonsense." His physical appearance contributes to his acceptability for he is a good-looking man, tall and slightly heavy. His build is somewhat athletic but his muscles seem to lack the firmness and tonus of a well-developed athlete. He is light-hearted, pleasure-oriented, and loves to eat.[29]

[29] In terms of the Freudian stages of development, this aspect of his personality would place him at the "oral" stage, the stage at which the infant, for example, does little more than *receive* love, care, and attention from the mother. The passive-

To achieve acceptance Frank relates to people in a deferent manner. He consciously admires and accepts his allies almost uncritically. He shows no tendency to control them nor to compel them to do what he wants; in other words, he reveals no need to dominate. Though he likes very much to have the spotlight himself, he is willing to share it with others and even to conceded it without resentment to people who are better attention-getters than he. He tends to establish friendships with such persons and to identify with them. Thus he receives vicarious gratification for his own need for recognition. This is illustrated in the fact that he joined the fraternity to which most of the "big wheels" on campus belonged though he himself was not a big wheel. Merely through association he felt he was able to share in their glory.

It is interesting to note that Frank does not limit his struggle for recognition to a few fields or a select group of persons as mature adults generally do. He is almost child-like in his willingness to perform. Once when drunk, he paid the singer in a night club twenty-five dollars to let him sing with her in front of the microphone. He still wears the badge that he received when he was deputized a sheriff for a week in his hometown. The importance of this incident was shown when Frank flipped his lapel so the interviewer could see the badge.

In addition to recognition, Frank seems to want love and affection. He tells that he was the "mascot" of a sorority at the first college he attended, and he was chosen "king of the prom" one season. If he feels blue, which he says is rare, he can be cheered by having women, peers or the mothers of peers, tell him how handsome he is.

Apparently since high school Frank always got along well with women because he always had a girl. He tended to date one girl at a time and to go with her pretty "seriously." He expected the same of her, and as a result most of these relationships broke up by his becoming jealous when the girl would date another fellow. He became jealous he says because he wanted "all her attention." The girls he dated were all short and very attractive. They conformed to his "ideal" of "one other fellows thought highly of, a popular girl in other words." Apparently a girl of this type brought vicarious recognition to Frank in the same manner as did the "big wheels" in the fraternity.

Now let us consider Frank's background. Frank was the third son in a family of four boys, all of whom were born during a period of eight years. His father, who was 57 when Frank was born, was a successful

---

dependent trends which we note in Frank's personality are considered the psychological counterparts of this stage of development. We shall note, however, that this characteristic is by no means the whole picture and that he is considerably more active than is implied for this stage.

salesman until the depression. After losing everything in the depression, the father stopped working. The major burden of supporting the family then fell upon his mother who was about 28 years younger than the father. In time this responsibility was shared by the oldest son. The mother was a petite and good-looking woman.[30] She was a very hard-working, efficient sort of person who, besides working at a full-time job, kept her house, herself, and her sons immaculately neat and also found time to participate in a few club activities. She had considerably more education than her husband in that she had a B.A. degree whereas he completed only the eighth grade. Frank remembers her as being undemonstrative in her affections and as a reasonably impartial judge in the children's quarrels but with a tendency to side with the underdog. Frank had little to say about his father's personality. Though the man had died only two years before the interview, Frank gave the impression that his father had participated little in family affairs. Frank's few descriptive comments portrayed an opinionated man, harsh in his judgments.

Among the seemingly more important aspects of this family is the absence of daughters. Having two sons already, both parents had desired that the next children be girls. Indeed Frank can remember the time when his mother gave him a girl's haircut. It would appear therefore that this attitude on the part of his parents, and especially his mother, laid the groundwork for the passive-dependent trends we have noted in his personality. It seems logical that Frank wanted the love and attention that is given to the baby. At the age of two years, however, he could no longer be gratified in these desires because of the arrival of the fourth and final brother. It appears that Frank resented this brother greatly. In one two-hour interview he mentioned both of the older brothers but not this one. Undoubtedly as a consequence of this situation Frank has developed a fear of rejection to which he has responded by always doing what is expected of him and by endeavoring to please people in order not to be rejected by them. Frank did not react to his feeling of rejection by rebellion. Perhaps this was because the mother never actually rejected him; she just did not give him all the affection he desired. To avoid losing what he did receive and to try to get more he reacted by being a "good boy."

But Frank was not a sissy in the common use of the term. He was interested in athletics and became captain of his high school football team. He liked mechanics and cars. Currently he is studying mechanical engineering and hopes someday to become a salesman for some large engineering firm.[31]

[30] It will be recalled that the girls he dated were of similar stature.

[31] It is not surprising that Frank wants to become a salesman because he enjoys so much talking with people and feels certain that he is able to get along with them well.

These masculine interests are very important for understanding Frank's personality. We have shown the tendency towards dependency in his personality which culturally is considered "feminine." Generally, males in our culture who tend to be passive experience some conflict if they are not able to live up to the cultural imperatives that they be assertive and "masculine." Frank shows little anxiety on this score, however, and appears to be very well adjusted. His not having developed a conflict on this score may be due to his having achieved such successful identifications with male authority figures that he consciously never questions his "maleness."

Undoubtedly, the oldest brother is a significant figure in understanding these identifications with males. Very early this brother became a counsellor to the mother. Frank felt ambivalent towards him. He was jealous because this brother played such an important role with the mother. On the other hand, if he hated his brother, then the mother would reject him completely; but if he were like his brother, he would get his mother's attention and at the same time establish a good relationship with the brother, who was moderately successful in his own business and popular with people. Thus, the brother became an ego-model for him and at the same time was a person who could meet some of Frank's dependent needs.

Thus, we now see Frank as an amiable, non-anxious person who does not have a great deal of ambition but who has the knack of relating himself to people who can do things for him.

Up to this point we have attempted to describe both Anne and Frank with very little reference to each other. Now we shall discuss their case with relation to complementary need theory.

Frank says that he was attracted to Anne because "she's probably the smartest woman I've run into, and I admired her a great deal I think before I truly loved her." On the other hand, Anne admired his easy-going manner and his ability to get along with people. Knowing what we do about each of them individually, we can see in these two remarks alone some ground for their complementary matching. First of all, we have pointed out that Anne has had some difficulty in getting along with people and that she would like to be able to do so more easily. Frank's ability to attract friends and to keep them facilitates Anne's social relationships in that he attracts their mutual friends. For Frank, Anne's initiative and her ability to attain the financial and other goals she sets for herself complements his lack of drive. The question is open, however, whether or not this particular pattern of interaction which is now mutually gratifying will continue to be so if Frank becomes a successful salesman.

In their interaction with each other we note that Anne has the authority. She handles their finances, and she decided that he should go back to school. As we have seen, this is the way she likes to do things and we have also noted that Frank shows little need to dominate and he accedes quite willingly to her plans.

Anne tends to be a very emotional person who is easily aroused and upset. At such times Frank's calm and easy-going manner is consoling to her. He has a good shoulder to cry on and he is willing to listen to her problems. She feels that he is helping to calm her down.

About the only thing that disturbs Anne about Frank's personality is that he does not have as much ambition as she would like to see. Indeed she has been somewhat bothered by his rather lethargic attitude towards school work. She would prefer to see him as excited about it as she has always been, but she feels that she is learning to accept his attitude that graduation is the important thing and that the level of one's performance in school is soon forgotten.

Occasionally Frank is a little perturbed by Anne for sometimes he is embarrassed when she pushes ahead in a crowd and drags him along with her, but he goes along and says nothing about it. Undoubtedly he is ambivalent about her aggressiveness. On the one hand, her behavior and her drive facilitate the realization of such desires as the new car which they recently bought. On the other hand, Frank fears that the same aspects of Anne's personality may put him in a position of stepping on other people which might result in their rejecting him. However, this aggressiveness does not constitute one of the things he would change about her if he could push a button to change anything. He would want to modify only her quick temper and her heaviness.

Anne is very different from the girls that Frank dated. The other girls were like his mother in physical characteristics in that they were all short and attractive. Anne has none of these physical characteristics, but does resemble Frank's mother in her efficiency. Although very different from Anne's father, Frank tends to be more like Anne's uncle and Anne's mother who are calm, easy-going, and dependent.

Both Anne and Frank desire considerable recognition from other people. Frank is attentive to Anne and considerate of her. She undoubtedly regards his submissiveness to her as admiration. Anne does not pay as much attention to Frank as he would like. It would seem that although Frank would like more in the way of demonstrated "hero-worship," he does not feel too deprived because she facilitates his getting the symbols (*e.g.,* the new sports car) which enable him to attract attention from other persons.

There is one other thing about Frank which Anne finds gratifying and which is worthy of mention here. Frank's attractive appearance and engaging manner enable Anne to compete successfully on a feminine basis with other women. Although this appeal on his part is gratifying to her in one sense, in another sense it threatens her. She mentioned that she is jealous if he pays too much attention to other women at parties. He also becomes jealous when she has occasion to lunch with another man. This mutual jealousy is understandable in terms of the marked need for recognition which each of them exhibits. On Frank's part, it undoubtedly is a manifestation of his fear of rejection; and from Anne's point of view, the insecurity stems from doubts about her feminine ability "to hold a man."

The complementariness that is described in this couple can be summarized generally as a case of a passive-dependent male finding gratification in relationship with a striving aggressive woman (and vice versa). Indeed, they are not complementary on all counts, *e.g.*, neither is willing to surrender his own desire for recognition in favor of the other. However, it would seem that the mutual choice that has been made satisfies the major, predominating trends within the personalities of each.

AN EXPERIMENT IN SCIENTIFIC MATCHMAKING

Karl Miles Wallace | Adapted from *Marriage and Family Living,* 21 (1959), 342-348.

This study had its beginning more than ten years ago, at the time the author was a graduate student, engaged in research for his doctoral dissertation on the prediction of marital adjustment. The study covers a span of ten years, 1947 through 1957. It was motivated by the thought that a research laboratory, where mates were actually being selected in a real life situation, would provide an opportunity to continue research already underway, and, at the same time, to investigate the behavior of members of introduction clubs (commonly dubbed "lonely hearts clubs").

The Research Design

A preliminary investigation clearly indicated that it was not only impracticable to utilize any existing introduction club or clubs for the project, but that even if such a procedure were possible it would not prove satisfactory. The procedures necessary for fruitful research could be initiated only in a new club of radical design, where the profit motive was of secondary importance.

In view of these obstacles, a research correspondence club, incorporated in California, under the name of Personal Acquaintance Service, was established.[1] It began business January 1, 1948, and the services of the club were terminated April 1 1954. To insure anonymity, the director of the club used only his first two names, Karl Miles, though he openly advertised that he was a college professor and marriage counselor, and that Personal Acquaintance Service was engaged in scientific matchmaking.

During the seventy-five month period that the Research Club was in operation, 59,417 inquiries from prospective members were received, primarily from newspaper and magazine advertising. Of this number, 12,421 (20.9 per cent) were considered undesirable for various reasons and were discarded. Of the remaining 46,966, 12.1 per cent were persuaded to register for membership, making a total of 5,670. Direct mail letters to more than 20,000 unmarried persons brought in an additional 353 members, making a total sample of 6,023 persons.

Both a dual-purpose personality inventory and a background data sheet were filled out by members when they joined. The background and personality data were punched on Remington cards to facilitate analysis. Methods for keeping detailed records were set up from the beginning, and every client was sent a questionaire after his membership was completed, asking for information about his experience in the club. Follow-up studies were made, where possible, by questionnaire and interview, of the 320 couples that reported marriage resulting from the club's introductions.

This was considered to be an exploratory study, and due to the comprehensive nature of the project only a few of the major findings regarding the social and personality characteristics of members are included in this report.[2]

Limitations of the Study

This study, of course, suffers from many limitations. Two major and unavoidable conditions added tremendously to the burden of the research

---

[1] This name was adapted from a suggestion in the writings of Joseph K. Folsom, who has referred to the lack of improvement of mate finding devices as a serious cultural lag. He suggested that a "Bureau of Personal Acquaintance" be established. See *The Family: Its Sociology and Social Psychiatry,* New York: John Wiley and Sons, 1934, p. 344.

[2] It was planned from the beginning that the study would be reported in both popular and professional publications. This report is adapted from a monograph to be published in the near future. The popular publications have already appeared. They were prepared in collaboration with Eve Odel, and were based on the active membership of 1,254 subjects participating in March, 1954. See "We Ran A Lonely Hearts Club," *McCall's,* 83 (March, 1956), pp. 43, 149-164; and *Love Is More Than Luck: An Experiment In Scientific Matchmaking,* New York: Wilfred Funk, Inc., 1957, 237 pp.

and also limited its effectiveness: (1) the necessity of using a pseudonym and maintaining secrecy of the project; and (2) the tremendous amount of extra work involved in operating an unprofitable business which was laden with tedious details, headaches, and problems not incident to other businesses. During the entire seventy-five months that the Research Club was in operation only a few trusted associates in the academic world knew of the project.[3] Secrecy in planning and executing the research deprived the author of valuable discussion and suggestions from colleagues, which undoubtedly would have improved the methods and procedures.

Two other major limitations should be mentioned. Many of the data were secured by questionnaire, and the inadequacies of this type of research are well known. The stigma in our society attached to forthright methods of seeking mates made it difficult to follow through on members' experiences and marriages, and this factor undoubtedly affected the validity of the data. It is also possible, as in most research of this type, that the participants gave biased answers to many of the questions asked.

Significance of the Study

The study is unique in several respects: (1) it is the first study of an attempt to match mates scientifically on a wide scale; (2) it fills a gap in our knowledge about behavior in introduction clubs since it is the first comprehensive study in this area; (3) the study utilized an unusually large and varied sample of the general population; and (4) it was conducted in a realistic setting, utilizing mate-seeking subjects, most of whom were not aware that they were being investigated.

The Research Club

It was decided that a correspondence club would be used as the Research Club because of its many advantages over other types of clubs. The three most apparent advantages were: (1) it would provide a larger and more representative sample; (2) it could be operated entirely by mail, making it possible for the researcher to maintain anonymity; and (3) the work involved could be done at irregular hours.

It was possible to attract a more representative clientele with Personal Acquaintance Service because it attracted clients from every state in the nation and from seventeen foreign countries, and it appealed to larger numbers of middle and upper-middle class people than other types of clubs. It selectively attracted people of higher socio-economic status because participation placed a premium on verbal ability, initiative, and in-

[3] One of these, Harvey J. Locke, was provided a preliminary report in 1952. See Ernest W. Burgess and Harvey J. Locke, *The Family,* New York: American Book Company, 1953, pp. 354-359.

dependence in selecting a mate, and it provided complete anonymity for those who wanted it.

The correspondence club also eliminates, to a great extent, the thrill seeker who wants to make an immediate local contact, and it discourages many of the emotionally maladjusted who shrink from the thought of being thrown on their own in social interaction with a strange person via correspondence. The rigorous procedures of the Research Club were designed for persons seeking marriage.

A highly complicated and effective matchmaking system was worked out in order to enhance opportunities for the clients to make successful marriages. Upon joining the club each client was provided a handbook, titled *Love Is More Than Luck,* which attempted to provide a rudimentary education in mate selection and to provide guidance regarding conduct in the club. It included practical information ranging from *Ten Do's* and *Ten Don'ts* in letter writing to advice in selecting a mate and making a marriage successful. A classification system was worked out whereby each client was assigned a socio-economic index score, ranging from 0 to 9. This score gave an indication of his cultural and educational level, or social class status. Each client was also tabbed with five personality trait scores, and he was given a profile and interpretation of these when he joined. The socio-economic index score, personality trait scores, and an average of twenty-nine other items of information for each client were punched on Remington cards, and a Remington electric card sorting machine was used in selecting the introductions.

In matching two persons for an introduction, the electric card sorting machine selected for any of the thirty-five characteristics. For example, if a thirty-five-year-old engineer wished to meet women between the ages of thirty and thirty-seven, who were Catholic, with college education, living in California, under 5' 6" tall, not divorced, who had outdoor life recreational interests, and who were compatible with him on the basis of the five personality traits measured, the machine could sort out all women in the active membership meeting these specifications in approximately fifteen minutes. An attempt was made to discourage the concept of the *one-and-only* or *perfect* mate. The objective was merely to enhance mate selection opportunities.

Cultural Background of Members

It was found that the cultural background of members varied with a number of factors inherent in the operation of the club, such as type and source of advertising, amount of membership fee, quality of promotion materials, screening of clients, and refinement of introduction procedures. Over the years, as Personal Acquaintance Service matured and improved its procedures, the socio-economic status of persons attracted to membership gradually rose. Information regarding the educational background of the director encouraged many college educated people who stated that they were skeptical of the integrity of other club directors, and the protection of the identity of clients by providing pen name and mail service also attracted members from prestige occupations that no other type of club could reach.

Education. In 1949, after one year of operation, there was an active membership of 881, 9.3 per cent of whom reported that they had completed four years college or more, 14.5 per cent reported some college, and 25.9 per cent only eight grades or less. In 1954, after six years of operation, there was an active membership of 1,254, 17.8 per cent of whom had completed four years college or more, 25.3 per cent some college, and only 11.1 per cent a grade school education or less. At the time Personal Acquaintance Service was dissolved approximately one in every five members being enrolled was a college graduate, and one-fourth of the graduates listed graduate training or graduate degrees.

For the total membership, 13.5 per cent reported a bachelor's degree or higher education, 16.8 per cent some college, 49.0 per cent nine to twelve grades, and 17.6 per cent eight grades or less.

It is understandable that the Research Club membership would be superior to the general population in education, since the uneducated could not ordinarily appreciate and utilize the scientific mate selection techniques that were provided. Census data for 1950 indicate that nearly half (47 per cent) of the general population over twenty-five years of age had completed only an eighth grade education or less, and that only 6 per cent had completed four years college or more.

Occupation. The occupational level of Personal Acquaintance Service members was comparable to their education. Nineteen and five-tenths per cent were classified as executive, managerial, professional, and semi-professional workers. In this group, teachers and nurses seemed to be unusually well represented among the women—engineers among the men.

A great variety of occupations was represented in the professional and semi-professional group, including ministers, college professors, school administrators, psychologists, social workers, sociologists, dentists, physicians and surgeons, chemists, accountants, magazine editors, publishers and many others.

About one in four of the men members reported that he was a skilled tradesman, such as a watchmaker, plumber, auto mechanic, or carpenter; 15.9 per cent were classified as semi-skilled and unskilled workers, 10.8 per cent were in military service, 6.0 per cent were farmers and ranchers, and 6.5 per cent did office or secretarial work.

The predominant occupational groups for women were homemaking and office or secretarial work—19.3 per cent and 18.2 per cent were employed in these groups respectively. The majority of the homemakers were women over fifty, most of whom were widows.

Sex Ratio. Despite constant and persistent efforts to enroll more women members, Personal Acquaintance Service was able to recruit only 2,054 women out of a total of 6,023; thus the men outnumbered the women approximately two to one. During the research period, 50,174 letters of inquiry were purchased from fourteen introduction clubs across the nation. Of this number, 38,081 were from men; only 12,093 were from women. The conscientious club operator everywhere was asking: How can I enroll more women? The need, however, was for women members under fifty years of age. There was a shortage of men and a surplus of women in the age group over fifty.

In Personal Acquaintance Service the pattern in the various age groups was approximately as follows:

18-26	5 men to every woman
27-32	4½ men to every woman
33-38	3 men to every woman
39-44	2 men to every woman
51-56	1½ women to every man
57-65	2 women to every man
Over 65	1 man to every woman

As is apparent, the predominant group in the Research Club was men under thirty-five; they comprised half of the male membership, and nearly a third of the total membership. Personal Acquaintance Service was able to serve best those members between the ages of thirty-five and fifty-five, since the sex ratio and cross-sex age composition for this group were in better balance.

It should be noted that only 1.3 per cent of the membership was over sixty-five years of age, though about 7 per cent was sixty years of age and over, and that in the age group over sixty the sexes were in reasonable balance. The age range for men was eighteen to eighty; for women it was eighteen to eighty-four. The median age for men was thirty-five, and for women it was forty-six.

There are perhaps several conditions in our society that are responsible for this unusual sex ratio and age distribution in introduction clubs. First, men are more aggressive, matter-of-fact, and less romantic in courtship and marriage than women. In joining a club they are merely exercising their prerogative—pursuit of the female. Women are more protected, longer under parental influence, and are less likely to engage in behavior that is disapproved by society.

Second, the popular press has done a good job of indoctrinating women against introduction clubs. It has depicted club membership rolls as over-loaded with lonely, love-starved women, who are often exploited by the few undesirable men who join. Women are therefore more fearful of ex-ploitation than men.

Third, there is a surplus of men in the marriageable age group under thirty-five in our population, and they suffer a numerical disadvantage in competition for mates. Census reports have not adequately clarified these data in the past; they do not ordinarily include young men in military service, and they do not adjust the data for errors in reporting age. A better method for determining the sex ratio for this group is through male-female birth and death rate differentials. Louis I. Dublin, Metropolitan Life Insurance Company vice-president and statistician, points out that there are 106 males born to every 100 females, and that although the mortality rate is higher for men at every age than women, men predominate in number up to about the age of fifty, and that after age fifty the sex ratio begins a continuous and rapid decline in deference to the female's greater durability.[4]

Fourth, in addition to the numerical disadvantage in their own age group, young men must compete with older men for the young women. Sometimes the older man is better established and more affluent, and the younger man loses out. Since many of the older unmarried men seek mates considerably younger than themselves, this also adds to the numer-ical disadvantage the older women suffer in their competition for mates.

Marital Status and Dependent Children. It was found that marital status and age composition were somewhat related. The abundant young men of Personal Acquaintance Service were predominantly single (never married), and the older women were predominantly widowed. Marital status for men was: single, 51.6 per cent; widowed, 15.4 per cent; di-vorced, 33.0 per cent; and for women it was 18.5 per cent, 47.3 per cent, and 34.2 per cent, respectively. Although more than half the men were single, less than one-fifth of the women had never been married, and while nearly half of the women were widowed, only 15.4 per cent of the men were widowers. Approximately a third of both sexes were divorced; three out of five members were widowed or divorced.

About three-fourths (76.6 per cent) of the total membership reported that they had no dependent children; 10.6 per cent had one dependent child, 6.9 per cent had two, and 4.1 per cent had three or more. The relatively small percentage of members reporting dependent children is to be expected, considering the age composition and marital status of Per-sonal Acquaintance Service members.

[4] Louis I. Dublin, *The Facts of Life from Birth to Death,* New York: The Mac-millan Company, 1951, p. 10.

Personality Characteristics of Members

It is commonly assumed that all introduction club members are either neurotic, physically inadequate, or socially rejected persons. An attempt was made in this study to assess the validity of this assumption by measuring and evaluating, so far as possible, the personality and the physical characteristics of Personal Acquaintance Service clients.

Since there was no adequate personality test available that was short and easy to fill out and to score, and that would function effectively for both research and scientific matchmaking, the only alternative was to construct one. Brevity was of great practical importance, so the test was designed to measure only five traits and only ten items were utilized for each trait. The traits were designed as *neurotic tendency, sociability, conformity, attitude toward sex,* and *religious orthodoxy.* Among the members they were known as traits *A, B, C, D,* and *E,* respectively, and the more euphemistic terms *temperament conformity to social standards,* and *religion and philosophy of life* were used for traits *A, C,* and *E,* respectively,

The subtest for *neurotic tendency* attempted to measure emotional stability, freedom from moodiness, depression, nervous tension, and various forms of emotional insecurity. *Sociability* was defined as interest and participation in group activities. The *conformity* subtest attempted to measure the degree to which a person conformed to majority attitudes on social issues. *Attitude toward sex* was considered to be a measure of inhibition and antagonism in relationships with the opposite sex, and *religious orthodoxy* was defined as the extent to which a person subscribed to commonly held beliefs supported by the major organized religious groups.

To compensate for the brevity of the test, and to increase both reliability and validity, a five point answer scale followed each test question, allowing the respondent to register his feelings with greater accuracy. The short test seemed to measure with a high degree of consistency; the test-retest reliability coefficients for the five subtests were: .91, .90, .86, .85 and .93, respectively. These were derived from a sample of fifty college students—the test-retest interval was two weeks.

The test was constructed in 1947, and it underwent two revisions during the life of the Research Club. A number of the items selected for the test had been used successfully in the previous research of Terman and his associates, Burgess and Cottrell, and other social scientists.[5] Some of the

---

[5] See Lewis M. Terman, *et al., Psychological Factors In Marital Happiness,* New York: McGraw-Hill Book Company, 1938, Chapter VI; Ernest W. Burgess and Leonard S. Cottrell, *Predicting Success or Failure In Marriage,* New York: Prentice-Hall, 1939, Chapter IV.

items had also been used in a test previously validated by the author.[6]

An attempt was made to further validate the test by eliminating the least discriminating and adding new and promising experimental items with each of the two revisions. Two methods were utilized to determine which items should be replaced: (1) the criterion of internal consistency, and (2) correlation of test results with case data that were gradually accumulating in the Personal Acquaintance Service files.

The first method involved the use of two criterion groups, selected from a sample of fifteen hundred cases in each instance. These were comprised of the two hundred persons making the highest scores on each subtest and the two hundred making the lowest. The performance of these two groups was compared for each item on the test and those items which differentiated least were considered to be the least valid and were eliminated.

The second method involved the use of detailed case data derived from letters, the application blank, and other sources on persons who would be expected, on the basis of such information, to make extreme positive and negative scores on the relevant traits. With each revision of the test those items which discriminated unsatisfactorily, according to the case data criterion, were also eliminated.

Another evidence of validity of the test was the relatively low intercorrelations between the various subtests. Five of these were .10 or lower, and the highest was .31, which was the correlation between *sociability* and *attitude toward sex*.

Over the years it was observed that a great many of the Research Club members were in occupations and environments that afforded little normal contact or social life with eligible members of the opposite sex. An analysis of their reported recreational interests indicated that their leisure time pursuits were predominantly of a nonsociable nature. Half were outdoor life enthusiasts and lovers of nature. Nearly a third listed home activities, such as gardening, homemaking, mechanical hobbies, pets, and work around the house. A third listed reading, intellectual and creative activities as leisure time pursuits. Only a fourth listed social events such as night clubbing, house parties, card playing, or dancing. It was therefore hypothesized that *sociability* test scores of members of Personal Acquaintance Service would be significantly lower than for persons of comparable cultural background in the general population.

It also became apparent early in the experiment that the Research Club clients were not predominantly neurotic, and that their social atti-

[6] Karl Miles Wallace, *Construction and Validation of Marital Adjustment and Prediction Scales,* Unpublished Doctoral Dissertation, The University of Southern California Library, 1947.

tudes were quite conservative and conventional. It was readily observed that for many of the neurotic members the Personal Acquaintance Service venture was a difficult experience. They were frequently sensitive, fearful, suspicious, and filled with anxiety at the thought of being thrown on their own in correspondence with a strange person. They rarely married, and they more often dropped out because of discouragement from lack of success. It therefore seemed reasonable to assume that even though a club such as Personal Acquaintance Service might be appealing to some types of neurotic persons for various reasons, this selective factor would discourage a sufficient number to make the Research Club sample approximately normal or average on this trait.

In view of these random observations, the null hypothesis was ventured for all of the personality traits except sociability—the assumption that the scores of Research Club members would be comparable to those of a general population sample, and that any obtained differences would be due to sampling variability.

In order to test these hypotheses, the personality test was administered to a sample of two-hundred persons secured from the general population in the Los Angeles area by research assistants and students. The sample was a predominantly Protestant, middle class group, equally divided between the sexes. A statistical comparison of the personality scores for each of the five traits was then made with scores taken from a random sample of Personal Acquaintance Service members that were matched for age, sex, religious preference, and socio-economic status.

Although the resulting score differentials suggested that the Research Club group was actually less neurotic, more conforming, more religious in attitude, and slightly less inhibited in attitude toward sex than the nonclub sample, these differences were not sufficiently great to give adequate assurance that they were not due to chance. For *neurotic tendency* and *conformity* the critical ratios were 1.69 and 1.83, respectively, and they could be considered significant only at the 10 per cent level of confidence. For *religious orthodoxy* the critical ratio was 1.95, barely under the 5 per cent level, which gives considerable assurance that the Research Club members were more religious than the nonmembers. Scores for the two groups were more nearly equal on *attitude toward sex*. The critical ratio was .99.

Personal Acquaintance Service members made much lower *sociability* scores, however, than persons in the nonclub sample, and this difference was statistically significant at the .001 per cent level of confidence; the critical ratio was 6.5.

It is possible that the Research Club members were not only less sociable, less neurotic, more religious, and more conforming than persons of like cultural background in the general population, but that they were also more rigid and inflexible, more intolerant, more demanding, and harder to please—that it was more often that they had rejected others instead of being rejected themselves. This was impossible to determine in the present study, however, because of insufficient data.[7]

In a society such as ours, where the courtship system is highly competitive, and where many cultural barriers operate to obstruct mate selection opportunities, the quiet, nonaggressive, and shy person is tremendously handicapped, no matter how stable his emotional adjustment or how little he expects in a mate. The problem is especially acute for the middle-aged who are seeking remarriage, many of whom are lost in a sea of married people and children. For these persons a correspondence club has great appeal because it allows one to join on his own, quietly, and even anonymously, in his own home. In this type of club members usually exchange letters for a time before meeting, assuring themselves that they have sufficient in common to make the meeting mutually interesting, and many of the awkward rebuffs, personality clashes, and other embarrassments common to other types of introduction clubs are avoided.

This experiment indicates that the correspondence club method, intelligently utilized, offers great promise of ameliorating many of the cultural barriers to intelligent mate selection in the United States, especially for middle-aged people.

[7] Data accrued in the present study suggest that a correspondence club selectively eliminates many types of neurotic and emotionally maladjusted persons because of the demands placed on the individual, whereas the personal-contact type of club is more appealing to the weak, fearful, and emotionally maladjusted person because he thinks he can dump his problems into the protective lap of a kindly club director or counselor and solve them with little or no effort on his part. Our data also suggest, however, that due to the stigma attached to forthright methods of seeking mates, and various other factors, the introduction club may selectively attract the perfectionist, the supercritical, the intolerant and inflexible, who expect too much from other human beings, and are therefore socially isolated and lonely. Further research will be required to determine the validity of these observations.

18. MARITAL ADJUSTMENT AND SUCCESS: MEASUREMENT AND STANDARDS

The general approach to the measurement of marital adjustment and success is described by Kirkpatrick, who adds a list of criticisms of published instruments. He asserts that two kinds of correlates of marital adjustment should be distinguished: Those that are clearly premarital and therefore are eligible to be considered as causes, and those that come up during the marriage. Hill has summarized Kirkpatrick's analysis of the results of a number of studies of marital happiness; some of Hill's summary is reproduced here in a table. What do we mean by marital success? What should we mean by it? A second excerpt from Kirkpatrick's book considers these questions.

ONLY NINE?

Reprinted from *Marriage and Family Living*, 18 (May, 1956), 113.

Those who believe that mass marital infelicity is a modern phenomenon may be interested in the following column, unsigned, which appeared in the September 1, 1785, issue of the New York *Daily Advertiser,* printed and edited by Francis Childs, whose print shop was on "Water Street, between the Coffee House and Fly-Market." The article was titled, "On the State of Marriage in South Britain."

If you see a man and woman, with little or no occasion, often finding fault, and correcting one another in company, you may be sure they are man and wife.

If you see a gentleman and lady in the same coach, in profound silence, the one looking out at one side, never imagine they mean any harm to one another; they are already honestly married. If you see a lady accidentally let fall a glove or handkerchief, and a gentleman that is next to her kindly telling her of it, that she might gather it up, man and wife.

If you see a lady presenting a gentleman with something sideways, at arm's length, with her head turned another way, speaking to him with a look and accent different from that she uses to others, it is her husband. If you see a man and woman walking in the fields in a direct line twenty yards distance from one another, the man strides over a stile, and goes on sans ceremonie, you may swear they are man and wife, without fear of perjury.

If you see a lady whose beauty and carriage attract the eyes, and engage the respect of all the company, except a certain gentleman, who speaks to her in a rough accent, not at all affected with her charms, you may be sure it is her husband, who married her for love, and now slights her.

If you see a gentleman that is courteous, obliging, and good-natured to every body, except a certain female that lives under the same roof with him, to whom he is unreasonably cross and ill-natured, it is his wife. If you see a male and female continually jarring, checking, and thwarting each other, yet under the kindest terms and appellations imaginable, as my dear, etc., man and wife.

The present state of Matrimony in South Britain:

Wives eloped from husbands	2,361
Husbands run away from their wives	1,362
Married pairs in a state of separation from each other	4,120
Married pairs living in a state of open war, under the same roof	191,023
Married pairs living in a state of inward hatred for each other though concealed from the world	162,320
Married pairs living in a state of coldness and indifference for each other	510,123
Married reputed happily in the esteem of the world	1,102
Married pairs comparatively happy	135
Married pairs absolutely and entirely happy	9
Married pairs in South Britain	872,564

MEASURING MARITAL ADJUSTMENT

Clifford Kirkpatrick

Adapted from Clifford Kirkpatrick, *The Family: As Process and Institution*, pp. 340-346. Copyright 1955, The Ronald Press Company, New York.

Pioneer Attempts to Measure Marital Adjustment

Simple Classifications and Measurements of Marital Adjustment. Ever since there has been marriage and giving in marriage, there have been estimates of marital adjustment in terms of amount or degree. In the modern scientific studies of factors related to success in marriage, various measures of marital adjustment are used. Very commonly there has been use of a happiness scale on which the marriage is rated by either a participant, an outsider, or both. The classification often runs (1) very unhappy, (2) unhappy, (3) average, (4) happy, and (5) very happy, although occasionally a sevenfold classification is used.

Bernard, in an interesting pioneer study, presented her subjects with terms or traits representing virtues and defects and scored adjustment in terms of attributing favorable rather than unfavorable traits to the spouse. [1] Hamilton, in the course of long interviews, asked his subjects many questions concerning dissatisfactions, desire to continue the relationship, willingness to press a magic button and thus never to have been married, rating of adjustment and desired changes in the mate, and the like. By assigning points to various answers to such questions, he derived a fourteen-point scale with score intervals corresponding to five categories of success in marriage. [2]

More Elaborate Scales for Measuring Marital Adjustment. On the basis of earlier efforts, Burgess and Cottrell were encouraged in the construction of one of the more widely used scales. Their scale differs from most previous scales in that there is (*a*) a systematic combination of various items, (*b*) an objective procedure for the selection and weighting of items, and, finally, (*c*) a greater number of points or degrees on the scale. Furthermore, the experimenters were concerned with the consistency between their scores and other measures or indices of marital adjustment.

They started with a five-point happiness-in-marriage scale and found considerable agreement in ratings between persons in a position to have knowledge of a particular marriage. They then selected aspects of marriage which seemed likely to be favorably associated with high happiness ratings, including agreement on various matters, common interests and activities, demonstration of affection, lack of dissatisfaction with the marriage, and lack of feelings of unhappiness and loneliness. Weights were assigned to the responses and made proportionate to the association of the response with a favorable self-rating of marital happiness. Thus by a summing of weights or points assigned to favorable responses, a score up to 194 could be derived to describe the marital adjustment of a given individual. Given this method of test construction, it is not surprising that the total scores of the subjects correlated highly with self-ratings of marital happiness, since the item weights were assigned on the basis of such a correlation.

Terman employed the term "happiness in marriage" rather than "marital adjustment," but used some of the Burgess-Cottrell items. Into his total scale he incorporated a seven-point rating scale of marital happiness, together with more specific complaints, items concerned with present un-happiness, and contemplation of marital disruption. The assignment of weights to items was based first on a high correlation of response on a given item with favorable responses on other items, and second on hus-band-wife agreement in responding to a particular item. His item weights were smaller than those of Burgess and Cottrell, and therefore the maxi-mum score on the Terman scale is 87, as compared with 194.

Persons with a high score on the Terman test, by virtue of test con-struction, consistently say nice things about their marriage in terms of responses which are associated with consistency of responses between hus-band and wife. Whether partners tell the complete truth about their mar-riage or not is another story. There was a great advantage in the Terman effort in that the information was gained independently from husbands and wives. Spouses returned their separately sealed questionnaires, placed in a covering envelope, by dropping them in a basket containing other covering envelopes.

The more recent Locke marital-adjustment scale was constructed chiefly with the aid of 201 divorced couples and 200 couples rated by out-siders as unusually happily married. [3] In most cases, Locke obtained in-formation from both husband and wife, and had the added advantages of interview methods and a more representative cross section of the popula-tion. Furthermore, he had a more objective basis for selecting and weigh-ing items in terms of their discrimination between the happily married and the divorced group. His approach was similar in many respects to that of Burgess-Cottrell, and of his 29 questions, 19 were taken from the Burgess-Cottrell scale. The weight given to a particular response varied from 0 to 8 in proportion as the response was differentially characteristic of the hap-pily married group. For men the maximum score was 157, and for women, 154. [4]

In view of common items and a similar conception of marital adjust-ment, it is not surprising that the Locke marital-adjustment scale corre-lated highly with marital adjustment as measured by Burgess-Cottrell items. By virtue of the construction method it was also to be expected that the marital-adjustment scale would permit inferences as to divorce or non-divorce.

At first glance it might seem that Locke had proved that divorced couples were extremely maladjusted. The mean adjustment score for mar-ried men in his group was 138.5 as compared with 100.8 for divorced men. In the case of the married women the score was 137.4 as compared with 102.4 for the divorced women. [5] But it should be noted that the happily married group were recommended for interviewing by a more or less random sample of the Bloomington population just because they were regarded as unusually happy in their marriages. Furthermore some of the happily married were recommended by divorced persons, who may have been biased in their ratings because of their own mental status. The happily married may have scored higher on the Locke test, not because they were married rather than divorced, but because they were selected as unusually happy in the first place. On the other hand the divorced persons may have scored low, not altogether because their marriages were extremely unhappy as compared with the average for married couples, but because of the divorce experience which created bitterness concerning their former marriages and, hence, low scores. It is unproved from Locke's study that persons *prior* to divorce are vastly more maladjusted than average persons not immediately inclined to divorce.[1]

Criticism of Attempts to Measure Marital Adjustment. Seven criticisms of attempts to measure marital adjustment will be offered.

1. Responses to questions recognizable as designed to measure marital adjustment tend to be biased by the desire to appear respectable according to the dominant values of the group. It is important in the United States to find happiness in marriage, and to admit failure is almost as shameful as to admit poverty in this land of opportunity. The majority of persons queried in the studies of marital adjustment write or say in one way or another that they are happy, adjusted, and successful in marriage. The skewed distribution of adjustment scores may be due to prior elimination of pairings which did not work out, but they may also reflect a tendency to put up a conformist "good front."

2. Devices for classifying and measuring marital adjustment tend to reflect American middle-class values. This type of critical comment has been made effectively by Kolb, Hill, and others. [6] This means that a person scoring high on a measurement of marital success or adjustment may be essentially a conventional person who claims to do that which is expected of a married person in terms of agreement, affection, harmony, comradeship, responsibility, and stability. Deviant values of persons, classes, or culture groups may not be fully recognized, and hence scores may be a measure of deviation from middle-class norms rather than of marital adjustment in some absolute sense.

3. There is a halo effect in the reaction of human beings such that the response to one stimulus in a pattern tends to carry over to other stimuli without a critical differentiation of response. This halo effect, stressed many years ago by Hollingworth, is a "monkey wrench" in the machinery for measuring and predicting marital adjustment. [7] A person who indicates in one way that he is happily or unhappily married is likely to give

---

[1] Numerous other attempts to measure marital adjustment have been made. See Clifford Kirkpatrick, *op. cit.*, p. 621.

the same verdict by other responses which are not really confirming evidence but rather different ways of saying the same thing, irrespective of the literal meaning of the questions which are put to him. The mood of the moment tends to dominate the words which are said or the marks which are put down on paper. Consistency may be due to diffusion of emotion.

4. Evaluation of a marriage from the report of one party to the marriage is a questionable procedure. The Burgess-Cottrell scale for measuring marital adjustment, both the original and the revised form, is really a scale for measuring the personal adjustment of the individual in marriage. Even personal adjustment may not be accurately inferable from a single report without consideration of the opinions of the other party to the marriage. While Burgess and Cottrell claimed a correlation of .88 between the adjustment scores of 66 couples with responses from both husband and wife, there is every reason to think that there was collaboration. [8] It could be that a husband with a wife leaning over his shoulder would hesitate to risk her wrath by complaints about his marriage. Terman, with a controlled separation of husbands and wives, obtained a correlation of only .59 between the happiness scores of husbands and happiness scores of wives. [9] Burgess and Wallin found a correlation of .57 between the adjustment scores of engaged persons and .41 for the marital adjustment scores of those whose engagements had culminated in marriage. A recent study by Terman and Oden revealed a correlation of .52. There is special interest in the finding of Locke concerning the agreement of husbands and wives in their marital adjustment scores as measured by the Locke scale. For the happily married group the correlation was .36, but for the divorced group it was only .04. It would seem that the divorced persons could not even agree upon their disagreements as husbands and wives. [10]

5. A marriage relationship is something more than the parties to the relationship, and hence it may be that measurement of marital adjustment should be based on the interrelated evidence from husband and wife. It is now realized that fertility in marriage is not individual fertility or even the sum of male and female fertilities, but a unique interrelationship between the biological capacities of a particular husband and a particular wife. The same principle applies, and even more strongly, to marital adjustment if it be distinguished from personal adjustment to marriage.[2] Perhaps both should be measured, but they should not be confused in the minds of investigators.

6. Doubts are appropriate concerning the reliability of various measures of marital adjustment. We accept the evidence that a person indicat-

[2] A combining of data from husband and wife has been attempted. [11]

ing a certain degree of adjustment on a scale is likely to reveal similar adjustment on a similar scale, as is shown by high correlations. Changing the weights attached to items does not seem to affect correlations greatly. It may even be true that applying a scale to a group of people and then giving them the same test over again would result in similar scores.[3] But all this means is that people say the same thing in different ways and that having said it they repeat what they have said. Perhaps this indicates reliability in the sense that a set of questions brings consistency of response, but the consistency may be due to a halo effect and to an individual's constant concern for "appearances" as compared with the concern of others.

It is certainly hard to distinguish the problem of reliability from that of validity. Burgess and Cottrell found a consistency in the case of 38 individuals who gave self-rating of marriage happiness once and then again after an interval of from 8 to 24 months. The correlation for this small group was .86 (tetrachoric). [13] This might mean that the scale was so reliable in the sense of yielding repeatedly the same result that it did not record a change in "real degree of adjustment." The couples may have put up "the same old front" at constant levels in spite of high points and low points in the marriages. Certainly a thermometer which always gave the same reading would not be a very good instrument for measuring the temperature of an individual. . . .

7. The validity of devices for measuring marital adjustment are open to friendly question on grounds similar to those already mentioned. Do marital adjustment scales measure marital adjustment? Halo effects operating on verbal responses may conceal an uneven profile of adjustment. The high correlations between different measures of adjustment (.90 for Burgess, Cottrell and Terman scales) may be based on common items and halo effects of the moment if both scales are taken at the same time. Both may reflect degree of concern for respectability, even though the scales are labeled "marital-adjustment scales." The correlation of a score with the rating of an outsider may mean only that the outsider got his impression from the kind of verbalizations which produced the score. A woman might falsely tell her daughter that she was happily married. The agreement of the daughter's husband with the daughter and with his mother-in-law would prove merely that three people can tell the same lie. An interesting study by Harter checked upon the degree to which family members saw things the same way within their common family situation. The fathers and mothers agreed in their ratings of their marital happiness to the extent of .65. The daughters agreed with their mothers concerning the mother's marital happiness to the extent of .58. The father-daughter agree-

[3] The test-retest reliability of the Burgess-Wallin engagement success scale was .75 for men and .71 for women. [12]

ment concerning his marriage was .48, the mother-son agreement was .56, and the father-son agreement concerning the father's marital adjustment was .56. [14]

Enthusiasts for the measurement of marital adjustment may argue that divorce is good external, objective evidence of poor marital adjustment, and that a scale on which the divorced get relatively low scores is a valid score. There is some truth in this claim, but prior marital adjustment may be underestimated because of bitterness caused by the divorce rather than by the marriage. To some extent maladjustment is due to divorce rather than divorce to maladjustment. Furthermore, given a scale constructed with reference to middle-class values, one of which is family stability, lower scores would be expected for divorced persons who by their action of divorce had flouted the family stability value. Some independence of American values is demonstrated by the fact that the Locke scale differentiated between Swedish groups which included 205 couples. The findings of Locke and Karlsson are given in Table 1.

TABLE 1.   **Mean Marital-Adjustment Scores of Four Swedish Groups, with Critical Ratios of the Differences Between Happily Married and the Other Groups**

Groups	Mean Scores of Men	C.R.	Mean Scores of Women	C.R.
Happily married	134.8		154.4	
General population	130.7	2.0	144.8	3.0
Unhappily married	117.4	5.0	124.6	6.1
Separated	88.6	11.9	86.8	17.0

Source: Harvey J. Locke and Georg Karlsson, "Marital Adjustment and Prediction in Sweden and the United States," *American Sociological Review,* XVII (1952), 12.

There is no reason to conclude that attempts to measure marital adjustment have resulted in total failure. It is merely suggested that there is danger in respectable responses, halo effects, and methodological circles. For the moment it may be assumed that there are some rough measures of marital adjustment. We ask, "What factors in the sense of traits, situations, and circumstances are associated with marital adjustment?"

Factors Associated with Marital Adjustment

In a large number of statistical studies an attempt has been made to find factors associated beyond chance with marital adjustment. Yet these researches certainly do not isolate "causative" factors as compared with factors merely associated with good marital adjustment. At risk of oversimplification it may be stated that a factor is more properly regarded as "causative" of marital adjustment if (*a*) it is prior in time, (*b*) it is highly associated or correlated, (*c*) if it is persistently associated in spite of absence of or variation in other factors, and (d) if it is distinct from marital adjustment rather than being merely an aspect of marital adjustment. Complaining about a spouse, for example, or getting a divorce could hardly be regarded as a cause of prior marital difficulties.

Convergence of Research Evidence. As research developed relating various factors to adjustment in marriage as measured by objective criteria the findings were widely quoted in the literature of the family. There was a marked tendency, however, to select the evidence which supported the opinions of the writer and to ignore contrary evidence. It seemed desirable, therefore, to compile evidence in condensed quantitative form and in a manner which would reveal precisely convergence and disagreement. This effort was made by the author. [15]

Hill was impressed by the lack of convergence of significant evidence in the Kirkpatrick compilation when he applied the test of labeling a finding significant only if associated with either a critical ratio of 2.8 or higher or a correlation of 49 or above. He prepared a table showing that of the 152 factors found significant, by the above-stated statistical test, in *at least* one study, there were 40 which were found insignificant by one or more studies. Of the 114 factors found significant by *only* one study, there were 29 factors which were found insignificant in one or more other studies. Of the 26 factors confirmed as significant by two studies, there were 4 which were found insignificant in one or more studies. There were only 9 factors confirmed as significant by three studies, 2 factors confirmed by four studies, and only 1 factor confirmed by five studies when the stated test of satistical significance was applied to each research. Even these significant factors were occasionally unconfirmed by other studies. [16] This is no very complete or unanimous verdict by science on what makes for happiness in marriage.

A Summary of Factors Associated with Marital Adjustment. Any condensation distorts to some extent the whole truth, especially when figures are replaced by words. After scrutinizing marriage studies published through 1954 it is the writer's impression that those studies most strongly support the following factors as being associated with marital adjustment. Premarital factors are listed separately from those operating during marriage. Both lists are roughly in the descending order of scientific verification.

A. Premarital Factors
    Happiness of parents' marriage
    Adequate length of acquaintance, courtship, and engagement
    Adequate sex information in childhood
    Personal happiness in childhood
    Approval of the marriage by parents and others

Engagement adjustment and normal motivation toward marriage
Ethnic and religious similarity
Higher social and educational status
Mature and similar chronological age
Harmonious affection with parents during childhood

B. Factors Operating during Marriage
Early and adequate orgasm capacity
Confidence in the marriage affection and satisfaction with affection shown
An equalitarian rather than a patriarchal marital relationship, with special
reference to the husband role
Mental and physical health
Harmonious companionship based on common interests and accompanied
by a favorable attitude toward the marriage and spouse

REFERENCES

1. Jessie Bernard, "An Instrument for the Measurement of Success in Marriage," *Publication of the American Sociological Society*, XXVII (1933), 94-106.

2. G. V. Hamilton, *A Research in Marriage* (New York: Albert & Charles Boni, Inc., 1929).

3. Harvey J. Locke, *Predicting Adjustment in Marriage: A Comparison of a Divorced and a Happily Married Group* (New York: Henry Holt & Co., 1951), pp. 21-23.

4. *Ibid.*, p. 47.

5. *Ibid.*, pp. 52-53.

6. Willard Waller and Reuben Hill, *The Family* (rev. ed.; New York: The Dryden Press, Inc., 1951), p. 353; William L. Kolb, "Sociologically Established Family Norms and Democratic Values," *Social Forces*, XXVI (1948), 452-53, quoted in Waller and Hill, *The Family*, pp. 353-56; John Sirjamaki, "Eight Cultural Configurations in the American Family," *American Journal of Sociology*, LIII (1948), 464-70.

7. H. L. Hollingworth, "Psychological Factors in Marital Happiness," *Psychological Bulletin*, XXXVI (1939), 191-97. A review of L. M. Terman, *Psychological Factors in Marital Happiness* (New York: McGraw-Hill Book Co., Inc., 1938).

8. Locke, *Predicting Adjustment*, p. 58.

9. Terman, *Psychological Factors*, p. 82.

10. Locke, *Predicting Adjustment*, pp. 58-59.

11. See Ernest W. Burgess and Paul Wallin, *Engagement and Marriage* (Philadelphia: J. B. Lippincott Co., 1953), pp. 551-54.

12. *Ibid.*, pp. 315-16.

13. Ernest W. Burgess and Leonard S. Cottrell, Jr., "The Prediction of Adjustment in Marriage," *American Sociological Review*, I (1936), 743.

14. Aubrey B. Harter, *Adjustment of High-School Seniors and the Marital Adjustment of Their Parents in a Southern California City* (Los Angeles: University of Southern California Library, 1950), pp. 65-71, quoted in Locke, *Predicting Adjustment*, p. 60.

15. Clifford Kirkpatrick, *What Science Says About Happiness in Marriage* (Minneapolis: Burgess Publishing Co., 1947), pp. 11-50.

16. Waller and Hill, *The Family*, p. 358.

FACTORS ASSOCIATED WITH MARITAL ADJUSTMENT

Reuben Hill

Adapted from Willard Waller, *The Family: A Dynamic Interpretation,* New York, Holt, Rinehart and Winston, 1951 (revised by Reuben Hill), p. 358.

Adaptation of Reuben Hill's Summary of Clifford Kirkpatrick's Analysis of Findings from Studies of Marital Happiness

Factor Studied	Number of Times		
	Studied	Confirmed	Not Confirmed
Happiness of husband's parents' marriage	6	5	1
Length of acquaintance before marriage	5	4	1
Duration of engagement	5	4	1
Husband's close attachment to father	3	3	0
Husband's childhood residence in country	3	3	0
Spouses' satisfaction with marriage	3	3	0
Spouses' willingness to relive life with same spouse	3	3	0
Spouses' ranking on Kirkpatrick's Community of Interests Scale	3	3	0
Participation in and sharing of outside interests together	4	3	1
Husband's sex drive (average or higher)	5	3	2
Husband's age at marriage	5	3	2
Husband's educational level (college or more)	6	3	3

Summary: Number of factors receiving as many as:

5 confirmations	1
4 confirmations	2
3 confirmations	9
Total	12

STANDARDS OF MARITAL SUCCESS

Clifford Kirkpatrick

Adapted from Clifford Kirkpatrick, *The Family: As Process and Institution*, pp. 358-362. Copyright 1955, The Ronald Press Company, New York.

Some Viewpoints. Burgess, a distinguished pioneer in the measurement of marital adjustment, has no illusions about a single standard of marital success. He points out that the middle class tends to think of marital success in terms of affectionate compatibility, child bearing and rearing, economic support by the husband, good housekeeping by the wife, a balanced budget, democracy in marriage, social participation, an outside interest for the wife, family objectives, and permanent monogamous marriage which gives the appearance of being happy. [1] He distinguishes eight criteria of marital success—namely, permanence, happiness, fulfilling expectations of the community, personality development, companionship, satisfaction with the marriage, integration in family life, and adjustment. Adjustment for him is indicated by agreement on the chief issues of marriage, consensus as to objectives, harmony in emotional intimacy, lack of complaints, and absence of emotional isolation. [2] From such an analysis Burgess is led to conceive of description of success in marriage in terms of a profile rather than a single score. Thus success on one criterion may be associated with relative failure with respect to others. Burgess, collaborating with Wallin, further developed the concept of multiple criteria. They constructed schedules to measure evaluation of permanence, conceptions of spouse satisfaction, marital happiness, general satisfaction, specific satisfaction, consensus, love, sexual satisfaction, companionship, and compatibility of personality. Specific scores correlate with general scores .44 to .65. [3]

The Burgess-Locke warning as to an uneven pattern of success is not fully heeded by Hill in his gallant effort to weave together the criteria of integration, adjustment, companionship, and personality development into a concept of developmental adjustment. For Hill, marriage is successful in proportion as there is love of mate, dynamic accommodations between mates, solidarity through common experience, compatible and mutually satisfactory roles, collective security, a wholesome atmosphere for child rearing, satisfaction of ego demands, a handling of economic problems, and free areas for self-expression. [4] The desire for such marital success is a splendid aspiration, but attempts at empirical measurement might result in the frustrating revelation of success profiles rather than uniform achievement of all worthy goals. Is there anything to be added to the thoughtful viewpoints of Burgess-Locke and of Hill?

Some Reflections on Success in Marriage. Dilemmas are often ignored in the writing on marital success. . . . Can there be, for example, both success based on order and success based on freedom? Each criterion of success in marriage implies a price, if a certain aspect of marital success cannot be pursued save by the sacrifice of some other aspect.

Social expectations tend to be opposed to personal expectations. This is merely an implication of the dilemma concept. It can be argued that the Burgess-Locke criteria of marital success—social expectations, permanence, integration, and adjustment—are opposed in varying degrees to happiness, personality development, satisfaction, and companionship. On the one side are criteria of order, and on the other side criteria of freedom. Many a marriage which is permanent in accordance with social expectation is stable because of a sacrifice of happiness, satisfaction, and personality development. Often a great effort with apparent success is associated with inner frustration deepened by awareness of the hollowness of the pretense. That is why psychiatrists have customers from some nice families.

There should be a distinction between the short-run and the long-run verdict in regard to the success of marriage. As in the case of parenthood, it takes a long time to accumulate the total evidence. An aged couple sitting by the fire engaging in collective introspection could consider the matter with full perspective. It may be that there were strange peaks and troughs in their forty-year curve of marital experience. The renounced lover may seem ultimately no great loss. Even shared misery may be the basis of a kind of comradeship. With recognition of the time element, we ask whether a short marriage is successful as compared with a long one, given equal adjustment. Marriage to a combat pilot should perhaps get a negative score on a scale for predicting duration of success in marriage.

There are varied implications of the happiness criterion. . . . One can ask, "Is marital happiness merely a matter of being a happy person? Is the success of a particular marriage in terms of moderate happiness for each spouse modified by the fact that each would have been still happier with another mate? What of ecstasy as compared with dull contentment?" A divorced couple say, "We had two wonderful years before we broke up, and it was worth a lifetime of placid routine." It is doubtful that people know their own state of happiness. In fooling others, people can fool themselves. A smile does not rule out the possibility of an ulcer, nor does a stolid face deny inner contentment.

There should be a sharper distinction between a relationship and the individuals who are part of that relationship. A relationship adjustment may be achieved by the subordination of one party to the relationship. Regardless of this possibility, there is a very real difference between the adjustment of an individual in marriage and the degree of adjustment inherent in the interaction pattern between two persons. An ordered relationship tends to put some limit upon freedom of the individual, although by a profound paradox there may be fulfillment through indentification with an achieved relationship. Family members can sometimes solace their personal frustrations with the thought that they are part of a continuing family process. Yet to predict there must be measurement or other evaluation of marital success, and precise prognostication demands a distinction between the individual and the marriage relationship.

Personality development is relative to personal values, and hence by this standard there may be many kinds of marital success. A man in the twilight of his life may say that by virtue of marriage to his particular mate he has become more the kind of person who in a moment of calm reflection he would like to be. Pasteur could say that of his devoted wife, and so could Robert Schumann speak of Klara. But on the other hand a safecracker could praise the contribution of his wife to his criminal achievements and extol her devotion in polishing his pistol, mixing nitroglycerin, driving the escape car from the bank entrance, and encouraging his struggle to the top of his profession. If in addition to rendering him professional assistance she found time to bear and rear four expert pickpockets and one big-time gambling racketeer he might feel that his cup of satisfaction was indeed well filled. Success as measured by self-estimated personality development could take countless forms, many of which would be at variance with the moral ideals of others. The above examples further illuminate the contrast between social and personal expectations.

Marital success could be regarded as a courtship bargain for the individual who got a desirable partner with reference to his own needs and his own desirability in the marriage market. The popular verdict so often heard is, "He did well for himself, but I can't see what she sees in him." The point here discussed reinforces the distinction between individual success and relationship success, for often a good bargain for one person is a bad bargain for the other party to the transaction. It is granted of course that, by virtue of obscure and complementary values, what looks like a bargain for one party might really be a bargain for both.

Marital success can be regarded as relative to obstacles. Certainly adjustment in marriage is harder for some people than for others, and perhaps there should be some evaluation comparable to the "A for effort" of the school. Perhaps a couple whose marriage ultimately ends in divorce should receive credit for having tried hard for adjustment in spite of the final failure. One hears the comment, "They got on mightly well considering what they are, what they were, and how much they differed."

Success in marriage may be usefully related to the concept of family roles and to the so-called basic hypothesis [set forth in the next paragraph]. It is plausible to think of marital success as a matter of harmonious interaction of complementary roles. The husband may play the man for the woman and the wife the woman for the man. Each in playing parental roles toward children may be playing complementary co-parent roles with respect to spouse. The harmonious playing of happy family dramas is furthered by convergence of family cultures which provide compatible scripts for family dramas which follow the mating of the two chief actors. Furthermore, one could conceive of success in proportion as the man and the woman make a smooth transition to more advanced roles. The man of the house who remains the baby of the family creates a family problem, and such family problems spell out failure rather than success in marriage.

According to the basic hypothesis a person tends to seek the continuance of satisfactions known in prior family experience and to strive with special intensity to make good the dissatisfactions of childhood family experience. This puts a strain upon maturity in role transition and may hamper satisfaction in the roles which are actually played. There is greater success, therefore, if each partner in the course of role-playing gives the other party the satisfaction of needs which exist, in accordance with the basic hypothesis, and receives in turn the satisfaction of such needs.

Marital success can be analyzed with the aid of the concept of personality adjustment [has been] defined in terms of clear-cut aspirations and the potentiality of fulfilling them in ways commanding approval by those persons who matter to the person whose adjustment is being considered . . . the types of maladjustment [which can be] derived from this definition [are] mental conflict, frustration, disapproval, and deprivation. It could be argued that success in marriage is essentially mutual personality adjustment. It is favorable for A and B in the marriage relationship each to have aspirations which are compatible or at least arranged in some similar order of relative importance. In the second place, it would be favorable for each to receive fulfillment or aid in fulfillment of aspiration from the other person. Finally, in this brand of success there is approval by each of the way in which the other party fulfills his aspirations. In other words, A would approve of the way in which B got satisfaction from A, and B would approve of the way in which A got satisfaction from B. This mutual approval would imply the compatibility of A's aspirations with reference to those of B, mutual aid, and communication of approval. Any breakdown in such a delightful pattern of adjustment would mean severe deprivation.

It might seem that such a subtle type of success in marriage is merely a complex ideal resulting from intellectualized wishful thinking. But given proper schedules, answered independently by husband and wife, yet interrelated, a rough measure of mutual personality adjustment might be achieved.*

REFERENCES

1. Ernest W. Burgess and Harvey J. Locke, *The Family* (New York: American Book Co., 1945), p. 435.

2. *Ibid.*, pp. 432-44. Sex adjustment is added in the revised edition (1953), p. 391.

3. Ernest W. Burgess and Paul Wallin, *Engagement and Marriage* (Philadelphia: J. B. Lippincott Co., 1953), p. 504.

4. Willard Waller and Reuben Hill, *The Family* (rev. ed., New York: The Dryden Press, Inc., 1951), pp. 368-69.

* The recent empirical work of Winch and Ktsanes is promising.

19. SOCIAL STRUCTURE, DOMESTIC POWER, AND MARITAL STABILITY

Performing a secondary analysis of the original 526 cases of the Burgess-Cottrell study, Roth and Peck find (a) that the higher scores on marital adjustment tend to be in the higher social classes and (b) that marital adjustment scores are higher on the average when both spouses are from the same social class rather than from different social classes.

Such findings raise the question as to what aspects of social class may be correlated with marital adjustment. To get some feel for the question we present Hollingshead's impressions of the relation between the class structure of small- and medium-sized communities and the stability of the marital dyad. Here again the conclusion is that there is a generally positive correlation between the socio-economic status of families and their stability. (More systematic evidence on this point is presented by Hillman in the next chapter.

Wolfe finds that marriages in which the husband is reported as dominant tend to report high annual income and high social status. (It should be borne in mind that the data in Wolfe's study come from wives only.) The highest levels of marital satisfaction were expressed by wives in marriages categorized as husband-dominant or as shared-authority (Wolfe's term is "syncratic").

SOCIAL CLASS AND SOCIAL MOBILITY FACTORS RE-
LATED TO MARITAL ADJUSTMENT

Julius Roth and Robert F. Peck | Adapted from the *American Sociological Review*, 16 (1951), 478-87.

In any stratified social order, marriage usually implies equality of status. Although the American class structure considered in this paper is not so rigid as the caste system discussed by Kingsley Davis in "Intermarriage in Caste Societies," [1] the fact remains that strata exist and the family is placed in the class structure as a unit. This means that if a man and woman of different social levels marry, there must generally be a shift of status for one or both of them.

The question arises: Does this necessary shift of status affect the subsequent adjustment of the spouses? Ruesch, Jacobson, and Loeb point out the feelings of stress, frustration, and confusion which often accompany the acculturation of immigrants.[2] The greater the degree of culture difference, the greater the maladjustment of the individual is likely to be. The differences in the characteristics of the social classes in our society indicate that these classes are different cultural groups. A person moving from one class to another must go through a process of acculturation similar to, though perhaps not so extreme as, that of an immigrant.

In some cases a shift of status of one or both spouses has taken place before marriage. A certain amount of stress and insecurity in social situations may thus be brought into the marriage relationship by one or both partners. This possibility suggests a second major problem: Do the premarital mobility patterns of the spouses affect the subsequent marital adjustment?

A fairly large group of cases which had some rating of marital adjustment and some good indication of the subject's social class was needed in order to examine these problems. The schedules Burgess and Cottrell used in their study of marital prediction provided such a group of cases.[3] Each of their 526 cases had a marital adjustment score which they had

[1] *American Anthropologist*, 43 (1941), 376-395.

[2] Jurgen Ruesch, Annemarie Jacobson, and Martin B. Loeb, *Acculturation and Illness,* Psychological Monographs, vol. LXII, No. 2, 1948.

[3] Ernest W. Burgess and Leonard S. Cottrell, *Predicting Success or Failure in Marriage,* New York, Prentice-Hall, Inc., 1939.

worked out with the help of statistical techniques. Most of the cases had data which made possible an estimate of the social class level of the subjects.

No detailed account of the collection of the data and the characteristics of the group will be given here, since Burgess and Cottrell have already presented this information in Chapter II of their book.[4] It is important to note that the great majority of the subjects were in their twenties or early thirties and that the time since marriage was fairly short (less than seven years) in all cases.

Social Class Ratings and Adjustment Index

The McGuire-Loeb modification of W. L. Warner's Index of Status Characteristics (I. S. C.) was used to determine the class status of each subject.[5] The schedules of 523 couples were consulted. Only those cases were used where a single, definite class placement could be made with reasonable confidence.[6] This provided a maximum total of 428 husbands and 417 wives. Some of the tables included here have a smaller number of subjects, since they pertain to questions or classifications which did not apply to all the subjects.

Four social class levels—upper-middle (UM) lower-middle (LM), upper-lower (UL), and lower-lower (LL)—had a sufficient number of cases to be useful in most of the tabulations.

The concept and measurement of marital adjustment used by Burgess and Cottrell, as well as the adjustment scores which they derived for each of their cases, were assumed without modification. The scores were likewise classified into the same GOOD (160-199), FAIR (120-159), and POOR (20-119) adjustment categories which Burgess and Cottrell used in almost all of their tables in *Predicting Success or Failure in Marriage*.[7]

Social Class and Marital Adjustment

Table I shows an evident trend in the case of both the husbands and wives for the marital adjustment score to increase as we move up the so-

[4] *Ibid.*

[5] The calculation of Warner's I. S. C. and the rationale on which it is based are presented in Part III of W. Lloyd Warner, Marchia Meeker, and Kenneth Eells, *Social Class in America,* Chicago, Science Research Associates, Inc., 1949. The McGuire-Loeb version may be found in Carson McGuire, "Social Status, Peer Status, and Social Mobility," Memorandum for the Committee on Human Development and supplement to *Social Class in America,* 1949, pp. 7, 8.

[6] This rigorous approach reduced the size of the sample; however, it appeared that the results of the analysis could be stated with more confidence if the composition of the groups begin compared was fairly uniform.

[7] Burgess and Cottrell, *op. cit.*

cial class scale. Testing this trend by the Chi-square method in comparison with a hypothesis of no relationship between social class and adjustment shows a significance at the 1 per cent level.

This finding is essentially in agreement with some of those of Burgess and Cottrell. Although they did not use any over-all measure of social class for comparison with adjustment, Burgess and Cottrell did examine many of the individual factors which are known to be related to class level. Education,[8] amount of organizational membership,[9] character of neighborhood,[10] and degree of economic security[11] are all positively correlated with class level. That is, in each case the part of the scale related to higher class status (e.g., more advanced education) included more couples who are well-adjusted than appear in the lower end of the scale (e.g., little education). The relationship of occupation to adjustment is less clear, because Burgess and Cottrell used an occupational classification which is only loosely related to the prestige value of the occupations. Executives and managers of large businesses and professional people, especially school teachers, show slightly better scores than clerical, sales,

TABLE 1. Distribution of the Husbands and Wives According to Social Class and Adjustment Index

Social Class at Marriage		Adjustment Scores of Couples							
		Number				Percentage			
		Good	Fair	Poor	Total	Good	Fair	Poor	Total
Husbands	UM	98	58	32	188	52.1	30.9	17.0	100.0
	LM	62	44	49	155	40.0	28.4	31.6	100.0
	UL*	27	17	28	72	37.5	23.6	38.9	100.0
	LL*	3	5	5	13	23.0	38.5	38.5	100.0
Total		190	124	114	428‡				
Wives	UM	63	32	25	120	52.5	26.7	20.8	100.0
	LM	106	71	53	230	46.1	30.9	23.0	100.0
	UL†	16	15	28	59	27.1	25.4	47.5	100.0
	LL†	1	2	5	8	12.5	25.0	62.5	100.0
Total		186	120	111	417‡				

* The cells in these two rows were combined in the Chi-square analysis.
† The cells in these two rows were combined in the Chi-square analysis.
‡ In this and in subsequent tables the total number of cases is less than 523 because the "unknown" and "indeterminate" cases are omitted.

[8] *Ibid.*, pp. 121, 391.
[9] *Ibid.*, p. 126.
[10] *Ibid.*, p. 132.
[11] *Ibid.*, p. 261.

small business, and skilled trades people.[12] Terman also found a slight positive relationship of marital happiness to education,[13] but no certain relationship to occupation[14] (again occupation was not classified according to prestige level).

When the social class background of each individual subject (that is, the social class of the spouse's parents) is tabulated, no class level trend in relation to adjustment is apparent (Table 2). The only fairly sharp difference lies between the lower-lower class and the remainder of the group. Lower-lower status has a slight negative relationship with marital adjustment. It thus appears that the social class of the spouses' parents *per se* has little relationship to the adjustment of the spouses.

Social Class Difference and Marital Adjustment

In over half the cases in this study in which the difference or similarity of the social class of the spouses was established, the spouses were of the

TABLE 2. Relationship of the Social Class of the Parental Family to the Adjustment Index of Each Subject

Social Class at Marriage	Adjustment Score of Each Subject							
	Number				Percentage			
	Good	Fair	Poor	Total	Good	Fair	Poor	Total
Subject's parents								
UM	67	43	40	150	44.7	28.6	26.7	100.0
LM	106	70	74	250	42.4	28.0	29.6	100.0
UL	87	54	45	186	46.8	29.0	24.2	100.0
LL	17	5	25	47	36.2	10.6	53.2	100.0
Total	277	172	184	633*				

* Of the 1,046 subjects (523 husbands and 523 wives) only 633 gave sufficient information about their parents to permit class ratings to be made. Each parental family was status-identified by the man's occupation, source of income, and education.

same social class at the time of marriage. (See Table 3.) Burgess and Cottrell state[15] that marriage tends to take place within a given cultural group and this has probably been the common finding in studies of marriage.[16]

[12] *Ibid.*, p. 136.

[13] Lewis M. Terman, *Psychological Factors in Marital Happiness,* New York, McGraw-Hill Book Co., 1938.

[14] *Ibid.*, p. 169.

[15] Burgess and Cottrell, *op. cit.*, p. 77.

[16] Robert T. McMillan, "Farm Ownership Status of Parents as a Determinant of Socio-economic Status of Farmers," *Rural Sociology,* 9 (June, 1944), 151-160. Mc-

Although most of the marriages take place within a given class, a substantial number of cross-class marriages are represented in Table 3. How do they compare in adjustment to same-class marriages? According to the hypothesis previously discussed, the cross-class marriages are likely to cause greater stress to the persons involved. How does this reflect on their relative adjustment? The percentage distribution in Table 3 shows that the adjustment scores tend to be higher in the case of same-class marriages. Testing this relationship by the Chi-square method[17] shows a significance at the 1 per cent level. This result suggests that the stress of a rapid shift in class values required by a cross-class marriage has a negative influence on the adjustment of that marriage.

TABLE 3. Distribution of Total Cases According to Adjustment Index and Similarity or Difference of the Social Class of the Spouses

Social Class at Marriage	Adjustment Score of Each Subject							
	Number				Percentage			
	Good	Fair	Poor	Total	Good	Fair	Poor	Total
Spouses of same class at time of marriage	115	56	44	215	53.5	26.0	20.5	100.0
Spouses 1 class apart at time of marriage	56	50	54	160	35.0	31.2	33.8	100.0
Spouses more than 1 class apart at time of marriage	3	8	10	21	14.3	38.1	47.6	100.0
Total	174	114	108	396				

In a cross-class marriage either the husband or the wife may be of the higher social class at the time of marriage. Does the effect on marital adjustment differ with the sex of the spouse of superior status? In Table 4 the direct comparison is made of all the cases in which the husband was of the higher class at marriage, with all those in which the wife was of the higher class at marriage. The wife-high cases seem to be more unfavorable to marital adjustment than the husband-high cases, although both show a tendency to lower scores than same-class marriages. Using the Chi-square technique we may test the hypothesis: Cross-class marriages in which the husband is of the higher social class generally show better adjustment than those in which the wife is of the higher class. The level of signifi-

---

Millan reports a strong tendency of farmers to marry within their own status group (defined by farm ownership) even where a marked disparity of the sex ratio made such pairing difficult.

[17] The corresponding cells for "1 class apart" and "more than 1 class apart" had to be combined because of the very small number of cases in the "more than 1 class apart" category.

TABLE 4. **Comparison of the Husband-High Cross-Class Marriages with Those in Which the Wife Is of the Higher Class**

Social Class at Marriage	Adjustment Score of Each Subject							
	Number				Percentage			
	Good	Fair	Poor	Total	Good	Fair	Poor	Total
Husband 1 or more classes higher than the wife at marriage	41	38	37	116	35.3	32.8	31.9	100.0
Wife 1 or more classes higher than the husband at marriage	18	20	27	65	27.7	30.8	41.5	100.0
Total	59	58	64	181				

cance proves to be very low $(0.30 < P < 0.50)$. The relationship is obscured by the fact that the class differences represented can have different origins. For example, a woman may have been mobile before marriage past the level of her future husband or she may have acquired her higher status from her parents. This problem will be further discussed in the section on mobility patterns.

Despite the low statistical significance of the relationship between wife-high and husband-high cross-class marriage, the relationship is in keeping with findings on this point in other studies. McMillan finds that in the case of marriages across class lines, it was most often the wife who "married up" the status ladder.[18] To put it another way, the men seemed to be more willing than the women to accept a lower status spouse. Terman found that in his subjects the wives who were markedly superior to their husbands in education[19] had low happiness scores, while the scores were much higher in cases where the husbands were markedly superior in education.[20] James West points out that in all of the cross-class dating in "Plainville" the boy is of the higher class. A boy dating a girl of a lower class is frowned upon, but "for an upper-class girl to have a date with a lower-class boy would be inconceivable."[21]

Why is adjustment smoother when the man enters marriage at the higher status than when the woman does so? An important finding of Burgess and Cottrell gives a clue for further study on this point. The

[18] McMillan, *op. cit.*

[19] Education may serve as a crude index of class level.

[20] Terman, *op. cit.*, p. 191.

[21] James West, *Plainville, U.S.A.*, New York, Columbia University Press, 1945. In this rural village dating is generally looked upon as a forerunner to marriage.

major adjustment, in fact almost the entire adjustment in most marriages, is made by the wife.[22] Since an upward shift in class status carries some rewards and also entails fewer punishments than a downward shift, we would expect less stress in those cases where the wife had to move upward (that is, the husband-high marriages) than in those where she was expected to shift her values downward (that is, the wife-high marriages).

TABLE 5.　Distribution of Adjustment Scores of Couples in Terms of the Social Class Difference of Their Parents

Social Class at Marriage	Adjustment Score of Each Subject							
	Number				Percentage			
	Good	Fair	Poor	Total	Good	Fair	Poor	Total
Couples' parents of same class	44	24	32	100	44.0	24.0	32.0	100.0
Couples' parents 1 class apart	48	30	30	108	44.4	27.8	27.8	100.0
Couples' parents more than 1 class apart	25	16	20	61	41.0	26.2	32.8	100.0
Total	117	70	82	269				

In the case of differences in class background the results were unexpected. Table 5 shows no relationship of the adjustment scores of the spouses to the social distance of their respective parents. That is, the adjustment of the spouses does not seem to be affected by the fact that their parental background is of the same or different class level.

The results presented in Table 5 indicate that the difference in the parental social status *per se* does not affect the marital adjustment of the spouses. This is a direct contradiction of the finding of Burgess and Cottrell that the closer the similarity of the family background, the better the marital adjustment of the spouses.[23] Why the difference? It is important to note that Burgess and Cottrell used a method of estimating parental status level which is different from the one used in this study. The latter uses Warner's social class concept and relies largely on the Index of Status Characteristics with the items: occupation, source of income, and education. Burgess and Cottrell used a numerical index of similarity in family backgrounds based on the weighted items: parents' religious preference, their church participation, their education, the father's occupation, the respondent's rating of their economic status, and the respondent's rating of their social status. The last two items are subjective ratings which appeared rather unreliable when compared to the objective data provided

22 Burgess and Cottrell, *op. cit.*, p. 341.
23 Burgess and Cottrell, *op. cit.*, pp. 82-85.

by the schedules. Some attempt was made to examine the biases in the methods of determining status level to account for the apparently contradictory results. This examination was inconclusive. The difference in the results of the two studies probably lies in a difference of classification of occupation, education, and religion and a different weighting of these factors.

Social Mobility and Marital Adjustment

Social mobility is defined in this study as the distance which a subject moved from his parents in terms of social class level. Thus, if the subject were of the same social class as his parents at the time of his marriage, he is classed as "non-mobile," if he is of a higher social class he is "upward mobile," and if he is of a lower social class he is "downward mobile." In over half the cases in which the mobility pattern could be identified one or both of the spouses were upward mobile. This high proportion probably results from the fact that Burgess and Cottrell included in their study a large number of subjects with advanced education.

The high proportion of cases including downward mobile subjects is harder to explain. Although no accurate estimates of downward mobility have been made, the 20 per cent of cases involving downward mobility in this study seems exceptionally high. It is likely that some of these are cases of "age-graded" mobility.[24] An example of age-graded mobility is the situation in which a man has established himself in a high status level, but his son is forced to start his business or professional career at a lower occupational level. The schedule data were not adequate for distinguishing between age-graded and permanently downgraded persons, so they had to be lumped together in the "downward mobile" category. This fact should be kept in mind wherever the downward mobile group is used in later analysis.

Earlier in this paper the problem was posed: How does mobility prior to marriage affect later adjustment? Does the stress of this earlier culture shift make marital adjustment more difficult? In order to examine this problem the major mobility groups—non-mobile, upward mobile, and downward mobile—were compared to the total group.

The non-mobile group has almost the same proportional distribution as the total group. The upward mobile group actually shows better adjustment than the total group, although this difference proves to be nonsignificant when the Chi-square test is applied. The stress of the earlier upward culture shift does not appear to affect marital adjustment adversely. It must be remembered that at the time of marriage an upward movement in status has already been achieved. The upward mobile per-

[24] McGuire, *op. cit.*, pp. 3, 19.

son has largely or entirely assimilated the values of his new position. If he (or she) marries a person of this new social level, he is likely to make his relatively new position more secure and thus improve his general social adjustment. If he has moved up quickly through the educational and occupational ladder, but has not yet assimilated the values of his new position, his marriage to a person at this level may facilitate the learning of new behavior. In their discusstion of the acculturation of imigrants, Ruesch, Jacobson, and Loeb point out that the migrant who marries a native partner acculturates much faster than his fellow migrants. "Constant exposure to a model, and reward in terms of affection, apparently acclerate the acculturation process." [25]

The downward mobile group, on the other hand, shows a distribution of significantly lower scores ($P < 0.02$) than the total group, despite the probable dilution of "age-graded mobility" cases.[26]

The tables in the previous section indicated that spouses of the same social class level at the time of marriage scored higher in adjustment than those who were at different levels at marriage. Does this hold true regardless of the spouses' mobility patterns or do the mobility patterns contribute unequally to this relationship? To examine this question the cases within each of the major mobility groups were divided into those who were at the same level at marriage and those who were at different levels.

The "non-mobile" and "one spouse upward mobile" categories show a marked tendency for same-class marriages to have good adjustment compared with cross-class marriages. The Chi-square test shows that this relationship is significant at the 1 per cent level. The "both spouses upward mobile" group shows the same relationship, but the number of cases is too small for statistical analysis. The "downward mobile" group, on the other hand, shows no marked difference between the same-class and cross-class marriages. The slight trend toward better adjustment favoring the same-class marriages proves to be of low significance ($0.30 < P < 0.50$). It appears that the unfavorable influence of downward mobility upon marital adjustment is so strong that the relationship of the class levels of the spouses is relatively unimportant.

Breaking the total group into a large number of mobility patterns left so few cases in each category that a further breakdown into "husband

[25] Ruesch, Jacobson, and Loeb, *op. cit.*, p. 23.

[26] The strong social disapproval of downward mobility may make the intimate relationship of marriage less stable. More likely, perhaps, the downward mobile person is apt to be a rebel against convention, and to be a person who rejects responsibilities. His downward mobility would then be a symptom of difficulties in maintaining stable emotional relationships with others. In this case, the unsatisfactory marital relationship would be only one reflection of a general personality pattern.

high" and "wife high" or "husband mobile" and "wife mobile" groups would make statistical analysis impossible. Nevertheless, some of the figures suggest explanations of relationships discussed earlier.

In the previous section it was noted that in cross-class marriages adjustment generally appeared poorer when the wife was of the higher class, than when the husband was of the higher class. The different mobility types are found to contribute very unequally to this relationship.

In the category "one spouse upward mobile, passing level of the other" the distribution is as follows:

	Good	Fair	Poor
Husband mobile	12	7	7
Wife mobile	2	2	8

Although the numbers are small the difference in distribution of "husband mobile" and "wife mobile" cases is extreme. When the wife is upward mobile and of higher class at marriage, the adjustment is markedly poor. That this poor adjustment is due primarily to the class difference rather than the mobility is shown by comparison with the tabulation of the category "one spouse upward mobile, reaching the level of the other":

	Good	Fair	Poor
Husband mobile	25	8	4
Wife mobile	15	5	0

In this tabulation the cases where the wife alone was upward mobile show no skewing toward poor adjustment; in fact, not a single one of the twenty cases falls in the "poor" classification. Also, in those cross-class marriages where there was no upward mobility on the part of either spouse before marriage, the distribution of adjustment scores is about the same for the husband-high and wife-high cases.[27]

Speculations and Suggestions for Future Research

In this paper marital adjustment has been examined in terms of certain relationships with the social class levels and mobility status of the spouses at marriage. Of these factors the one that appears to be of primary importance is the similarity or dissimilarity of the social class level of the

[27] This might suggest that women who were upward mobile before marriage find the adjustment to a lower status husband more difficult than those who were not upward mobile. We may speculate that a young woman who moves upward in class status before marriage by means of the educational and occupational ladder is likely to desire a role other than (or in addition to) that of housewife and mother. If she marries a man of lower status, her position will appear to be one of superiority over her husband. Since most men in our culture find such a position ego-shattering and since a downward shift in status on the part of the originally upward-mobile wife is similarly difficult, such a marriage is likely to prove unsatisfactory.

spouses at marriage. In other words the chief question is: Are the spouses of the same social class or of different class levels at marriage? The former case is favorable to good adjustment, the latter unfavorable, in the population studied here. Whether the spouses are non-mobile or one or both are upward mobile, whether they are of the higher or lower classes, whether their parents are of the same or different social class levels, the most important factor is still the similarity of the class level of husband and wife at the time of marriage.

This does not mean that the other factors can be ignored. The social class level of the husband and wife in itself is related to adjustment. The subjects show progressively higher adjustment scores as we go up the social class scale. The values of the various social classes concerning marriage and family life may differ in such a way as to make for better adjustment between the spouses at the higher levels. Perhaps Burgess's concept of "companionship" [28] in marriage is found more often at the higher class levels. Downward mobility has a strong unfavorable influence on marital adjustment which tends to obscure other factors. Since in these cases the person was already downward mobile at the time of marriage, it is possible that the person is rebellious and rejects responsibility. His marriage relationships will therefore be unstable. Since the downward mobile group in this study is almost certainly diluted with cases of age-graded mobility, the unfavorable effect of downward mobility is probably even greater than appears in the analysis presented in this paper.

Surprisingly, difference in the social class background of the spouses (that is, the social class of their respective parents), does not appear to affect their adjustment. Neither does the husband's or wife's social class background *per se*, except for a possible unfavorable effect in the case of lower-lower class parents. Of course, the class level of the parents has an indirect influence insofar as it affects the class levels of the spouses. But the actual levels of the spouses at marriage, whether they have been inherited directly from the parents or have been moved into through some mobility route, seem to determine the success of the marriage.

Our data show this pattern, but do not explain it. Why should it exist? In attempting to account for it, it seems reasonable to assume that it is the present, operating values of the husband and the wife which determine how they behave and how they evaluate each other's behavior. Thus, it would not be the sociological fact of their social status which makes them happy or unhappy. Rather, it would be the class-typical day-to-day behavior which would tend to harmonize in the case of same-class marriages, and conflict in cross-class marriages.

[28] Ernest W. Burgess and Harvey J. Locke, *The Family,* New York, American Book Co., 1945.

These data further suggest that there is no necessary, mechanical inheritance of values from the parental family. Instead, they suggest that it is possible to learn a whole way of life which is different from that of the family one is born into; and to do it successfully, by the standards of the new social group. Thus it is not some mysterious, automatic reproduction of a sociological pattern; not some inexplicable, but inescapable, "background factor" which determines adult behavior and adjustment. Rather, the evidence indicates, as does so much other evidence, that the socially significant aspects of human behavior are largely learned. While the impress of childhood training is a powerful influence (and many of our mobile subjects may have been trained by a father or mother who envisioned and encouraged upward mobility) it remains that much of the acquisition of values and behaviors could and probably did occur outside the parental home. In the case of our well-adjusted subjects with disparate backgrounds but similar class-status at the time of marriage, some such process seems to be the most likely explanation.

Thus, a fatalistic prediction that the children of Park Avenue and Railroad Street could not marry happily appears untrue. Some of them can and do learn to be a different kind of person than their parents were, and live the new role successfully with a marriage partner of the new class.

On the other hand, it may be that the ability or willingness to learn a new pattern of life declines after marriage. Those subjects who had been mobile, but married a person of a different social class (as of the time of marriage), did not show the same success in adapting to their marriage that they displayed in adapting to their new social position.

It may be that one's pattern of behavior is largely set by the time one marries. *On the average* (for this is a statistical deduction that should not be applied uncritically to a specific case), it may be that it is not nearly so easy to learn new behavior patterns after marriage as it was in the earlier years. This might be a function of increasing psychological rigidity with age.

There is another explanation, however, which may fit the facts better. We know that there are cases in which husband and wife jointly move up the social ladder *after* they are married. The ability to achieve the complex learning this requires does not disappear at the point of the nuptial ceremony. Acculturation to a new group can still occur. In a sense, the adjustment to a marriage partner might also be considered an acculturation process, insofar as it involves the modification of behaviors and attitudes. Yet, our data point out that the same people who successfully acculturate to a new social class find it harder to adapt to a spouse of a different social status. Cross-class marriages, even among the mobile, are a poorer risk than same-class marriages. It may be that the motivation to adapt to a higher social class is actually stronger, and that the rewards appear more desirable to the individual, than is the case in adapting to the way of life of one's spouse.* This is an unromantic explanation, but it seems very possible. Put bluntly, people may be more willing to change themselves in order to be successfully mobile than to make the changes necessary for a satisfying, stable marriage.

If nothing more, these data suggest certain deep differences among the different social classes; deep enough to make it relatively hard for two people of different classes to live together happily as man and wife. To illustrate, a person who believes in accumulating property and providing economically for the future (middle-class, especially upper-middle) would scarcely be able to agree in money matters with a person who prefers to spend all his money for immediate satisfactions (lower class, especially lower-lower). A person who seeks rather compulsively to impose "proper" behavior and attitudes on his children (middle class) would have difficulty in agreeing on child rearing problems with a person who had a more indulgent approach (lower class).[29] *

As usual, a relatively small study such as this, using only statistical comparisons of groups, raises more questions than it answers. One hopes that they are different questions, based on a larger and clearer fund of knowledge than one had at the outset. The next step would preferably be an intensive study of married individuals, gathering the fullest possible information about childhood experiences and training, and the later influences and events that ultimately produce the adult behavior-value pattern present at the time of marriage. Further, we would need to know the crucial behaviors and attitudes that each partner shows within the successful and unsuccessful marriage. . . .

* This point is considered in greater detail by Whyte in "The Wife Problem"—*Eds.*

[29] Of course, middle-class parents often hold a conscious philosophy of permissiveness, but the weight they give to "proper training," in practice, still contrasts markedly with lower-class parents.

* On this topic see chapter 11 above—*Eds.*

CLASS DIFFERENCES IN FAMILY STABILITY

August B. Hollingshead

Adapted from *The Annals of the American Academy of Political and Social Science,* 272 (1950), 39-46.

Sociologists in recent years have become aware of the interdependence that exists between the family and status systems in American society, but no studies have been focused on the analysis of the problem of class differences in family stability. Consequently, there is no comprehensive body of either quantitative or qualitative data that we may draw upon for a statement of similarities and differences in family stability and instability in the several classes found in our society. Official city, county, state, and national statistics on marriage and divorce do not recognize the existence of social classes, so these data are not appropriate for our purposes. In view of these limitations this paper will merely outline some of the major differences in family stability revealed by studies of social stratification at the community level.[1] However, before we turn to a discussion of the problem of family stability and the status structure, a few paragraphs of theoretical orientation are in order.

Relation of Family and Class Systems

The nexus between the family and class systems arises from the fact that every individual is simultaneously a member of both systems. He is created in the family and placed in the class system whether he wills it or not. However, the functions of the two systems are essentially different; the family is the procreative and primary training institution, whereas the class system functions as a ranking device. The two systems are interwoven at many points in ways that are too intricate for us to unravel here. It is sufficient for present purposes to point out that each individual's original position in the class system is ascribed to him on the basis of a combination of social and biological characteristics inherited from his family

---

[1] W. Lloyd Warner and P. S. Lunt, *The Social Life of a Modern Community,* New Haven, Yale University Press, 1941, pp. 60-61, 92-104; James West, *Plainville, U.S.A.,* New York, Columbia University Press, 1945, pp. 57-69, 115-41; Allison Davis, Burleigh B. Gardner, and Mary R. Gardner, *Deep South,* Chicago, University of Chicago Press, 1941, pp. 59-136; August B. Hollingshead, *Elmtown's Youth,* New York, John Wiley and Sons, 1949, pp. 66-126, 335-88, 414-36; August B. Hollingshead, "Class and Kinship in a Middle Western Community," *American Sociological Review,* vol. 14 (Aug. 1949), pp. 469-75.

through genetic and social processes. This position may be modified, and in some cases changed sharply, during the course of the individual's life; but the point of origin in the status system for every individual is the family into which he is born.

The nuclear group of husband, wife, and dependent children constitues the primary family and common household unit throughout our society. This group normally passes through a family cycle[2] which begins with marriage and extends through the childbearing and childrearing years and on into the old age of the parental pair. It is the maintenance of the family cycle from marriage to old age that we will take as our criterion of a stable family. Each marriage of a man and a woman brings into being a new family cycle. Upon the birth of their first child the nuclear pair becomes a family of procreation, but for the child this family of origin is his family of orientation.* Thus, each individual who marries and rears children has a family of orientation and a family of procreation. He also has an ascribed status which he inherits from his family of orientation, and an achieved status which he acquires in the course of his life. His achieved status may be different from his ascribed status, but not necessarily, particularly from the viewpoint of class position; but his family of procreation, of necessity, is different from his family of orientation. In the case of a man, his achieved status normally becomes the status of his wife and of his children during their early years.

Each nuclear family is related to a number of other nuclear families by consanguineal and affinal ties.† Also, each family in the kin group occupies a position in the status system. All nuclear families in a kin group may be in the same class or may be in different class positions from others. The latter situation is produced by mobility on the part of some individual families, while other families remain in the approximate status position ascribed to them by their family of orientation. This movement of the individual nuclear family in the status system, while it is approved, and often lauded as "the American way," has important effects on kin group relations.[3]

With these considerations in mind, we turn to the discussion of class and family stability. We wish to warn the reader, however, that the statements presented in the following analysis are based on a few community

---

[2] For a discussion of this concept see Paul C. Glick, "The Life Cycle of the Family."

  * See George Peter Murdock, "Structure and Functions of the Family"—*Eds.*

  † See George Peter Murdock, *op. cit.*—*Eds.*

[3] The effects of a nuclear family's mobility, both upward and downward, on its relations to the kin group will be explored in a forthcoming paper by the author, published elsewhere.

studies in different parts of the Nation, and therefore the bases of the generalizations are fragmentary; heuristic observations are made in the hope that they will draw attention to this area of the social structure, and that they will give readers new insight into these facets of our society.

The Upper Class

Families in the upper class may be divided into two categories on the basis of the length of time they have occupied upper-class position: (1) *established* families, which have been in the upper class for two or more generations; and (2) *new* families, which have achieved their position through the success of the present adult generation.

Who one's ancestors were, and who one's relatives are, count for more in the established family group than what one has achieved in one's own lifetime. "Background" is stressed most heavily when it comes to the crucial question of whom a member may or may not marry, for marriage is the institution that determines membership in the family group. Indeed, one of the perennial problems of the established family is the control of the marriage choices of its young men. Young women can be controlled more easily than young men, because of the sheltered life they lead and their passive role in courtship. The passivity of the upper-class female, coupled with sex exploitation of females from lower social positions by upper-class males that sometimes leads to marriage, results in a considerable number of old maids in established upper-class families. Strong emphasis on family background is accompanied by the selection of marriage mates from within the old-family group in an exceptionally high percentage of cases, and if not from the old-family group, then from the new-family segment of the upper class. The degree of kinship solidarity, combined with intraclass marriages, found in this level results in a high order of stability in the upper class, in the extended kin group, and in the nuclear family within it.

The stablished upper-class family is basically an extended kin group, solidified by lineage and a heritage of common experience in a communal setting. A complicated network of consanguineal and affinal ties unites nuclear families of orientation and procreation into an in-group that rallies when its position is threatened by the behavior of one of its members, particularly where out-marriage is involved. Each nuclear family usually maintains a separate household, but it does not conceive of itself as a unit apart from the larger kin group. The nuclear family is viewed as only a part of a broader kin group that includes the consanguineal descendants of a known ancestral pair, plus kin that have been brought into the group by marriage.

An important factor in the extended established family's ability to maintain its position through several generations is its economic security. Usually a number of different nuclear families within a kin group are supported, in part at least, by income from a family estate held in trust. Also, because of the practice of intramarriage it is not unusual for a family to be the beneficiary. . . .

The In-Group Marriage Test. The tradition relative to Protestant intra-upper-class marriages had a severe test in recent years. A son in one family, who had spent four years in the armed services in the late war, asked a middle-class Catholic girl to marry him. The engagement was announced by the girl's family, to the consternation of the Scotts.[4] The Scotts immediately brought pressure on the boy to "break off the affair." His mother "bristled" at the very idea of her son's marriage; his father "had a talk with him"; his 84-year-old paternal grandmother snorted, "A Scott marry a Flaherty, never!" A great-aunt remarked icily, "No Scott is dissolute enough to *have* to marry a Flaherty." After the first shock of indignation had passed, the young man was told he was welcome in "any Scott home" without that "Flaherty flip." A few weeks later his maternal grandfather told him he would be disinherited if he "demeaned" himself by marrying "that girl."

After several months of family and class pressure against the marriage, the young man "saw his error" and broke the engagement. A year later he married a family-approved "nice" girl from one of the other "old" families in the city. Today he is assistant cashier in his wife's family's bank, and his father is building him a fine suburban home.

Nancy Flaherty, when the storm broke over her engagement, quit her job as a secretary in an insurance office. A few weeks later she left home to seek a job in another city. After the engagement was broken she quit this job and went to New York City. Today she is unmarried, living alone and working in New York.

This case illustrates a number of characteristics typical of the established upper-class family. It is stable, extended, tends to pull together when its position is threatened—in this instance by an out-marriage—exerts powerful controls on its members to ensure that their behavior conforms to family and class codes, and provides for its members economically by trust funds and appropriate positions.

The New Family. The new upper-class family is characterized most decisively by phenomenal economic success during a short interval of time. Its meteoric rise in the economic system is normally the personal triumph of the money-maker. While its head is busy making a "million bucks," the family acquires the purchasable symbols associated with the wealthy

[4] All names are pseudonyms; they are used because some of the quotations have meaning only in terms of them.

American family: a large house, fine furniture, big automobiles, and expensive clothes. The new tycoon knows the power of money in the market place, and he often attempts to buy a high position in the status system. The new family is able to meet *the means test,* but not *the lineage test* of the established families. Consequently, it is generally systematically excluded from membership in the most prestigeful cliques and associations in the community. This is resented, especially by the wife and children; less often by the tycoon.

The new family is very unstable in comparison with the established family. It lacks the security of accepted position at the top of the local status system—a position that will come only through time; it cannot be purchased. The stabilizing influence exerted on the deviant individual by an extended family group, as well as friends, is absent. (Many upwardly mobile families break with their kin group as part of the price they pay for their mobility.) Then, too, the new family is composed of adults who are self-directing, full of initiative, believe in the freedom of the individual, and rely upon themselves rather than upon a kin group. The result is, speaking broadly, conspicuous expenditure, fast living, insecurity, and family instability. Thus, we find divorces broken homes, alcoholism, and other symptoms of disorganization in a large number of new families. Because new families are so conspicuous in their consumption and behavior they become, in the judgment of the general population, symbolic of upper-class actions and values, much to the detriment, and resentment, of established families.

The Middle Classes*

The nuclear upper-middle-class family, composed of husband, wife, and two or three dependent children during the major years of the family cycle, is a very stable unit in comparison with the new upper-class family and the working-class family. Divorce is rare, desertion by the husband or wife is most infrequent, and premature death rates are low.

During the past half-century changes that have taken place in American society have created a demand for technically trained personnel in such large numbers that the old middle class could not provide enough recruits to fill the new positions. Concomitantly, our educational institutions expanded enormously to meet the need for professionally, scientifically, and administratively trained personnel. A vast area of opportunity opened for boys and girls in the lower-middle and working classes to move upward in the economic and status structures. Thus, the majority of upper-middle-class persons now above thirty-five years of age are upward mo-

* Whyte discusses the effects of social and spatial mobility on the role of the wife in up-moving middle-class families. See William F. Whyte, Jr., "The Wife Problem"—*Eds.*

bile. Their mobility has been made possible by education, self-discipline, and opportunity in the professional and administrative channels of our economic system.

Geographic mobility has been a second concomitant in this process. The man—or woman—who is now in the upper-middle class more often than not left his home community as a young adult to attend college. After his formal schooling was completed he generally took a job in a different community from the one where he was trained, and oftentimes it was in a different one from his home town. If he began his adult work career with a national business firm, the chances are high that he was transferred from one city to another as he moved up the job ladder.

Geographic movement is typical of an upward mobile family, even when it lives out the family cycle in its home community. In a large number of cases, when a mobile couple is newly married, both partners work. The couple often lives in an apartment or flat in a residential area that is not desirable as a permanent residence. As the husband achieves a higher economic status, the new family generally moves to a small single-family house, or a two-family one, farther from the center of the city, where there are yards and trees. Often about this time the wife quits work and the first of two or three children is born. A third or fourth move, some years later, into a six- to eight-room single-family house on a well landscaped lot in the better residential areas of the city or the suburbs normally completes the family's odyssey. While it is moving from house to house, many of its social contacts change as the husband passes through the successive stages of his business or professional career.

Even though there is a high prevalence of social and geographic mobility, and no extended kin group to bring pressure on the family, there is a negligible amount of instability. Self-discipline, the demands of the job, and the moral pressures exerted by friends and associates keep the nuclear family together. The principal family goals are success in business or a profession, a good college or university education for the children, and economic security for the parents in their old age. These goals are realized in the vast majority of cases, and the family is generally a happy, well-knit group.

The Lower-Middle Class. The lower-middle-class family, like the upper-middle, is a stable unit for the most part. In fact, there is no essential difference between these two levels of the status system in so far as family stability is concerned. In Elmtown 85 per cent of the upper-middle (class II) and 82 per cent of the lower-middle families (class III) were intact after fifteen and more years of marriage.[5] Oren found in an industrial city in Connecticut that 93 per cent of the lower-middle families with adoles-

[5] Hollingshead, *Elmtown's Youth, op. cit.,* note 1 *supra,* p. 99.

cent children were unbroken after eighteen and more years of marriage.[6]

Probably a higher proportion of lower-middle-class individuals have achieved their positions through their own efforts than is true of any other status level except the new family group in the upper class. The majority of lower-middle-class adults have come from a working-class background; many have an ethnic background of recent immigrant origin. Through ability, hard work, and an element of luck they have founded small businesses, operated by the family members and a few employees, or acquired some technical training which has enabled them to obtain clerical, sales, and minor administrative posts in industry and government.

The major problems of the lower-middle-class family are connected with the security of its economic position and the education of its children. Parents generally have high educational aspirations for their children, but income limitations often compel them to compromise with less education than they desire, and possibly a different kind from what they would choose. Parents acutely see the need for a good formal education, and they make heavy sacrifices to give their children the educational training that will enable them to take over positions held by persons in the upper-middle class. By stressing education for the child, parents many times unwittingly create conflicts for themselves and their children, because the educational goals they set for the child train him in values that lead him away from his family. This process, while it does not have a direct bearing on the stability of the nuclear family, acts as a divisive factor that splits parents and children apart, as well as brothers and sisters who have received different amounts of education and follow different job channels.

The Working Class

The family cycle is broken prematurely in the working class about twice as frequently as it is in the middle classes. Community studies indicate that from one-fourth to one-third of working-class families are broken by divorce, desertion, and death of a marital partner, after a family of procreation has been started but before it is reared. This generalization does not include families broken before the birth of children or after they leave the parental home. In Elmtown I found that 33 per cent of the working-class families (class IV) had been broken after fifteen and more years of marriage;[7] Oren[8] reported that 29 per cent of his working-class families with adolescent children were broken ones. The norm and the ideal in the working class are a stable family, but broken homes occur

[6] Paul Oren, "Becoming a Milltown American," unpublished doctoral dissertation, Yale University, 1950.

[7] Hollingshead, *Elmtown's Youth, op. cit.,* note 1 *supra,* p. 106.

[8] *Op. cit.,* note 6 *supra.*

with such frequency that most parents realize that they are, along with unemployment, a constant hazard.

Family instability is a product of the conditions under which most working-class families live. In the first place, they are completely dependent on the swings of the business cycle in our wage-price-profits system, for the working-class family is almost invariably supported by wages earned by the hour, the piece, the day, or the week. Ideally its wages are earned by the male head, but in a considerable proportion of families the wife too is employed as a wage earner outside the home. When a working-class wife takes a job it is for a substantial reason, usually necessity, rather than the desire for "a career." [9]

Factors of Stress. The home is the center of family life, and the hope of most working-class families is a single-family dwelling with a yard; but from a fifth to one-half are forced to live in multiple dwelling units with inadequate space for family living. Added to this is the working-class *mos* that one is obligated to give shelter and care in a crisis to a husband's or wife's relatives or to a married child. Thus, in a considerable percentage of these families the home is shared with some relative. Then, too, resources are stringently limited, so when a family is faced with unemployment, illness, and death it must turn to someone for help. In such crises, a relative is called upon in most instances before some public agency. The relative normally has little to offer, but in most cases that little is shared with the family in need, even though grudgingly.

While crises draw family members together, they also act as divisive agents; for when a family has to share its limited living space and meager income with relatives, kin ties are soon strained, often to the breaking point. One family is not able to give aid to another on an extensive scale without impairing its own standard of living; possibly its own security may be jeopardized. In view of this risk, some persons do everything short of absolute refusal to aid a relative in distress; some even violate the "blood is thicker than water" *mos* and refuse to give help when it is requested. This ordinarily results in the permanent destruction of kin ties, but it is justified by the belief that one's own family's needs come first.

Although the principle is stressed here that the working-class family lives very close to the limits of its economic resources at all times, and when a crisis comes its effects upon family stability are profound, we should not overlook the fact that moral, personal, and emotional factors contribute to family instability. It is possible that these factors are as important as the economic ones, but this and other observations made here

[9] See Floyd Dotson, "The Associations of Urban Workers," unpublished doctoral dissertation, Yale University, 1950, for an excellent analysis of urban working-class culture.

need to be verified by field studies. Actually, while we know that the family at this level of the status structure is susceptible to instability, we have little knowledge derived from systematic research to tell us what cultural conditions are associated with unstable, in contrast to stable, families. A carefully planned series of studies of stable and unstable families with class level held constant is needed. Until this is done, we can only guess about the factors which condition stability and instability in family life.

The Lower Class

Lower-class families exhibit the highest prevalence of instability of any class in the status structure. If we view the lower-class family in terms of a continuum, we find at one end stable families throughout the family cycle; at the other end, the nuclear family of a legally wedded husband and wife and dependent children has given way to a reciprocal companionate relationship between a man and a woman. This latter relationship, in most cases, is the result of their personal desire to live together; it is not legally sanctioned. A companionate family is often a complicated one. It may include the natural children of the couple, plus the woman's children from a previous legal or companionate relationship; also there may be dependent children of the man living with the woman. Normally, when the lower-class family is broken, as in the higher classes, the mother keeps the children. However, the mother may desert her "man" for another man, and leave her children with him, her mother or sister, or a social agency. In the Deep South and Elmtown, from 50 to 60 per cent of lower-class family groups are broken once, and often more, by desertion, divorce, death, or separation, often due to imprisonment of the man, between marriage legal or companionate, and its normal dissolution through the marriage of adult children and the death of aged parents.[10]

Economic insecurity is but one of a number of factors that give rise to this amount of instability. Lower-class people are employed in the most menial, the poorest-paid, and the dirtiest jobs; these jobs also tend to be seasonal and cyclical, and of short duration. Moreover, from one-half to two-thirds of the wives are gainfully employed outside the home; in many cases they are the sole support of the family. However, the problem of economic insecurity does not account for amoral behavior that ranges from the flagrant violation of conventional sex mores to open rebellion against formal agencies of social control.

The very nature of our society may be responsible, in large part, for the number, the intensity, and the variety of social problems associated with the lower class. Such cultural values as individualism, wealth, posi-

[10] Davis, Gardner, and Gardner, *Deep South, op. cit.,* note 1 *supra,* pp. 118-36; Hollingshead, *Elmtown's Youth, op. cit.,* note 1 *supra,* pp. 116-20.

tion, and power must be considered in an analysis of social problems from the viewpoint of the class system. Ours is a competitive, acquisitive society where individuals successful in the competitive arena are admired by most other Americans; they achieve positions of prestige and of power desired by many and attained by few. Less successful individuals may struggle as hard but not be able to do more than hold the status in which they were born; their goal may be to avoid the sorry drift toward lower-class existence. Other individuals may fail in the struggle and sink to the bottom. To be sure, some were born there and failed to rise from the unenviable position they inherited at birth. . . .

POWER AND AUTHORITY IN THE FAMILY

Donald M. Wolfe

Adapted from Dorwin Cartwright (ed.), *Studies in Social Power,* Ann Arbor, Michigan, The Institute for Social Research, Univ. of Michigan, 1959, pp. 99-117.

The purpose of this study is to investigate several possible sources of power in the husband-wife relationship and their effects on the family authority structure.[1] This report is divided into four sections. The first offers a conceptual analysis of power and authority in general, and suggests ways in which these concepts may be used in the study of the family. The second presents a diagrammatic model by which to compare various family authority structures. The third section describes procedures employed in an empirical study of family authority structure. The fourth presents the results of that study.

Theoretical Approach

The theoretical orientation upon which this analysis is based is primarily the group dynamics enlargement upon the Lewinian conceptualization of social power [1, 6, 7], and the development of the concept "authority" by Dubin [2].

Definition 1: Power is the potential ability of one person, O, to induce

[1] The data for this study were provided by the Detroit Area Survey of 1954-1955. The Detroit Area Study is an ongoing research program of the Department of Sociology and the Survey Research Center, University of Michigan. . . . Additional findings from this survey relevant to family life may be found in *Husbands and Wives: the Dynamics of Married Living* by R. O. Blood and D. M. Wolfe.

forces on another person, P, toward (or against) movement or change in a given direction, within a given behavior region, at a given time.

In general, O and P may be either persons or groups, but here we shall be concerned only with persons, i.e., the husband and wife in the nuclear family. Implicit in this definition is that this potential—power of O over P—is made up of the maximum forces O can induce on P, over and above the maximum forces P can exert in resistance to O's inductions. In the family setting, the husband may be able to influence his wife with respect to caring for his clothes and use of the car, but she may be better able to influence him with respect to shopping or the use of leisure time. Thus, power is a concept of potential behavior in a social relationship, and allows for variation from one behavioral region to another from time to time.[2]

A number of assumptions are made about the nature of the individual and of interpersonal relations: (a) Every individual is continually attempting to satisfy his needs and desires and to attain his goals. (b) Most of the individual's needs are satisfied and goals attained through social interaction with other persons or groups. (c) During this interaction, there is a continual exchange of "resources" which contribute to the satisfaction of individual needs and to the attainment of individual or group goals.

These resources are personal properties or attributes of the individual which may be physical, intellectual, or emotional in nature (e.g., physical strength, personal appearance, knowledge or skills, or "capacities" for giving and receiving love); or they may be material possessions or status distinctions (intrinsically valuable or instrumental objects, or rights and privileges concomitant with holding a special social position—perhaps an office).

Definition 2: A *resource* is a property of a person or group which can be made available to others as instrumental to the satisfaction of their needs or the attainment of their goals.

A resource then is a property of one person which can be utilized by others and transferred socially to others. The more resources one person

[2] Inherent in this approach is the fact that the power of O over P and the power of P over O are conceptually independent of each other. Both O and P may have considerable power over each other in the same region, or the power of one may be greater than the power of the other in some regions and vice versa in other regions. Or, in some power relationships, the power of O over P and the power of P over O may be small in all regions. In a single power relationship, the power of O over P may differ in magnitude from region to region and from time to time within the same region.

Since power is defined as a social relationship, and not a personal attribute, we should expect the source or basis for this power to be found in both entities or in the interaction between them.

has under his control, the more he can contribute to the need-gratification or goal-attainment of others. Implicit in this definition is the idea that a resource may be instrumental negatively as well as positively, i.e., a resource may also contribute to a deprivation of needs or may act as a barrier to the attainment of goals. Thus a person who is very strong is capable of inflicting bodily injury on others, depriving them of their need for safety from harm. If a person, O, actually makes his resources available to P, then O is said to be "actualizing" his resources. If P uses O's resources to satisfy his own needs, then P is "utilizing" O's resources.

Two conditions are necessary for O to have power over P: (a) P must have needs or goals which he feels can be satisfied or attained with the help of another's resources but not without such help, and (b) P must perceive O as having resources which might be made available to him. O's power over P is based on potential rewards when P perceives that O has resources that P needs, desires, or values, or which have a positive utility for P. It is based on potential punishments when P perceives that O might actualize resources which are negatively valent for P, or which have a negative utility with respect to P's needs, desires, or goals.

We now come to the question of resources which may be a source of power for the husband or wife in the family. The assumption is made that people participate in a marriage relationship in order to satisfy certain needs and desires and to attain certain goals, not necessarily selfish in nature. In this relationship a number of common goals develop (e.g., procreation and socialization of offspring) but these are based on the needs and values of the individual members. So power may be based on resources of the husband or the wife which contribute to movement toward joint goals as well as toward individual goals. These resources may be skills, competence, or knowledge with respect to regions of common goal-directed behavior. No doubt some of the increased power of the wife in American families in recent years is based on the increase of education for the wife in various regions of behavior.

Actually, any conceivable property which can be utilized by other persons may be a source of power. It must be noted, however, that this theoretical approach to power in the family views it as being based on an exchange of needed or desired resources, and not just as a fulfillment of social norms. Norms may enter the picture, however, with respect to the values and needs of the husband and wife. Family goals may be based on a culturally derived value system, and certain resources, especially expertness in certain regions or capacities for love and affection, may be derived from the culture through a socialization process.

Power and authority are similar concepts in that they both deal with the ability of one social entity to influence or affect the behavior of another.

However, authority is a special case having to do with the decision-making process in social groups.

Definition 3: Authority is the ability of one person (or group), O, to make decisions which guide the behavior of another person, P, in a given behavioral region at a given time, where both O and P perceive this ability as O's right.

Authority, as power, is a concept of potential ability involving two social entities, and is limited to a specific region and time. There are two aspects of this conception which need to be stressed. First, it pertains to the making of decisions which have interpersonal implications, and second, it involves the perception of legitimacy and propriety. The term *decision* is taken to mean a process whereby a person considers a set of alternative courses of action in a given situation and selects one of them as the course of action to be followed. In the exercise of authority, O selects a course of action which P is to follow, and communicates this to P. That is, P's behavior is determined by a decision made by O.

In an authority relationship, not only can O make decisions concerning P's behavior, but both P and O perceive that O has a legitimate right to do so and that P has an obligation to comply with these decisions. The subordinate, P, "holds in abeyance his own critical faculties for choosing between (behavioral) alternatives and uses the formal criterion of the receipt of a command or signal as his basis for choice" (8, p. 189). The concept of authority as used here is very nearly the same as one which French and Raven call "legitimate power." "Legitimate power of O/P is defined as that power which stems from internalized values in P which dictate that O has a legitimate right to influence P and that P has an obligation to accept his influence." . . . They conceive of legitimate power as being (a) culturally derived, (b) situationally required, (c) determined by social structure, or (d) designated by a legitimizing agent.

Power is seen here as an aspect of an informal social relationship based on the ability of one person to contribute to the gratification or deprivation of another's needs. But authority is an aspect of the formal structure of a group based on the role prescriptions and founded in the norm system of the group.

In some families a certain pattern of authority may be based on the mores of the society. It is "traditional" for the husband to make decisions in certain areas. In other families the wife may be more competent to move the family toward its goals, and therefore both she and her husband feel that she should make the decisions in the regions pertinent to those goals; thus her authority is legitimized by family standards or norms. Still other families may set up an authority relationship which they feel is legitimized by the state in giving a marriage license or by the church official who married them. It seems likely that when a couple first marries, societal norms will dictate which of them will make decisions in which regions. That is, there are certain pressures toward establishing a pattern of authority similar to an "ideal pattern" for that society.

From one family to another, roles and (internal) norms may differ, and even within the same family the norms may change and role prescriptions may be altered. For example, the goals and values of the family may change from the time of marriage to the time the children are grown and leave home. With these changes may come a change in the authority relationship of the husband and wife, but at any one time the authority relationship is supported by the family's own set of norms—the family "sees" the authority pattern as being "right."

Power, on the other hand, is seen as being independent of roles. Anyone may have needs or goals or may have resources valued by others regardless of his roles. However, the norms of the family may set some limits as to the kinds of influence attempts which are acceptable.

The traditional cultural norms may prescribe that the husband shall make most of the decisions in most family regions and that the wife shall behave in accordance with these decisions, and such a pattern may at first be established in the family. However, it may be that the husband has certain strong needs for which the wife has resources for gratification, thus giving her increased power. For example, the husband may be seen as the one who should make the final decisions with respect to expenditures for major purchases for the family, yet because of his insecurity in financial matters and because of her greater competence (derived from training or experience) she may be better able to advance the family toward its financial goals and she may derive more power than he in this region where he has the authority.

In order to maintain authority in an ongoing group, the authority figure must have a certain minimum of power with which to enforce conformity with his decisions—i.e., he must control certain resources which are necessary to need-gratification of those under him. In most formal organizations, such resources are made available to persons in positions of authority so that they will have power to back up their decisions. In the modern American family there is considerable variation in role structure and a high degree of ambiguity regarding the norms of authority relationships. Adequate rationale is readily available to support nearly any pattern of authority between husband and wife. Therefore we shall assume that if either the wife or the husband has a high degree of power over the other, the power figure will become accepted as the decision-maker in his regions of power and will achieve dominance in the authority relationship.

Thus, various sources of power may become bases of authority if the family norms regulating the authority structure change.

Types of Family Authority Relationship

In the family behavioral field, the husband has a range of authority, the wife has a range of authority, and a range of authority is shared. There may also be regions in which neither the husband nor the wife has authority over the other. Authority relationships may differ from family to family in at least two respects: the extent of the ranges of authority of the husband and wife, and the extent of the shared range of authority. The husband's range of authority may be larger than, equal to, or smaller than the wife's range.

Definition 4: The *relative authority* (RA) of husband and wife is the ratio of the wife's range of authority to the husband's range of authority.

If the wife's range of authority is larger than her husband's, she has more relative authority than he. If his range is larger, he has more relative authority. If the two ranges are of equal size, there is a balance of authority in the family.

Definition 5: The *degree of shared authority* (DS) in the family is the proportion of behavioral regions of the family field which are in the shared range of authority.

DS may vary from zero to unity—no regions of the family field in the shared range to all regions in the shared range.

Diagram 1 offers a graphic model of the possible distribution of authority relationships on these two dimensions. The ordinate represents the possible RA values, with wife dominant in all regions at the top, and husband dominant in all regions at the bottom. At the center, the wife's and husband's ranges of authority are of equal size. The abscissa indicates the degree of shared authority, with shared authority in all regions on the left extreme, and with the right extreme indicating no shared range of authority. Any point within Diagram 1 represents an authority relationship with a given RA and a given DS.

The triangular shape of the diagram has theoretical significance. A family field in which all regions are in the husband's range of authority must, by definition, have no regions in a shared range. The same is true of a family in which the wife is dominant in all regions. These two cases are represented by the lower right and upper right corners of the triangle, respectively. In like manner, the family field in which all regions are in a shared range of authority must have a balance of relative authority between husband and wife, and in Diagram 1 would be placed at the left extreme of the DS dimension.

   The dotted lines arbitrarily divide the distribution of authority relationships into four types: (a) Wife Dominant is made up of those families in
which the wife's range of authority is considerably larger than her husband's. (b) The Syncratic type is made up of families in which there is
nearly a balance of relative authority and the shared range is equal to or
greater than the combined ranges of the husband and wife. (c) The Autonomic type also has nearly a balance of RA but the husband's and wife's
ranges together are greater than the shared range. (d) The Husband
Dominant type is an authority relationship in which his range of authority
is considerably larger than his wife's.

Method

   The data used in this study are based on personal interviews with wives
in Metropolitan Detroit as part of the Detroit Area Study of the University of Michigan for 1954-1955. Interviews were taken with 731 wives
who are living with their husbands, and thus belong to ongoing family

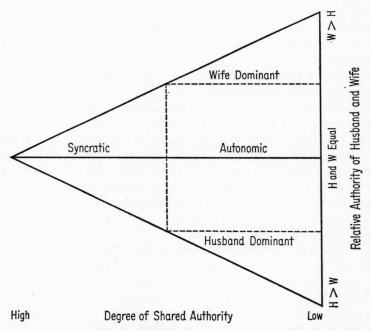

Diagram 1. Theoretical distribution of husband-wife authority relationships. The higher the degree of shared authority the more equal the relative authority of the husband and wife. The broken lines divide the distribution into four authority types: Wife Dominant, Syncratic, Autonomic,
and Husband Dominant.

groups. This represents a multi-stage probability sample of dwelling units in Wayne, Oakland, and Macomb counties. Each interview took approximately one hour and was conducted in the home of the respondent by a trained interviewer, either a graduate student in the Department of Sociology or a member of the staff of the Survey Research Center of the Institute for Social Research.

The dimensions or variables making up Diagram 1 were measured by the use of eight questions concerning who makes the final decision about a variety of problems in various family activities. (See Appendix.) These are in no sense exhaustive of all activity regions in any one family. The eight questions, however, were thought to be pertinent to almost all urban families and sufficiently representative to give an indication of the relative authority of the husband and wife and the degree of shared authority in the family.

The relative authority index (RA) is based on the sum of the numerical codes across the eight questions on decision-making; the lower the RA score, the greater the husband's range; the higher the RA score the greater the wife's range of authority.

The Degree of Shared Authority index (DS) is based on the number of regions in which husband and wife share the authority—scores are determined by the number of questions in the decision set which are answered "husband and wife exactly the same." A DS score of 8 would indicate complete sharing of authority in all regions—completely syncratic. The lower the DS score, the more autocratic or autonomic the authority relationship.

The RA and DS indices together determine the Authority Type. The Wife Dominant group includes all families with an RA score of 29 or more (where the range is from 8 to 40); those in the Husband Dominant group have an RA score of 19 or less. The Syncratic group is made up of all families having an RA score from 20 to 28 inclusive and a DS score of 4 to 8; the Autonomic group have RA scores from 20 to 28 and DS scores of 3 or less.

Note that these are types only in the statistical sense, and are not meant to be "ideal types" in the sense of configurations of behavior. They are determined by the two variables which define the triangular model, and no other variables are inherent in them. The purpose of this research is to explore the relations between this conceptualization and other variables which may be pertinent to family structure and interaction.[3]

[3] This approach differs from that of Herbst (3, 4) in that authority as used here is based only on questions of "who makes decisions" and not on "who carries them out" as well. The considerable influence of the Herbst studies on the present analysis is obvious and deeply appreciated.

Results and Discussion

Several possible sources of authority for the husband or wife and some of the effects of different authority relationships have been explored and will be discussed in the following sections: (a) authority and demographic characteristics of the husband, (b) the effects of the wife's employment on the authority relationship, (c) control of family money and bills, (d) emotional factors in family authority structure, (e) age of wife and authority structure, and (f) authority and marital satisfaction.

Authority and Some Demographic Characteristics of the Husband. A man who is successful in the business world must have resources for goal attainment, some of which could presumably be utilized in the home as well. His wife is more likely to perceive him as a person who is generally competent and who has resources for need-satisfaction and goal-attainment which he can readily actualize within the home as well as elsewhere. Thus she will attribute more power to him and will accept his influence attempts. In families in which the husband is not so successful, the wife is apt to perceive him as being less competent; she will feel that he is less capable of helping the family move toward its goals, and she will therefore attribute less power to him. There are of course many circumstances under which this would not be true; for example, when the wife's needs are very strong, when she perceives success and competence somewhat differently than most, or when the husband has and actualizes resources with respect to his wife but not when away from home. Other things being equal, however, we should expect that husbands who have resources of value in the business world probably control resources of value to the home and will have more power in the home than will less successful husbands.

There are three variables—income, occupational level, and general social status—which are estimates of "success" of the husband in society at large and which probably influence his wife's perception of his general competence. A person with a large income not only has financial resources which may be sources of power but also prove that he is a capable person in general. Occupational level and income are certainly related, but being high in the occupational scale is also evidence of general competence. Professional and managerial workers are generally thought of as being quite capable of handling their affairs—they probably have intellectual and skill resources of high utility for locomotion toward individual and group goals. At least it is likely that their wives so perceive them. If so, these husbands should have relatively high power, and, in turn, considerable authority in the home.

Hypothesis 1: Husbands who are generally successful and prestigeful will have more power and will therefore derive more authority in the home than husbands who are less successful.

Table 1 shows the mean annual income and percentage of families having high (above median) social status for each authority type. The average income ($6437) for Husband Dominant families is significantly higher than the total mean income for all others ($t = 3.1$, $p < .001$), and the Wife Dominant mean income ($4692) is lower than the mean for all other groups combined, just missing the .05 criterion of significance $t = 1.5, p < .07$).

Turning now to occupational level, we find that nearly a third of the husbands in Husband Dominant families are classified as professional, technical, or kindred workers, or as managers, officials and proprietors, a disproportionately high number when the total sample contains less than a quarter in these occupational classes. This goes along with the finding above with respect to income and lends general support to Hypothesis 1. Oddly enough, the husbands in Wife Dominant families have more than the expected number in the managers, officials, and proprietors class. They tend, however, to be self-employed and to have lower incomes and thus perhaps do not contribute to a perception of general competence.

The social status index is an indicator of the husband's position on a general prestige dimension. It is a composite of his position (based on percentile rankings) on four variables—education, occupation, income, and ethnic background. Social status in an upward mobile society is certainly an important factor in the perception of one's general competence. In a sense, high social status and "success" are almost synonymous. Table 1 gives the percentage of families in each authority group who are above the median for social status. The Husband Dominant group is significantly higher than expected by chance, and only 36 per cent of the Wife Dominant families have social status scores above the median for the Detroit Area population. Thus, higher-status husbands tend to have more authority in their homes than lower-status husbands.

TABLE 1.   Mean Annual Income and Social Status by Authority Type

Authority Type	Mean Income	Percentage High Social Status*	N
Husband Dominant	$6437	58%	166
Syncratic	5490	49	201
Autonomic	5378	46	267
Wife Dominant	4692	36	22
Total Sample	5646	50	656

* Families have High Social Status if their index score is above the median for the total sample (see text).

The results of this analysis of income and social status strongly support Hypothesis 1, and there is a tendency in that direction with respect to occupational level. Husbands who are successful in the business world do tend to have more power and authority in the home than do less successful men.

The Effects of the Wife's Employment on Family Authority Relationship. The wife gains in power in at least two ways when she works outside the home. First, she develops resources (financial, intellectual, skill, etc.) which usually are not readily developed in the home. Second, she becomes less dependent upon her husband for the satisfaction of her needs. The contacts she makes outside the home provide resources for satisfying some of her affiliational need, her own success and prestige are less contingent upon those of her husband, and she becomes aware of other sources of aid for her goal-attainment. The husband also finds that his wife now controls many resources which he could utilize for his need-satisfaction and goal-attainment, and he will attribute more power to her. In many homes, the right to make decisions is almost an aspect of the role of provider. "He who pays the piper calls the tune."

Hypothesis 2: Wives who are working or who have worked outside the home have more power and will derive more authority than wives who have not worked.

Putting Hypothesis 2 somewhat differently, more wives in Husband Dominant homes will have never worked outside the home, and fewer will be working now than in any other group. Wives who are now working will be more likely to belong to families with a balance of power and authority or with the relative power in favor of the wife. The results in Table 2 strongly support this prediction. Significantly more wives in Husband Dominant homes than in all others have never worked. Wives who are now working tend to participate in Syncratic or Autonomic authority relationships.

Control of Family Money and Bills. The control of financial resources in the family presents some theoretical problems. On the one hand, the person who controls the purse strings has a resource which may be a basis for power in the home. But it is equally likely that the person who has the greater power and authority, from whatever source, will be able to gain and maintain control over this important resource. As Lasswell and Kaplan (5) hypothesize: "Forms of power and influence are agglutinative: those with some forms tend to acquire other forms also" (p. 97). Since many of the decisions which must be made in carrying out the affairs of the family involve money, it is likely that the norms which determine who will make the major decisions will also establish that spouse as the controller of family finances. Perhaps differential control of other-than-financial resources lies behind the authority relationship in this area.

TABLE 2. Employment Experience of Wife by Authority Type

Employment Experience	Authority Type				
	Husband Dominant	Syncratic	Autonomic	Wife Dominant	All Groups
Wife never employed	40%	25%	21%	27%	27%
Wife previously employed	48	48	55	46	51
Wife now employed	10	25	22	18	20
Not ascertained	2	2	2	9	2
Total	100%	100%	100%	100%	100%
N	166	201	267	22	656

Hypothesis 3: Control of financial resources will be in the hands of the dominant authority figure in the family.

Although Hypothesis 3 predicts that the person in greater authority will also control the financial matters of the family, the direction of causation is dubious. Correlational data, as we have here, can shed little light on this problem. However, Table 3 presents the results of this cross-tabulation. As was predicted, in Husband Dominant homes the husband is much more apt to handle the money and bills, and in Wife Dominant homes the wife is likely to control these resources. In Syncratic homes this responsibility is more often shared than handled by either spouse alone, but the wife often handles this function.

These results confirm Hypothesis 3. Although the relationship is not perfect, the authority figure is much more apt to maintain control over family financial resources than the nondominant spouse. This seems to be one of the sources of power by which the authority figure can enforce his decisions. Since in Autonomic families each member is apt to carry out his own decisions, it is likely that the wife has the authority and is active in those regions which call for handling most of the financial matters.

TABLE 3. Pattern of Handling Money and Bills by Authority Type

Spouse Controlling Financial Resources	Authority Type				
	Husband Dominant	Syncratic	Autonomic	Wife Dominant	Total Sample
Husband more than wife	43%	20%	20%	18%	26%
Husband and wife equal	36	41	30	5	34
Wife more than husband	21	39	50	77	40
Total	100%	100%	100%	100%	100%
N	166	201	267	22	656

Some Emotional Factors in Family Authority Structure. In the theoretical discussion of sources of power, two conditions were specified for the power of O over P: O must have resources of utility to P, and P must have needs or goals which can be satisfied through the utilization of O's resources. So far, we have been looking primarily at possible resources which may be a basis for power, and therefore authority, in the family. Now we shall turn to the second condition and look at some of the needs (especially of the wife) which may be a source of power and authority for the other spouse.

The respondents were asked to rank in importance to them several aspects of marriage: companionship, chance to have children, husband's understanding, love and affection, and standard of living. The average ranking for the total sample was in the order given. In every group, companionship was ranked first, on the average. Standard of living had the lowest mean rank in every group except Wife Dominant, in which case it was ranked second from the bottom.

It is assumed that those wives who rank love and affection first or second have a high need for love and affection. It is also assumed that it is usually up to the husband to satisfy this need. This would be indicative of a situation in which the wife has a strong need and the husband has certain resources, emotional in nature, which can be utilized for the satisfaction of that need. From this situation the husband should derive power which may bring about a change in the authority relationship, especially if his expression of affection is in any way contingent upon her compliant behavior.

Hypothesis 4: Wives who have a strong need for love and affection will have relatively less power, and will have less relative authority, than wives with a weaker need.

Table 4 gives the percentage of wives in each authority type who have a strong need for love and affection (who rank love and affection either first or second). Although there is a slight trend in the predicted direction, the proportion of Husband Dominant wives who are high in this need is not significantly above that of the total population. On the other hand, Wife Dominant wives tend to be very low in their need for love and affection. While the data are consistent with Hypothesis 4, it appears that other factors also need to be considered. Perhaps a strong need for affection is a source of power for the other spouse only under special circumstances, but a weak need enables the person to develop and exercise his own power. That is, a wife who is not seeking emotional support from her husband may feel quite free to exercise her power stemming from other need-resource bases, while a wife who needs continual affectional gratification may fear that this gratification will be threatened if she over-extends her authority. Any such attempt to refine Hypothesis 4 of course needs subsequent empirical validation which cannot be supplied in the present study.

What of the other aspects of marriage which were ranked by the wife? It is assumed that each of them represents a need or goal. Ranking any of them first or second would indicate that it is an important need or goal for the wife. Companionship is an important need to 68 per cent of the wives in the total sample, and all four groups fluctuate within six percentage points of this figure. This seems to be an important need for most wives, but is not an important source of power or authority. This need, unlike many others, may be best satisfied under conditions of equal power. Therefore, a strong imbalance of power might tend to deprive one of satisfaction. If this is true, it would be unlikely that the husband could derive much power from his wife's need for companionship.

TABLE 4. Wife's Need for Love and Affection by Authority Type

Wife's Need for Love and Affection	Authority Type				
	Husband Dominant	Syncratic	Autonomic	Wife Dominant	Total Sample
High need	37%	34%	30%	9%	32%
Low need	61	64	69	86	66
Not ascertained	2	2	1	5	2
Total	100%	100%	100%	100%	100%
N	166	201	267	22	656

Having children is a primary goal for 54 per cent of the wives in the sample, but 68 per cent of the wives in Wife Dominant authority relationships rank this as first or second in importance. Since they play the most active part in having and rearing children, it may be that wives derive more power for themselves from this than do their husbands: they are more capable of attaining this goal without continual help from their husbands than is true for other goals. Wife's need for her husband's understanding is moderate in all groups, and no group deviates from the average for the total population by more than seven percentage points. A high standard of living is stated as an important goal for only 13 per cent of the wives in this sample, with no group differing from this significantly. Thus, a desire for a high standard of living is a seldom expressed need.

Age of Wife and Authority Structure. When children first come to a young couple, the responsibilities of the wife increase and at first she is quite dependent upon her husband for help and for emotional support. However, the husband usually cannot be present in the home during much of the day, and the wife must satisfy her needs and those of her children (this being a goal of hers) without the resources he might contribute.

She must develop resources of her own to cope with the exigencies of child care and development. Therefore, the power of the husband should be high at first, but the wife's relative power should increase as she becomes more and more competent. By the time the children are grown and about to leave the home, the wife's power (from this source) should be at a peak.

The presence of growing children can contribute to the relative power of the wife in another way as well. As we have seen, emotional resources are important to the wife. Since the husband is not always present to actualize his resources in this direction, many mothers turn to their children for partial satisfaction of this need. Love and affection from her children replaces some of that which formerly came from her husband. The wife's need for her husband's emotional resources decreases for a time.

When a wife passes the age of fifty, and her children are about to leave the home or have already left, she has developed her resources to their highest peak. She may have become accustomed to the security and satisfaction of relatively high power and authority in the home. With the

TABLE 5.   Age of Wife by Authority Type

Age of Wife	Authority Type				
	Husband Dominant	Syncratic	Autonomic	Wife Dominant	Total Sample
Under 30 years	32%	20%	27%	9%	26%
30 to 39 years	33	46	30	18	35
40 to 49 years	18	15	24	23	20
50 years and over	17	20	19	50	20
Total	100%	101%	100%	100%	101%
N	166	201	267	22	656

maturation of her children, she has lost what is perhaps her most important function, child rearing, as well as a source of her own need-gratification. At this time, we might expect two contrary trends: first, a return to dependency on her husband for the satisfaction of her emotional needs, and, second, an effort to find a replacement for her lost function and to redirect her resources, and thus power, toward other family regions. The former trend will be successful only if the husband is able and willing to fill her needs. (If not, she may seek satisfaction in her mother-in-law and grandmother roles.) With reference again to the Lasswell and Kaplan hypothesis, we should expect that the wife will have gained power and authority in most family regions and therefore that the second trend will be predominant. Most wives should tend to maintain and increase their authority at least up to the time of their husbands' retirement.

These two changes—an increase in the wife's own resources, and a decrease in her need for those of her husband—should bring an increase in the relative power of the wife. Since this increase in power should in turn bring an increase in her relative authority, we should expect fewer Husband Dominant wives and more Wife Dominant wives in the older age categories.

Hypothesis 5: The relative power of the wife will increase with age, leading to an increase in her authority relative to her husband.

Table 5 shows the age distribution for wives in the various authority groups. The Husband Dominant type shows a predominance of wives under thirty years of age and between thirty and thirty-nine. Syncratic wives are typically in the thirty to thirty-nine category. Autonomic family wives are more evenly distributed with respect to age, with a tendency toward the forty to forty-nine category. And half of the Wife Dominant wives are over fifty years of age.

Authority Patterns and Marital Satisfaction. In predicting the marital satisfaction of the wife, several factors must be taken into account—her ability to satisfy her needs and attain her goals, and her adherence to social norms—to mention only two. Both of these factors are involved in the power and authority relationships between husband and wife. Generally speaking, the more power the wife has, the greater her ability to attain her goals, for she can influence others to this effect. Therefore, we might predict that the more power the wife has, the greater her satisfaction in marriage. However, it is also important for her to integrate with society, to behave within the norms of society as a whole, and to have a marital relationship which is advantageous to both herself and her family. Thus, we should expect the wife, if she wishes to be happy, to desire an authority relationship close to that prescribed by the social norms—either a democratic one or one in which the husband is dominant. This need to conform to the social norm might conflict with the prediction suggested above.

An index of marital satisfaction was constructed on a basis of the ranking of the wife's needs and goals and her satisfaction with her need gratification in the family. From this index it is possible to get an estimate of how well the wife's needs and goals are being attained. The index shows the wife's position along a dimension from low to high satisfaction. The respondents are then coded as either above or below the median score for the whole sample. Table 6 gives the percentage in each authority group who are above the median for marital satisfaction.

Syncratic relationships, in which authority is shared by husband and wife, are most conducive to satisfaction for the wife. In these cases, the wife's power is fairly high, but the authority pattern conforms to societal norms. Autonomic wives are less satisfied than the average, perhaps because there is less interaction between husband and wife, and less opportunity for need-gratification. The Wife Dominant authority relationship coincides with low marital satisfaction, perhaps because of societal pressures, but more likely because lack of satisfaction over long periods of time may lead to a striving on the part of the wife for more power and authority. Thus in this case, low satisfaction probably came before the increase of authority for the wife. A Husband Dominant relationship may lead to high or low satisfaction for the wife, as this group shows a greater average deviation than any other.

TABLE 6. Marital Satisfaction of Wife by Authority Type

Authority Type	Percentage with High Satisfaction	Number of Cases*
Husband dominant	51%	153
Syncratic	61	186
Autonomic	43	256
Wife dominant	26	19
Total sample	50%	614

* Respondents for whom marital satisfaction scores were not ascertained have been excluded from this table so that 50 per cent could be used as a base for comparison. High marital satisfaction is defined here as above the median score for the total sample.

From this we can conclude that power is important but not "all important" in bringing general satisfaction for the wife. Freedom from conflict is also important, as is close adherence to societal norms. Certainly many other factors, not related to power or authority, are also very important in determining satisfaction in marriage for the wife. However, these data suggest an interesting direction for further study.

Summary and Conclusions

The concepts of social power (as based on differential control of resources of value to others for need-satisfaction) and authority (an aspect of role-prescriptions and group norms) were discussed as they pertain to the urban family. Several sources of authority for the husband or wife were explored using questionnaire data from a cross-sectional sample of the Detroit area population. A conceptual scheme for comparing various authority relationships between husband and wife was presented, allowing for a division of the sample into four "types" of family authority structure —Husband Dominant, Syncratic, Autonomic, and Wife Dominant.

1. Husband Dominant families are generally high on mean annual income and on general social status; Wife Dominant families are generally low on these two variables.

2. Wives in Husband Dominant families are less likely ever to have worked

or to be currently working outside the home than wives in any other authority type.

3. The spouse who is the dominant authority figure in the home is generally most apt also to handle the family money and bills.

4. A strong need for love and affection on the part of the wife is a source of power and authority for the husband.

5. The authority of the wife increases over the years of marriage so that Husband Dominant wives are apt to be younger, and Wife Dominant wives are older (half are over fifty years of age).

6. Wives in Syncratic families are likely to be well satisfied with their marriage, while Autonomic and especially Wife Dominant wives are more likely to be low on marital satisfaction.

Appendix

Decision Questions Upon Which RA and DS Scores Are Based

54. In every family somebody has to decide such things as where the family will live and so on. Many couples talk such things over first, but the *final* decision often has to be made by the husband or the wife. For instance, who usually makes the final decision about what car to get?
(SHOW CARD VI)

(CARD VI)	husband always	husband more than wife	husband and wife exactly the same	wife more than husband	wife always
	1	2	3	4	5

55. . . . about whether or not to buy some life insurance

	1	2	3	4	5

56. . . . about what house or apartment to take?

	1	2	3	4	5

57. Who usually makes the final decision about what job your husband should take?

	1	2	3	4	5

58. . . . about whether or not you should go to work or quit work?

	1	2	3	4	5

59. . . . about how much money your family can afford to spend per week on food?

	1	2	3	4	5

60. . . . about what doctor to have when someone is sick?

	1	2	3	4	5

61. . . . and, about where to go on a vacation?

	1	2	3	4	5

REFERENCES

1. Cartwright, D., & Zander, A., Group Dynamics: Research and Theory (Evanston: Row, Peterson, 1953).

2. Dubin, R. (Ed.), Human Relations in Administration (New York: Prentice-Hall, 1951).

3. Herbst, P. G., "The Measurement of Family Relationships," Hum. Relat., 1952, 5, 3-35.

4. Herbst, P. G., & Middleton, Margaret, "The Members of a Family," in Oeser, O. A., & Hammond, S. B., eds., Social

Structure and Personality in a City (New York: Macmillan, 1954).

5. Lasswell, H. D., & Kaplan, A., *Power and Society: A Framework for Political Inquiry* (New Haven: Yale Univer. Press, 1950).

6. Lewin, K., *Field Theory in Social Science* (New York: Harper, 1951).

7. Rosen, S., Lippitt, R., & Levinger, G. K., *The Development and Maintenance of Power in Children's Groups*, a monograph.

8. Simon, H. A., "Authority," in R. Dubin, ed., *Human Relations in Administration* (New York: Prentice-Hall, 1951).

20. MARITAL DISORGANIZATION AND REORGANIZATION: DEATH, DIVORCE, SEPARATION, AND REMARRIAGE

Jacobson has estimated that the United States divorce rate rose gradually from 1860 to 1940, jumped to a peak during the first year after World War II, and then dropped back to a level about eight times higher than it was in 1860.[1] Over the same period he has estimated that the proportion of marriages broken by the deaths of spouses dropped by more than a third.[2] He concludes that the resultant of an eightfold rise in the divorce rate and a one-third reduction in the proportion of marriages broken by death is a decrease over the period 1860-1956 of somewhat under ten percent in the rate of marital dissolutions from both causes.[3] These trends suggest that there should be a decrease in the proportion of brides and grooms who have been widowed and an increase in those who have been divorced. On this point Jacobson's estimates cover the period 1900-1950. He finds the proportion widowed among marrying men fell from 10.3 percent in 1900 to 5.9 in 1950, and among women from 8.3 percent to 6.5. The proportion of newlyweds who had been divorced rose sharply over the same period from 2.3 percent for men and 3.3 percent for women to 16.6 and 16.7 percent respectively.[4]

It appears that for some time there was a popular impression that divorce

[1] Jacobson estimates that the number of divorces per 1000 existing marriages rose from 1.2 in 1860 to 4.0 in 1900, 8.7 in 1940, peaked at 18.2 in 1946, and dropped thereafter to a level of around 10. See Paul H. Jacobson, *American Marriage and Divorce*, New York, Holt, Rinehart and Winston, 1959, p. 142.

[2] His estimates run from 28.4 per 1000 existing marriages in 1860 to 27.7 in 1900, 20.9 in 1940, 19.0 in 1946 and around 17.5 for the period 1950-1956. *Ibid.*

[3] *Ibid.*

[4] *Ibid.*, p. 71.

was a luxury that could be indulged in only by the rich. A parallel impression was that desertion was "the poor man's divorce." More recently the sociological literature has been informing us of a negative correlation between divorce rates and socioeconomic status. Such studies have perforce been based on small sets of local data. The data of the census of 1950 have enabled Karen Hillman to check this relationship on a national basis. With respect to white males, her analysis turns up a clear-cut trend for marital instability (divorces and separations) to correlate negatively with socioeconomic status. However, she sets forth a series of qualifications concerning this conclusion and shows that the relationship is less clear-cut for females and for nonwhites.

One of the most recurrent questions of college students has to do with the outlook for marriages wherein the spouses are of different religious faiths. Landis has studied several thousand marriages of parents of students at a large state school. His analysis concerns the categories of Protestant and Catholic. He concludes that mixed Protestants and Catholic marriages are more likely to run into trouble than are marriages wherein the spouses are of the same faith. The religious training of children is seen as a focus of conflict in mixed marriages. It should be emphasized that, since his data are based on the parents of college students, the findings are biased in not representing childless marriages at all and in underrepresenting families of low socioeconomic status.

Do "forced marriages" account for more than their share of divorces? By his method of "record linkage" Christensen has devised a procedure for identifying couples wherein conception occurred before the wedding ceremony.

Another point of popular concern has been the question as to whether or not the marriages of persons under twenty-one have a greater than average probability of breaking up. Glick's analysis of the ages of women at the time of their first marriages leads to the conclusion that the proportion remarried among women married for the first time at eighteen or nineteen was "about half again as high as that for women who first married at twenty or twenty-one." For women who first married at ages below eighteen the remarriage rate was still higher. In the light of the Hillman and Christensen-Meissner papers, however, it is possible that the correlation between marital instability and age at marriage might disappear if allowance could be made in the analysis for socioeconomic status[5] and premarital pregnancies.

Finally, the question frequently arises as to the outlook for remarriages. Monahan remarks that our literature seems to hold two conflicting views—that remarriages tend to be less stable than first marriages and that they tend to be equally stable. Using records for a three-year period in Iowa he shows that remarriages of divorced persons are much more likely to end in divorce than

[5] There is evidence that average age at first marriage is positively correlated with socioeconomic status. Cf. Robert F. Winch, *The Modern Family*, New York, Holt, Rinehart and Winston, 1952, and the forthcoming revision.

are first marriages. Remarriages of widowed persons, however, are less likely than first marriages to lead to divorce.

MARITAL INSTABILITY AND ITS RELATION TO EDUCATION, INCOME, AND OCCUPATION: AN ANALYSIS BASED ON CENSUS DATA

Karen G. Hillman | Original manuscript.

Introduction

The relationship between socioeconomic status and marital instability has been the subject of considerable discussion among sociologists. Various writers have found an inverse relationship between income levels, educational background and occupational status on the one hand, and marital instability on the other.[1] Goode "supposes" that in the upper socioeconomic strata there are more "external supports" to marriage than in the lower levels. Among these external supports he lists a greater continuity of social relations, more long term investments, greater difference between the husband's and the wife's earning power, and less anonymity in the event of a marital rupture.[2]

It was not until the publication of the census of 1950 that any substantial body of data on this topic became available. Thereupon it became possible to study the problem on a more comprehensive basis than the quite limited and local evidence used in previous studies. The purpose of this paper is to use the United States population as reported in the 1950 census in order to examine the relationship between socioeconomic status and marital instability.

In terms of the operations employed in this study, a person whose marriage has proved unstable is one who was recorded as either "divorced" or "separated" on the day in 1950 when the census-taker entered that person on the record. For fairly obvious reasons education is an index of socioeconomic status of both sexes, irrespective of marital status, but oc-

---

[1] Studies in which this inverse relationship is shown most clearly are the following: H. Ashley Weeks, "Differential Divorce Rates by Occupation," *Social Forces,* 21, 1943, 334-337; August B. Hollingshead, "Class Differences in Family Stability," *American Academy of Political and Social Science,* 272, 1950, 39-46; William A. Kephart, "Occupational Level and Marital Disruption," *American Sociological Review,* 20, 1955, 456-465; William J. Goode, *After Divorce,* Glencoe, Free Press, 1956.

[2] Goode, *op. cit.,* chapter 5.

TABLE 1. Number of Persons in U.S. 25 Years of Age and Over Ever Married,

White Males

Education	Ever Married	Separated	Divorced	Separated per 1000 of Ever Married	Divorced per 1000 of Ever Married	Both Divorced and Separated per 1000 of Ever Married
No school	681,150	18,510	15,510	27.17	22.77	49.94
Elem. 1-4 years	2,516,700	56,550	71,820	22.47	28.54	51.01
5-7 years	5,317,290	102,060	155,610	19.19	29.26	48.45
8 years	7,561,710	106,560	206,610	14.09	27.32	41.41
High School 1-3 yrs.	5,885,820	76,500	166,590	13.00	28.30	41.30
4 yrs.	6,387,240	63,300	156,360	9.91	24.48	34.39
College 1-3 years	2,451,030	22,890	63,450	9.34	25.89	35.23
4 years or more	2,543,760	15,900	44,040	6.25	17.31	23.56
School yrs. not known	922,890	32,760	38,970	35.50	42.22	77.72
Total	34,267,590	495,030	918,960	14.45	26.26	41.47

White Females

Education	Ever Married	Separated	Divorced	Separated per 1000 of Ever Married	Divorced per 1000 of Ever Married	Both Divorced and Separated per 1000 of Ever Married
No school	776,640	16,860	11,220	21.71	14.45	36.16
Elem. 1-4 years	2,245,590	48,840	45,960	21.75	20.47	42.22
5-7 years	5,401,560	114,780	135,390	21.25	25.06	46.31
8 years	7,786,620	129,090	207,570	16.58	26.66	43.24
High School 1-3 yrs.	6,748,380	115,440	239,850	17.11	35.54	52.65
4 yrs.	8,711,670	108,960	289,140	12.51	33.19	45.70
College 1-3 yrs.	2,823,570	28,110	104,700	9.96	37.08	47.04
4 years or more	1,647,060	14,310	58,500	8.69	35.52	44.21
School yrs. not known	778,681	18,330	28,650	23.54	36.79	60.33
Total	36,919,770	594,720	1,120,980	16.11	30.36	46.47

* Compiled from U. S. Bureau of the Census, *U. S. Census of Population: 1950, Special Reports,*

cupation and income are available as indexes only for males. With respect to education and income it is possible to present the data for whites and nonwhites separately. This breakdown is not available for the data on occupations.

Results

Table 1 shows the number of divorced and separated per 1000 persons ever married, by sex and race for several levels of education. The most obvious fact in the table is that the rates of marital instability for nonwhites run between two and three times as high as for whites. Almost

Separated, and Divorced, 1950, by Education, Sex, and Race*

Nonwhite Males

Ever Married	Separated	Divorced	Separated per 1000 of Ever Married	Divorced per 1000 of Ever Married	Both Divorced and Separated per 1000 of Ever Married
266,520	20,340	4,080	76.32	15.30	91,62
1,006,440	84,510	18,750	83.99	18.63	102.62
917,670	78,690	23,040	85.74	25.11	110.85
379,320	32,040	12,120	84.45	31.95	116.40
393,120	34,020	14,010	86.54	35.64	122.18
240,390	17,220	9,180	71.63	38.19	109.82
89,670	5,310	3,630	59.22	40.48	99.70
67,320	2,640	2,040	39.22	30.30	69.52
130,770	14,010	4,200	107.13	32.12	139.25
3,491,220	288,780	91,050	82.71	26.08	108.79

Nonwhite Females

Ever Married	Separated	Divorced	Separated per 1000 of Ever Married	Divorced per 1000 of Ever Married	Both Divorced and Separated per 1000 of Ever Married
214,980	13,920	3,210	64.75	14.93	79.68
891,420	87,090	17,910	97.70	20.09	97.79
1,137,870	123,960	34,980	108.94	30.74	139.68
478,350	55,140	21,240	115.27	44.40	159.67
557,430	69,480	28,950	124.64	51.93	176.57
330,150	33,330	19,020	100.95	57.61	158.56
111,870	8,490	6,480	75.89	57.92	133.81
79,020	4,500	5,100	56.95	64.54	121.49
101,280	11,760	3,750	116.11	37.02	153.13
3,902,370	407,670	140,640	104.47	36.04	140.51

Education, table 8, U .S. Government Printing Office, Washington, D. C., 1953.

equally obvious is the fact that among whites divorces account for well over half of the marital instability whereas among the nonwhites divorces account for only about a third. Table 2 presents divorce and separation rates for white and nonwhite men by income, and Table 3 does so by occupational grouping.

To determine whether or not these data show a negative correlation between socioeconomic status and marital instability rank correlations were computed. With respect to education eight ranks were used; those reported as "school years not known" were omitted. With respect to occupation nine ranks were used in the order shown in Table 3 from "Profes-

TABLE 2.   Number of Men 14 Years of Age and Older in U. S. Population Ever

| | White Men | | | | | |
Income	Ever Married	Separated	Divorced	Separated per 1000 of Ever Married	Divorced per 1000 of Ever Married	Both Divorced and Separated per 1000 of Ever Married
No income	1,794,750	46,950	64,800	26.16	36.10	62.26
$1-$999	4,242,940	119,940	188,580	28.27	44.44	72.71
$1000-$1999	5,211,410	98,130	167,190	18.82	32.08	50.90
$2000-$2999	8,334,000	131,220	255,150	15.74	30.62	46.36
$3000-$3999	7,320,240	70,740	167,610	9.66	22.90	32.56
$4000 and over	7,261,920	5,100	18,090	.70	2.49	3.19
Total	34,165,260	472,080	861,420	13.82	25.21	39.03

* Compiled from: *U. S. Bureau of the Census, U. S. Census of Population: 1950, Special Re-*

sional . . ." to "Laborers . . ." The farm occupations and "occupations not listed" (last three rows) were omitted. The results appear in Table 4.

It is seen that in general the data support the negative relationship. For white men all the correlations are negative, irrespective of the index of socio-economic status or of the nature of the instability. For separations all the correlations are negative, irrespective of sex or the index of socio-economic status. In addition to negative coefficients, however, Table 4 has some in the high positive range and also two fairly close to zero. (The values of the latter two are actually —.17.) One of these, for nonwhite

TABLE 3.   Number of Men 14 Years of Age and Over in U. S. Population, 1950, Ever Married and Divorced, by Occupational Grouping*

Occupational Grouping	Ever Married	Divorced	Divorced per 1000 of Ever Married
Professional, technical, & kindred workers	2,493,870	46,110	18.49
Managers, officials, & proprietors exc. farm	3,979,230	66,030	16.59
Sales workers	2,074,230	50,250	24.22
Clerical & kindred workers	1,991,190	51,180	25.70
Craftsmen, foremen, & kindred workers	6,945,840	167,610	24.15
Operatives & kindred workers	6,859,560	179,550	26.18
Service workers exc. private household	1,933,680	81,900	42.35
Private household workers	57,150	2,700	47.24
Laborers exc. farm & mine	2,630,340	86,280	32.80
Farmers & farm managers	3,735,150	29,370	7.86
Farm laborers & foremen	923,070	37,620	40.76
Occupations not reported	516,030	23,850	46.22
Total	34,139,340	822,450	24.09

* Compiled from: U. S. Bureau of the Census, *U. S. Census of Population, 1950: Special Reports, Occupational Characteristics,* Table 8, Washington, Government Printing Office, 1953.

Married, Divorced, and Separated, by Income Received in 1949 and Race*

		Nonwhite Men				
Income	Ever Married	Separated	Divorced	Separated per 1000 of Ever Married	Divorced per 1000 of Ever Married	Both Divorced and Separated per 1000 of Ever Married
No income	254,660	28,590	7,770	112.27	30.51	142.78
$1-$999	1,096,740	98,130	25,440	89.47	23.20	112.67
$1000-$1999	1,022,010	86,190	23,700	84.33	23.19	107.52
$2000-$2999	802,860	57,120	19,650	71.14	24.48	95.62
$3000-$3999	239,670	12,870	6,840	53.70	28.54	82.24
$4000 and over	63,390	210	210	3.31	3.31	6.62
Total	3,479,330	283,110	83,610	81.37	24.03	105.40

ports, *Marital Status,* Tables 6 and 7, U. S. Government Printing Office, Washington, D. C., 1953.

males, represents two contradictory trends—a negative correlation between separation rate and educational level, and a positive correlation between divorce rate and educational level. The other, for nonwhite women, reflects a curvilinear relationship wherein the highest rate of separations occurs near the middle of the range of educational level.

Discussion

Let us notice some of the limitations and implications of the operations used:

1. Since persons with annulled marriages are classified as single, they cannot enter into this analysis of marital instability.
2. Under the census definition, the category "separated" includes not only persons with legal separations but also those who live, permanently or temporarily apart from their spouses because of marital discord but who have not obtained legal separations.

TABLE 4. Rank Correlations between Indexes of Marital Instability and Indexes of Socioeconomic Status, by Race and Sex

Index of Socioeconomic Status	Sex	Race	Index of Marital Instability		
			Separations	Divorces	Divorces & Separations
Education	Male	White	−1.00	− .43	− .95
		Nonwhite	− .55	+ .76	− .17
	Female	White	− .95	+ .88	+ .57
		Nonwhite	− .17	+1.00	+ .33
Income	Male	White	− .94	− .94	− .94
		Nonwhite	− .49	−1.00	−1.00
Occupational grouping (except farms)	Male	All		− .88	

3. The Bureau of the Census indicates that the number of divorced persons is underestimated.[3] Jacobson estimates that as many as one fifth of the divorced persons are not enumerated as divorced in the census. He suggests that a number of divorced men are probably reported as single, as are women without children. Divorced women with children probably are frequently recorded as widowed.[4]

4. The census takes its data as of a single day in the respondent's life. That is, the records show the marital status of the respondent on that one day rather than any such marital status as "ever divorced." Accordingly divorced persons who have remarried do not enter into our category of having unstable marriages.

5. There is a correlation between age and rate of remarriage. Glick suggests that "under current conditions . . . about two-thirds of the divorced women and three-fourths of the divorced men will eventually remarry." [5] He found a high positive correlation between income and remarriage rates. Before the data were standardized for age, he discovered a strong negative correlation between educational attainment and number of times married for both sexes. After such standardization, however, the relationship becomes positive for women. As Glick indicates, the age factor is important because older persons are the ones most likely to remarry, and they received their education at a time when the average level of education was considerably lower than it is today. Thus when figures are not standardized for age, the remarriages of older persons with preponderantly lower education influence the trend so as to show an inverse relationship between education and remarriage. But if age is controlled, a different trend is discovered—a direct relationship between remarriage rate and education. The data in this paper have not been standardized for age.

Taking the foregoing considerations into account and qualifying our findings accordingly, we conclude that for American white males there was as of 1950 a negative correlation between socioeconomic status (whether measured by education, income, or occupation) and marital dissolution (whether separation or divorce). For nonwhite males and for females, both white and nonwhite, no such clear-cut relationship emerged. Some relationships were curvilinear and some showed positive correlations. According to this evidence, then, any general statement concerning the negative relationship between socioeconomic status and marital instability will have to be confined to white males.

[3] U. S. Bureau of the Census, *U. S. Census of Population: 1950. Special Reports, Education,* Washington, Government Printing Office, 1953, p. 5B-9.

[4] Paul H. Jacobson, *American Marriage and Divorce,* New York, Holt, Rinehart and Winston, 1959, pp. 6-7.

[5] Paul C. Glick, *American Families,* New York, Wiley, 1957, p. 199.

MARRIAGES OF MIXED AND NON-MIXED RELIGIOUS FAITH

Judson T. Landis

Adapted from the *American Sociological Review,* 14 (1949), pp. 401-407.

. . . Catholics, Protestants, and Jews have frowned upon mixed marriages and have done much to discourage their young people from entering mixed unions. Some young people feel that the discouraging of mixed unions among the followers of the different faiths is largely a battle for souls and that there is no practical reason why they should not enter mixed marriages. Many young people today are probably not much interested in the struggle for souls, but they are interested in knowing whether a mixed marriage has less chance for success than marriage within a faith.

In an attitude study among 2,000 students 50 per cent of them said that other things being equal with respect to the prospective spouse, they would marry into a different faith. There was little divergence between the responses of Catholics and of Protestants. One-third of those who would marry outside their faith would be willing to change to the faith of the partner. Protestant students were more willing to change than Catholics. The attitude expression of students on inter-faith marriages is quite in contrast to the feeling of most church leaders, both Catholic and Protestant. Edgar Schmiedeler, director of the family life bureau of the National Catholic Welfare Conference, says, "Since courtship is the beginning which leads ultimately to a marriage contract, the sound starting point toward this goal will be to avoid courtship with any and all non-Catholics.[1]

In spite of the strong Catholic stand on the matter, Clement S. Mihanovich of St. Louis University recently reported, after a survey of all dioceses, that 25 per cent of Roman Catholics married outside the faith.[2]

To gain more information on the success or failure of marriages of mixed and non-mixed religious faith we collected, for 3 years, information on their parents' marriages from the students in our marriage lecture sections. Early in the course and before the subject of mixed marriages had been discussed, each student was asked to complete a questionnaire which gave several facts about his parents' marriage, such as age when

[1] Edgar Schmiedeler, *Marriage and the Family,* New York: McGraw-Hill Book Company, 1946, p. 111.
[2] *Family Life,* December, 1948, p. 6.

married, occupation, education, religion, present marital status, whether either parent changed religious faith at or after marriage, who took the responsibility for giving the religious training, how much conflict over religion had been evident to the children, and the eventual faith chosen by the children.

This discussion is a summary of the information gained on religion in the parent families. We have to date the histories of 4,108 families. Of these, almost two thirds, 2,794, were both Protestant. Almost one-third were either Catholic (573) or Catholic non-Catholic mixed (346). Of the 1,492 individual Catholics in the study, 346 or 23 per cent had married outside the Catholic faith. In 192 of these 346 marriages, each spouse maintained his or her own religion after marriage; in 113 either the Catholic or the Protestant changed to the faith of the other; and in 41 marriages the Catholic had married a person with no religious faith.

This method of doing research has advantages as well as disadvantages. The chief advantage is that a mass of data can be collected in a short time and at little cost. The results of research are more meaningful to students when information is collected from among their own families. The research information may or may not apply to a cross section of the population. It represents the background of the young people in college classes in the Mid-west. They are upper middle class in the main. However, its results agree with those of more representative cross sectional studies.

Because of the method used in collecting the information, the evidence given will shed light only upon mixed marriages in which there are children. Approximately two out of three couples who divorce do not have children. A study of childless mixed marriages might show entirely different results than are revealed in this study. In fact, the results of our study among couples with children would lead us to believe that if there are not children, the Catholic-Protestant marriage has few elements which would make marital adjustment difficult. It is the presence of children in the home which makes for marital conflict in the Catholic-Protestant marriage.

Table I summarizes the divorce and separation rates in our study of marriages of mixed and non-mixed religious faiths and compares the findings with the findings of the studies made in a similar manner in different regions of the country. H. Ashley Weeks[3] made an analysis of 6,548 families of public and parochial school children in Spokane, Washington and Howard Bell[4] analyzed 13,528 families in Maryland. Weeks and Bell

[3] H. Ashley Weeks, "Differential Divorce Rates by Occupation," *Social Forces,* Vol. 21, no. 3, March 1943, p. 336.

[4] Howard M. Bell, *Youth Tell Their Story*, Washington, D.C.: American Council on Education, 1938, p. 21.

did not make all of the breakdowns that we have in our study. These more detailed breakdowns have revealed some interesting material on mixed marriages. Our information on the divorce and separation rate in Catholic, Jewish, Protestant, and mixed marriages confirms the two earlier studies. This might not be expected, since our sample was taken from people who have children in college, while the other samples were more representative of cross sections of whole communities.

TABLE I. **Percentage of Marriages of Mixed and Non-mixed Religious Faiths Ending in Divorce or Separation as Revealed by Studies of Marriages in Michigan, Maryland, and Washington**

Religious Categories		Land's study in Michigan (N-4, 108)	Bell study in Maryland (N-13, 528)	Weeks study in Washington (N-6548)
	Number	Percent	Percent	Percent
Both Catholic	573	4.4	6.4	3.8
Both Jewish	96	5.2	4.6	
Both Protestant	2794	6.0	6.8	10.0
Mixed, Catholic-Protestant	192	14.1	15.2	17.4
Both none	39	17.9	16.7	23.9
Protestant changed to Catholic	56	10.7		
Catholic changed to Protestant	57	10.6		
Protestant father-Catholic mother	90	6.7		
Catholic father-Protestant mother	102	20.6		
Father none-Mother Catholic	41	9.8		
Father none-Mother Protestant	84	19.0		

Approximately 5 per cent of the Catholic and Jewish marriages had ended in divorce or separation in all three studies, 8 per cent of the Protestant marriages, 15 per cent of mixed Catholic-Protestant, and 18 per cent of the marriages in which there was no religious faith. On further analysis, we found that it makes a difference whether the mother is Catholic or Protestant in the mixed marriage. The divorce rate had been highest of all in marriages of a Catholic man to a Protestant women. Twenty-one per cent of these marriages had ended in divorce, while only seven per cent of the marriages in which a Protestant man was married to a Catholic wife had ended in divorce. It is also interesting to note that when a Catholic woman marries a man who has no religious faith the divorce rate is relatively low, 10 per hundred, when compared with other types of mixed marriages. However, when the Protestant woman married a man with no religion the divorce rate was higher, 19 per one hundred. More research is needed on the latter type of marriage. It is our theory that this type of marriage would more nearly fall within the classification of both no religion.

We recognize that using the divorce rate as an index of success or failure in the mixed religious marriage does not necessarily give a true picture of marital happiness. To illustrate, the divorce rate could be low in cases in which the wife is Catholic, and yet the marriages could be unhappy. In the United States three out of four divorces are granted to women. If the wife is a good Catholic she cannot take the initiative toward divorce, although the marriage may be unhappy. On the other hand,

TABLE II.   Percentages of 392 Children Brought Up in Each Faith in 192 Completed Families of Protestant-Catholic Faiths

Faith of Children	Father Protestant Mother Catholic N-90				Father Catholic Mother Protestant N-102			
	Boys		Girls		Boys		Girls	
	No.	Percent	No.	Percent	No.	Percent	No.	Percent
Catholic	70	58.8	58	75.3	31	27.2	18	22.0
Protestant	35	29.4	18	23.4	80	70.2	63	76.8
None	14	11.8	1	1.3	3	2.6	1	1.2
Total	119	100.0	77	100.0	114	100.0	82	100.0

the Protestant wife is free to ask for a divorce if the marriage is unsatisfactory. We found, however, that there are fewer factors making for disharmony in marriages in which the mother is Catholic than there are if she is Protestant.

Factors Making for Disharmony in Catholic-Protestant Marriages

When a Protestant marries a Catholic the Protestant must sign the Ante-Nuptial Agreement in which he promises that the marriage contract can be broken only by death, that all children shall be baptized and educated in the Catholic faith even though the Catholic spouse dies, that no other marriage ceremony than that by the Catholic priest shall take place, and further he shall have respect for the religious principles and convictions of the Catholic partner on birth control.

Of all the differences between Catholics and Protestants the one making for the greatest conflict seems to be the one over the religious training of the children. There is considerable publicity about the use of birth control but this does not seem to be a very serious problem. Studies have revealed that some Catholics do not follow the teachings of the church on this point. If the wife is Catholic and opposes the use of birth control, this may not result in serious conflict since she is the one who has to bear the children. In the families studied, Catholic women married to Protestants had had 2.2 children; Protestant women married to Catholics, 1.9 children; both Catholics, 3.6 children; both Protestant, 2.7; and both Jewish, 2.1. If the use of birth control is related to the size of the family we might conclude from the above that the Catholic woman married to a Protestant uses birth control since she has a much smaller family on the average than when two Catholics marry.

Our research thus supports Baber's conclusion, after his case studies of mixed marriages, that the chief source of friction centers around the religious training of the children. Although the young couple agree before marriage that the children will be baptized in the Catholic faith, they may find they cannot follow through on this agreement in marriage. It is impossible for the Protestant member to project himself into the future and to know how he will feel as a parent. When the children arrive the couple must then face the issue in the light of their present feelings, and they often break the agreements which they sincerely made during the period of courtship. In support of this statement we would refer to Table II, which gives the religious faith of 392 children who were reared in families of mixed Protestant-Catholic faiths. Half of the children had been reared in the Protestant faith, 45 per cent in the Catholic faith and 5 per cent had no faith. The most common tendency seems to be that the children, especially the daughters, follow the faith of the mother. Approximately 65 per cent of the boys and 75 per cent of the girls follow the faith of the mother.

TABLE III.　Parental Policy on Religious Instruction in 192 Catholic-Protestant Families (Percent distribution)

Policy	Father Protestant Mother Catholic N-90	Father Catholic Mother Protestant N-102
Mother took all responsibility for the religious training	42.2	33.7
Our parents told us about both faiths but let us decide for ourselves when we were old enough	22.7	33.7
Responsibility was equally divided	22.7	19.1
We took turns going to both the church of my father and my mother	6.8	6.8
Father took all responsibility for the religious training	1.1	5.6
Some of us went with my father to his church and some went with my mother to her church	4.5	1.1
Total	100.0	100.0

The children who were the products of these mixed marriages were asked to check one of six statements which best described who took the responsibility for religious training in their homes. Table III presents a description of the policies and the percentages following each policy. It will be observed that the most common policy was that "mother took all responsibility for the religious training." The next most common policy was that "our parents told us about both faiths but let us decide for ourselves when we were old enough." Both of these policies are contrary to the agreement made before marriage. When parents face reality they are more likely to follow one of these courses.

Table IV summarizes the children's beliefs as to how much of a handicap differences in religious beliefs had been to their parents' marriages. It will be observed that the greatest handicap had occurred in marriages in which a Catholic man married a Protestant woman.

In the American home the mother is more likely to be a church member and is more apt to take the responsibility for the religious instruction of the children. When a man who has no faith or is a Protestant marries a Catholic woman, he signs the ante-nuptial agreement and does not find it difficult to abide by the agreement when his children are born. He expects his wife to be responsible for their religious training. There is then no great cause for conflict in this type of a mixed marriage.

TABLE IV. **Children's Statements of the Degree to Which Religious Differences Had Handicapped the Parents' Marriage in Mixed and Non-mixed Marriage of Religion (Percent distribution)**

Degree of Handicap	Both Protestant N-721	Both Catholics N-103	Father Protestant Mother Catholic N-90	Father Catholic Mother Protestant N-102
Not at all	85.7	87.4	59.2	45.2
Very little	11.4	8.8	21.0	20.5
Somewhat	2.7	2.8	13.2	23.3
Great	.2	1.0	5.3	5.5
Very great			1.3	5.5

If the mother is Protestant the marriage seems to have many more serious problems. The Protestant mother has agreed that the children will be baptized Catholic, and yet she can hardly bring up her children in a faith which she herself does not accept. Since the major responsibility for religious training falls upon her, she will probably bring the children up in the only faith she knows and believes in. This means that the agreement made before marriage must be scrapped. The Catholic husband is more apt to be a church member than the Protestant husband who marries a Catholic. It may be quite a blow to him to find that his wife will not have the children baptized into his faith. Conflict results since many Catholic fathers cannot give up without a struggle. The Catholic father not only has his own conscience to live with but he is constantly aware of the attitude of his church and of his family when they see his children being brought up in the Protestant faith.

An additional source of trouble in the mixed religious marriage is the serious and sometimes emotional interest which the inlaws take in the marriage. The two grandmothers are on the alert to see which faith will claim the grandchildren. The Protestant grandmother sometimes gives up when the son marries a Catholic because she figures that the children are lost from the beginning. On the other hand, the Catholic grandmother does all she can to see that her son's children are brought up in the Catholic faith. A careful study of in-law frictions in marriage would probably find more intense friction in cases where the children have made a mixed religious marriage.

In general, priests try to discourage mixed marriages in which the non-Catholic is strong in his faith. They may exert less pressure against the marriage if the non-Catholic is weak in his faith. The highest divorce rate was in marriages in which the largest percentage of husbands and wives were church members, i.e., the Catholic father-Protestant mother combinations. Seventy-five per cent of these husbands and wives were church members, while only 48 per cent of the husbands and 91 per cent of the wives were church members in Protestant father-Catholic mother marriages.

Change to Faith of Spouse

In two out of three of the mixed marriages the spouses maintained their own religious faith. In the other third of the mixed marriages one partner changed to the faith of the spouse. In 56 of the marriages the Protestant member had changed to the Catholic faith, and in 57 marriages the Catholic member had changed to the Protestant faith. What evidence we have shows that the marriage has a better chance of success when both spouses accept the same faith, that is, when one changes to the other's faith. The divorce rate had been 10.6 per cent in this type of marriage, but higher when the Catholic wife had changed to the Protestant faith (15.6) per cent and when the Protestant husband had changed to the Catholic faith (16.7 per cent).[5] Part of the explanation for the lower divorce rate is the lack of conflict over the religious training of the children. In these marriages, from 90 to 95 per cent of the children followed the religion of the faith adopted by the couple.

Conclusions

Marriages between Catholics and Protestants entail more hazards than do those between members of one faith. Although couples discuss before marriage the problems arising from religious differences, they can find no final solution to the problems and the differences do not usually decrease with the passing of time after marriage.

[5] When a Protestant wife changed to the Catholic faith the divorce rate was 7.9 per cent; when a Catholic husband changed to Protestant it was 4.0 per cent.

The divorce rate varies according to whether the wife is Catholic or Protestant in the mixed Protestant-Catholic marriage.

Children are important factors in the conflict in mixed marriages. Children tend to follow the religious faith of the mother.

Needed Research

Studies of mixed religious marriages among those who do not have children.

Study of a larger sample conducted in a manner similar to the present study to get larger numbers of all possible combinations.

Interview studies of mixed marriages to check on the present statistical analyses.

Studies of mixed marriages which have not ended in divorce to determine how well adjusted these couples are as compared with couples who do not make mixed marriages.

Research among marriages of different combinations of Protestants to determine whether certain of these marriages have contrasts which hinder marital adjustment.

More research to determine the chances for success in marriages of religiously oriented and non-religiously oriented persons.

PREMARITAL PREGNANCY AS A FACTOR IN DIVORCE

Harold T. Christensen and

Hanna H. Meissner

Adapted from the *American Sociological Review*, 18 (1953), pp. 641-644.

Research relevant to the question of whether sexual intimacies before marriage are related to the success or failure of marriage, is scarce and somewhat inconclusive. Studies reported by Hamilton,[1] Davis,[2] Terman,[3] Locke,[4] and Burgess & Wallin,[5] all tend to support the hypothesis that mar-

[1] Gilbert V. Hamilton, *A Research in Marriage*, New York: Albert & Charles Boni Inc., 1929, pp. 393-95.

[2] Katharine B. Davis, *Factors in the Sex Life of Twenty-Two Hundred Women*, New York: Harper & Bros., 1929, p. 59.

[3] Lewis M. Terman, *Psychological Factors in Marital Happiness*, New York: McGraw-Hill Book Co., 1938, pp. 324-25.

[4] Harvey J. Locke, *Predicting Adjustment in Marriage*, New York: Henry Holt and Company, 1951, p. 156.

[5] Ernest W. Burgess and Paul Wallin, *Engagement and Marriage*, Chicago: J. B. Lippincott Co., 1953, pp. 353-390.

riage is more successful where there has been no premarital sexual intercourse. Yet, most of these authors express certain cautions and qualifications. Furthermore, the Burgess & Wallin data "do not support the theory that coitus before marriage has an adverse effect on the *sexual relationships* after marriage." [6]

It was for the purpose of throwing additional light on this problem that the present study was undertaken. Our focus has been upon the phenomenon of premarital pregnancy as a factor in divorce, viewing the latter as an index of marriage failure.

In an earlier article by the senior author, it was reported that premarital conception was involved in approximately twenty per cent of all first births within marriage, and that disproportionately higher percentages were associated with the depression marriages of 1929-31, a young age at marriage, a non-religious wedding, and a laboring occupation.[7] These generalizations were based upon analyses of official marriage and birth data for Tippecanoe County, Indiana, for the years 1919-1921, 1929-31 and 1939-41.[8]

However, these earlier data provided no basis for testing the possibility of a relationship between premarital pregnancy and the success or failure of marriage. Since that time, we have secured from the files of the County Clerk's office available data for every divorce occurring in Tippecanoe County from the beginning of 1919 through the end of 1952. Names from this list were then carefully checked against those of the earlier marriage and birth lists to determine which, from our original samples, were also divorce cases. In this manner we were able to assemble complete marriage, birth, and divorce information on 137 cases.[9]

Preliminary Description of Divorce Data. Divorce (137 cases) constituted 8.95 per cent of the total sample (1531 cases).[10] Undoubtedly, this percentage, as found, is lower than one might expect if he were considering all present and future divorces within the sample. But, since most of these marriages are still in process, and since there is no good

[6] *Ibid.,* p. 366. Italics supplied.

[7] Harold T. Christensen, "Studies in Child Spacing: I—Premarital Pregnancy as Measured by the Spacing of the First Birth From Marriage," *American Sociological Review,* 18 (February 1953), pp. 53-59.

[8] Tippecanoe County's population is fairly homogeneous as to race and nativity, with less than 1 per cent Negroes and about two per cent foreign born.

[9] Help in recording and checking was given by Witold Krassowski, Sybille Fritzsche, Marilyn Swihart, Delores Aurenz, Nancy Falk, and Lenore Miller. The earlier marriage and birth data had been gathered entirely by Olive P. Bowden.

[10] Since our present concern is with the relationship between divorce and the timing of conception, our overall sample is the 1531 marriages for which we found evidence of the birth of a first child. For further detail see Christensen, *loc. cit.*

way of following our couples once they have left the county, such a figure is unobtainable.[11]

So far as our data reveal, there has been a general increase in divorce rate over time, with the depression marriages showing the highest rate of all. Considering, for this comparison, only those divorces occurring within eleven years from marriage,[12] we found rates of 4.52, 7.10, and 5.35 for the 1919-21, 1929-31, and 1939-41 year groups respectively.

Positive relationships were found between divorce and the following factors: young age at marriage, wide age differences between the mates, urban residence, unskilled occupation, nonreligious wedding, and remarriage. . . .

Testing the Relationships. In Table 1 are presented distributions of total cases and divorce cases according to the time interval between marriage and birth of a child. The first two columns show cases in which there is a reasonable presumption of premarital pregnancy.[13]

Divorce and interval to first birth are shown to be negatively related. Proportionately, there were over three times as many divorces in the shortest interval group as in the longest interval group, with the graduations in between running in a consistent direction. A chi square test revealed these differences to be significant beyond the .001 level of confidence.[14]

---

[11] Our method has the limitation of missing those divorces which took place outside of Tippecanoe County and which occurred (or will still occur) in years later than the ones searched. Some estimation of this loss can be obtained by comparing all marriages with all divorces in Tippecanoe County during the nine sample years of our study; thus considered, divorces (1042) amounted to 21.72 per cent of the marriages (4798). This means that we have located less than half of all actual and potential divorces in our sample (assuming that divorce in a parent-group such as ours is in as high a proportion as with all marriages, an assumption which we do not believe, but which we grant here in the interest of being conservative). However, since there is no known bias affecting these omissions, we feel justified in assuming that they will not substantially affect the comparisons of this paper.

[12] This was for the purpose of making the data comparable, since, for example, one would expect to find a larger number of divorces in the 1919 marriage group during its 33 years of exposure than in the 1941 marriage group during its 11 years of exposure.

[13] The normal period of uterogestation is 266 days. Premature birth would make a few postmarital pregnancy cases appear as premarital, just as late birth would make a few premarital pregnancy cases appear as postmarital, but it seems likely that these discrepancies would not be many. Cf. Christensen *loc. cit.*

[14] It is possible that real differences are even greater than those found here due to differential migration away from the county. If, as is reasonable to suppose, disproportionately more of the premarital pregnancy cases migrated, in order to keep their problem secret, we would have missed disproportionately more of them in our search of the divorce records.

There are at least three important conclusions to be drawn from the above observation pertaining to our sample: (1) Disproportionately high divorce rates are found to be associated with the condition of premarital pregnancy. (2) Within the premarital pregnancy group, rates are higher (highest of any in the sample) for those who delay marriage until shortly before the expected birth. (3) Within the postmarital pregnancy group, rates are highest when conception takes place rather soon after marriage, lowest when it is delayed for several months or years.

What about duration of marriage as possibly affected by premarital pregnancy? Though our data are not entirely satisfactory with respect to this factor,[15] they at least tentatively suggest a relationship; namely, the higher the incidence of premarital pregnancy, the shorter the duration of marriage. There were eleven divorces in our sample where marriage duration was less than two years. Of these, seven were premarital pregnancy cases and the remaining four involved early conception following marriage. Apparently, our premaritally pregnant couples get divorced, not only at a higher rate than others, but within a shorter time.

Controlling Other Variables. We have already reported association between premarital pregnancy on the one hand and such factors as young age, non-religious wedding, and a laboring occupation on the other. Similar relationships were found to exist between divorce and these very

TABLE 1. **Divorces Distributed According to the Spacing of the First Birth from Marriage**

Year Groups	Intervals between Marriage and Birth of First Child (in days)					
	0-139	140-265	266-391	392-531	532-1819	Total
1919-21						
Total cases	25	80	151	88	143	487
Number of divorces	3	11	17	5	6	42
1929-31						
Total cases	23	77	85	48	119	352
Number of divorces	5	15	9	8	10	47
1939-41						
Total cases	23	119	139	102	309	692
Number of divorces	6	13	8	4	17	48
All sample years						
Total cases	71	276	375	238	571	1531
Number of divorces	14	39	34	17	33	137
Per cent divorces	19.72	14.13	9.07	7.14	5.78	8.95

[15] Some of the marriages of our sample, now intact, will later end in divorce. Without these, calculations on mean duration would be meaningless. Furthermore, our figures on cases with short duration are extremely small.

same factors. What we do not know at this point, therefore, is whether the disproportionately high divorce rate in the premarital pregnancy group is due to the reaction of the couples to premarital pregnancy, or to the prevalence in the premarital pregnancy group of other factors known to be associated with high divorce rate.

To find an answer, we first eliminated from consideration all cases with time-intervals of 196-335 days—cases, in other words, where the placing of conception either before or after marriage might be held in some doubt. This left 178 cases where conception was rather definitely premarital and 989 where it was definitely postmarital. Divorce percentages within these two groups were found to be 18.54 and 6.27 respectively.

The second step was to match the two groups on the following factors: wife's age at marriage, age differences between the mates, place of residence, type of occupation, type of wedding, and number of times married.[16] This, of course, had the effect of controlling the named variables. Divorce percentages now became 18.54 and 8.62 respectively for the experimental (premarital pregnancy) and control (postmarital pregnancy) groups. Thus, it would seem that, though the disproportionately high divorce rate is to be partially explained by the presence there of other divorce-favoring factors, a large part may also be accounted for by the fact of premarital pregnancy itself, or in other words, by the attitudes and reactions of couples with reference to this condition.

Summary and Interpretations.   This study has been based upon a sample drawn from Tippecanoe County, Indiana, using data from official marriage, birth and divorce records. Premarital pregnancy has been found to be associated with disproportionately high divorce rates. This was true even after the control, through matching, of interfering variables, and was especially true for cases in which couples waited until late pregnancy before marrying. There was also an indication that those premaritally pregnant couples who got divorced, did so sooner than others. Within marriage, early conceivers showed higher rates of divorce than did late conceivers. All of this is suggestive of the following typology:

(1) "Delayed marriage following pregnancy." This group is the most disposed toward divorce of any. For one thing, it is disproportionately loaded with divorce-disposing factors—young age at marriage, wide age differences between the mates, urban residence, unskilled occupation, nonreligious wedding, and second or subsequent marriages. These factors do not account for all of the difference, however. In a culture such as ours, where premarital pregnancy is generally disapproved, one might expect

---

[16] Matching dichotomies were as follows: wife 20 or under, other; mates 4 or more years different in age, other; urban residence, other; unskilled occupation, other; religious ceremony, other; both first marriage, other.

those who "get caught" to be relatively more anxious and hence less adjustable. Furthermore, within this particular group where marriage is delayed until the last minute, so to speak, one might expect to find disproportionately more who are resentful over being forced into a relationship and determined to get out of it just as soon as they can.

(2) "Early marriage following pregnancy." This group has the second highest divorce rate in the sample. It, also, is rather heavily loaded with the divorce-disposing factors listed in the above paragraph. In addition, there would likely be some of the same anxieties surrounding a disapproved act as described above. There would be fewer forced marriages in this group, however, or, putting it another way, more of these had been planning marriage anyway and were in better positions to adjust to married life because of more favorable attitudes as well as previous preparation. Hence, the lower divorce rate than in group one.

(3) "Early pregnancy following marriage." This group has a lower divorce rate than those involving premarital pregnancy, but higher than those in which pregnancy is delayed for awhile after marriage. It is lower than the premarital pregnancy groups, both because of the presence of more favorable factors, and because of the relative absence of either anxiety or coercion. It is higher than the delayed pregnancy group, probably because of the frustrations coming from unsuccessful birth control [17] and from having the responsibilities of a child before the early marriage adjustments have become established.

(4) "Delayed pregnancy following marriage." This group has the lowest divorce rate of all. In addition to the advantages named for group three, it is largely free of the frustrations peculiar to that group. Here, pregnancy is kept within reasonable control, both as to the demands of the culture, and as to needs and desires of the couple.

Admittedly, some of these interpretations are mere speculations—hypotheses for future research. Beyond the statistical generalizations which we have been able to make for our relatively small group, lie other, more subtle, reasons for the behavior under study. While the explanations suggested above for our four types seem reasonable they have not as yet been sufficiently tested. Especially needed are follow-up case studies and analyses of clinical data as applied to this problem in this particular

[17] Cf. Shirley and Thomas Poffenberger and Judson T. Landis, "Intent Toward Conception and the Pregnancy Experience," *American Sociological Review,* 17 (October 1952), pp. 616-620; also Harold T. Christensen and Robert E. Philbrick, "Family Size as a Factor in the Marital Adjustments of College Couples," *American Sociological Review,* 17 (June 1952), pp. 306-312. Poffenberger and Landis found larger proportions of unplanned pregnancies among couples who conceived early in marriage (p. 617). Christensen and Philbrick found lower marital adjustment scores for couples who had unplanned children (p. 309).

sample. It would also be important to extend the testing of these relationships to other populations and the various subcultures of each.

STABILITY OF MARRIAGE IN RELATION TO AGE AT MARRIAGE

Paul C. Glick

Adapted by permission from Paul C. Glick, *American Families,* New York, John Wiley & Sons, Inc., 1957, pp. 56-58, 110-112.

Women who were still in their first marriage in 1954 were about two years older at first marriage, on the average, than women who had remarried, regardless of the period of the first marriage. This fact is consistent with the hypothesis that marriages of young persons are less stable than those of more mature persons.

Further evidence of the greater instability of marriages where age at first marriage is relatively young can be gathered from a comparison of the interquartile ranges of age at first marriage of those with and without broken marriages. (The interquartile range contains the central half of the cases; one-fourth fall below the first quartile and one-fourth fall above the third quartile of age at first marriage.) Thus, for women who had remarried, the range between the first and third quartiles (17.3 years to 21.4 years) is only about four years as compared with about five years (18.9 years to 24.0 years) for women in first marriages. These figures indicate that a large proportion of women with remarriages had married for the first time at a relatively young age.

Distributions of age at first marriage shown in table 1 bear testimony of the relatively narrow range of ages at first marriage. Among men who married for the first time during the early 1950's, about two-thirds were within an age range of seven years, 19 to 25. Among women, the range was smaller; two-thirds became brides for the first time within the six-year span for 17 to 22 years. The most common ages at first marriage for men were 21 and 22 years and for women, 18 and 19 years; about 20 to 30 percent of all first marriages occurred at these particular ages. At the younger extreme of age at first marriage, less than 10 percent of the men married before they were 19 years of age and less than 10 percent of the women married before they were 17 years of age. At the older extreme, less than 15 percent of the men and less than 10 percent of the women entered their first marriage after they had reached 30 years of age.

TABLE 2. Ever-Married Women Married Once or Remarried, by Age at First Marriage, Period of First Marriage, and Presence of Husband: April 1954 [Numbers in thousands]

Period of first marriage and age at first marriage	All ever-married women in 1954		Married once			Married more than once		
	Number	Percent	Total	Husband present	Other	Total	Husband present	Other
1940 TO 1949								
Total	12,727		11,357	10,270	1,087	1,370	1,232	138
Percent		100.0	89.2	80.7	8.5	10.8	9.7	1.1
14 to 17 years	1,956	100.0	78.1	67.3	10.8	21.9	19.2	2.7
18 and 19 years	3,017	100.0	86.6	79.1	7.5	13.4	11.7	1.7
20 and 21 years	2,646	100.0	91.4	84.1	7.3	8.6	8.2	0.4
22 to 24 years	2,472	100.0	93.5	84.3	9.3	6.5	5.9	0.6
25 to 29 years	1,731	100.0	94.2	86.4	7.7	5.8	5.5	0.3
30 years and over	905	100.0	94.7	84.1	10.6	5.3	5.0	0.3
1930 TO 1939								
Total	9,901		8,244	7,164	1,080	1,657	1,389	268
Percent		100.0	83.3	72.4	10.9	16.7	14.0	2.7
14 to 17 years	1,640	100.0	68.6	58.3	10.3	31.4	27.3	4.1
18 and 19 years	2,296	100.0	78.8	69.3	9.5	21.2	17.2	4.0
20 and 21 years	2,092	100.0	85.7	74.6	11.1	14.3	11.9	2.4
22 to 24 years	2,020	100.0	89.4	78.3	11.0	10.6	8.9	1.8
25 to 29 years	1,332	100.0	91.0	80.9	10.2	8.9	7.4	1.6
30 years and over	521	100.0	95.8	76.4	19.4	4.2	3.6	0.6
1920 TO 1929								
Total	8,602		6,931	5,461	1,470	1,671	1,255	416
Percent		100.0	80.6	63.5	17.1	19.4	14.6	4.8
14 to 17 years	1,626	100.0	61.9	49.6	12.2	38.1	29.7	8.4
18 and 19 years	1,943	100.0	77.4	65.1	12.2	22.6	18.0	4.7
20 and 21 years	1,841	100.0	84.0	65.9	18.0	16.0	10.8	5.2
22 to 24 years	1,699	100.0	89.5	69.7	19.8	10.5	7.8	2.6
25 to 29 years	1,010	100.0	89.3	69.6	19.7	10.7	7.4	3.3
30 years and over	483	100.0	93.8	59.6	34.2	6.2	3.3	2.9

Source: Derived from U. S. Bureau of the Census, *Current Population Reports*, Series P-20, No. 67, table 13, and unpublished tabulation of Current Population Survey data.

cycles much different from those of other women. Furthermore, the remarriage rate among these women is quite low, as a rule. Perhaps the characteristics of a woman that make her slow to enter first marriage likewise minimize her chances of remarriage.

As a concluding observation on this phase of the discussion of remarriage, it is appropriate to call attention to the fact that the great majority of people go through life with their original spouse. Among women who entered their first marriages during the 1920's and who had, by 1954, already passed through about 30 years of exposure to the risk of losing their husbands by divorce or death, four out of every five had married only once. Moreover, among those in this group who had been above the age of 21 at the time of first marriage, fully nine out of every ten had not remarried by 1954. These facts point to the essential stability of marriages in the United States, especially of those contracted after the woman has passed her teens.

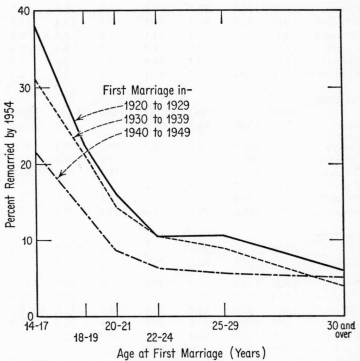

Figure 1.   Percent remarried by 1954, among women with first marriages between 1920 and 1949, by age at first marriage.

Note: Based on data in table 2.

THE CHANGING NATURE AND INSTABILITY OF RE-MARRIAGES

Thomas P. Monahan

Adapted from *Eugenics Quarterly*, 5:2 (1958), 73-85.

Changes in Remarriage

. . . Until recently, although there are no reliable statistics on the subject in this country, it had been the consensus that most divorcees did not remarry. Many well-known writers subscribed to the guess that only about one-third remarried, while some others estimated it to be about 50 per cent and as high as 88 per cent.[1] Available statistics do not easily lend themselves to an exact determination of the proportion of the divorced or widowed who ultimately remarry, now or in the past. In spite of the uncertainty of the information on "divorce," the data we do have, however, give considerable support to the new pronouncements that a large proportion of the divorcees remarry, perhaps as high as three-fourths.

In contrast to colonial times when widowhood accounted for nearly all remarriages, in the last few generations a new element has interposed itself into the American marital pattern—namely, the gradually swelling tide of remarrying divorcees. This drastic change is illustrated in the charts showing the nature of remarriages in the contiguous states of New York and Connecticut.[2] Data from a number of other states and cities ex-

[1] Alfred Cahen, *Statistical Analysis of American Divorce* (New York: Columbia University Press, 1932), pp. 98-109; Popenoe, Paul, *Modern Marriage*, New York, Macmillan, 1943, p. 118; Willard Waller, *The Family, A Dynamic Interpretation* (New York: The Cordon Co., 1938), p. 531; E. R. Groves and W. F. Ogburn, *American Marriage and Family Relationships* (New York: Henry Holt and Co., 1928), pp. 363-64; Ray E. Baber, *Marriage and the Family* (New York: McGraw-Hill Book Co., 1939), pp. 475-79; Paul C. Glick, "First Marriages and Remarriages," *American Sociological Review*, XIV (December, 1949), 730, and *American Families* (New York: John Wiley and Sons, 1957), p. 139; and United States Bureau of the Census, *Population—Special Reports*, P-46, No. 4, June 1, 1946.

[2] *Census of the State of New York for 1855* (and *for 1865*), Secretary of State, Albany, New York, 1865, p. xcv, and *Annual Reports of the Division of Vital Statistics*, State Department of Health, Albany, New York; *Annual Registration Reports, Division of Vital Statistics*, Connecticut State Department of Health, Hartford, Connecticut. According to some data in the early 1900's, New York City, with its large Catholic and Jewish population, seems to have lagged behind up-state New York in the extent of divorce among persons remarrying. However, by 1940 the two jurisdictions were not far apart (cf., National Office of Vital Statistics, *Vital Statistics—*

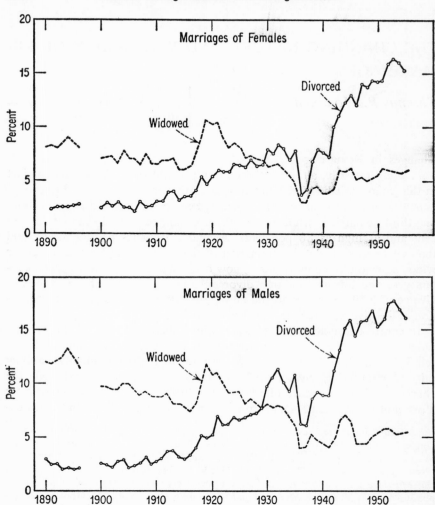

Remarriages in Connecticut: The divorced and widowed as a percent of total marriages.

tending over a period of years further substantiate the pattern of change.[3]
Some *population* survey data on marriages in all of the United States

---

Special Reports, Vol. 15, No. 19, and Vol. 17, Nos. 9 and 13). New York City currently publishes no statistics. Because of migratory-marriage effects state data cannot be accepted as being precisely definitive.

[3] For example, Maine, New Hampshire, Rhode Island (and Providence), Wisconsin (and Milwaukee), and others. See Ruby Jo Reeves, *Marriage in New Haven Since 1870,* unpublished Ph.D. thesis, Yale University, 1938, pp. 1-17, IV—1 to 13.

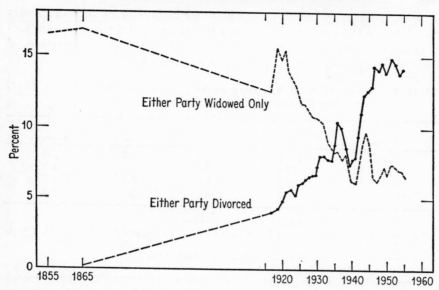

Remarriages in New York State (excluding New York City)*: Divorced and widowed combinations as a percent of total marriages.

* Includes New York City in 1855 and 1865 only.

gathered in 1953 disclosed that 30 per cent of all marriages from January 1950 to April 1953 were remarriages for either party and in 13 per cent of all marriages both parties had been married before. Seventy-five per cent of the remarrying males and 70 per cent of the remarrying females had been *previously* divorced (Table 1).

TABLE 1. Persons Married during January 1950 to April 1953 by Number of Times Married, Percent Distribution*

Husband	Wife		
	Total	Once Only	Twice or More
Total	100.0	77.7	22.3
Once Only	79.6	70.1	9.5
Twice or More	20.4	7.6	12.8

	Previous Marital Status of Remarried		
	Total	Divorced	Widowed
Grooms	100.0	74.6	25.4
Brides	100.0	69.9	30.1

* National Office of Vital Statistics, *Vital Statistics-Special Reports*, Vol. 39, Nos. 3 and 5, pp. 108, 210. Compare to Glick, *op. cit.*, 1957, pp. 141 and 144.

A compilation of statistics from *marriage* records in 21 reporting states for the year 1954 indicated the same thing: about three out of four persons who remarried had been divorced, and in a higher proportion of cases at least one of the parties had been divorced (Table 2).

TABLE 2. Marriages in 21 Reporting States by Previous Marital Status, 1954, Percentage Distribution*

Husband	Wife			
	Total	Single	Widowed	Divorced
Total	100.0	75.9	7.0	17.1
Single	77.1	68.4	1.8	6.9
Widowed	6.2	1.4	3.1	1.7
Divorced	16.7	6.1	2.1	8.5

* Eliminating cases of marital status not reported, one or both.

Source: National Office of Vital Statistics, *Vital Statistics—Special Reports,* Vol. 44, No. 6, p. 130.

Approximately the same relationship holds for 23 states in 1955.

Explanation for the change in the nature of remarriage is two-fold. With the increase in life expectancy widowhood has been occurring at older ages when eventual remarriage is less likely and, at the same time, divorce has been becoming more common for American couples at the younger marriageable ages. As a result divorcees predominate among those who remarry. Actually, if it had not been for the rise in divorce in the earlier years of life, the over-all proportion of remarriages would have been on the decline in the past 50 years in the United States.[4]

The Success of Remarriages

Present Interest. Americans are not merely interested in achieving the state of marriage; they are also eager to be successfully married, with divorce being an unintentional and unpremeditated outcome. The allegation that the prospects of a successful or a happy marriage for divorcees were poor (according to some cogent reasoning and selected case studies) was an acceptable conclusion in sociological and popular literature at one time—when divorces were fewer and held in low esteem.[5] In view of the swelling tide of remarriages of divorcees and the apparent success of many of them, and the more tolerant public attitude nowadays, sociologists are reporting more favorably upon this class of remarriages, in support of which viewpoint some rather selected kinds of data are offered in evi-

[4] Paul H. Jacobson, "Differentials in Divorce by Duration of Marriage and Size of Family," *American Sociological Review,* XV (April, 1950), 235-44.

[5] W. Waller, *The Old Love and the New* (New York: Horace Liveright, Inc., 1930), pp. 157-58, and *op. cit.,* 1938, p. 566.

dence. The push to make divorce acceptable is growing, but, as yet, there is no espousal of the novel idea expressed years ago (1924) by a judge who averred that "Every divorce makes four people happy." [6] One authority on divorce, however, was recently credited with the statement, "There is no question now that second marriages are happier than first marriages . . . (and) the percentage of failures is less among all second marriages than among all first marriages." [7] This conclusion is shared by others but is in direct opposition to certain facts and needs careful evaluation.

The magnitude of the problem and the intricate psychological and sociological character of present-day remarriages will soon lead to an exploration and development of this field by social scientists. Some recent family textbooks, it is true, give it hardly any treatment at all; on the other hand, one monograph published in 1956 deals entirely with this subject. Unfortunately, we have very little refined *and* comprehensive data on remarriage and the literature contains many contradictions which have yet to be resolved. . . .

Iowa Data

Only two or three states gather records showing the marital background of couples entering into marriage or being divorced. In Iowa a rather full marital history of the parties is asked for on the new (1953) marriage and divorce forms for state registration. With the help of Mr. L. E. Chancellor, the Iowa State Director of Vital Statistics, it was possible to obtain tabulation sheets on both marriages and divorces according to prior marital history of the couples. Whereas, heretofore, we have been limited in our analysis to remarriages by the number of the marriage without regard to type, now we have for the *first time* information detailed according to widowhood or divorce. In the period 1953-1955 Iowa marriages totaled 70,901; divorces, 15,502.

Representativeness. Iowa is a fairly representative area in the mid-Western United States, with a strong agricultural base and a small-industry economy. An outward migration state, its population is over 95 per cent white, about one-fifth Catholic and less than one per cent Jewish; and Des Moines, with a population of 178,000 in 1950, is the only city

[6] R. L. Hartt, "The Habit of Getting Divorces," *World's Work,* XLVIII (August, 1924), 407. Cf., A. Q. Maisel, "What Divorce Crisis?", *McCall's,* LXXV (August, 1948), 18ff.

[7] William J. Goode cited in E. M. Duvall and R. Hill, *When You Marry* (New York: Association Press, 1953), p. 291—a widely used pre-college text. Cf., Bernard, Goode, Monahan, Popenoe, cited herein; J. H. S. Bossard, *Parent and Child* (Philadelphia: University of Pennsylvania Press, 1953), pp. 149-50; and M. F. Nimkoff in *Conference on Divorce* (Chicago: University of Chicago Law School, 1952), Conference Series No. 9, p. 61.

with over 100,000 inhabitants. There seems to be a liberal attitude toward divorce and no special legal impediments are imposed against it, although remarriage within a year without court permission is technically prohibited by law. In spite of Iowa's agricultural emphasis and non-urban nature, the divorce rate per 1,000 population in 1954 was 8.7 compared to the national figure of 9.2. In 1954 the duration-specific divorce rates indicated the marriage fatality rate in Iowa to be about 21 per cent of all marriages.[8] All in all, even though it may not represent the condition prevailing in highly urbanized, industrial areas, or in other places characterized by a large nonwhite population, in its geographic region Iowa may be taken as a typical middle-American state.

The nature and accuracy of all data on marriage and divorce (and marital status) need careful appraisal, and familiarity with registration records gives one reason to be most cautious about small differences in findings. This matter and other methodological perplexities are variously discussed in the literature cited. . . .

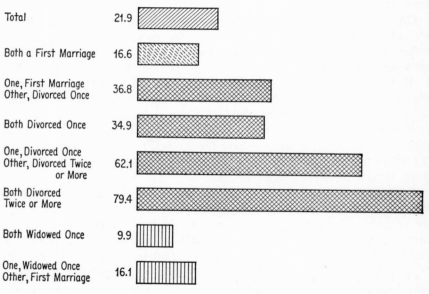

Ratio of divorces per 100 marriages of selected type of marital background,* Iowa, 1953-1955.

* Divorced-widowed combinations excluded.

[8] See, Iowa State Department of Health, *Annual Report of the Division of Vital Statistics, 1951* (Des Moines, Iowa), pp. 83-86.

Divorce Stability by Marital Background

Ratio of Divorces per 100 Marriages by Marital Type. The overall ratio of divorces per 100 marriages in the 3-year period was 21.9.

Primary marriages (a first marriage for both parties) show a ratio of only 16.6 divorces per 100 marriages in that category, but where both parties had been divorced *once* before the figure doubles to 34.9, and where both parties had been divorced *twice or more* times before the ratio climbs to 79.4. Not all marital combinations are shown in the chart; some have too few cases per cell. One of the ratios obtained (a first-married bride with a twice-or-more divorced husband) appears extremely high. In this regard it must be remembered that marital types which are relatively infrequent may not be accurately gauged in this 3-year compilation; and also, in view of the nature of the information and the method, the ratios can only be accepted as approximate. A certain variability appears in the set of figures, but one must hesitate about drawing any inference from them regarding any unequal fragility on the male or female sides which have a similar prior marital history. Allowances for a later age at marriage for divorcees as compared to first marriages (and mortality reduction before divorce could occur) would only accentuate the divorce ratios found. The duration of these remarriages before divorce takes place a second time is shorter than for the first marriages, and this would tend to exaggerate their "rate" of failure. Nevertheless, everything considered, the differences in the "divorce ratios" obtained need not be disqualified to any appreciable extent on these scores.

As to the widowed, they show abnormally low ratios: 9.9 for the both widowed *once* only, and 2.0 for the both widowed *twice or more* only. One very important element, which it is not yet possible to eliminate from the results for the widowed classes, is their much older age at remarriage. In a group of 14 reporting states in 1954, divorced grooms (last or prior status) remarried at about age 36, whereas widowed males remarried at age 57—and brides of corresponding classes about 5 to 9 years earlier, respectively.[9] The low divorce ratios of the widowed may be due largely to the effect of joint-mortality, such that previously *widowed* couples do not survive as a unit to endure the rigors of divorce or make it necessary. The divorce ratio of first-married with widowed spouses bears out this surmise of age bias, as witnessed by the ratios of 15.1 for the widower with a first-married wife, and 16.8 for the widow with a first married husband. . . .

[9] Supra, *Vital Statistics—Special Reports,* Vol. 44, No. 6, p. 108.

Conclusion

Marriages which are successful must certainly survive the social storms of life. Those which do not can hardly be called a "success." One author in this field, who compared a divorced and a happily married group in great detail, based his thesis squarely on the point that "divorce is a criterion of marital adjustment." [10] However, he and others have produced results which purport to show that remarried divorcees "on the whole are happy and adjusted in second marriages," [11] and a new school of thought seems to be developing around this belief. Whatever the reason for it, there is a fundamental contradiction between the two viewpoints: that remarriages of the divorced are more likely to end in divorce *and* that divorced persons are as successful in their remarriages as once-married persons.

For the first time in the United States we have been able to gather data on a broad basis to show the incidence of divorce among remarrying classes. According to this information for one state, Iowa, as for marriage survival alone, it seems that the widowed who remarry fare as well as do the *primary* marriages, after making considerable allowance for the differential mortality effects upon the widowed group. With the divorced it is the opposite story: a divorce for one party weakens the strength of the marriage bond, and a second divorce experience greatly lessens the chances of survival of the marriage. It may well be that the twice-divorced, who are sometimes referred to as the "neurotics," [12] have only about 20 to 25 chances in 100 for a lasting marriage.

Such are the hard demographic facts for the general population of Iowa, at least. Incidental information from a few other cities confirms this finding. The qualifications to the Iowa data given above should not be overlooked, however; and similar studies should be made to test the general or special validity of these findings.

All this is not to deny that there are other elements to a truly happy

[10] H. J. Locke, *Predicting Adjustment in Marriage* (New York: Holt, 1951), p. 3.

[11] See feature article on the Goode study (*supra*) in *This Week* newspaper magazine, January 8, 1956, p. 8, which states, "Nine out of 10 divorcees who remarry claim that their second marriages are better." Cf., footnotes 14, 16, 18.

[12] Waller, *op. cit.*, 1930, p. 140; Edmund Bergler, *Unhappy Marriage and Divorce* (New York: International Universities Press, 1947); Shirley Raines, "Personality Traits of Persons Who Divorced and Remarried," M.A. Thesis, University of Southern California, 1949, as reported in *Family Life* X (June, 1950), 1-2; Kingsley Davis in M. Fishbein and R. Kennedy, eds., *Modern Marriage and Family Living* (New York: Oxford University Press, 1957), p. 111 (1947 edition, p. 467); and Rev. H. Loomis, "Divorce Legislation in Connecticut," *New Englander and Yale Review*, XXV (July, 1866), 450-51.

marriage; and the fact of mere legal endurance is not enough. Some marriages are indeed an endless round of tensions and conflict, until "death do them part." On the other hand some divorced persons do achieve a successful remarriage, and perhaps for the 50 per cent or so which manage to survive there is a high level of happiness.

If we take into account the higher proportion of married persons in the population of late, the divorce rate in the United States has been noticeably receding since the all-time high (1946) of the post-war readjustment period. From the demographic point of view, however, it seems we have gradually arrived at a stage in our American divorce history when the eventuality anticipated by Willcox and the statistical warning voiced by him can no longer be ignored.[13] Our divorce rate is being compounded by the repetitiousness of divorce among a divorce-prone population group,[14] and there seems to exist in our American system a dual pattern of marriage, with "sequential polygamy" being the practice of a substantial minority of the population, entirely "within the law." [15] According to one well-known family sociologist, "We are, in effect, operating a type of trial marriage system." [16] A realization of these facts should arouse a host of theoretical speculations and perhaps some more extensive research on the subject of remarriage.[17]

[13] Walter F. Willcox, *Studies in American Demography* (New York: Cornell University Press, 1940), p. 350; Cf., Inter-Agency Committee, National Conference on Family Life, May 1948, *The American Family, A Factual Background* (Washington, D. C.: Government Printing Office, 1949), p. 21.

[14] If we assume that 18 per cent of all first marriages are divorced, following which 75 per cent remarry and one-third of these fail a second time, and 50 per cent of the twice-divorced remarry with two-thirds of these failing again, a rough calculation comparable to Willcox's (*op. cit.*, 1940, p. 350) would show the following: the initial rate of marriage failure being only 18 per 100 marriages, the crude overall rate because of repetitiousness of divorces would be 24 per 100 marriages.

[15] P. H. Landis, "Sequential Marriage," *Journal of Home Economics,* XLII (October, 1950), 625-28; Jessie Bernard, *Remarriage,* N. Y., Dryden, 1956, Ch. 3; Morris Ploscowe, *Sex and the Law* (New York: Prentice-Hall, Inc., 1951), p. 63; and E. E. Eubank, "Why Marriages Fail," *World Tomorrow,* X (June, 1927), 274.

[16] Report of an address by Reuben Hill to the National Conference on Social Welfare, *Philadelphia Evening Bulletin,* May 22, 1957.

[17] For other references, background material, and qualifying or methodological considerations the reader should consult the originally published source.

NAME INDEX

Bold face references indicate selections contained in this volume.

SUBJECT INDEX